Iterative Methods for Solving Nonlinear Equations and Systems

.

Iterative Methods for Solving Nonlinear Equations and Systems

Special Issue Editors

Juan R. Torregrosa
Alicia Cordero
Fazlollah Soleymani

MDPI • Basel • Beijing • Wuhan • Barcelona • Belgrade

MDPI

Special Issue Editors

Juan R. Torregrosa
Polytechnic University of
Valencia
Spain

Alicia Cordero
Polytechnic University of
Valencia
Spain

Fazlollah Soleymani
Institute for Advanced Studies in
Basic Sciences (IASBS)
Iran

Editorial Office
MDPI
St. Alban-Anlage 66
4052 Basel, Switzerland

This is a reprint of articles from the Special Issue published online in the open access journal *Mathematics* (ISSN 2227-7390) from 2018 to 2019 (available at: https://www.mdpi.com/journal/mathematics/special_issues/Iterative_Methods_Solving_Nonlinear_Equations_Systems).

For citation purposes, cite each article independently as indicated on the article page online and as indicated below:

LastName, A.A.; LastName, B.B.; LastName, C.C. Article Title. *Journal Name* **Year**, *Article Number*, Page Range.

ISBN 978-3-03921-940-7 (Pbk)
ISBN 978-3-03921-941-4 (PDF)

Contents

About the Special Issue Editors

Juan Ramón Torregrosa, Dr., has a Bachelor in Mathematical Sciences (Universitat de València) and obtained his PhD (1990, Universitat de València) defending his thesis "Algunas propiedades geométricas uniformes y no uniformes de un espacio de Banach." He is Full Professor of Applied Mathematics in the Institute for Multidisciplinary Mathematics of the Polytechnical University of València.

He published several papers about locally convex spaces and Banach spaces in the 1990s. Afterwards, he launched new research projects in linear algebra, matrix analysis, and combinatorics. He has supervised several PhD theses on these topics. He also has published a significant number of papers in related journals: *Linear Algebra and Its Applications*, *Applied Mathematics Letters*, and *SIAM Journal Matrix Analysis.*

His current research is in numerical analysis. He focuses on different problems related to the solution of nonlinear equations and systems, matrix equations, and dynamical analysis of rational functions involved in iterative methods. He has published more than 200 papers in JCR journals and he also has presented numerous communications in international conferences.

Alicia Cordero, Dr., has a Bachelor in Mathematic Sciences (1995, Universitat de València). She obtained her PhD in Mathematics (2003, Universitat Jaume I) defending her PhD Thesis "Cadenas de órbitas periódicas en la variedad S2xS1," which was supervised by José Martínez and Pura Vindel.

Through the years, she has published many papers about the decomposition into round loops of three-dimensional varieties, links of periodic orbits of non-singular Morse-Smale fluxes, and their applications to celestial mechanics.

She is Full Professor of Applied Mathematics in the Institute for Multidisciplinary Mathematics of the Polytechnical University of València. Her current research is focused on dynamical systems and numerical analysis, highlighting the iterative methods for solving nonlinear equations and systems as well as the dynamical study of the rational functions involved in these processes. She has published more than 150 papers in JCR Journals. She also has presented numerous communications in international conferences.

Fazlollah Soleymani, Dr., acquired his PhD degree in numerical analysis at Ferdowsi University of Mashhad in Iran. He also accomplished his postdoctoral fellowship at the Polytechnic University of Valencia in Spain. Soleymani's main research interests are in the areas of computational mathematics. Recently, he has been working on a number of different problems, which fall under semi-discretized and (localized) RBF(–(H)FD) meshfree schemes for financial partial (integro–) differential equations, high-order iterative methods for nonlinear systems, numerical solution of stochastic differential equations, and iteration methods for generalized inverses.

.

Preface to "Iterative Methods for Solving Nonlinear Equations and Systems"

Solving nonlinear equations in any Banach space (real or complex nonlinear equations, nonlinear systems, and nonlinear matrix equations, among others) is a non-trivial task that involves many areas of science and technology. Usually the solution is not directly affordable and requires an approach utilizing iterative algorithms. This is an area of research that has grown exponentially over the last few years.

This Special Issue focuses mainly on the design, analysis of convergence, and stability of new iterative schemes for solving nonlinear problems and their application to practical problems. Included papers study the following topics: Methods for finding simple or multiple roots, either with or without derivatives, iterative methods for approximating different generalized inverses, and real or complex dynamics associated with the rational functions resulting from the application of an iterative method on a polynomial function. Additionally, the analysis of the convergence of the proposed methods has been carried out by means of different sufficient conditions assuring the local, semilocal, or global convergence.

This Special issue has allowed us to present the latest research results in the area of iterative processes for solving nonlinear equations as well as systems and matrix equations. In addition to the theoretical papers, several manuscripts on signal processing, nonlinear integral equations, partial differential equations, or convex programming reveal the connection between iterative methods and other branches of science and engineering.

Juan R. Torregrosa, Alicia Cordero, Fazlollah Soleymani

Special Issue Editors

mathematics **MDPI**

Article

A Few Iterative Methods by Using [1, n]-Order Padé Approximation of Function and the Improvements

Shengfeng Li [1,*] **, Xiaobin Liu** [2] **and Xiaofang Zhang** [1]

[1] Institute of Applied Mathematics, Bengbu University, Bengbu 233030, China; zhangxiaofangah@126.com
[2] School of Computer Engineering, Bengbu University, Bengbu 233030, China; 18807925677@139.com
* Correspondence: lsf@bbc.edu.cn; Tel.: +86-552-317-5158

Received: 15 November 2018; Accepted: 30 December 2018; Published: 7 January 2019

Abstract: In this paper, a few single-step iterative methods, including classical Newton's method and Halley's method, are suggested by applying [1, n]-order Padé approximation of function for finding the roots of nonlinear equations at first. In order to avoid the operation of high-order derivatives of function, we modify the presented methods with fourth-order convergence by using the approximants of the second derivative and third derivative, respectively. Thus, several modified two-step iterative methods are obtained for solving nonlinear equations, and the convergence of the variants is then analyzed that they are of the fourth-order convergence. Finally, numerical experiments are given to illustrate the practicability of the suggested variants. Henceforth, the variants with fourth-order convergence have been considered as the imperative improvements to find the roots of nonlinear equations.

Keywords: nonlinear equations; Padé approximation; iterative method; order of convergence; numerical experiment

1. Introduction

It is well known that a variety of problems in different fields of science and engineering require to find the solution of the nonlinear equation $f(x) = 0$ where $f : I \rightarrow D$, for an interval $I \subseteq \mathbb{R}$ and $D \subseteq \mathbb{R}$, is a scalar function. In general, iterative methods, such as Newton's method, Halley's method, Cauchy's method, and so on, are the most used techniques. Hence, iterative algorithms based on these iterative methods for finding the roots of nonlinear equations are becoming one of the most important aspects in current researches. We can see the works, for example, [1–22] and references therein. In the last few years, some iterative methods with high-order convergence have been introduced to solve a single nonlinear equation. By using various techniques, such as Taylor series, quadrature formulae, decomposition techniques, continued fraction, Padé approximation, homotopy methods, Hermite interpolation, and clipping techniques, these iterative methods can be constructed. For instance, there are many ways of introducing Newton's method. Among these ways, using Taylor polynomials to derive Newton's method is probably the most widely known technique [1,2]. By considering different quadrature formulae for the computation of the integral, Weerakoon and Fernando derive an implicit iterative scheme with cubic convergence by the trapezoidal quadrature formulae [4], while Cordero and Torregrosa develope some variants of Newton's method based in rules of quadrature of fifth order [5]. In 2005, Chun [6] have presented a sequence of iterative methods improving Newton's method for solving nonlinear equations by applying the Adomian decomposition method. Based on Thiele's continued fraction of the function, Li et al. [7] give a fourth-order convergent iterative method. Using Padé approximation of the function, Li et al. [8] rederive the Halley's method and by the divided differences to approximate the derivatives, they arrive at some modifications with third-order convergence. In [9], Abbasbandy et al. present an efficient numerical algorithm for solving nonlinear algebraic equations based on Newton–Raphson method and homotopy analysis method. Noor and Khan suggest and analyze a new class of iterative methods by using the homotopy perturbation method in [10]. In 2015, Wang et al. [11] deduce a general family of n-point Newton type iterative methods for solving nonlinear equations by using direct Hermite interpolation. Moreover, for a particular class of functions, for instance, if f is a polynomial, there exist some efficient univariate root-finding algorithms to compute all solutions of the polynomial equation (see [12,13]). In the literature [13], Bartoň et al. present an algorithm for computing all roots of univariate polynomial based on degree reduction, which has the higher convergence rate than Newton's method. In this article, we will mainly solve more general nonlinear algebraic equations.

Newton's method is probably the best known and most widely used iterative algorithm for root-finding problems. By applying Taylor's formula for the function $f(x)$, let us recall briefly how to derive Newton iterative method. Suppose that $f(x) \in C^n[I], n = 1, 2, 3, \ldots$, and $\eta \in I$ is a single root of the nonlinear equation $f(x) = 0$. For a given guess value $x_0 \in I$ and a $\delta \in \mathbb{R}$, assume that $f'(x) \neq 0$ for each x belongs to the neighborhood $(x_0 - \delta, x_0 + \delta)$. For any $x \in (x_0 - \delta, x_0 + \delta)$, we expand $f(x)$ into the following Taylor's formula about x_0:

$$f(x) = f(x_0) + f'(x_0)(x - x_0) + \frac{1}{2!}f''(x_0)(x - x_0)^2 + \cdots + \frac{1}{k!}(x - x_0)^k f^{(k)}(x_0) + \cdots,$$

where $k = 0, 1, 2, \cdots$. Let $|\eta - x_0|$ be sufficiently small. Then the terms involving $(\eta - x_0)^k, k = 2, 3, \ldots$, are much smaller. Hence, we think the fact that the first Taylor polynomial is a good approximation to the function near the point x_0 and give that

$$f(x_0) + f'(x_0)(\eta - x_0) \approx 0.$$

Notice the fact $f'(x_0) \neq 0$, and solving the above equation for η yields

$$\eta \approx x_0 - \frac{f(x_0)}{f'(x_0)},$$

which follows that we can construct the Newton iterative scheme as below

$$x_{k+1} = x_k - \frac{f(x_k)}{f'(x_k)}, k = 0, 1, 2, \ldots.$$

It has been known that Newton iterative method is a celebrated one-step iterative method. The order of convergence of Newton's method is quadratic for a simple zero and linear for multiple root.

Motivated by the idea of the above technique, in this paper, we start with using Padé approximation of a function to construct a few one-step iterative schemes which includes classical Newton's method and Halley's method to find roots of nonlinear equations. In order to avoid calculating the high-order derivatives of the function, then we employ the approximants of the higher derivatives to improve the presented iterative method. As a result, we build several two-step iterative formulae, and some of them do not require the operation of high-order derivatives. Furthermore, it is shown that these modified iterative methods are all fouth-order convergent for a simple root of the equation. Finally, we give some numerical experiments and comparison to illustrate the efficiency and performance of the presented methods.

The rest of this paper is organized as follows. we introduce some basic preliminaries about Padé approximation and iteration theory for root-finding problem in Section 2. In Section 3, we firstly construct several one-step iterative schemes based on Padé approximation. Then, we modify the presented iterative method to obtain a few iterative formulae without calculating the high-order derivatives. In Section 4, we show that the modified methods have fourth-order convergence at least for a simple root of the equation. In Section 5 we give numerical examples to show the performance of the presented methods and compare them with other high-order methods. At last, we draw conclusions from the experiment results in Section 6.

2. Preliminaries

In this section, we briefly review some basic definitions and results for Padé approximation of function and iteration theory for root-finding problem. Some surveys and complete literatures about iteration theory and Padé approximation could be found in Alfio [1], Burden et al. [2], Wuytack [23], and Xu et al. [24].

Definition 1. *Assume that $f(x)$ is a function whose $(n+1)$-st derivative $f^{(n+1)}(x), n = 0, 1, 2, \ldots$, exists for any x in an interval I. Then for each $x \in I$, we have*

$$f(x) = f(x_0) + f'(x_0)(x - x_0) + \frac{f''(x_0)}{2!}(x - x_0)^2 + \cdots + \frac{f^{(n)}(x_0)}{n!}(x - x_0)^n + o[(x - x_0)^n], \quad (1)$$

which is called the Taylor's formula with Peano remainder term of order n based at x_0, and the error $o[(x - x_0)^n]$ is called the Peano remainder term or the Peano truncation error.

Definition 2. *If $P(x)$ is a polynomial, the accurate degree of the polynomial is $\partial(P)$, and the order of the polynomial is $\omega(P)$, which is the degree of the first non-zero term of the polynomial.*

Definition 3. *If it can be found two ploynomials*

$$P(x) = \sum_{i=0}^{m} a_i(x - x_0)^i \quad and \quad Q(x) = \sum_{i=0}^{n} b_i(x - x_0)^i$$

such that

$$\partial(P(x)) \le m, \quad \partial(Q(x)) \le n, \quad \omega(f(x)Q(x) - P(x)) \ge m + n + 1,$$

then we have the following incommensurable form of the rational fraction $\frac{P(x)}{Q(x)}$:

$$R_{m,n}(x) = \frac{P_0(x)}{Q_0(x)} = \frac{P(x)}{Q(x)},$$

which is called $[m, n]$-order Padé approximation of function $f(x)$.

We give the computational formula of Padé approximation of function $f(x)$ by the use of determinant, as shown in the following lemma [23,24].

Lemma 1. *Assume that $R_{m,n}(x) = \frac{P_0(x)}{Q_0(x)}$ is Padé approximation of function $f(x)$. If the matrix*

$$A_{m,n} = \begin{pmatrix} a_m & a_{m-1} & \cdots & a_{m+1-n} \\ a_{m+1} & a_m & \cdots & a_{m+2-n} \\ \vdots & \vdots & \ddots & \vdots \\ a_{m+n-1} & a_{m+n-2} & \cdots & a_m \end{pmatrix}$$

is nonsingular, that is the determinant $|A_{m,n}| = d \neq 0$, then $P_0(x), Q_0(x)$ can be written by the following determinants

$$P_0(x) = \frac{1}{d} \begin{vmatrix} T_m(x) & (x-x_0)T_{m-1}(x) & \cdots & (x-x_0)^n T_{m+1-n}(x) \\ a_{m+1} & a_m & \cdots & a_{m+1-n} \\ \vdots & \vdots & \ddots & \vdots \\ a_{m+n} & a_{m+n-1} & \cdots & a_m \end{vmatrix}$$

and

$$Q_0(x) = \frac{1}{d} \begin{vmatrix} 1 & (x-x_0) & \cdots & (x-x_0)^n \\ a_{m+1} & a_m & \cdots & a_{m+1-n} \\ \vdots & \vdots & \ddots & \vdots \\ a_{m+n} & a_{m+n-1} & \cdots & a_m \end{vmatrix},$$

where $a_n = \frac{f^{(n)}(x_0)}{n!}, n = 0, 1, 2, \ldots$, and we appoint that

$$T_k(x) = \begin{cases} \sum_{i=0}^{k} a_i(x-x_0)^i, & for \quad k \geq 0, \\ 0, & for \quad k < 0. \end{cases}$$

Next, we recall the speed of convergence of an iterative scheme. Thus, we give the following definition and lemma.

Definition 4. *Assume that a sequence $\{x_i\}_{i=0}^{\infty}$ converges to η, with $x_i \neq \eta$ for all $i, i = 0, 1, 2, \ldots$. Let the error be $e_i = x_i - \eta$. If there exist two positive constants α and β such that*

$$\lim_{i \to \infty} \frac{|e_{i+1}|}{|e_i|^{\alpha}} = \beta,$$

then $\{x_i\}_{i=0}^{\infty}$ converges to the constant η of order α. When $\alpha = 1$, the sequence $\{x_i\}_{i=0}^{\infty}$ is linearly convergent. When $\alpha > 1$, the sequence $\{x_i\}_{i=0}^{\infty}$ is said to be of higher-order convergence.

For a single-step iterative method, sometimes it is convenient to use the following lemma to judge the order of convergence of the iterative method.

Lemma 2. *Assume that the equation $f(x) = 0, x \in I$, can be rewritten as $x = \varphi(x)$, where $f(x) \in C[I]$ and $\varphi(x) \in C^{\gamma}[I], \gamma \in \mathbb{N}^{+}$. Let η be a root of the equation $f(x) = 0$. If the iterative function $\varphi(x)$ satisfies*

$$\varphi^{(j)}(\eta) = 0, j = 1, 2, \ldots, \gamma - 1, \varphi^{(\gamma)}(\eta) \neq 0,$$

then the order of convergence of the iterative scheme $x_{i+1} = \varphi(x_i), i = 0, 1, 2, \ldots$, is γ.

3. Some Iterative Methods

Let η be a simple real root of the equation $f(x) = 0$, where $f : I \to D, I \subseteq \mathbb{R}, D \subseteq \mathbb{R}$. Suppose that $x_0 \in I$ is an initial guess value sufficiently close to η, and the function $f(x)$ has n-th derivative $f^{(n)}(x), n = 1, 2, 3, \ldots$, in the interval I. According to Lemma 1, $[m, n]$-order Padé approximation of function $f(x)$ is denoted by the following rational fraction:

$$f(x) \approx R_{m,n}(x) = \frac{\begin{vmatrix} T_m(x) & (x-x_0)T_{m-1}(x) & \cdots & (x-x_0)^n T_{m+1-n}(x) \\ a_{m+1} & a_m & \cdots & a_{m+1-n} \\ \vdots & \vdots & \ddots & \vdots \\ a_{m+n} & a_{m+n-1} & \cdots & a_m \end{vmatrix}}{\begin{vmatrix} 1 & (x-x_0) & \cdots & (x-x_0)^n \\ a_{m+1} & a_m & \cdots & a_{m+1-n} \\ \vdots & \vdots & \ddots & \vdots \\ a_{m+n} & a_{m+n-1} & \cdots & a_m \end{vmatrix}}. \tag{2}$$

Recall Newton iterative method derived by Taylor's series in Section 1. The first Taylor polynomial is regarded as a good approximation to the function $f(x)$ near the point x_0. Solving the linear equation denoted by $f(x_0) + f'(x_0)(\eta - x_0) \approx 0$ for η gives us the stage for Newton's method. Then, we think whether or not a novel or better linear function is selected to approximate the function $f(x)$ near the point x_0. Maybe Padé approximation can solve this question. In the process of obtaining new iterative methods based on Padé approximation of function, on the one hand, we consider that the degree of the numerator of Equation (2) is always taken as 1, which guarantees to obtain the different linear function. On the other hand, we discuss the equations are mainly nonlinear algebraic equations, which differ rational equations and have not the poles. Clearly, as n grows, the poles of the denominator of Equation (2) do not affect the linear functions that we need. These novel linear functions may be able to set the stage for new methods. Next, let us start to introduce a few iterative methods by using $[1, n]$-order Padé approximation of function.

3.1. Iterative Method Based on $[1, 0]$-Order Padé Approximation

Firstly, when $m = 1, n = 0$, we consider $[1, 0]$-order Padé approximation of function $f(x)$. It follows from the expression (2) that

$$f(x) \approx R_{1,0}(x) = T_1(x) = a_0 + a_1(x - x_0).$$

Let $R_{1,0}(x) = 0$, then we have

$$a_0 + a_1(x - x_0) = 0. \tag{3}$$

Due to the determinant $|A_{1,0}| \neq 0$, i.e., $f'(x_0) \neq 0$, we obtain the following equation from Equation (3).

$$x = x_0 - \frac{a_0}{a_1}.$$

In view of $a_0 = f(x_0), a_1 = f'(x_0)$, we reconstruct the Newton iterative method as below.

Method 1. *Assume that the function $f : I \to D$ has its first derivative at the point $x_0 \in I$. Then we obtain the following iterative method based on $[1, 0]$-order Padé approximation of function $f(x)$:*

$$x_{k+1} = x_k - \frac{f(x_k)}{f'(x_k)}, k = 0, 1, 2, \ldots. \tag{4}$$

Starting with an initial approximation x_0 that is sufficiently close to the root η and using the above scheme (4), we can get the iterative sequence $\{x_i\}_{i=0}^{\infty}$.

Remark 1. *Method 1 is the well-known Newton's method for solving nonlinear equation [1,2].*

3.2. Iterative Method Based on [1,1]-Order Padé Approximation

Secondly, when $m = 1, n = 1$, we think about $[1,1]$-order Padé approximation of function $f(x)$. Similarly, it follows from the expression (2) that

$$f(x) \approx R_{1,1}(x) = \frac{\begin{vmatrix} T_1(x) & (x - x_0)T_0(x) \\ a_2 & a_1 \end{vmatrix}}{\begin{vmatrix} 1 & (x - x_0) \\ a_2 & a_1 \end{vmatrix}}.$$

Let $R_{1,1}(x) = 0$, then we get

$$a_0 a_1 + a_1^2 (x - x_0) - a_0 a_2 (x - x_0) = 0. \tag{5}$$

Due to the determinant $|A_{1,1}| \neq 0$, that is,

$$\begin{vmatrix} a_1 & a_0 \\ a_2 & a_1 \end{vmatrix} = \begin{vmatrix} f'(x_0) & f(x_0) \\ \frac{f''(x_0)}{2} & f'(x_0) \end{vmatrix} = f'^2(x_0) - f(x_0)\frac{f''(x_0)}{2} \neq 0.$$

Thus, we obtain the following equality from Equation (5):

$$x = x_0 - \frac{a_0 a_1}{a_1^2 - a_0 a_2}.$$

Combining $a_0 = f(x_0)$, $a_1 = f'(x_0)$, $a_2 = \frac{1}{2}f''(x_0)$, gives Halley iterative method as follows.

Method 2. *Assume that the function $f : I \to D$ has its second derivative at the point $x_0 \in I$. Then we obtain the following iterative method based on $[1,1]$-order Padé approximation of function $f(x)$:*

$$x_{k+1} = x_k - \frac{2f(x_k)f'(x_k)}{2f'^2(x_k) - f(x_k)f''(x_k)}, k = 0, 1, 2, \ldots. \tag{6}$$

Starting with an initial approximation x_0 that is sufficiently close to the root η and applying the above scheme (6), we can obtain the iterative sequence $\{x_i\}_{i=0}^{\infty}$.

Remark 2. *Method 2 is the classical Halley's method for finding roots of nonlinear equation [1,2], which converges cubically.*

3.3. Iterative Method Based on [1,2]-Order Padé Approximation

Thirdly, when $m = 1, n = 2$, we take into account $[1,2]$-order Padé approximation of function $f(x)$. By the same manner, it follows from the expression (2) that

$$f(x) \approx R_{1,2}(x) = \frac{\begin{vmatrix} T_1(x) & (x - x_0)T_0(x) & 0 \\ a_2 & a_1 & a_0 \\ a_3 & a_2 & a_1 \end{vmatrix}}{\begin{vmatrix} 1 & x - x_0 & (x - x_0)^2 \\ a_2 & a_1 & a_0 \\ a_3 & a_2 & a_1 \end{vmatrix}}.$$

Let $R_{1,2}(x) = 0$, then one has

$$a_0 a_1^2 + a_0^2 a_2 + (a_1^3 - 2a_0 a_1 a_2 + a_0^2 a_3)(x - x_0) = 0. \tag{7}$$

Due to the determinant $|A_{1,2}| \neq 0$, that is,

$$
\begin{vmatrix} a_1 & a_0 & 0 \\ a_2 & a_1 & a_0 \\ a_3 & a_2 & a_1 \end{vmatrix} = \begin{vmatrix} f'(x_0) & f(x_0) & 0 \\ \frac{f''(x_0)}{2} & f'(x_0) & f(x_0) \\ \frac{f'''(x_0)}{6} & \frac{f''(x_0)}{2} & f'(x_0) \end{vmatrix} = f'^3(x_0) - f(x_0)f'(x_0)f''(x_0) + \frac{f^2(x_0)f'''(x_0)}{6} \neq 0.
$$

Thus, we gain the following equality from Equation (7):

$$
x = x_0 - \frac{a_0 a_1^2 - a_0^2 a_2}{a_1^3 - 2a_0 a_1 a_2 + a_0^2 a_3}.
$$

Substituting $a_0 = f(x_0), a_1 = f'(x_0), a_2 = \frac{1}{2}f''(x_0)$, and $a_3 = \frac{1}{6}f'''(x_0)$ into the above equation gives a single-step iterative method as follows.

Method 3. *Assume that the function $f : I \rightarrow D$ has its third derivative at the point $x_0 \in I$. Then we obtain the following iterative method based on $[1,2]$-order Padé approximation of function $f(x)$:*

$$
x_{k+1} = x_k - \frac{3f(x_k)\left(2f'^2(x_k) - f(x_k)f''(x_k)\right)}{6f'^3(x_k) - 6f(x_k)f'(x_k)f''(x_k) + f^2(x_k)f'''(x_k)}, k = 0, 1, 2, \ldots. \tag{8}
$$

Starting with an initial approximation x_0 that is sufficiently close to the root η and applying the above scheme (8), we can receive the iterative sequence $\{x_i\}_{i=0}^{\infty}$.

Remark 3. *Method 3 could be used to find roots of nonlinear equation. Clearly, for the sake of applying this iterative method, we must compute the second derivative and the third derivative of the function $f(x)$, which may generate inconvenience. In order to overcome the drawback, we suggest approximants of the second derivative and the third derivative, which is a very important idea and plays a significant part in developing some iterative methods free from calculating the higher derivatives.*

3.4. Modified Iterative Method Based on Approximant of the Third Derivative

In fact, we let $z_k = x_k - \frac{f(x_k)}{f'(x_k)}$. Then expanding $f(z_k)$ into third Taylor's series about the point x_k yields

$$
f(z_k) \approx f(x_k) + f'(x_k)(z_k - x_k) + \frac{1}{2!}f''(x_k)(z_k - x_k)^2 + \frac{1}{3!}f'''(x_k)(z_k - x_k)^3,
$$

which follows that

$$
f'''(x_k) \approx \frac{3f^2(x_k)f'(x_k)f''(x_k) - 6f(z_k)f'^3(x_k)}{f^3(x_k)}. \tag{9}
$$

Substituting (9) into (8), we can have the following iterative method.

Method 4. *Assume that the function $f : I \rightarrow D$ has its second derivative about the point $x_0 \in I$. Then we possess a modified iterative method as below:*

$$
\begin{cases} z_k = x_k - \frac{f(x_k)}{f'(x_k)}, \\ x_{k+1} = x_k - \frac{x_k - z_k}{1 + 2f(z_k)f'^2(x_k)L^{-1}(x_k)}, \end{cases} k = 0, 1, 2, \ldots, \tag{10}
$$

where $L(x_k) = f(x_k)\left(f(x_k)f''(x_k) - 2f'^2(x_k)\right)$. Starting with an initial approximation x_0 that is sufficiently close to the root η and using the above scheme (10), we can have the iterative sequence $\{x_i\}_{i=0}^{\infty}$.

Remark 4. *Methods 4 is a two-step iterative method free from third derivative of the function.*

3.5. Modified Iterative Method Based on Approximant of the Second Derivative

It is obvious that the iterative method (10) requires the operation of the second derivative of the function $f(x)$. In order to avoid computing the second derivative, we introduce an approximant of the second derivative by using Taylor's series.

Similarly, expanding $f(z_k)$ into second Taylor's series about the point x_k yields

$$f(z_k) \approx f(x_k) + (z_k - x_k)f'(x_k) + \frac{1}{2!}(z_k - x_k)^2 f''(x_k),$$

which means

$$f''(x_k) \approx \frac{2f(z_k)f'^2(x_k)}{f^2(x_k)}. \tag{11}$$

Using (11) in (10), we can get the following modified iterative method without computing second derivative.

Method 5. *Assume that the function $f : I \to D$ has its first derivative about the point $x_0 \in I$. Then we have a modified iterative method as below:*

$$\begin{cases} z_k = x_k - \frac{f(x_k)}{f'(x_k)}, \\ x_{k+1} = x_k - \frac{f(x_k)-f(z_k)}{f(x_k)-2f(z_k)}(x_k - z_k), \quad k = 0, 1, 2, \ldots. \end{cases} \tag{12}$$

Starting with an initial approximation x_0 that is sufficiently close to the root η and using the above scheme (12), we can obtain the iterative sequence $\{x_i\}_{i=0}^{\infty}$.

Remark 5. *Method 5 is another two-step iterative method. It is clear that Method 5 does not require to calculate the high-order derivative. But more importantly, the characteristic of Method 5 is that per iteration it requires two evaluations of the function and one of its first-order derivative. The efficiency of this method is better than that of the well-known other methods involving the second-order derivative of the function.*

4. Convergence Analysis of Iterative Methods

Theorem 1. *Suppose that $f(x)$ is a function whose n-th derivative $f^{(n)}(x), n = 1, 2, 3, \ldots$, exists in a neighborhood of its root η with $f'(\eta) \neq 0$. If the initial approximation x_0 is sufficiently close to η, then the Method 3 defined by (8) is fourth-order convergent.*

Proof of Theorem 1. By the hypothesis $f(\eta) = 0$ and $f'(\eta) \neq 0$, we know that η is an unique single root of the equation $f(x) = 0$. So, for each positive integer $n \geq 1$, we have that the derivatives of high orders $f^{(n)}(\eta) \neq 0$. Considering the iterative scheme (8) in Method 3, we denote its corresponding iterative function as shown below:

$$\varphi(x) = x - \frac{3f(x)\left(2f'^2(x) - f(x)f''(x)\right)}{6f'^3(x) - 6f(x)f'(x)f''(x) + f^2(x)f'''(x)}. \tag{13}$$

By calculating the first and high-order derivatives of the iterative function $\varphi(x)$ with respect to x at the point η, we verify that

$$\varphi'(\eta) = 0, \quad \varphi''(\eta) = 0, \quad \varphi'''(\eta) = 0$$

and

$$\varphi^{(4)}(\eta) = \frac{3f''^3(\eta) - 4f'(\eta)f''(\eta)f'''(\eta) + f'^2(\eta)f^{(4)}(\eta)}{f'^3(\eta)} \neq 0.$$

Thus, it follows from Lemma 2 that Method 3 defined by (8) is fourth-order convergent. This completes the proof. □

Theorem 2. *Suppose that $f(x)$ is a function whose n-th derivative $f^{(n)}(x), n = 1,2,3,\ldots$, exists in a neighborhood of its root η with $f'(\eta) \neq 0$. If the initial approximation x_0 is sufficiently close to η, then the Method 4 defined by (10) is at least fourth-order convergent with the following error equation*

$$e_{k+1} = (b_2^3 - 2b_2b_3)e_k^4 + O(e_k^5),$$

where $e_k = x_k - \eta, k = 1,2,3,\ldots$, and the constants $b_n = \frac{a_n}{f'(\eta)}, a_n = \frac{f^{(n)}(\eta)}{n!}, n = 1,2,3,\ldots.$

Proof of Theorem 2. By the hypothesis, it is clear to see that η is an unique single root of the equation $f(x) = 0$. By expanding $f(x_k)$, $f'(x_k)$ and $f''(x_k)$ into Taylor's series about η, we obtain

$$\begin{aligned} f(x_k) &= e_k f'(\eta) + \frac{e_k^2}{2!}f''(\eta) + \frac{e_k^3}{3!}f'''(\eta) + \frac{e_k^4}{4!}f^{(4)}(\eta) + \frac{e_k^5}{5!}f^{(5)}(\eta) + \frac{e_k^6}{6!}f^{(6)}(\eta) + O(e_k^7) \\ &= f'(\eta)\left(b_1 e_k + b_2 e_k^2 + b_3 e_k^3 + b_4 e_k^4 + b_5 e_k^5 + b_6 e_k^6 + O(e_k^7)\right), \end{aligned} \tag{14}$$

$$f'(x_k) = f'(\eta)\left(b_1 + 2b_2 e_k + 3b_3 e_k^2 + 4b_4 e_k^3 + 5b_5 e_k^4 + 6b_6 e_k^5 + O(e_k^6)\right) \tag{15}$$

and

$$f''(x_k) = f'(\eta)\left(2b_2 + 6b_3 e_k + 12b_4 e_k^2 + 20b_5 e_k^3 + 30b_6 e_k^4 + O(e_k^5)\right), \tag{16}$$

where $b_n = \frac{1}{n!}\frac{f^{(n)}(\eta)}{f'(\eta)}, n = 1,2,\cdots$. Clearly, $b_1 = 1$. Dividing (14) by (15) directly, gives us

$$\begin{aligned} \frac{f(x_k)}{f'(x_k)} &= x_k - z_k = e_k - b_2 e_k^2 - 2(b_3 - b_2^2)e_k^3 - (4b_2^3 + 3b_4 - 7b_2b_3)e_k^4 \\ &\quad - 2(10b_2^2 b_3 - 2b_5 + 5b_2 b_4 + 4b_2^4 + 3b_3^2)e_k^5 \\ &\quad - (16b_2^5 + 28b_2^2 b_4 + 33b_2 b_3^2 + 5b_6 - 52b_2^3 b_3 - 17b_3 b_4 - 13b_2 b_5)e_k^6 + O(e_k^7). \end{aligned} \tag{17}$$

By substituting (17) into (10) in Method 4, one has

$$\begin{aligned} z_k &= \eta + b_2 e_k^2 + 2(b_3 - b_2^2)e_k^3 + (4b_2^3 + 3b_4 - 7b_2b_3)e_k^4 \\ &\quad + 2(10b_2^2 b_3 - 2b_5 + 5b_2 b_4 + 4b_2^4 + 3b_3^2)e_k^5 \\ &\quad + (16b_2^5 + 28b_2^2 b_4 + 33b_2 b_3^2 + 5b_6 - 52b_2^3 b_3 - 17b_3 b_4 - 13b_2 b_5)e_k^6 + O(e_k^7). \end{aligned} \tag{18}$$

Again, expanding $f(z_k)$ by Taylor's series about η, we have

$$\begin{aligned} f(z_k) &= f'(\eta)\left(b_2 e_k^2 - 2(b_2^2 - b_3)e_k^3 - (7b_2b_3 - 5b_2^3 - 3b_4)e_k^4\right. \\ &\quad - 2(5b_2 b_4 + 6b_2^4 + 3b_3^2 - 12b_2^2 b_3 - 2b_5)e_k^5 \\ &\quad \left. + (28b_2^5 + 34b_2^2 b_4 + 37b_2 b_3^2 + 5b_6 - 73b_2^3 b_3 - 17b_3 b_4 - 13b_2 b_5)e_k^6 + O(e_k^7)\right). \end{aligned} \tag{19}$$

Hence, from (15) and (19), we have

$$\begin{aligned} f(z_k)f'^2(x_k) &= f'^3(\eta)\left(b_2 e_k^2 + 2(b_2^2 + b_3)e_k^3 + (7b_2b_3 + b_2^3 + 3b_4)e_k^4 + 2(5b_2 b_4 + 2b_2^2 b_3 + 3b_3^2 + 2b_5)e_k^5\right. \\ &\quad \left. + (4b_2 b_3^2 + 6b_2^2 b_4 + b_2^3 b_3 + 13b_2 b_5 + 17b_3 b_4 + 5b_6)e_k^6 + O(e_k^7)\right). \end{aligned} \tag{20}$$

Also, from (14), (15), and (16), one has

$$
\begin{aligned}
L(x_k) = \quad & -2f'^3(\eta)\left(e_k + 4b_2 e_k^2 + 2(3b_2^2 + 2b_3)e_k^3 + (14b_2 b_3 + 3b_2^3 + 3b_4)e_k^4\right. \\
& + (14b_2 b_4 + 11b_2^2 b_3 + 9b_3^2 + b_5)e_k^5 \\
& \left. + 2(7b_2 b_3^2 + 6b_2^2 b_4 + 6b_2 b_5 + 10b_3 b_4 + b_6)e_k^6 + O(e_k^7)\right).
\end{aligned}
\tag{21}
$$

Therefore, combining (20) and (21), one can have

$$
\begin{aligned}
\frac{2f(z_k)f'^2(x_k)}{L(x_k)} = \quad & -b_2 e_k - 2(b_3 - b_2^2)e_k^2 - (3b_2^3 + 3b_4 - 5b_2 b_3)e_k^3 \\
& - (6b_2^2 b_3 + 4b_5 - 5b_2 b_4 - 3b_2^4 - 2b_3^2)e_k^4 + O(e_k^5).
\end{aligned}
\tag{22}
$$

Furthermore, from (17) and (22), we get

$$
\frac{x_k - y_k}{1 + 2f(z_k)f'^2(x_k)L^{-1}(x_k)} = e_k - (b_2^3 - 2b_2 b_3)e_k^4 - (12b_2^2 b_3 - 5b_2 b_4 - 4b_2^4 - 4b_3^2)e_k^5 + O(e_k^6).
\tag{23}
$$

So, substituting (23) into (10) in Method 4, one obtains

$$
x_{k+1} = \eta + (b_2^3 - 2b_2 b_3)e_k^4 + O(e_k^5).
\tag{24}
$$

Noticing that the $(k+1)$-st error $e_{k+1} = x_{k+1} - \eta$, from (24) we have the following error equation

$$
e_{k+1} = (b_2^3 - 2b_2 b_3)e_k^4 + O(e_k^5),
\tag{25}
$$

which shows that Method 4 defined by (10) is at least fourth-order convergent according to Definition 4. We have shown Theorem 2. □

Theorem 3. *Suppose that $f(x)$ is a function whose n-th derivative $f^{(n)}(x), n = 1, 2, 3, \ldots$, exists in a neighborhood of its root η with $f'(\eta) \neq 0$. If the initial approximation x_0 is sufficiently close to η, then the Method 5 defined by (12) is also at least fourth-order convergent with the following error equation*

$$
e_{k+1} = (b_2^3 - b_2 b_3)e_k^4 + O(e_k^5),
$$

where $e_k = x_k - \eta, k = 1, 2, 3, \ldots$, and the constants $b_n = \frac{a_n}{f'(\eta)}, a_n = \frac{f^{(n)}(\eta)}{n!}, n = 1, 2, 3, \ldots$.

Proof of Theorem 3. Referring to (14) and (19) in the proof of Theorem 2, then, dividing $f(z_k)$ by $f(x_k) - f(z_k)$ we see that

$$
\begin{aligned}
\frac{f(z_k)}{f(x_k) - f(z_k)} = \quad & b_2 e_k - 2(b_2^2 - b_3)e_k^2 - 3(2b_2 b_3 - b_2^3 - b_4)e_k^3 - (3b_2^4 + 4b_3^2 + 8b_2 b_4 - 11b_2^2 b_3 - 4b_5)e_k^4 \\
& - (10b_2^3 b_3 + 10b_3 b_4 + 10b_2 b_5 - 11b_2^2 b_3^2 - 14b_2^2 b_4 - 5b_6)e_k^5 \\
& - (221b_2^4 b_3 + 16b_3^3 + 78b_2 b_3 b_4 + 27b_2^2 b_5 - 73b_2^6 \\
& - 158b_2^2 b_3^2 - 91b_3^3 b_4 - 6b_4^2 - 10b_3 b_5 - 4b_2 b_6)e_k^6 + O(e_k^7).
\end{aligned}
\tag{26}
$$

From (26), we obtain

$$
\begin{aligned}
\frac{f(x_k) - f(z_k)}{f(x_k) - 2f(z_k)} = \quad & \frac{1}{1 - \frac{f(z_k)}{f(x_k) - f(z_k)}} = 1 + b_2 e_k + (2b_3 - b_2^2)e_k^2 + (3b_4 - 2b_2 b_3)e_k^3 \\
& + (2b_2^4 + 4b_5 - 3b_2^2 b_3 - 2b_2 b_4)e_k^4 + (14b_2^3 b_3 + 2b_3 b_4 + 5b_6 - 5b_2^5 \\
& - 9b_2 b_3^2 - 5b_2^2 b_4 - 2b_2 b_5)e_k^5 + (77b_2^6 + 192b_2^2 b_3^2 + 121b_2^3 b_4 + 15b_4^2 \\
& + 26b_3 b_5 + 14b_2 b_6 - 240b_2^4 b_3 - 24b_3^3 - 130b_2 b_3 b_4 - 51b_2^2 b_5)e_k^6 + O(e_k^7).
\end{aligned}
\tag{27}
$$

Multiplying (27) by (17) yields that

$$\frac{f(x_k) - f(y_k)}{f(x_k) - 2f(y_k)}(x_k - z_k) = e_k - (b_2^3 - b_2 b_3)e_k^4 + 2(2b_2^4 - 4b_2^2 b_3 + b_3^2 + b_2 b_4)e_k^5 + O(e_k^6). \tag{28}$$

Consequently, from (12) and the above Equation (28), and noticing the error $e_{k+1} = x_{k+1} - \eta$, we get the error equation as below:

$$e_{k+1} = (b_2^3 - b_2 b_3)e_k^4 + O(e_k^5). \tag{29}$$

Thus, according to Definition 4, we have shown that Method 5 defined by (12) has fourth-order convergence at least. This completes the proof of Theorem 3. □

Remark 6. *Per iteration, Method 5 requires two evaluations of the function and one of its first-order derivative. If we consider the definition of efficiency index [3] as $\sqrt[\tau]{\lambda}$, where λ is the order of convergence of the method and τ is the total number of new function evaluations (i.e., the values of f and its derivatives) per iteration, then Method 5 has the efficiency index equal to $\sqrt[3]{4} \approx 1.5874$, which is better than the ones of Halley iterative method $\sqrt[3]{3} \approx 1.4423$ and Newton iterative method $\sqrt{2} \approx 1.4142$.*

5. Numerical Results

In this section, we present the results of numerical calculations to compare the efficiency of the proposed iterative methods (Methods 3–5) with Newton iterative method (Method 1, NIM for short), Halley iterative method (Method 2, HIM for short) and a few classical variants defined in literatures [19–22], such as the next iterative schemes with fourth-order convergence:

(i)   Kou iterative method (KIM for short) [19].

$$x_{k+1} = x_k - \frac{2}{1 + \sqrt{1 - 2\bar{L}_f(x_k)}}\frac{f(x_k)}{f'(x_k)}, \qquad k = 0, 1, 2, \ldots,$$

where $\bar{L}_f(x_k)$ is defined by the equation as follows:

$$\bar{L}_f(x_k) = \frac{f''\left(x_k - f(x_k)/(3f'(x_k))\right)f(x_k)}{f'^2(x_k)}.$$

(ii) Double-Newton iterative method (DNIM for short) [20].

$$\begin{cases} z_k = x_k - \frac{f(x_k)}{f'(x_k)}, \\ x_{k+1} = x_k - \frac{f(x_k)}{f'(x_k)} - \frac{f(z_k)}{f'(z_k)}, \qquad k = 0, 1, 2, \ldots. \end{cases}$$

(iii) Chun iterative method (CIM for short) [21].

$$\begin{cases} z_k = x_k - \frac{f(x_k)}{f'(x_k)}, \\ x_{k+1} = x_k - \frac{f(x_k)}{f'(x_k)} - \left(1 + 2\frac{f(z_k)}{f(x_k)} + \frac{f^2(z_k)}{f^2(x_k)}\right)\frac{f(z_k)}{f'(x_k)}, \qquad k = 0, 1, 2, \ldots. \end{cases}$$

(iv) Jarratt-type iterative method (JIM for short) [22].

$$\begin{cases} z_k = x_k - \frac{2}{3}\frac{f(x_k)}{f'(x_k)}, \\ x_{k+1} = x_k - \frac{4f(x_k)}{f'(x_k) + 3f'(z_k)}\left(1 + \frac{9}{16}\left(\frac{f'(z_k)}{f'(x_k)} - 1\right)^2\right), \qquad k = 0, 1, 2, \ldots. \end{cases}$$

In iterative process, we use the following stopping criteria for computer programs:

$$|x_{k+1} - x_k| < \varepsilon \quad and \quad |f(x_{k+1})| < \varepsilon,$$

where the fixed tolerance ε is taken as 10^{-14}. When the stopping criteria are satisfied, x_{k+1} can be regarded as the exact root η of the equation. Numerical experiments are performed in Mathematica 10 environment with 64 digit floating point arithmetics (Digits: =64). Different test equations $f_i = 0$, $i = 1, 2, \ldots, 5$, the initial guess value x_0, the number of iterations $k + 1$, the approximate root x_{k+1}, the values of $|x_{k+1} - x_k|$ and $|f(x_{k+1})|$ are given in Table 1. The following test equations are used in the numerical results:

$$f_1(x) = x^3 - 11 = 0,$$
$$f_2(x) = \cos x - x = 0,$$
$$f_3(x) = x^3 + 4x^2 - 25 = 0,$$
$$f_4(x) = x^2 - e^x - 3x + 2 = 0,$$
$$f_5(x) = (x + 2)e^x - 1 = 0.$$

Table 1. Numerical results and comparison of various iterative methods.

Methods	Equation	x_0	$k+1$	x_{k+1}	$\lvert x_{k+1} - x_k\rvert$	$\lvert f(x_{k+1})\rvert$
NIM	$f_1 = 0$	1.5	7	2.2239800905693155211653633767221 5719652	1.1×10^{-25}	4.1×10^{-47}
HIM	$f_1 = 0$	1.5	5	2.2239800905693155211653633767221 5719652	1.7×10^{-41}	1.0×10^{-46}
Method 3	$f_1 = 0$	1.5	4	2.2239800905693155211653633767221 5719652	8.3×10^{-40}	1.6×10^{-48}
Method 4	$f_1 = 0$	1.5	4	2.2239800905693155211653633767221 5719652	8.3×10^{-22}	1.9×10^{-47}
Method 5	$f_1 = 0$	1.5	4	2.2239800905693155211653633767221 5719652	7.5×10^{-30}	7.4×10^{-45}
KIM	$f_1 = 0$	1.5	4	2.2239800905693155211653633767221 5719652	8.5×10^{-38}	3.9×10^{-48}
DNIM	$f_1 = 0$	1.5	4	2.2239800905693155211653633767221 5719652	1.1×10^{-25}	1.1×10^{-47}
CIM	$f_1 = 0$	1.5	5	2.2239800905693155211653633767221 5719652	1.5×10^{-41}	6.6×10^{-45}
JIM	$f_1 = 0$	1.5	5	2.2239800905693155211653633767221 5719652	1.2×10^{-45}	4.3×10^{-47}
NIM	$f_2 = 0$	1	5	0.7390851332151606416553120876738 7340401	6.4×10^{-21}	1.5×10^{-41}
HIM	$f_2 = 0$	1	4	0.7390851332151606416553120876738 7340401	3.4×10^{-29}	5.1×10^{-49}
Method 3	$f_2 = 0$	1	3	0.7390851332151606416553120876738 7340401	8.2×10^{-19}	7.5×10^{-49}
Method 4	$f_2 = 0$	1	3	0.7390851332151606416553120876738 7340401	1.4×10^{-17}	9.4×10^{-48}
Method 5	$f_2 = 0$	1	3	0.7390851332151606416553120876738 7340401	1.1×10^{-18}	7.5×10^{-47}
KIM	$f_2 = 0$	1	3	0.7390851332151606416553120876738 7340401	1.5×10^{-20}	8.3×10^{-49}
DNIM	$f_2 = 0$	1	3	0.7390851332151606416553120876738 7340401	6.4×10^{-21}	9.5×10^{-48}
CIM	$f_2 = 0$	1	3	0.7390851332151606416553120876738 7340401	2.2×10^{-17}	9.4×10^{-48}
JIM	$f_2 = 0$	1	3	0.7390851332151606416553120876738 7340401	7.4×10^{-18}	8.3×10^{-49}
NIM	$f_3 = 0$	3.5	7	2.0352684811819591535475504154736 1249916	6.4×10^{-28}	2.9×10^{-47}
HIM	$f_3 = 0$	3.5	5	2.0352684811819591535475504154736 1249916	2.0×10^{-39}	5.8×10^{-47}
Method 3	$f_3 = 0$	3.5	4	2.0352684811819591535475504154736 1249916	2.0×10^{-33}	6.0×10^{-47}
Method 4	$f_3 = 0$	3.5	4	2.0352684811819591535475504154736 1249916	2.0×10^{-33}	8.0×10^{-46}
Method 5	$f_3 = 0$	3.5	4	2.0352684811819591535475504154736 1249916	3.4×10^{-30}	3.1×10^{-45}
KIM	$f_3 = 0$	3.5	4	2.0352684811819591535475504154736 1249916	4.3×10^{-33}	8.6×10^{-47}
DNIM	$f_3 = 0$	3.5	4	2.0352684811819591535475504154736 1249916	6.4×10^{-28}	9.9×10^{-46}
CIM	$f_3 = 0$	3.5	4	2.0352684811819591535475504154736 1249916	1.1×10^{-20}	9.6×10^{-46}
JIM	$f_3 = 0$	3.5	4	2.0352684811819591535475504154736 1249916	1.9×10^{-22}	1.1×10^{-49}

Table 1. *Cont.*

Methods	Equation	x_0	$k+1$	x_{k+1}	$|x_{k+1} - x_k|$	$|f(x_{k+1})|$
NIM	$f_4 = 0$	3.6	8	0.257530285439860760455367304937724178138	6.5×10^{-29}	3.5×10^{-46}
HIM	$f_4 = 0$	3.6	6	0.257530285439860760455367304937724178138	4.8×10^{-37}	1.6×10^{-46}
Method 3	$f_4 = 0$	3.6	4	0.257530285439860760455367304937724178138	9.6×10^{-14}	2.7×10^{-46}
Method 4	$f_4 = 0$	3.6	5	0.257530285439860760455367304937724178138	1.1×10^{-36}	3.3×10^{-44}
Method 5	$f_4 = 0$	3.6	4	0.257530285439860760455367304937724178138	2.5×10^{-19}	4.9×10^{-44}
KIM	$f_4 = 0$	3.6	5	0.257530285439860760455367304937724178138	2.1×10^{-14}	8.9×10^{-44}
DNIM	$f_4 = 0$	3.6	4	0.257530285439860760455367304937724178138	2.6×10^{-14}	3.5×10^{-46}
CIM	$f_4 = 0$	3.6	4	0.257530285439860760455367304937724178138	2.8×10^{-12}	2.8×10^{-46}
JIM	$f_4 = 0$	3.6	5	0.257530285439860760455367304937724178138	9.7×10^{-38}	9.7×10^{-46}
NIM	$f_5 = 0$	3.5	11	$-0.442854401002388583141327999999933681972$	8.2×10^{-22}	7.7×10^{-43}
HIM	$f_5 = 0$	3.5	7	$-0.442854401002388583141327999999933681972$	2.2×10^{-37}	6.1×10^{-45}
Method 3	$f_5 = 0$	3.5	5	$-0.442854401002388583141327999999933681972$	1.8×10^{-24}	3.4×10^{-45}
Method 4	$f_5 = 0$	3.5	5	$-0.442854401002388583141327999999933681972$	5.3×10^{-37}	7.9×10^{-44}
Method 5	$f_5 = 0$	3.5	6	$-0.442854401002388583141327999999933681972$	2.0×10^{-42}	3.9×10^{-42}
KIM	$f_5 = 0$	3.5	7	$-0.442854401002388583141327999999933681972$	3.6×10^{-23}	2.7×10^{-42}
DNIM	$f_5 = 0$	3.5	6	$-0.442854401002388583141327999999933681972$	8.2×10^{-22}	4.9×10^{-45}
CIM	$f_5 = 0$	3.5	7	$-0.442854401002388583141327999999933681972$	3.3×10^{-37}	8.6×10^{-44}
JIM	$f_5 = 0$	3.5	6	$-0.442854401002388583141327999999933681972$	9.3×10^{-13}	6.6×10^{-46}

6. Conclusions

In Section 3 of the paper, it is evident that we have obtained a few single-step iterative methods including classical Newton's method and Halley's method, based on $[1, n]$-order Padé approximation of a function for finding a simple root of nonlinear equations. In order to avoid calculating the higher derivatives of the function, we have tried to improve the proposed iterative method by applying approximants of the second derivative and the third derivative. Hence, we have gotten a few modified two-step iterative methods free from the higher derivatives of the function. In Section 4, we have given theoretical proofs of the several methods. It is seen that any modified iterative method reaches the convergence order 4. However, it is worth mentioning that Method 5 is free from second order derivative and its efficiency index is 1.5874. Furthermore, in Section 5, numerical examples are employed to illustrate the practicability of the suggested variants for finding the approximate roots of some nonlinear scalar equations. The computational results presented in Table 1 show that in almost all of the cases the presented variants converge more rapidly than Newton iterative method and Halley iterative method, so that they can compete with Newton iterative method and Halley iterative method. Finally, for more nonlinear equations we tested, the presented variants have at least equal performance compared to the other existing iterative methods that are of the same order.

Author Contributions: The contributions of all of the authors have been similar. All of them have worked together to develop the present manuscript.

Funding: This research was funded by the Natural Science Key Foundation of Education Department of Anhui Province (Grant No. KJ2013A183), the Project of Leading Talent Introduction and Cultivation in Colleges and Universities of Education Department of Anhui Province (Grant No. gxfxZD2016270) and the Incubation Project of National Scientific Research Foundation of Bengbu University (Grant No. 2018GJPY04).

Acknowledgments: The authors are thankful to the anonymous reviewers for their valuable comments.

Conflicts of Interest: The authors declare no conflict of interest.

References

1. Alfio, Q.; Riccardo, S.; Fausto, S. Rootfinding for nonlinear equations. In *Numerical Mathematics*; Springer: New York, NY, USA, 2000; pp. 251–285, ISBN 0-387-98959-5.
2. Burden, A.M.; Faires, J.D.; Burden, R.L. Solutions of equations in one variable. In *Numerical Analysis*, 10th ed.; Cengage Learning: Boston, MA, USA, 2014; pp. 48–101, ISBN 978-1-305-25366-7.

3. Gautschi, W. Nonlinear equations. In *Numerical Analysis*, 2nd ed.; Birkhäuser: Boston, MA, USA, 2011; pp. 253–323, ISBN 978-0-817-68258-3.
4. Weerakoon, S.; Fernando, T.G.I. A variant of Newton's method with accelerated third-order convergence. *Appl. Math. Lett.* **2000**, *13*, 87–93. [CrossRef]
5. Cordero, A.; Torregrosa, I.R. Variants of Newton's method using fifth-order quadrature formulas. *Appl. Math. Comput.* **2007**, *190*, 686–698. [CrossRef]
6. Chun, C. Iterative methods improving Newton's method by the decomposition method. *Comput. Math. Appl.* **2005**, *50*, 1559–1568. [CrossRef]
7. Li, S.; Tan, J.; Xie, J.; Dong, Y. A new fourth-order convergent iterative method based on Thiele's continued fraction for solving equations. *J. Inform. Comput. Sci.* **2011**, *8*, 139–145.
8. Li, S.; Wang, R.; Zhang, X.; Xie, J.; Dong, Y. Halley's iterative formula based on Padé approximation and its modifications. *J. Inform. Comput. Sci.* **2012**, *9*, 997–1004.
9. Abbasbandy, S.; Tan, Y.; Liao, S.J. Newton-homotopy analysis method for nonlinear equations. *Appl. Math. Comput.* **2007**, *188*, 1794–1800. [CrossRef]
10. Noor, M.A.; Khan, W.A. New iterative methods for solving nonlinear equation by using homotopy perturbation method. *Appl. Math. Comput.* **2012**, *219*, 3565–3574. [CrossRef]
11. Wang, X.; Qin, Y.; Qian, W.; Zhang, S.; Fan, X. A family of Newton type iterative methods for solving nonlinear equations. *Algorithms* **2015**, *8*, 786–798. [CrossRef]
12. Sederberg, T.W.; Nishita, T. Curve intersection using Bézier clipping. *Comput.-Aided Des.* **1990**, *22*, 538–549. [CrossRef]
13. Bartoň, M.; Jüttler, B. Computing roots of polynomials by quadratic clipping. *Comput. Aided Geom. Des.* **2007**, *24*, 125–141. [CrossRef]
14. Morlando, F. A class of two-step Newton's methods with accelerated third-order convergence. *Gen. Math. Notes* **2015**, *29*, 17–26.
15. Thukral, R. New modification of Newton method with third-order convergence for solving nonlinear equations of type $f(0) = 0$. *Am. J. Comput. Appl. Math.* **2016**, *6*, 14–18. [CrossRef]
16. Rafiq, A.; Rafiullah, M. Some multi-step iterative methods for solving nonlinear equations. *Comput. Math. Appl.* **2009**, *58*, 1589–1597. [CrossRef]
17. Ali, F.; Aslam, W.; Ali, K.; Anwar, M.A.; Nadeem, A. New family of iterative methods for solving nonlinear models. *Disc. Dyn. Nat. Soc.* **2018**, 1–12. [CrossRef]
18. Qureshi, U.K. A new accelerated third-order two-step iterative method for solving nonlinear equations. *Math. Theory Model.* **2018**, *8*, 64–68.
19. Kou, J.S. Some variants of Cauchy's method with accelerated fourth-order convergence. *J. Comput. Appl. Math.* **2008**, *213*, 71–78. [CrossRef]
20. Traub, J.F. *Iterative Methods for the Solution of Equations*; Chelsea Publishing Company: New York, NY, USA, 1977; pp. 1–49.
21. Chun, C. Some fourth-order iterative methods for solving nonlinear equations. *Appl. Math. Comput.* **2008**, *195*, 454–459. [CrossRef]
22. Sharifi, M.; Babajee, D.K.R.; Soleymani, F. Finding the solution of nonlinear equations by a class of optimal methods. *Comput. Math. Appl.* **2012**, *63*, 764–774. [CrossRef]
23. Wuytack, L. Padé approximants and rational functions as tools for finding poles and zeros of analytical functions measured experimentally. In *Padé Approximation and its Applications*; Springer: New York, NY, USA, 1979; pp. 338–351, ISBN 0-387-09717-1.
24. Xu, X.Y.; Li, J.K.; Xu, G.L. Definitions of Padé approximation. In *Introduction to Padé Approximation*; Shanghai Science and Technology Press: Shanghai, China, 1990; pp. 1–8, ISBN 978-7-532-31887-2.

mathematics

MDPI

Article

Improving the Computational Efficiency of a Variant of Steffensen's Method for Nonlinear Equations

Fuad W. Khdhr [1], **Rostam K. Saeed** [1] and **Fazlollah Soleymani** [2,*]

[1] Department of Mathematics, College of Science, Salahaddin University, Erbil, Iraq;
     fuad.khdhr@su.edu.krd (F.W.K.); rostam.saeed@su.edu.krd (R.K.S.)
[2] Department of Mathematics, Institute for Advanced Studies in Basic Sciences (IASBS),
     Zanjan 45137-66731, Iran
*   Correspondence: fazlollah.soleymani@gmail.com

Received: 21 January 2019; Accepted: 18 March 2019; Published: 26 March 2019

Abstract: Steffensen-type methods with memory were originally designed to solve nonlinear equations without the use of additional functional evaluations per computing step. In this paper, a variant of Steffensen's method is proposed which is derivative-free and with memory. In fact, using an acceleration technique via interpolation polynomials of appropriate degrees, the computational efficiency index of this scheme is improved. It is discussed that the new scheme is quite fast and has a high efficiency index. Finally, numerical investigations are brought forward to uphold the theoretical discussions.

Keywords: iterative methods; Steffensen's method; R-order; with memory; computational efficiency

1. Introduction

One of the commonly encountered topics in computational mathematics is to tackle solving a nonlinear algebraic equation. The equation can be presented as in the scalar case $f(x) = 0$, or more complicated as a system of nonlinear algebraic equations. The procedure of finding the solutions (if it exists) cannot be done analytically. In some cases, the analytic techniques only give the real result while its complex zeros should be found and reported. As such, numerical techniques are a viable choice for solving such nonlinear problems. Each of the existing computational procedures has their own domain of validity with some pros and cons [1,2].

Two classes of methods with the use of derivatives and without the use of derivatives are known to be useful depending on the application dealing with [3]. In the derivative-involved methods, a larger attraction basin along with a simple coding effort for higher dimensional problems is at hand which, in derivative-free methods, the area of choosing the initial approximations is smaller and extending to higher dimensional problems is via the application of a divided difference operator matrix, which is basically a dense matrix. However, the ease in not computing the derivative and, subsequently, the Jacobians, make the application of derivative-free methods more practical in several problems [4–7].

Here, an attempt is made at developing a computational method which is not only efficient in terms of the computational efficiency index, but also in terms of larger domains for the choice of the initial guesses/approximations for starting the proposed numerical method.

The Steffensen's method [8] for solving nonlinear scalar equations has quadratic convergence for simple zeros and given by:

$$
\begin{cases}
x_{k+1} = x_k - \dfrac{f(x_k)}{f[x_k, w_k]}, \\
w_k = x_k + \beta f(x_k),\ \beta \in R\backslash\{0\},\ k \geq 0,
\end{cases}
\tag{1}
$$

where the two-point divided difference is defined by:

$$f[x_k, w_k] = \frac{f(x_k) - f(w_x)}{x_k - w_k},$$

This scheme needs two function evaluations per cycle. Scheme (1) shows an excellent tool for constructing efficient iterative methods for nonlinear equations. This is because it is derivative-free with a free parameter. This parameter can, first of all, enlarge the attraction basins of Equation (1) or any of its subsequent methods and, second, can directly affect the improvement of the R-order of convergence and the efficiency index.

Recalling that Kung and Traub conjectured that the iterative method without memory based on m functions evaluation per iteration attain the optimal convergence of order 2^{m-1} [9,10].

The term "with memory" means that the values of the function associated with the computed approximations of the roots are used in subsequent iterations. This is unlike the term "without memory" in which the method only uses the current values to find the next estimate. As such, in a method with memory, the calculated results up to the desired numbers of iterations should be stored and then called to proceed.

Before proceeding the given idea to improve the speed of convergence, efficiency index, and the attraction basins, we provide a short literature by reviewing some of the existing methods with accelerated convergence order. Traub [11] proposed the following two-point method with memory of order 2.414:

$$\begin{cases} x_{k+1} = x_k - \dfrac{f(x_k)}{f[x_k, x_k + \beta_k f(x_k)]}, \\ \beta_k = \dfrac{-1}{f[x_k, z_{k-1}]}, \end{cases} \tag{2}$$

where $z_{k-1} = x_{k-1} + \beta_{k-1} f(x_{k-1})$, and $\beta_0 = -\mathrm{sign}(f'(x_0))$ or $-\dfrac{1}{f[x_0, x_0 + f(x_0)]}$. This is one of the pioneering and fundamental methods with memory for solving nonlinear equations.

Džunić in [12] suggested an effective bi-parametric iterative method with memory of $\frac{1}{2}\left(3 + \sqrt{17}\right)$ R-order of convergence as follows:

$$\begin{cases} w_k = x_k + \beta_k f(x_k), \\ \beta_k = -\dfrac{1}{N_2'(x_k)}, \quad \zeta_k = -\dfrac{N_3''(w_k)}{2N_3'(w_k)}, \quad k \geq 1, \\ x_{k+1} = x_k - \dfrac{f(x_k)}{f[x_k, w_k] + \zeta_k f(w_k)} \quad k \geq 0. \end{cases} \tag{3}$$

Moreover, Džunić and Petković [13] derived the following cubically convergent Steffensen-like method with memory:

$$\begin{cases} x_{k+1} = x_k - \dfrac{f(x_k)}{f[x_k, x_k + \beta_k f(x_k)]}, \\ \beta_k = \dfrac{-1}{f[x_k, z_{k-1}] + f[x_k, x_{k-1}] + f[x_{k-1}, z_{k-1}]}, \end{cases} \tag{4}$$

where $z_{k-1} = x_{k-1} + \beta_{k-1} f(x_{k-1})$ depending on the second-order Newton interpolation polynomial. Various Steffensen-type methods are proposed in [14–17].

In fact, it is possible to improve the performance of the aforementioned method by considering several more sub-steps and improve the computational efficiency index via multi-step iterative methods. However, this procedure is more computational burdensome. Thus, the motivation here is to know that is it possible to improve the performance of numerical methods in terms of the computational efficiency index, basins of attraction, and the rate of convergence without adding more sub-steps and propose a numerical method as a one-step solver.

Hence, the aim of this paper is to design a one-step method with memory which is quite fast and has an improved efficiency index, based on the modification of the one-step method of Steffensen (Equation (1)) and increase the convergence order to 3.90057 without any additional functional evaluations.

The rest of this paper is ordered as follows: In Section 2, we develop the one-point Steffensen-type iterative scheme (Equation (1)) with memory which was proposed by [18]. We present the main goal in Section 3 by approximating the acceleration parameters involved in our contributed scheme by Newton's interpolating polynomial and, thus, improve the convergence R-order. The numerical reports are suggested in Section 4 to confirm the theoretical results. Some discussions are given in Section 5.

2. An Iterative Method

The following iterative method without memory was proposed by [18]:

$$
\begin{cases}
w_k = x_k - \beta f(x_k), \\
x_{k+1} = x_k - \dfrac{f(x_k)}{f[x_k, w_k]}\left(1 + \zeta \dfrac{f(w_k)}{f[x_k, w_k]}\right), \ \zeta \in \mathbb{R},
\end{cases}
\tag{5}
$$

with the following error equation to improve the performance of (1) in terms of having more free parameters:

$$
e_{k+1} = -(-1 + \beta f'(\alpha))(c_2 - \zeta)e_k^2 + O\left(e_k^3\right),
\tag{6}
$$

where $c_i = \frac{1}{i!}\frac{f^{(i)}(\alpha)}{f'(\alpha)}$. Using the error Equation (6), to derive Steffensen-type iterative methods with memory, we calculate the following parameters: $\beta = \beta_k$, $\zeta = \zeta_k$, by the formula:

$$
\begin{cases}
\beta_k = \dfrac{1}{\overline{f}'(x_\alpha)}, \\
\zeta_k = \overline{c}_2,
\end{cases}
\tag{7}
$$

for $k = 1, 2, 3, \cdots$, while $\overline{f}'(x_\alpha)$, $\overline{c}_2$ are approximations to $f'(\alpha)$ and c_2, respectively; where α is a simple zero of $f(x)$. In fact, Equation (7) shows a way to minimize the asymptotic error constant of Equation (6) by making this coefficient closer and closer to zero when the iterative method is converging to the true solution.

The initial estimates β_0 and ζ_0 must be chosen before starting the process of iterations. We state the Newton's interpolating polynomial of fourth and fifth-degree passing through the saved points as follows:

$$
\begin{cases}
N_4(t) = N_4(t; x_k, w_{k-1}, x_{k-1}, w_{k-2}, x_{k-2}), \\
N_5(t) = N_5(t; w_k, x_k, w_{k-1}, x_{k-1}, w_{k-2}, x_{k-2}).
\end{cases}
\tag{8}
$$

Recalling that $N(t)$ is an interpolation polynomial for a given set of data points also known as the Newton's divided differences interpolation polynomial because the coefficients of the polynomial are calculated using Newton's divided differences method. For instance, here the set of data points for $N_4(t)$ are $\{\{x_k, f(x_k)\}, \{w_{k-1}, f(w_{k-1})\}, \{x_{k-1}, f(x_{k-1})\}, \{w_{k-2}, f(w_{k-2})\}, \{x_{k-2}, f(x_{k-2})\}\}$.

Now, using some modification on Equation (5) we present the following scheme:

$$
\begin{cases}
w_k = x_k - \beta_k f(x_k), \\
\beta_k = \dfrac{1}{N_4'(x_k)}, \ \zeta_k = \dfrac{N_5''(w_k)}{2N_5'(w_k)}, \ k \geq 2, \\
x_{k+1} = x_k - \dfrac{f(x_k)}{f[x_k, w_k]}\left(1 + \zeta_k \dfrac{f(w_k)}{f[x_k, w_k]}\right), \ k \geq 0.
\end{cases}
\tag{9}
$$

Noting that the accelerator parameters β_k, ζ_k are getting updated and then used in the iterative method right after the second iterations, viz, $k \geq 2$. This means that the third line of Equation (9) is imposed at the beginning and after that the computed values are stored and used in the subsequent iterates. For $k = 1$, the degree of Newton interpolation polynomials would be two and three. However, for $k \geq 2$, interpolations of degrees four and five as given in Equation (8) can be used to increase the convergence order.

Additionally speaking, this acceleration of convergence would be attained without the use any more functional evaluations as well as imposing more steps. Thus, the proposed scheme with memory (Equation (9)) can be attractive for solving nonlinear equations.

3. Convergence Analysis

In this section, we show the convergence criteria of Equation (9) using Taylor's series expansion and several extensive symbolic computations.

Theorem 1. *Let the function $f(x)$ be sufficiently differentiable in a neighborhood of its simple zero α. If an initial approximation x_0 is necessarily close to α. Then, R-order of convergence for the one-step method (Equation (9)) with memory is 3.90057.*

Proof. The proof is done using the definition of the error equation as the difference between the k-estimate and the exact zero along with symbolic computations. Let the sequence $\{x_k\}$ and $\{w_k\}$ have convergence orders r and p, respectively. Namely,

$$e_{k+1} \sim e_k^r, \tag{10}$$

and:

$$e_{w,k} \sim e_k^p, \tag{11}$$

Therefore, using Equations (10) and (11), we have:

$$e_{k+1} \sim e_k^r \sim e_{k-1}^{r^2} \sim e_{k-2}^{r^3}, \tag{12}$$

and:

$$e_{w,k} \sim e_k^p \sim (e_{k-1}^r)^p \sim e_{k-2}^{pr^2}. \tag{13}$$

The associated error equations to the accelerating parameters β_k and ζ_k for Equation (9) can now be written as follows:

$$e_{w,k} \sim (-1 + \beta_k f'(\alpha))e_k, \tag{14}$$

and:

$$e_{k+1} \sim -(-1 + \beta_k f'(\alpha))(c_2 - \zeta_k)e_k^2. \tag{15}$$

On the other hand, by using a symbolic language and extensive computations one can find the following error terms for the involved terms existing in the fundamental error Equation (6):

$$-1 + \beta_k f'(\alpha) \sim c_5 e_{k-2} e_{k-1} e_{w,k-1} e_{w,k-2}, \tag{16}$$

$$c_2 - \zeta_k \sim c_6 e_{k-2} e_{k-1} e_{w,k-1} e_{w,k-2} \tag{17}$$

Combining Equations (14)–(17), we get that:

$$e_{w,k} \sim e_{k-2}^{r^2 + pr + r + p + 1}, \tag{18}$$

$$e_{k+1} \sim e_{k-2}^{2(r^2 + pr + r + p + 1)}. \tag{19}$$

We now compare the left and right hand side of Equations (12)–(19) and Equations (13)–(18), respectively. Thus, we have the following nonlinear system of equations in order to find the final R-orders:

$$\begin{cases} r^2 p - (r^2 + pr + r + p + 1) = 0, \\ r^3 - 2(r^2 + pr + r + p + 1) = 0. \end{cases} \tag{20}$$

The positive real solution of (20) is $r = 3.90057$ and $p = 1.9502$. Therefore, the convergence R-order for Equation (9) is 3.90057. □

Since improving the convergence R-order is useless if the whole computational method is expensive, basically researcher judge on a new scheme based upon its computational efficiency index which is a tool in order to provide a trade-off between the whole computational cost and the attained R-order. Assuming the cost of calculating each functional evaluation is one, we can use the definition of efficiency index as $EI = p^{1/\theta}$, θ is the whole computational cost [19].

The computational efficiency index of Equation (9) is $3.90057^{\frac{1}{2}} \approx 1.97499 \approx 2$, which is clearly higher than efficiency index $2^{\frac{1}{2}} \approx 1.4142$ of Newton's and Steffensen's methods, $3.56155^{\frac{1}{2}} \approx 1.8872$ of (3) $3^{1/2} \approx 1.73205$ of Equation (4).

However, this improved computational efficiency is reported by ignoring the number of multiplication and division per computing cycle. By imposing a slight weight for such calculations one may once again obtain the improved computational efficiency of (9) in contrast to the existing schemes of the same type.

4. Numerical Computations

In this section, we compare the convergence performance of Equation (9), with three well-known iterative methods for solving four test problems numerically carried out in Mathematica 11.1. [20].

We denote Equations (1), (3), (5) and (9) with SM, DZ, PM, M4, respectively. We compare the our method with different methods, using $\beta_0 = 0.1$ and $\zeta_0 = 0.1$. Here, the computational order of convergence (coc) has been computed by the following formula [21]:

$$coc = \frac{\ln|(f(x_k)/f(x_{k-1})|}{\ln|(f(x_{k-1})/f(x_{k-2})|} \tag{21}$$

Recalling that using a complex initial approximation, one is able to find the complex roots of the nonlinear equations using (9).

Experiment 1. *Let us consider the following nonlinear test function:*

$$f_1(x) = (x - 2\tan(x))(x^3 - 8), \tag{22}$$

where $\alpha = 2$ and $x_0 = 1.7$.

Experiment 2. *We take into account the following nonlinear test function:*

$$f_2(x) = (x - 1)(x^{10} + x^3 + 1)\sin(x), \tag{23}$$

where $\alpha = 1$ and $x_0 = 0.7$.

Experiment 3. *We consider the following test problem now:*

$$f_3(x) = \frac{-x^3}{2} + 2\tan^{-1}(x) + 1, \tag{24}$$

where $\alpha \approx 1.8467200$ and $x_0 = 4$.

Experiment 4. *The last test problem is taken into consideration as follows:*

$$f_4(x) = \tan^{-1}(\exp(x+2)+1) + \tanh(\exp(-x\cos(x))) - \sin(\pi x),\qquad(25)$$

where $\alpha \approx -3.6323572\cdots$ *and* $x_0 = -4.1$.

Tables 1–4 show that the proposed Equation (9) is of order 3.90057 and it is obviously believed to be of more advantageous than the other methods listed due to its fast speed and better accuracy.

For better comparisons, we present absolute residual errors $|f(x)|$, for each test function which are displayed in Tables 1–4. Additionally, we compute the computational order of convergence. Noting that we have used multiple precision arithmetic considering 2000 significant digits to observe and the asymptotic error constant and the coc as obviously as possible.

The results obtained by our proposed Equation (M4) are efficient and show better performance than other existing methods.

A significant challenge of executing high-order nonlinear solvers is in finding initial approximation to start the iterations when high accuracy calculating is needed.

Table 1. Result of comparisons for the function f_1.

Methods	$|f_1(x_3)|$	$|f_1(x_4)|$	$|f_1(x_5)|$	$|f_1(x_6)|$	coc
SM	4.1583	3.0743	1.4436	0.25430	2.00
DZ	0.13132	2.0026×10^{-7}	1.0181×10^{-27}	7.1731×10^{-99}	3.57
PM	1.8921×10^{-6}	4.5864×10^{-24}	1.0569×10^{-88}	7.5269×10^{-318}	3.55
M4	9.1741×10^{-6}	3.3242×10^{-26}	4.4181×10^{-103}	1.1147×10^{-404}	3.92

Table 2. Result of comparisons for the function f_2.

Methods	$|f_2(x_5)|$	$|f_2(x_6)|$	$|f_2(x_7)|$	$|f_2(x_8)|$	coc
SM	—	—	—	—	—
DZ	0.14774	0.0016019.	1.3204×10^{-10}	1.5335×10^{-35}	3.56
PM	2.1191×10^{-10}	8.0792×10^{-35}	1.9037×10^{-121}	3.7062×10^{-430}	3.56
M4	5.9738×10^{-15}	4.1615×10^{-57}	1.7309×10^{-220}	1.8231×10^{-857}	3.90

Table 3. Result of comparisons for the function f_3.

Methods	$|f_3(x_3)|$	$|f_3(x_4)|$	$|f_3(x_5)|$	$|f_3(x_6)|$	coc
SM	0.042162	0.00012627	1.1589×10^{-9}	9.7638×10^{-20}	2.00
DZ	1.0219×10^{-11}	4.4086×10^{-44}	1.6412×10^{-157}	1.5347×10^{-562}	3.57
PM	7.9792×10^{-8}	3.712×10^{-30}	4.9556×10^{-108}	2.9954×10^{-386}	3.57
M4	4.4718×10^{-6}	2.9187×10^{-25}	4.7057×10^{-101}	1.0495×10^{-395}	3.89

To discuss further, mostly based on interval mathematics, one can find a close enough guess to start the process. There are some other ways to determine the real initial approximation to start the process. An idea of finding such initial guesses given in [22] is based on the useful commands in Mathematica 11.1 NDSolve [] for the nonlinear function on the interval $D = [a, b]$.

Following this the following piece of Mathematica code could give a list of initial approximations in the working interval for Experiment 4:

```
ClearAll["Global`*"]

(*Defining the nonlinear function.*)
f[x_]:=ArcTan[Exp[x+2]+1]+Tanh[Exp[-x Cos[x]]]-Sin[Pi x];

(*Defining the interval.*)
a=-4.; b=4.;

(*Find the list of initial estimates.*)
Zeros = Quiet@Reap[soln=y[x]/.First[NDSolve[{y'[x]
==Evaluate[D[f[x],x]],y[b]==(f[b])},y[x],{x,a,b},
Method->{"EventLocator","Event"->y[x], "EventAction":>Sow[{x,y[x]}]}]]]][[2,1]];
initialPoints = Sort[Flatten[Take[zeros,Length[zeros],1]]]
```

To check the position of the zero and the graph of the function, we can use the following code to obtain Figure 1.

```
Length[initialPoints]
Plot[f[x],{x,a,b}, Epilog->{PointSize[Medium], Red, Point[zeros]},PlotRange->All,  PerformanceGoal-
>"Quality", PlotStyle->{Thick, Blue}]
```

Table 4. Result of comparisons for the function f_4.

Methods	$|f_4(x_3)|$	$|f_4(x_4)|$	$|f_4(x_5)|$	$|f_4(x_6)|$	coc
SM	0.00001166	3.7123×10^{-10}	3.7616×10^{-19}	3.8622×10^{-37}	2.00
DZ	1.6×10^{-13}	6.9981×10^{-47}	1.0583×10^{-164}	7.0664×10^{-585}	3.57
PM	3.0531×10^{-11}	3.2196×10^{-38}	3.7357×10^{-134}	6.5771×10^{-476}	3.56
M4	2.5268×10^{-13}	1.5972×10^{-49}	2.8738×10^{-191}	1.6018×10^{-744}	3.90

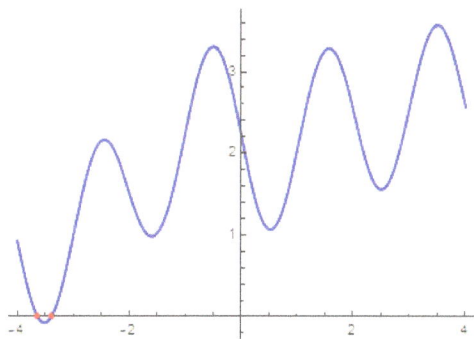

Figure 1. The plot of the nonlinear function in Experiment 4 along with its roots colored in red.

As a harder test problem, for the nonlinear function $g(x) = 2x + 0.5\sin(20\pi x) - x^2$, we can simply find a list of estimates as initial guesses using the above piece of codes as follows: {-0.185014, -0.162392, -0.0935912, -0.0535277, 6.73675×10^{-9}, 0.0533287, 0.0941576, 0.160021, 0.188066, 0.269075, 0.279428, 1.76552, 1.78616, 1.8588, 1.89339, 1.95294, $2.$, 2.04692, 2.10748, 2.13979, 2.2228, 2.22471}. The plot of the function in this case is brought forward in Figure 2.

We observe that the two self-accelerating parameters β_0 and ζ_0 have to be selected before the iterative procedure is started. That is, they are calculated by using information existing from the present and previous iterations (see, e.g., [23]). The initial estimates β_0 and ζ_0 should be preserved as precise small positive values. We use $\beta_0 = \zeta_0 = 0.1$ whenever required.

After a number of iterates, the (nonzero) free parameters start converging to a particular value which makes the coefficient of Equation (6) zero as well as make the numerical scheme to converge with high R-order.

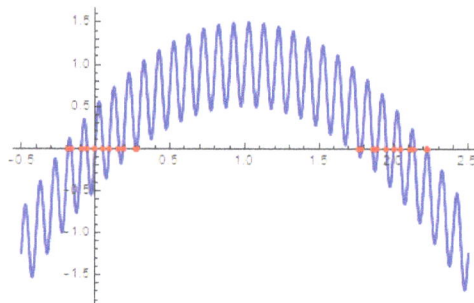

Figure 2. The behavior of the function g and the position of its roots (the red dots show the location of the zeros of the nonlinear functions).

5. Ending Comments

In this paper, we have constructed a one-step method with memory to solve nonlinear equations. By using two self-accelerator parameters our scheme equipped with Newton's interpolation polynomial without any additional functional calculation possesses the high computational efficiency index 1.97499, which is higher than many of the existing methods.

The efficacy of our scheme is confirmed by some of numerical examples. The results in Tables 1–4 shows that our method (Equation (M4)) is valuable to find an adequate estimate of the exact solution of nonlinear equations.

Author Contributions: The authors contributed equally to this paper.

Funding: This research received no external funding.

Acknowledgments: The authors are thankful to two anonymous referees for careful reading and valuable comments which improved the quality of this manuscript.

Conflicts of Interest: The authors declare no conflict of interest.

References

1. Cordero, A.; Hueso, J.L.; Martinez, E.; Torregrosa, J.R. Steffensen type methods for solving nonlinear equations. *J. Comput. Appl. Math.* **2012**, *236*, 3058–3064. [CrossRef]
2. Soleymani, F. Some optimal iterative methods and their with memory variants. *J. Egypt. Math. Soc.* **2013**, *21*, 133–141. [CrossRef]
3. Praks, P.; Brkić, D. Choosing the optimal multi-point iterative method for the Colebrook Flow friction equation. *Processes* **2018**, *6*, 130. [CrossRef]
4. Zafar, F.; Cordero, A.; Torregrosa, J.R. An efficient family of optimal eighth-order multiple root finders. *Mathematics* **2018**, *6*, 310. [CrossRef]
5. Saeed, R.K.; Aziz, K.M. An iterative method with quartic convergence for solving nonlinear equations. *Appl. Math. Comput.* **2008**, *202*, 435–440. [CrossRef]
6. Saeed, R.K. Six order iterative method for solving nonlinear equations. *World Appl. Sci. J.* **2010**, *11*, 1393–1397.
7. Torkashvand, V.; Lotfi, T.; Araghi, M.A.F. A new family of adaptive methods with memory for solving nonlinear equations. *Math. Sci.* **2019**, 1–20.
8. Noda, T. The Steffensen iteration method for systems of nonlinear equations. *Proc. Jpn. Acad.* **1987**, *63*, 186–189.
9. Kung, H.T.; Traub, J.F. Optimal order of one-point and multipoint iteration. *J. Assoc. Comput. Math.* **1974**, *21*, 634–651. [CrossRef]

10. Ahmad, F. Comment on: On the Kung-Traub conjecture for iterative methods for solving quadratic equations. *Algorithms* **2016**, *9*, 30. [CrossRef]
11. Traub, J.F. *Iterative Methods for the Solution of Equations*; Prentice-Hall: Englewood Cliffs, NJ, USA, 1964.
12. Džunić, J. On efficient two-parameter methods for solving nonlinear equations. *Numer. Algorithms* **2013**, *63*, 549–569. [CrossRef]
13. Džunić, J.; Petković, M.S. On generalized biparametric multipoint root finding methods with memory. *J. Comput. Appl. Math.* **2014**, *255*, 362–375. [CrossRef]
14. Zheng, O.; Wang, J.; Zhang, P.L. A Steffensen-like method and its higher-order variants. *Appl. Math. Comput.* **2009**, *214*, 10–16. [CrossRef]
15. Lotfi, T.; Tavakoli, E. On a new efficient Steffensen-like iterative class by applying a suitable self-accelerator parameter. *Sci. World J.* **2014**, *2014*, 769758. [CrossRef]
16. Zheng, O.P.; Zhao, L.; Ma, W. Variants of Steffensen-Secant method and applications. *Appl. Math. Comput.* **2010**, *216*, 3486–3496. [CrossRef]
17. Petković, M.S.; Ilić, S.; Džunić, J. Derivative free two-point methods with and without memory for solving nonlinear equations. *Appl. Math. Comput.* **2010**, *217*, 1887–1895.
18. Khaksar Haghani, F. A modiffied Steffensen's method with memory for nonlinear equations. *Int. J. Math. Model. Comput.* **2015**, *5*, 41–48.
19. Howk, C.L.; Hueso, J.L.; Martinez, E.; Teruel, C. A class of efficient high-order iterative methods with memory for nonlinear equations and their dynamics. *Math. Meth. Appl. Sci.* **2018**, 1–20. [CrossRef]
20. Cliff, H.; Kelvin, M.; Michael, M. *Hands-on Start to Wolfram Mathematica and Programming with the Wolfram Language*, 2nd ed.; Wolfram Media, Inc.: Champaign, IL, USA, 2016; ISBN 9781579550127.
21. Weerakoon, S.; Fernando, T.G.I. A variant of Newton's method with accelerated third-order convergence. *Appl. Math. Lett.* **2000**, *13*, 87–93. [CrossRef]
22. Soleymani, F.; Shateyi, S. Two optimal eighth-order derivative-free classes of iterative methods. *Abstr. Appl. Anal.* **2012**, *2012*, 318165. [CrossRef]
23. Zaka, M.U.; Kosari, S.; Soleymani, F.; Khaksar, F.H.; Al-Fhaid, A.S. A super-fast tri-parametric iterative method with memory. *Appl. Math. Comput.* **2016**, *289*, 486–491.

mathematics

MDPI

Article

Optimal Fourth, Eighth and Sixteenth Order Methods by Using Divided Difference Techniques and Their Basins of Attraction and Its Application

Yanlin Tao [1],[†] and Kalyanasundaram Madhu [2],*,[†]

[1]   School of Computer Science and Engineering, Qujing Normal University, Qujing 655011, China;
     taoyanl@126.com
[2]   Department of Mathematics, Saveetha Engineering College, Chennai 602105, India
*   Correspondence: kalyan742@pec.edu; Tel.: +91-9840050042
†   These authors contributed equally to this work.

Received: 26 February 2019; Accepted: 26 March 2019; Published: 30 March 2019

Abstract: The principal objective of this work is to propose a fourth, eighth and sixteenth order scheme for solving a nonlinear equation. In terms of computational cost, per iteration, the fourth order method uses two evaluations of the function and one evaluation of the first derivative; the eighth order method uses three evaluations of the function and one evaluation of the first derivative; and sixteenth order method uses four evaluations of the function and one evaluation of the first derivative. So these all the methods have satisfied the Kung-Traub optimality conjecture. In addition, the theoretical convergence properties of our schemes are fully explored with the help of the main theorem that demonstrates the convergence order. The performance and effectiveness of our optimal iteration functions are compared with the existing competitors on some standard academic problems. The conjugacy maps of the presented method and other existing eighth order methods are discussed, and their basins of attraction are also given to demonstrate their dynamical behavior in the complex plane. We apply the new scheme to find the optimal launch angle in a projectile motion problem and Planck's radiation law problem as an application.

Keywords: non-linear equation; basins of attraction; optimal order; higher order method; computational order of convergence

MSC: 65H05, 65D05, 41A25

---

1. Introduction

One of the most frequent problems in engineering, scientific computing and applied mathematics, in general, is the problem of solving a nonlinear equation $f(x) = 0$. In most of the cases, whenever real problems are faced, such as weather forecasting, accurate positioning of satellite systems in the desired orbit, measurement of earthquake magnitudes and other high-level engineering problems, only approximate solutions may get resolved. However, only in rare cases, it is possible to solve the governing equations exactly. The most familiar method of solving non linear equation is Newton's iteration method. The local order of convergence of Newton's method is two and it is an optimal method with two function evaluations per iterative step.

In the past decade, several higher order iterative methods have been developed and analyzed for solving nonlinear equations that improve classical methods such as Newton's method, Chebyshev method, Halley's iteration method, etc. As the order of convergence increases, so does the number of function evaluations per step. Hence, a new index to determine the efficiency called the efficiency index is introduced in [1] to measure the balance between these quantities. Kung-Traub [2] conjectured that the order of convergence of any multi-point without memory method with d function evaluations cannot exceed the bound 2^{d-1}, the optimal order. Thus the optimal order for three evaluations per iteration would be four, four evaluations per iteration would be eight, and so on. Recently, some fourth and eighth order optimal iterative methods have been developed (see [3–14] and references therein). A more extensive list of references as well as a survey on the progress made in the class of multi-point methods is found in the recent book by Petkovic et al. [11].

This paper is organized as follows. An optimal fourth, eighth and sixteenth order methods are developed by using divided difference techniques in Section 2. In Section 3, convergence order is analyzed. In Section 4, tested numerical examples to compare the proposed methods with other known optimal methods. The problem of Projectile motion is discussed in Section 5 where the presented methods are applied on this problem with some existing ones. In Section 6, we obtain the conjugacy maps of these methods to make a comparison from dynamical point of view. In Section 7, the proposed methods are studied in the complex plane using basins of attraction. Section 8 gives concluding remarks.

2. Design of an Optimal Fourth, Eighth and Sixteenth Order Methods

Definition 1 ([15]). *If the sequence $\{x_n\}$ tends to a limit x^* in such a way that*

$$\lim_{n \to \infty} \frac{x_{n+1} - x^*}{(x_n - x^*)^p} = C$$

for $p \geq 1$, then the order of convergence of the sequence is said to be p, and C is known as the asymptotic error constant. If $p = 1$, $p = 2$ or $p = 3$, the convergence is said to be linear, quadratic or cubic, respectively. Let $e_n = x_n - x^*$, then the relation

$$e_{n+1} = C\, e_n^p + O\left(e_n^{p+1}\right) = O\left(e_n^p\right). \tag{1}$$

is called the error equation. The value of p is called the order of convergence of the method.

Definition 2 ([1]). *The Efficiency Index is given by*

$$EI = p^{\frac{1}{d}}, \tag{2}$$

where d is the total number of new function evaluations (the values of f and its derivatives) per iteration.

Let $x_{n+1} = \psi(x_n)$ define an Iterative Function (IF). Let x_{n+1} be determined by new information at $x_n, \phi_1(x_n), ..., \phi_i(x_n), i \geq 1$. No old information is reused. Thus,

$$x_{n+1} = \psi(x_n, \phi_1(x_n), ..., \phi_i(x_n)). \tag{3}$$

Then ψ is called a multipoint IF without memory.
The Newton (also called Newton-Raphson) IF ($2^{nd}NR$) is given by

$$\psi_{2^{nd}NR}(x) = x - \frac{f(x)}{f'(x)}. \tag{4}$$

The $2^{nd}NR$ IF is one-point IF with two function evaluations and it satisfies the Kung-Traub conjecture with $d = 2$. Further, $EI_{2^{nd}NR} = 1.414$.

2.1. An Optimal Fourth Order Method

We attempt to get a new optimal fourth order IF as follows, let us consider two step Newton's method

$$\psi_{4^{th}NR}(x) = \psi_{2nd\,NR}(x) - \frac{f(\psi_{2nd\,NR}(x))}{f'(\psi_{2nd\,NR}(x))}. \tag{5}$$

The above one is having fourth order convergence with four function evaluations. But, this is not an optimal method. To get an optimal, need to reduce a function and preserve the same convergence order, and so we estimate $f'(\psi_{2nd\,NR}(x))$ by the following polynomial

$$q(t) = a_0 + a_1(t - x) + a_2(t - x)^2, \tag{6}$$

which satisfies

$$q(x) = f(x),\, q'(x) = f'(x),\, q(\psi_{2nd\,NR}(x)) = f(\psi_{2nd\,NR}(x)).$$

On implementing the above conditions on Equation (6), we obtain three unknowns a_0, a_1 and a_2. Let us define the divided differences

$$f[y, x] = \frac{f(y) - f(x)}{y - x},\, f[y, x, x] = \frac{f[y, x] - f'(x)}{y - x}.$$

From conditions, we get $a_0 = f(x)$, $a_1 = f'(x)$ and $a_2 = f[\psi_{2nd\,NR}(x), x, x]$, respectively, by using divided difference techniques. Now, we have the estimation

$$f'(\psi_{2nd\,NR}(x)) \approx q'(\psi_{2nd\,NR}(x)) = a_1 + 2a_2(\psi_{2^{th}NR}(x) - x).$$

Finally, we propose a new optimal fourth order method as

$$\psi_{4^{th}YM}(x) = \psi_{2nd\,NR}(x) - \frac{f(\psi_{2nd\,NR}(x))}{f'(x) + 2f[\psi_{2nd\,NR}(x), x, x](\psi_{2^{th}NR}(x) - x)}. \tag{7}$$

The efficiency of the method (7) is $EI_{4thYM} = 1.587$.

2.2. An Optimal Eighth Order Method

Next, we attempt to get a new optimal eighth order IF as following way

$$\psi_{8^{th}YM}(x) = \psi_{4^{th}YM}(x) - \frac{f(\psi_{4^{th}YM}(x))}{f'(\psi_{4^{th}YM}(x))}. $$

The above one is having eighth order convergence with five function evaluations. But, this is not an optimal method. To get an optimal, need to reduce a function and preserve the same convergence order, and so we estimate $f'(\psi_{4^{th}YM}(x))$ by the following polynomial

$$q(t) = b_0 + b_1(t - x) + b_2(t - x)^2 + b_3(t - x)^3, \tag{8}$$

which satisfies

$$q(x) = f(x),\, q'(x) = f'(x),\, q(\psi_{2nd\,NR}(x)) = f(\psi_{2nd\,NR}(x)),\, q(\psi_{4^{th}YM}(x)) = f(\psi_{4^{th}YM}(x)).$$

On implementing the above conditions on (8), we obtain four linear equations with four unknowns b_0, b_1, b_2 and b_3. From conditions, we get $b_0 = f(x)$ and $b_1 = f'(x)$. To find b_2 and b_3, we solve the following equations:

$$
\begin{aligned}
f(\psi_{2nd\,NR}(x)) &= f(x) + f'(x)(\psi_{2nd\,NR}(x) - x) + b_2(\psi_{2nd\,NR}(x) - x)^2 + b_3(\psi_{2nd\,NR}(x) - x)^3\\
f(\psi_{4^{th}YM}(x)) &= f(x) + f'(x)(\psi_{4^{th}YM}(x) - x) + b_2(\psi_{4^{th}YM}(x) - x)^2 + b_3(\psi_{4^{th}YM}(x) - x)^3.
\end{aligned}
$$

Thus by applying divided differences, the above equations simplifies to

$$b_2 + b_3(\psi_{2nd\,NR}(x) - x) = f[\psi_{2nd\,NR}(x), x, x] \tag{9}$$
$$b_2 + b_3(\psi_{4th\,YM}(x) - x) = f[\psi_{4th\,YM}(x), x, x] \tag{10}$$

Solving Equations (9) and (14), we have

$$b_2 = \frac{f[\psi_{2nd\,NR}(x), x, x](\psi_{4th\,PM}(x) - x) - f[\psi_{4th\,YM}(x), x, x](\psi_{2nd\,NR}(x) - x)}{\psi_{4th\,YM}(x) - \psi_{2nd\,NR}(x)},$$
$$b_3 = \frac{f[\psi_{4th\,YM}(x), x, x] - f[\psi_{2nd\,NR}(x), x, x]}{\psi_{4th\,YM}(x) - \psi_{2nd\,NR}(x)}. \tag{11}$$

Further, using Equation (11), we have the estimation

$$f'(\psi_{4th\,YM}(x)) \approx q'(\psi_{4th\,YM}(x)) = b_1 + 2b_2(\psi_{4th\,YM}(x) - x) + 3b_3(\psi_{4th\,YM}(x) - x)^2.$$

Finally, we propose a new optimal eighth order method as

$$\psi_{8th\,YM}(x) = \psi_{4th\,YM}(x) - \frac{f(\psi_{4th\,YM}(x))}{f'(x) + 2b_2(\psi_{4th\,YM}(x) - x) + 3b_3(\psi_{4th\,YM}(x) - x)^2}. \tag{12}$$

The efficiency of the method (12) is $EI_{8thYM} = 1.682$. Remark that the method is seems a particular case of the method of Khan et al. [16], they used weight function to develop their methods. Whereas we used finite difference scheme to develop proposed methods. We can say the methods $4^{th}YM$ and $8^{th}YM$ are reconstructed of Khan et al. [16] methods.

2.3. An Optimal Sixteenth Order Method

Next, we attempt to get a new optimal sixteenth order IF as following way

$$\psi_{16th\,YM}(x) = \psi_{8th\,YM}(x) - \frac{f(\psi_{8th\,YM}(x))}{f'(\psi_{8th\,YM}(x))}.$$

The above one is having eighth order convergence with five function evaluations. However, this is not an optimal method. To get an optimal, need to reduce a function and preserve the same convergence order, and so we estimate $f'(\psi_{8th\,YM}(x))$ by the following polynomial

$$q(t) = c_0 + c_1(t - x) + c_2(t - x)^2 + c_3(t - x)^3 + c_4(t - x)^4, \tag{13}$$

which satisfies

$$q(x) = f(x), \ q'(x) = f'(x), \ q(\psi_{2nd\,NR}(x)) = f(\psi_{2nd\,NR}(x)),$$
$$q(\psi_{4th\,YM}(x)) = f(\psi_{4th\,YM}(x)), q(\psi_{8th\,YM}(x)) = f(\psi_{8th\,YM}(x)).$$

On implementing the above conditions on (13), we obtain four linear equations with four unknowns c_0, c_1, c_2 and c_3. From conditions, we get $c_0 = f(x)$ and $c_1 = f'(x)$. To find c_2, c_3 and c_4, we solve the following equations:

$$f(\psi_{2nd\,NR}(x)) = f(x) + f'(x)(\psi_{2nd\,NR}(x) - x) + c_2(\psi_{2nd\,NR}(x) - x)^2 + c_3(\psi_{2nd\,NR}(x) - x)^3 + c_4(\psi_{2nd\,NR}(x) - x)^4$$
$$f(\psi_{4th\,YM}(x)) = f(x) + f'(x)(\psi_{4th\,YM}(x) - x) + c_2(\psi_{4th\,YM}(x) - x)^2 + c_3(\psi_{4th\,YM}(x) - x)^3 + c_4(\psi_{4th\,YM}(x) - x)^4$$
$$f(\psi_{8th\,YM}(x)) = f(x) + f'(x)(\psi_{8th\,YM}(x) - x) + c_2(\psi_{8th\,YM}(x) - x)^2 + c_3(\psi_{8th\,YM}(x) - x)^3 + c_4(\psi_{8th\,YM}(x) - x)^4.$$

Thus by applying divided differences, the above equations simplifies to

$$
\begin{aligned}
c_2 + c_3(\psi_{2nd\,NR}(x) - x) + c_4(\psi_{2nd\,NR}(x) - x)^2 &= f[\psi_{2nd\,NR}(x), x, x] \\
c_2 + c_3(\psi_{4th\,YM}(x) - x) + c_4(\psi_{4th\,YM}(x) - x)^2 &= f[\psi_{4th\,YM}(x), x, x] \\
c_2 + c_3(\psi_{8th\,YM}(x) - x) + c_4(\psi_{8th\,YM}(x) - x)^2 &= f[\psi_{8th\,YM}(x), x, x]
\end{aligned}
\tag{14}
$$

Solving Equation (14), we have

$$
\begin{aligned}
c_2 &= \frac{\left(f[\psi_{2nd\,NR}(x), x, x]\left(-S_2^2 S_3 + S_2 S_3^2\right) + f[\psi_{4th\,YM}(x), x, x]\left(S_1^2 S_3 - S_1 S_3^2\right) + f[\psi_{8th\,YM}(x), x, x]\left(-S_1^2 S_2 + S_1 S_2^2\right)\right)}{-S_1^2 S_2 + S_1 S_2^2 + S_1^2 S_3 - S_2^2 S_3 - S_1 S_3^2 + S_2 S_3^2}, \\[4pt]
c_3 &= \frac{\left(f[\psi_{2nd\,NR}(x), x, x]\left(S_2^2 - S_3^2\right) + f[\psi_{4th\,YM}(x), x, x]\left(-S_1^2 + S_3^2\right) + f[\psi_{8th\,YM}(x), x, x]\left(S_1^2 - S_2^2\right)\right)}{-S_1^2 S_2 + S_1 S_2^2 + S_1^2 S_3 - S_2^2 S_3 - S_1 S_3^2 + S_2 S_3^2}, \\[4pt]
c_4 &= \frac{\left(f[\psi_{2nd\,NR}(x), x, x]\left(-S_2 + S_3\right) + f[\psi_{4th\,YM}(x), x, x]\left(S_1 - S_3\right) + f[\psi_{8th\,YM}(x), x, x]\left(-S_1 + S_2\right)\right)}{-S_1^2 S_2 + S_1 S_2^2 + S_1^2 S_3 - S_2^2 S_3 - S_1 S_3^2 + S_2 S_3^2},
\end{aligned}
\tag{15}
$$

$$S_1 = \psi_{2nd\,NR}(x) - x, \; S_2 = \psi_{4th\,YM}(x) - x, \; S_3 = \psi_{8th\,YM}(x) - x.$$

Further, using Equation (15), we have the estimation

$$f'(\psi_{8th\,YM}(x)) \approx q'(\psi_{8th\,YM}(x)) = c_1 + 2c_2(\psi_{8th\,YM}(x) - x) + 3c_3(\psi_{8th\,YM}(x) - x)^2 + 4c_4(\psi_{8th\,YM}(x) - x)^3.$$

Finally, we propose a new optimal sixteenth order method as

$$
\psi_{16th\,YM}(x) = \psi_{8th\,YM}(x) - \frac{f(\psi_{8th\,YM}(x))}{f'(x) + 2c_2(\psi_{8th\,YM}(x) - x) + 3c_3(\psi_{8th\,YM}(x) - x)^2 + 4c_4(\psi_{8th\,YM}(x) - x)^3}.
\tag{16}
$$

The efficiency of the method (16) is $EI_{16thYM} = 1.741$.

3. Convergence Analysis

In this section, we prove the convergence analysis of proposed *IFs* with help of Mathematica software.

Theorem 1. *Let $f : D \subset \mathbb{R} \to \mathbb{R}$ be a sufficiently smooth function having continuous derivatives. If $f(x)$ has a simple root x^* in the open interval D and x_0 chosen in sufficiently small neighborhood of x^*, then the method $4^{th}YM$ IFs (7) is of local fourth order convergence, and the $8^{th}YM$ IFs (12) is of local eighth order convergence.*

Proof. Let $e = x - x^*$ and $c[j] = \dfrac{f^{(j)}(x^*)}{j! f'(x^*)}$, $j = 2, 3, 4, \dots$. Expanding $f(x)$ and $f'(x)$ about x^* by Taylor's method, we have

$$
f(x) = f'(x^*)\left(e + e^2 c[2] + e^3 c[3] + e^4 c[4] + e^5 c[5] + e^6 c[6] + e^7 c[7] + e^8 c[8] + \dots\right)
\tag{17}
$$

and

$$
f'(x) = f'(x^*)\left(1 + 2e\,c[2] + 3e^2 c[3] + 4e^3 c[4] + 5e^4 c[5] + 6e^5 c[6] + 7e^6 c[7] + 8e^7 c[8] + 9e^8 c[9] + \dots\right)
\tag{18}
$$

Thus,

$$
\begin{aligned}
\psi_{2^{nd}NR}(x) = {}& x^* + c[2]e^2 + \left(-2c[2]^2 + 2c[3]\right)e^3 + \left(4c[2]^3 - 7c[2]c[3] + 3c[4]\right)e^4 + \left(-8c[2]^4\right. \\
& + 20c[2]^2c[3] - 6c[3]^2 - 10c[2]c[4] + 4c[5]\right)e^5 + \left(16c[2]^5 - 52c[2]^3c[3] + 28c[2]^2c[4] - 17c[3]c[4]\right. \\
& + c[2](33c[3]^2 - 13c[5]) + 5c[6]\right)e^6 - 2\left(16c[2]^6 - 64c[2]^4c[3] - 9c[3]^3 + 36c[2]^3c[4] + 6c[4]^2 + 9c[2]^2(7c[3]^2\right. \\
& - 2c[5]) + 11c[3]c[5] + c[2](-46c[3]c[4] + 8c[6]) - 3c[7]\right)e^7 + \left(64c[2]^7 - 304c[2]^5c[3]\right. \\
& + 176c[2]^4c[4] + 75c[3]^2c[4] + c[2]^3(408c[3]^2 - 92c[5]) - 31c[4]c[5] - 27c[3]c[6] \\
& + c[2]^2(-348c[3]c[4] + 44c[6]) + c[2](-135c[3]^3 + 64c[4]^2 + 118c[3]c[5] - 19c[7]) + 7c[8]\right)e^8 + \dots.
\end{aligned}
\tag{19}
$$

Expanding $f(\psi_{2^{nd}NR}(x))$ about x^* by Taylor's method, we have

$$
\begin{aligned}
f(\psi_{2^{nd}NR}(x)) = {}& f'(x^*)\left(c[2]e^2 + \left(-2c[2]^2 + 2c[3]\right)e^3 + \left(5c[2]^3 - 7c[2]c[3] + 3c[4]\right)e^4 - 2\left(6c[2]^4\right.\right. \\
& - 12c[2]^2c[3] + 3c[3]^2 + 5c[2]c[4] - 2c[5]\right)e^5 + \left(28c[2]^5 - 73c[2]^3c[3] + 34c[2]^2c[4] - 17c[3]c[4]\right. \\
& + c[2](37c[3]^2 - 13c[5]) + 5c[6]\right)e^6 - 2\left(32c[2]^6 - 103c[2]^4c[3] - 9c[3]^3 + 52c[2]^3c[4] + 6c[4]^2\right. \\
& + c[2]^2(80c[3]^2 - 22c[5]) + 11c[3]c[5] + c[2](-52c[3]c[4] + 8c[6]) - 3c[7]\right)e^7 \\
& + \left(144c[2]^7 - 552c[2]^5c[3] + 297c[2]^4c[4] + 75c[3]^2c[4] + 2c[2]^3(291c[3]^2 - 67c[5])\right. \\
& - 31c[4]c[5] - 27c[3]c[6] + c[2]^2(-455c[3]c[4] + 54c[6]) + c[2](-147c[3]^3 + 73c[4]^2 \\
& \left.\left. + 134c[3]c[5] - 19c[7]) + 7c[8]\right)e^8 + \dots.\right)
\end{aligned}
\tag{20}
$$

Using Equations (17)–(20) in divided difference techniques. We have

$$
\begin{aligned}
f[\psi_{2^{nd}NR}(x), x, x] = {}& f'(x^*)\left(c[2] + 2c[3]e + \left(c[2]c[3] + 3c[4]\right)e^2 + 2\left(-c[2]^2c[3] + c[3]^2\right.\right. \\
& + c[2]c[4] + 2c[5]\right)e^3 + \left(4c[2]^3c[3] - 3c[2]^2c[4] + 7c[3]c[4] + c[2](-7c[3]^2 + 3c[5]) + 5c[6]\right)e^4 \\
& + \left(-8c[2]^4c[3] - 6c[3]^3 + 4c[2]^3c[4] + 4c[2]^2(5c[3]^2 - c[5]) + 10c[3]c[5]\right. \\
& + 4c[2](-5c[3]c[4] + c[6]) + 6(c[4]^2 + c[7])\right)e^5 + \left(16c[2]^5c[3] - 4c[2]^4c[4]\right. \\
& - 25c[3]^2c[4] + 17c[4]c[5] + c[2]^3(-52c[3]^2 + 5c[5]) + c[2]^2(46c[3]c[4] - 5c[6]) \\
& \left.\left. + 13c[3]c[6] + c[2](33c[3]^3 - 14c[4]^2 - 26c[3]c[5] + 5c[7]) + 7c[8]\right)e^6 + \dots.\right)
\end{aligned}
\tag{21}
$$

Substituting Equations (18)–(21) into Equation (7), we obtain, after simplifications,

$$
\begin{aligned}
\psi_{4^{th}YM}(x) = {}& x^* + \left(c[2]^3 - c[2]c[3]\right)e^4 - 2\left(2c[2]^4 - 4c[2]^2c[3] + c[3]^2 + c[2]c[4]\right)e^5 + \left(10c[2]^5 - 30c[2]^3c[3]\right. \\
& + 12c[2]^2c[4] - 7c[3]c[4] + 3c[2](6c[3]^2 - c[5])\right)e^6 - 2\left(10c[2]^6 - 40c[2]^4c[3] - 6c[3]^3\right. \\
& + 20c[2]^3c[4] + 3c[4]^2 + 8c[2]^2(5c[3]^2 - c[5]) + 5c[3]c[5] + c[2](-26c[3]c[4] + 2c[6])\right)e^7 + \left(36c[2]^7\right. \\
& - 178c[2]^5c[3] + 101c[2]^4c[4] + 50c[3]^2c[4] + 3c[2]^3(84c[3]^2 - 17c[5]) - 17c[4]c[5] - 13c[3]c[6] \\
& + c[2]^2(-209c[3]c[4] + 20c[6]) + c[2](-91c[3]^3 + 37c[4]^2 + 68c[3]c[5] - 5c[7])\right)e^8 + \dots.
\end{aligned}
\tag{22}
$$

Expanding $f(\psi_{4^{th}YM}(x))$ about x^* by Taylor's method, we have

$$
\begin{aligned}
f(\psi_{4^{th}YM}(x)) = f'(x^*)\Big(& \big(c[2]^3 - c[2]c[3]\big)e^4 - 2\big(2c[2]^4 - 4c[2]^2c[3] + c[3]^2 + c[2]c[4]\big)e^5 + \Big(10c[2]^5 \\
& - 30c[2]^3c[3] + 12c[2]^2c[4] - 7c[3]c[4] + 3c[2]\big(6c[3]^2 - c[5]\big)\Big)e^6 - 2\big(10c[2]^6 - 40c[2]^4c[3] \\
& - 6c[3]^3 + 20c[2]^3c[4] + 3c[4]^2 + 8c[2]^2(5c[3]^2 - c[5]) + 5c[3]c[5] + c[2](-26c[3]c[4] + 2c[6])\big)e^7 \\
& + \Big(37c[2]^7 - 180c[2]^5c[3] + 101c[2]^4c[4] + 50c[3]^2c[4] + c[2]^3(253c[3]^2 - 51c[5]) - 17c[4]c[5] \\
& - 13c[3]c[6] + c[2]^2(-209c[3]c[4] + 20c[6]) + c[2](-91c[3]^3 + 37c[4]^2 + 68c[3]c[5] - 5c[7])\Big)e^8 + \ldots\Big)
\end{aligned}
\tag{23}
$$

Now,

$$
\begin{aligned}
f[\psi_{4^{th}YM}(x), x, x] = f'(x^*)\Big(& c[2] + 2c[3]e + 3c[4]e^2 + 4c[5]e^3 + \big(c[2]^3c[3] - c[2]c[3]^2 + 5c[6]\big)e^4 \\
& + \big(-4c[2]^4c[3] + 8c[2]^2c[3]^2 - 2c[3]^3 + 2c[2]^3c[4] - 4c[2]c[3]c[4] + 6c[7]\big)e^5 \\
& + \big(10c[2]^5c[3] - 8c[2]^4c[4] + 28c[2]^2c[3]c[4] - 11c[3]^2c[4] + c[2]^3(-30c[3]^2 + 3c[5]) + \\
& 2c[2](9c[3]^3 - 2c[4]^2 - 3c[3]c[5]) + 7c[8]\big)e^6 + \ldots\Big)
\end{aligned}
\tag{24}
$$

Substituting Equations (19)–(21), (23) and (24) into Equation (12), we obtain, after simplifications,

$$
\psi_{8^{th}YM}(x) - x^* = c[2]^2\Big(c[2]^2 - c[3]\Big)\Big(c[2]^3 - c[2]c[3] + c[4]\Big)e^8 + O(e^9)
\tag{25}
$$

Hence from Equations (22) and (25), we concluded that the convergence order of the $4^{th}YM$ and $8^{th}YM$ are four and eight respectively. □

The following theorem is given without proof, which can be worked out with the help of Mathematica.

Theorem 2. *Let $f : D \subset \mathbb{R} \to \mathbb{R}$ be a sufficiently smooth function having continuous derivatives. If $f(x)$ has a simple root x^* in the open interval D and x_0 chosen in sufficiently small neighborhood of x^*, then the method (16) is of local sixteenth order convergence and and it satisfies the error equation*
$$
\psi_{16^{th}YM}(x) - x^* = \Big((c[2]^4)((c[2]^2 - c[3])^2)(c[2]^3 - c[2]c[3] + c[4])(c[2]^4 - c[2]^2c[3] + c[2]c[4] - c[5])\Big)e^{16} + O(e^{17}).
$$

4. Numerical Examples

In this section, numerical results on some test functions are compared for the new methods $4^{th}YM$, $8^{th}YM$ and $16^{th}YM$ with some existing eighth order methods and Newton's method. Numerical computations have been carried out in the MATLAB software with 500 significant digits. We have used the stopping criteria for the iterative process satisfying $error = |x_N - x_{N-1}| < \epsilon$, where $\epsilon = 10^{-50}$ and N is the number of iterations required for convergence. The computational order of convergence is given by ([17])

$$
\rho = \frac{\ln|(x_N - x_{N-1})/(x_{N-1} - x_{N-2})|}{\ln|(x_{N-1} - x_{N-2})/(x_{N-2} - x_{N-3})|}.
$$

We consider the following iterative methods for solving nonlinear equations for the purpose of comparison: $\psi_{4^{th}SB}$, a method proposed by Sharma et al. [18]:

$$
y = x - \frac{2f(x)}{3f'(x)}, \quad \psi_{4^{th}SB}(x) = x - \left(-\frac{1}{2} + \frac{9}{8}\frac{f'(x)}{f'(y)} + \frac{3}{8}\frac{f'(y)}{f'(x)}\right)\frac{f(x)}{f'(x)}.
\tag{26}
$$

$\psi_{4^{th}CLND}$, a method proposed by Chun et al. [19]:

$$
y = x - \frac{2f(x)}{3f'(x)}, \quad \psi_{4^{th}CLND}(x) = x - \frac{16f(x)f'(x)}{-5f'(x)^2 + 30f(x)f'(y) - 9f'(y)^2}.
\tag{27}
$$

$\psi_{4^{th}SJ}$, a method proposed by Singh et al. [20]:

$$y = x - \frac{2}{3}\frac{f(x)}{f'(x)}, \psi_{4^{th}SJ}(x) = x - \left(\frac{17}{8} - \frac{9}{4}\frac{f'(y)}{f'(x)} + \frac{9}{8}\left(\frac{f'(y)}{f'(x)}\right)^2\right)\left(\frac{7}{4} - \frac{3}{4}\frac{f'(y)}{f'(x_n)}\right)\frac{f(x)}{f'(x)}. \quad (28)$$

$\psi_{8^{th}KT}$, a method proposed by Kung-Traub [2]:

$$y = x - \frac{f(x)}{f'(x)}, z = y - \frac{f(y) * f(x)}{(f(x) - f(y))^2}\frac{f(x)}{f'(x)},$$

$$\psi_{8^{th}KT}(x) = z - \frac{f(x)}{f'(x)}\frac{f(x)f(y)f(z)}{(f(x) - f(y))^2}\frac{f(x)^2 + f(y)(f(y) - f(z))}{(f(x) - f(z))^2(f(y) - f(z))}. \quad (29)$$

$\psi_{8^{th}LW}$, a method proposed by Liu et al. [8]

$$y = x - \frac{f(x)}{f'(x)}, z = y - \frac{f(x)}{f(x) - 2f(y)}\frac{f(y)}{f'(x)},$$

$$\psi_{8^{th}LW}(x) = z - \frac{f(z)}{f'(x)}\left(\left(\frac{f(x) - f(y)}{f(x) - 2f(y)}\right)^2 + \frac{f(z)}{f(y) - f(z)} + \frac{4f(z)}{f(x) + f(z)}\right). \quad (30)$$

$\psi_{8^{th}PNPD}$, a method proposed by Petkovic et al. [11]

$$y = x - \frac{f(x)}{f'(x)}, z = x - \left(\left(\frac{f(y)}{f(x)}\right)^2 - \frac{f(x)}{f(y) - f(x)}\right)\frac{f(x)}{f'(x)}, \psi_{8^{th}PNPD}(x) = z - \frac{f(z)}{f'(x)}\left(\varphi(t) + \frac{f(z)}{f(y) - f(z)} + \frac{4f(z)}{f(x)}\right),$$

$$where \; \varphi(t) = 1 + 2t + 2t^2 - t^3 \; and \; t = \frac{f(y)}{f(x)}. \quad (31)$$

$\psi_{8^{th}SA1}$, a method proposed by Sharma et al. [12]

$$y = x - \frac{f(x)}{f'(x)}, z = y - \left(3 - 2\frac{f[y, x]}{f'(x)}\right)\frac{f(y)}{f'(x)}, \psi_{8^{th}SA1}(x) = z - \frac{f(z)}{f'(x)}\left(\frac{f'(x) - f[y, x] + f[z, y]}{2f[z, y] - f[z, x]}\right). \quad (32)$$

$\psi_{8^{th}SA2}$, a method proposed by Sharma et al. [13]

$$y = x - \frac{f(x)}{f'(x)}, z = y - \frac{f(y)}{2f[y, x] - f'(x)}, \psi_{8^{th}SA2}(x) = z - \frac{f[z, y]}{f[z, x]}\frac{f(z)}{2f[z, y] - f[z, x]} \quad (33)$$

$\psi_{8^{th}CFGT}$, a method proposed by Cordero et al. [6]

$$y = x - \frac{f(x)}{f'(x)}, z = y - \frac{f(y)}{f'(x)}\frac{1}{1 - 2t + t^2 - t^3/2}, \psi_{8^{th}CFGT}(x) =$$

$$z - \frac{1 + 3r}{1 + r}\frac{f(z)}{f[z, y] + f[z, x, x](z - y)}, r = \frac{f(z)}{f(x)}. \quad (34)$$

$\psi_{8^{th}CTV}$, a method proposed by Cordero et al. [7]

$$y = x - \frac{f(x)}{f'(x)}, z = x - \frac{1 - t}{1 - 2t}\frac{f(x)}{f'(x)}, \psi_{8^{th}CTV}(x) = z - \left(\frac{1 - t}{1 - 2t} - v\right)^2\frac{1}{1 - 3v}\frac{f(z)}{f'(x)}, v = \frac{f(z)}{f(y)}. \quad (35)$$

Table 1 shows the efficiency indices of the new methods with some known methods.

Table 1. Comparison of Efficiency Indices.

Methods	p	d	*EI*
2^{nd}NR	2	2	1.414
4^{th}SB	4	3	1.587
4^{th}CLND	4	3	1.587
4^{th}SJ	4	3	1.587
4^{th}YM	4	3	1.587
8^{th}KT	8	4	1.682
8^{th}LW	8	4	1.682
8^{th}PNPD	8	4	1.682
8^{th}SA1	8	4	1.682
8^{th}SA2	8	4	1.682
8^{th}CFGT	8	4	1.682
8^{th}CTV	8	4	1.682
8^{th}YM	8	4	1.682
16^{th}YM	16	5	1.741

The following test functions and their simple zeros for our study are given below [10]:

$$f_1(x) = \sin(2\cos x) - 1 - x^2 + e^{\sin(x^3)}, \qquad x^* = -0.7848959876612125352...$$

$$f_2(x) = xe^{x^2} - \sin^2 x + 3\cos x + 5, \qquad x^* = -1.2076478271309189270...$$

$$f_3(x) = x^3 + 4x^2 - 10, \qquad x^* = 1.3652300134140968457...$$

$$f_4(x) = \sin(x) + \cos(x) + x, \qquad x^* = -0.4566247045676308244...$$

$$f_5(x) = \frac{x}{2} - \sin x, \qquad x^* = 1.8954942670339809471...$$

$$f_6(x) = x^2 + \sin(\frac{x}{5}) - \frac{1}{4}, \qquad x^* = 0.4099920179891371316...$$

Table 2, shows that corresponding results for $f_1 - f_6$. We observe that proposed method $4^{th}YM$ is converge in a lesser or equal number of iterations and with least error when compare to compared methods. Note that $4^{th}SB$ and $4^{th}SJ$ methods are getting diverge in f_5 function. Hence, the proposed method $4^{th}YM$ can be considered competent enough to existing other compared equivalent methods.

Also, from Tables 3–5 are shows the corresponding results for $f_1 - f_6$. The computational order of convergence agrees with the theoretical order of convergence in all the functions. Note that $8^{th}PNPD$ method is getting diverge in f_5 function and all other compared methods are converges with least error. Also, function f_1 having least error in $8^{th}CFGT$, function f_2 having least error in $8^{th}CTV$, functions f_3 and f_4 having least error in $8^{th}YM$, function f_5 having least error in $8^{th}SA2$, function f_6 having least error in $8^{th}CFGT$. The proposed $16^{th}YM$ method converges less number of iteration with least error in all the tested functions. Hence, the $16^{th}YM$ can be considered competent enough to existing other compared equivalent methods.

Table 2. Numerical results for nonlinear equations.

Methods	$f_1(x), x_0 = -0.9$				$f_2(x), x_0 = -1.6$			
	N	$|x_1 - x_0|$	$|x_N - x_{N-1}|$	ρ	N	$|x_1 - x_0|$	$|x_N - x_{N-1}|$	ρ
$2^{nd}NR$ (4)	7	0.1080	7.7326×10^{-74}	1.99	9	0.2044	9.2727×10^{-74}	1.99
$4^{th}SB$ (26)	4	0.1150	9.7275×10^{-64}	3.99	5	0.3343	1.4237×10^{-65}	3.99
$4^{th}CLND$ (27)	4	0.1150	1.4296×10^{-64}	3.99	5	0.3801	1.1080×10^{-72}	3.99
$4^{th}SJ$ (28)	4	0.1150	3.0653×10^{-62}	3.99	5	0.3190	9.9781×10^{-56}	3.99
$4^{th}YM$ (7)	4	0.1150	6.0046×10^{-67}	3.99	5	0.3737	7.2910×10^{-120}	4.00
Methods	$f_3(x), x_0 = 0.9$				$f_4(x), x_0 = -1.9$			
$2^{nd}NR$ (4)	8	0.6263	1.3514×10^{-72}	2.00	8	1.9529	1.6092×10^{-72}	1.99
$4^{th}SB$ (26)	5	0.5018	4.5722×10^{-106}	3.99	5	1.5940	6.0381×10^{-92}	3.99
$4^{th}CLND$ (27)	5	0.5012	4.7331×10^{-108}	3.99	5	1.5894	2.7352×10^{-93}	3.99
$4^{th}SJ$ (28)	5	0.4767	3.0351×10^{-135}	3.99	5	1.5776	9.5025×10^{-95}	3.99
$4^{th}YM$ (7)	5	0.4735	2.6396×10^{-156}	3.99	5	1.5519	1.4400×10^{-102}	3.99
Methods	$f_5(x), x_0 = 1.2$				$f_6(x), x_0 = 0.8$			
$2^{nd}NR$ (4)	9	2.4123	1.3564×10^{-83}	1.99	8	0.3056	3.2094×10^{-72}	1.99
$4^{th}SB$ (26)			Diverge		5	0.3801	2.8269×10^{-122}	3.99
$4^{th}CLND$ (27)	14	0.0566	6.8760×10^{-134}	3.99	5	0.3812	7.8638×10^{-127}	3.99
$4^{th}SJ$ (28)			Diverge		5	0.3780	1.4355×10^{-114}	3.99
$4^{th}YM$ (7)	6	1.2887	2.3155×10^{-149}	3.99	5	0.3840	1.1319×10^{-143}	3.99

Table 3. Numerical results for nonlinear equations.

Methods	$f_1(x), x_0 = -0.9$				$f_2(x), x_0 = -1.6$			
	N	$|x_1 - x_0|$	$|x_N - x_{N-1}|$	ρ	N	$|x_1 - x_0|$	$|x_N - x_{N-1}|$	ρ
$8^{th}KT$ (29)	3	0.1151	1.6238×10^{-61}	7.91	4	0.3876	7.2890×10^{-137}	7.99
$8^{th}LW$ (30)	3	0.1151	4.5242×10^{-59}	7.91	4	0.3904	1.1195×10^{-170}	8.00
$8^{th}PNPD$ (31)	3	0.1151	8.8549×10^{-56}	7.87	4	0.3734	2.3461×10^{-85}	7.99
$8^{th}SA1$ (32)	3	0.1151	3.4432×10^{-60}	7.88	4	0.3983	8.4343×10^{-121}	8.00
$8^{th}SA2$ (33)	3	0.1151	6.9371×10^{-67}	7.99	4	0.3927	5.9247×10^{-225}	7.99
$8^{th}CFGT$ (34)	3	0.1151	1.1715×10^{-82}	7.77	5	0.1532	2.0650×10^{-183}	7.99
$8^{th}CTV$ (35)	3	0.1151	4.4923×10^{-61}	7.94	4	0.3925	2.3865×10^{-252}	7.99
$8^{th}YM$ (12)	3	0.1151	1.1416×10^{-70}	7.96	4	0.3896	8.9301×10^{-163}	8.00
$16^{th}YM$ (16)	3	0.1151	0	15.99	3	0.3923	3.5535×10^{-85}	16.20

Table 4. Numerical results for nonlinear equations.

Methods	$f_3(x), x_0 = 0.9$				$f_4(x), x_0 = -1.9$			
	N	$|x_1 - x_0|$	$|x_N - x_{N-1}|$	ρ	N	$|x_1 - x_0|$	$|x_N - x_{N-1}|$	ρ
$8^{th}KT$ (29)	4	0.4659	5.0765×10^{-216}	7.99	4	1.4461	5.5095×10^{-204}	8.00
$8^{th}LW$ (30)	4	0.4660	2.7346×10^{-213}	7.99	4	1.4620	3.7210×10^{-146}	8.00
$8^{th}PNPD$ (31)	4	0.3821	9.9119×10^{-71}	8.02	4	1.3858	2.0603×10^{-116}	7.98
$8^{th}SA1$ (32)	4	0.4492	1.5396×10^{-122}	8.00	4	1.4170	2.2735×10^{-136}	7.99
$8^{th}SA2$ (33)	4	0.4652	4.1445×10^{-254}	7.98	4	1.4339	2.5430×10^{-166}	7.99
$8^{th}CFGT$ (34)	4	0.4654	2.4091×10^{-260}	7.99	4	1.4417	4.7007×10^{-224}	7.99
$8^{th}CTV$ (35)	4	0.4652	3.8782×10^{-288}	8.00	4	1.3957	3.7790×10^{-117}	7.99
$8^{th}YM$ (12)	4	0.4653	3.5460×10^{-309}	7.99	4	1.4417	2.9317×10^{-229}	7.99
$16^{th}YM$ (16)	3	0.4652	3.6310×10^{-154}	16.13	3	1.4434	1.8489×10^{-110}	16.36

Table 5. Numerical results for nonlinear equations.

Methods		$f_5(x)$, $x_0 = 1.2$				$f_6(x)$, $x_0 = 0.8$		
	N	$|x_1 - x_0|$	$|x_N - x_{N-1}|$	ρ	N	$|x_1 - x_0|$	$|x_N - x_{N-1}|$	ρ
$8^{th} KT$ (29)	5	1.8787	2.6836×10^{-182}	7.99	4	0.3898	6.0701×10^{-234}	7.99
$8^{th} LW$ (30)	6	40.5156	4.6640×10^{-161}	7.99	4	0.3898	6.1410×10^{-228}	7.99
$8^{th} PNPD$ (31)		Diverge			4	0.3894	3.6051×10^{-190}	7.99
$8^{th} SA1$ (32)	7	891.9802	2.1076×10^{-215}	9.00	4	0.3901	5.9608×10^{-245}	8.00
$8^{th} SA2$ (33)	4	0.7161	5.3670×10^{-128}	7.99	4	0.3900	8.3398×10^{-251}	8.61
$8^{th} CFGT$ (34)	5	2.8541	0	7.99	4	0.3900	0	7.99
$8^{th} CTV$ (35)	5	0.6192	1.6474×10^{-219}	9.00	4	0.3901	1.0314×10^{-274}	8.00
$8^{th} YM$ (12)	4	0.7733	1.3183×10^{-87}	7.98	4	0.3900	1.2160×10^{-286}	7.99
$16^{th} YM$ (16)	4	0.6985	0	16.10	3	0.3900	1.1066×10^{-143}	15.73

5. Applications to Some Real World Problems

5.1. Projectile Motion Problem

We consider the classical projectile problem [21,22] in which a projectile is launched from a tower of height $h > 0$, with initial speed v and at an angle θ with respect to the horizontal onto a hill, which is defined by the function ω, called the impact function which is dependent on the horizontal distance, x. We wish to find the optimal launch angle θ_m which maximizes the horizontal distance. In our calculations, we neglect air resistances.

The path function $y = P(x)$ that describes the motion of the projectile is given by

$$P(x) = h + x \tan \theta - \frac{gx^2}{2v^2} \sec^2 \theta \tag{36}$$

When the projectile hits the hill, there is a value of x for which $P(x) = \omega(x)$ for each value of x. We wish to find the value of θ that maximize x.

$$\omega(x) = P(x) = h + x \tan \theta - \frac{gx^2}{2v^2} \sec^2 \theta \tag{37}$$

Differentiating Equation (37) implicitly w.r.t. θ, we have

$$\omega'(x) \frac{dx}{d\theta} = x \sec^2 \theta + \frac{dx}{d\theta} \tan \theta - \frac{g}{v^2} \left(x^2 \sec^2 \theta \tan \theta + x \frac{dx}{d\theta} \sec^2 \theta \right) \tag{38}$$

Setting $\frac{dx}{d\theta} = 0$ in Equation (38), we have

$$x_m = \frac{v^2}{g} \cot \theta_m \tag{39}$$

or

$$\theta_m = \arctan \left(\frac{v^2}{g \, x_m} \right) \tag{40}$$

An enveloping parabola is a path that encloses and intersects all possible paths. Henelsmith [23] derived an enveloping parabola by maximizing the height of the projectile for a given horizontal distance x, which will give the path that encloses all possible paths. Let $w = \tan \theta$, then Equation (36) becomes

$$y = P(x) = h + xw - \frac{gx^2}{2v^2} (1 + w^2) \tag{41}$$

Differentiating Equation (41) w.r.t. w and setting $y' = 0$, Henelsmith obtained

$$y' = x - \frac{xg^2}{v^2}(w) = 0$$
$$w = \frac{v^2}{g\,x} \tag{42}$$

so that the enveloping parabola defined by

$$y_m = \rho(x) = h + \frac{v^2}{2g} - \frac{gx^2}{2v^2} \tag{43}$$

The solution to the projectile problem requires first finding x_m which satisfies $\rho(x) = w(x)$ and solving for θ_m in Equation (40) because we want to find the point at which the enveloping parabola ρ intersects the impact function w, and then find θ that corresponds to this point on the enveloping parabola. We choose a linear impact function $w(x) = 0.4x$ with $h = 10$ and $v = 20$. We let $g = 9.8$. Then we apply our IFs starting from $x_0 = 30$ to solve the non-linear equation

$$f(x) = \rho(x) - w(x) = h + \frac{v^2}{2g} - \frac{gx^2}{2v^2} - 0.4x$$

whose root is given by $x_m = 36.102990117.....$ and

$$\theta_m = \arctan\left(\frac{v^2}{g\,x_m}\right) = 48.5°.$$

Figure 1 shows the intersection of the path function, the enveloping parabola and the linear impact function for this application. The approximate solutions are calculated correct to 500 significant figures. The stopping criterion $|x_N - x_{N-1}| < \epsilon$, where $\epsilon = 10^{-50}$ is used. Table 6 shows that proposed method 16^{th}YM is converging better than other compared methods. Also, we observe that the computational order of convergence agrees with the theoretical order of convergence.

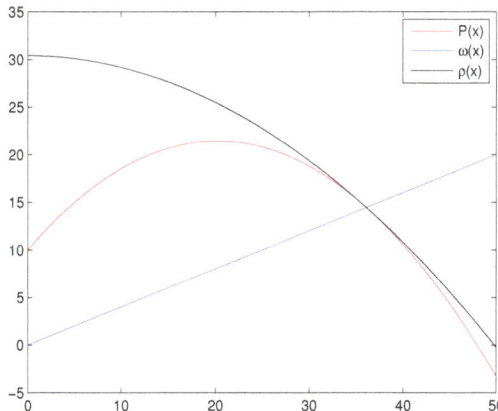

Figure 1. The enveloping parabola with linear impact function.

Table 6. Results of projectile problem.

IF	N	Error	cpu Time(s)	ρ
2^{nd}NR	7	4.3980×10^{-76}	1.074036	1.99
4^{th}YM	4	4.3980×10^{-76}	0.902015	3.99
8^{th}KT	3	1.5610×10^{-66}	0.658235	8.03
8^{th}LW	3	7.8416×10^{-66}	0.672524	8.03
8^{th}PNPD	3	4.2702×10^{-57}	0.672042	8.05
8^{th}SA1	3	1.2092×10^{-61}	0.654623	8.06
8^{th}CTV	3	3.5871×10^{-73}	0.689627	8.02
8^{th}YM	3	4.3980×10^{-76}	0.618145	8.02
16^{th}YM	3	0	0.512152	16.01

5.2. Planck's Radiation Law Problem

We consider the following Planck's radiation law problem found in [24]:

$$\varphi(\lambda) = \frac{8\pi ch\lambda^{-5}}{e^{ch/\lambda kT} - 1},\tag{44}$$

which calculates the energy density within an isothermal blackbody. Here, λ is the wavelength of the radiation, T is the absolute temperature of the blackbody, k is Boltzmann's constant, h is the Planck's constant and c is the speed of light. Suppose, we would like to determine wavelength λ which corresponds to maximum energy density $\varphi(\lambda)$. From (44), we get

$$\varphi'(\lambda) = \left(\frac{8\pi ch\lambda^{-6}}{e^{ch/\lambda kT} - 1}\right)\left(\frac{(ch/\lambda kT)e^{ch/\lambda kT}}{e^{ch/\lambda kT} - 1} - 5\right) = A \cdot B.$$

It can be checked that a maxima for φ occurs when $B = 0$, that is, when

$$\left(\frac{(ch/\lambda kT)e^{ch/\lambda kT}}{e^{ch/\lambda kT} - 1}\right) = 5.$$

Here putting $x = ch/\lambda kT$, the above equation becomes

$$1 - \frac{x}{5} = e^{-x}.\tag{45}$$

Define

$$f(x) = e^{-x} - 1 + \frac{x}{5}.\tag{46}$$

The aim is to find a root of the equation $f(x) = 0$. Obviously, one of the root $x = 0$ is not taken for discussion. As argued in [24], the left-hand side of (45) is zero for $x = 5$ and $e^{-5} \approx 6.74 \times 10^{-3}$. Hence, it is expected that another root of the equation $f(x) = 0$ might occur near $x = 5$. The approximate root of the Equation (46) is given by $x^* \approx 4.96511423174427630369$ with $x_0 = 3$. Consequently, the wavelength of radiation (λ) corresponding to which the energy density is maximum is approximated as

$$\lambda \approx \frac{ch}{(kT)4.96511423174427630369}.$$

Table 7 shows that proposed method 16^{th}YM is converging better than other compared methods. Also, we observe that the computational order of convergence agrees with the theoretical order of convergence.

Table 7. Results of Planck's radiation law problem.

IF	N	Error	*cpu Time*(s)	ρ
2^{nd}NR	7	1.8205×10^{-70}	0.991020	2.00
4^{th}YM	5	1.4688×10^{-181}	0.842220	4.00
8^{th}KT	4	4.0810×10^{-288}	0.808787	7.99
8^{th}LW	4	3.1188×10^{-268}	0.801304	7.99
8^{th}PNPD	4	8.0615×10^{-260}	0.800895	7.99
8^{th}SA1	4	1.9335×10^{-298}	0.791706	8.00
8^{th}CTV	4	5.8673×10^{-282}	0.831006	8.00
8^{th}YM	4	2.5197×10^{-322}	0.855137	8.00
16^{th}YM	3	8.3176×10^{-153}	0.828053	16.52

Hereafter, we will study the optimal fourth and eighth order methods along with Newton's method.

6. Corresponding Conjugacy Maps for Quadratic Polynomials

In this section, we discuss the rational map R_p arising from $2^{nd}NR$, proposed methods $4^{th}YM$ and $8^{th}YM$ applied to a generic polynomial with simple roots.

Theorem 3. *($2^{nd}NR$) [18] For a rational map $R_p(z)$ arising from Newton's method (4) applied to $p(z) = (z - a)(z - b)$, $a \neq b$, $R_p(z)$ is conjugate via the a Möbius transformation given by $M(z) = (z - a)/(z - b)$ to*

$$S(z) = z^2.$$

Theorem 4. *($4^{th}YM$) For a rational map $R_p(z)$ arising from Proposed Method (7) applied to $p(z) = (z - a)(z - b)$, $a \neq b$, $R_p(z)$ is conjugate via the a Möbius transformation given by $M(z) = (z - a)/(z - b)$ to*

$$S(z) = z^4.$$

Proof. Let $p(z) = (z - a)(z - b)$, $a \neq b$, and let M be Möbius transformation given by $M(z) = (z - a)/(z - b)$ with its inverse $M^{-1}(z) = \frac{(zb-a)}{(z-1)}$, which may be considered as map from $\mathbb{C} \cup \{\infty\}$. We then have

$$S(z) = M \circ R_p \circ M^{-1}(z) = M\left(R_p\left(\frac{zb - a}{z - 1}\right)\right) = z^4.$$

$\square$

Theorem 5. *($8^{th}YM$) For a rational map $R_p(z)$ arising from Proposed Method (12) applied to $p(z) = (z - a)(z - b)$, $a \neq b$, $R_p(z)$ is conjugate via the a Möbius transformation given by $M(z) = (z - a)/(z - b)$ to*

$$S(z) = z^8.$$

Proof. Let $p(z) = (z - a)(z - b)$, $a \neq b$, and let M be Möbius transformation given by $M(z) = (z - a)/(z - b)$ with its inverse $M^{-1}(z) = \frac{(zb-a)}{(z-1)}$, which may be considered as map from $\mathbb{C} \cup \{\infty\}$. We then have

$$S(z) = M \circ R_p \circ M^{-1}(z) = M\left(R_p\left(\frac{zb - a}{z - 1}\right)\right) = z^8.$$

$\square$

Remark 1. *The methods (29)–(35) are given without proof, which can be worked out with the help of Mathematica.*

Remark 2. *All the maps obtained above are of the form $S(z) = z^p R(z)$, where $R(z)$ is either unity or a rational function and p is the order of the method.*

7. Basins of Attraction

The study of dynamical behavior of the rational function associated to an iterative method gives important information about convergence and stability of the method. The basic definitions and dynamical concepts of rational function which can found in [4,25].

We take a square $\mathbb{R} \times \mathbb{R} = [-2,2] \times [-2,2]$ of 256×256 points and we apply our iterative methods starting in every $z^{(0)}$ in the square. If the sequence generated by the iterative method attempts a zero z_j^* of the polynomial with a tolerance $|f(z^{(k)})| < \times 10^{-4}$ and a maximum of 100 iterations, we decide that $z^{(0)}$ is in the basin of attraction of this zero. If the iterative method starting in $z^{(0)}$ reaches a zero in N iterations ($N \leq 100$), then we mark this point $z^{(0)}$ with colors if $|z^{(N)} - z_j^*| < \times 10^{-4}$. If $N > 50$, we conclude that the starting point has diverged and we assign a dark blue color. Let N_D be a number of diverging points and we count the number of starting points which converge in 1, 2, 3, 4, 5 or above 5 iterations. In the following, we describe the basins of attraction for Newton's method and some higher order Newton type methods for finding complex roots of polynomials $p_1(z) = z^2 - 1$, $p_2(z) = z^3 - 1$ and $p_3(z) = z^5 - 1$.

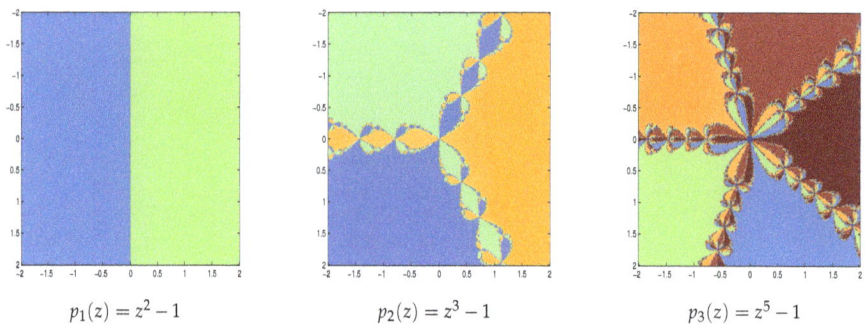

$$p_1(z) = z^2 - 1 \qquad p_2(z) = z^3 - 1 \qquad p_3(z) = z^5 - 1$$

Figure 2. Basins of attraction for $2^{nd}NR$ for the polynomial $p_1(z)$, $p_2(z)$, $p_3(z)$.

Figures 2 and 3 shows the polynomiographs of the methods for the polynomial $p_1(z)$. We can see that the methods 2^{nd}NR, 4^{th}YM, 8^{th}SA2 and 8^{th}YM performed very nicely. The methods 4^{th}SB, 4^{th}SJ, 8^{th}KT, 8^{th}LW, 8^{th}PNPD, 8^{th}SA1, 8^{th}CFGT and 8^{th}CTV are shows some chaotic behavior near the boundary points. The method 4^{th}CLND have sensitive to the choice of initial guess in this case.

Figures 2 and 4 shows the polynomiographs of the methods for the polynomial $p_2(z)$. We can see that the methods 2^{nd}NR, 4^{th}YM, 8^{th}SA2 and 8^{th}YM performed very nicely. The methods 4^{th}SB, 8^{th}KT, 8^{th}LW and 8^{th}CTV are shows some chaotic behavior near the boundary points. The methods 4^{th}CLND, 4^{th}SJ, 8^{th}PNPD, 8^{th}SA1, and 8^{th}CFGT have sensitive to the choice of initial guess in this case.

Figures 2 and 5 shows the polynomiographs of the methods for the polynomial $p_3(z)$. We can see that the methods 4^{th}YM, 8^{th}SA2 and 8^{th}YM are shows some chaotic behavior near the boundary points. The methods 2^{nd}NR, 4^{th}SB, 4^{th}CLND, 4^{th}SJ, 8^{th}KT, 8^{th}LW, 8^{th}PNPD, 8^{th}SA1, 8^{th}CFGT and 8^{th}CTV have sensitive to the choice of initial guess in this case. In Tables 8–10, we classify the number of converging and diverging grid points for each iterative method.

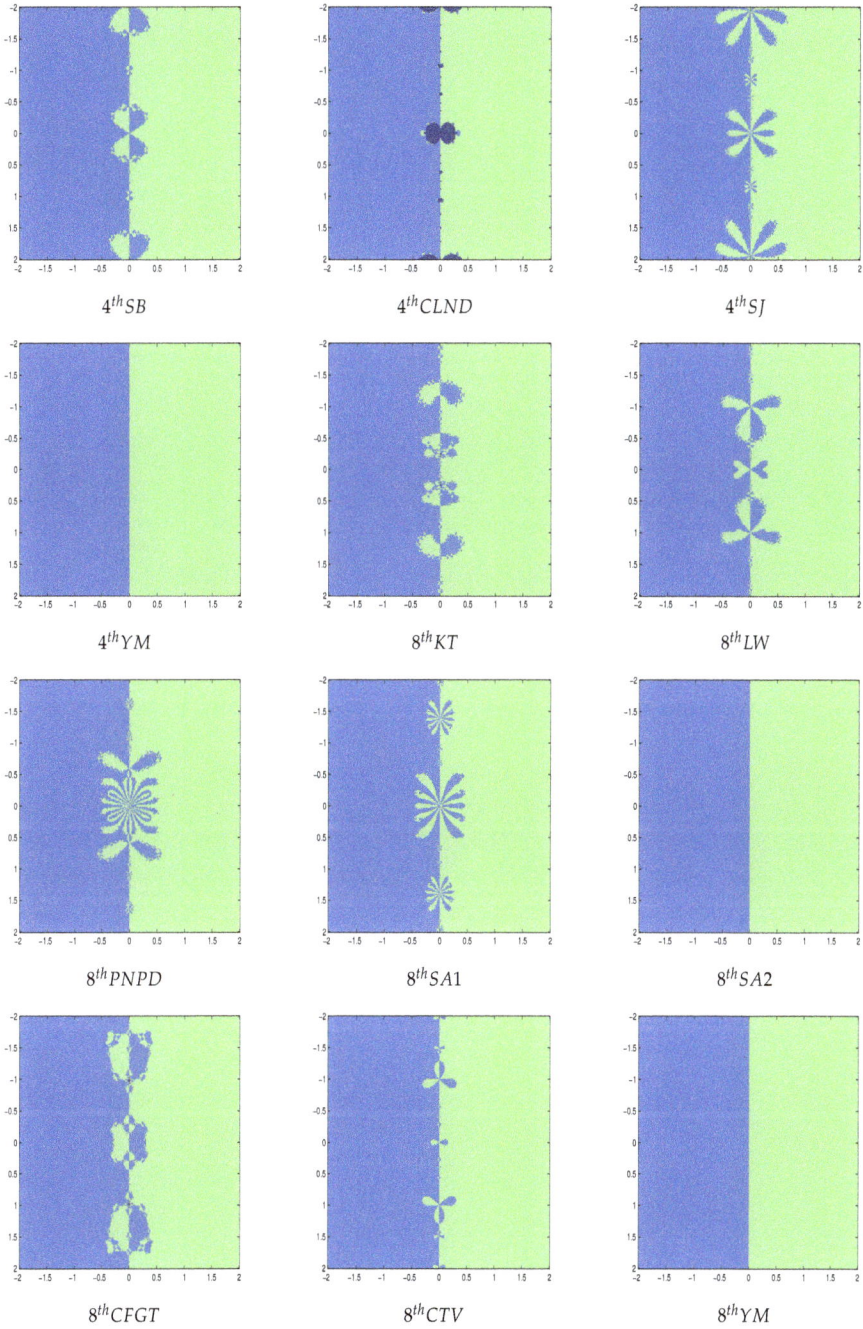

Figure 3. Basins of attraction for $p_1(z) = z^2 - 1$.

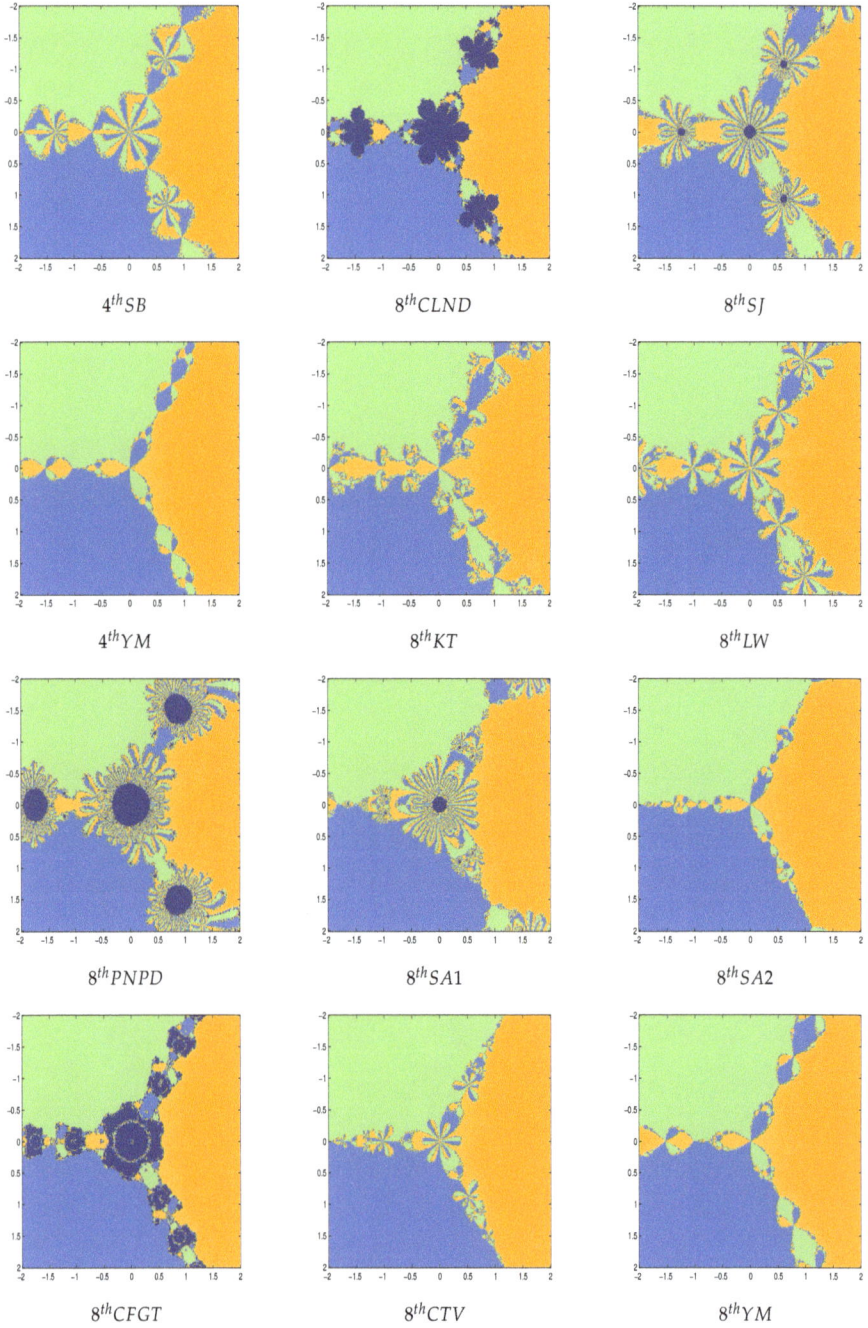

$4^{th}SB$

$8^{th}CLND$

$8^{th}SJ$

$4^{th}YM$

$8^{th}KT$

$8^{th}LW$

$8^{th}PNPD$

$8^{th}SA1$

$8^{th}SA2$

$8^{th}CFGT$

$8^{th}CTV$

$8^{th}YM$

Figure 4. Basins of attraction for $p_2(z) = z^3 - 1$.

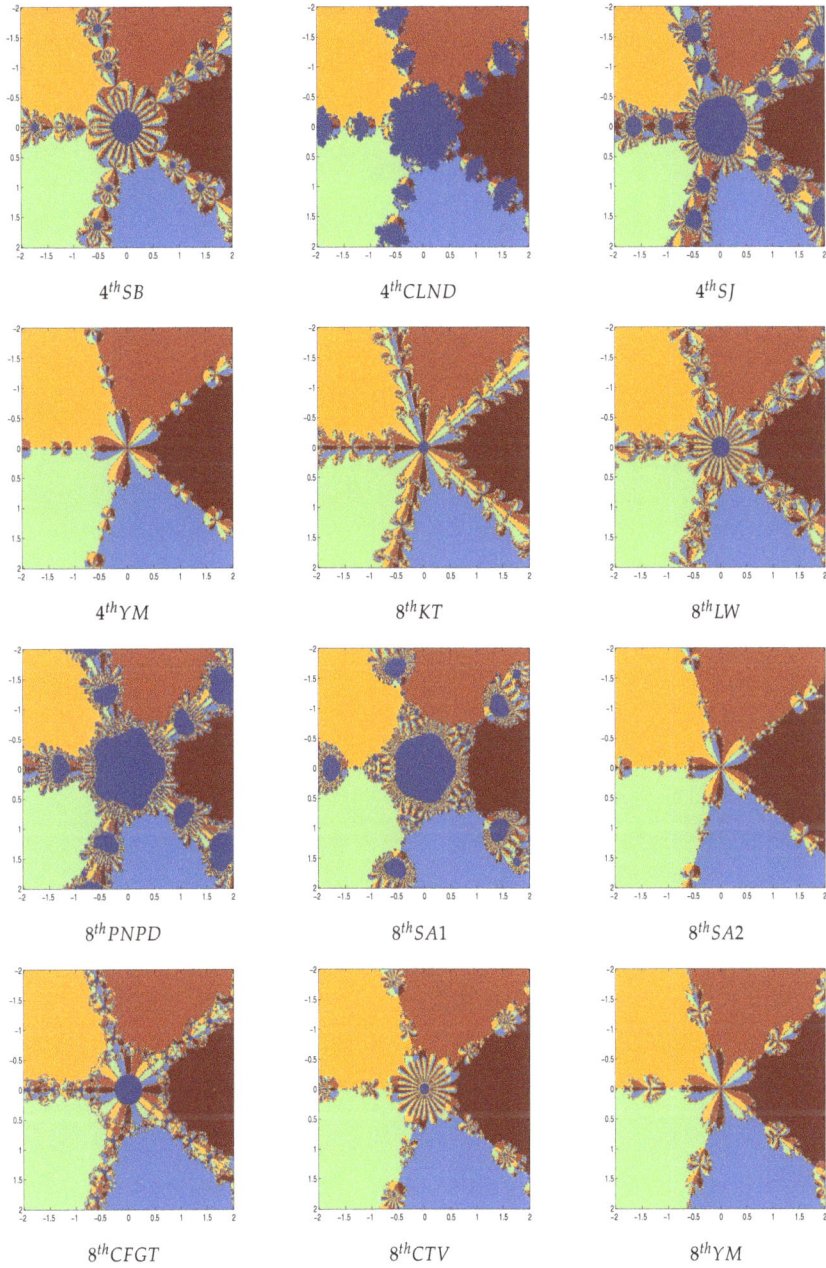

$4^{th}SB$ $4^{th}CLND$ $4^{th}SJ$

$4^{th}YM$ $8^{th}KT$ $8^{th}LW$

$8^{th}PNPD$ $8^{th}SA1$ $8^{th}SA2$

$8^{th}CFGT$ $8^{th}CTV$ $8^{th}YM$

Figure 5. Basins of attraction for $p_3(z) = z^5 - 1$.

Table 8. Results of the polynomials $p_1(z) = z^2 - 1$.

IF	$N = 1$	$N = 2$	$N = 3$	$N = 4$	$N = 5$	$N > 5$	N_D
2^{nd}NR	4	516	7828	23,272	20,548	13,368	0
4^{th}SB	340	22,784	29,056	6836	2928	3592	0
4^{th}CLND	372	24,600	29,140	6512	2224	2688	1076
4^{th}SJ	300	19,816	28,008	5844	2968	8600	0
4^{th}YM	520	31,100	27,520	4828	1208	360	0
8^{th}KT	4684	44,528	9840	3820	1408	1256	24
8^{th}LW	4452	43,236	11,408	3520	1540	1380	0
8^{th}PNPD	2732	39,768	13,112	3480	1568	4876	16
8^{th}SA1	4328	45,824	8136	2564	1484	3200	0
8^{th}SA2	15,680	45,784	3696	376	0	0	0
8^{th}CFGT	9616	43,716	7744	2916	980	564	64
8^{th}CTV	7124	48,232	7464	1892	632	192	0
8^{th}YM	8348	50,792	5572	824	0	0	0

Table 9. Results of the polynomials $p_2(z) = z^3 - 1$.

IF	$N = 1$	$N = 2$	$N = 3$	$N = 4$	$N = 5$	$N > 5$	N_D
2^{nd}NR	0	224	2908	11,302	19,170	31,932	0
4^{th}SB	160	9816	27,438	9346	5452	13,324	6
4^{th}CLND	170	11,242	28,610	9984	4202	11,328	7176
4^{th}SJ	138	7760	25,092	8260	5058	19,228	1576
4^{th}YM	270	18,064	30,374	9862	3688	3278	0
8^{th}KT	2066	34,248	11,752	6130	4478	6862	0
8^{th}LW	2092	33,968	12,180	4830	3030	9436	0
8^{th}PNPD	1106	25,712	11,258	3854	1906	21,700	10,452
8^{th}SA1	1608	36,488	12,486	3718	1780	9456	872
8^{th}SA2	6432	46,850	9120	2230	640	264	0
8^{th}CFGT	3688	40,740	13,696	4278	1390	1744	7395
8^{th}CTV	3530	43,554	11,724	3220	1412	2096	0
8^{th}YM	3816	43,596	12,464	3636	1302	722	0

Table 10. Results of the polynomials $p_3(z) = z^5 - 1$.

IF	$N = 1$	$N = 2$	$N = 3$	$N = 4$	$N = 5$	$N > 5$	N_D
2^{nd}NR	2	100	1222	4106	7918	52,188	638
4^{th}SB	76	3850	15,458	18,026	5532	22,594	5324
4^{th}CLND	86	4476	18,150	17,774	5434	19,616	12,208
4^{th}SJ	62	3094	11,716	16,840	5682	28,142	19,900
4^{th}YM	142	7956	27,428	15,850	5726	8434	0
8^{th}KT	950	17,884	20,892	5675	4024	16,111	217
8^{th}LW	1032	18,764	20,622	5056	3446	16,616	1684
8^{th}PNPD	496	12,770	21,472	6576	2434	21,788	14,236
8^{th}SA1	692	26,212	15,024	4060	1834	17,714	8814
8^{th}SA2	2662	41,400	12,914	4364	1892	2304	0
8^{th}CFGT	2008	21,194	23,734	6180	3958	8462	1953
8^{th}CTV	1802	36,630	13,222	4112	2096	7674	350
8^{th}YM	1736	27,808	21,136	5804	2704	6348	0

We note that a point z_0 belongs to the Julia set if and only if the dynamics in a neighborhood of z_0 displays sensitive dependence on the initial conditions, so that nearby initial conditions lead to wildly different behavior after a number of iterations. For this reason, some of the methods are getting divergent points. The common boundaries of these basins of attraction constitute the Julia set of the iteration function. It is clear that one has to use quantitative measures to distinguish between the methods, since we have a different conclusion when just viewing the basins of attraction.

In order to summarize the results, we have compared mean number of iteration and total number of functional evaluations (TNFE) for each polynomials and each methods in Table 11. The best method based on the comparison in Table 11 is 8^{th}SA2. The method with the fewest number of functional evaluations per point is 8^{th}SA2 followed closely by 4^{th}YM. The fastest method is 8^{th}SA2 followed closely by 8^{th}YM. The method with highest number of functional evaluation and slowest method is 8^{th}PNPD.

Table 11. Mean number of iteration (N_μ) and TNFE for each polynomials and each methods.

IF	N_μ for $p_1(z)$	N_μ for $p_2(z)$	N_μ for $p_3(z)$	Average	TNFE
2^{nd}NR	4.7767	6.4317	9.8531	7.0205	14.0410
4^{th}SB	3.0701	4.5733	9.2701	5.6378	16.9135
4^{th}CLND	3.6644	8.6354	12.8612	8.3870	25.1610
4^{th}SJ	3.7002	7.0909	14.5650	8.4520	25.3561
4^{th}YM	2.6366	3.1733	4.0183	3.2760	9.8282
8^{th}KT	2.3647	3.1270	4.4501	3.3139	13.2557
8^{th}LW	2.3879	3.5209	6.3296	4.0794	16.3178
8^{th}PNPD	2.9959	10.5024	12.3360	8.6114	34.4457
8^{th}SA1	2.5097	4.5787	9.7899	5.6262	22.5044
8^{th}SA2	1.8286	2.1559	2.5732	2.1859	8.7436
8^{th}CFGT	2.1683	2.8029	3.4959	2.8223	11.2894
8^{th}CTV	2.1047	2.4708	3.9573	2.8442	11.3770
8^{th}YM	1.9828	2.3532	3.3617	2.5659	10.2636

8. Concluding Remarks and Future Work

In this work, we have developed optimal fourth, eighth and sixteenth order iterative methods for solving nonlinear equations using the divided difference approximation. The methods require the computations of three functions evaluations reaching order of convergence is four, four functions evaluations reaching order of convergence is eight and five functions evaluations reaching order of convergence is sixteen. In the sense of convergence analysis and numerical examples, the Kung-Traub's conjecture is satisfied. We have tested some examples using the proposed schemes and some known schemes, which illustrate the superiority of the proposed method 16^{th}YM. Also, proposed methods and some existing methods have been applied on the Projectile motion problem and Planck's radiation law problem. The results obtained are interesting and encouraging for the new method 16^{th}YM. The numerical experiments suggests that the new methods would be valuable alternative for solving nonlinear equations. Finally, we have also compared the basins of attraction of various fourth and eighth order methods in the complex plane.

Future work includes:

- Now we are investigating the proposed scheme to develop optimal methods of arbitrarily high order with Newton's method, as in [26].
- Also, we are investigating to develop derivative free methods to study dynamical behavior and local convergence, as in [27,28].

Author Contributions: The contributions of both the authors have been similar. Both of them have worked together to develop the present manuscript.

Funding: This paper is supported by three project funds: 1. National College Students Innovation and entrepreneurship training program of Ministry of Education of the People's Republic of China in 2017: Internet Animation Company in Minority Areas–Research Model of "Building Dream" Animation Company. (Project number: 201710684001). 2. Yunnan Provincial Science and Technology Plan Project University Joint Project 2017: Research on Boolean Satisfiability Dividing and Judging Method Based on Clustering and Partitioning (Project number: 2017FH001-056). 3. Qujing Normal college scientific research fund special project (Project number: 2018zx003).

Acknowledgments: The authors would like to thank the editors and referees for the valuable comments and for the suggestions to improve the readability of the paper.

Conflicts of Interest: The authors declare no conflict of interest.

References

1. Ostrowski, A.M. *Solutions of Equations and System of Equations*; Academic Press: New York, NY, USA, 1960.
2. Kung, H.T.; Traub, J.F. Optimal order of one-point and multipoint iteration. *J. Assoc. Comput. Mach.* **1974**, *21*, 643–651.
3. Amat, S.; Busquier, S.; Plaza, S. Dynamics of a family of third-order iterative methods that do not require using second derivatives. *Appl. Math. Comput.* **2004**, *154*, 735–746.
4. Amat, S.; Busquier, S.; Plaza, S. Review of some iterative root-finding methods from a dynamical point of view. *Scientia* **2004**, *10*, 3–35.
5. Babajee, D.K.R; Madhu, K.; Jayaraman, J. A family of higher order multi-point iterative methods based on power mean for solving nonlinear equations. *Afrika Matematika* **2016**, *27*, 865–876. [CrossRef]
6. Cordero, A.; Fardi, M.; Ghasemi, M.; Torregrosa, J.R. Accelerated iterative methods for finding solutions of nonlinear equations and their dynamical behavior. *Calcolo* **2014**, *51*, 17–30.
7. Cordero, A.; Torregrosa, J.R.; Vasileva, M.P. A family of modified ostrowski's methods with optimal eighth order of convergence. *Appl. Math. Lett.* **2011**, *24*, 2082–2086.
8. Liu, L.; Wang, X. Eighth-order methods with high efficiency index for solving nonlinear equations. *Appl. Math. Comput.* **2010**, *215*, 3449–3454.
9. Madhu, K. Some New Higher Order Multi-Point Iterative Methods and Their Applications to Differential and Integral Equation and Global Positioning System. Ph.D. Thesis, Pndicherry University, Kalapet, India, June 2016.
10. Madhu, K.; Jayaraman, J. Higher order methods for nonlinear equations and their basins of attraction. *Mathematics* **2016**, *4*, 22.
11. Petkovic, M.S.; Neta, B.; Petkovic, L.D.; Dzunic, J. *Multipoint Methods for Solving Nonlinear Equations*; Elsevier: Amsterdam, The Netherlands, 2012.
12. Sharma, J.R.; Arora, H. An efficient family of weighted-newton methods with optimal eighth order convergence. *Appl. Math. Lett.* **2014**, *29*, 1–6.
13. Sharma, J.R.; Arora, H. A new family of optimal eighth order methods with dynamics for nonlinear equations. *Appl. Math. Comput.* **2016**, *273*, 924–933.
14. Soleymani, F.; Khratti, S.K.; Vanani, S.K. Two new classes of optimal Jarratt-type fourth-order methods. *Appl. Math. Lett.* **2011**, *25*, 847–853.
15. Wait, R. *The Numerical Solution of Algebraic Equations*; John Wiley and Sons: Hoboken, NJ, USA, 1979.
16. Khan, Y.; Fardi, M.; Sayevand, K. A new general eighth-order family of iterative methods for solving nonlinear equations. *Appl. Math. Lett.* **2012**, *25*, 2262–2266.
17. Cordero, A.; Torregrosa, J.R. Variants of Newton's method using fifth-order quadrature formulas. *Appl. Math. Comput.* **2007**, *190*, 686–698.
18. Sharma, R.; Bahl, A. An optimal fourth order iterative method for solving nonlinear equations and its dynamics. *J. Complex Anal.* **2015**, *2015*, 259167.
19. Chun, C; Lee, M. Y.; Neta, B; Dzunic, J. On optimal fourth-order iterative methods free from second derivative and their dynamics. *Appl. Math. Comput.* **2012**, *218*, 6427–6438. [CrossRef]
20. Singh, A.; Jaiswal, J.P. Several new third-order and fourth-order iterative methods for solving nonlinear equations. *Int. J. Eng. Math.* **2014**, *2014*, 828409.
21. Babajee, D.K.R.; Madhu, K. Comparing two techniques for developing higher order two-point iterative methods for solving quadratic equations. *SeMA J.* **2018**, 1–22. [CrossRef]
22. Kantrowitz, R.; Neumann, M.M. Some real analysis behind optimization of projectile motion. *Mediterr. J. Math.* **2014**, *11*, 1081–1097.
23. Henelsmith, N. *Finding the Optimal Launch Angle*; Whitman College: Walla Walla, WA, USA, 2016.
24. Bradie, B. *A Friendly Introduction to Numerical Analysis*; Pearson Education Inc.: New Delhi, India, 2006.
25. Scott, M.; Neta, B.; Chun, C. Basin attractors for various methods. *Appl. Math. Comput.* **2011**, *218*, 2584–2599.

26. Cordero, A.; Hueso, J. L.; Martinez, E.; Torregrosa, J.R. Generating optimal derivative free iterative methods for nonlinear equations by using polynomial interpolation. *Appl. Math. Comput.* **2013**, *57* 1950–1956. [CrossRef]

27. Argyros, I.K.; Magrenan, A.A.; Orcos, L. Local convergence and a chemical application of derivative free root finding methods with one parameter based on interpolation. *J. Math. Chem.* **2016**, *54*, 1404–1416.

28. Zafar, F.; Cordero, A.; Torregrosa, J.R. An efficient family of optimal eighth-order multiple root finders. *Mathematics* **2018**, *6*, 310.

![Σ mathematics logo] *mathematics*

MDPI

Article

Design and Complex Dynamics of Potra–Pták-Type Optimal Methods for Solving Nonlinear Equations and Its Applications

Prem B. Chand [1,†], Francisco I. Chicharro [2,†], Neus Garrido [3,*,†] and Pankaj Jain [1,†]

1   Department of Mathematics, South Asian University, Akbar Bhawan, Chanakya Puri, New Delhi 110021, India; premchand13@gmail.com (P.B.C.); pankaj.jain@sau.ac.in (P.J.)
2   Escuela Superior de Ingeniería y Tecnología, Universidad Internacional de La Rioja, Av. La Paz 137, 26006 Logroño, Spain; francisco.chicharro@unir.net
3   Institute for Multidisciplinary Mathematics, Universitat Politècnica de València, Cno. de Vera s/n, 46022 València, Spain
*   Correspondence: neugarsa@upvnet.upv.es
†   These authors contributed equally to this work.

Received: 10 September 2019; Accepted: 9 October 2019; Published: 11 October 2019

Abstract: In this paper, using the idea of weight functions on the Potra–Pták method, an optimal fourth order method, a non optimal sixth order method, and a family of optimal eighth order methods are proposed. These methods are tested on some numerical examples, and the results are compared with some known methods of the corresponding order. It is proved that the results obtained from the proposed methods are compatible with other methods. The proposed methods are tested on some problems related to engineering and science. Furthermore, applying these methods on quadratic and cubic polynomials, their stability is analyzed by means of their basins of attraction.

Keywords: nonlinear equations; Potra–Pták method; optimal methods; weight function; basin of attraction

PACS: 65H05

---

1. Introduction

For solving nonlinear equations iteratively, the Newton's method given by

$$x_{n+1} = x_n - \frac{f(x_n)}{f'(x_n)}$$

is one of the most commonly used methods. The efficiency index as defined by Ostroswki in [1], which relates the order of convergence of a method p with the number of function evaluations per iteration d, is given by the expression $p^{1/d}$. Newton's method is quadratically convergent and requires two function evaluations per iteration and, thereby, has the efficiency index value of $2^{1/2} \approx 1.414$. Numerous methods have appeared giving higher order of convergence or better efficiency. One of the recent strategies to increase the order of the methods is the use of weight functions [2–5]. In this regard, Sharma and Behl [6] presented the fourth order method:

$$
\begin{aligned}
y_n &= x_n - \frac{2}{3}\frac{f(x_n)}{f'(x_n)}, \\
x_{n+1} &= x_n - \left(-\frac{1}{2} + \frac{3}{8}\frac{f'(y_n)}{f'(x_n)} + \frac{9}{8}\frac{f'(x_n)}{f'(y_n)}\right)\frac{f(x_n)}{f'(x_n)}.
\end{aligned}
\tag{1}
$$

Similarly, Sharifi et al. [7] used weight functions on the third order Heun's method and proposed the fourth order method

$$
\begin{aligned}
y_n &= x_n - \frac{2}{3}\frac{f(x_n)}{f'(x_n)},\\
x_{n+1} &= x_n - \frac{f(x_n)}{4}\left(\frac{1}{f'(x_n)} + \frac{3}{f'(y_n)}\right)\left(1 + \frac{3}{8}\left(\frac{f'(y_n)}{f'(x_n)} - 1\right)^2 - \frac{69}{64}\left(\frac{f'(y_n)}{f'(x_n)} - 1\right)^3 + \frac{f(x_n)}{f'(y_n)}\right).
\end{aligned}
\tag{2}
$$

According to Kung and Traub [8], an iterative method is said to be optimal if its order is 2^{d-1}, where d is the number of function evaluations per iteration. Notice that Newton's method as well as (1) and (2) are all optimal.

Potra and Pták [9], as an attempt to improve Newton's method, gave the method

$$
\begin{aligned}
y_n &= x_n - \frac{f(x_n)}{f'(x_n)},\\
x_{n+1} &= x_n - \frac{f(x_n)+f(y_n)}{f'(x_n)}.
\end{aligned}
\tag{3}
$$

This method is cubically convergent but is not optimal, as it requires three function evaluations per iteration.

The aim, in the present paper, is to further investigate the method (3). Precisely, we use weight functions and improve the order of convergence of (3). We do it in three ways which correspond to the methods of orders 4, 6 and 8. Out of these, the methods with orders 4 and 8 are optimal.

Dynamics of a rational operator give important information about the convergence, efficiency and stability of the iterative methods. During the last few decades, many researchers, e.g., [10–16] and references therein, study the dynamical behavior of rational operators associated with iterative methods. Furthermore, there is an extensive literature [17–21] to understand and implement further results on the dynamics of rational functions. In this paper, we also analyze the dynamical behavior of the methods that we have developed in this paper. Furthermore, at the end of this work, the basins of attraction are also presented and compared among the proposed and other methods of the corresponding order.

The remaining part of the paper is organized as follows. In Section 2, the development of the methods and their convergence analysis are given. In Section 3, the proposed methods are tested on some functions, and the results are compared with other methods in the head of Numerical Examples. In Section 4, the proposed methods are tested on some engineering and science related designs. Section 5 is devoted to analyze the stability of the introduced methods by means of complex dynamics. In this sense, the study of the rational function resulting from the application of the methods to several nonlinear functions is developed, and their basins of attraction are represented. Finally, Section 6 covers the conclusions of the research.

2. Development of Methods and Their Convergence Analysis

In this section, the methods of order four, six and eight are introduced, and its convergence is analyzed.

2.1. Optimal Fourth Order Method

Based on the Potra–Pták method (3), we propose the following two-step method using a weight function, whose iterative expression is

$$
\begin{aligned}
y_n &= x_n - \frac{f(x_n)}{f'(x_n)},\\
x_{n+1} &= x_n - w(t_n)\frac{f(x_n)+f(y_n)}{f'(x_n)},
\end{aligned}
\tag{4}
$$

where $w(t_n) = a_1 + a_2 t_n + a_3 t_n^2$ and $t_n = \frac{f(y_n)}{f(x_n)}$. The convergence of (4) is proved in the following theorem.

Theorem 1. *Let f be a real or complex valued function defined in the interval I having a sufficient number of smooth derivatives. Let α be a simple root of the equation $f(x) = 0$ and the initial point x_0 is close enough to α. Then, the method (4) is fourth order of convergence if $a_1 = 1$, $a_2 = 0$ and $a_3 = 2$.*

Proof. We denote $c_j = \frac{f^{(j)}(\alpha)}{j! \, f'(\alpha)}$. Let $e_n = x_n - \alpha$ be the error in x_n. Then, Taylor's series expansion of $f(x_n)$ and $f'(x_n)$ about α gives

$$f(x_n) = f'(\alpha) \left(e_n + c_2 e_n^2 + c_3 e_n^3 + c_4 e_n^4 + c_5 e_n^5 + c_6 e_n^6 + c_7 e_n^7 + c_8 e_n^8 + O(e_n^9) \right) \tag{5}$$

and

$$f'(x_n) = f'(\alpha) \left(1 + 2c_2 e_n + 3c_3 e_n^2 + 4c_4 e_n^3 + 5c_5 e_n^4 + 6c_6 e_n^5 + 7c_7 e_n^6 + 8c_8 e_n^7 + O(e_n^8) \right). \tag{6}$$

Let $d_n = y_n - \alpha$, then, from the first equation of (4), we get

$$
\begin{aligned}
d_n &= c_2 e_n^2 - 2\left(c_2^2 - c_3\right) e_n^3 + \left(4c_2^3 - 7c_2 c_3 + 3c_4\right) e_n^4 + \left(-8c_2^4 + 20c_3 c_2^2 - 10c_4 c_2 - 6c_3^2 + 4c_5\right) e_n^5 \\
&\quad + \left(16c_2^5 - 52c_3 c_2^3 + 28c_4 c_2^2 + \left(33c_3^2 - 13c_5\right) c_2 - 17c_3 c_4 + 5c_6\right) e_n^6 \\
&\quad - 2\left(16c_2^6 - 64c_3 c_2^4 + 36c_4 c_2^3 + 9\left(7c_3^2 - 2c_5\right) c_2^2 + \left(8c_6 - 46c_3 c_4\right) c_2 - 9c_3^3 + 6c_4^2 + 11c_3 c_5 - 3c_7\right) e_n^7 \\
&\quad + \left(64c_2^7 - 304c_3 c_2^5 + 176c_4 c_2^4 + \left(408c_3^2 - 92c_5\right) c_2^3 + \left(44c_6 - 348c_3 c_4\right) c_2^2 \right. \\
&\quad \left. + 75c_3^2 c_4 - 31c_4 c_5 - 27c_3 c_6 + c_2\left(-135c_3^3 + 118c_5 c_3 + 64c_4^2 - 19c_7\right) + 7c_8\right) e_n^8 + O(e_n^9)
\end{aligned}
$$

so that, using Taylor's series expansion of $f(y_n)$ about α, we get

$$
\begin{aligned}
f(y_n) &= f(d_n + \alpha) \\
&= f'(\alpha) \left[c_2 e_n^2 - 2(c_2^2 - c_3) e_n^3 + \left(5c_2^3 - 7c_2 c_3 + 3c_4\right) e_n^4 - 2\left(6c_2^4 - 12c_3 c_2^2 + 5c_4 c_2 + 3c_3^2 - 2c_5\right) e_n^5 \right. \\
&\quad + \left(28c_2^5 - 73c_3 c_2^3 + 34c_4 c_2^2 + \left(37c_3^2 - 13c_5\right) c_2 - 17c_3 c_4 + 5c_6\right) e_n^6 \\
&\quad - 2\left(32c_2^6 - 103c_3 c_2^4 + 52c_4 c_2^3 + \left(80c_3^2 - 22c_5\right) c_2^2 + \left(8c_6 - 52c_3 c_4\right) c_2 \right. \\
&\quad \left. - 9c_3^3 + 6c_4^2 + 11c_3 c_5 - 3c_7\right) e_n^7 + \left(144c_2^7 - 552c_3 c_2^5 + 297c_4 c_2^4 + 2\left(291c_3^2 - 67c_5\right) c_2^3 \right. \\
&\quad + \left(54c_6 - 455c_3 c_4\right) c_2^2 + 75c_3^2 c_4 - 31c_4 c_5 - 27c_3 c_6 + c_2\left(-147c_3^3 + 134c_5 c_3 + 73c_4^2 - 19c_7\right) \\
&\quad \left. + 7c_8\right) e_n^8 + O(e_n^9) \right].
\end{aligned}
\tag{7}
$$

Now, from (5) and (7), we get

$$
\begin{aligned}
t_n &= \frac{f(y_n)}{f(x_n)} \\
&= c_2 e_n + \left(-3c_2^2 + 2c_3\right) e_n^2 + \left(8c_2^3 - 10c_2 c_3 + 3c_4\right) e_n^3 + \left(-20c_2^4 + 37c_2^2 c_3 - 14c_2 c_4 - 8c_3^2 + 4c_5\right) e_n^4 \\
&\quad + \left(48c_2^5 - 118c_2^3 c_3 + 51c_2^2 c_4 + 55c_2 c_3^2 - 18c_2 c_5 - 22c_3 c_4 + 5c_6\right) e_n^5 \\
&\quad + \left(-112c_2^6 + 344c_3 c_2^4 - 163c_4 c_2^3 + \left(65c_5 - 252c_3^2\right) c_2^2 + 2\left(75c_3 c_4 - 11c_6\right) c_2 \right. \\
&\quad \left. + 26c_3^3 - 15c_4^2 - 28c_3 c_5 + 6c_7\right) e_n^6 + O(e_n^7).
\end{aligned}
\tag{8}
$$

Therefore, using the results obtained above in the second equation of (4), we get

$$
\begin{aligned}
e_{n+1} &= (1 - a_1) e_n - a_2 c_2 e_n^2 + \left(2a_1 c_2^2 + 3a_2 c_2^2 - 2a_2 c_3 - a_3 c_2^2\right) e_n^3 \\
&\quad + \left(-9a_1 c_2^3 + 7a_1 c_2 c_3 - 6a_2 c_2^3 + 10a_2 c_2 c_3 - 3a_2 c_4 + 6a_3 c_2^3 - 4a_3 c_2 c_3\right) e_n^4 + O(e_n^5).
\end{aligned}
\tag{9}
$$

In order to obtain fourth order of convergence, in view of (9), we must have

$$\begin{aligned} 1 - a_1 &= 0, \\ a_2 &= 0, \\ 2a_1 c_2^2 + 3a_2 c_2^2 - 2a_2 c_3 - a_3 c_2^2 &= 0, \end{aligned}$$

which gives $a_1 = 1$, $a_2 = 0$ and $a_3 = 2$. Therefore, from (9), the error equation of the method (4) becomes

$$e_{n+1} = (3c_2^3 - c_2 c_3)e_n^4 + \mathcal{O}\left(e_n^5\right),$$

and the assertion follows. $\square$

In view of Theorem 1, the proposed fourth order method is

$$\begin{aligned} y_n &= x_n - \frac{f(x_n)}{f'(x_n)}, \\ x_{n+1} &= x_n - \left(1 + 2\left(\frac{f(y_n)}{f(x_n)}\right)^2\right)\frac{f(x_n)+f(y_n)}{f'(x_n)}, \end{aligned} \tag{10}$$

which requires three function evaluations per iteration and consequently is optimal. In addition, the efficiency index of (10) is 1.5874, which is higher than that of (3) having an efficiency index of 1.442.

2.2. Sixth Order Method

Using the results obtained in (10), we propose a new method defined by

$$\begin{aligned} y_n &= x_n - \frac{f(x_n)}{f'(x_n)}, \\ z_n &= x_n - \left(1 + 2\left(\frac{f(y_n)}{f(x_n)}\right)^2\right)\frac{f(x_n)+f(y_n)}{f'(x_n)}, \\ x_{n+1} &= z_n - w_1(t_n)\frac{f(z_n)}{f'(x_n)}, \end{aligned} \tag{11}$$

where $w_1(t_n) = b_1 + b_2 t_n$ is a new weight function and t_n is as in (4). The order of convergence is shown in the following result.

Theorem 2. *Let f be a real or complex valued function defined in an interval I having a sufficient number of smooth derivatives. Let α be a simple root of the equation $f(x) = 0$ and the initial point x_0 is close enough to α. Then, (11) has a sixth order of convergence if $b_1 = 1$ and $b_2 = 2$.*

Proof. Let $\theta_n = z_n - \alpha$. Then, from second equation of (11), we obtain

$$\begin{aligned} \theta_n &= (3c_2^3 - c_2 c_3)e_n^4 - 2\left(8c_2^4 - 10c_2^2 c_3 + c_3^2 + c_2 c_4\right)e_n^5 \\ &\quad + (46c_2^5 - 114c_2^3 c_3 + 30c_2^2 c_4 + 42c_2 c_3^2 - 3c_2 c_5 - 7c_3 c_4)e_n^6 + O(e_n^7). \end{aligned} \tag{12}$$

Now, by expanding $f(z_n)$ about α using Equation (12), we obtain

$$\begin{aligned} f(z_n) &= f(\theta_n + \alpha) \\ &= f'(\alpha)[(3c_2^3 - c_2 c_3)\,e_n^4 - 2\left(8c_2^4 - 10c_2^2 c_3 + c_3^2 + c_2 c_4\right)e_n^5 \\ &\quad + \left(46c_2^5 - 114c_2^3 c_3 + 30c_2^2 c_4 + 42c_2 c_3^2 - 3c_2 c_5 - 7c_3 c_4\right)e_n^6 + O(e_n^7)]. \end{aligned} \tag{13}$$

Therefore, using (6), (8) and (13) in the third equation of (11), we obtain

$$\begin{aligned} e_{n+1} &= (1 - b_1)c_2(3c_2^2 - c_3)e_n^4 \\ &\quad + \left(c_2^4(22b_1 - 3b_2 - 16) + c_2^2 c_3(-22b_1 + b_2 + 20) + 2(b_1 - 1)c_2 c_4 + 2(b_1 - 1)c_3^2\right)e_n^5 \\ &\quad + (c_2^5(-90b_1 + 31b_2 + 46) + c_2^3 c_3(167b_1 - 31b_2 - 114) + 2c_2^2 c_4(-17b_1 + b_2 + 15) \\ &\quad + c_2(c_3^2(-49b_1 + 4b_2 + 42) + 3(b_1 - 1)c_5) + 7(b_1 - 1)c_3 c_4)e_n^6 + O(e_n^7). \end{aligned} \tag{14}$$

In order to obtain sixth order of convergence, the coefficients of e_n^4 and e_n^5 must vanish in (14), i.e., $b_1 = 1$ and $b_2 = 2$. Therefore, the error equation of the method (11) becomes

$$e_{n+1} = c_2 \left(18c_2^4 - 9c_2^2 c_3 + c_3^2 \right) e_n^6 + \mathcal{O}\left(e_n^7 \right),$$

and the assertion follows. □

In view of Theorem 2, the following is the sixth order method

$$\begin{aligned} y_n &= x_n - \frac{f(x_n)}{f'(x_n)}, \\ z_n &= x_n - \left(1 + 2 \left(\frac{f(y_n)}{f(x_n)} \right)^2 \right) \frac{f(x_n) + f(y_n)}{f'(x_n)}, \\ x_{n+1} &= z_n - \left(1 + 2 \frac{f(y_n)}{f(x_n)} \right) \frac{f(z_n)}{f'(x_n)}. \end{aligned} \tag{15}$$

2.3. Optimal Eighth Order Method

Notice that the method (15) is not optimal as it requires four function evaluation per iteration to achieve sixth order of convergence. Its efficiency index is 1.5651, which is less than that of the fourth order method (10). However, an eighth order method is obtained by (10) using an additional Newton step. The resulting iterative scheme is

$$\begin{aligned} y_n &= x_n - \frac{f(x_n)}{f'(x_n)}, \\ z_n &= x_n - \left(1 + 2 \left(\frac{f(y_n)}{f(x_n)} \right)^2 \right) \frac{f(x_n) + f(y_n)}{f'(x_n)}, \\ x_{n+1} &= z_n - \frac{f(z_n)}{f'(z_n)}. \end{aligned} \tag{16}$$

Nevertheless, this method requires five function evaluation per iteration, so that its efficiency index reduces to 1.5157, and, moreover, it is not optimal. Towards making the method (16) more efficient and optimal, we approximate $f'(z)$ as

$$f'(z_n) \approx \frac{f'(x_n)}{J(t_n, u_n) \cdot G(s_n)}, \tag{17}$$

where

$$t_n = \frac{f(y_n)}{f(x_n)}, \quad u_n = \frac{f(z_n)}{f(x_n)}, \quad s_n = \frac{f(z_n)}{f(y_n)}.$$

Here, J and G are some appropriate weight functions of two variables and one variable, respectively. This type of approximations was done by Matthies et al. in [22]. Accordingly, we propose the following method:

$$\begin{aligned} y_n &= x_n - \frac{f(x_n)}{f'(x_n)}, \\ z_n &= x_n - \left(1 + 2 \left(\frac{f(y_n)}{f(x_n)} \right)^2 \right) \frac{f(x_n) + f(y_n)}{f'(x_n)}, \\ x_{n+1} &= z_n - \frac{f(z_n)}{f'(x_n)} \cdot J(t_n, u_n) \cdot G(s_n), \end{aligned} \tag{18}$$

where t_n, u_n, and s_n, are as in (17). For the method (18), we take the functions J and G as

$$J(t_n, u_n) = \frac{1 + 2t_n + (\beta + 2)u_n + 3t_n^2}{1 + \beta u_n} \tag{19}$$

and

$$G(s_n) = \frac{1 + \lambda s_n}{1 + (\lambda - 1)s_n}, \tag{20}$$

where β and λ belong to $\mathbb{C}$. We prove the following result.

Theorem 3. *Let f be a real or complex valued function defined on some interval I having a sufficient number of smooth derivatives. Let α be a simple root of the equation $f(x) = 0$ and the initial point x_0 is close enough to α. Then, (18) is an eighth order of convergence for the functions J and G given by (19) and (20), respectively.*

Proof. In view of (5) and (13), we obtain

$$
\begin{aligned}
u_n &= \frac{f(z_n)}{f(x_n)} \\
&= \left(3c_2^3 - c_2 c_3\right) e_n^3 + \left(-19c_2^4 + 21c_2^2 c_3 - 2c_2 c_4 - 2c_3^2\right) e_n^4 \\
&\quad + \left(65c_2^5 - 138c_2^3 c_3 + 32c_2^2 c_4 + 45c_2 c_3^2 - 3c_2 c_5 - 7c_3 c_4\right) e_n^5 + O(e_n^6).
\end{aligned}
$$

Similarly, (7) and (13) yield

$$
\begin{aligned}
s_n &= \frac{f(z_n)}{f(y_n)} \\
&= \left(3c_2^2 - c_3\right) e_n^2 - 2\left(5c_2^3 - 6c_2 c_3 + c_4\right) e_n^3 + \left(11c_2^4 - 44c_2^2 c_3 + 17c_2 c_4 + 11c_3^2 - 3c_5\right) e_n^4 \\
&\quad + \left(56c_2^5 + 28c_2^3 c_3 - 56c_2^2 c_4 - 60c_2 c_3^2 + 22c_2 c_5 + 30c_3 c_4 - 4c_6\right) e_n^5 + O(e_n^6).
\end{aligned}
$$

Consequently, (19) gives

$$
\begin{aligned}
J(t_n, u_n) &= 1 + 2c_2 e_n + \left(4c_3 - 3c_2^2\right) e_n^2 + \left(4c_2^3 - 10c_2 c_3 + 6c_4\right) e_n^3 \\
&\quad + \left(-3(2\beta + 1)c_2^4 + 2(\beta + 10)c_2^2 c_3 - 14c_2 c_4 - 8c_3^2 + 8c_5\right) e_n^4 \\
&\quad + \left((47\beta - 38)c_2^5 - (57\beta + 14)c_2^3 c_3 + 4(\beta + 7)c_2^2 c_4 \right. \\
&\qquad \left. + 2c_2\left(4(\beta + 4)c_3^2 - 9c_5\right) - 22c_3 c_4 + 10c_6\right) e_n^5 + O(e_n^6),
\end{aligned}
\tag{21}
$$

and (20) gives

$$
\begin{aligned}
G(s_n) &= 1 + \left(3c_2^2 - c_3\right) e_n^2 - 2\left(5c_2^3 - 6c_2 c_3 + c_4\right) e_n^3 \\
&\quad + \left((20 - 9\lambda)c_2^4 + 2(3\lambda - 25)c_2^2 c_3 - (\lambda - 12)c_3^2 + 17c_2 c_4 - 3c_5\right) e_n^4 \\
&\quad + 2\left((30\lambda - 2)c_2^5 + (60 - 46\lambda)c_2^3 c_3 + 2(3\lambda - 17)c_2^2 c_4 + c_2\left(6(2\lambda - 7)c_3^2 + 11c_5\right)\right. \\
&\qquad \left. + (17 - 2\lambda)c_3 c_4 - 2c_6\right) e_n^5 + O(e_n^6).
\end{aligned}
\tag{22}
$$

Now, using the values from (6), (12), (13), (21), and (22) in (18), the error equation of the method is

$$
e_{n+1} = c_2\left(3c_2^2 - c_3\right)\left(c_2^4(6\beta + 9\lambda + 9) - 2c_2^2 c_3(\beta + 3\lambda + 4) + c_2 c_4 + c_3^2 \lambda\right) e_n^8 + \mathcal{O}\left(e_n^9\right),
$$

which gives the eighth order of convergence. □

3. Numerical Examples

In this section, we test the performance of the methods proposed in Section 2 with the help of some numerical examples. We compare the results obtained with the known methods of the corresponding order. We consider the following nonlinear equations and initial guesses:

- $f_1(x) = \sin^2 x - x^2 + 1, x_0 = 2,$
- $f_2(x) = \ln(1 + x^2) + \exp(x^2 - 3x)\sin x, x_0 = 2,$
- $f_3(x) = x^2 - (1 - x)^5, x_0 = 1,$
- $f_4(x) = x^2 - \exp(x) - 3x + 2, x_0 = 1,$
- $f_5(x) = \sqrt{x^2 + 2x + 5} - 2\sin x - x^2 + 3, x_0 = 2.$

In the previous section, we have proved the theoretical order of convergence of various methods. For practical purposes, we can test numerically the order of convergence of these methods by using

Approximated Computational Order of Convergence (or ACOC), defined by Cordero and Torregrosa [23]. They defined the ACOC of a sequence $\{x_k\}, k \geq 0$ as

$$ACOC = \frac{\log\left(|x_{k+1} - x_k| \,/\, |x_k - x_{k-1}|\right)}{\log\left(|x_k - x_{k-1}| \,/\, |x_{k-1} - x_{k-2}|\right)}. \tag{23}$$

The use of $ACOC$, given by (23), serves as a practical check on the theoretical error calculations.

We apply our proposed methods and other existing methods as discussed in the following subsections on each of the test functions. Various results of up to four iterations are observed, and we compare the results obtained at the 4th iteration among different methods of the corresponding order and shown in Tables 1–3. For a particular test function, we take the same initial guess x_0 for each of the methods under consideration. We compare the approximate error $\Delta x_n \equiv |x_n - x_{n-1}|$, the approximate solution x_n, the absolute value of corresponding functional value $|f(x_n)|$, and approximated computational order of convergence $(ACOC)$ at $n = 4$. In the tables, "NC" stands for no convergence of the method. We use Mathematica 9.0 for the calculations.

3.1. Comparison of the Fourth Order Method

Let us denote our method (10) by M_{41}. We shall compare this method with

- Sharma and Behl method (1), denoted by M_{42},
- Sharifi et al. method (2), denoted by M_{43},
- Jarratt's method [24], denoted by M_{44} and given by

$$
\begin{aligned}
y_n &= x_n - \frac{2}{3}\frac{f(x_n)}{f'(x_n)}, \\
x_{n+1} &= x_n - \left(\frac{3f'(y_n)+f'(x_n)}{6f'(y_n)-2f'(x_n)}\right)\frac{f(x_n)}{f'(x_n)},
\end{aligned}
$$

- Kung–Traub [8] method, denoted by M_{45}, and given by

$$
\begin{aligned}
y_n &= x_n - \frac{f(x_n)}{f'(x_n)}, \\
x_{n+1} &= y_n - \left(\frac{f(x_n)\cdot f(y_n)}{(f(x_n)-f(y_n))^2}\right)\frac{f(x_n)}{f'(x_n)}.
\end{aligned}
$$

All the methods M_{4i}, $i = 1, 2, 3, 4, 5$ are optimal. Table 1 records the performance of all these methods.

Table 1. Comparison of numerical results of fourth order methods at the 4th iteration.

		f_1	f_2	f_3	f_4	f_5
Δx_n	M_{41}	8.7309×10^{-26}	2.7730×10^{-55}	9.9454×10^{-30}	1.2399×10^{-65}	9.2139×10^{-82}
	M_{42}	1.1188×10^{-27}	2.9815×10^{-28}	1.0915×10^{-24}	7.7434×10^{-72}	3.5851×10^{-101}
	M_{43}	1.1523×10^{-23}	NC	6.1887×10^{-13}	1.3049×10^{-15}	3.6376×10^{-49}
	M_{44}	2.0493×10^{-32}	2.0594×10^{-31}	1.1971×10^{-20}	1.5448×10^{-71}	1.1488×10^{-97}
	M_{45}	4.0043×10^{-28}	2.8464×10^{-57}	2.4018×10^{-30}	4.7295×10^{-65}	2.8215×10^{-81}
x_n	M_{41}	1.4045	-7.8835×10^{-218}	0.3460	0.2575	2.3320
	M_{42}	1.4045	-6.9805×10^{-110}	0.3460	0.2575	2.3320
	M_{43}	1.4045	NC	0.3460	0.2575	2.3320
	M_{44}	1.4045	3.2977×10^{-123}	0.3460	0.2575	2.3320
	M_{45}	1.4045	-3.5010×10^{-226}	0.3460	0.2575	2.3320
$|f(x_n)|$	M_{41}	1.9828×10^{-100}	7.8835×10^{-218}	1.9230×10^{-116}	2.5756×10^{-262}	1.1861×10^{-326}
	M_{42}	4.0436×10^{-108}	6.9805×10^{-110}	1.1758×10^{-96}	6.8107×10^{-287}	1.9034×10^{-404}
	M_{43}	3.6237×10^{-93}	NC	6.4877×10^{-49}	7.5782×10^{-62}	2.9990×10^{-196}
	M_{44}	1.7439×10^{-127}	3.2977×10^{-123}	4.4608×10^{-80}	1.3131×10^{-285}	2.5652×10^{-390}
	M_{45}	5.7027×10^{-110}	3.5010×10^{-226}	9.4841×10^{-120}	6.9959×10^{-260}	1.1952×10^{-324}
ACOC	M_{41}	3.9919	4.0000	4.0184	4.0000	4.0000
	M_{42}	3.9935	3.9953	4.0646	4.0000	4.0000
	M_{43}	4.1336	NC	3.5972	4.6265	4.0214
	M_{44}	3.9978	4.0069	3.9838	4.0000	4.0000
	M_{45}	3.9946	4.0001	3.9878	4.0000	4.0000

3.2. Comparison of Sixth Order Methods

We denote our sixth order method (15) by M_{61}. We shall compare this method with

- M_{62} : Method of Neta [25] with $a = 1$, given by

$$
\begin{aligned}
y_n &= x_n - \frac{f(x_n)}{f'(x_n)}, \\
z_n &= y_n - \frac{f(x_n)+af(y_n)}{f(x_n)+(a-2)f(y_n)} \frac{f(y_n)}{f'(x_n)}, \\
x_{n+1} &= z_n - \frac{f(x_n)-f(y_n)}{f(x_n)-3f(y_n)} \frac{f(z_n)}{f'(x_n)},
\end{aligned}
$$

- M_{63} : Method of Grau et al. [26] given by

$$
\begin{aligned}
y_n &= x_n - \frac{f(x_n)}{f'(x_n)}, \\
z_n &= y_n - \frac{f(y_n)}{f'(x_n)} \frac{f(x_n)}{f(x_n)-2f(y_n)}, \\
x_{n+1} &= z_n - \frac{f(z_n)}{f'(x_n)} \frac{f(x_n)}{f(x_n)-2f(y_n)}.
\end{aligned}
$$

- M_{64} : Method of Sharma and Guha [27] with $a = 2$, given by

$$
\begin{aligned}
y_n &= x_n - \frac{f(x_n)}{f'(x_n)}, \\
z_n &= y_n - \frac{f(y_n)}{f'(x_n)} \frac{f(x_n)}{f(x_n)-2f(y_n)}, \\
x_{n+1} &= z_n - \frac{f(z_n)}{f'(x_n)} \frac{f(x_n)+af(y_n)}{f(x_n)+(a-2)f(y_n)},
\end{aligned}
$$

- M_{65} : Method of Chun and Neta [28] given by

$$
\begin{aligned}
y_n &= x_n - \frac{f(x_n)}{f'(x_n)}, \\
z_n &= y_n - \frac{f(y_n)}{f'(x_n)} \frac{1}{\left(1-\frac{f(y_n)}{f(x_n)}\right)^2}, \\
x_{n+1} &= z_n - \frac{f(y_n)}{f'(x_n)} \frac{1}{\left(1-\frac{f(y_n)}{f(x_n)}-\frac{f(z_n)}{f(x_n)}\right)^2}.
\end{aligned}
$$

The comparison of the methods M_{6i}, $i = 1, 2, 3, 4, 5$ is tabulated in Table 2. From the table, we observe that the proposed method M_{61} is compatible with the other existing methods. We can see that method M_{63} gives different results for the test functions f_2 and f_5 with given initial guesses.

Table 2. Comparison of numerical results of sixth order methods at the 4th iteration.

		f_1	f_2	f_3	f_4	f_5
	M_{61}	1.8933×10^{-73}	1.8896×10^{-148}	5.1627×10^{-90}	1.3377×10^{-199}	9.5891×10^{-261}
	M_{62}	1.6801×10^{-106}	2.9382×10^{-152}	2.4137×10^{-64}	1.7893×10^{-191}	3.75383×10^{-255}
Δx_n	M_{63}	2.9803×10^{-95}	2.9803×10^{-95}	2.9815×10^{-82}	2.9815×10^{-82}	2.9803×10^{-95}
	M_{64}	5.0012×10^{-85}	2.4246×10^{-153}	4.9788×10^{-69}	4.6397×10^{-198}	4.0268×10^{-259}
	M_{65}	9.9516×10^{-88}	2.1737×10^{-154}	3.3993×10^{-86}	2.7764×10^{-193}	3.4903×10^{-256}
	M_{61}	1.4045	-1.1331×10^{-884}	0.3460	0.2575	2.3320
	M_{62}	1.4045	4.5753×10^{-908}	0.3460	0.2575	2.3320
x_n	M_{63}	1.4045	1.4045	0.3460	0.2575	1.4045
	M_{64}	1.4045	1.0114×10^{-914}	0.3460	0.2575	2.3320
	M_{65}	1.4045	-3.7511×10^{-921}	0.3460	0.2575	2.3320

Table 2. *Cont.*

		f_1	f_2	f_3	f_4	f_5
	M_{61}	5.6523×10^{-436}	1.1331×10^{-884}	1.8046×10^{-535}	0.0	0.0
	M_{62}	6.7308×10^{-636}	4.5753×10^{-908}	1.0347×10^{-381}	0.0	0.0
$\|f(x_n)\|$	M_{63}	8.1802×10^{-568}	8.1802×10^{-568}	8.2004×10^{-490}	8.2004×10^{-490}	8.1802×10^{-568}
	M_{64}	5.7605×10^{-506}	1.0114×10^{-914}	1.8726×10^{-409}	0.0	0.0
	M_{65}	3.7794×10^{-522}	3.7511×10^{-921}	4.8072×10^{-514}	0.0	0.0
	M_{61}	5.9980	6.0000	5.9980	6.0000	6.0000
	M_{62}	5.9992	6.0000	5.9854	6.0000	6.000
ACOC	M_{63}	5.9997	5.9997	5.9992	5.9992	5.9997
	M_{64}	5.9991	6.0000	5.9984	6.0000	6.0000
	M_{65}	5.9993	6.0000	6.0088	6.0000	6.0000

3.3. Comparison of Eighth Order Methods

Consider the eighth order method (18), which involves the parameter pair (β, λ). We denote

- M_{81} the case where $(\beta, \lambda) = (0,0)$, whose iterative expression results in

$$
\begin{aligned}
y_n &= x_n - \frac{f(x_n)}{f'(x_n)}, \\
z_n &= x_n - \frac{f(x_n)+f(y_n)}{f'(x_n)}\left(1 + 2\left(\frac{f(y_n)}{f(x_n)}\right)^2\right), \\
x_{n+1} &= z_n - \frac{f(z_n)}{f'(x_n)}\left(\frac{1+2t_n+2u_n+3t_n^2}{1-s_n}\right),
\end{aligned}
$$

- M_{82} for $(\beta, \lambda) = (1,1)$, resulting in the iterative scheme given by M_{81} :

$$
\begin{aligned}
y_n &= x_n - \frac{f(x_n)}{f'(x_n)}, \\
z_n &= x_n - \frac{f(x_n)+f(y_n)}{f'(x_n)}\left(1 + 2\left(\frac{f(y_n)}{f(x_n)}\right)^2\right), \\
x_{n+1} &= z_n - \frac{f(z_n)}{f'(x_n)}\left(\frac{1+2t_n+3u_n+3t_n^2}{1+u_n}(1+s_n)\right),
\end{aligned}
$$

- M_{83} for $(\beta, \lambda) = (0,1)$, whose iterative method is

$$
\begin{aligned}
y_n &= x_n - \frac{f(x_n)}{f'(x_n)}, \\
z_n &= x_n - \frac{f(x_n)+f(y_n)}{f'(x_n)}\left(1 + 2\left(\frac{f(y_n)}{f(x_n)}\right)^2\right), \\
x_{n+1} &= z_n - \frac{f(z_n)}{f'(x_n)}\left((1+2t_n+2u_n+3t_n^2)(1+s_n)\right).
\end{aligned}
$$

Along with these, we take the following methods for the comparison of numerical results:

- Matthies et al. in [22] presented an optimal class of 8*th* order method from the Kung–Traub method [8]. For some particular values of the parameters, one of the methods denoted by M_{84} is given by

$$
\begin{aligned}
y_n &= x_n - \frac{f(x_n)}{f'(x_n)}, \\
z_n &= y_n - \left(\frac{f(x_n)f(y_n)}{(f(x_n)-f(y_n))^2}\right)\frac{f(x_n)}{f'(x_n)}, \\
x_{n+1} &= z_n - \frac{f(z_n)}{f'(x_n)}\left(\frac{2+t_n+5u_n+4t_n^2+4t_n^3}{2-3t_n+u_n+2t_n^2} \cdot \frac{2+s_n}{2-s_n}\right).
\end{aligned}
$$

- Babajee et al. in [11] presented a family of eighth order methods. For some fixed values of parameters, the method denoted by M_{85} is given by

$$
\begin{aligned}
y_n &= x_n - \frac{f(x_n)}{f'(x_n)}\left(1 + \left(\frac{f(x_n)}{f'(x_n)}\right)^5\right), \\
z_n &= y_n - \frac{f(y_n)}{f'(x_n)}\left(1 - \frac{f(y_n)}{f(x_n)}\right)^{-2}, \\
x_{n+1} &= z_n - \frac{f(z_n)}{f'(x_n)}\left(\frac{1 + t_n^2 + 5t_n^4 + s_n}{(1 - t_n - u_n)^2}\right).
\end{aligned}
$$

- Chun and Lee in [29] presented a family of optimal eighth order methods. For some particular values of parameters, the method denoted by M_{86} is given by

$$
\begin{aligned}
y_n &= x_n - \frac{f(x_n)}{f'(x_n)}, \\
z_n &= y_n - \frac{f(y_n)}{f'(x_n)}\frac{1}{\left(1 - \frac{f(y_n)}{f(x_n)}\right)^2}, \\
x_{n+1} &= z_n - \frac{f(z_n)}{f'(x_n)}\frac{1}{\left(1 - t_n - \frac{t_n^2}{2} + \frac{t_n^3}{2} - \frac{u_n}{2} - \frac{s_n}{2}\right)^2}.
\end{aligned}
$$

In all the above methods, t_n, u_n and s_n are as given in (17). The performance of the methods M_{8i}, $i = 1, 2, 3, 4, 5, 6$ are recorded in Table 3.

Table 3. Comparison of numerical results of eighth order methods at the 4th iteration.

		f_1	f_2	f_3	f_4	f_5
Δx_n	M_{81}	5.8768×10^{-187}	1.5404×10^{-393}	2.5345×10^{-165}	6.1099×10^{-495}	4.4344×10^{-658}
	M_{82}	2.0563×10^{-165}	9.0158×10^{-321}	1.1101×10^{-167}	5.4494×10^{-421}	4.0437×10^{-598}
	M_{83}	4.5429×10^{-170}	1.5139×10^{-324}	2.9710×10^{-168}	2.8838×10^{-421}	2.9107×10^{-604}
	M_{84}	2.4469×10^{-187}	4.9438×10^{-351}	4.3825×10^{-171}	1.8592×10^{-438}	4.3404×10^{-614}
	M_{85}	2.6744×10^{-204}	NC	1.7766×10^{-177}	6.5231×10^{-192}	9.8976×10^{-553}
	M_{86}	4.1482×10^{-235}	1.3271×10^{-380}	5.6991×10^{-175}	2.5934×10^{-455}	7.1011×10^{-617}
x_n	M_{81}	1.4045	0.0	0.3460	0.2575	2.3320
	M_{82}	1.4045	0.0	0.3460	0.2575	2.3320
	M_{83}	1.4045	0.0	0.3460	0.2575	2.3320
	M_{84}	1.4045	0.0	0.3460	0.2575	2.3320
	M_{85}	1.4045	NC	0.3460	0.2575	2.3320
	M_{86}	1.4045	0.0	0.3460	0.2575	2.3320
$|f(x_n)|$	M_{81}	0.0	0.0	0.0	0.0	0.0
	M_{82}	0.0	0.0	0.0	0.0	0.0
	M_{83}	0.0	0.0	0.0	0.0	0.0
	M_{84}	0.0	0.0	0.0	0.0	0.0
	M_{85}	0.0	0.0	0.0	0.0	0.0
	M_{86}	0.0	0.0	0.0	0.0	0.0
ACOC	M_{81}	7.9999	8.0000	7.9993	8.0000	8.0000
	M_{82}	7.9996	8.0000	8.0000	8.0000	8.0000
	M_{83}	7.9997	8.0000	7.9996	8.0000	8.0000
	M_{84}	7.9998	8.0000	8.0047	8.0000	8.0000
	M_{85}	7.9995	NC	8.0020	8.0004	8.0000
	M_{86}	8.0000	8.0000	8.0023	8.0000	8.0000

From Tables 1–3, we observe that the proposed methods are compatible with other existing methods (and sometimes perform better than other methods) of the corresponding order. Not any particular method is superior to others for all examples. Among the family of eighth order methods (18), from Table 3, we observe that the method M_{81} performs better than other two. For more understanding about the iterative methods, we study the dynamics of these methods in the next section.

4. Applications

The applications discussed in Sections 4.1–4.3 are based on standard engineering examples, and we refer to [30]. We use the proposed methods M_{41}, M_{61}, and M_{8i}, $i = 1, 2, 3$ to obtain the various results from the first three iterations of these examples. In particular, we compute the value of the unknowns x_{n-1} and x_n, absolute value of the function $f(x_n)$ and absolute value of the difference d of unknown in two consecutive iterations, i.e., $d = |x_n - x_{n-1}|$, $n = 1, 2, 3$.

4.1. Pipe Friction Problem

Determining fluid flow through pipes and tubes has great relevance in many areas of engineering and science. In engineering, typical applications include the flow of liquids and gases through pipelines and cooling systems. Scientists are interested in topics ranging from flow in blood vessels to nutrient transmission through a plant's vascular system. The resistance to flow in such conduits is parameterized by a dimensionless number called the friction factor f. For a flow with turbulence, the Colebrook equation [31] provides a means to calculate the friction factor:

$$0 = \frac{1}{\sqrt{f}} + 2.0 \log\left(\frac{\epsilon}{3.7D} + \frac{2.51}{Re\sqrt{f}}\right), \tag{24}$$

where ϵ is the roughness (m), D is the diameter (m) and Re is the Reynolds number

$$Re = \frac{\rho v D}{\mu}.$$

Here, ρ denotes the fluid density (kg/m^3), v the velocity of the fluid (m/s) and μ the dynamical viscosity (N·s/m^2). A flow is said to be turbulent if $Re > 4000$.

To determine f for air flow through a smooth and thin tube, the parameters are taken to be $\rho = 1.23$ kg/m^3, $\mu = 1.79 \times 10^{-5}$ N·s/m^2, $D = 0.005$ m, $V = 40$ m/s and $\epsilon = 0.0000015$ m. Since the friction factors range from about 0.008 to 0.08, we choose initial guess $f_0 = 0.023$. To determine the approximate value of f, we use the function

$$g(f) = \frac{1}{\sqrt{f}} + 2.0 \log\left(\frac{\epsilon}{3.7D} + \frac{2.51}{Re\sqrt{f}}\right). \tag{25}$$

The results obtained by the various methods are presented in Table 4.

Table 4. Results of pipe friction problem.

# Iter	Value	M_{41}	M_{61}	M_{81}	M_{82}	M_{83}
1	f	0.0169	0.0170	0.0170	0.0170	0.0170
	$g(f)$	0.0240	0.0104	0.0009	0.0005	0.0008
	d	0.0061	0.0060	0.0060	0.0060	0.0060
2	f	0.0170	0.0170	0.0170	0.0170	0.0170
	$g(f)$	3.0954×10^{-9}	2.6645×10^{-15}	8.8818×10^{-16}	8.8818×10^{-16}	8.8818×10^{-16}
	d	0.0001	4.1700×10^{-5}	3.7223×10^{-6}	2.0962×10^{-6}	3.3172×10^{-6}
3	f	0.0170	0.0170	0.0170	0.0170	0.0170
	$g(f)$	8.8818×10^{-16}	8.8818×10^{-16}	8.8818×10^{-16}	8.8818×10^{-16}	8.8818×10^{-16}
	d	1.2442×10^{-11}	1.0408×10^{-17}	6.9389×10^{-18}	0.0	0.0

4.2. Open-Channel Flow

An open problem in civil engineering is to relate the flow of water with other factors affecting the flow in open channels such as rivers or canals. The flow rate is determined as the volume of water passing a particular point in a channel per unit time. A further concern is related to what happens when the channel is slopping.

Under uniform flow conditions, the flow of water in an open channel is given by Manning's equation

$$Q = \frac{\sqrt{S}}{n} A R^{2/3}, \tag{26}$$

where S is the slope of the channel, A is the cross-sectional area of the channel, R is the hydraulic radius of the channel and n is the Manning's roughness coefficient. For a rectangular channel having the width B and the defth of water in the channel y, it is known that

$$A = By$$

and

$$R = \frac{By}{B + 2y}.$$

With these values, (26) becomes

$$Q = \frac{\sqrt{S}}{n} By \left(\frac{By}{B + 2y} \right)^{2/3}. \tag{27}$$

Now, if it is required to determine the depth of water in the channel for a given quantity of water, (27) can be rearranged as

$$f(y) = \frac{\sqrt{S}}{n} By \left(\frac{By}{B + 2y} \right)^{2/3} - Q. \tag{28}$$

In our work, we estimate y when the remaining parameters are assumed to be given as $Q = 14.15 \, \text{m}^3/\text{s}$, $B = 4.572 \, \text{m}$, $n = 0.017$ and $S = 0.0015$. We choose as an initial guess $y_0 = 4.5 \, \text{m}$. The results obtained by the various methods are presented in Table 5.

Table 5. Results of an open channel problem.

# Iter	Value	M_{41}	M_{61}	M_{81}	M_{82}	M_{83}
	y	1.4804	1.4666	1.4652	1.4653	1.4653
1	$f(y)$	0.2088	0.0204	0.0016	0.0029	0.0028
	d	3.0200	3.0334	3.0348	3.0347	3.0347
	y	1.4651	1.4651	1.4651	1.4651	1.4651
2	$f(y)$	4.5027×10^{-9}	1.7764×10^{-15}	$\times 10^{-15}$	3.5527×10^{-15}	3.5527×10^{-15}
	d	0.0154	0.0015	0.0001	0.0002	0.0002
	y	1.4651	1.4651	1.4651	1.4651	1.4651
3	$f(y)$	3.5527×10^{-15}	7.1054×10^{-14}	6.5725×10^{-14}	5.3291×10^{-15}	1.7764×10^{-15}
	d	3.3152×10^{-10}	5.1070×10^{-15}	5.1070×10^{-15}	6.6613×10^{-16}	2.2204×10^{-16}

4.3. Ideal and Non-Ideal Gas Laws

The ideal gas law is

$$PV = nRT,$$

where P is the absolute pressure, V is the volume, n is the number of moles, R is the universal gas constant and T is the absolute temperature. Due to its limited use in engineering, an alternative equation of state for gases is the given van der Waals equation [32–35]

$$\left(P + \frac{a}{v^2} \right) (v - b) = RT,$$

where $v = \frac{V}{n}$ is the molal volume and a, b are empirical constants that depend on the particular gas. The computation of the molal volume is done by solving

$$f(v) = \left(P + \frac{a}{v^2}\right)(v - b) - RT. \tag{29}$$

We take the remaining parameters as $R = 0.082054$ L atm/(mol K), for carbon dioxide $a = 3.592$, $b = 0.04267$, $T = 300$ K, $p = 1$ atm, and the initial guess for the molal volume is taken as $v_0 = 3$. The results obtained by the various methods are presented in Table 6. In this table, *IND* stands for indeterminate form.

Table 6. Numerical results of ideal and non-ideal gas law.

# Iter	Value	M_{41}	M_{61}	M_{81}	M_{82}	M_{83}
	v	26.4881	27.0049	23.9583	24.1631	24.0274
1	$f(v)$	1.9647	2.4788	0.5509	0.3474	0.4823
	d	23.4881	24.0049	20.9583	21.1631	21.0274
	v	24.5126	24.5126	24.5126	24.5126	24.5126
2	$f(v)$	2.7340×10^{-8}	3.3573×10^{-12}	0.0	0.0	0.0
	d	1.9756	2.4923	0.5543	0.3495	0.4852
	v	24.5126	24.5126	*IND*	*IND*	*IND*
3	$f(v)$	0.0	0.0	*IND*	*IND*	*IND*
	d	2.7503×10^{-8}	3.3786×10^{-12}	*IND*	*IND*	*IND*

5. Dynamical Analysis

The stability analysis of the methods M_{41}, M_{61} and M_{8i}, $i = 1, 2, 3$, is performed in this section. The dynamics of the proposed methods on a generic quadratic polynomial will be studied, analyzing the associated rational operator for each method. This analysis shows their performance depending on the initial estimations. In addition, method M_{41} is analyzed for cubic polynomials. First, we recall some basics on complex dynamics.

5.1. Basics on Complex Dynamics

Let $R : \hat{\mathbb{C}} \longrightarrow \hat{\mathbb{C}}$ be a rational function defined on the Riemann sphere. Let us recall that every holomorphic function from the Riemann sphere to itself is in fact a rational function $R(z) = \frac{P(z)}{Q(z)}$, where P and Q are complex polynomials (see [36]). For older work on dynamics on the Riemann sphere, see, e.g., [37].

The orbit of a point $z_0 \in \hat{\mathbb{C}}$ is composed by the set of its images by R, i.e.,

$$\{z_0, R(z_0), R^2(z_0), \dots, R^n(z_0), \dots\}.$$

A point $z^F \in \hat{\mathbb{C}}$ is a fixed point if $R(z^F) = z^F$. Note that the roots z^* of an equation $f(z) = 0$ are fixed points of the associated operator of the iterative method. Fixed points that do not agree with a root of $f(x) = 0$ are strange fixed points.

The asymptotical behavior of a fixed point z^F is determined by the value of its multiplier $\mu = |R'(z^F)|$. Then, z^F is attracting, repelling or neutral if μ is lower, greater or equal to 1, respectively. In addition, it is superattracting when $\mu = 0$.

For an attracting fixed point z^F, its basin of attraction is defined as the set of its pre-images of any order:

$$\mathcal{A}(z^F) = \{z_0 \in \hat{\mathbb{C}} : R^n(z_0) \longrightarrow z^F, n \to \infty\}.$$

The dynamical plane represents the basins of attraction of a method. By iterating a set of initial guesses, their convergence is analyzed and represented. The points $z^C \in \hat{\mathbb{C}}$ that satisfy $R'(z^C) = 0$ are called critical points of R. When a critical point does not agree with a solution of $f(x) = 0$, it is a free critical point. A classical result [21] establishes that there is at least one critical point associated with each immediate invariant Fatou component.

5.2. Rational Operators

Let $p(z)$ be a polynomial defined on $\hat{\mathbb{C}}$. Corresponding to the methods developed in this paper, i.e., methods (10), (15) and family (18), we define the operators $R_4(z)$, $R_6(z)$ and $R_8(z)$, respectively, in $\hat{\mathbb{C}}$ as follows:

$$R_4(z) = z - \left(1 + 2\left(\frac{p(y(z))}{p(z)}\right)^2\right)\frac{p(z) + p(y(z))}{p'(z)}, \tag{30}$$

$$R_6(z) = R_4(z) - \left(1 + 2\frac{p(y(z))}{p(z)}\right)\frac{p(R_4(z))}{p'(z)},$$

$$R_8(z) = R_4(z) - \frac{p(R_4(z))}{p'(z)}J(z)G(z),$$

where $y(z) = z - \frac{p(z)}{p'(z)}$ and

$$J(z) = \frac{1 + 2\frac{p(y(z))}{p(z)} + (\beta+2)\frac{p(R_4(z))}{p(z)} + 3\left(\frac{p(y(z))}{p(z)}\right)^2}{1 + \beta\frac{p(R_4(z))}{p(z)}},$$

$$G(z) = \frac{1 + \lambda\frac{p(R_4(z))}{p(y(z))}}{1 + (\lambda-1)\frac{p(R_4(z))}{p(y(z))}}.$$

First, we recall the following result for the generalization of the dynamics of M_{41}.

Theorem 4 (Scaling Theorem for method M_{41}). *Let $f(z)$ be an analytic function in the Riemann sphere and let $A(z) = \eta z + \sigma$, with $\eta \neq 0$, be an affine map. Let $h(z) = \mu(f \circ A)(z)$ with $\mu \neq 0$. Then, the fixed point operator R_4^f is affine conjugated to R_4^h by A, i.e.,*

$$(A \circ R_4^h \circ A^{-1})(z) = R_4^f(z).$$

Proof. From (30), let the fixed point operators associated with f and h be, respectively,

$$R_4^f(z) = z - \left(1 + 2\left(\frac{f(y(z))}{f(z)}\right)^2\right)\frac{f(z) + f(y(z))}{f'(z)},$$

$$R_4^h(z) = z - \left(1 + 2\left(\frac{h(y(z))}{h(z)}\right)^2\right)\frac{h(z) + h(y(z))}{h'(z)}.$$

Thus, we have

$$(R_4^f \circ A)(z) = A(z) - \left(1 + 2\frac{f^2(A(y))}{f^2(A(z))}\right)\frac{f(A(z)) + f(A(y))}{f'(A(z))}. \tag{31}$$

Being $h'(z) = \eta\mu f'(A(z))$, we obtain

$$\begin{aligned}R_4^h(z) &= z - \left(1 + 2\frac{\mu^2 f^2(A(y))}{\mu^2 f^2(A(z))}\right)\frac{\mu f(A(z)) + \mu f(A(y))}{\eta\mu f'(A(z))}\\ &= z - \left(1 + 2\frac{f^2(A(y))}{f^2(A(z))}\right)\frac{f(A(z)) + f(A(y))}{\eta f'(A(z))}.\end{aligned}$$

The affine map A satisfies $A(z_1 - z_2) = A(z_1) - A(z_2) + \sigma$, $\forall z_1, z_2$. Then, from (32), we have

$$
\begin{aligned}
(A \circ R_4^h)(z) &= A(z) - A\left(\left(1 + 2\frac{f^2(A(y))}{f^2(A(z))}\right)\frac{f(A(z)) + f(A(y))}{\eta f'(A(z))}\right) + \sigma \\
&= A(z) - \left(\eta\left(1 + 2\frac{f^2(A(y))}{f^2(A(z))}\right)\frac{f(A(z)) + f(A(y))}{\eta f'(A(z))} + \sigma\right) + \sigma \\
&= A(z) - \left(1 + 2\frac{f^2(A(y))}{f^2(A(z))}\right)\frac{f(A(z)) + f(A(y))}{f'(A(z))}.
\end{aligned}
$$

Thus, it proves that $(R_4^f \circ A)(z) = (A \circ R_4^h)(z)$ and then method M_{41} satisfies the Scaling Theorem. $\square$

Theorem 4 allows for generalizing the dynamical study of a specific polynomial to a generic family of polynomials by using an affine map. Analogous to the way we proved the Scaling Theorem for the operator R_4, it also follows that the fixed point operators R_6 and R_8 obey the Scaling Theorem.

5.3. Dynamics on Quadratic Polynomials

The application of the rational functions on a generic quadratic polynomial $p(z) = (z - a)(z - b)$, $a, b \in \hat{\mathbb{C}}$ is studied below. Let $R_{4,a,b}$ be the rational operator associated with method M_{41} on $p(z)$. When the Möbius transformation $h(u) = \frac{a-u}{b-u}$ is applied to $R_{4,a,b}$, we obtain

$$
S_4(z) = (h \circ R_{4,a,b} \circ h^{-1})(z) = \frac{z^4 \left(z^4 + 6z^3 + 14z^2 + 14z + 3\right)}{3z^4 + 14z^3 + 14z^2 + 6z + 1}. \tag{32}
$$

The rational operator associated with M_{41} on $p(z)$ does not depend on a and b. Then, the dynamical analysis of the method on all quadratic polynomials can be studied through the analysis of (32). In addition, the Möbius transformation h maps its roots a and b to $z_1^* = 0$ and $z_2^* = \infty$, respectively.

The fixed point operator $S_4(z)$ has nine fixed points: $z_1^F = 0$ and $z_2^F = \infty$, which are superattracting, and $z_3^F = 1$, $z_{4,5}^F = \frac{1}{2}(-3 \pm \sqrt{5})$, $z_{6-7}^F = \frac{-2+\sqrt{2}}{2} \pm i\sqrt{\frac{3}{2} - \sqrt{2}}$, $z_{8-9}^F = \frac{-2-\sqrt{2}}{2} \pm i\sqrt{\frac{3}{2} + \sqrt{2}}$, all of them being repelling. Computing $S_4'(z) = 0$, 5 critical points can be found. $z_{1,2}^C = z_{1,2}^*$ and the free critical points $z_3^C = -1$ and $z_{4,5}^C = \frac{1}{6}(-13 \pm \sqrt{133})$.

Following the same procedure, when Möbius transformation is applied to methods M_6 and M_{8i}, $i = 1, 2, 3$, on polynomial $p(z)$, the respective fixed point operators turn into

$$
S_6(z) = \frac{z^6\left(z^{12} + 16z^{11} + 119z^{10} + 544z^9 + 1700z^8 + 3808z^7 + 6206z^6 + 7288z^5 + 5973z^4 + 3248z^3 + 1111z^2 + 216z + 18\right)}{18z^{12} + 216z^{11} + 1111z^{10} + 3248z^9 + 5973z^8 + 7288z^7 + 6206z^6 + 3808z^5 + 1700z^4 + 544z^3 + 119z^2 + 16z + 1},
$$

$$
S_{81}(z) = \frac{P_{30}(z)}{P_{22}(z)}, \qquad S_{82}(z) = \frac{P_{42}(z)}{P_{34}(z)}, \qquad S_{83}(z) = \frac{Q_{42}(z)}{Q_{34}(z)},
$$

where P_k and Q_k denote polynomials of degree k.

The fixed point operator S_6 has 19 fixed points: the two superattracting fixed points $z_{1,2}^F = z_{1,2}^*$, the repelling fixed point $z_3^F = 1$ and the repelling fixed points $z_4^F, \ldots, z_{19}^F$, which are the roots of a sixteenth-degree polynomial.

Regarding the critical points of S_6, the roots of $p(z)$ are critical points, and S_6 has the free critical points $z_3^C = -1$ and the roots of a tenth-degree polynomial, $z_4^C, \ldots, z_{11}^C$.

The dynamical planes are a useful tool in order to analyze the stability of an iterative method. Taking each point of the plane as initial estimation to start the iterative process, they represent the convergence of the method depending on the initial guess. In this sense, the dynamical planes show the basins of attraction of the attracting points.

Figure 1 represents the dynamical planes of the methods S_4 and S_6. The generation of the dynamical planes follows the guidelines established in [38]. A mesh of 500×500 complex values has been set as initial guesses in the intervals $-5 < \Re\{z\} < 5$, $-5 < \Im\{z\} < 5$. The roots $z_1^* = 0$ and $z_2^* = \infty$ are mapped with orange and blue colors, respectively. The regions where the colors are darker represent that more iterations are necessary to converge than with the lighter colors, with a maximum of 40 iterations of the methods and a stopping criteria of a difference between two consecutive iterations lower than 10^{-6}.

As Figure 1 illustrates, there is convergence to the roots for every initial guess. Let us remark that, when the order of the method increases, the basin of attraction of $z_1^* = 0$ becomes more intricate.

Finally, for the fixed point operators associated with family M_8, the solutions of $S_{8i}(z) = z$ for $i = 1, 2, 3$ give the superattracting fixed points $z_{1,2}^F = z_{1,2}^*$ and the repelling point $z_3^F = 1$. In addition, S_{81} has 28 repelling points. S_{82} and S_{83} have 38 repelling points, corresponding to the roots of polynomials of 28 and 38 degree, respectively, and the strange fixed points $z_{4,5}^F = \frac{1}{2}(-1 \pm \sqrt{5})$.

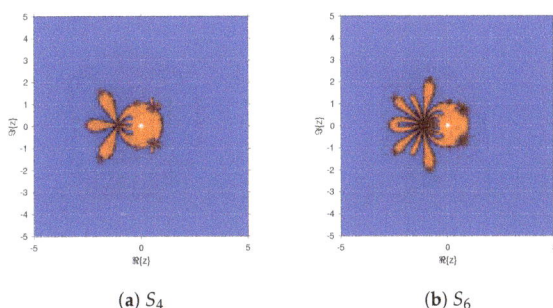

Figure 1. Dynamical planes of methods S_4 and S_6.

The number of critical points of the fixed point operators S_{8i} are collected in Table 7. In addition, the number of strange fixed points and free critical points are also included in the table for all of the methods.

Table 7. Number of strange fixed points (SFP) and free critical points (FCP) for the methods on quadratic polynomials.

	S_4	S_6	S_{81}	S_{82}	S_{83}
Strange fixed points	7	17	29	41	41
Free critical points	3	29	29	43	29

Figure 2 represents the dynamical planes of the methods S_{81}, S_{82} and S_{83}. Since the original methods satisfy the Scaling Theorem, the generation of one dynamical plane involves the study of every quadratic polynomial.

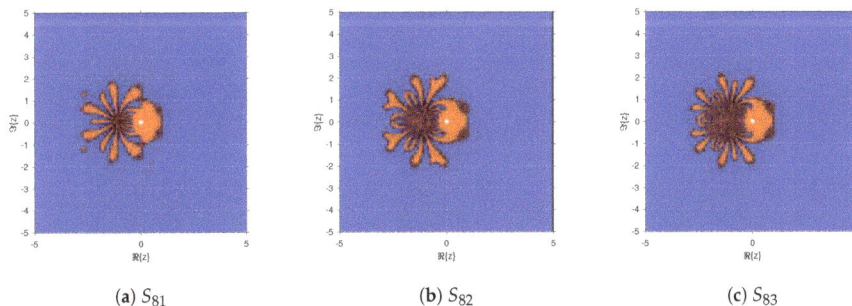

Figure 2. Dynamical planes of methods $S_{8i}, i = 1, 2, 3$.

There is an intricate region around $z = -1$ in Figure 2a, becoming wider in Figure 2b,c around $z = -1.5$. However, for every initial guess in the three dynamical planes of Figure 2, there is convergence to the roots.

5.4. Dynamics on Cubic Polynomials

The stability of method M_{41} on cubic polynomials is analyzed below. As stated by the authors in [39], the Scaling Theorem reduces the dynamical analysis on cubic polynomials to the study of dynamics on the cubic polynomials $p_0(z) = z^3$, $p_+(z) = z^3 + z$, $p_-(z) = z^3 - z$ and the family of polynomials $p_\gamma(z) = z^3 + \gamma z + 1$. Let us recall that the first one only has the root $z_1^* = 0$, while $p_+(z)$ and $p_-(z)$ have three simple roots: $z_1^* = 0$ and $z_{2,3}^* = \mp i$ or $z_{2,3}^* = \mp 1$, respectively. For each $\gamma \in \mathbb{C}$, the polynomial $p_\gamma(z)$ also has three simple roots that depend on the value of γ. They will be denoted by $z_{1,2,3}^*(\gamma)$.

By applying method M_{41} to polynomials $p_0(z)$, $p_+(z)$ and $p_-(z)$, the fixed point operators obtained are, respectively,

$$S_{4,0}(z) = \frac{46z}{81}, \quad S_{4,+}(z) = \frac{6z^5 + 36z^7 + 46z^9}{(1+3z^2)^4}, \quad S_{4,-}(z) = \frac{6z^5 - 36z^7 + 46z^9}{(1-3z^2)^4}.$$

The only fixed point of $S_{4,0}(z)$ agrees with the root of the polynomial, so it is superattracting, and the operator does not have critical points.

The rest of the fixed point operators have six repelling fixed points, in addition to the roots of the corresponding polynomials: $z_{4,5}^F = \pm\frac{i\sqrt{5}}{5}$ and $z_{6-9}^F = \pm i\sqrt{\frac{1}{7}(3\pm\sqrt{2})}$ for $S_{4,+}(z)$, and $z_{4,5}^F = \pm\frac{\sqrt{5}}{5}$ and $z_{6-9}^F = \pm\sqrt{\frac{1}{7}(3\pm\sqrt{2})}$ for $S_{4,-}(z)$.

Regarding the critical points of $S_{4,+}(z)$ and $S_{4,-}(z)$, they match with the roots of the polynomials. Moreover, there is the presence of free critical points with values $z_{4,5}^C = \pm i\sqrt{\frac{5}{23}}$ for $S_{4,+}(z)$ and $z_{4,5}^C = \pm\sqrt{\frac{5}{23}}$ for $S_{4,-}(z)$.

As for quadratic polynomials, the dynamical planes of method M_{41} when it is applied to the cubic polynomials have been represented in Figure 3. Depending on the roots of each polynomial, the convergence to $z_1^* = 0$ is represented in orange, while the convergence to z_2^* and z_3^* is represented in blue and green, respectively. It can be see in Figure 3 that there is full convergence to a root in the three cases. However, there are regions with darker colors that indicate a higher number of iterations until the convergence is achieved.

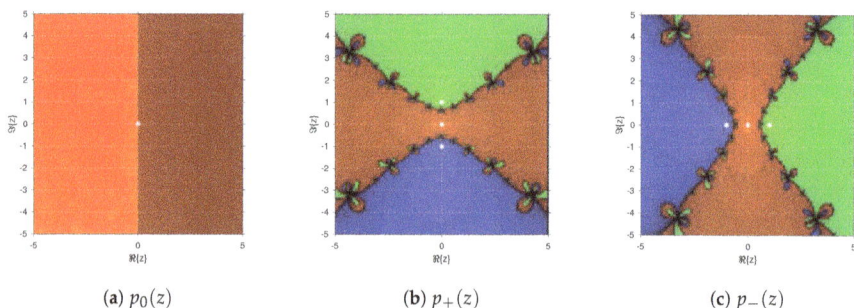

(a) $p_0(z)$ (b) $p_+(z)$ (c) $p_-(z)$

Figure 3. Dynamical planes of method M_{41} on polynomials $p_0(z)$, $p_+(z)$ and $p_-(z)$.

When method M_{41} is applied on $p_\gamma(z)$, the fixed point function turns into

$$S_{4,\gamma}(z) = -\frac{\gamma^3 - 46z^9 - 36\gamma z^7 + 42z^6 - 6\gamma^2 z^5 + 45\gamma z^4 + 6z^3 + 12\gamma^2 z^2 - 1}{(\gamma + 3z^2)^4}.$$

The fixed points of $S_{4,\gamma}(z)$ are the roots of the polynomial $z_{1,2,3}^*(\gamma)$, being superattracting, and the strange fixed points $z_{4-9}^F(\gamma)$ that are the roots of the sixth-degree polynomial $q(z,\gamma) = 35z^6 + 37\gamma z^4 + 7z^3 + 11\gamma^2 z^2 + \gamma z + \gamma^3 - 1$.

As the asymptotical behavior of $z_4^F(\gamma), \ldots, z_9^F(\gamma)$ depends on the value of γ, the stability planes corresponding to these points are represented in Figure 4. For each strange fixed point, a mesh of 100×100 points covers the values of $\Re(\gamma) \in [-5,5]$ and $\Im(\gamma) \in [-5,5]$. The stability plane shows the values for the parameter where $|S_{4,\gamma}'(z^F)|$ is lower or greater than 1, represented in red or green, respectively.

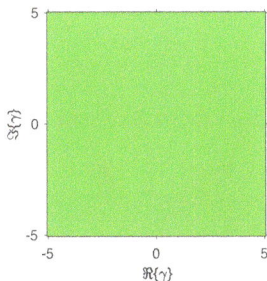

Figure 4. Stability planes of $z_{4-9}^F(\gamma)$.

From Figure 4, the strange fixed points are always repelling for $(\Re(\gamma), \Im(\gamma)) \in [-5,5] \times [-5,5]$. Then, the only attracting fixed points are the roots of the polynomial. This fact guarantees a better stability of the method.

The solutions of $S_{4,\gamma}'(z) = 0$ are the critical points $z_{1,2,3}^C(\gamma) = z_{1,2,3}^*(\gamma)$ and the free critical points $z_4^C = 0$ and

$$z_5^C(\gamma) = \frac{\left(\sqrt{69}\sqrt{125\gamma^3+2484}+414\right)^{2/3}-5\sqrt[3]{69}\gamma}{69^{2/3}\sqrt[3]{\sqrt{69}\sqrt{125\gamma^3+2484}+414}},$$

$$z_{6,7}^C(\gamma) = \frac{\left(-1\pm i\sqrt{3}\right)\left(\sqrt{69}\sqrt{125\gamma^3+2484}+414\right)^{2/3}+5\sqrt[3]{69}\left(1\pm i\sqrt{3}\right)\gamma}{2\,69^{2/3}\sqrt[3]{\sqrt{69}\sqrt{125\gamma^3+2484}+414}}.$$

When the fixed point function has dependence on a parameter, another useful representation is the parameters' plane. This plot is generated in a similar way to the dynamical planes, but, in this case, by iterating the method taking as an initial guess a free critical point and varying the value of γ in a complex mesh of values, so each point in the plane represents a method of the family. The parameters' plane helps to select the values for the parameter that give rise to the methods of the family with more stability.

The parameters' planes of the four free critical points are shown in Figure 5. Parameter γ takes the values of 500×500 points in a complex mesh in the square $[-5,5] \times [-5,5]$. Each point is represented in orange, green or blue when the corresponding method converges to an attracting fixed point. The iterative process ends when the maximum number of 40 iterations is reached, in which case the point is represented in black, or when the method converges as soon as, by the stopping criteria, a difference between two consecutive iterations lower than 10^{-6} is reached.

For the parameters' planes in Figure 5, there is not any black region. This guarantees that the corresponding iterative schemes converge to a root of $p_\gamma(z)$ for all the values of γ.

In order to visualize the basins of attraction of the fixed points, several values of γ have been chosen to perform the dynamical planes of method M_{41}. These values have been selected from the different regions of convergence observed in the parameters planes. Figure 6, following the same code of colours and stopping criteria as in the other representations, shows the dynamical planes obtained when these values of γ are fixed.

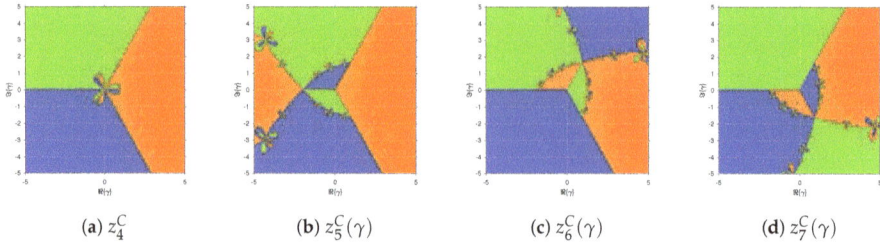

(a) z_4^C **(b)** $z_5^C(\gamma)$ **(c)** $z_6^C(\gamma)$ **(d)** $z_7^C(\gamma)$

Figure 5. Parameter planes of the critical points of method M_{41} on $p_\gamma(z)$.

As Figure 6 shows, there is not any initial guess that tends to a point different than the roots. This fact guarantees the stability of these methods on the specific case of any cubic polynomial.

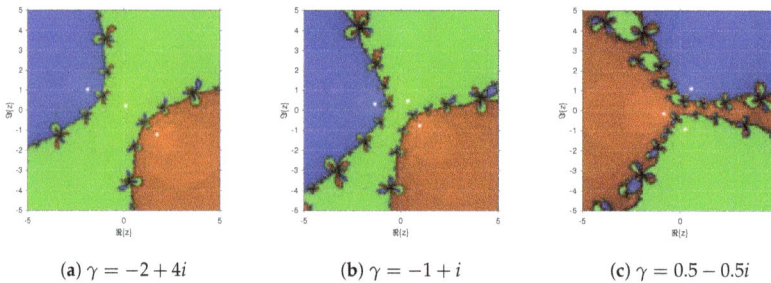

(a) $\gamma = -2 + 4i$ **(b)** $\gamma = -1 + i$ **(c)** $\gamma = 0.5 - 0.5i$

Figure 6. Dynamical planes for method M_{41} on $p_\gamma(z)$ for different values of γ.

6. Conclusions

Two iterative schemes of orders of convergence four and six, and a family of methods of order eight have been introduced. The method of order four and the family of order eight are optimal in the sense of Kung–Traub's conjecture. The development of the order of convergence of every method has been performed. For every method, we have made a numerical experiment, over both test functions and real engineering problems. In order to analyze the stability of the introduced methods, the dynamical behavior of them has been studied. The results confirm that the methods have wide basins of attraction, guaranteeing the stability over some nonlinear problems.

Author Contributions: The individual contributions of the authors are as follows: conceptualization, P.J.; validation, P.B.C. and P.J.; formal analysis, F.I.C.; writing, original draft preparation, N.G. and F.I.C.; numerical experiments, N.G. and P.B.C.

Funding: This research received no external funding.

Acknowledgments: The second and third authors have been partially supported by PGC2018-095896-B-C22 (MCIU/AEI/FEDER/UE) and Generalitat Valenciana PROMETEO/2016/089.

Conflicts of Interest: The authors declare no conflict of interest.

References

1. Ostrowski, A.M. *Solution of Equations and Systems of Equations*; Prentice-Hall: Englewood Cliffs, NJ, USA, 1964.
2. Cordero, A.; Fardi, M.; Ghasemi, M.; Torregrosa, J.R. Accelerated iterative methods for finding solutions of nonlinear equations and their dynamical behavior. *Calcolo* **2012**, *51*, 17–30. [CrossRef]
3. Lotfi, T.; Sharifi, S.; Salimi, M.; Siegmund, S. A new class of three-point methods with optimal convergence order eight and its dynamics. *Numer. Algor.* **2015**, *68*, 261–288. [CrossRef]
4. Chicharro, F.I.; Cordero, A.; Garrido, N.; Torregrosa, J.R. Wide stability in a new family of optimal fourth-order iterative methods. *Comput. Math. Meth.* **2019**, *1*, e1023. [CrossRef]

5.   Chicharro, F.I.; Cordero, A.; Torregrosa, J.R. Dynamics of iterative families with memory based on weight functions procedure. *J. Comput. Appl. Math.* **2019**, *354*, 286–298. [CrossRef]

6.   Sharma, R.; Bahl, A. An optimal fourth order iterative method for solving nonlinear equations and its dynamics. *J. Complex Anal.* **2015**, *2015*, 259167. [CrossRef]

7.   Sharifi, M.; Babajee, D.K.R.; Soleymani, F. Finding solutions of nonlinear equations by a class of optimal methods. *Comput. Math. Appl.* **2012**, *63*, 764–774. [CrossRef]

8.   Kung, H.T.; Traub, J.F. Optimal order of one-point and multipoint iterations. *J. Assoc. Comput. Mach.* **1974**, *21*, 643–651. [CrossRef]

9.   Potra, F.A.; Pták, V. Nondiscrete induction and iterative processes. *Res. Notes Math.* **1984**, *103*, 112–119.

10.   Amat, S.; Busquier, S.; Plaza, S. Review of some iterative root finding methods from a dynamical point of view. *Scientia* **2004**, *10*, 3–35.

11.   Babajee, D.K.R.; Cordero, A.; Soleymani, F.; Torregrosa, J.R. On improved three-step schemes with high efficiency index and their dynamics. *Numer. Algor.* **2014**, *65*, 153–169. [CrossRef]

12.   Chicharro, F.I.; Cordero, A.; Torregrosa, J.R.; Vassileva, M.P. King-type derivative-free iterative families: Real and memory dynamics. *Complexity* **2017**, *2017*, 2713145. [CrossRef]

13.   Cordero, A.; Feng, L.; Magreñán, Á.A.; Torregrosa, J.R. A new fourth order family for solving nonlinear problems and its dynamics. *J. Math. Chem.* **2015**, *53*, 893–910. [CrossRef]

14.   Scott, M.; Neta, B.; Chun, C. Basin attractors for various methods. *Appl. Math. Comput.* **2011**, *218*, 2584–2599. [CrossRef]

15.   Varona, J.L. Graphic and numerical comparison between iterative methods. *Math. Intell.* **2002**, *24*, 37–46. [CrossRef]

16.   Vrscay, E.R.; Gilbert, W.J. Extraneous fixed points, basin boundaries and chaotic dynamics for Schroder and Konig rational iteration functions. *Numer. Math.* **1988**, *52*, 1–16. [CrossRef]

17.   Alexander, D.S. *A History of Complex Dynamics: From Schröder to Fatou and Julia*; Vieweg & Teubner Verlag: Wiesbaden, Germany, 1994.

18.   Beardon, A.F. *Iteration of Rational Functions*; Springer: New York, NY, USA, 2000.

19.   Blanchard, P. The dynamics of Newton's method. *Proc. Sympos. Appl. Math.* **1994**, *49*, 139–154.

20.   Carleson, L.; Gamelin, T.W. *Complex Dynamics*; Springer-Verlag: New York, NY, USA, 1992.

21.   Devaney, R.L. *An Introduction to Chaotic Dynamical Systems*; Addison-Wesley: Redwood City, CA, USA, 1989.

22.   Matthies, G.; Salimi, M.; Sharifi, S.; Varona, J.L. An optimal class of eighth-order iterative methods based on Kung and Traub's method with its dynamics. *arXiv* **2015**, arXiv:1508.01748v1.

23.   Cordero, A.; Torregrosa, J.R. Variants of Newton's method using fifth order quadrature formulas. *Appl. Math. Comput.* **2007**, *190*, 686–698. [CrossRef]

24.   Jarratt, P. Some efficient fourth order multipoint methods for solving equations. *BIT* **1969**, *9*, 119–124. [CrossRef]

25.   Neta, B. A sixth-order family of methods for nonlinear equations. *Int. J. Comput. Math.* **1979**, *7*, 157–161. [CrossRef]

26.   Grau, M.; Díaz-Barrero, J.L. An improvement to Ostrowski root-finding method. *Appl. Math. Comput.* **2006**, *173*, 450–456. [CrossRef]

27.   Sharma, J.R.; Guha, R.K. A family of modified Ostrowski's methods with accelerated sixth order convergence. *Appl. Math. Comput.* **2007**, *190*, 111–115.

28.   Chun, C.; Neta, B. A new sixth-order scheme for nonlinear equations. *Appl. Math. Lett.* **2012**, *25*, 185–189. [CrossRef]

29.   Chun, C.; Lee, M.Y. A new optimal eighth-order family of iterative methods for the solution of nonlinear equations. *Appl. Math. Comput.* **2013**, *223*, 506–519. [CrossRef]

30.   Chapra, S.C.; Canale, R.C. *Numerical Methods for Engineers*; McGraw Hills Education: New York, NY, USA, 2015.

31.   Brkić, D. A note on explicit approximations to Colebrook's friction factor in rough pipes under highly turbulent cases. *Int. J. Heat Mass tramsf.* **2016**, *93*, 513–515. [CrossRef]

32.   Wang, J.; Pang, Y.; Zhang, Y. Limits of solutions to the isentropic Euler equations for van der Waals gas. *Int. J. Nonlinear Sci. Numer. Simul.* **2019**, *20*, 461–473. [CrossRef]

33.   Gates, D.J.; Penrose, O. The van der Waals limit for classical systems. I. A variational principle. *Comm. Math. Phys.* **1969**, *15*, 255–276. [CrossRef]

34. Gates, D.J.; Penrose, O. The van der Waals limit for classical systems. II. Existence and continuity of the canonical pressure. *Comm. Math. Phys.* **1970**, *16*, 231–237. [CrossRef]

35. Gates, D.J.; Penrose, O. The van der Waals limit for classical systems. III. Deviation from the van der Waals-Maxwell theory. *Comm. Math. Phys.* **1970**, *17*, 194–209. [CrossRef]

36. Blanchard, P. Complex analytic dynamics on the Riemann sphere. *Bull. Am. Math. Soc.* **1984**, *11*, 85–141. [CrossRef]

37. Schlag, W. *A Course in Complex Analysis and Riemann Surfaces*; American Mathematical Society: Providence, RI, USA, 2014.

38. Chicharro, F.I.; Cordero, A.; Torregrosa, J.R. Drawing dynamical and parameters planes of iterative families and methods. *Sci. World J.* **2013**, *2013*, 780153. [CrossRef] [PubMed]

39. Amat, S.; Busquier, S.; Plaza, S. Chaotic dynamics of a third-order Newton-type method. *J. Math. Anal. Appl.* **2010**, *366*, 24–32. [CrossRef]

![mathematics logo] *mathematics*

MDPI

Article

Higher-Order Derivative-Free Iterative Methods for Solving Nonlinear Equations and Their Basins of Attraction

Jian Li [1],*, Xiaomeng Wang [1] and Kalyanasundaram Madhu [2],* ⓘ

[1] Inner Mongolia Vocational College of Chemical Engineering, Hohhot 010070, China; xiaomengw081@gmail.com
[2] Department of Mathematics, Saveetha Engineering College, Chennai 602105, India
*   Correspondence: ljian4064@gmail.com (J.L.); kalyan742@pec.edu (K.M.)

Received: 08 September 2019; Accepted: 16 October 2019; Published: 4 November 2019

Abstract: Based on the Steffensen-type method, we develop fourth-, eighth-, and sixteenth-order algorithms for solving one-variable equations. The new methods are fourth-, eighth-, and sixteenth-order converging and require at each iteration three, four, and five function evaluations, respectively. Therefore, all these algorithms are optimal in the sense of Kung–Traub conjecture; the new schemes have an efficiency index of 1.587, 1.682, and 1.741, respectively. We have given convergence analyses of the proposed methods and also given comparisons with already established known schemes having the same convergence order, demonstrating the efficiency of the present techniques numerically. We also studied basins of attraction to demonstrate their dynamical behavior in the complex plane.

Keywords: Kung–Traub conjecture; multipoint iterations; nonlinear equation; optimal order; basins of attraction

MSC: 65H05; 65D05; 41A25

1. Introduction

Finding faster and exact roots of scalar nonlinear equations is the most important problem in engineering, scientific computing, and applied mathematics. In general, this is the problem of solving a nonlinear equation $f(x) = 0$. Analytical methods for finding solutions of such problems are almost nonavailable, so the only way to get appropriate solutions by numerical methods is based on iterative algorithms. Newton's method is one of the well-known and famous methods for finding solutions of nonlinear equations or local minima in problems of optimization. Despite its nice properties, it will often not work efficiently in some real-life practical applications. Ill conditioning of the problems, the computational expense of functional derivative, accurate initial guesses, and a late convergence rate generally lead to difficulties in its use. Nevertheless, many advantages in all of these drawbacks have been found and led to efficient algorithms or codes that can be easily used (see References [1,2] and references therein). Hence, Steffensen developed a derivative-free iterative method ($SM2$) (see References [3]):

$$w^{(n)} = x^{(n)} + f(x^{(n)}), \ x^{(n+1)} = x^{(n)} - \frac{f(x^{(n)})}{f[x^{(n)}, w^{(n)}]}, \tag{1}$$

where $f[x^{(n)}, w^{(n)}] = \frac{f(x^{(n)}) - f(w^{(n)})}{x^{(n)} - w^{(n)}}$, which preserves the convergence order and efficiency index of Newton's method.

The main motivation of this work is to implement efficient derivative-free algorithms for finding the solution of nonlinear equations. We obtained an optimal iterative method that will support the conjecture [4]. Kung–Traub conjectured that multipoint iteration methods without memory based on d functional evaluations could achieve an optimal convergence order 2^{d-1}. Furthermore, we studied the behavior of iterative schemes in the complex plane.

Let us start a short review of the literature with some of the existing methods with or without memory before proceeding to the proposed idea. Behl et al. [5] presented an optimal scheme that does not need any derivative evaluations. In addition, the given scheme is capable of generating new optimal eighth-order methods from the earlier optimal fourth-order schemes in which the first sub-step employs Steffensen's or a Steffensen-type method. Salimi et al. [6] proposed a three-point iterative method for solving nonlinear equations. The purpose of this work is to upgrade a fourth-order iterative method by adding one Newton step and by using a proportional approximation for the last derivative. Salimi et al. [7] constructed two optimal Newton–Secant-like iterative methods for finding solutions of nonlinear equations. The classes have convergence orders of four and eight and cost only three and four function evaluations per iteration, respectively. Matthies et al. [8] proposed a three-point iterative method without memory for solving nonlinear equations with one variable. The method provides a convergence order of eight with four function evaluations per iteration. Sharifi et al. [9] presented an iterative method with memory based on the family of King's methods to solve nonlinear equations. The method has eighth-order convergence and costs only four function evaluations per iteration. An acceleration of the convergence speed is achieved by an appropriate variation of a free parameter in each step. This self-accelerator parameter is estimated using Newton's interpolation fourth degree polynomial. The order of convergence is increased from eight to 12 without any extra function evaluation. Khdhr et al. [10] suggested a variant of Steffensen's iterative method with a convergence order of 3.90057 for solving nonlinear equations that are derivative-free and have memory. Soleymani et al. [11] presented derivative-free iterative methods without memory with convergence orders of eight and sixteen for solving nonlinear equations. Soleimani et al. [12] proposed a optimal family of three-step iterative methods with a convergence order of eight by using a weight function alongside an approximation for the first derivative. Soleymani et al. [13] gave a class of four-step iterative schemes for finding solutions of one-variable equations. The produced methods have better order of convergence and efficiency index in comparison with optimal eighth-order methods. Soleymani et al. [14] constructed a class of three-step eighth order iterative methods by using an interpolatory rational function in the third step. Each method of the class reaches the optimal efficiency index according to the Kung–Traub conjecture concerning multipoint iterative methods without memory. Kanwar et al. [15] suggested two new eighth-order classes of Steffensen–King-type methods for finding solutions of nonlinear equations numerically. Cordero et al. [1] proposed a general procedure to obtain derivative-free iterative methods for finding solutions of nonlinear equations by polynomial interpolation. In addition, many authors have worked with these ideas on different iterative schemes [16–24], describing the basin of attraction of some well-known iterative scheme. In this work, we developed a novel fourth-order iterative scheme, eighth-order iterative scheme, and sixteenth-order iterative scheme, that are without memory, are derivative-free, and are optimal.

The rest of this paper is ordered as follows. In Section 2, we present the proposed fourth-, eighth-, and sixteenth-order methods that are free from derivatives. Section 3 presents the convergence order of the proposed scheme. In Section 4, we discuss some well-known iterative methods for the numerical and effectiveness comparison of the proposed schemes. In Section 5, we display the performance of proposed methods and other compared algorithms described by problems. The respective graphical fractal pictures obtained from each iteration scheme for test problems are given in Section 6 to show the consistency of the proposed methods. Finally, Section 7 gives concluding remarks.

2. Development of Derivative-Free Scheme

2.1. Optimal Fourth-Order Method

Let us start from Steffensen's method and explain the procedure to get optimal methods of increasing order. The idea is to compose a Steffensen's iteration with Newton's step as follows:

$$
\begin{cases}
w^{(n)} = x^{(n)} + f(x^{(n)})^3, \\
y^{(n)} = x^{(n)} - \dfrac{f(x^{(n)})^4}{f(w^{(n)}) - f(x^{(n)})}, \\
z^{(n)} = y^{(n)} - \dfrac{f(y^{(n)})}{f'(y^{(n)})}.
\end{cases}
\tag{2}
$$

The resulting iteration has convergence order four, with the composition of two second-order methods, but the method is not optimal because it uses four function evaluations. In order to get an optimality, we need to reduce a function and to preserve the same convergence order, and so, we estimate $f'(y^{(n)})$ by the following polynomial:

$$
\mathcal{N}_2(t) = f(y^{(n)}) + (t - y^{(n)})f[y^{(n)}, w^{(n)}] + (t - y^{(n)})(t - w^{(n)})f[y^{(n)}, w^{(n)}, x^{(n)}],
\tag{3}
$$

where

$$
f[x^{(0)}, x^{(1)}, x^{(2)}, ..., x^{(k-1)}, x^{(k)}] = \frac{f[x^{(1)}, x^{(2)}, ..., x^{(k-1)}, x^{(k)}] - f[x^{(0)}, x^{(1)}, x^{(2)}, ..., x^{(k-1)}]}{x^{(k)} - x^{(0)}}, \quad x^{(k)} \neq x^{(0)},
$$

is the generalized divided differences of kth order at $x^{(0)} \leq x^{(1)} \leq x^{(2)} \leq ... \leq x^{(k-1)} \leq x^{(k)}$. It is noted that $\mathcal{N}_2(y^{(n)}) = f(y^{(n)})$. Differentiating Equation (3) and putting $t = y^{(n)}$, we get

$$
\mathcal{N}_2'(y^{(n)}) = f[y^{(n)}, w^{(n)}] + (y^{(n)} - w^{(n)})f[y^{(n)}, w^{(n)}, x^{(n)}].
\tag{4}
$$

Now, approximating $f'(y^{(n)}) \approx \mathcal{N}_2'(y^{(n)})$ in Equation (2), we get a new derivative-free optimal fourth-order method (*PM4*) given by

$$
\begin{cases}
w^{(n)} = x^{(n)} + f(x^{(n)})^3, \\
y^{(n)} = x^{(n)} - \dfrac{f(x^{(n)})^4}{f(w^{(n)}) - f(x^{(n)})}, \\
z^{(n)} = y^{(n)} - \dfrac{f(y^{(n)})}{f[y^{(n)}, w^{(n)}] + (y^{(n)} - w^{(n)})f[y^{(n)}, w^{(n)}, x^{(n)}]}.
\end{cases}
\tag{5}
$$

2.2. Optimal Eighth-Order Method

Next, we attempt to get a new optimal eighth-order method in the following way:

$$
\begin{cases}
w^{(n)} = x^{(n)} + f(x^{(n)})^3, \\
y^{(n)} = x^{(n)} - \dfrac{f(x^{(n)})^4}{f(w^{(n)}) - f(x^{(n)})}, \\
z^{(n)} = y^{(n)} - \dfrac{f(y^{(n)})}{f[y^{(n)}, w^{(n)}] + (y^{(n)} - w^{(n)})f[y^{(n)}, w^{(n)}, x^{(n)}]}, \\
p^{(n)} = z^{(n)} - \dfrac{f(z^{(n)})}{f'(z^{(n)})}.
\end{cases}
\tag{6}
$$

The above has eighth-order convergence with five function evaluations, but this is not an optimal method. To get an optimal, we need to reduce a function and to preserve the same convergence order, and so, we estimate $f'(z^{(n)})$ by the following polynomial:

$$
\begin{aligned}
\mathcal{N}_3(t) = {}& f(z^{(n)}) + (t - z^{(n)})f[z^{(n)}, y^{(n)}] + (t - z^{(n)})(t - y^{(n)})f[z^{(n)}, y^{(n)}, w^{(n)}] \\
& + (t - z^{(n)})(t - y^{(n)})(t - w^{(n)})f[z^{(n)}, y^{(n)}, w^{(n)}, x^{(n)}].
\end{aligned}
\tag{7}
$$

It is clear that $\mathcal{N}_3(z^{(n)}) = f(z^{(n)})$. Differentiating Equation (7) and setting $t = z^{(n)}$, we get

$$\mathcal{N}_3'(z^{(n)}) = f[z^{(n)}, y^{(n)}] + (z^{(n)} - y^{(n)})f[z^{(n)}, y^{(n)}, w^{(n)}] + (z^{(n)} - y^{(n)})(z^{(n)} - w^{(n)})f[z^{(n)}, y^{(n)}, w^{(n)}, x^{(n)}]. \qquad (8)$$

Now, approximating $f'(z^{(n)}) \approx \mathcal{N}_3'(z^{(n)})$ in (6), we get a new derivative-free optimal eighth-order method (*PM8*) given by

$$\begin{cases} w^{(n)} = x^{(n)} + f(x^{(n)})^3, \\ y^{(n)} = x^{(n)} - \dfrac{f(x^{(n)})^4}{f(w^{(n)}) - f(x^{(n)})}, \\ z^{(n)} = y^{(n)} - \dfrac{f(y^{(n)})}{f[y^{(n)}, w^{(n)}] + (y^{(n)} - w^{(n)})f[y^{(n)}, w^{(n)}, x^{(n)}]}, \\ p^{(n)} = z^{(n)} - \dfrac{f(z^{(n)})}{f[z^{(n)}, y^{(n)}] + (z^{(n)} - y^{(n)})f[z^{(n)}, y^{(n)}, w^{(n)}] + (z^{(n)} - y^{(n)})(z^{(n)} - w^{(n)})f[z^{(n)}, y^{(n)}, w^{(n)}, x^{(n)}]}. \end{cases} \qquad (9)$$

2.3. Optimal Sixteenth-Order Method

Next, we attempt to get a new optimal sixteenth-order method in the following way:

$$\begin{cases} w^{(n)} = x^{(n)} + f(x^{(n)})^3, \\ y^{(n)} = x^{(n)} - \dfrac{f(x^{(n)})^4}{f(w^{(n)}) - f(x^{(n)})}, \\ z^{(n)} = y^{(n)} - \dfrac{f(y^{(n)})}{f[y^{(n)}, w^{(n)}] + (y^{(n)} - w^{(n)})f[y^{(n)}, w^{(n)}, x^{(n)}]}, \\ p^{(n)} = z^{(n)} - \dfrac{f(z^{(n)})}{f[z^{(n)}, y^{(n)}] + (z^{(n)} - y^{(n)})f[z^{(n)}, y^{(n)}, w^{(n)}] + (z^{(n)} - y^{(n)})(z^{(n)} - w^{(n)})f[z^{(n)}, y^{(n)}, w^{(n)}, x^{(n)}]}, \\ x^{(n+1)} = p^{(n)} - \dfrac{f(p^{(n)})}{f'(p^{(n)})}. \end{cases} \qquad (10)$$

The above has sixteenth-order convergence with six function evaluations, but this is not an optimal method. To get an optimal, we need to reduce a function and to preserve the same convergence order, and so, we estimate $f'(p^{(n)})$ by the following polynomial:

$$\begin{aligned} \mathcal{N}_4(t) = {} & f(p^{(n)}) + (t - p^{(n)})f[p^{(n)}, z^{(n)}] + (t - p^{(n)})(t - z^{(n)})f[p^{(n)}, z^{(n)}, y^{(n)}] \\ & + (t - p^{(n)})(t - z^{(n)})(t - y^{(n)})f[p^{(n)}, z^{(n)}, y^{(n)}, w^{(n)}] \\ & + (t - p^{(n)})(t - z^{(n)})(t - y^{(n)})(t - w^{(n)})f[p^{(n)}, z^{(n)}, y^{(n)}, w^{(n)}, x^{(n)}]. \end{aligned} \qquad (11)$$

It is clear that $\mathcal{N}_4(p^{(n)}) = f(p^{(n)})$. Differentiating Equation (11) and setting $t = p^{(n)}$, we get

$$\begin{aligned} \mathcal{N}_4'(p^{(n)}) = {} & f[p^{(n)}, z^{(n)}] + (p^{(n)} - z^{(n)})f[p^{(n)}, z^{(n)}, y^{(n)}] + (p^{(n)} - z^{(n)})(p^{(n)} - y^{(n)})f[p^{(n)}, z^{(n)}, y^{(n)}, w^{(n)}] \\ & + (p^{(n)} - z^{(n)})(p^{(n)} - y^{(n)})(p^{(n)} - w^{(n)})f[p^{(n)}, z^{(n)}, y^{(n)}, w^{(n)}, x^{(n)}]. \end{aligned} \qquad (12)$$

Now, approximating $f'(p^{(n)}) \approx \mathcal{N}_4'(p^{(n)})$ in Equation (10), we get a new derivative-free optimal sixteenth-order iterative method (*PM16*) given by

$$\begin{cases} w^{(n)} = x^{(n)} + f(x^{(n)})^3, \\ y^{(n)} = x^{(n)} - \dfrac{f(x^{(n)})^4}{f(w^{(n)}) - f(x^{(n)})}, \\ z^{(n)} = y^{(n)} - \dfrac{f(y^{(n)})}{f[y^{(n)}, w^{(n)}] + (y^{(n)} - w^{(n)})f[y^{(n)}, w^{(n)}, x^{(n)}]}, \\ p^{(n)} = z^{(n)} - \dfrac{f(z^{(n)})}{f[z^{(n)}, y^{(n)}] + (z^{(n)} - y^{(n)})f[z^{(n)}, y^{(n)}, w^{(n)}] + (z^{(n)} - y^{(n)})(z^{(n)} - w^{(n)})f[z^{(n)}, y^{(n)}, w^{(n)}, x^{(n)}]}, \\ x^{(n+1)} = p^{(n)} - \dfrac{f(p^{(n)})}{\mathcal{N}_4'(p^{(n)})}, \end{cases} \qquad (13)$$

where $\mathcal{N}_4'(p^{(n)})$ given in Equation (12).

3. Convergence Analysis

In this part, we will derive the convergence analysis of the proposed schemes in Equations (5), (9), and (13) with the help of MATHEMATICA software.

Theorem 1. *Let $f : D \subset \mathbb{R} \to \mathbb{R}$ be a sufficiently smooth function having continuous derivatives. If $f(x)$ has a simple root x^* in the open interval D and $x^{(0)}$ is chosen in a sufficiently small neighborhood of x^*, then the method of Equation (5) is of local fourth-order convergence and and it satisfies the error equation*

$$e_{n+1} = (c[2]^3 - c[2]c[3])e_n^4 + O(e_n^5).$$

Proof. Let $e_n = x^{(n)} - x^*$ and $c[j] = \frac{f^{(j)}(x^*)}{j!f'(x^*)}$, $j = 2, 3, 4, \dots$. Expanding $f(x^{(n)})$ and $f(w^{(n)})$ about x^* by Taylor's method, we have

$$f(x^{(n)}) = f'(x^*)[e_n + c[2]e_n^2 + c[3]e_n^3 + c[4]e_n^4 + \dots], \tag{14}$$

$$w^{(n)} = e_n + f'(x^*)^3[e_n + c[2]e_n^2 + c[3]e_n^3 + c[4]e_n^4 + \dots]^3, \tag{15}$$

$$f(w^{(n)}) = f'(x^*)[e_n + c[2]e_n^2 + (f'(x^*)^3 + c[3])e_n^3 + (5f'(x^*)^3c[2] + c[4])e_n^4 + \dots]. \tag{16}$$

Then, we have

$$y^{(n)} = x^* + c[2]e_n^2 + (-2c[2]^2 + 2c[3])e_n^3 + (4c[2]^3 - 7c[2]c[3] + 3c[4] + f'(x^*)^3c[2])e_n^4 + \dots . \tag{17}$$

Expanding $f(y^{(n)})$ about x^*, we have

$$f(y^{(n)}) = f'(x^*)[c[2]e_n^2 - 2(c[2]^2 - c[3])e_n^3 + (5c[2]^3 - 7c[2]c[3] + 3c[4] + f'(x^*)^3c[2])e_n^4 + \dots]. \tag{18}$$

Now, we get the Taylor's expansion of $f[y^{(n)}, w^{(n)}] = \frac{f(y^{(n)}) - f(w^{(n)})}{y^{(n)} - w^{(n)}}$ by replacing Equation (15)–(18).

$$f[y^{(n)}, w^{(n)}] = f'(x^*)[1 + c[2]e_n + (c[2]^2 + c[3])e_n^2 + (f'(x^*)^3c[2] - 2c[2]^3 + c[2]c[3] + c[4])e_n^3 + \dots]. \tag{19}$$

Also, we have

$$f[y^{(n)}, w^{(n)}, x^{(n)}] = f'(x^*)[c[2] + 2c[3]e_n + (c[2]c[3] + c[4])e_n^2 + \dots] \tag{20}$$

Using Equations (14)–(20) in the scheme of Equation (5), we obtain the following error equation:

$$e_{n+1} = (c[2]^3 - c[2]c[3])e_n^4 + \dots . \tag{21}$$

This reveals that the proposed method *PM4* attains fourth-order convergence. $\square$

Theorem 2. *Let $f : D \subset \mathbb{R} \to \mathbb{R}$ be a sufficiently smooth function having continuous derivatives. If $f(x)$ has a simple root x^* in the open interval D and $x^{(0)}$ is chosen in a sufficiently small neighborhood of x^*, then the method of Equation (9) is of local eighth-order convergence and and it satisfies the error equation*

$$e_{n+1} = c[2]^2(c[2]^2 - c[3])(c[2]^3 - c[2]c[3] + c[4])e_n^8 + O(e_n^9).$$

Theorem 3. *Let $f : D \subset \mathbb{R} \to \mathbb{R}$ be a sufficiently smooth function having continuous derivatives. If $f(x)$ has a simple root x^* in the open interval D and $x^{(0)}$ is chosen in a sufficiently small neighborhood of x^*, then the method of Equation* (13) *is of local sixteenth-order convergence and and it satisfies the error equation*

$$e_{n+1} = c[2]^4 \left(c[2]^2 - c[3]\right)^2 \left(c[2]^3 - c[2]c[3] + c[4]\right)\left(c[2]^4 - c[2]^2 c[3] + c[2]c[4] - c[5]\right) e_n^{16} + O(e_n^{17}).$$

4. Some Known Derivative-Free Methods

Let us consider the following schemes for the purpose of comparison. Derivative-free Kung–Traub's two-step method (KTM4) [4] is as follows:

$$y^{(n)} = x^{(n)} - \frac{f(x^{(n)})}{f[x^{(n)}, w^{(n)}]}, \ w^{(n)} = x^{(n)} + f(x^{(n)}), \ x^{(n+1)} = y^{(n)} - \frac{f(y^{(n)})f(w^{(n)})}{[f(w^{(n)}) - f(y^{(n)})]f[x^{(n)}, y^{(n)}]}. \tag{22}$$

Derivative-free Argyros et al. two-step method (AKKB4) [25] is as follows:

$$y^{(n)} = x^{(n)} - \frac{f(x^{(n)})}{f[x^{(n)}, w^{(n)}]}, \ w^{(n)} = x^{(n)} + f(x^{(n)}), \ x^{(n+1)} = y^{(n)} - \frac{f(y^{(n)})}{[f(x^{(n)}) - 2f(y^{(n)})]} \frac{f(x^{(n)})}{f[y^{(n)}, w^{(n)}]} \left(1 - \frac{f(y^{(n)})}{f(x^{(n)})}\right). \tag{23}$$

Derivative-free Zheng et al. two-step method (ZLM4) [26] is as follows:

$$y^{(n)} = x^{(n)} - \frac{f(x^{(n)})}{f[x^{(n)}, w^{(n)}]}, \ w^{(n)} = x^{(n)} + f(x^{(n)}), \ x^{(n+1)} = y^{(n)} - \frac{f(y^{(n)})}{f[x^{(n)}, y^{(n)}] + (y^{(n)} - x^{(n)})f[x^{(n)}, w^{(n)}, y^{(n)}]}. \tag{24}$$

Derivative-free Argyros et al. three-step method (AKKB8) [25] is as follows:

$$\begin{cases} y^{(n)} = x^{(n)} - \frac{f(x^{(n)})}{f[x^{(n)}, w^{(n)}]}, w^{(n)} = x^{(n)} + f(x^{(n)}), z^{(n)} = y^{(n)} - \frac{f(y^{(n)})}{[f(x^{(n)}) - 2f(y^{(n)})]} \frac{f(x^{(n)})}{f[y^{(n)}, w^{(n)}]} \left(1 - \frac{f(y^{(n)})}{f(x^{(n)})}\right), \\ x^{(n+1)} = z^{(n)} - \frac{f(z^{(n)})}{f[z^{(n)}, y^{(n)}] + (z^{(n)} - y^{(n)})f[z^{(n)}, y^{(n)}, x^{(n)}] + (z^{(n)} - y^{(n)})(z^{(n)} - x^{(n)})f[z^{(n)}, y^{(n)}, x^{(n)}, w^{(n)}]}. \end{cases} \tag{25}$$

Derivative-free Kanwar et al. three-step method (KBK8) [15] is as follows:

$$\begin{cases} y^{(n)} = x^{(n)} - \frac{f(x^{(n)})}{f[x^{(n)}, w^{(n)}]}, \ w^{(n)} = x^{(n)} + f(x^{(n)})^3, \ z^{(n)} = y^{(n)} - \frac{f(y^{(n)})}{2f[y^{(n)}, x^{(n)}] - f[x^{(n)}, w^{(n)}]}, \\ x^{(n+1)} = z^{(n)} - \frac{f(z^{(n)})}{f[y^{(n)}, z^{(n)}] + f[w^{(n)}, y^{(n)}, z^{(n)}](z^{(n)} - y^{(n)})} \left(1 - \left(\frac{f(y^{(n)})}{f(x^{(n)})}\right)^3 - 8\frac{f(y^{(n)})f(z^{(n)})}{f(x^{(n)})^2} + \frac{f(z^{(n)})}{f(x^{(n)})} + 5\left(\frac{f(z^{(n)})}{f(y^{(n)})}\right)^2\right). \end{cases} \tag{26}$$

Derivative-free Soleymani three-step method (SM8) [2] is as follows:

$$\begin{cases} w^{(n)} = x^{(n)} + f(x^{(n)}), y^{(n)} = x^{(n)} - \frac{f(x^{(n)})}{f[x^{(n)}, w^{(n)}]}, z^{(n)} = y^{(n)} - \frac{f(y^{(n)})}{f[x^{(n)}, w^{(n)}]} \phi_n, \\ x^{(n+1)} = z^{(n)} - \frac{f(z^{(n)})}{f[x^{(n)}, w^{(n)}]} \phi_n \psi_n, \text{ where } \phi_n = \frac{1}{1 - f(y^{(n)})/f(x^{(n)}) - f(y^{(n)})/f(w^{(n)})}, \\ \psi_n = 1 + \frac{1}{1 + f[x^{(n)}, w^{(n)}]} \left(\frac{f(y^{(n)})}{f(x^{(n)})}\right)^2 + \left((1 + f[x^{(n)}, w^{(n)}])(2 + f[x^{(n)}, w^{(n)}])\right)\left(\frac{f(y^{(n)})}{f(w^{(n)})}\right)^3 + \frac{f(z^{(n)})}{f(y^{(n)})} + \frac{f(z^{(n)})}{f(x^{(n)})} + \frac{f(z^{(n)})}{f(w^{(n)})}. \end{cases} \tag{27}$$

Derivative-free Zheng et al. four-step method (ZLM16) [26] is as follows:

$$\begin{cases} y^{(n)} = x^{(n)} - \frac{f(x^{(n)})^2}{f(w^{(n)}) - f(x^{(n)})}, w^{(n)} = x^{(n)} + f(x^{(n)}), z^{(n)} = y^{(n)} - \frac{f(y^{(n)})}{f[y^{(n)}, w^{(n)}] + (y^{(n)} - w^{(n)})f[y^{(n)}, w^{(n)}, x^{(n)}]}, \\ p^{(n)} = z^{(n)} - \frac{f(z^{(n)})}{f[z^{(n)}, y^{(n)}] + (z^{(n)} - y^{(n)})f[z^{(n)}, y^{(n)}, w^{(n)}] + (z^{(n)} - y^{(n)})(z^{(n)} - w^{(n)})f[z^{(n)}, y^{(n)}, w^{(n)}, x^{(n)}]}, \\ x^{(n+1)} = p^{(n)} - \frac{f(p^{(n)})}{f'(p^{(n)})}, \\ \text{where } f'(p^{(n)}) \approx f[p^{(n)}, z^{(n)}] + (p^{(n)} - z^{(n)})f[p^{(n)}, z^{(n)}, y^{(n)}] + (p^{(n)} - z^{(n)})(p^{(n)} - y^{(n)})f[p^{(n)}, z^{(n)}, y^{(n)}, w^{(n)}] \\ + (p^{(n)} - z^{(n)})(p^{(n)} - y^{(n)})(p^{(n)} - w^{(n)})f[p^{(n)}, z^{(n)}, y^{(n)}, w^{(n)}, x^{(n)}]. \end{cases} \tag{28}$$

5. Test Problems

We compare the performance of the proposed methods along with some existing methods for test problems by using MATLAB. We use the conditions for stopping criteria for $|f(x^{(N)})| < \epsilon$ where

$\epsilon = 10^{-50}$ and N is the number of iterations needed for convergence. The computational order of convergence (*coc*) is given by

$$\rho = \frac{\ln\left|(x^{(N)} - x^{(N-1)})/(x^{(N-1)} - x^{(N-2)})\right|}{\ln\left|(x^{(N-1)} - x^{(N-2)})/(x^{(N-2)} - x^{(N-3)})\right|}.$$

The test problems and their roots are given below:

$$f_1(x) = \sin(2\cos x) - 1 - x^2 + e^{\sin(x^3)}, \qquad x^* = -0.7848959876612125352...$$
$$f_2(x) = x^3 + 4x^2 - 10, \qquad x^* = 1.36523001341140968457...$$
$$f_3(x) = \sqrt{x^2 + 2x + 5} - 2\sin x - x^2 + 3, \qquad x^* = 2.3319676558839640103...$$
$$f_4(x) = e^{-x}\sin x + \log(1 + x^2) - 2, \qquad x^* = 2.4477482864524245021...$$
$$f_5(x) = \sin(x) + \cos(x) + x, \qquad x^* = -0.4566247045676308244...$$

Tables 1–5 show the results of all the test functions with a given initial point. The computational order of convergence conforms with theoretical order of convergence. If the initial points are close to the zero, then we obtain less number of iterations with least error. If the initial points are away from the zero, then we will not obtained the least error. We observe that the new methods in all the test function have better efficiency as compared to other existing methods of the equivalent methods.

Table 1. Comparisons between different methods for $f_1(x)$ at $x^{(0)} = -0.9$.

Methods	N	$|x^{(1)} - x^{(0)}|$	$|x^{(2)} - x^{(1)}|$	$|x^{(3)} - x^{(2)}|$	$|x^{(N)} - x^{(N-1)}|$	coc
SM2 (1)	8	0.0996	0.0149	6.1109×10^{-4}	1.0372×10^{-89}	1.99
KTM4 (22)	5	0.1144	6.7948×10^{-4}	3.4668×10^{-12}	5.1591×10^{-178}	4.00
AKKB4 (23)	4	0.1147	3.6299×10^{-4}	9.5806×10^{-14}	4.6824×10^{-52}	3.99
ZLM4 (24)	5	0.1145	6.1744×10^{-4}	1.5392×10^{-12}	1.3561×10^{-184}	4.00
PM4 (5)	4	0.1150	1.3758×10^{-4}	2.6164×10^{-16}	3.4237×10^{-63}	3.99
AKKB8 (25)	3	0.1151	1.2852×10^{-8}	3.7394×10^{-62}	3.7394×10^{-62}	7.70
KBK8 (26)	3	0.1151	8.1491×10^{-8}	1.5121×10^{-56}	1.5121×10^{-56}	7.92
SM8 (27)	4	0.1151	1.8511×10^{-6}	1.0266×10^{-43}	0	7.99
PM8 (9)	3	0.1151	7.1154×10^{-9}	9.3865×10^{-67}	9.3865×10^{-67}	8.02
ZLM16 (28)	3	0.1151	5.6508×10^{-15}	1.4548×10^{-225}	1.4548×10^{-225}	15.82
PM16 (13)	3	0.1151	5.3284×10^{-17}	1.2610×10^{-262}	1.2610×10^{-262}	16.01

Table 2. Comparisons between different methods for $f_2(x)$ at $x^{(0)} = 1.6$.

Methods	N	$|x^{(1)} - x^{(0)}|$	$|x^{(2)} - x^{(1)}|$	$|x^{(3)} - x^{(2)}|$	$|x^{(N)} - x^{(N-1)}|$	coc
SM2 (1)	12	0.0560	0.0558	0.0520	1.7507×10^{-83}	1.99
KTM4 (22)	5	0.2184	0.0163	3.4822×10^{-6}	4.7027×10^{-79}	3.99
AKKB4 (23)	33	0.0336	0.0268	0.0171	2.4368×10^{-52}	0.99
ZLM4 (24)	5	0.2230	0.0117	4.4907×10^{-7}	3.9499×10^{-95}	3.99
PM4 (5)	5	0.2123	0.0224	2.3433×10^{-7}	4.3969×10^{-112}	4.00
AKKB8 (25)	4	0.2175	0.0173	1.2720×10^{-9}	1.0905×10^{-66}	8.00
KBK8 (26)	D	D	D	D	D	D
SM8 (27)	4	0.2344	4.1548×10^{-4}	9.5789×10^{-24}	7.7650×10^{-181}	7.89
PM8 (9)	4	0.2345	2.4307×10^{-4}	4.6428×10^{-32}	8.2233×10^{-254}	8.00
ZLM16 (28)	3	0.2348	2.2048×10^{-7}	1.9633×10^{-124}	1.9633×10^{-124}	15.57
PM16 (13)	3	0.2348	2.8960×10^{-8}	1.7409×10^{-126}	1.7409×10^{-126}	17.11

Table 3. Comparisons between different methods for $f_3(x)$ at $x^{(0)} = 2.7$.

Methods	N	$\|x^{(1)} - x^{(0)}\|$	$\|x^{(2)} - x^{(1)}\|$	$\|x^{(3)} - x^{(2)}\|$	$\|x^{(N)} - x^{(N-1)}\|$	coc
SM2 (1)	7	0.3861	0.0180	4.6738×10^{-05}	1.0220×10^{-82}	1.99
KTM4 (22)	4	0.3683	2.8791×10^{-4}	1.0873×10^{-16}	2.2112×10^{-66}	3.99
AKKB4 (23)	4	0.3683	2.5241×10^{-4}	5.2544×10^{-17}	9.8687×10^{-68}	3.99
ZLM4 (24)	4	0.3683	3.1466×10^{-4}	1.7488×10^{-16}	1.6686×10^{-65}	4.00
PM4 (5)	4	0.3683	2.2816×10^{-4}	2.3732×10^{-17}	2.7789×10^{-69}	3.99
AKKB8 (25)	3	0.3680	1.7343×10^{-8}	3.8447×10^{-67}	3.8447×10^{-67}	8.00
KBK8 (26)	4	0.3680	4.2864×10^{-5}	1.8700×10^{-38}	2.4555×10^{-305}	7.99
SM8 (27)	3	0.3680	7.8469×10^{-8}	2.9581×10^{-61}	2.9581×10^{-61}	8.00
PM8 (9)	3	0.3680	9.7434×10^{-9}	1.0977×10^{-69}	1.0977×10^{-69}	8.04
ZLM16 (28)	3	0.3680	1.4143×10^{-16}	6.3422×10^{-240}	6.3422×10^{-240}	16.03
PM16 (13)	3	0.3680	3.6568×10^{-17}	7.4439×10^{-274}	7.4439×10^{-274}	16.04

Table 4. Comparisons between different methods for $f_4(x)$ at $x^{(0)} = 1.9$.

Methods	N	$\|x^{(1)} - x^{(0)}\|$	$\|x^{(2)} - x^{(1)}\|$	$\|x^{(3)} - x^{(2)}\|$	$\|x^{(N)} - x^{(N-1)}\|$	coc
SM2 (1)	7	0.4975	0.0500	2.5378×10^{-4}	1.9405×10^{-73}	2.00
KTM4 (22)	4	0.2522	1.7586×10^{-6}	1.5651×10^{-26}	9.8198×10^{-107}	3.99
AKKB4 (23)	4	0.5489	0.0011	3.8305×10^{-15}	5.5011×10^{-61}	3.99
ZLM4 (24)	4	0.5487	9.0366×10^{-4}	1.4751×10^{-15}	1.0504×10^{-62}	3.99
PM4 (5)	4	0.5481	3.0864×10^{-4}	8.0745×10^{-18}	3.7852×10^{-72}	3.99
AKKB8 (25)	3	0.5477	5.4938×10^{-7}	4.9628×10^{-56}	4.9628×10^{-56}	8.17
KBK8 (26)	3	0.5477	4.1748×10^{-7}	5.8518×10^{-59}	5.8518×10^{-59}	8.47
SM8 (27)	3	0.5477	5.4298×10^{-7}	4.1081×10^{-56}	4.1081×10^{-56}	8.18
PM8 (9)	3	0.5477	5.8222×10^{-8}	1.1144×10^{-64}	1.1144×10^{-64}	8.13
ZLM16 (28)	3	0.5477	2.7363×10^{-14}	7.2982×10^{-229}	7.2982×10^{-229}	16.13
PM16 (13)	3	0.5477	5.6240×10^{-16}	1.9216×10^{-257}	1.9216×10^{-257}	16.11

Table 5. Comparisons between different methods for $f_5(x)$ at $x^{(0)} = -0.2$.

Methods	N	$\|x^{(1)} - x^{(0)}\|$	$\|x^{(2)} - x^{(1)}\|$	$\|x^{(3)} - x^{(2)}\|$	$\|x^{(N)} - x^{(N-1)}\|$	coc
SM2 (1)	7	0.3072	0.0499	6.4255×10^{-4}	4.1197×10^{-59}	2.00
KTM4 (22)	5	0.2585	0.0019	1.5538×10^{-12}	3.4601×10^{-194}	4.00
AKKB4 (23)	4	0.2571	4.4142×10^{-4}	3.4097×10^{-15}	1.2154×10^{-59}	3.99
ZLM4 (24)	4	0.2580	0.0013	3.5840×10^{-13}	1.8839×10^{-51}	3.99
PM4 (5)	4	0.2569	2.8004×10^{-4}	6.2960×10^{-17}	1.6097×10^{-67}	3.99
AKKB8 (25)	3	0.2566	4.1915×10^{-8}	6.3444×10^{-65}	6.3444×10^{-65}	8.37
KBK8 (26)	4	0.2566	4.0069×10^{-6}	5.1459×10^{-47}	0	7.99
SM8 (27)	4	0.2566	2.9339×10^{-6}	1.0924×10^{-46}	0	7.99
PM8 (9)	3	0.2566	3.7923×10^{-11}	9.0207×10^{-90}	9.0207×10^{-90}	7.99
ZLM16 (28)	3	0.2566	5.3695×10^{-16}	7.0920×10^{-252}	7.0920×10^{-252}	16.06
PM16 (13)	3	0.2566	1.1732×10^{-19}	1.2394×10^{-314}	1.2394×10^{-314}	16.08

6. Basins of Attraction

The iterative scheme gives information about convergence and stability by studying basins of attraction of the rational function. The basic definitions and dynamical concepts of rational function can found in References [17,27,28]. Let us consider a region $\mathbb{R} \times \mathbb{R} = [-2, 2] \times [-2, 2]$ with 256×256 grids. We test iterative methods in all the grid point $z^{(0)}$ in the square. The iterative algorithms attempt roots z_j^* of the equation with condition $|f(z^{(k)})| < \times 10^{-4}$ and a maximum of 100 iterations; we conclude that $z^{(0)}$ is in the basin of attraction of this zero. If the iterative method starting in $z^{(0)}$ reaches a zero in N iterations, then we mark this point $z^{(0)}$ with colors if $|z^{(N)} - z_j^*| < \times 10^{-4}$. If $N > 50$, then we assign a dark blue color for diverging grid points. We describe the basins of attraction for finding complex roots of $p_1(z) = z^2 - 1$, $p_2(z) = z^3 - 1$, $p_3(z) = (z^2 + 1)(z^2 - 1)$, and $p_4(z) = z^5 - 1$ for proposed methods and some higher-order iterative methods.

In Figures 1–5, we have given the basins of attraction for new methods with some existing methods. We confirm that a point z_0 containing the Julia set whenever the dynamics of point shows sensitivity to the conditions. The neighbourhood of initial points leads to the slight variation in behavior after some iterations. Therefore, some of the compared algorithms obtain more divergent initial conditions.

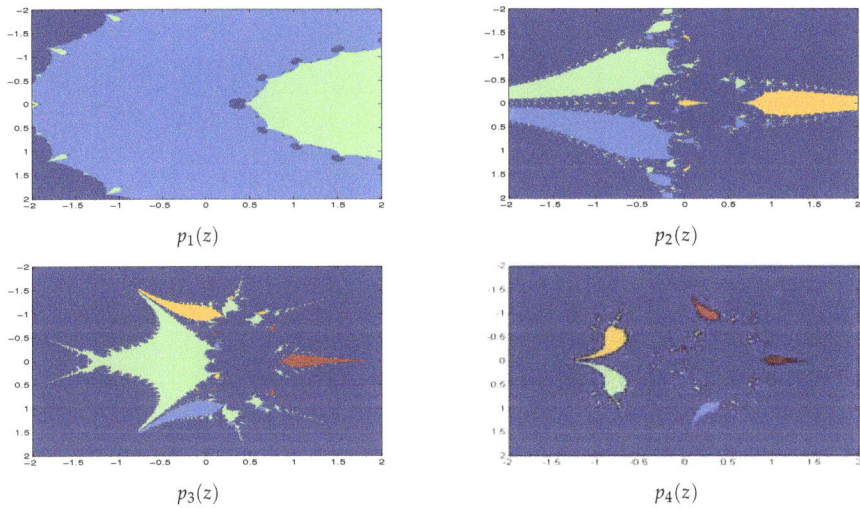

$p_1(z)$

$p_2(z)$

$p_3(z)$

$p_4(z)$

Figure 1. Basins of attraction for *SM2* for the polynomial.

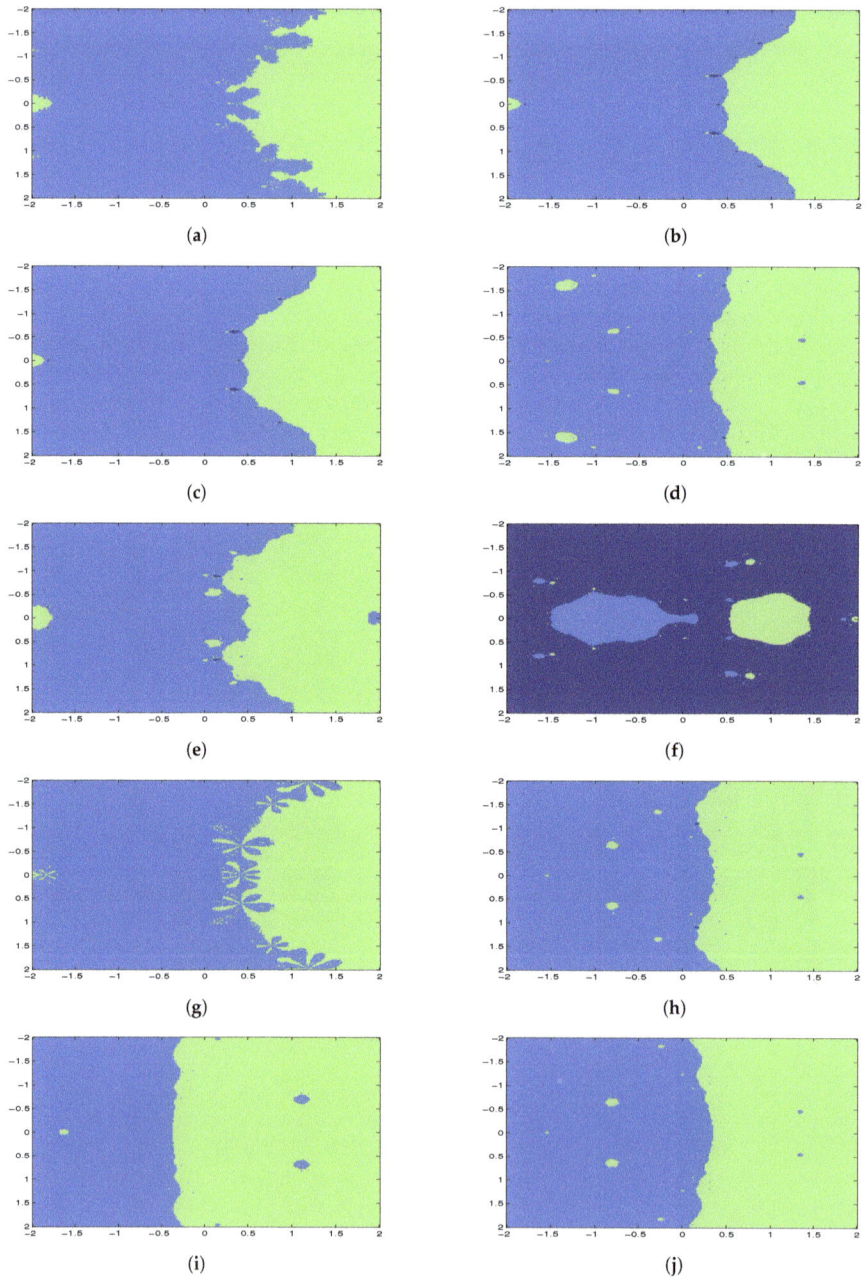

Figure 2. Polynomiographs of $p_1(z)$: (**a**) $KTM4$; (**b**) $AKKB4$; (**c**) $ZLM4$; (**d**) $PM4$; (**e**) $AKKB8$; (**f**) $KBK8$; (**g**) $SM8$; (**h**) $PM8$; (**i**) $ZLM16$; and (**j**) $PM16$.

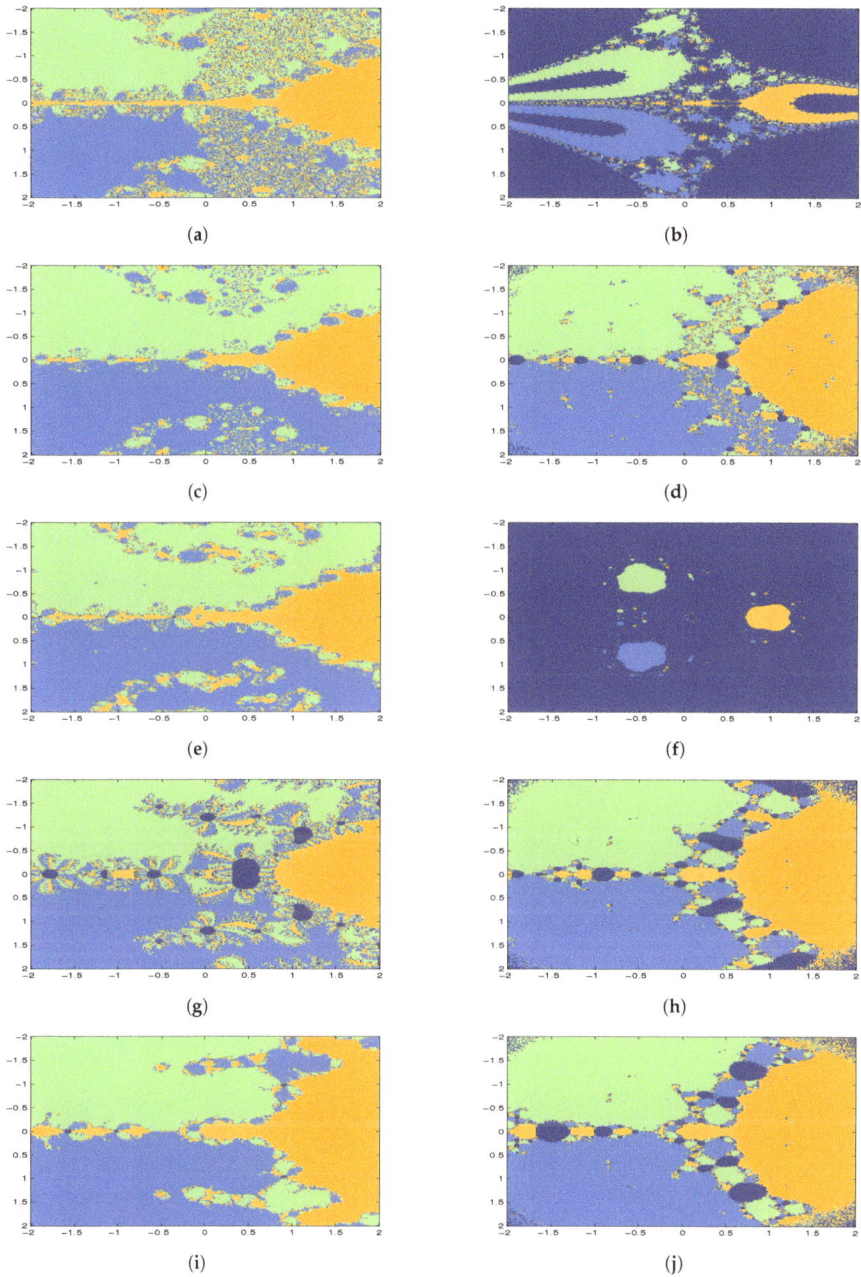

Figure 3. Polynomiographs of $p_2(z)$: (**a**) $KTM4$; (**b**) $AKKB4$; (**c**) $ZLM4$; (**d**) $PM4$; (**e**) $AKKB8$; (**f**) $KBK8$; (**g**) $SM8$; (**h**) $PM8$; (**i**) $ZLM16$; and (**j**) $PM16$.

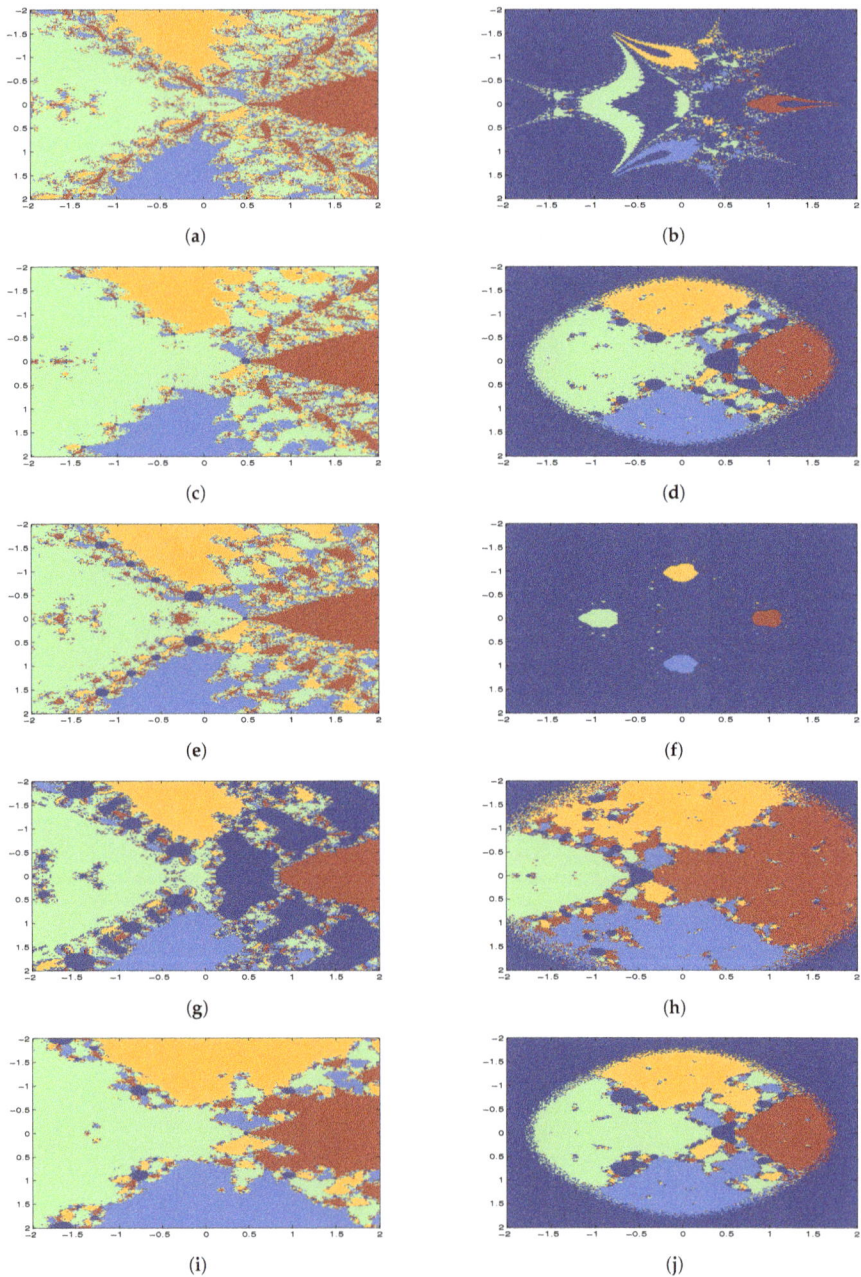

Figure 4. Polynomiographs of $p_3(z)$: (**a**) *KTM*4; (**b**) *AKKB*4; (**c**)*ZLM*4; (**d**) *PM*4; (**e**) *AKKB*8; (**f**) *KBK*8; (**g**) *SM*8; (**h**) *PM*8; (**i**) *ZLM*16; and (**j**) *PM*16.

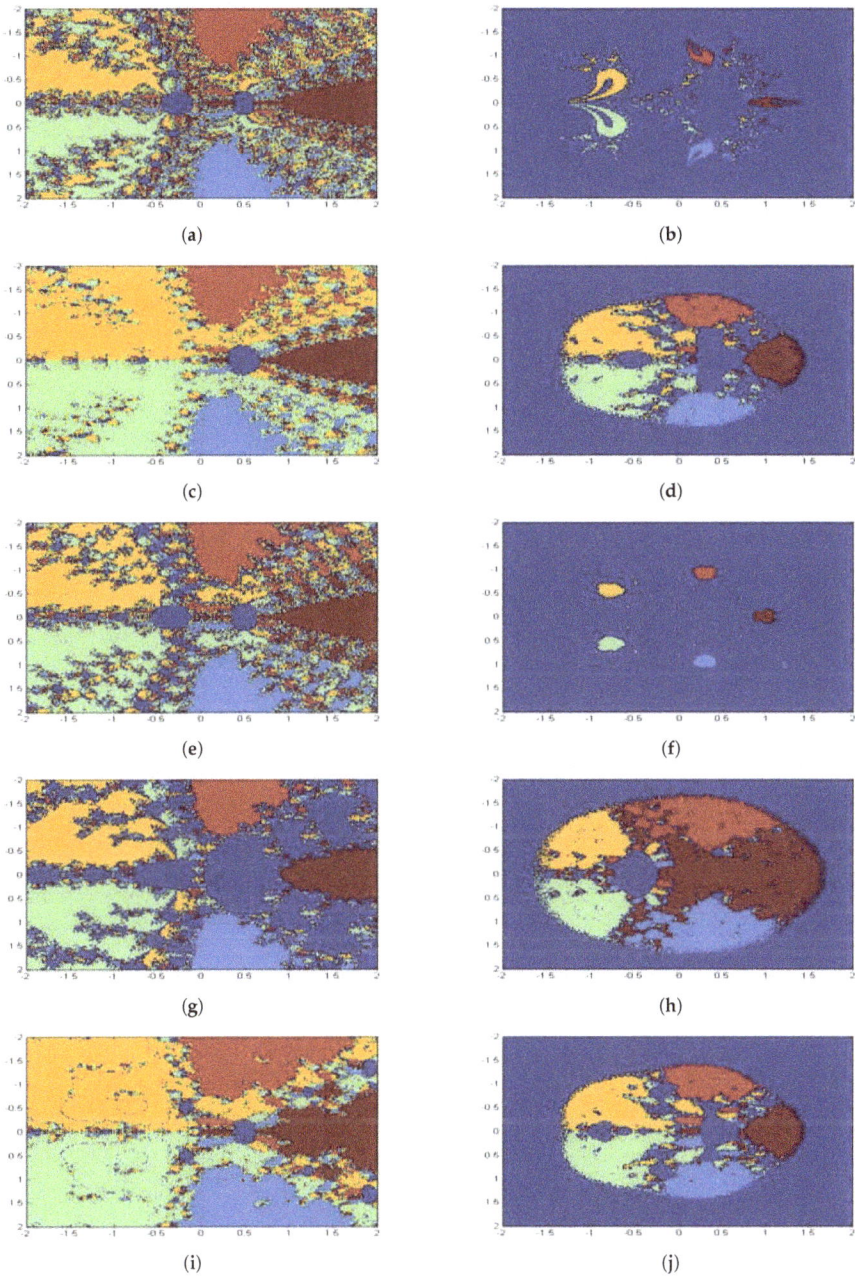

Figure 5. Polynomiographs of $p_4(z)$: (**a**) $KTM4$; (**b**) $AKKB4$; (**c**)$ZLM4$; (**d**) $PM4$; (**e**) $AKKB8$; (**f**) $KBK8$; (**g**) $SM8$; (**h**) $PM8$; (**i**) $ZLM16$; and (**j**) $PM16$.

7. Concluding Remarks

We have proposed fourth-, eighth-, and sixteenth-order methods using finite difference approximations. Our proposed new methods requires 3 functions to get the 4th-order method, 4 functions to obtain the 8th-order method, and 5 functions to get the 16th-order one. We have increased the convergence order of the proposed method, respectively, to four, eight, and sixteen with efficiency indices 1.587, 1.565, and 1.644 respectively. Our new proposed schemes are better than the Steffensen method in terms of efficiency index (1.414). Numerical solutions are tested to show the performance of the proposed algorithms. Also, we have analyzed on the complex region for iterative methods to study their basins of attraction. Hence, we conclude that the proposed methods are comparable to other well-known existing equivalent methods.

Author Contributions: Conceptualization, K.M.; Funding acquisition, J.L. and X.W.; Methodology, K.M.; Project administration, J.L. and X.W.; Resources, J.L. and X.W.; Writing—original draft, K.M.

Funding: This research received no external funding.

Acknowledgments: The authors would like to thank the editors and referees for the valuable comments and for the suggestions to improve the readability of the paper.

Conflicts of Interest: The authors declare no conflict of interest.

References

1. Cordero, A.; Hueso, J.L.; Martinez, E.; Torregrosa, J.R. Generating optimal derivative free iterative methods for nonlinear equations by using polynomial interpolation. *Appl. Math. Comp.* **2013**, *57*, 1950–1956. [CrossRef]
2. Soleymani, F. Efficient optimal eighth-order derivative-free methods for nonlinear equations. *Jpn. J. Ind. Appl. Math.* **2013**, *30*, 287–306. [CrossRef]
3. Steffensen, J.F. Remarks on iteration. *Scand. Aktuarietidskr.* **1933**, *16*, 64–72. [CrossRef]
4. Kung, H.; Traub, J. Optimal order of one-point and multi-point iteration. *J. Assoc. Comput. Math.* **1974**, *21*, 643–651. [CrossRef]
5. Behl, R.; Salimi, M.; Ferrara, M.; Sharifi, S.; Alharbi, S.K. Some Real-Life Applications of a Newly Constructed Derivative Free Iterative Scheme. *Symmetry* **2019**, *11*, 239. [CrossRef]
6. Salimi, M.; Long, N.M.A.N.; Sharifi, S.; Pansera, B.A. A multi-point iterative method for solving nonlinear equations with optimal order of convergence. *Jpn. J. Ind. Appl. Math.* **2018**, *35*, 497–509. [CrossRef]
7. Salimi, M.; Lotfi, T.; Sharifi, S.; Siegmund, S. Optimal Newton-Secant like methods without memory for solving nonlinear equations with its dynamics. *Int. J. Comput. Math.* **2017**, *94*, 1759–1777. [CrossRef]
8. Matthies, G.; Salimi, M.; Sharifi, S.; Varona, J.L. An optimal three-point eighth-order iterative method without memory for solving nonlinear equations with its dynamics. *Jpn. J. Ind. Appl. Math.* **2016**, *33*, 751–766. [CrossRef]
9. Sharifi, S.; Siegmund, S.; Salimi, M. Solving nonlinear equations by a derivative-free form of the King's family with memory. *Calcolo* **2016**, *53*, 201–215. [CrossRef]
10. Khdhr, F.W.; Saeed, R.K.; Soleymani, F. Improving the Computational Efficiency of a Variant of Steffensen's Method for Nonlinear Equations. *Mathematics* **2019**, *7*, 306. [CrossRef]
11. Soleymani, F.; Babajee, D.K.R.; Shateyi, S.; Motsa, S.S. Construction of Optimal Derivative-Free Techniques without Memory. *J. Appl. Math.* **2012**, *2012*, 24. [CrossRef]
12. Soleimani, F.; Soleymani, F.; Shateyi, S. Some Iterative Methods Free from Derivatives and Their Basins of Attraction for Nonlinear Equations. *Discret. Dyn. Nat. Soc.* **2013**, *2013*, 10. [CrossRef]
13. Soleymani, F.; Sharifi, M. On a General Efficient Class of Four-Step Root-Finding Methods. *Int. J. Math. Comp. Simul.* **2011**, *5*, 181–189.
14. Soleymani, F.; Vanani, S.K.; Paghaleh, M.J. A Class of Three-Step Derivative-Free Root Solvers with Optimal Convergence Order. *J. Appl. Math.* **2012**, *2012*, 15. [CrossRef]
15. Kanwar, V.; Bala, R.; Kansal, M. Some new weighted eighth-order variants of Steffensen-King's type family for solving nonlinear equations and its dynamics. *SeMA J.* **2016**. [CrossRef]

16. Amat, S.; Busquier, S.; Plaza, S. Dynamics of a family of third-order iterative methods that do not require using second derivatives. *Appl. Math. Comput.* **2004**, *154*, 735–746. [CrossRef]

17. Amat, S.; Busquier, S.; Plaza, S. Review of some iterative root-finding methods from a dynamical point of view. *SCIENTIA Ser. A Math. Sci.* **2004**, *10*, 3–35.

18. Babajee, D.K.R.; Madhu, K. Comparing two techniques for developing higher order two-point iterative methods for solving quadratic equations. *SeMA J.* **2019**, *76*, 227–248. [CrossRef]

19. Cordero, A.; Hueso, J.L.; Martinez, E.; Torregrosa, J.R. A family of iterative methods with sixth and seventh order convergence for nonlinear equations. *Math. Comput. Model.* **2010**, *52*, 1490–1496. [CrossRef]

20. Curry, J.H.; Garnett, L.; Sullivan, D. On the iteration of a rational function: computer experiments with Newton's method. *Commun. Math. Phys.* **1983**, *91*, 267–277. [CrossRef]

21. Soleymani, F.; Babajee, D.K.R.; Sharifi, M. Modified Jarratt Method Without Memory With Twelfth-Order Convergence. *Ann. Univ. Craiova Math. Comput. Sci. Ser.* **2012**, *39*, 21–34.

22. Tao, Y.; Madhu, K. Optimal Fourth, Eighth and Sixteenth Order Methods by Using Divided Difference Techniques and Their Basins of Attraction and Its Application. *Mathematics* **2019**, *7*, 322. [CrossRef]

23. Vrscay, E.R. Julia sets and mandelbrot-like sets associated with higher order Schroder rational iteration functions: a computer assisted study. *Math. Comput.* **1986**, *46*, 151–169.

24. Vrscay, E.R.; Gilbert, W.J. Extraneous fxed points, basin boundaries and chaotic dynamics for Schroder and Konig rational iteration functions. *Numer. Math.* **1987**, *52*, 1–16. [CrossRef]

25. Argyros, I.K.; Kansal, M.; Kanwar, V.; Bajaj, S. Higher-order derivative-free families of Chebyshev-Halley type methods with or without memory for solving nonlinear equations. *Appl. Math. Comput.* **2017**, *315*, 224–245. [CrossRef]

26. Zheng, Q.; Li, J.; Huang, F. An optimal Steffensen-type family for solving nonlinear equations. *Appl. Math. Comput.* **2011**, *217*, 9592–9597. [CrossRef]

27. Madhu, K. Some New Higher Order Multi-Point Iterative Methods and Their Applications to Differential and Integral Equation and Global Positioning System. Ph.D. Thesis, Pndicherry University, Pondicherry, India, June 2016.

28. Scott, M.; Neta, B.; Chun, C. Basin attractors for various methods. *Appl. Math. Comput.* **2011**, *218*, 2584–2599. [CrossRef]

![Sigma logo] *mathematics*

MDPI

Article

A Generic Family of Optimal Sixteenth-Order Multiple-Root Finders and Their Dynamics Underlying Purely Imaginary Extraneous Fixed Points

Min-Young Lee [1], Young Ik Kim [1,*] and Beny Neta [2]

[1] Department of Applied Mathematics, Dankook University, Cheonan 330-714, Korea; leemy@dankook.ac.kr
[2] Naval Postgraduate School, Department of Applied Mathematics, Monterey, CA 93943, USA; bneta@nps.edu
* Correspondence: yikbell79@yahoo.com; Tel.: +82-41-550-3415

Received: 26 April 2019; Accepted: 18 June 2019; Published: 20 June 2019

Abstract: A generic family of optimal sixteenth-order multiple-root finders are theoretically developed from general settings of weight functions under the known multiplicity. Special cases of rational weight functions are considered and relevant coefficient relations are derived in such a way that all the extraneous fixed points are purely imaginary. A number of schemes are constructed based on the selection of desired free parameters among the coefficient relations. Numerical and dynamical aspects on the convergence of such schemes are explored with tabulated computational results and illustrated attractor basins. Overall conclusion is drawn along with future work on a different family of optimal root-finders.

Keywords: sixteenth-order optimal convergence; multiple-root finder; asymptotic error constant; weight function; purely imaginary extraneous fixed point; attractor basin

MSC: 65H05; 65H99

1. Introduction

Many nonlinear equations governing real-world natural phenomena cannot be solved exactly by virtue of their intrinsic complexities. It would be certainly an important matter to discuss methods for approximating such solutions of the nonlinear equations. The most widely accepted method under general circumstances is Newton's method, which has quadratic convergence for a simple-root and linear convergence for a multiple-root. Other higher-order root-finders have been developed by many researchers [1–9] with optimal convergence satisfying Kung–Traub's conjecture [10]. Several authors [10–14] have proposed optimal sixteenth-order simple-root finders, although their applications to real-life problems are limited due to the high degree of their algebraic complexities. Optimal sixteenth-order multiple-root finders are hardly found in the literature to the best of our knowledge at the time of writing this paper. It is not too much to emphasize the theoretical importance of developing optimal sixteenth-order multiple root-finders as well as to apply them to numerically solve real-world nonlinear problems.

In order to develop an optimal sixteenth-order multiple-root finders, we pursue a family of iterative methods equipped with generic weight functions of the form:

$$
\begin{cases}
y_n = x_n - m\frac{f(x_n)}{f'(x_n)}, \\
z_n = y_n - mQ_f(s)\frac{f(y_n)}{f'(x_n)} = x_n - m\left[1 + sQ_f(s)\right]\frac{f(x_n)}{f'(x_n)}, \\
w_n = z_n - mK_f(s,u)\frac{f(z_n)}{f'(x_n)} = x_n - m\left[1 + sQ_f(s) + suK_f(s,u)\right]\frac{f(x_n)}{f'(x_n)}, \\
x_{n+1} = w_n - mJ_f(s,u,v)\frac{f(w_n)}{f'(x_n)} = x_n - m\left[1 + sQ_f(s) + suK_f(s,u) + suvJ_f(s,u,v)\right]\frac{f(x_n)}{f'(x_n)},
\end{cases}
\tag{1}
$$

where $s = \left(\frac{f(y_n)}{f(x_n)}\right)^{1/m}$, $u = \left(\frac{f(z_n)}{f(y_n)}\right)^{1/m}$, $v = \left(\frac{f(w_n)}{f(z_n)}\right)^{1/m}$; $Q_f : \mathbb{C} \to \mathbb{C}$ is analytic [15] in a neighborhood of 0, $K_f : \mathbb{C}^2 \to \mathbb{C}$ holomorphic [16,17] in a neighborhood of $(0,0)$, and $J_f : \mathbb{C}^3 \to \mathbb{C}$ holomorphic in a neighborhood of $(0,0,0)$. Since s, u and v are respectively one-to-m multiple-valued functions, their principal analytic branches [15] are considered. Hence, for instance, it is convenient to treat s as a principal root given by $s = \exp[\frac{1}{m}\text{Log}(\frac{f(y_n)}{f(x_n)})]$, with $\text{Log}(\frac{f(y_n)}{f(x_n)}) = \text{Log}|\frac{f(y_n)}{f(x_n)}| + i\,\text{Arg}(\frac{f(y_n)}{f(x_n)})$ for $-\pi < \text{Arg}(\frac{f(y_n)}{f(x_n)}) \leq \pi$; this convention of $\text{Arg}(z)$ for $z \in \mathbb{C}$ agrees with that of $\text{Log}[z]$ command of Mathematica [18] to be employed later in numerical experiments.

The case for $m = 1$ has been recently developed by Geum–Kim–Neta [19]. Many other existing cases for $m = 1$ are special cases of (1) with appropriate forms of weight functions Q_f, K_f, and J_f; for example, the case developed in [10] uses the following weight functions:

$$\begin{cases} Q_f(s) = \frac{1}{(1-s)^2}, \\ K_f(s,u) = \frac{1+(1-u)s^2}{(1-s)^2(1-u)(1-su)^2}, \\ J_f(s,u,v) = \frac{-1+2su^2(v-1)+s^4(u-1)u^2(v-1)(uv-1)+s^2[uv-1-u^3(v^2-1)]}{(1-s)^2(u-1)(su-1)^2(v-1)(uv-1)(suv-1)^2}. \end{cases} \qquad (2)$$

One goal of this paper is to construct a family of optimal sixteenth-order multiple-root finders by characterizing the generic forms of weight functions $Q_f(s)$, $K_f(s,u)$, and $J_f(s,u,v)$. The other goal is to investigate the convergence behavior by exploring their numerical behavior and dynamics through basins of attractions [20] underlying the extraneous fixed points [21] when $f(z) = (z-a)^m(z-b)^m$ is applied. In view of the right side of final substep of (1), we can conveniently locate extraneous fixed points from the roots of the weight function $m[1 + sQ_f(s) + suK_f(s,u) + suvJ_f(s,u,v)]$.

A motivation undertaking this research is to investigate the local and global characters on the convergence of proposed family of methods (1). The local convergence of an iterative method for solving nonlinear equations is usually guaranteed with an initial guess taken in a sufficiently close neighborhood of the sought zero. On the other hand, effective information on its global convergence is hardly achieved under general circumstances. We can obtain useful information on the global convergence from attractor basins through which relevant dynamics is worth exploring. Especially the dynamics underlying the extraneous fixed points (to be described in Section 3) would influence the dynamical behavior of the iterative methods by the presence of possible attractive, indifferent, repulsive, and other chaotic orbits. One way of reducing such influence is to control the location of the extraneous fixed points. We prefer the location to be the imaginary axis that divides the entire complex plane into two symmetrical half-planes. The dynamics underlying the extraneous fixed points on the imaginary axis would be less influenced by the presence of the possible periodic or chaotic attractors.

The main theorem is presented in Section 2 with required constraints on weight functions, Q_f, K_f, and J_f to achieve the convergence order of 16. Section 2 discusses special cases of rational weight functions. Section 3 extensively investigates the purely imaginary extraneous fixed points and investigates their stabilities. Section 4 presents numerical experiments as well as the relevant dynamics, while Section 5 states the overall conclusions along with the short description of future work.

2. Methods and Special Cases

A main theorem on the convergence of (1) is established here with the error equation and relationships among generic weight functions $Q_f(s)$, $K_f(s,u)$, and $J_f(s,u,v)$:

Theorem 1. *Suppose that* $f : \mathbb{C} \to \mathbb{C}$ *has a multiple root* α *of multiplicity* $m \geq 1$ *and is analytic in a neighborhood of* α. *Let* $c_j = \frac{m!}{(m-1+j)!}\frac{f^{(m-1+j)}(\alpha)}{f^{(m)}(\alpha)}$ *for* $j = 2,3,\cdots$. *Let* x_0 *be an initial guess selected in a sufficiently small region containing* α. *Assume* $L_f : \mathbb{C} \to \mathbb{C}$ *is analytic in a neighborhood of 0. Let* $Q_i = \frac{1}{i!}\frac{d^i}{ds^i}Q_f(s)|_{(s=0)}$ *for* $0 \leq i \leq 6$. *Let* $K_f : \mathbb{C}^2 \to \mathbb{C}$ *be holomorphic in a neighborhood of* $(0,0)$. *Let* $J_f : \mathbb{C}^3 \to \mathbb{C}$ *be holomorphic in a neighborhood of* $(0,0,0)$. *Let* $K_{ij} = \frac{1}{i!j!}\frac{\partial^{i+j}}{\partial s^i \partial u^j}K_f(s,u)|_{(s=0,u=0)}$ *for*

$0 \leq i \leq 12$ *and* $0 \leq j \leq 6$. *Let* $J_{ijk} = \frac{1}{i!j!k!} \frac{\partial^{i+j+k}}{\partial s^i \partial u^j \partial v^j} J_f(s,u,v)\big|_{(s=0,u=0,v=0)}$ *for* $0 \leq i \leq 8$, $0 \leq j \leq 4$ *and* $0 \leq k \leq 2$. *If* $Q_0 = 1, Q_1 = 2, K_{00} = 1, K_{10} = 2, K_{01} = 1, K_{20} = 1 + Q_2, K_{11} = 4, K_{30} = -4 + 2Q_2 + Q_3,$ $J_{000} = 1, J_{100} = 2, J_{200} = 1 + Q_2, J_{010} = 1, J_{110} = 4, J_{300} = -4 + 2Q_2 + Q_3, J_{001} = 1, J_{020} = K_{02}, J_{210} = 1 + K_{21}, J_{400} = K_{40}, J_{101} = 2, J_{120} = 2 + K_{12}, J_{310} = -4 + K_{31} + 2Q_2, J_{500} = K_{50}, J_{011} = 2, J_{201} = 1 + Q_2, J_{030} = -1 + K_{02} + K_{03}, J_{220} = 1 + K_{21} + K_{22} - Q_2, J_{410} = -3 + K_{40} + K_{41} + Q_2 - Q_4, J_{600} = K_{60}, J_{111} = 8, J_{301} = -4 + 2Q_2 + Q_3, J_{130} = -4 + 2K_{02} + K_{12} + K_{13}, J_{320} = -6 + 2K_{21} + K_{31} + K_{32} - 2Q_2 - Q_3, J_{510} = 6 + 2K_{40} + K_{50} + K_{51} - 3Q_3 - 2Q_4 - Q_5, J_{700} = K_{70}$ *are fulfilled, then Scheme* (1) *leads to an optimal class of sixteenth-order multiple-root finders possessing the following error equation: with* $e_n = x_n - \alpha$ *for* $n = 0, 1, 2, \cdots,$

$$e_{n+1} = \frac{1}{3456m^{15}} c_2(\rho c_2^2 - 2mc_3)\left[\beta_0 c_2^4 + \beta_1 c_2^2 c_3 + 12m^2(K_{02} - 1)c_3^2 - 12m^2 c_2 c_4\right]\Psi \, e_n^{16} + O(e_n^{17}), \quad (3)$$

where $\rho = 9 + m - 2Q_2$, $\beta_0 = (-431 + 12K_{40} - 7m^2 + 6m(-17 + Q_2) + 102Q_2 - 24Q_3 - 12Q_4 + 6K_{21}\rho + 3K_{02}\rho^2)$, $\beta_1 = -12m(-17 + K_{21} - 2m + Q_2 + K_{02}\rho)$, $\Psi = \Delta_1 c_2^8 + \Delta_2 c_2^6 c_3 + \Delta_3 c_2^5 c_4 + \Delta_4 c_2^3 c_3 c_4 + \Delta_5 c_2^4 + \Delta_6 c_2^2 + \Delta_7 c_3^4 + \Delta_8 c_2 c_3^2 c_4$,

$\Delta_1 = (-255124 + 144J_{800} - 144K_{80} - 122577m - 23941m^2 - 2199m^3 - 79m^4 + 24K_{40}(472 + 93m + 5m^2) - 72(17 + m)Q_3^3 - 576Q_3^2 + Q_3(48(-566 + 6K_{40} - 117m - 7m^2) - 576Q_4) + 24(-485 + 6K_{40} - 108m - 7m^2)Q_4 - 144Q_4^2 + Q_2^2(36(-87 + 14m + m^2) + 288Q_3 + 144Q_4) + Q_2(18(5300 + 1529m + 172m^2 + 7m^3 - 8K_{40}(18 + m)) + 144(35 + m)Q_3 + 72(29 + m)Q_4) + 18\rho^3\sigma + 6\rho(12J_{610} - 12(2K_{50} + K_{60} + K_{61} - 2Q_5 - Q_6) + J_{211}(-12K_{40} + \sigma_2) + 2K_{21}(-J_{002}\sigma_2 + \sigma_3 + 6\eta_0)) + \rho^2(36J_{420} - 36J_{211}K_{21} + 36J_{002}K_{21}^2 - 72K_{31} - 36J_{021}K_{40} - 36K_{41} - 36K_{42} + 3J_{021}\sigma_2 + 6K_{02}(-6J_{401} + 12J_{002}K_{40} - J_{002}\sigma_2 + \sigma_3)) + 12J_{401}\sigma_7 + J_{002}\sigma_7^2 + 9\rho^4\tau)$,

$\Delta_2 = m(144(Q_2^3 - 2Q_5 - Q_6) + 288Q_3(-K_{21} + (39 + 4m) - K_{02}\rho) + 144Q_2^2(-2K_{21} - (7 + m) - K_{02}\rho) + 144Q_4(-K_{21} + 4(9 + m) - K_{02}\rho) + Q_2(144K_{21}(58 + 5m) - 36(1529 - 8K_{40} + m(302 + 17m)) - 288Q_3 - 144Q_4 + 144K_{02}(38 + 3m)\rho) - 108\rho^2\sigma + 6(40859 - 24J_{610} + 48K_{50} + 24K_{60} + 24K_{61} + 24K_{40}(-31 + J_{211} - 3m) + m(14864 + m(1933 + 88m)) - 2J_{211}\sigma_8) - 72\rho^3\tau - 24\rho_2(J_{002}\sigma_2 - 6\eta_0) - 24K_{21}(1309 + m(267 + 14m) - J_{002}\sigma_8 + 6\eta_0) + \rho(144J_{211}K_{21} - 144J_{002}K_{21}^2 + 12(-12J_{420} + 12(2K_{31} + J_{021}K_{40} + K_{41} + K_{42}) - J_{021}\sigma_4) - 12K_{02}(1781 + m(360 + 19m) - 2J_{002}\sigma_4 + 12\eta_0)))$,

$\Delta_3 = 12m^2(6(63 + 5m)Q_2 - 48Q_3 - 24Q_4 + 6(J_{211} + K_{21} - 2J_{002}K_{21})\rho - (1645 - 12J_{401} + 12(-1 + 2J_{002})K_{40} + 372m + 23m^2 - 2J_{002}\sigma_2) + 3\rho^2\sigma_5)$,

$\Delta_4 = 144m^3(-3Q_2 + (53 - J_{211} + (-1 + 2J_{002})K_{21} + 6m - 2J_{002}\rho_2) - \rho\sigma_5)$,

$\Delta_5 = 72\rho m^3 c_5 + 12m^2 c_3^2((-12K_{21}(J_{211} - 4(10 + m)) + K_{02}(4778 - 12J_{401} + m(990 + 52m)) + 3(-1929 + 4J_{401} + 4J_{420} - 8K_{31} + 8K_{40} - 4K_{41} - 4K_{42} - 476m - 31m^2 + 2J_{211}(43 + 5m) + J_{021}(115 + 22m + m^2))) - 6(-88 + 121J_{021} + 4J_{211} + 232K_{02} + 8K_{21} + (-10 + 13J_{021} + 22K_{02})m + 2J_{002}(51 - 4K_{21} + 5m))Q_2 + 12(1 + J_{002} + 3J_{021} + 6K_{02})Q_2^2 + 18\rho\sigma + 18\rho^2\tau + Q_4(12(-2 + K_{02}) + 12\eta_1) + Q_3(24(-2 + K_{02}) + 24\eta_1) - J_{021}\eta_2 + \eta_1((2165 - 12K_{40} + m(510 + 31m)) + \eta_2))$,

$\Delta_6 = 72m^3(2(-1 + J_{002})mc_4^2 - 2mc_3 c_5 + c_3^3(2Q_2(-1 + 6K_{02}) - 2K_{02}(49 - J_{211} + 5m + 2J_{002}\rho_3) + (85 - 2J_{211} - 2J_{230} + 4K_{12} - 4K_{21} + 4J_{002}(K_{21} - \rho_2) + 2K_{22} + 2K_{23} + 11m + 2J_{021}\rho_3) - 4\rho\tau))$,

$\Delta_7 = 144m^4(-1 + J_{002} + J_{021} + J_{040} - K_{03} - K_{04} + (2 - \eta_1)K_{02} + J_{002}K_{02}^2)$, $\Delta_8 = 144m^4(-3 + \eta_1 + (1 - 2J_{002})K_{02})$,

$\tau = J_{040} - J_{021}K_{02} + J_{002}K_{02}^2 - K_{03} - K_{04}$, $\rho_2 = 17 + 2m - Q_2$, $\rho_3 = -26 + K_{21} - 3m + 3Q_2$, $\sigma = J_{230} - J_{211}K_{02} - 2K_{12} - J_{021}K_{21} + 2J_{002}K_{02}K_{21} - K_{22} - K_{23}$, $\sigma_2 = 431 + 7m^2 - 6m(-17 + Q_2) - 102Q_2 + 24Q_3 + 12Q_4$, $\sigma_3 = 472 + 5m^2 + m(93 - 6Q_2) - 108Q_2 + 12Q_3 + 6Q_4$, $\sigma_4 = 890 + 13m^2 - 231Q_2 + 6Q_2^2 - 3m(-69 + 7Q_2) + 24Q_3 + 12Q_4$, $\sigma_5 = J_{021} + K_{02} - 2J_{002}K_{02}$, $\sigma_6 = -1255 + 6K_{40} - 288m - 17m^2 + 363Q_2 + 39mQ_2 - 18Q_2^2 - 12Q_3 - 6Q_4$, $\sigma_7 = 431 - 12K_{40} + 7m^2 - 6m(-17 + Q_2) - 102Q_2 + 24Q_3 + 12Q_4$, $\sigma_8 = 1349 + 19m^2 + m(312 - 36Q_2) - 360Q_2 + 12Q_2^2 + 24Q_3 + 12Q_4$, $\eta_0 = -J_{401} + 2J_{002}K_{40}$, $\eta_1 = 2J_{002} + J_{021}$, $\eta_2 = 6K_{21}^2 - 6K_{21}(43 + 5m) + 2\sigma_6 K_{02}$.

Proof. Since Scheme (1) employs five functional evaluations, namely, $f'(x_n)$, $f(x_n)$, $f(y_n)$, $f(z_n)$, and $f(w_n)$, optimality can be achieved if the corresponding convergence order is 16. In order to induce the desired order of convergence, we begin by the 16th-order Taylor series expansion of $f(x_n)$ about α:

$$f(x_n) = \frac{f'(\alpha)}{m!} e_n^m \{1 + \sum_{i=2}^{17} c_i \, e_n^{i-1} + O(e_n^{17})\}. \tag{4}$$

It follows that

$$f'(x_n) = \frac{f'(\alpha)}{(m-1)!} e_n^{m-1} \{1 + \sum_{i=2}^{16} i \, \frac{m+i-1}{m} c_i \, e_n^{i-1} + O(e_n^{16})\}. \tag{5}$$

For brevity of notation, we abbreviate e_n as e. Using Mathematica [18], we find:

$$y_n = x_n - m\frac{f(x_n)}{f'(x_n)} = \alpha + \frac{c_2}{m} e^2 + \frac{(-(m+1)c_2^2 + 2mc_3)}{m^2} e^3 + \frac{Y_4}{m^3} e^4 + \sum_{i=5}^{16} \frac{Y_i}{m^{i-1}} e_n^i + O(e^{17}), \tag{6}$$

where $Y_4 = (1+m)^2 c_2^3 - m(4+3m)c_2c_3 + 3m^2c_4$ and $Y_i = Y_i(c_2, c_3, \cdots, c_{16})$ for $5 \le i \le 16$.
After a lengthy computation using the fact that $f(y_n) = f(x_n)|_{e_n \to (y_n - \alpha)}$, we get:

$$s = \left(\frac{f(y_n)}{f(x_n)}\right)^{1/m} = \frac{c_2}{m} e + \frac{(-(m+2)c_2^2 + 2mc_3)}{m^2} e^2 + \frac{\gamma_3}{2m^3} e^3 + \sum_{i=4}^{15} E_i \, e^i + O(e^{16}), \tag{7}$$

where $\gamma_3 = (7+7m+2m^2)c_2^3 - 2m(7+3m)c_2c_3 + 6m^2c_4$, $E_i = E_i(c_2, c_3, \cdots, c_{16})$ for $4 \le i \le 15$.
In the third substep of Scheme (1), $w_n = O(e^8)$ can be achieved based on Kung–Traub's conjecture. To reflect the effect on w_n from z_n in the second substep, we need to expand z_n up to eighth-order terms; hence, we carry out a sixth-order Taylor expansion of $Q_f(s)$ about 0 by noting that $s = O(e)$ and $\frac{f(y_n)}{f'(x_n)} = O(e^2)$:

$$Q_f(s) = Q_0 + Q_1 s + Q_2 s^2 + Q_3 s^3 + Q_4 s^4 + Q_5 s^5 + Q_6 s^6 + O(e^7), \tag{8}$$

where $Q_j = \frac{1}{j!} \frac{d^j}{ds^j} Q_f(s)$ for $0 \le j \le 6$. As a result, we come up with:

$$z_n = x_n - mQ_f(s)\frac{f(y_n)}{f'(x_n)} = \alpha + \frac{(1-Q_0)}{m} e^2 + \frac{\mu_3}{m^2} e^3 + \sum_{i=4}^{16} W_i \, e^i + O(e^{17}), \tag{9'}$$

where $\mu_3 = (-1 + m(Q_0 - 1) + 3Q_0 - Q_1)c_2^2 - 2m(Q_0 - 1)c_3$ and $W_i = W_i(c_2, c_3, \cdots, c_{16}, Q_0, \cdots, Q_6)$ for $4 \le i \le 16$. Selecting $Q_0 = 1$ and $Q_1 = 2$ leads us to an expression:

$$z_n = \alpha + \frac{c_2(\rho c_2^2 - 2mc_3)}{m^2} e^4 + \sum_{i=5}^{16} W_i \, e^i + O(e^{17}). \tag{9}$$

By a lengthy computation using the fact that $f(z_n) = f(x_n)|_{e_n \to (z_n - \alpha)}$, we deduce:

$$u = \left(\frac{f(z_n)}{f(y_n)}\right)^{1/m} = \frac{(\rho c_2^2 - 2mc_3)}{2m^2} e^2 + \frac{\delta_3}{3m^3} e^3 + \sum_{i=4}^{16} G_i \, e^i + O(e^{17}), \tag{10}$$

where $\delta_3 = (49 + 2m^2 + m(27 - 6Q_2) - 18Q_2 + 3Q_3)c_2^3 - 6m\rho c_2c_3 + 6m^2c_4$ and $G_i = G_i(c_2, c_3, \cdots, c_{16}, Q_2, \cdots, Q_6)$ for $4 \le i \le 16$.
In the last substep of Scheme (1), $x_{n+1} = O(e^{16})$ can be achieved based on Kung-Traub's conjecture. To reflect the effect on x_{n+1} from w_n in the third substep, we need to expand w_n up to sixteenth-order

terms; hence, we carry out a 12th-order Taylor expansion of $K_f(s,u)$ about $(0,0)$ by noting that: $s = O(e)$, $u = O(e^2)$ and $\frac{f(z_n)}{f'(x_n)} = O(e^4)$ with $K_{ij} = 0$ satisfying $i + 2j > 12$ for all $0 \leq i \leq 12, 0 \leq j \leq 6$:

$$
\begin{aligned}
K_f(s,u) = & \, K_{00} + K_{10}s + K_{20}s^2 + K_{30}s^3 + K_{40}s^4 + K_{50}s^5 + K_{60}s^6 + K_{70}s^7 + K_{80}s^8 + K_{90}s^9 + K_{100}s^{10} + K_{110}s^{11} + \\
& \, K_{120}s^{12} + (K_{01} + K_{11}s + K_{21}s^2 + K_{31}s^3 + K_{41}s^4 + K_{51}s^5 + K_{61}s^6 + K_{71}s^7 + K_{81}s^8 + K_{91}s^9 + K_{101}s^{10})u + \\
& \, (K_{02} + K_{12}s + K_{22}s^2 + K_{32}s^3 + K_{42}s^4 + K_{52}s^5 + K_{62}s^6 + K_{72}s^7 + K_{82}s^8)u^2 + \\
& \, (K_{03} + K_{13}s + K_{23}s^2 + K_{33}s^3 + K_{43}s^4 + K_{53}s^5 + K_{63}s^6)u^3 + \\
& \, (K_{04} + K_{14}s + K_{24}s^2 + K_{34}s^3 + K_{44}s^4)u^4 + (K_{05} + K_{15}s + K_{25}s^2)u^5 + K_{06}u^6 + O(e^{13}).
\end{aligned}
\tag{11}
$$

Substituting $z_n, f(x_n), f(y_n), f(z_n), f'(x_n)$, and $K_f(s,u)$ into the third substep of (1) leads us to:

$$
w_n = z_n - mK_f(s,u) \cdot \frac{f(z_n)}{f'(x_n)} = \alpha + \frac{(1 - K_{00})c_2(\rho c_2^2 - 2mc_3)}{2m^3}e^4 + \sum_{i=5}^{16} \Gamma_i \, e^i + O(e^{17}),
\tag{12}
$$

where $\Gamma_i = \Gamma_i(c_2, c_3, \cdots, c_{16}, Q_2, \cdots, Q_6, K_{j\ell})$, for $5 \leq i \leq 16, 0 \leq j \leq 12$ and $0 \leq \ell \leq 6$. Thus $K_{00} = 1$ immediately annihilates the fourth-order term. Substituting $K_{00} = 1$ into $\Gamma_5 = 0$ and solving for K_{10}, we find:

$$
K_{10} = 2.
\tag{13}
$$

Continuing the algebraic operations in this manner at the i-th ($6 \leq i \leq 7$) stage with known values of $K_{j\ell}$, we solve $\Gamma_i = 0$ for remaining $K_{j\ell}$ to find:

$$
K_{20} = 1 + Q_2, \ K_{01} = 1.
\tag{14}
$$

Substituting $K_{00} = 1, K_{10} = 2, K_{20} = 1 + Q_2, K_{01} = 1$ into (12) and simplifying we find:

$$
v = \left(\frac{f(w_n)}{f(z_n)}\right)^{1/m} = -\frac{[\beta_0 c_2^4 + \beta_1 c_2^2 c_3 + 12m^2(K_{02} - 1)c_3^2 - 12m^2 c_2 c_4]}{12m^4}e^4 + \sum_{i=5}^{16} T_i \, e^i + O(e^{17}),
\tag{15}
$$

where β_0 and β_1 are described in (3) and $T_i = T_i(c_2, c_3, \cdots, c_{16}, Q_2, \cdots, Q_6)$ for $5 \leq i \leq 16$.

To compute the last substep of Scheme (1), it is necessary to have an eighth-order Taylor expansion of $J_f(s,u,v)$ about $(0,0,0)$ due to the fact that $\frac{f(w_n)}{f'(x_n)} = O(e^8)$. It suffices to expand J_f up to eighth-, fourth-, and second-order terms in s, u, v in order, by noting that $s = O(e)$, $u = O(e^2)$, $v = O(e^4)$ with $J_{ijk} = 0$ satisfying $i + 2j + 4k > 8$ for all $0 \leq i \leq 8, 0 \leq j \leq 4, 0 \leq k \leq 2$:

$$
\begin{aligned}
J_f(s,u,v) = & \, J_{000} + J_{100}s + J_{200}s^2 + J_{300}s^3 + J_{400}s^4 + J_{500}s^5 + J_{600}s^6 + J_{700}s^7 + J_{800}s^8 + (J_{010} + J_{110}s + J_{210}s^2 + \\
& \, J_{310}s^3 + J_{410}s^4 + J_{510}s^5 + J_{610}s^6)u + (J_{020} + J_{120}s + J_{220}s^2 + J_{320}s^3 + J_{420}s^4)u^2 + (J_{030} + J_{130}s + J_{230}s^2)u^3 + \\
& \, J_{040}u^4 + (J_{001} + J_{101}s + J_{201}s^2 + J_{301}s^3 + J_{401}s^4) + (J_{011} + J_{111}s + J_{211}s^2)u + J_{021}u^2)v + J_{002}v^2.
\end{aligned}
\tag{16}
$$

Substituting $w_n, f(x_n), f(y_n), f(z_n), f(w_n), f'(x_n)$ and $J_f(s,u,v)$ in (1), we arrive at:

$$
x_{n+1} = w_n - mJ_f(s,u,v) \cdot \frac{f(w_n)}{f'(x_n)} = \alpha + \phi e^8 + \sum_{i=9}^{16} \Omega_i \, e^i + O(e^{17}),
\tag{17}
$$

where $\phi = \frac{1}{24m^7}(1 - J_{000})c_2(\rho c_2^2 - 2mc_3)[\beta_0 c_2^4 + \beta_1 c_2^2 c_3 + 12m^2(K_{02} - 1)c_3^2 - 12m^2 c_2 c_4]$ and $\Omega_i = \Omega_i(c_2, c_3, \cdots, c_{16}, Q_2, \cdots, Q_6, K_{\delta\theta}, J_{jk\ell})$, for $9 \leq i \leq 16, 0 \leq \delta \leq 12, 0 \leq \theta \leq 6, 0 \leq j \leq 8, 0 \leq k \leq 4, 0 \leq \ell \leq 2$.

Since $J_{000} = 1$ makes $\phi = 0$, we substitute $J_{000} = 1$ into $\Omega_9 = 0$ and solve for J_{100} to find:

$$
J_{100} = 2.
\tag{18}
$$

Continuing the algebraic operations in the same manner at the *i*-th ($10 \leq i \leq 15$) stage with known values of $J_{jk\ell}$, we solve $\Omega_i = 0$ for remaining $J_{jk\ell}$ to find:

$$
\begin{cases}
J_{200} = 1 + Q_2, J_{010} = 1, J_{110} = 4, J_{300} = -4 + 2Q_2 + Q_3, J_{001} = 1, J_{020} = K_{02}, J_{210} = 1 + K_{21}, \\
J_{400} = K_{40}, J_{101} = 2, J_{120} = 2 + K_{12}, J_{310} = -4 + K_{31} + 2Q_2, J_{500} = K_{50}, J_{011} = 2, J_{201} = 1 + Q_2, \\
J_{111} = 8, J_{030} = -1 + K_{02} + K_{03}, J_{220} = 1 + K_{21} + K_{22} - Q_2, J_{410} = -3 + K_{40} + K_{41} + Q_2 - Q_4, \\
J_{301} = -4 + 2Q_2 + Q_3, J_{130} = -4 + 2K_{02} + K_{12} + K_{13}, J_{320} = -6 + 2K_{21} + K_{31} + K_{32} - 2Q_2 - Q_3, \\
J_{600} = K_{60}, J_{510} = 6 + 2K_{40} + K_{50} + K_{51} - 3Q_3 - 2Q_4 - Q_5, J_{700} = K_{70}.
\end{cases}
\tag{19}
$$

Upon substituting Relation (19) into Ω_{16}, we finally obtain:

$$
\Omega_{16} = \frac{1}{3456m^{15}} c_2(\rho c_2^2 - 2mc_3)\left[\beta_0 c_2^4 + \beta_1 c_2^2 c_3 + 12m^2(K_{02} - 1)c_3^2 - 12m^2 c_2 c_4\right]\Psi,
\tag{20}
$$

where ρ, β_0, β_1, and Ψ as described in (3). This completes the proof. $\square$

Remark 1. *Theorem 1 clearly reflects the case for $m = 1$ with the same constraints on weight functions Q_f, K_f, J_f studied in [19].*

Special Cases of Weight Functions

Theorem 1 enables us to obtain $Q_f(s)$, $K_f(s, u)$, and $J_f(s, u, v)$ by means of Taylor polynomials:

$$
\begin{cases}
Q_f(s) = 1 + 2s + Q_2 s^2 + Q_3 s^3 + Q_4 s^4 + Q_5 s^5 + Q_6 s^6 + O(e^7), \\
K_f(s, u) = 1 + 2s + (1 + Q_2)s^2 + (2Q_2 + Q_3 - 4)s^3 + K_{40}s^4 + K_{50}s^5 + K_{60}s^6 + K_{70}s^7 + K_{80}s^8 \\
\quad + K_{90}s^9 + K_{100}s^{10} + K_{110}s^{11} + K_{120}s^{12} + (1 + 4s + K_{21}s^2 + K_{31}s^3 + K_{41}s^4 + K_{51}s^5 + K_{61}s^6 \\
\quad + K_{71}s^7 + K_{81}s^8 + K_{91}s^9 + K_{101}s^{10})u + (K_{02} + K_{12}s + K_{22}s^2 + K_{32}s^3 + K_{42}s^4 + K_{52}s^5 \\
\quad + K_6 s^6 + K_{72}s^7 + K_{82}s^8)u^2 + (K_{03} + K_{13}s + K_{23}s^2 + K_{33}s^3 + K_{43}s^4 + K_{53}s^5 + K_{63}s^6)u^3 \\
\quad + (K_{04} + K_{14}s + K_{24}s^2 + K_{34}s^3 + K_{44}s^4)u^4 + (K_{05} + K_{15}s + K_{25}s^2)u^5 + K_{06}u^6 + O(e^{13}), \\
J_f(s, u, v) = 1 + 2s + (1 + Q_2)s^2 + (2Q_2 + Q_3 - 4)s^3 + K_{40}s^4 + K_{50}s^5 + K_{60}s^6 + K_{70}s^7 + J_{800}s^8 \\
\quad + (1 + 4s + (1 + K_{21})s^2 + (K_{31} + 2Q_2 - 4)s^3 + (K_{40} + K_{41} - 3 + Q_2 - Q_4)s^4 + (2K_{40} + K_{50} + K_{51} + 6 \\
\quad - 3Q_3 - 2Q_4 - Q_5)s^5 + J_{610}s^6)u + (K_{02} + (2 + K_{12})s + (K_{21} + K_{22} - Q_2 + 1)s^2 + (2K_{21} + K_{31} + K_{32} - 6 \\
\quad - 2Q_2 - Q_3)s^3 + J_{420}s^4)u^2 + (K_{02} + K_{03} - 1 + (2K_{02} + K_{12} + K_{13} - 4)s + J_{230}s^2)u^3 + J_{040}u^4 \\
\quad + (1 + 2s + (1 + Q_2)s^2 + (2Q_2 + Q_3 - 4)s^3 + J_{401}s^4 + (2 + 8s + J_{211}s^2)u + J_{021}u^2)v + J_{002}v^2 + O(e^9),
\end{cases}
\tag{21}
$$

where parameters Q_2–Q_6, $K_{40}, K_{50}, K_{60}, K_{70}, K_{80}, K_{90}, K_{100}, K_{110}, K_{120}, K_{21}, K_{31}, K_{41}, K_{51}, K_{61}, K_{71}, K_{81}, K_{91}, K_{101}, K_{02}, K_{12}, K_{22}, K_{32}, K_{42}, K_{52}, K_{62}, K_{72}, K_{82}, K_{03}, K_{13}, K_{23}, K_{33}, K_{13}, K_{53}, K_{63}, K_{04}, K_{14}, K_{24}, K_{34}, K_{44}, K_{05}, K_{15}, K_{25}, K_{06}$ and $J_{040}, J_{002}, J_{021}, J_{211}, J_{230}, J_{401}, J_{420}, J_{610}, J_{800}$ may be free.

Although various forms of weight functions $Q_f(s)$, $K_f(s, u)$ and $J_f(s, u, v)$ are available, in the current study we limit ourselves to all three weight functions in the form of rational functions, leading us to possible purely imaginary extraneous fixed points when $f(z) = (z^2 - 1)^m$ is employed. In the current study, we will consider two special cases described below:

The first case below will represent the best scheme, W3G7, studied in [19] only for $m = 1$.

Case 1:

$$
\begin{cases}
Q_f(s) & = \frac{1}{1 - 2s}, \\
K_f(s, u) & = Q_f(s) \cdot \frac{(s-1)^2}{1 - 2s - u + 2s^2 u}, \\
J_f(s, u, v) & = K_f(s, u) \cdot \frac{1 + \sum_{i=1}^{3} q_i s^i + u \sum_{i=4}^{8} q_i s^{i-4} + u^2 \sum_{i=9}^{14} q_i s^{i-9} + u^3 \sum_{i=15}^{21} q_i s^{i-15}}{\mathcal{A}_1(s, u) + v \cdot (\sum_{i=22}^{25} r_i s^{i-22} + u(r_{26} + r_{27}s + \lambda s^2))},
\end{cases}
\tag{22}
$$

where

$$
\begin{cases}
q_9 = q_{12} = q_{13} = q_{17} = q_{18} = q_{19} = q_{20} = r_9 = r_{10} = r_{19} = r_{20} = 0, \\
q_1 = \frac{-3055820263252 - 76497245\lambda}{142682111242}, q_2 = \frac{56884034112404 + 44614515451\lambda}{285364222484}, \\
q_3 = \frac{-45802209949332 - 44308526471\lambda}{142682111242}, q_4 = -\frac{3(17778426888128 + 67929066997\lambda)}{1426821112420}, \\
q_5 = \frac{2(21034820227211 + 132665343294\lambda)}{356705278105}, q_6 = \frac{-1589080655012451 + 134087681464\lambda}{142682111242}, \\
q_7 = \frac{2(-780300304419180 + 71852971399\lambda)}{71341055621}, q_8 = \frac{12288(-727219117761 + 128167952\lambda)}{71341055621}, \\
q_{10} = 2, q_{11} = \frac{2(-741727036224277 + 126275739062\lambda)}{71341055621}, \\
q_{14} = -\frac{8192(-3964538065856 + 615849113\lambda)}{71341055621}, q_{15} = \frac{8(-226231159891830 + 34083208621\lambda)}{71341055621}, \\
q_{16} = -\frac{24(-908116719056544 + 136634733499\lambda)}{356705278105}, q_{21} = \frac{131072(-918470889768 + 136352293\lambda)}{356705278105}, \\
r_1 = q_1, r_2 = q_2, r_3 = q_3, r_4 = q_4, r_5 = q_5, r_6 = q_6 - 1, r_7 = q_7 - q_1 - 2, r_8 = q_8 + \frac{q_3}{2}, \\
r_{11} = \frac{-29558910226378916 + 5256346708371\lambda}{1426821112420}, r_{12} = \frac{-55018830261476 - 109759858153\lambda}{142682111242}, \\
r_{13} = \frac{25(-75694849962572 + 11301475999\lambda)}{71341055621}, r_{14} = -\frac{4096(-1500792372416 + 228734011\lambda)}{15508925135}, \\
r_{15} = q_{15}, r_{16} = \frac{43641510974266076 - 6354680006961\lambda}{713410556210}, r_{17} = -\frac{2(-1060205894022116 + 202907726307\lambda)}{71341055621}, \\
r_{18} = \frac{2(-2870055173156756 + 475573395275\lambda)}{71341055621}, r_{21} = \frac{q_{21}}{2}, r_{22} = -1, r_{23} = -q_1, r_{24} = -q_2, \\
r_{25} = -q_3, r_{26} = -1 - q_4, r_{27} = -2 - q_1 - q_5, \lambda = \frac{1353974063793787}{212746858830},
\end{cases}
\tag{23}
$$

and $A_1(s, u) = 1 + \sum_{i=1}^{3} r_i s^i + u \sum_{i=4}^{8} r_i s^{i-4} + u^2 \sum_{i=9}^{14} r_i s^{i-9} + u^3 \sum_{i=15}^{21} r_i s^{i-15}$.

As a second case, we will consider the following set of weight functions:

Case 2:

$$
\begin{cases}
Q_f(s) &= \frac{1}{1-2s}, \\
K_f(s, u) &= Q_f(s) \cdot \frac{(s-1)^2}{1-2s-u+2s^2 u}, \\
J_f(s, u, v) &= \frac{1 + \sum_{i=1}^{3} q_i s^i + u \sum_{i=4}^{8} q_i s^{i-4} + u^2 \sum_{i=9}^{14} q_i s^{i-9} + u^3 \sum_{i=15}^{19} q_i s^{i-15}}{A_0(s, u) + v \cdot (\sum_{i=20}^{23} r_i s^{i-20} + u \sum_{i=24}^{28} r_i s^{i-24} + r_{29} u^2)},
\end{cases}
\tag{24}
$$

where $A_0(s, u) = 1 + \sum_{i=1}^{3} r_i s^i + u \sum_{i=4}^{8} r_i s^{i-4} + u^2 \sum_{i=9}^{14} r_i s^{i-9} + u^3 \sum_{i=15}^{19} r_i s^{i-15}$ and determination of the 48 coefficients q_i, r_i of J_f is described below. Relationships were sought among all free parameters of $J_f(s, u, v)$, giving us a simple governing equation for extraneous fixed points of the proposed family of methods (1).

To this end, we first express s, u and v for $f(z) = (z^2 - 1)^m$ as follows with $t = z^2$:

$$
s = \frac{1}{4}\left(1 - \frac{1}{t}\right), \; u = \frac{1}{4} \cdot \frac{(t-1)^2}{(t+1)^2}, \; v = \frac{(t-1)^4}{4(1 + 6t + t^2)^2}.
\tag{25}
$$

In order to obtain a simple form of $J_f(s, u, v)$, we needed to closely inspect how it is connected with $K_f(s, u)$. When applying to $f(z) = (z^2 - 1)^m$, we find $K_f(s, u)$ with $t = z^2$ as shown below:

$$
K_f(s, u) = \frac{4t(1 + t)}{t^2 + 6t + 1}.
\tag{26}
$$

Using the two selected weight functions Q_f, K_f, we continue to determine coefficients q_i, r_i of J_f yielding a simple governing equation for extraneous fixed points of the proposed methods when $f(z) = (z^2 - 1)^m$ is applied. As a result of tedious algebraic operations reflecting the 25 constraints (with possible rank deficiency given by (18) and (19)), we find only 23 effective relations, as follows:

$$
\begin{cases}
q_1 = \frac{1}{4}(-8 - r_{23}), q_2 = -3 - 2q_1, q_3 = 2 + q_1, q_4 = -r_{24}, q_5 = 2q_4 - r_{25}, \\
q_6 = 5 + \frac{7q_3}{4} + \frac{13q_4}{4} + \frac{9q_5}{8} - \frac{5q_9}{4} - \frac{25q_{10}}{8} + \frac{5q_{12}}{4} + \frac{r_8}{4} - \frac{5r_{11}}{4} - \frac{5r_{12}}{8}, \\
q_7 = -\frac{4q_3}{5} + \frac{4q_4}{5} + \frac{q_5}{5} - \frac{2q_6}{5} - \frac{2r_8}{5}, q_8 = q_4 + \frac{q_5}{2} - \frac{q_7}{2}, q_9 = q_{15} - r_{15}, \\
q_{10} = -2q_4 - 2q_{15} + q_{16} - r_{16}, q_{11} = -6 + \frac{q_5}{2} - \frac{5q_7}{2} + q_9 + 2q_{10} - r_8 + r_{11}, \\
r_1 = -2 + q_1, r_2 = 2(1 + q_2), r_3 = 4q_3, r_4 = -1 + q_4, r_5 = 2 - q_3 - 2q_4 + q_5, \\
r_6 = 1 + 2q_3 - q_4 - 2q_5 + q_6, r_7 = -2 + 5q_3 - 4q_4 - 2q_5 + 6q_7 + 2r_8, r_9 = -q_4 + q_9, \\
r_{10} = -2 - q_5 - 2q_9 + q_{10}, r_{20} = -1, r_{21} = 4 - q_3, r_{22} = -4(1 - q_3).
\end{cases}
\tag{27}
$$

The three relations, $J_{500} = K_{50}$, $J_{600} = K_{60}$, and $J_{700} = K_{70}$ give one relation $r_{22} = -4(1 - q_3)$.

Due to 23 constraints in Relation (27), we find that 18 free parameters among 48 coefficients of J_f in (24) are available. We seek relationships among the free parameters yielding purely imaginary extraneous fixed points of the proposed family of methods when $f(z) = (z^2 - 1)^m$ is applied.

To this end, after substituting the 23 effective relations given by (27) into J_f in (24) and by applying to $f(z) = (z^2 - 1)^m$, we can construct $H(z) = 1 + sQ_f(s) + suK_f(s, u) + suvJ_f(s, u, v)$ in (1) and seek its roots for extraneous fixed points with $t = z^2$:

$$H(z) = \frac{\mathcal{A} \cdot G(t)}{t(1 + t)^2(1 + 6t + t^2) \cdot W(t)}, \tag{28}$$

where $\mathcal{A}$ is a constant factor, $G(t) = \sum_{i=0}^{20} g_i t^i$, with $g_0 = -q_{14}$, $g_1 = -16 - 2q_{12} - 4q_{13} - 8q_{14} - 16q_{15} + 10q_{16} - 4r_8 + 4r_{11} + 2r_{12} - 16_{14} - 4r_{15} - 10r_{16} + 3r_{20} + 60r_{21} + 10r_{22}$, $g_i = g_i(q_{12}, r_{12}, \cdots, r_{25})$, for $2 \leq i \leq 20$ and $W(t) = \sum_{i=0}^{15} w_i t^i$, with $w_0 = -r_{14}$, $w_1 = 16r_8 + 4r_{13} - 5r_{14} + 4r_{25}$, $w_i = w_i(q_{12}, r_{12}, \cdots, r_{25})$, for $2 \leq i \leq 15$. The coefficients of both polynomials, $G(t)$ and $W(t)$, contain at most 18 free parameters.

We first observe that partial expressions of $H(z)$ with $t = z^2$, namely, $1 + sQ_f(s) = \frac{1+3t}{2(1+t)}$, $1 + sQ_f(s) + suK_f(s, u) = \frac{1+21t+35t^2+7t^3}{4(1+t)(1+6t+t^2)}$ and the denominator of (28) contain factors t, $(1+3t)$, $(1+t)$, $(1+6t+t^2)$, $(1+21t+35t^2+7t^3)$ when $f(z) = (z^2-1)^m$ is applied. With an observation of presence of such factors, we seek a special subcase in which $G(t)$ may contain all the interested factors as follows:

$$G(t) = t(1+3t)(1+t)^\lambda(1+6t+t^2)^\beta(1+21t+35t^2+7t^3) \cdot (1+10t+5t^2)(1+92t+134t^2+28t^3+t^4) \cdot \Phi(t), \tag{29}$$

where $\Phi(t)$ is a polynomial of degree $(9 - (\lambda + 2\beta))$, with $\lambda \in \{0, 1, 2\}$, $\beta \in \{1, 2, 3\}$ and $1 \leq \lambda + \beta \leq 3$; two polynomial factors $(1 + 10t + 5t^2)$ and $(1 + 92t + 134t^2 + 28t^3 + t^4)$ were found in Case 3G of the previous study done by Geum–Kim–Neta [19]. Notice that factors $(1 + 6t + t^2)$, $(1 + 21t + 35t^2 + 7t^3)$, $(1 + 10t + 5t^2)$ and $(1 + 92t + 134t^2 + 28t^3 + t^4)$ of $G(t)$ are all negative, i.e., the corresponding extraneous fixed points are all purely imaginary.

In fact, the degree of $\Phi(t)$ will be decreased by annihilating the relevant coefficients containing free parameters to make all its roots negative. We take the 6 pairs of $(\lambda, \beta) \in \{(0,1), (0,2), (0,3), (1,1), (1,2), (2,1)\}$ to form 6 subcases named as Case 2A–2F in order. The lengthy algebraic process eventually leads us to additional constraints to each subcase described below:

Case 2A: $(\lambda, \beta) = (0, 1)$

$$\begin{pmatrix} q_{12} \\ q_{15} \\ r_8 \\ r_{11} \\ r_{12} \\ r_{13} \\ r_{14} \\ r_{15} \\ r_{16} \\ r_{18} \\ r_{23} \\ r_{24} \end{pmatrix} = \begin{bmatrix} -3 & 9 & 9 & \frac{13}{2} & \frac{7}{2} & -13 & 0 & 0 & -12 & -\frac{23}{2} & -13 & -134 \\ 0 & -\frac{1}{2} & -\frac{1}{4} & -\frac{1}{8} & -\frac{1}{16} & 0 & 0 & 0 & 0 & 0 & 0 & 0 \\ -\frac{7}{4} & \frac{17}{4} & 5 & \frac{59}{16} & 2 & -\frac{1}{4} & \frac{1}{16} & \frac{181}{28} & \frac{699}{112} & \frac{99}{16} & \frac{1807}{224} & \frac{1163}{16} \\ -\frac{11}{4} & 34 & \frac{171}{4} & \frac{125}{4} & \frac{159}{8} & -3 & \frac{3}{4} & \frac{291}{7} & \frac{274}{7} & \frac{75}{2} & \frac{629}{14} & -454 \\ -\frac{11}{2} & 14 & \frac{23}{2} & \frac{35}{4} & \frac{9}{2} & 1 & -\frac{1}{4} & \frac{255}{14} & \frac{425}{28} & -14 & \frac{821}{56} & \frac{693}{4} \\ 7 & -30 & -33 & -\frac{49}{2} & -\frac{29}{2} & 1 & -\frac{3}{7} & \frac{297}{7} & \frac{537}{14} & 37 & \frac{1199}{28} & \frac{937}{2} \\ 0 & 1 & 1 & \frac{3}{4} & \frac{1}{2} & 0 & 0 & -\frac{9}{14} & -\frac{15}{14} & -1 & -\frac{27}{28} & -\frac{27}{2} \\ \frac{1}{2} & -\frac{21}{4} & -\frac{23}{4} & -\frac{67}{16} & -\frac{21}{8} & \frac{1}{4} & -\frac{1}{16} & \frac{165}{28} & \frac{599}{112} & \frac{79}{224} & \frac{1291}{224} & \frac{967}{16} \\ -\frac{3}{2} & 16 & \frac{35}{4} & \frac{51}{4} & 8 & -1 & \frac{1}{4} & \frac{241}{14} & \frac{439}{28} & -\frac{29}{2} & \frac{947}{56} & -\frac{707}{4} \\ 2 & -21 & -23 & -\frac{67}{4} & -\frac{21}{2} & 0 & -1 & \frac{152}{7} & \frac{279}{14} & \frac{37}{2} & \frac{603}{28} & \frac{447}{2} \\ 0 & 0 & 0 & 0 & 0 & 0 & 0 & -\frac{2}{7} & -\frac{4}{7} & -1 & -\frac{12}{7} & 0 \\ 0 & 0 & 0 & 0 & 0 & 0 & 0 & -\frac{2}{7} & -\frac{1}{14} & 0 & \frac{1}{28} & -\frac{1}{2} \end{bmatrix} \begin{pmatrix} q_{13} \\ q_{16} \\ q_{17} \\ q_{18} \\ q_{19} \\ r_{17} \\ r_{19} \\ r_{25} \\ r_{26} \\ r_{27} \\ r_{28} \\ r_{29} \end{pmatrix} + \begin{pmatrix} -74 \\ 0 \\ -\frac{2207}{56} \\ -\frac{1900}{7} \\ -\frac{1355}{14} \\ \frac{1929}{7} \\ -\frac{61}{7} \\ \frac{1987}{56} \\ -\frac{1453}{14} \\ \frac{919}{7} \\ \frac{16}{7} \\ \frac{2}{7} \end{pmatrix} \tag{30}$$

and $q_{14} = 0$. These 12 additional constraints $q_{12}, q_{15}, r_8, r_{11}, r_{12}, r_{13}, r_{14}, r_{15}, r_{16}, r_{18}, r_{23}, r_{24}$ are expressed in terms of 12 parameters $q_{13}, q_{16}, q_{17}, q_{18}, q_{19}, r_{17}, r_{19}, r_{25}, r_{26}, r_{27}, r_{28}, r_{29}$ that are arbitrarily free for the purely imaginary extraneous fixed points. Those 12 free parameters are chosen at our disposal. Then, using Relations (27) and (30), the desired form of $J_f(s, u, v)$ in (24) can be constructed.

Case 2B: $(\lambda, \beta) = (0, 2)$

$$
\begin{pmatrix} q_{12} \\ q_{15} \\ r_8 \\ r_{11} \\ r_{12} \\ r_{13} \\ r_{14} \\ r_{15} \\ r_{16} \\ r_{18} \\ r_{23} \\ r_{24} \\ r_{25} \\ r_{26} \end{pmatrix}
=
\begin{bmatrix}
-3 & 9 & 9 & \frac{13}{2} & \frac{7}{2} & 0 & 0 & \frac{1}{6} & -\frac{1}{3} & -\frac{178}{3} \\
0 & -\frac{1}{2} & -\frac{1}{4} & -\frac{1}{8} & -\frac{1}{16} & 0 & 0 & 0 & 0 & 0 \\
-\frac{7}{4} & \frac{17}{4} & 5 & \frac{59}{16} & 2 & -\frac{1}{4} & \frac{1}{16} & -\frac{1}{48} & \frac{161}{96} & -\frac{917}{24} \\
-\frac{11}{4} & 34 & \frac{171}{4} & \frac{125}{4} & \frac{159}{8} & -3 & \frac{3}{4} & \frac{5}{6} & -\frac{25}{6} & -\frac{668}{3} \\
-\frac{11}{2} & 14 & \frac{23}{2} & \frac{35}{4} & \frac{9}{2} & 1 & -\frac{1}{4} & \frac{1}{6} & \frac{61}{24} & \frac{317}{6} \\
7 & -30 & -33 & -\frac{49}{2} & -\frac{29}{2} & 1 & -\frac{1}{2} & 0 & \frac{7}{4} & 217 \\
0 & 1 & 1 & \frac{3}{4} & \frac{1}{2} & 0 & 0 & 0 & \frac{1}{4} & -5 \\
\frac{1}{2} & -\frac{21}{4} & -\frac{23}{4} & -\frac{67}{16} & -\frac{21}{8} & \frac{1}{4} & -\frac{1}{16} & -\frac{11}{48} & \frac{5}{96} & \frac{617}{24} \\
-\frac{3}{2} & 16 & \frac{35}{2} & \frac{51}{4} & 8 & -1 & \frac{1}{4} & \frac{2}{3} & -\frac{5}{24} & -\frac{455}{6} \\
2 & -21 & -23 & -\frac{67}{4} & -\frac{21}{2} & 0 & -1 & -\frac{5}{6} & \frac{5}{12} & \frac{293}{3} \\
0 & 0 & 0 & 0 & 0 & 0 & 0 & -\frac{1}{3} & -\frac{4}{3} & -\frac{4}{3} \\
0 & 0 & 0 & 0 & 0 & 0 & 0 & 0 & \frac{1}{4} & 3 \\
0 & 0 & 0 & 0 & 0 & 0 & 0 & \frac{1}{3} & -\frac{2}{3} & -\frac{44}{3} \\
0 & 0 & 0 & 0 & 0 & 0 & 0 & -\frac{4}{3} & -\frac{1}{3} & \frac{29}{3}
\end{bmatrix}
\begin{pmatrix} q_{13} \\ q_{16} \\ q_{17} \\ q_{18} \\ q_{19} \\ r_{17} \\ r_{19} \\ r_{27} \\ r_{28} \\ r_{29} \end{pmatrix}
+
\begin{pmatrix} -10 \\ 0 \\ -\frac{87}{8} \\ -76 \\ \frac{25}{2} \\ 57 \\ -1 \\ \frac{43}{8} \\ -\frac{33}{2} \\ 23 \\ 0 \\ 4 \\ -16 \\ 12 \end{pmatrix}
\tag{31}
$$

and $q_{14} = 0$. These 14 additional constraints are expressed in terms of 10 parameters $q_{13}, q_{16}, q_{17}, q_{18}, q_{19}, r_{17}, r_{19}, r_{27}, r_{28}, r_{29}$ that are arbitrarily free for the purely imaginary extraneous fixed points. Those 10 free parameters are chosen at our disposal. Then, using Relations (27) and (31), the desired form of $J_f(s, u, v)$ in (24) can be constructed.

Case 2C: $(\lambda, \beta) = (0, 3)$

$$
\begin{pmatrix} q_{12} \\ q_{15} \\ q_{16} \\ r_8 \\ r_{11} \\ r_{12} \\ r_{13} \\ r_{14} \\ r_{15} \\ r_{16} \\ r_{18} \\ r_{23} \\ r_{24} \\ r_{25} \\ r_{26} \\ r_{27} \end{pmatrix}
=
\begin{bmatrix}
-3 & 0 & -\frac{1}{4} & -1 & 0 & 0 & -\frac{23}{16} & -\frac{29}{4} \\
0 & \frac{1}{4} & \frac{1}{4} & \frac{3}{16} & 0 & 0 & \frac{1}{32} & -\frac{21}{8} \\
0 & -1 & -\frac{3}{4} & -\frac{1}{2} & 0 & 0 & -\frac{1}{16} & \frac{21}{4} \\
-\frac{7}{4} & \frac{3}{4} & \frac{1}{2} & -\frac{1}{8} & -\frac{1}{4} & \frac{1}{16} & -\frac{15}{8} & -\frac{33}{2} \\
-\frac{11}{4} & \frac{35}{4} & \frac{23}{4} & \frac{23}{8} & -3 & \frac{3}{4} & -9 & -20 \\
-\frac{11}{2} & -\frac{5}{2} & -\frac{7}{4} & -\frac{5}{4} & 1 & -\frac{1}{4} & \frac{9}{8} & \frac{51}{2} \\
7 & -3 & -2 & \frac{1}{2} & 1 & -\frac{1}{2} & \frac{29}{8} & \frac{119}{2} \\
0 & 0 & 0 & 0 & 0 & 0 & \frac{3}{16} & \frac{1}{4} \\
\frac{1}{2} & -\frac{1}{2} & -\frac{1}{4} & 0 & \frac{1}{4} & -\frac{1}{16} & \frac{9}{8} & -\frac{17}{2} \\
-\frac{3}{2} & \frac{3}{2} & \frac{3}{4} & 0 & -1 & \frac{1}{4} & -\frac{27}{8} & \frac{55}{2} \\
2 & -2 & -1 & 0 & 0 & -1 & \frac{71}{16} & -\frac{147}{4} \\
0 & 0 & 0 & 0 & 0 & 0 & -\frac{1}{4} & -11 \\
0 & 0 & 0 & 0 & 0 & 0 & \frac{1}{4} & 3 \\
0 & 0 & 0 & 0 & 0 & 0 & -\frac{7}{4} & -5 \\
0 & 0 & 0 & 0 & 0 & 0 & 4 & -29 \\
0 & 0 & 0 & 0 & 0 & 0 & -\frac{13}{4} & 29
\end{bmatrix}
\begin{pmatrix} q_{13} \\ q_{17} \\ q_{18} \\ q_{19} \\ r_{17} \\ r_{19} \\ r_{28} \\ r_{29} \end{pmatrix}
+
\begin{pmatrix} -\frac{57}{4} \\ -\frac{1}{8} \\ \frac{1}{4} \\ -9 \\ -100 \\ \frac{19}{2} \\ \frac{99}{2} \\ -\frac{3}{4} \\ 13 \\ -\frac{77}{2} \\ \frac{201}{4} \\ 13 \\ 4 \\ -29 \\ 64 \\ -39 \end{pmatrix}
\tag{32}
$$

and $q_{14} = 0$. These 16 additional constraints are expressed in terms of 8 parameters $q_{17}, q_{18}, q_{19}, r_{17}, r_{18}, r_{19}, r_{28}, r_{29}$ that are arbitrarily free for the purely imaginary extraneous fixed points. Those 8 free parameters are chosen at our disposal. Then, using Relations (27) and (32), the desired form of $J_f(s, u, v)$ in (24) can be constructed.

Case 2D: $(\lambda, \beta) = (1, 1)$, being identical with Case 2A.

Case 2E: $(\lambda, \beta) = (1, 2)$, being identical with Case 2B.

Case 2F: $(\lambda, \beta) = (2,1)$,

$$
\begin{pmatrix} q_{12} \\ q_{15} \\ q_{16} \\ r_8 \\ r_{11} \\ r_{12} \\ r_{13} \\ r_{14} \\ r_{15} \\ r_{16} \\ r_{18} \\ r_{23} \\ r_{24} \end{pmatrix}
=
\begin{bmatrix}
-3 & 0 & -\frac{1}{4} & -1 & 0 & 0 & -13 & -12 & -\frac{23}{2} & -13 & -134 \\
0 & \frac{1}{4} & \frac{1}{4} & \frac{3}{16} & 0 & 0 & 0 & 0 & 0 & 0 & 0 \\
0 & -1 & -\frac{3}{4} & -\frac{1}{2} & 0 & 0 & 0 & 0 & 0 & 0 & 0 \\
-\frac{7}{4} & \frac{3}{4} & \frac{1}{2} & -\frac{1}{8} & -\frac{1}{4} & \frac{1}{16} & -\frac{181}{28} & -\frac{699}{112} & -\frac{99}{16} & -\frac{1807}{224} & -\frac{1163}{16} \\
-\frac{11}{4} & \frac{35}{4} & \frac{23}{4} & \frac{23}{8} & -3 & \frac{3}{4} & -\frac{291}{7} & -\frac{274}{7} & -\frac{75}{2} & -\frac{629}{14} & -454 \\
-\frac{11}{2} & -\frac{5}{2} & -\frac{7}{4} & -\frac{5}{2} & 1 & -\frac{1}{4} & -\frac{255}{14} & -\frac{425}{28} & -14 & -\frac{821}{56} & -\frac{693}{4} \\
7 & -3 & -2 & \frac{1}{2} & 1 & -\frac{1}{2} & \frac{297}{7} & \frac{537}{14} & 37 & \frac{1199}{28} & \frac{937}{2} \\
0 & 0 & 0 & 0 & 0 & 0 & -\frac{9}{7} & -\frac{15}{14} & -1 & -\frac{27}{28} & -\frac{27}{2} \\
\frac{1}{2} & -\frac{1}{2} & -\frac{1}{4} & 0 & \frac{1}{4} & -\frac{1}{16} & \frac{165}{28} & \frac{599}{112} & \frac{79}{16} & \frac{1291}{224} & \frac{967}{16} \\
-\frac{3}{2} & \frac{3}{2} & \frac{3}{4} & 0 & -1 & \frac{1}{4} & -\frac{241}{14} & -\frac{439}{28} & -\frac{29}{2} & -\frac{947}{56} & -\frac{707}{4} \\
2 & -2 & -1 & 0 & 0 & -1 & \frac{152}{7} & \frac{279}{14} & \frac{37}{2} & \frac{603}{28} & \frac{447}{2} \\
0 & 0 & 0 & 0 & 0 & 0 & -\frac{2}{7} & -\frac{4}{7} & -1 & -\frac{12}{7} & 0 \\
0 & 0 & 0 & 0 & 0 & 0 & -\frac{2}{7} & -\frac{1}{14} & 0 & \frac{1}{28} & -\frac{1}{2}
\end{bmatrix}
\begin{pmatrix} q_{13} \\ q_{17} \\ q_{18} \\ q_{19} \\ r_{17} \\ r_{19} \\ r_{25} \\ r_{26} \\ r_{27} \\ r_{28} \\ r_{29} \end{pmatrix}
+
\begin{pmatrix} -74 \\ 0 \\ 0 \\ -\frac{2207}{56} \\ -\frac{1900}{7} \\ -\frac{1355}{14} \\ \frac{1929}{7} \\ -\frac{61}{7} \\ \frac{1987}{56} \\ \frac{1453}{14} \\ \frac{919}{7} \\ \frac{16}{7} \\ \frac{2}{7} \end{pmatrix}
\tag{33}
$$

and $q_{14} = 0$. These 13 additional constraints are expressed in terms of 11 parameters $q_{13}, q_{17}, q_{18}, q_{19}$, $r_{17}, r_{19}, r_{25}, r_{26}, r_{27}, r_{28}, r_{29}$ that are arbitrarily free for the purely imaginary extraneous fixed points. Those 11 free parameters are chosen at our disposal. Then, using Relations (27) and (33), the desired form of $J_f(s, u, v)$ in (24) can be constructed. After a process of careful factorization, we find the expression for $H(z)$ in (28) stated in the following lemma.

Proposition 1. *The expression $H(z)$ in (28) is identical in each subcase of 2A–2F and given by a unique relation below:*

$$
H(z) = \frac{(1+3t)(1+10t+5t^2)(1+92t+134t^2+28t^3+t^4)}{8(1+t)(1+6t+t^2)(1+28t+70t^2+28t^3+t^4)}, \quad t = z^2,
\tag{34}
$$

despite the possibility of different coefficients in each subcase.

Proof. Let us write $G(t)$ in (28) as $G(t) = t(1 + 3t) \cdot \psi_1(t) \cdot \psi_2(t) \cdot \Phi(t) \cdot (1 + t)^\lambda (1 + 6t + t^2)^{\beta-1}$ with $\psi_1(t) = (1 + 6t + t^2)(1 + 21t + 35t^2 + 7t^3)$ and $\psi_2(t) = (1 + 10t + 5t^2)(1 + 92t + 134t^2 + 28t^3 + t^4)$. Then after a lengthy process of a series of factorizations with the aid of Mathematica symbolic ability, we find $\Phi(t)$ and $W(t)$ in each subcase as follows.

(1) Case 2A: with $\lambda = 0$ and $\beta = 1$, we get

$$
\begin{cases}
\Phi(t) = -\frac{2}{7}(1 + t) \cdot \Gamma_1(t), \\
W(t) = -\frac{16}{7}\psi_1(t)(1 + 28t + 70t^2 + 28t^3 + t^4)\Gamma_1(t),
\end{cases}
\tag{35}
$$

where $\Gamma_1(t) = -244 + 28q_{16} + 28q_{17} + 21q_{18} + 14q_{19} - 36r_{25} - 30r_{26} - 28r_{27} - 27r_{28} - 378r_{29} + 14t(-72 + 4q_{16} + 4q_{17} + 3q_{18} + 2q_{19} - 4r_{25} + 2r_{27} + 4r_{28} - 72r_{29}) + t^2(1692 - 476q_{16} - 476q_{17} - 357q_{18} - 238q_{19} + 548r_{25} + 578r_{26} + 672r_{27} + 957r_{28} + 6006r_{29}) + 4t^3(-2288 + 196q_{16} + 196q_{17} + 147q_{18} + 98q_{19} - 148r_{25} - 100r_{26} - 42r_{27} - 6r_{28} - 1540r_{29}) - 7t^4(1636 + 68q_{16} + 68q_{17} + 51q_{18} + 34q_{19} + 4r_{25} + 22r_{26} + 36r_{27} + 55r_{28} + 386r_{29}) + t^5(-4176 + 56q_{16} + 56q_{17} + 42q_{18} + 28q_{19} + 648r_{25} + 400r_{26} + 140r_{27} - 32r_{28} + 7168r_{29}) + t^6(-4332 + 28q_{16} + 28q_{17} + 21q_{18} + 14q_{19} - 484r_{25} - 394r_{26} - 392r_{27} - 545r_{28} - 2926r_{29})$.

(2) Case 2B: with $\lambda = 0$ and $\beta = 2$, we get

$$
\begin{cases}
\Phi(t) = -\frac{2}{3}(1 + t) \cdot \Gamma_2(t), \\
W(t) = -\frac{16}{3}(1 + 6t + t^2)\psi_1(t)(1 + 28t + 70t^2 + 28t^3 + t^4)\Gamma_2(t),
\end{cases}
\tag{36}
$$

where $\Gamma_2(t) = (1 + 6t + t^2)(3(-4 + 4q_{16} + 4q_{17} + 3q_{18} + 2q_{19} + r_{28} - 20r_{29}) + t(24 - 48q_{16} - 48q_{17} - 36q_{18} - 24q_{19} + 4r_{27} + 22r_{28} + 280r_{29}) + 6t^2(-32 + 12q_{16} + 12q_{17} + 9q_{18} + 6q_{19} + 2r_{27} + 6r_{28} - 16r_{29}) - 2t^3(396 + 24q_{16} + 24q_{17} + 18q_{18} + 12q_{19} + 2r_{27} + 11r_{28} + 140r_{29}) + 3t^4(-188 + 4q_{16} + 4q_{17} + 3q_{18} + 2q_{19} - 4r_{27} - 13r_{28} + 52r_{29}))$.

(3) Case 2C: with $\lambda = 0$ and $\beta = 3$, we get

$$\begin{cases} \Phi(t) = \frac{1}{2}(1+t) \cdot \Gamma_3(t), \\ W(t) = 4(1+6t+t^2)^2 \psi_1(t)(1+28t+70t^2+28t^3+t^4)\Gamma_3(t), \end{cases} \tag{37}$$

where $\Gamma_3(t) = (1+6t+t^2)^2(12 - 3r_{28} + 2t(60 + r_{28} - 84r_{29}) - 4r_{29} + t^2(124 + r_{28} + 172r_{29}))$.

(4) Case 2D: with $\lambda = 1$ and $\beta = 1$, we get

$$\begin{cases} \Phi(t) = -\frac{2}{7} \cdot \Gamma_1(t), \\ W(t) = -\frac{16}{7}\psi_1(t)(1+28t+70t^2+28t^3+t^4)\Gamma_1(t). \end{cases} \tag{38}$$

(5) Case 2E: with $\lambda = 1$ and $\beta = 2$, we get

$$\begin{cases} \Phi(t) = -\frac{2}{3} \cdot \Gamma_2(t), \\ W(t) = -\frac{16}{3}(1+6t+t^2)\psi_1(t)(1+28t+70t^2+28t^3+t^4)\Gamma_2(t), \end{cases} \tag{39}$$

(6) Case 2F: with $\lambda = 2$ and $\beta = 1$, we get

$$\begin{cases} \Phi(t) = \frac{2}{7} \cdot \Gamma_4(t), \\ W(t) = \frac{2}{7}(1+t)\psi_1(t)(1+28t+70t^2+28t^3+t^4)\Gamma_4(t), \end{cases} \tag{40}$$

where $\Gamma_4(t) = 244 + 36r_{25} + 30r_{26} + 28r_{27} + 27r_{28} + 378r_{29} + t(764 + 20r_{25} - 30r_{26} - 56r_{27} - 83r_{28} + 630r_{29}) - 2t^2(1228 + 284r_{25} + 274r_{26} + 308r_{27} + 437r_{28} + 3318r_{29}) + 2t^3(5804 + 580r_{25} + 474r_{26} + 392r_{27} + 449r_{28} + 6398r_{29}) - t^4(156 + 1132r_{25} + 794r_{26} + 532r_{27} + 513r_{28} + 10094r_{29}) + t^5(4332 + 484r_{25} + 394r_{26} + 392r_{27} + 545r_{28} + 2926r_{29})$.

Substituting each pair of $(\Phi(t), W(t))$ into (28) yields an identical Relation (34) as desired. □

Remark 2. *The factorization process in the above proposition yields the additional constraints given by* (30)–(33) *for subcases 2A–2F, after a lengthy computation. Case 2D and Case 2E are found to be identical with Case 2A and Case 2B, respectively, by direct computation.*

In Table 1, we list free parameters selected for typical subcases of 2A–2F. Combining these selected free parameters with Relations (27) and (30)–(33), we can construct special iterative schemes named as W2A1, W2A2, $\cdots$, W2F3, W2F4. Such schemes together with W3G7 for Case 1 shall be used in Section 4 to display results on their numerical and dynamical aspects.

Table 1. Free parameters selected for typical subcases of 2A1–2F4.

SCN	q_{13}	q_{16}	q_{17}	q_{18}	q_{19}	r_{17}	r_{19}	r_{25}	r_{26}	r_{27}	r_{28}	r_{29}
2A1	0	0	0	0	0	0	0	0	0	0	0	0
2A2	20	0	0	0	0	0	1012	4	2	-8	0	0
2A3	$-\frac{89}{26}$	0	0	0	0	0	0	$-\frac{149}{26}$	0	0	0	0
2A4	0	0	$\frac{711}{26}$	$-\frac{622}{13}$	$\frac{222}{13}$	0	0	$-\frac{149}{26}$	0	0	0	0
2B1	0	0	0	0	0	0	0	-	-	0	0	0
2B2	0	0	0	0	0	0	96	-	-	-31	-1	$-\frac{1}{4}$
2B3	-19	0	0	0	0	0	0	-	-	-18	0	0
2B4	0	0	45	-52	-12	0	0	-	-	-18	0	0
2C1	0	-	0	0	0	0	0	-	-	-	0	0
2C2	-34	-	0	0	0	174	0	-	-	-	4	0
2C3	0	-	0	0	0	0	280	-	-	-	4	0
2C4	$-\frac{39}{4}$	-	$\frac{375}{4}$	$-\frac{627}{4}$	0	0	0	-	-	-	0	0
2F1	0	-	0	0	0	-40	0	$-\frac{122}{3}$	$\frac{290}{3}$	$-\frac{184}{3}$	0	0
2F2	16	-	0	0	0	0	0	1	0	-10	0	0
2F3	$\frac{138}{7}$	-	$-\frac{356}{7}$	$\frac{3963}{14}$	0	0	0	0	0	0	0	0
2F4	0	-	0	0	0	-32	0	-33	78	-46	-4	0

3. The Dynamics behind the Extraneous Fixed Points

The dynamics behind the extraneous fixed points [21] of iterative map (1) have been investigated by Stewart [20], Amat et al. [22], Argyros–Magreñan [23], Chun et al. [24], Chicharro et al. [25], Chun–Neta [26], Cordero et al. [27], Geum et al. [14,19,28–30], Rhee at al. [9], Magreñan [31], Neta et al. [32,33], and Scott et al. [34].

We locate a root α of a given function $f(x)$ as a fixed point ξ of the iterative map R_f:

$$x_{n+1} = R_f(x_n), n = 0, 1, \cdots, \tag{41}$$

where R_f is the iteration function associated with f. Typically, R_f is written in the form: $R_f(x_n) = x_n - \frac{f(x_n)}{f'(x_n)} H_f(x_n)$, where H_f is a weight function whose zeros are other fixed points $\xi \neq \alpha$ called extraneous fixed points of R_f. The dynamics of R_f might be influenced by presence of possible attractive, indifferent, or repulsive, and other periodic or chaotic orbits underlying the extraneous fixed points. For ease of analysis, we rewrite the iterative map (41) in a more specific form:

$$x_{n+1} = R_f(x_n) = x_n - m \frac{f(x_n)}{f'(x_n)} H_f(x_n), \tag{42}$$

where $H_f(x_n) = 1 + sQ_f(s) + suK_f(s,u) + suvJ_f(s,u,v)$ can be regarded as a weight function in the classical modified Newton's method for a multiple root of integer multiplicity m. Notice that α is a fixed point of R_f, while $\xi \neq \alpha$ for which $H_f(\xi) = 0$ are extraneous fixed points of R_f.

The influence of extraneous fixed points on the convergence behavior was well demonstrated for simple zeros via König functions and Schröder functions [21] applied to a class of functions $\{f_k(x) = x^k - 1, k \geq 2\}$. The basins of attraction may be altered due to the trapped sequence $\{x_n\}$ by the attractive extraneous fixed points of R_f. An initial guess x_0 chosen near a desired root may converge to another unwanted remote root when repulsive or indifferent extraneous fixed points are present. These aspects of the Schröder functions were observed when applied to the same class of functions $\{f_k(x) = x^k - 1, k \geq 2\}$.

To simply treat dynamics underlying the extraneous fixed points of iterative map (42), we select a member $f(z) = (z^2 - 1)^m$. By a similar approach made by Chun et al. [35] and Neta et al. [33,36], we construct $H_f(x_n) = s \cdot Q_f(s) + s \cdot u \cdot K_f(s,u) + s \cdot u \cdot v \cdot J_f(s,u,v)$ in (42). Applying $f(z) = (z^2 - 1)^m$ to H_f, we find a rational function $H(z)$ with $t = z^2$:

$$H(z) = \frac{\mathcal{N}(t)}{\mathcal{D}(t)}, \tag{43}$$

where both $\mathcal{D}(t)$ and $\mathcal{N}(t)$ are co-prime polynomial functions of t. The underlying dynamics of the iterative map (42) can be favorably investigated on the Riemann sphere [37] with possible fixed points "0(zero)" and "∞". As can be seen in Section 5, the relevant dynamics will be illustrated in a 6×6 square region centered at the origin.

Indeed, the roots t of $\mathcal{N}(t)$ provide the extraneous fixed points ξ of R_f in Map (42) by the relation:

$$\xi = \begin{cases} t^{\frac{1}{2}}, & \text{if } t \neq 0, \\ 0(\text{double root}), & \text{if } t = 0. \end{cases} \tag{44}$$

Extraneous Fixed Points and their Stability

The following proposition describes the stability of the extraneous fixed points of (42).

Proposition 2. *Let $f(z) = (z^2 - 1)^m$. Then the extraneous fixed points ξ for Case 2 discussed earlier are all found to be repulsive.*

Proof. By direct computation of $R'_f(z)$ with $f(z) = (z^2 - 1)^m$, we write it as with $t = z^2$:

$$R'_f(z) = \frac{\Psi_n(t)}{\Psi_d(t)},$$

where $\Psi_n(t) = (-1 + t)^{15}$ and $\Psi_d(t) = 16t(1 + t)^2(1 + 6t + t^2)^2(1 + 28t + 70t^2 + 28t^3 + t^4)^2$. With the help of Mathematica, we are able to express $\Psi_n(t) = \frac{1}{61509375}(1 + 3t)(1 + 10t + 5t^2)(1 + 92t + 134t^2 + 28t^3 + t^4) \cdot Q_n(t) - 2097152 \cdot R_n(t)$ and $\Psi_d(t) = -\frac{1}{61509375}16(1 + 3t)(1 + 10t + 5t^2)(1 + 92t + 134t^2 + 28t^3 + t^4) \cdot Q_d(t) - 131072 \cdot R_d(t)$, with $Q_n(t)$ and $Q_d(t)$ as six- and eight-degree polynomials, while $R_n(t) = (327,923,929,643 + 34,417,198,067,010t + 446,061,306,116,505t^2 + 1621107643125740t^3 + 2,036,953,856,667,405t^4 + 892,731,761,917,554t^5 + 108,873,731,877,775t^6)$ and $R_d(t) = (327,923,929,643 + 34417198067010t + 446,061,306,116,505t^2 + 1621107643125740t^3 + 2,036,953,856,667,405t^4 + 892,731,761,917,554t^5 + 108,873,731,877,775t^6)$. Further, we express $R_n(t) = (1 + 10t + 5t^2)Q_v(t) + R_v(t)$ and $R_d(t) = (1 + 10t + 5t^2)Q_\delta(t) + R_\delta(t)$, with $R_v(t) = -\frac{10,077,696}{25}(36 + 341t) = R_\delta(t)$. Now let $t = \xi^2$, then

$$R'_f(\xi) = 16$$

using the fact that $(1 + 3t)(1 + 10t + 5t^2)(1 + 92t + 134t^2 + 28t^3 + t^4) = 0$. Hence ξ for Case 2 are all found to be repulsive. □

Remark 3. *Although not described here in detail due to limited space, by means of a similar proof as shown in Proposition 2, extraneous fixed points ξ for Case 1 was found to be indifferent in [19].*

If $f(z) = p(z)$ is a generic polynomial other than $(z^2 - 1)^m$, then theoretical analysis of the relevant dynamics may not be feasible as a result of the highly increased algebraic complexity. Nevertheless, we explore the dynamics of the iterative map (42) applied to $f(z) = p(z)$, which is denoted by R_p as follows:

$$z_{n+1} = R_p(z_n) = z_n - m\frac{p(z_n)}{p'(z_n)}H_p(z_n). \tag{45}$$

Basins of attraction for various polynomials are illustrated in Section 5 to observe the complicated dynamics behind the fixed points or the extraneous fixed points. The letter W was conveniently prefixed to each case number in Table 1 to symbolize a way of designating the numerical and dynamical aspects of iterative map (42).

4. Results and Discussion on Numerical and Dynamical Aspects

We first investigate numerical aspects of the local convergence of (1) with schemes W3G7 and W2A1–W2F4 for various test functions; then we explore the dynamical aspects underlying extraneous fixed points based on iterative map (45) applied to $f(z) = (z^2 - 1)^m$, whose attractor basins give useful information on the global convergence.

Results of numerical experiments are tabulated for all selected methods in Tables 2–4. Computational experiments on dynamical aspects have been illustrated through attractor basins in Figures 1–7. Both numerical and dynamical aspects have strongly confirmed the desired convergence.

Throughout the computational experiments with the aid of Mathematica, $MinPrecision = 400$ has been assigned to maintain 400 digits of minimum number of precision. If α is not exact, then it is given by an approximate value with 416 digits of precision higher than $MinPrecision$.

Limited paper space allows us to list x_n and α with up to 15 significant digits. We set error bound $\epsilon = \frac{1}{2} \times 10^{-360}$ to meet $|x_n - \alpha| < \epsilon$. Due to the high-order of convergence and root multiplicity, close initial guesses have been selected to achieve a moderate number of accurate digits of the asymptotic error constants.

Methods W3G7, W2A1, W2C2 and W2F2 successfully located desired zeros of test functions $F_1 - F_4$:

$$
\begin{cases}
\textbf{W3G7}: F_1(x) = [\cos\left(\frac{\pi x}{2}\right) + 2x^2 - 3\pi]^4, \ \alpha \approx 2.27312045629419, m = 4, \\
\textbf{W2A1}: F_2(x) = [\cos(x^2 + 1) - x\log(x^2 - \pi + 2) + 1]^4 \cdot (x^2 + 1 - \pi), \ \alpha = \sqrt{\pi - 1}, m = 5, \\
\textbf{W2C2}: F_3(x) = [\sin^{-1}(x^2 - 1) + 3e^x - 2x - 3]^2, \ \alpha \approx 0.477696831914490, m = 2, \\
\textbf{W2F2}: F_4(x) = (x^2 + 1)^4 + \log[1 + (x^2 + 1)^3], \ \alpha = i, m = 3, \\
\text{where } \log z (z \in \mathbb{C}) \text{ is a principal analytic branch with } -\pi < Im(\log z) \leq \pi.
\end{cases}
\tag{46}
$$

We find that Table 2 ensures sixteenth-order convergence. The computational asymptotic error constant $|e_n|/|e_{n-1}|^{16}$ is in agreement with the theoretical one $\eta = \lim_{n\to\infty} |e_n|/|e_{n-1}|^{16}$ up to 4 significant digits. The computational convergence order $p_n = \log|e_n/\eta|/\log|e_{n-1}|$ well approaches 16.

Additional test functions in Table 3 confirm the convergence of Scheme (1). The errors $|x_n - \alpha|$ are listed in Table 4 for comparison among the listed methods W3G7 and W2A1–W2F4. In the current experiments, W3G7 has slightly better convergence for f_5 and slightly poor convergence for all other test functions than the rest of the listed methods. No specific method performs better than the other among methods W2A1–W2F4 of Case 2.

According to the definition of the asymptotic error constant $\eta(c_i, Q_f, K_f, J_f) = \lim_{n\to\infty} |R_f(x_n) - \alpha|/|x_n - \alpha|^{16}$, the convergence is dependent on iterative map $R_f(x_n)$, $f(x)$, x_0, α and the weight functions Q_f, K_f and J_f. It is clear that no particular method always achieves better convergence than the others for any test functions.

Table 2. Convergence of methods W3G7, W2A1, W2C2, W2F2 for test functions $F_1(x) - F_4(x)$.

Method	F	n	x_n	$|F(x_n)|$	$|x_n - \alpha|$	$|e_n/e_{n-1}^{16}|$	η	p_n
W3G7	F_1	0	2.2735	1.873×10^{-10}	0.000379544			
		1	2.27312045629419	1.927×10^{-233}	6.798×10^{-60}	0.00003666355445	0.00003666729357	
		2	2.27312045629419	0.0×10^{-400}	1.004×10^{-237}			16.00000
W2A1	F_2	0	1.4634	1.93×10^{-21}	0.0000181404			
		1	1.46341814037882	3.487×10^{-366}	2.040×10^{-74}	148.4148965	148.4575003	
		2	1.46341814037882	0.0×10^{-400}	0.0×10^{-399}			16.00000
W2C2	F_3	0	0.4777	1.890×10^{-10}	3.168×10^{-6}			
		1	0.477696831914490	6.116×10^{-183}	1.802×10^{-92}	0.0001750002063	0.0001749999826	
		2	0.477696831914490	0.0×10^{-400}	8.522×10^{-367}			16.00000
W2F2	F_4	0	0.99995i	1.000×10^{-12}	0.00005			
		1	1.00000000000000 i	4.820×10^{-215}	1.391×10^{-72}	0.001037838436	0.001041219259	
		2	1.00000000000000 i	0.0×10^{-400}	0.0×10^{-400}			16.00030

$$i = \sqrt{-1}, \ \eta = \lim_{n\to\infty} \frac{|e_n|}{|e_{n-1}|^{16}}, \ p_n = \frac{\log|e_n/\eta|}{\log|e_{n-1}|}.$$

Table 3. Additional test functions $f_i(x)$ with zeros α and initial values x_0 and multiplicities.

i	$f_i(x)$	α	x_0	m
1	$[4 + 3\sin x - 2x^2]^4$	≈ 1.85471014256339	1.86	4
2	$[2x - Pi + x\cos x]^5$	$\frac{\pi}{2}$	1.5707	5
3	$[2x^3 + 3e^{-x} + 4\sin(x^2) - 5]^2$	$\approx -0.402282449584416$	-0.403	2
4	$[\sqrt{3}x^2 \cdot \cos\frac{\pi x}{6} + \frac{1}{x^3+1} - \frac{1}{28}] \cdot (x-3)^3$	3	3.0005	4
5	$(x-1)^2 + \frac{1}{12} - \log[\frac{25}{12} - 2x + x^2]$	$1 - i\frac{\sqrt{3}}{6}$	$0.99995 - 0.28i$	2
6	$[x\log x - \sqrt{x} + x^3]^3$	1	1.0001	3

Here, $\log z$ $(z \in \mathbb{C})$ represents a principal analytic branch with $-\pi \leq Im(\log z) < \pi$.

The proposed family of methods (1) has efficiency index *EI* [38], which is $16^{1/5} \approx 1.741101$ and larger than that of Newton's method. In general, the local convergence of iterative methods (45) is guaranteed with good initial values x_0 that are close to α. Selection of good initial values is a difficult task, depending on precision digits, error bound, and the given function $f(x)$.

Table 4. Comparison of $|x_n - \alpha|$ among selected methods applied to various test functions.

Method	$|x_n - \alpha|$	$f(x)$					
		f_1	f_2	f_3	f_4	f_5	f_6
W3G7	$|x_1 - \alpha|$	1.77×10^{-40} *	1.62×10^{-57}	4.89×10^{-51}	1.50×10^{-61}	5.76×10^{-7}	1.19×10^{-62}
	$|x_2 - \alpha|$	1.02×10^{-159}	1.13×10^{-225}	1.24×10^{-201}	3.27×10^{-245}	1.08×10^{-95}	2.40×10^{-247}
W2A1	$|x_1 - \alpha|$	2.83×10^{-42}	1.05×10^{-58}	1.92×10^{-52}	1.23×10^{-62}	1.29×10^{-6}	1.11×10^{-63}
	$|x_2 - \alpha|$	0.0×10^{-399}	4.24×10^{-230}	0.0×10^{-400}	0.0×10^{-399}	6.62×10^{-90}	3.61×10^{-251}
W2A2	$|x_1 - \alpha|$	1.63×10^{-41}	2.33×10^{-58}	1.45×10^{-51}	1.05×10^{-61}	1.32×10^{-6}	2.34×10^{-63}
	$|x_2 - \alpha|$	0.0×10^{-399}	1.11×10^{-228}	0.0×10^{-400}	0.0×10^{-399}	2.53×10^{-89}	7.39×10^{-250}
W2A3	$|x_1 - \alpha|$	2.53×10^{-43}	1.82×10^{-60}	4.56×10^{-54}	1.20×10^{-63}	4.43×10^{-6}	3.85×10^{-65}
	$|x_2 - \alpha|$	0.0×10^{-399}	1.40×10^{-236}	8.40×10^{-213}	0.0×10^{-399}	3.03×10^{-83}	1.53×10^{-256}
W2A4	$|x_1 - \alpha|$	1.24×10^{-42}	1.35×10^{-59}	1.53×10^{-52}	8.38×10^{-63}	1.58×10^{-4}	1.12×10^{-64}
	$|x_2 - \alpha|$	1.70×10^{-125}	5.96×10^{-424}	5.22×10^{-155}	8.16×10^{-187}	3.34×10^{-57}	2.79×10^{-424}
W2B1	$|x_1 - \alpha|$	2.39×10^{-42}	2.28×10^{-59}	1.30×10^{-52}	7.80×10^{-63}	2.14×10^{-6}	2.81×10^{-64}
	$|x_2 - \alpha|$	0.0×10^{-399}	1.57×10^{-232}	0.0×10^{-400}	0.0×10^{-399}	6.04×10^{-87}	2.23×10^{-253}
W2B2	$|x_1 - \alpha|$	4.44×10^{-42}	2.73×10^{-59}	3.03×10^{-52}	1.79×10^{-62}	4.30×10^{-6}	3.29×10^{-64}
	$|x_2 - \alpha|$	0.0×10^{-399}	6.69×10^{-232}	0.0×10^{-400}	0.0×10^{-399}	6.01×10^{-82}	$7.78 \times 10^{e-253}$
W2B3	$|x_1 - \alpha|$	9.85×10^{-43}	3.11×10^{-61}	4.26×10^{-53}	3.01×10^{-63}	4.46×10^{-6}	3.26×10^{-65}
	$|x_2 - \alpha|$	0.0×10^{-399}	1.17×10^{-239}	0.0×10^{-400}	0.0×10^{-399}	3.06×10^{-83}	7.91×10^{-257}
W2B4	$|x_1 - \alpha|$	1.04×10^{-42}	1.92×10^{-59}	1.77×10^{-52}	1.12×10^{-62}	1.77×10^{-4}	1.53×10^{-64}
	$|x_2 - \alpha|$	1.12×10^{-125}	4.68×10^{-405}	9.06×10^{-155}	2.23×10^{-186}	2.32×10^{-56}	0.0×10^{-400}
W2C1	$|x_1 - \alpha|$	4.87×10^{-42}	4.27×10^{-59}	2.95×10^{-52}	1.50×10^{-62}	1.08×10^{-6}	5.14×10^{-64}
	$|x_2 - \alpha|$	0.0×10^{-399}	0.0×10^{-399}	0.0×10^{-400}	0.0×10^{-399}	2.41×10^{-91}	2.22×10^{-191}
W2C2	$|x_1 - \alpha|$	9.31×10^{-42}	1.01×10^{-58}	5.94×10^{-52}	2.61×10^{-62}	1.47×10^{-6}	1.18×10^{-63}
	$|x_2 - \alpha|$	0.0×10^{-399}	4.23×10^{-230}	0.0×10^{-400}	0.0×10^{-399}	7.11×10^{-89}	4.95×10^{-251}
W2C3	$|x_1 - \alpha|$	9.17×10^{-42}	8.88×10^{-59}	6.85×10^{-52}	4.30×10^{-62}	4.36×10^{-6}	9.66×10^{-64}
	$|x_2 - \alpha|$	0.0×10^{-399}	7.84×10^{-230}	0.0×10^{-400}	0.0×10^{-399}	2.11×10^{-81}	6.05×10^{-251}
W2C4	$|x_1 - \alpha|$	5.36×10^{-42}	6.72×10^{-59}	6.55×10^{-52}	4.38×10^{-62}	1.02×10^{-4}	6.13×10^{-64}
	$|x_2 - \alpha|$	9.92×10^{-124}	5.96×10^{-424}	2.97×10^{-153}	8.55×10^{-185}	1.30×10^{-59}	0.0×10^{-400}
W2F1	$|x_1 - \alpha|$	4.36×10^{-42}	8.67×10^{-60}	2.55×10^{-52}	1.29×10^{-62}	4.34×10^{-6}	1.60×10^{-64}
	$|x_2 - \alpha|$	0.0×10^{-399}	7.12×10^{-234}	0.0×10^{-400}	0.0×10^{-399}	4.57×10^{-82}	4.61×10^{-254}
W2F2	$|x_1 - \alpha|$	1.20×10^{-42}	1.74×10^{-60}	5.52×10^{-53}	3.67×10^{-63}	4.08×10^{-6}	5.06×10^{-65}
	$|x_2 - \alpha|$	0.0×10^{-399}	1.16×10^{-236}	0.0×10^{-400}	0.0×10^{-399}	2.33×10^{-83}	4.56×10^{-256}
W2F3	$|x_1 - \alpha|$	1.08×10^{-41}	1.54×10^{-58}	1.55×10^{-51}	1.14×10^{-61}	8.67×10^{-5}	1.396×10^{-63}
	$|x_2 - \alpha|$	1.41×10^{-424}	5.27×10^{-172}	8.47×10^{-424}	0.0×10^{-399}	1.66×10^{-60}	5.22×10^{-188}
W2F4	$|x_1 - \alpha|$	3.80×10^{-42}	5.01×10^{-60}	2.15×10^{-52}	1.07×10^{-62}	4.35×10^{-6}	1.18×10^{-64}
	$|x_2 - \alpha|$	0.0×10^{-399}	1.16×10^{-235}	0.0×10^{-400}	0.0×10^{-399}	3.65×10^{-82}	1.38×10^{-254}

The global convergence with appropriate initial values x_0 is effectively described by means of a basin of attraction that is the set of initial values leading to long-time behavior approaching the attractors under the iterative action of R_f. Basins of attraction contain information about the region of convergence. A method occupying a larger region of convergence is likely to be a more robust method. A quantitative analysis will play the important role for measuring the region of convergence.

The basins of attraction, as well as the relevant statistical data, are constructed in a similar manner shown in the work of Geum–Kim–Neta [19]. Because of the high order, we take a smaller square $[-1.5, 1.5]^2$ and use 601×601 initial points uniformly distributed in the domain. Maple software has been used to perform the desired dynamics with convergence stopping criteria satisfying $|x_{n+1} - x_n| < 10^{-6}$ within the maximum number of 40 iterations. An initial point is painted with a color whose intensity measures the number of iterations converging to a root. The brighter color implies the faster convergence. The black point means that its orbit did not converge within 40 iterations.

Despite the limited space, we will explore the dynamics of all listed maps W3G7 and W2A1–W2F4, with applications to $p_k(z)$, $(1 \le k \le 7)$ through the following seven examples. In each example, we have shown dynamical planes for the convergence behavior of iterative map $x_{n+1} = R_f(x_n)$ (42)

with $f(z) = p_k(z)$ by illustrating the relevant basins of attraction through Figures 1–7 and displaying relevant statistical data in Tables 5–7 with colored fonts indicating best results.

Example 1. *As a first example, we have taken a quadratic polynomial raised to the power of two with all real roots:*

$$p_1(z) = (z^2 - 1)^2. \tag{47}$$

Clearly the roots are ± 1. Basins of attraction for W3G7, W2A1–W2F4 are given in Figure 1. Consulting Tables 5–7, we find that the methods **W2B2** and **W2F4** use the least number (2.71) of iterations per point on average (ANIP) followed by W2F1 with 2.72 ANIP, W2C3 with 2.73 and W2B1 with 2.74. The fastest method is W2A2 with 969.374 s followed closely by W2A3 with 990.341 s. The slowest is W2A4 with 4446.528 s. Method W2C4 has the lowest number of black points (601) and W2A4 has the highest number (78843). We will not include **W2A4** in the coming examples.

Table 5. Average number of iterations per point for each example (1–7).

Map	Example							
	1: m = 2	2: m = 3	3: m = 3	4: m = 4	5: m = 3	6: m = 3	7: m = 3	Average
W3G7	2.94	3.48	3.83	3.93	3.95	3.97	6.77	4.12
W2A1	2.84	3.50	3.70	4.04	6.84	3.74	5.49	4.31
W2A2	2.76	3.15	3.52	3.84	3.62	3.66	4.84	3.63
W2A3	2.78	3.21	3.61	3.89	3.70	3.74	4.98	3.70
W2A4	11.40	-	-	-	-	-	-	-
W2B1	2.74	3.25	3.70	3.88	3.67	3.72	5.01	3.71
W2B2	2.95	3.42	3.66	4.01	3.75	3.77	5.15	3.82
W2B3	2.78	3.28	3.64	3.89	3.69	4.65	5.13	3.86
W2B4	3.29	3.91	4.99	-	-	-	-	-
W2C1	2.88	3.66	3.87	4.08	3.89	5.45	6.25	4.30
W2C2	2.93	3.68	3.95	4.15	6.70	4.67	5.75	4.55
W2C3	2.73	3.22	3.53	3.98	3.60	3.62	4.94	3.66
W2C4	3.14	3.81	4.96	-	-	-	-	-
W2F1	2.72	3.24	3.55	3.84	3.49	3.57	5.41	3.69
W2F2	2.81	3.28	3.80	4.06	5.02	4.50	5.29	4.10
W2F3	2.91	3.54	4.36	4.41	-	-	-	-
W2F4	2.71	3.19	3.50	3.86	3.42	3.53	5.52	3.68

Example 2. *As a second example, we have taken the same quadratic polynomial now raised to the power of three:*

$$p_2(z) = (z^2 - 1)^3. \tag{48}$$

The basins for the best methods are plotted in Figure 2. This is an example to demonstrate the effect of raising the multiplicity from two to three. In one case, namely W3G7, we also have $m = 5$ with CPU time of 4128.379 s. Based on the figure we see that W2B4, W2C4 and W2F3 were chaotic. The worst are W2B4, W2C4 and W2F3. In terms of ANIP, the best was W2A2 (3.15) followed by W2F4 (3.19) and the worst was W2B4 (3.91). The fastest was W2B3 using (2397.111 s) followed by W2F1 using 2407.158 s and the slowest was W2C4 (4690.295 s) preceded by W3G7 (2983.035 s). Four methods have the highest number of black points (617). Those were W2A1, W2B4, W2C1 and W2F2. The lowest number was 601 for W2A2, W2C2, W2C4 and W2F1.

Comparing the CPU time for the cases $m = 2$ and $m = 3$ of W3G7, we find it is about doubled. But when increasing from three to five, we only needed about 50% more.

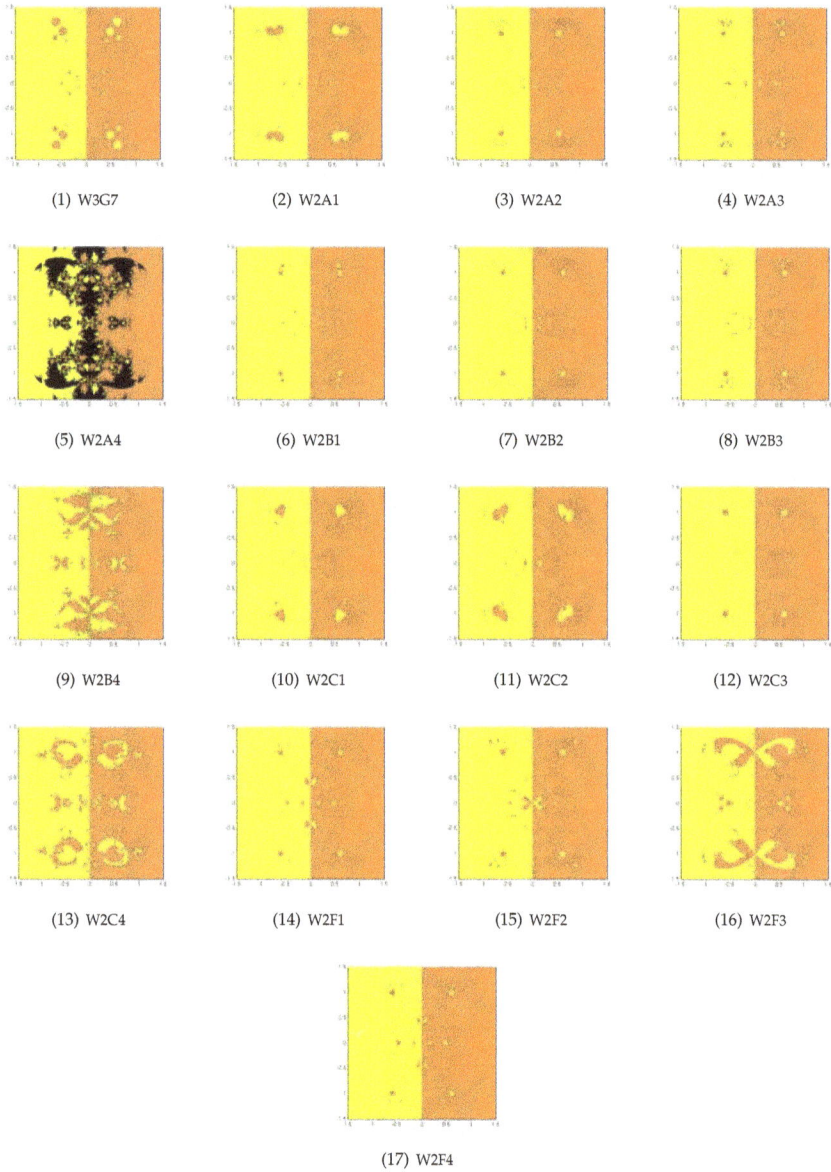

Figure 1. The top row for W3G7 (**left**), W2A1 (**center left**), W2A2 (**center right**) and W2A3 (**right**). The second row for W2A4 (**left**), W2B1 (**center left**), W2B2 (**center right**) and W2B3 (**right**). The third row for W2B4 (**left**), W2C1 (**center left**), W2C2 (**center right**) and W2C3 (**right**). The third row for W2C4 (**left**), W2F1 (**center left**), W2F2 (**center right**), and W2F3 (**right**). The bottom row for W2F4 (**center**), for the roots of the polynomial equation $(z^2 - 1)^2$.

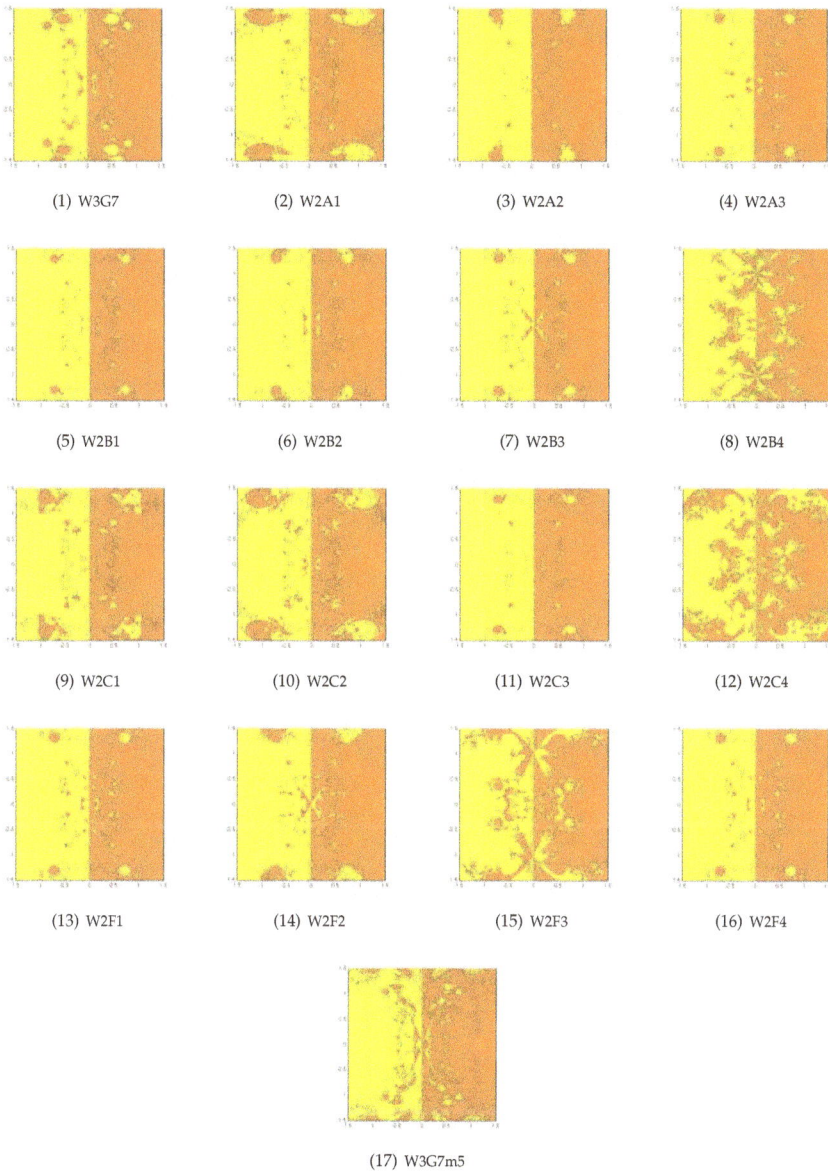

(1) W3G7

(2) W2A1

(3) W2A2

(4) W2A3

(5) W2B1

(6) W2B2

(7) W2B3

(8) W2B4

(9) W2C1

(10) W2C2

(11) W2C3

(12) W2C4

(13) W2F1

(14) W2F2

(15) W2F3

(16) W2F4

(17) W3G7m5

Figure 2. The top row for W3G7 (**left**), W2A1 (**center left**), W2A2 (**center right**) and W2A3 (**right**). The second row for W2B1 (**left**), W2B2 (**center left**), W2B3 (**center right**) and W2B4 (**right**). The third row for W2C1 (**left**), W2C2 (**center left**), W2C3 (**center right**) and W2C4 (**right**). The fourth row for W2F1 (**left**), W2F2 (**center left**), W2F3 (**center right**), and W2F4 (**right**). The bottom row for W3G7m5 (**center**), for the roots of the polynomial equation $(z^2 - 1)^k$, $k \in \{3, 5\}$.

Example 3. *In our third example, we have taken a cubic polynomial raised to the power of three:*

$$p_3(z) = (3z^3 + 4z^2 - 10)^3. \tag{49}$$

Basins of attraction are given in Figure 3. It is clear that W2B4, W2C4 and W2F3 were too chaotic and they should be eliminated from further consideration. In terms of ANIP, the best was W2F4 (3.50) followed by W2A2 (3.52), W2C3 (3.53) and W2F1 (3.55) and the worst were W2B4 and W2C4 with 4.99 and 4.96 ANIP, respectively. The fastest was W2C3 using 2768.362 s and the slowest was W2B3 (7193.034 s). There were 13 methods with only one black point and one with two points. The highest number of black points was 101 for W2F2.

Table 6. CPU time (in seconds) required for each example(1–7) using a Dell Multiplex-990.

Map	Example							
	1: m = 2	2: m = 3	3: m = 3	4: m = 4	5: m = 3	6: m = 3	7: m = 3	Average
W3G7	1254.077	2983.035	3677.848	3720.670	3944.937	3901.679	4087.102	3367.050
W2A1	1079.667	2694.537	3528.149	3119.911	5896.635	2938.747	3526.840	3254.927
W2A2	969.374	2471.727	3287.081	2956.702	3218.223	2891.478	2981.179	2682.252
W2A3	990.341	2843.789	2859.093	2999.712	3002.146	3074.811	3155.307	2703.600
W2A4	4446.528	-	-	-	-	-	-	-
W2B1	1084.752	2634.826	3295.162	3051.941	2835.755	3238.363	3272.667	2773.352
W2B2	1075.393	2429.996	3130.223	3051.192	2929.106	3581.456	3155.619	2764.712
W2B3	1180.366	2397.111	7193.034	3000.383	2970.711	3739.766	3139.084	3374.351
W2B4	1274.653	2932.008	4872.972	-	-	-	-	-
W2C1	1132.069	2685.355	3242.637	3287.066	3147.663	4080.019	4802.662	3196.782
W2C2	1112.162	2881.697	3189.706	3873.037	5211.619	3665.773	3950.896	3412.127
W2C3	1052.570	2421.026	2768.362	3014.033	2778.518	2914.941	3953.346	2700.399
W2C4	2051.710	4690.295	7193.034	-	-	-	-	-
W2F1	1071.540	2407.158	2909.965	3472.317	2832.230	3490.896	3246.584	2775.813
W2F2	1015.051	2438.483	3031.802	3061.270	3703.152	3737.394	3324.537	2901.670
W2F3	1272.188	2596.200	3603.655	4130.158	-	-	-	-
W2F4	1216.839	2620.052	3589.177	3233.168	3534.312	3521.660	3934.845	3092.865

Table 7. Number of points requiring 40 iterations for each example (1–7).

Map	Example							
	1: m = 2	2: m = 3	3: m = 3	4: m = 4	5: m = 3	6: m = 3	7: m = 3	Average
W3G7	677	605	1	250	40	1265	33,072	5130.000
W2A1	657	617	1	166	34,396	1201	18,939	7996.714
W2A2	697	601	1	162	1	1201	15,385	2578.286
W2A3	675	605	55	152	9	1221	14,711	2489.714
W2A4	78,843	-	-	-	-	-	-	-
W2B1	679	613	1	204	9	1201	13,946	2379.000
W2B2	635	609	1	116	1	1217	15,995	2653.429
W2B3	679	613	1	146	3	10,645	16,342	4061.286
W2B4	659	617	2	-	-	-	-	-
W2C1	689	617	1	400	20	18,157	24,239	6303.286
W2C2	669	601	1	174	17,843	1265	18,382	5562.143
W2C3	659	609	1	184	1	1201	14,863	2502.571
W2C4	601	601	1	-	-	-	-	-
W2F1	681	601	1	126	10	1225	18,772	3059.429
W2F2	679	617	101	614	3515	1593	17,469	3512.571
W2F3	663	605	1	78	-	-	-	-
W2F4	645	605	1	130	12	1217	20,020	3232.857

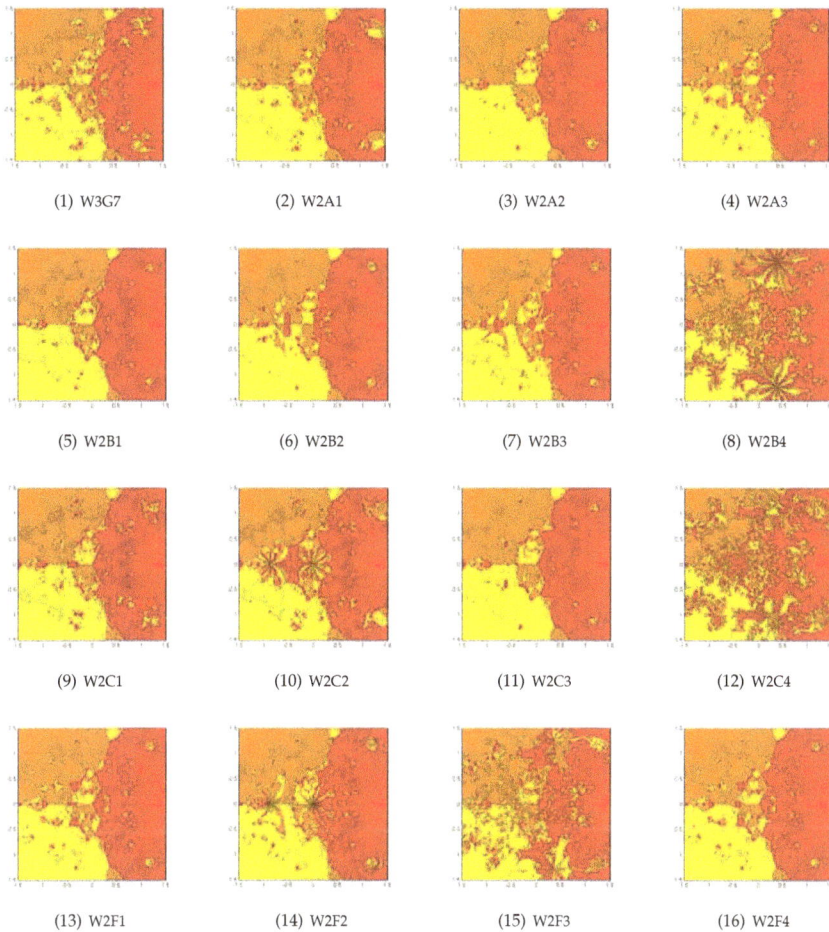

Figure 3. The top row for W3G7 (**left**), W2A1 (**center left**), W2A2 (**center right**) and W2A3 (**right**). The second row for W2B1 (**left**), W2B2 (**center left**), W2B3 (**center right**) and W2B4 (**right**). The third row for W2C1 (**left**), W2C2 (**center left**), W2C3 (**center right**) and W2C4 (**right**). The bottom row for W2F1 (**left**), W2F2 (**center left**), W2F3 (**center right**), and W2F4 (**right**), for the roots of the polynomial equation $(3z^3 + 4z^2 - 10)^3$.

Example 4. *As a fourth example, we have taken a different cubic polynomial raised to the power of four:*

$$p_4(z) = (z^3 - z)^4. \tag{50}$$

The basins are given in Figure 4. We now see that W2F3 is the worst. In terms of ANIP, W2A2 and W2F1 were the best (3.84 each) and the worst was W2F3 (4.41). The fastest was W2A2 (2956.702 s) and the slowest was W2F3 (4130.158 s). The lowest number of black points (78) was for method W2F3 and the highest number (614) for W2F2. We did not include W2F3 in the rest of the experiments.

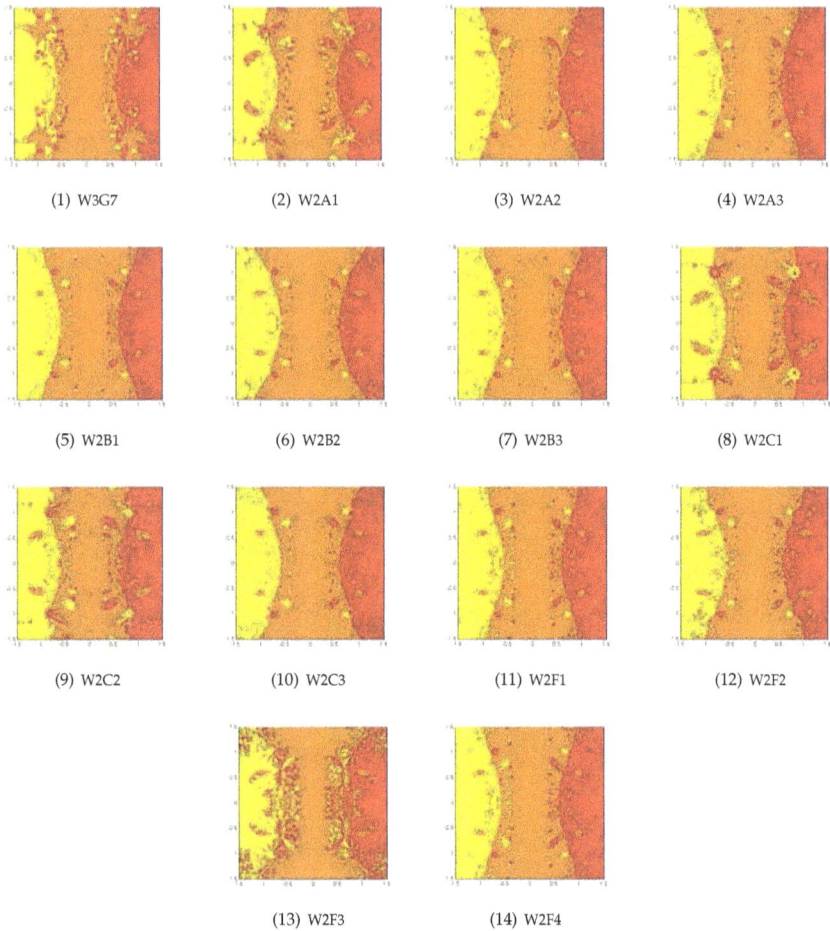

(1) W3G7   (2) W2A1   (3) W2A2   (4) W2A3

(5) W2B1   (6) W2B2   (7) W2B3   (8) W2C1

(9) W2C2   (10) W2C3   (11) W2F1   (12) W2F2

(13) W2F3   (14) W2F4

Figure 4. The top row for W3G7 (**left**), W2A1 (**center left**), W2A2 (**center right**) and W2A3 (**right**). The second row for W2B1 (**left**), W2B2 (**center left**), W2B3 (**center right**) and W2C1 (**right**). The third row for W2C2 (**left**), W2C3 (**center left**), W2F1 (**center right**) and W2F2 (**right**). The bottom row for W2F3 (**left**) and W2F4 (**right**), for the roots of the polynomial equation $(z^3 - z)^4$.

Example 5. *As a fifth example, we have taken a quintic polynomial raised to the power of three:*

$$p_3(z) = (z^5 - 1)^3. \tag{51}$$

The basins for the best methods left are plotted in Figure 5. The worst were W2A1 and W2C2. In terms of ANIP, the best was W2F4 (3.42) followed by W2F1 (3.49) and the worst were W2A1 (6.84) and W2C2 (6.70). The fastest was W2C3 using 2778.518 s followed by W2F1 using 2832.23 s and W2B1 using 2835.755 s. The slowest was W2A1 using 5896.635 s. There were three methods with one black point (W2A2, W2B2 and W2C3) and four others with 10 or less such points, namely W2B3 (3), W2A3 and W2B1 (9) and W2F1 (10). The highest number was for W2A1 (34,396) preceded by W2C2 with 17,843 black points.

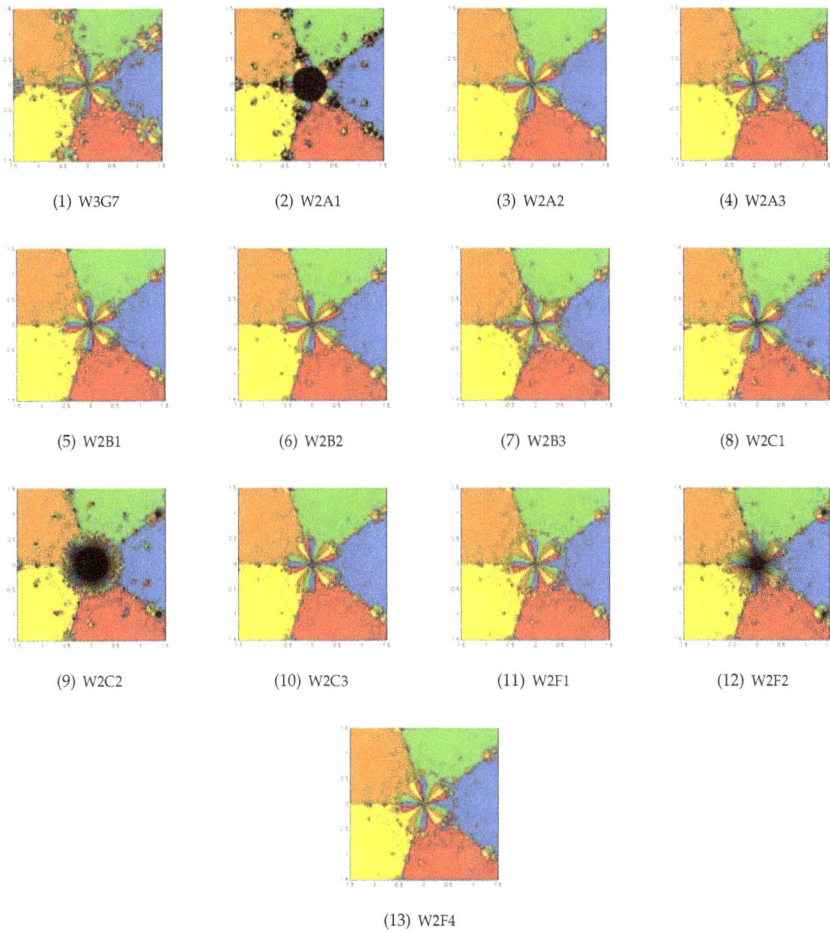

(1) W3G7   (2) W2A1   (3) W2A2   (4) W2A3

(5) W2B1   (6) W2B2   (7) W2B3   (8) W2C1

(9) W2C2   (10) W2C3   (11) W2F1   (12) W2F2

(13) W2F4

Figure 5. The top row for W3G7 (**left**), W2A1 (**center left**), W2A2 (**center right**) and W2A3 (**right**). The second row for W2B1 (**left**), W2B2 (**center left**), W2B3 (**center right**) and W2C1 (**right**). The third row for W2C2 (**left**), W2C3 (**center left**), W2F1 (**center right**) and W2F2 (**right**). The bottom row for W2F4 (**center**), for the roots of the polynomial equation $(z^5 - 1)^3$.

Example 6. *As a sixth example, we have taken a quartic polynomial raised to the power of three:*

$$p_6(z) = (z^4 - 1)^3. \tag{52}$$

The basins for the best methods left are plotted in Figure 6. It seems that most of the methods left were good except W2B3 and W2C1. Based on Table 5 we find that W2F4 has the lowest ANIP (3.53) followed by W2F1 (3.57). The fastest method was W2A2 (2891.478 s) followed by W2C3 (2914.941 s). The slowest was W2C1 (4080.019 s) preceded by W3G7 using 3901.679 s. The lowest number of black points was for W2A1, W2A2, W2B1 and W2C3 (1201) and the highest number was for W2C1 with 18,157 black points.

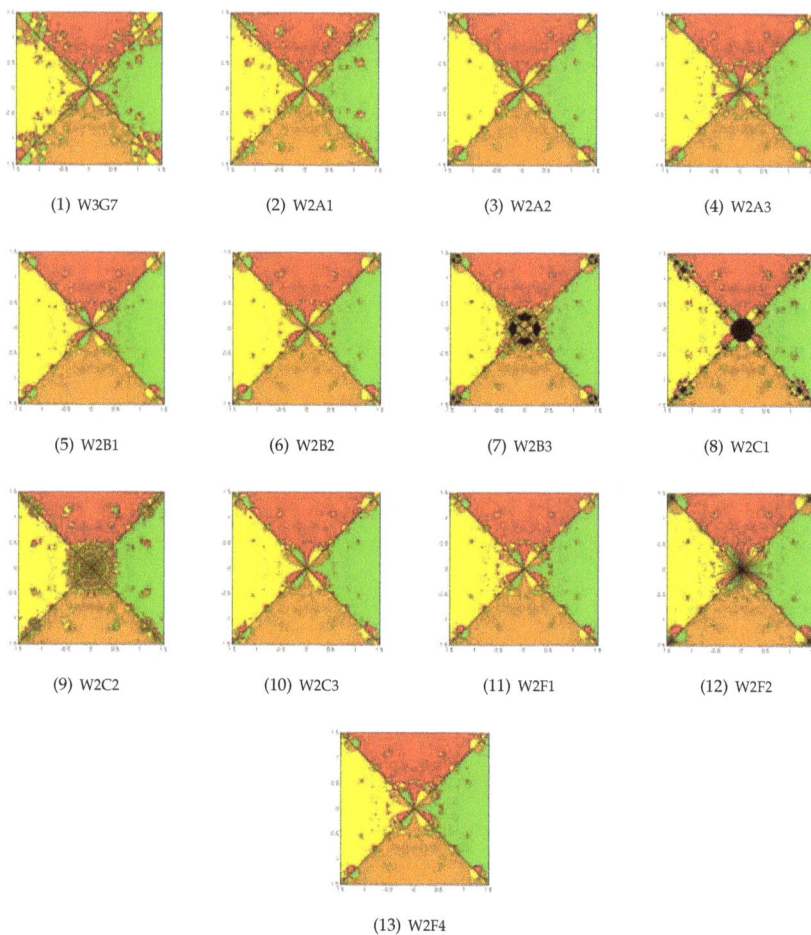

(1) W3G7   (2) W2A1   (3) W2A2   (4) W2A3

(5) W2B1   (6) W2B2   (7) W2B3   (8) W2C1

(9) W2C2   (10) W2C3   (11) W2F1   (12) W2F2

(13) W2F4

Figure 6. The top row for W3G7 (**left**), W2A1 (**center left**), W2A2 (**center right**) and W2A3 (**right**). The second row for W2B1 (**left**), W2B2 (**center left**), W2B3 (**center right**) and W2C1 (**right**). The third row for W2C2 (**left**), W2C3 (**center left**), W2F1 (**center right**) and W2F2 (**right**). The bottom row for W2F4 (**center**), for the roots of the polynomial equation $(z^4 - 1)^3$.

Example 7. *As a seventh example, we have taken a non-polynomial equation having $\pm i$ as its triple roots:*

$$p_6(z) = (z + i)^3 (e^{z-i} - 1)^3, \text{ with } i = \sqrt{-1}. \tag{53}$$

The basins for the best methods left are plotted in Figure 7. It seems that most of the methods left have a larger basin for the root $-i$, i.e., the boundary does not match the real line exactly. Based on Table 5 we find that W2A2 has the lowest ANIP (4.84) followed by W2C3 (4.94) and W2A3 (4.98). The fastest method was W2A2 (2981.179 seconds) followed by W2B3 (3139.084 s), W2A3 (3155.307 s) and W2B2 (3155.619 s). The slowest was W2C1 (4802.662 s). The lowest number of black points was for W2B1 (13,946) and the highest number was for W3G7 with 33,072 black points. In general all methods had higher number of black points compared to the polynomial examples.

We now average all these results across the seven examples to try and pick the best method. W2A2 had the lowest ANIP (3.63), followed by W2C3 with 3.66, W2F4 with 3.68 and W2F1 with 3.69. The fastest method was W2A2 (2682.252 seconds), followed by W2C3 (2700.399 s) and W2A3 using 2703.600 s of CPU. W2B1 has the lowest number of black points on average (2379), followed by W2A3 (2490 black points). The highest number of black points was for W2A1.

Based on these seven examples we see that **W2F4** has four examples with the lowest ANIP, W2A2 had three examples and W2F1 has one example. On average, though, W2A2 had the lowest ANIP. W2A2 was the fastest in four examples and on average. W2C3 was the fastest in two examples and W2B3 in one example. In terms of black points, W2A2, W2B1 and W2B3 had the lowest number in three examples and W2F1 in two examples. On average W2B1 has the lowest number. Thus, we recommend W2A2, since it is in the top in all categories.

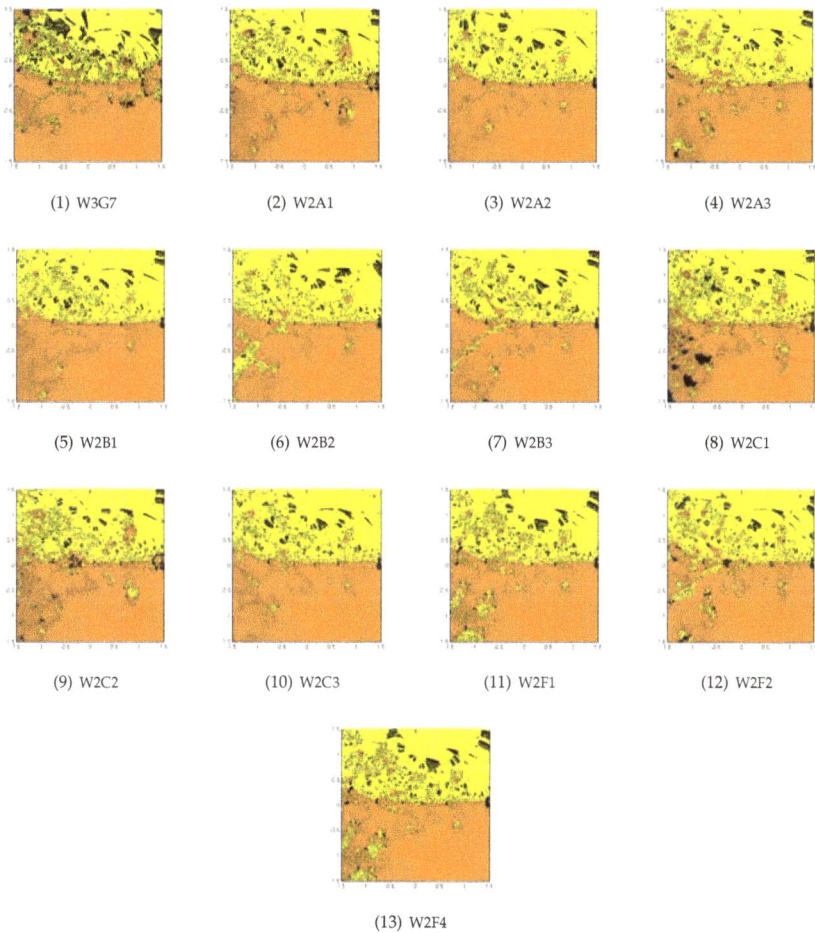

(1) W3G7 (2) W2A1 (3) W2A2 (4) W2A3

(5) W2B1 (6) W2B2 (7) W2B3 (8) W2C1

(9) W2C2 (10) W2C3 (11) W2F1 (12) W2F2

(13) W2F4

Figure 7. The top row for W3G7 (**left**), W2A1 (**center left**), W2A2 (**center right**) and W2A3 (**right**). The second row for W2B1 (**left**), W2B2 (**center left**), W2B3 (**center right**) and W2C1 (**right**). The third row for W2C2 (**left**), W2C3 (**center left**), W2F1 (**center right**) and W2F2 (**right**). The bottom row for W2F4 (**center**), for the roots of the non-polynomial equation $(z+i)^3(e^{z-i}-1)^3$.

5. Conclusions

Both numerical and dynamical aspects of iterative map (1) support the main theorem well through a number of test equations and examples. The W2C2 and W2B3 methods were observed to occupy relatively slower CPU time. Such dynamical aspects would be greatly strengthened if we could include a study of parameter planes with reference to appropriate parameters in Table 1.

The proposed family of methods (1) employing generic weight functions favorably cover most of optimal sixteenth-order multiple-root finders with a number of feasible weight functions. The dynamics behind the purely imaginary extraneous fixed points will choose best members of the family with improved convergence behavior. However, due to the high order of convergence, the algebraic difficulty might arise resolving its increased complexity. The current work is limited to univariate nonlinear equations; its extension to multivariate ones becomes another task.

Author Contributions: investigation, M.–Y.L.; formal analysis, Y.I.K.; supervision, B.N.

Conflicts of Interest: The authors have no conflict of interest to declare.

References

1. Bi, W.; Wu, Q.; Ren, H. A new family of eighth-order iterative methods for solving nonlinear equations. *Appl. Math. Comput.* **2009**, *214*, 236–245. [CrossRef]
2. Cordero, A.; Torregrosa, J.R.; Vassileva, M.P. Three-step iterative methods with optimal eighth-order convergence. *J. Comput. Appl. Math.* **2011**, *235*, 3189–3194. [CrossRef]
3. Geum, Y.H.; Kim, Y.I. A uniparametric family of three-step eighth-order multipoint iterative methods for simple roots. *Appl. Math. Lett.* **2011**, *24*, 929–935. [CrossRef]
4. Lee, S.D.; Kim, Y.I.; Neta, B. An optimal family of eighth-order simple-root finders with weight functions dependent on function-to-function ratios and their dynamics underlying extraneous fixed points. *J. Comput. Appl. Math.* **2017**, *317*, 31–54. [CrossRef]
5. Liu, L.; Wang, X. Eighth-order methods with high efficiency index for solving nonlinear equations. *Appl. Math. Comput.* **2010**, *215*, 3449–3454. [CrossRef]
6. Petković, M.S.; Neta, B.; Petković, L.D.; Džunić, J. *Multipoint Methods for Solving Nonlinear Equations*; Elsevier: New York, NY, USA, 2012.
7. Petković, M.S.; Neta, B.; Petković, L.D.; Džunić, J.; Multipoint methods for solving nonlinear equations: A survey. *Appl. Math. Comput.* **2014**, *226*, 635–660. [CrossRef]
8. Sharma, J.R.; Arora, H. A new family of optimal eighth order methods with dynamics for nonlinear equations. *Appl. Math. Comput.* **2016**, *273*, 924–933. [CrossRef]
9. Rhee, M.S.; Kim, Y.I.; Neta, B. An optimal eighth-order class of three-step weighted Newton's methods and their dynamics behind the purely imaginary extraneous fixed points. *Int. J. Comput. Math.* **2017**, *95*, 2174–2211. [CrossRef]
10. Kung, H.T.; Traub, J.F. Optimal order of one-point and multipoint iteration. *J. Assoc. Comput. Mach.* **1974**, *21*, 643–651. [CrossRef]
11. Maroju, P.; Behl, R.; Motsa, S.S. Some novel and optimal families of King's method with eighth and sixteenth-order of convergence. *J. Comput. Appl. Math.* **2017**, *318*, 136–148. [CrossRef]
12. Sharma, J.R.; Argyros, I.K.; Kumar, D. On a general class of optimal order multipoint methods for solving nonlinear equations. *J. Math. Anal. Appl.* **2017**, *449*, 994–1014. [CrossRef]
13. Neta, B. On a family of Multipoint Methods for Non-linear Equations. *Int. J. Comput. Math.* **1981**, *9*, 353–361. [CrossRef]
14. Geum, Y.H.; Kim, Y.I.; Neta, B. Constructing a family of optimal eighth-order modified Newton-type multiple-zero finders along with the dynamics behind their purely imaginary extraneous fixed points. *J. Comput. Appl. Math.* **2018**, *333*, 131–156. [CrossRef]
15. Ahlfors, L.V. *Complex Analysis*; McGraw-Hill Book, Inc.: New York, NY, USA, 1979.
16. Hörmander, L. *An Introduction to Complex Analysis in Several Variables*; North-Holland Publishing Company: Amsterdam, The Netherlands, 1973.

17. Shabat, B.V. *Introduction to Complex Analysis PART II, Functions of several Variables*; American Mathematical Society: Providence, RI, USA, 1992.

18. Wolfram, S. *The Mathematica Book*, 5th ed.; Wolfram Media: Champaign, IL, USA, 2003.

19. Geum, Y.H.; Kim, Y.I.; Neta, B. Developing an Optimal Class of Generic Sixteenth-Order Simple-Root Finders and Investigating Their Dynamics. *Mathematics* **2019**, *7*, 8. [CrossRef]

20. Stewart, B.D. Attractor Basins of Various Root-Finding Methods. Master's Thesis, Naval Postgraduate School, Department of Applied Mathematics, Monterey, CA, USA, June 2001.

21. Vrscay, E.R.; Gilbert, W.J. Extraneous Fixed Points, Basin Boundaries and Chaotic Dynamics for shröder and König rational iteration Functions. *Numer. Math.* **1988**, *52*, 1–16. [CrossRef]

22. Amat, S.; Busquier, S.; Plaza, S. Review of some iterative root-finding methods from a dynamical point of view. *Scientia* **2004**, *10*, 3–35.

23. Argyros, I.K.; Magreñán, A.Á. On the convergence of an optimal fourth-order family of methods and its dynamics. *Appl. Math. Comput.* **2015**, *252*, 336–346. [CrossRef]

24. Chun, C.; Lee, M.Y.; Neta, B.; Džunić, J. On optimal fourth-order iterative methods free from second derivative and their dynamics. *Appl. Math. Comput.* **2012**, *218*, 6427–6438. [CrossRef]

25. Chicharro, F.; Cordero, A.; Gutiérrez, J.M.; Torregrosa, J.R. Complex dynamics of derivative-free methods for nonlinear equations. *Appl. Math. Comput.* **2013**, *219*, 7023–7035. [CrossRef]

26. Chun, C.; Neta, B. Comparison of several families of optimal eighth order methods. *Appl. Math. Comput.* **2016**, *274*, 762–773. [CrossRef]

27. Cordero, A.; García-Maimó, J.; Torregrosa, J.R.; Vassileva, M.P.; Vindel, P. Chaos in King's iterative family. *Appl. Math. Lett.* **2013**, *26*, 842–848. [CrossRef]

28. Geum, Y.H.; Kim, Y.I.; Magreñán, Á.A. A biparametric extension of King's fourth-order methods and their dynamics. *Appl. Math. Comput.* **2016**, *282*, 254–275. [CrossRef]

29. Geum, Y.H.; Kim, Y.I.; Neta, B. A class of two-point sixth-order multiple-zero finders of modified double-Newton type and their dynamics. *Appl. Math. Comput.* **2015**, *270*, 387–400. [CrossRef]

30. Geum, Y.H.; Kim, Y.I.; Neta, B. A sixth-order family of three-point modified Newton-like multiple-root finders and the dynamics behind their extraneous fixed points. *Appl. Math. Comput.* **2016**, *283*, 120–140. [CrossRef]

31. Magreñan, Á.A. Different anomalies in a Jarratt family of iterative root-finding methods. *Appl. Math. Comput.* **2014**, *233*, 29–38.

32. Neta, B.; Scott, M.; Chun, C. Basin attractors for various methods for multiple roots. *Appl. Math. Comput.* **2012**, *218*, 5043–5066. [CrossRef]

33. Neta, B.; Chun, C.; Scott, M. Basins of attraction for optimal eighth order methods to find simple roots of nonlinear equations. *Appl. Math. Comput.* **2014**, *227*, 567–592. [CrossRef]

34. Scott, M.; Neta, B.; Chun, C. Basin attractors for various methods. *Appl. Math. Comput.* **2011**, *218*, 2584–2599. [CrossRef]

35. Chun, C.; Neta, B.; Basins of attraction for Zhou-Chen-Song fourth order family of methods for multiple roots. *Math. Comput. Simul.* **2015**, *109*, 74–91. [CrossRef]

36. Neta, B.; Chun, C. Basins of attraction for several optimal fourth order methods for multiple roots. *Math. Comput. Simul.* **2014**, *103*, 39–59. [CrossRef]

37. Beardon, A.F. *Iteration of Rational Functions*; Springer: New York, NY, USA, 1991.

38. Traub, J.F. *Iterative Methods for the Solution of Equations*; Chelsea Publishing Company: Chelsea, VT, USA, 1982.

mathematics

MDPI

Article

An Efficient Family of Optimal Eighth-Order Multiple Root Finders

Fiza Zafar [1,2,*] , Alicia Cordero [2] and Juan R. Torregrosa [2]

1   Centre for Advanced Studies in Pure and Applied Mathematics, Bahauddin Zakariya University, Multan 60800, Pakistan

2   Instituto de Matemáticas Multidisciplinar, Universitat Politènica de València, 46022 València, Spain; acordero@mat.upv.es (A.C.); jrtorre@mat.upv.es (J.R.T.)

*   Correspondence: fizazafar@gmail.com

Received: 13 November 2018 ; Accepted: 5 December 2018; Published: 7 December 2018

Abstract: Finding a repeated zero for a nonlinear equation $f(x) = 0, f : I \subseteq R \to R$ has always been of much interest and attention due to its wide applications in many fields of science and engineering. Modified Newton's method is usually applied to solve this kind of problems. Keeping in view that very few optimal higher-order convergent methods exist for multiple roots, we present a new family of optimal eighth-order convergent iterative methods for multiple roots with known multiplicity involving a multivariate weight function. The numerical performance of the proposed methods is analyzed extensively along with the basins of attractions. Real life models from life science, engineering, and physics are considered for the sake of comparison. The numerical experiments and dynamical analysis show that our proposed methods are efficient for determining multiple roots of nonlinear equations.

Keywords: nonlinear equations; multiple zeros; optimal iterative methods; higher order of convergence

1. Introduction

It is well-known that Newton's method converges linearly for non-simple roots of a nonlinear equation. For obtaining multiple roots of a univariate nonlinear equation with a quadratic order of convergence, Schröder [1] modified Newton's method with prior knowledge of the multiplicity $m \geq 1$ of the root as follows:

$$x_{n+1} = x_n - m\frac{f(x_n)}{f'(x_n)}. \tag{1}$$

Scheme (1) can determine the desired multiple root with quadratic convergence and is optimal in the sense of Kung-Traub's conjecture [2] that any multipoint method without memory can reach its convergence order of at most 2^{p-1} for p functional evaluations.

In the last few decades, many researchers have worked to develop iterative methods for finding multiple roots with greater efficiency and higher order of convergence. Among them, Li et al. [3] in 2009, Sharma and Sharma [4] and Li et al. [5] in 2010, Zhou et al. [6] in 2011, Sharifi et al. [7] in 2012, Soleymani et al. [8], Soleymani and Babajee [9], Liu and Zhou [10] and Zhou et al. [11] in 2013, Thukral [12] in 2014, Behl et al. [13] and Hueso et al. [14] in 2015, and Behl et al. [15] in 2016 presented optimal fourth-order methods for multiple zeros. Additionally, Li et al. [5] (among other optimal methods) and Neta [16] presented non-optimal fourth-order iterative methods. In recent years, efforts have been made to obtain an optimal scheme with a convergence order greater than four for multiple zeros with multiplicity $m \geq 1$ of univariate function. Some of them only succeeded in developing iterative schemes of a maximum of sixth-order convergence, in the case of multiple zeros; for example, see [17,18]. However, there are only few multipoint iterative schemes with optimal eighth-order convergence for multiple zeros which have been proposed very recently.

Behl et al. [19] proposed a family of optimal eighth-order iterative methods for multiple roots involving univariate and bivariate weight functions given as:

$$y_n = x_n - m\frac{f(x_n)}{f'(x_n)}, m \geq 1,$$

$$z_n = y_n - u_n Q(h_n)\frac{f(x_n)}{f'(x_n)}, \tag{2}$$

$$x_{n+1} = z_n - u_n t_n G(h_n, t_n)\frac{f(x_n)}{f'(x_n)},$$

where weight functions $Q : \mathbb{C} \to \mathbb{C}$ and $G : \mathbb{C}^2 \to \mathbb{C}$ are analytical in neighborhoods of (0) and $(0,0)$, respectively, with $u_n = \left(\frac{f(y_n)}{f(x_n)}\right)^{\frac{1}{m}}$, $h_n = \frac{u_n}{a_1 + a_2 u_n}$ and $t_n = \left(\frac{f(z_n)}{f(y_n)}\right)^{\frac{1}{m}}$, being a_1 and a_2 complex nonzero free parameters.

A second optimal eighth-order scheme involving parameters has been proposed by Zafar et al. [20], which is given as follows:

$$y_n = x_n - m\frac{f(x_n)}{f'(x_n)}, m \geq 1,$$

$$z_n = y_n - m u_n H(u_n)\frac{f(x_n)}{f'(x_n)}, \tag{3}$$

$$x_{n+1} = z_n - u_n t_n (B_1 + B_2 u_n) P(t_n) G(w_n)\frac{f(x_n)}{f'(x_n)},$$

where $B_1, B_2 \in \mathbb{R}$ are free parameters and the weight functions $H : \mathbb{C} \to \mathbb{C}$, $P : \mathbb{C} \to \mathbb{C}$ and $G : \mathbb{C} \to \mathbb{C}$ are analytic in the neighborhood of 0 with $u_n = \left(\frac{f(y_n)}{f(x_n)}\right)^{\frac{1}{m}}$, $t_n = \left(\frac{f(z_n)}{f(y_n)}\right)^{\frac{1}{m}}$ and $w_n = \left(\frac{f(z_n)}{f(x_n)}\right)^{\frac{1}{m}}$.

Recently, Geum et al. [21] presented another optimal eighth-order method for multiple roots:

$$y_n = x_n - m\frac{f(x_n)}{f'(x_n)}, n \geq 0$$

$$w_n = x_n - m L_f(s)\frac{f(x_n)}{f'(x_n)}, \tag{4}$$

$$x_{n+1} = x_n - m\left[L_f(s) + K_f(s,u)\right]\frac{f(x_n)}{f'(x_n)},$$

where $L_f : \mathbb{C} \to \mathbb{C}$ is analytic in the neighborhood of 0 and $K_f : \mathbb{C}^2 \to \mathbb{C}$ is holomorphic in the neighborhood of $(0,0)$ with $s = \left(\frac{f(y_n)}{f(x_n)}\right)^{\frac{1}{m}}$, $u = \left(\frac{f(w_n)}{f(y_n)}\right)^{\frac{1}{m}}$.

Behl et al. [22] also developed another optimal eighth-order method involving free parameters and a univariate weight function as follows:

$$y_n = x_n - m\frac{f(x_n)}{f'(x_n)},$$

$$z_n = y_n - m u\frac{f(x_n)}{f'(x_n)}\frac{1 + \beta u}{1 + (\beta - 2)u}, \beta \in \mathbb{R} \tag{5}$$

$$x_{n+1} = z_n - u v\frac{f(x_n)}{f'(x_n)}\left(\alpha_1 + (1 + \alpha_2 v)P_f(u)\right),$$

where α_1, $\alpha_2 \in \mathbb{R}$ are two free disposable parameters and the weight function $P_f : \mathbb{C} \to \mathbb{C}$ is an analytic function in a neighborhood of 0 with $u = \left(\frac{f(y_n)}{f(x_n)}\right)^{\frac{1}{m}}$, $v = \left(\frac{f(z_n)}{f(y_n)}\right)^{\frac{1}{m}}$.

Most recently, Behl at al. [23] presented an optimal eighth-order method involving univariate weight functions given as:

$$y_n = x_n - m\frac{f(x_n)}{f'(x_n)}, m \geq 1,$$

$$z_n = y_n - mu_n G_f(u_n)\frac{f(x_n)}{f'(x_n)}, \tag{6}$$

$$x_{n+1} = z_n + \frac{u_n w_n}{1 - w_n}\frac{f(x_n)}{f'(x_n)}\left(H_f(u_n) + K_f(v_n)\right),$$

where $B_1, B_2 \in \mathbb{R}$ are free parameters and the weight functions $G_f, H_f, K_f : \mathbb{C} \to \mathbb{C}$ are analytic in the neighborhood of 0 with $u_n = \left(\frac{f(y_n)}{f(x_n)}\right)^{\frac{1}{m}}$, $v_n = \left(\frac{f(z_n)}{f(x_n)}\right)^{\frac{1}{m}}$ and $w_n = \left(\frac{f(z_n)}{f(y_n)}\right)^{\frac{1}{m}}$.

Motivated by the research going on in this direction and with a need to give more stable optimal higher-order methods, we propose a new family of optimal eighth-order iterative methods for finding simple as well as multiple zeros of a univariate nonlinear function with multiplicity $m \geq 1$. The derivation of the proposed class is based on a univariate and trivariate weight function approach. In addition, our proposed methods not only give the faster convergence but also have smaller residual error. We have demonstrated the efficiency and robustness of the proposed methods by performing several applied science problems for numerical tests and observed that our methods have better numerical results than those obtained by the existing methods. Further, the dynamical performance of these methods on the above mentioned problems supports the theoretical aspects, showing a good behavior in terms of dependence on initial estimations.

The rest of the paper is organized as follows: Section 2 provides the construction of the new family of iterative methods and the analysis of convergence to prove the eighth order of convergence. In Section 3, some special cases of the new family are defined. In Section 4, the numerical performance and comparison of some special cases of the new family with the existing ones are given. The numerical comparisons is carried out using the nonlinear equations that appear in the modeling of the predator–prey model, beam designing model, electric circuit modeling, and eigenvalue problem. Additionally, some dynamical planes are provided to compare their stability with that of known methods. Finally, some conclusions are stated in Section 5.

2. Construction of the Family

This section is devoted to the main contribution of this study, the design and convergence analysis of the proposed scheme. We consider the following optimal eighth-order class for finding multiple zeros with multiplicity $m \geq 1$:

$$y_n = x_n - m\frac{f(x_n)}{f'(x_n)}, n \geq 0,$$

$$z_n = y_n - mu_n G(u_n)\frac{f(x_n)}{f'(x_n)}, \tag{7}$$

$$x_{n+1} = z_n - mu_n H(u_n, t_n, w_n)\frac{f(x_n)}{f'(x_n)},$$

where $G : \mathbb{C} \to \mathbb{C}$ and $H : \mathbb{C}^3 \to \mathbb{C}$ are analytical functions in a neighborhood of (0) and $(0,0,0)$, respectively being $u_n = \left(\frac{f(y_n)}{f(x_n)}\right)^{\frac{1}{m}}$, $t_n = \left(\frac{f(z_n)}{f(y_n)}\right)^{\frac{1}{m}}$ and $w_n = \left(\frac{f(z_n)}{f(x_n)}\right)^{\frac{1}{m}}$.

In the next result, we demonstrate that the order of convergence of the proposed family reaches optimal order eight.

Theorem 1. *Let us consider* $x = \xi$ *(say) is a zero with multiplicity* $m \geq 1$ *of the involved function* f. *In addition, we assume that* $f : \mathbb{C} \to \mathbb{C}$ *is an analytical function in the region enclosing the multiple zero* ξ. *The proposed class defined by Equation* (7) *has an optimal eighth order of convergence, when the following conditions are satisfied:*

$$G(0) = 1, \ G_1 = G'(0) = 2, \ G_2 = G''(0) = 4 - \frac{G_3}{6}, G_3 = G'''(0)$$

$$H_{000} = 0, H_{100} = 0, H_{010} = 1, H_{101} = 3 - \frac{G_3}{12}, H_{110} = 2 - H_{001}, \tag{8}$$

$$H_{011} = 4, H_{020} = 1, |G_3| < \infty, |H_{001}| < \infty,$$

where $H_{ijk} = \frac{1}{i!j!k!} \frac{\partial^{i+j+k}}{\partial u_n^i \partial t_n^j \partial w_n^k} H(u_n, t_n, w_n)|_{(u_n=0,t_n=0,w_n=0)}$ *for* $0 \leq i, k \leq 1, 0 \leq j \leq 2$.

Proof. Let us assume that $e_n = x_n - \xi$ is the error at nth step. By expanding $f(x_n)$ and $f'(x_n)$ about $x = \xi$ using Taylor series expansion, we have:

$$f(x_n) = \frac{f^{(m)}(\xi)}{m!} e_n^m \left(1 + c_1 e_n + c_2 e_n^2 + c_3 e_n^3 + c_4 e_n^4 + c_5 e_n^5 + c_6 e_n^6 + c_7 e_n^7 + c_8 e_n^8 + O(e_n^9)\right) \tag{9}$$

and:

$$f'(x_n) = \frac{f^{(m)}(\xi)}{m!} e_n^{m-1} \left(m + c_1(m+1)e_n + c_2(m+2)e_n^2 + c_3(m+3)e_n^3 + c_4(m+4)e_n^4 + c_5(m+5)e_n^5 \right.$$
$$\left. + c_6(m+6)e_n^6 + c_7(m+7)e_n^7 + c_8(m+8)e_n^8 + O(e_n^9)\right), \tag{10}$$

respectively, where $c_k = \frac{m!}{(m+k)!} \frac{f^{(m+k)}(\xi)}{f^{(m)}(\xi)}, \ k = 1, 2, 3, \ldots 8$.

By inserting the above Equations (9) and (10), in the first substep of Equation (7), we obtain:

$$y_n - \xi = \frac{c_1 e_n^2}{m} + \frac{\left(2mc_2 - (m+1)c_1^2\right)e_n^3}{m^2} + \sum_{k=0}^{4} A_k e_n^{k+4} + O(e_n^9), \tag{11}$$

where $A_k = A_k(m, c_1, c_2, \ldots, c_8)$ are given in terms of $m, c_1, c_2, c_3, \ldots, c_8$ with two explicitly written coefficients $A_0 = \frac{1}{m^3}\left\{3c_3m^2 + c_1^3(m+1)^2 - c_1c_2m(3m+4)\right\}$ and $A_1 = -\frac{1}{m^4}\left\{c_1^4(m+1)^3 - 2c_2c_1^2m(2m^2 + 5m + 3) + 2c_3c_1m^2(2m+3) + 2m^2(c_2^2(m+2) - 2c_4m)\right\}$, etc.

With the help of Taylor series expansion and Equation (11), we get:

$$f(y_n) = f^{(m)}(\xi)e_n^{2m}\left[\frac{(\frac{c_1}{m})^m}{m!} + \frac{\left(2mc_2 - (m+1)c_1^2\right)(\frac{c_1}{m})^m e_n}{m!c_1} + \sum_{k=0}^{6} \bar{A}_k e_n^{k+2} + O(e_n^9)\right]. \tag{12}$$

Using Equations (9) and (12), we have:

$$u_n = \left(\frac{f(y_n)}{f(x_n)}\right)^{\frac{1}{m}} = \frac{c_1 e_n}{m} + \frac{\left(2mc_2 - (m+2)c_1^2\right)e_n^2}{m^2} + \tau_1 e_n^3 + \tau_2 e_n^4 + \tau_3 e_n^5 + O(e_n^6), \tag{13}$$

where $\tau_1 = \frac{1}{2m^3}\left[c_1^3(2m^2 + 7m + 7) + 6c_3m^2 - 2c_1c_2m(3m+7)\right]$, $\tau_2 = -\frac{1}{6m^4}\left[c_1^4(6m^3 + 29m^2 + 51m + 34) - 6c_2c_1^2m(4m^2 + 16m + 17) + 12c_1c_3m^2(2m+5) + 12m^2(c_2^2(m+3) - 2c_4m)\right]$ and $\tau_3 = $

$\frac{1}{24m^5}\Big[-24m^3\left(c_2c_3(5m+17)-5c_5m\right)+12c_3c_1^2m^2(10m^2+43m+49)+12c_1m^2\Big\{c_2^2(10m^2+47m+53)-$

$2c_4m(5m+13)\Big\}-4c_2c_1^3m(30m^3+163m^2+306m+209)+c_1^5(24m^4+146m^3+355m^2+418m+209)\Big].$

It is clear from Equation (13) that u_n is of order one. Therefore, we can expand the weight function $G(u_n)$ in the neighborhood of origin by Taylor series expansion up to third-order terms for eighth-order convergence as follows:

$$G(u_n)\approx G(0)+u_nG'(0)+\frac{u_n^2}{2!}G''(0)+\frac{u_n^3}{3!}G'''(0). \tag{14}$$

Now, by inserting Equations (11)–(14) in the second substep of the proposed class (Equation (7)), we obtain:

$$z_n-\xi=-\frac{(G(0)-1)c_1}{m}e_n^2-\frac{((1+G'(0)+m-G(0)(m+3))c_1^2+2mc_2(G(0)-1))}{m^2}e_n^3$$
$$+\sum_{j=1}^{5}B_je_n^{j+3}+O(e_n^9), \tag{15}$$

where $B_j=B_j(G(0),G'(0),G''(0),G'''(0),m,c_1,c_2,\dots,c_7),j=1,2,3,4,5.$

In order to obtain fourth-order convergence, the coefficients of e_n^2 and e_n^3 must be simultaneously equal to zero. Thus, from Equation (15), we obtain the following values of $G(0)$ and $G'(0)$:

$$G(0)=1,G'(0)=2. \tag{16}$$

Using Equation (16), we have:

$$z_n-\xi=-\frac{((9-G''(0)+m)c_1^2-2mc_1c_2)}{2m^3}e_n^4+\sum_{j=1}^{4}P_je_n^{j+4}+O(e_n^9), \tag{17}$$

where $P_j=P_j(G''(0),G'''(0),m,c_1,c_2,\dots,c_7),j=1,2,3,4.$

With the help of Equation (17) and Taylor series expansion, we have:

$$f(z_n)=f^{(m)}(\xi)e_n^{4m}\left[\frac{2^{-m}\left(\frac{(9-G''(0)+m)c_1^3-2mc_1c_2}{m^3}\right)^m}{m!}-\frac{\left(2^{-m}\left(\frac{(9-G''(0)+m)c_1^3-2mc_1c_2}{m^3}\right)^{m-1}\eta_0\right)}{3(m^3m!)}e_n\right.$$
$$\left.+\sum_{j=1}^{7}\overline{P}_je_n^{j+1}+O(e_n^9)\right], \tag{18}$$

where:

$$\eta_0=\left(124+G'''(0)-3G''(0)(7+3m)c_1^4-6m(-3G''(0)+4(7+m)c_1^2c_2+12m^2c_2^2+12m^2c_1c_3)\right)$$

and $\overline{P}_j=\overline{P}_j(G''(0),G'''(0),m,c_1,c_2,\dots,c_7),j=1,2,\dots7.$

Using Equations (9), (12) and (18), we further obtain:

$$t_n=\left(\frac{f(z_n)}{f(y_n)}\right)^{\frac{1}{m}}=\frac{c_1^2(9-G''(0)+m)-2mc_2}{2m^2}e_n^2+\sum_{j=1}^{4}Q_je_n^{j+2}+O(e_n^7), \tag{19}$$

where $Q_j = Q_j(G''(0), G'''(0), m, c_1, c_2, \ldots, c_6), j = 1, 2, 3, 4$ and:

$$w_n = \left(\frac{f(z_n)}{f(x_n)}\right)^{\frac{1}{m}} = \frac{c_1^3(9 - G''(0) + m) - 2mc_1c_2}{2m^3}e_n^3 + \sum_{j=1}^{4}\overline{Q}_je_n^{j+3} + O(e_n^8), \tag{20}$$

where $\overline{Q}_j = \overline{Q}_j(G''(0), G'''(0), m, c_1, c_2, \ldots, c_6), j = 1, 2, 3, 4$.

Hence, it is clear from Equation (13) that t_n and w_n are of order 2 and 3, respectively. Therefore, we can expand weight function $H(u_n, t_n, w_n)$ in the neighborhood of $(0, 0, 0)$ by Taylor series expansion up to second-order terms as follows:

$$H(u_n, t_n, w_n) \approx H_{000} + u_n H_{100} + t_n H_{010} + w_n H_{001} + u_n t_n H_{110} + u_n w_n H_{101} + w_n t_n H_{110} + t_n^2 H_{020}, \tag{21}$$

where $H_{ijk} = \dfrac{1}{i!j!k!}\dfrac{\partial^{i+j+k}}{\partial u_n^i \partial t_n^j \partial w_n^k}H(u_n, t_n, w_n)|_{(0,0,0)}$, for $0 \leq i, k \leq 1, 0 \leq j \leq 2$.

Using Equations (9)–(21) in the last substep of proposed scheme (Equation (7)), we have:

$$e_{n+1} = -\frac{H_{000}c_1}{m}e_n^2 + \sum_{i=1}^{5}E_ie_n^{i+2} + O(e_n^8), \tag{22}$$

where $E_i = E_i(m, G''(0), G'''(0), H_{000}, H_{100}, H_{010}, H_{001}, H_{101}, H_{110}, H_{020}, c_1, c_2, \ldots, c_6), i = 1, 2, 3, 4$.

From Equation (22), it is clear that we can easily obtain at least cubic order of convergence, for:

$$H_{000} = 0. \tag{23}$$

Moreover, $E_1 = 0$ for $H_{100} = 0$, we also have:

$$E_2 = \frac{(-1 + H_{010})c_1(-9 + G''(0)c_1^2 + 2mc_2)}{2m^3}.$$

Thus, we take:

$$-1 + H_{010} = 0, \tag{24}$$

Thus, by inserting Equation (24), it results that $E_2 = 0$ and:

$$E_3 = \frac{(-2 + H_{001} + H_{110})c_1^2(-9 + G''(0)c_1^2 + 2mc_2)}{2m^4}. \tag{25}$$

Therefore, by taking:

$$H_{110} = 2 - H_{001}, \tag{26}$$

we have at least a sixth-order convergence. Additionally, for $H_{020} = 1$:

$$E_4 = \frac{(-2 + 2H_{101} - G''(0))c_1^3(-9 + G''(0)c_1^2 + 2mc_2)}{4m^5}, \tag{27}$$

which further yields:

$$H_{101} = 1 + \frac{G''(0)}{2}. \tag{28}$$

Finally, we take:

$$H_{011} = 4, \quad G''(0) = 4 - \frac{G_3}{6}.$$

where $G_3 = G'''(0)$.

Then, by substituting Equations (23), (24), (26) and (28) in Equation (22), we obtain the following optimal asymptotical error constant term:

$$
\begin{aligned}
e_{n+1} = {} & \frac{c_1}{288m^7}(G_3 + 6(5+m))c_1^2 - 12mc_2)((G_3(25+m) + 2(227 + 90m + 7m^2))c_1^4 \\
& - 2m(180 + G_3 + 24m)c_1^2 c_2 + 24m^2 c_2^2 + 24m^2 c_1 c_3)e_n^8 + O(e_n^9).
\end{aligned}
\tag{29}
$$

Equation (29) reveals that the proposed scheme (Equation (7)) reaches optimal eighth-order convergence using only four functional evaluations (i.e., $f(x_n), f'(x_n), f(y_n)$ and $f(z_n)$) per iteration. This completes the proof. □

3. Some Special Cases of Weight Function

In this section, we discuss some special cases of our proposed class (7) by assigning different kinds of weight functions. In this regard, please see the following cases, where we have mentioned some different members of the proposed family.

Case 1: Let us describe the following polynomial weight functions directly from the hypothesis of Theorem 1:

$$
\begin{aligned}
G(u_n) &= 1 + 2u_n + \left(2 - \frac{G_3}{12}\right)u_n^2 + \frac{1}{6}G_3 u_n^3, \\
H(u_n, t_n, w_n) &= t_n + \left(H_{001} + \left(3 - \frac{G_3}{12}\right)u_n\right)w_n + ((2 - H_{001})u_n + 4w_n + t_n)t_n,
\end{aligned}
\tag{30}
$$

where H_{001} and G_3 are free parameters.

Case 1A: When $H_{001} = 2, G_3 = 0$, we obtain the corresponding optimal eighth-order iterative method as follows:

$$
\begin{aligned}
y_n &= x_n - m\frac{f(x_n)}{f'(x_n)}, n \geq 0, \\
z_n &= y_n - mu_n(1 + 2u_n + 2u_n^2)\frac{f(x_n)}{f'(x_n)}, \\
x_{n+1} &= z_n - mu_n(t_n^2 + w_n(2 + 3u_n + 4t_n) + t_n)\frac{f(x_n)}{f'(x_n)}.
\end{aligned}
\tag{31}
$$

Case 2: Now, we suggest a mixture of rational and polynomial weight functions satisfying condition Equation (8) as follows:

$$
\begin{aligned}
G(u_n) &= \frac{1 + a_0 u_n}{1 + (a_0 - 2)u_n + a_3 u_n^2}, \\
H(u_n, t_n, w_n) &= t_n + \left(H_{001} + \left(3 - \frac{G_3}{12}\right)u_n\right)w_n + ((2 - H_{001})u_n + 4w_n + t_n)t_n,
\end{aligned}
\tag{32}
$$

where $a_3 = -2(a_0 - 1) + \frac{G_3}{12}$ and a_0, G_3 and H_{001} are free parameters.

Case 2A: When $a_0 = 2, H_{001} = 2, G_3 = 12$, the corresponding optimal eighth-order iterative scheme is given by:

$$
\begin{aligned}
y_n &= x_n - m\frac{f(x_n)}{f'(x_n)}, n \geq 0, \\
z_n &= y_n - mu_n\left(\frac{1 + 2u_n}{1 - u_n^2}\right)\frac{f(x_n)}{f'(x_n)}, \\
x_{n+1} &= z_n - mu_n(t_n + 2(1 + u_n)w_n + (t_n + 4w_n)t_n)\frac{f(x_n)}{f'(x_n)}.
\end{aligned}
\tag{33}
$$

Case 3: Now, we suggest another rational and polynomial weight function satisfying Equation (8) as follows:

$$G(u_n) = \frac{1 + a_0 u_n}{1 + (a_0 - 2) u_n + a_3 u_n^2 + a_4 u_n^3},$$

$$H(u_n, t_n, w_n) = t_n + \left(H_{001} + \left(3 - \frac{G_3}{12} \right) u_n \right) w_n + ((2 - H_{001}) u_n + 4w_n + t_n) t_n,$$

(34)

where $a_3 = -2(a_0 - 1) + \frac{G_3}{12}$, $a_4 = 2a_0 + (a_0 - 6) \frac{G_3}{12}$ and a_0, H_{001} and G_3 are free.

Case 3A: By choosing $a_0 = 4$, $a_3 = -5$, $a_4 = 6$, $H_{001} = 2$, $G_3 = 12$, the corresponding optimal eighth-order iterative scheme is given by:

$$y_n = x_n - m \frac{f(x_n)}{f'(x_n)}, n \geq 0,$$

$$z_n = y_n - m u_n \left(\frac{1 + 4u_n}{1 + 2u_n - 5u_n^2 + 6u_n^3} \right) \frac{f(x_n)}{f'(x_n)}$$

(35)

$$x_{n+1} = z_n - m u_n (t_n + 2(1 + u_n) w_n + (t_n + 4w_n) t_n) \frac{f(x_n)}{f'(x_n)}.$$

In a similar way, we can develop several new and interesting optimal schemes with eighth-order convergence for multiple zeros by considering new weight functions which satisfy the conditions of Theorem 1.

4. Numerical Experiments

This section is devoted to demonstrating the efficiency, effectiveness, and convergence behavior of the presented family. In this regard, we consider some of the special cases of the proposed class, namely, Equations (31), (33) and (35), denoted by *NS1*, *NS2*, and *NS3*, respectively. In addition, we choose a total number of four test problems for comparison: The first is a predator–prey model, the second is a beam designing problem, the third is an electric circuit modeling for simple zeros, and the last is an eigenvalue problem.

Now, we want to compare our methods with other existing robust schemes of the same order on the basis of the difference between two consecutive iterations, the residual errors in the function, the computational order of convergence ρ, and asymptotic error constant η. We have chosen eighth-order iterative methods for multiple zeros given by Behl et al. [19,23]. We take the following particular case (Equation (27)) for ($a_1 = 1$, $a_2 = -2$, $G_{02} = 2m$) of the family by Behl et al. [19] and denote it by *BM1* as follows:

$$y_n = x_n - m \frac{f(x_n)}{f'(x_n)},$$

$$z_n = y_n - m (1 + 2h_n) \frac{f(x_n)}{f'(x_n)} u_n,$$

(36)

$$x_{n+1} = z_n - m \left(1 + t_n + t_n^2 + 3h_n^2 + h_n(2 + 4t_n - 2h_n) \right) \frac{f(x_n)}{f'(x_n)} u_n t_n.$$

From the eighth-order family of Behl et al. [23], we consider the following special case denoted by *BM2*:

$$y_n = x_n - m\frac{f(x_n)}{f'(x_n)},$$

$$z_n = y_n - mu_n\left(1 + 2u_n\right)\frac{f(x_n)}{f'(x_n)},$$

$$x_{n+1} = z_n + m\frac{u_n w_n}{1 - w_n}\frac{f(x_n)}{f'(x_n)}\left(-1 - 2u_n - u_n^2 + 4u_n^3 - 2v_n\right). \tag{37}$$

Tables 1–4 display the number of iteration indices (n), the error in the consecutive iterations $|x_{n+1} - x_n|$, the computational order of convergence $\rho \approx \frac{\log|f(x_{n+1})/f(x_n)|}{\log|(f(x_n)/f(x_{n-1}))|}$, $n \geq 1$, (the formula by Jay [24]), the absolute residual error of the corresponding function $(|f(x_n)|)$, and the asymptotical error constant $\eta \approx \left|\frac{e_n}{e_{n-1}^8}\right|$. We did our calculations with 1000 significant digits to minimize the round-off error. We display all the numerical values in Tables 1–4 up to 7 significant digits with exponent. Finally, we display the values of approximated zeros up to 30 significant digits in Examples 1–4, although a minimum of 1000 significant digits are available with us.

For computer programming, all computations have been performed using the programming package *Maple* 16 with multiple precision arithmetics. Further, the meaning of $a(\pm b)$ is $a \times 10^{(\pm b)}$ in Tables 1–4.

Now, we explain the real life problems chosen for the sake of comparing the schemes as follows:

Example 1 (Predator-Prey Model). *Let us consider a predator-prey model with ladybugs as predators and aphids as preys [25]. Let x be the number of aphids eaten by a ladybug per unit time per unit area, called the predation rate, denoted by P(x). The predation rate usually depends on prey density and is given as:*

$$P(x) = K\frac{x^3}{a^3 + x^3}, \quad a, K > 0.$$

Let the growth of aphids obey the Malthusian model; therefore, the growth rate of aphids G per hour is:

$$G(x) = rx, \quad r > 0.$$

The problem is to find the aphid density x for which:

$$P(x) = G(x).$$

This gives:

$$rx^3 - Kx^2 + ra^3 = 0.$$

Let $K = 30$ aphids eaten per hour, $a = 20$ aphids and $r = 2^{-\frac{1}{3}}$ per hour. Thus, we are required to find the zero of:

$$f_1(x) = 0.7937005260x^3 - 30x^2 + 6349.604208$$

The desired zero of f_1 is 25.1984209978974632953442121455564 with $m = 2$. We choose $x_0 = 20$.

Table 1. Comparison of different multiple root finding methods for $f_1(x)$.

	BM1	BM2	NS1	NS2	NS3
$|x_1 - x_0|$	2.064550(1)	4.789445	1.219414(1)	1.214342(1)	1.213887(1)
$|f(x_1)|$	1.008384(4)	4.963523	1.739946(3)	1.712863(3)	1.710446(3)
$|x_2 - x_1|$	1.544682(1)	4.088744(−1)	6.995715	6.944984	6.940438
$|f(x_2)|$	1.967429(−6)	3.035927(−7)	3.672323(−9)	6.792230(−9)	4.951247(−9)
$|x_3 - x_2|$	2.560869(−4)	1.005971(−4)	1.106393(−5)	1.504684(−5)	1.284684(−5)
$|f(x_3)|$	5.685107(−81)	6.093227(−29)	1.223217(−100)	5.427728(−98)	2.522949(−99)
η	7.900841	0.1287852	1.928645(−12)	2.780193(−12)	2.386168(−12)
ρ	7.676751	3.0078946	7.834927	7.814388	7.825421

Example 2 (Beam Designing Model). *We consider a beam positioning problem (see [26]) where an r meter long beam is leaning against the edge of the cubical box with sides of length 1 m each, such that one of its ends touches the wall and the other touches the floor, as shown in Figure 1.*

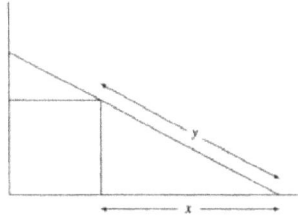

Figure 1. Beam positioning problem.

What should be the distance along the floor from the base of the wall to the bottom of the beam? Let y be the distance in meters along the beam from the floor to the edge of the box and let x be the distance in meters from the bottom of the box to the bottom of the beam. Then, for a given value of r, we have:

$$f_2(x) = x^4 + 4x^3 - 24x^2 + 16x + 16 = 0.$$

The positive solution of the equation is a double root x = 2. We consider the initial guess $x_0 = 1.7$.

Table 2. Comparison of different multiple root finding methods for $f_2(x)$.

	BM1	BM2	NS1	NS2	NS3
$|x_1 - x_0|$	1.288477	2.734437(−1)	7.427026(−1)	7.391615(−1)	7.388023(−1)
$|f(x_1)|$	35.99479	1.670143(−2)	5.783224	5.682280	5.672098
$|x_2 - x_1|$	9.884394(−1)	2.654643(−2)	4.427007(−1)	4.391589(−1)	4.388001(−1)
$|f(x_2)|$	3.566062(−8)	2.333107(−9)	8.652078(−11)	1.664205(−10)	1.1624462(−10)
$|x_3 - x_2|$	3.854647(−5)	9.859679(−6)	1.898691(−6)	2.633282(−7)	2.200800(−6)
$|f(x_3)|$	7.225712(−77)	5.512446(−30)	2.306147(−95)	1.620443(−92)	4.8729521(−94)
e η	4.230427(−1)	3.997726(7)	1.286982(−3)	1.903372(−3)	1.601202(−3)
ρ	7.629155	3.0090640	7.812826	7.7859217	7.800775

Example 3 (The Shockley Diode Equation and Electric Circuit). *Let us consider an electric circuit consisting of a diode and a resistor. By Kirchoff's voltage law, the source voltage drop V_S is equal to the sum of the voltage drops across the diode V_D and resistor V_R :*

$$V_s = V_R + V_D. \tag{38}$$

Let the source voltage be $V_S = 0.5 \, V$ and from Ohm's law:

$$V_R = RI. \tag{39}$$

Additionally, the voltage drop across the diode is given by the Shockley diode equation as follows:

$$I = I_S \left(e^{\frac{V_D}{nV_T}} - 1 \right), \tag{40}$$

where I is the diode current in amperes, I_S is saturation current (amperes), n is the emission or ideality constant ($1 \le n \le 2$ for silicon diode), and V_D is the voltage applied across the diode. Solving Equation (40) for V_D and using all the values in Equation (38), we obtain:

$$-0.5 + RI + nV_T \ln \left(\frac{I}{I_S} + 1 \right) = 0$$

Now, for the given values of n, V_T, R and I_S, we have the following equation [27]:

$$-0.5 + 0.1I + 1.4 \ln (I + 1) = 0.$$

Replacing I with x, we have

$$f_3 (x) = -0.5 + 0.1x + 1.4 \ln (x + 1).$$

The true root of the equation is $0.38997719839007758658645353 2646$. We take $x_0 = 0.5$.

Table 3. Comparison of different multiple root finding methods for $f_3(x)$.

	BM1	BM2	NS1	NS2	NS3
$\lvert x_1 - x_0 \rvert$	1.100228(−1)	1.100228(−1)	1.100228(−1)	1.100228(−1)	1.100228(−1)
$\lvert f(x_1) \rvert$	3.213611(−12)	1.902432(−10)	7.591378(−11)	4.728795(−10)	1.626799(−10)
$\lvert x_2 - x_1 \rvert$	2.902439(−12)	1.718220(−10)	6.856308(−11)	4.270907(−10)	1.469276(−10)
$\lvert f(x_2) \rvert$	9.512092(−97)	6.797214(−81)	2.215753(−84)	2.393956(−77)	1.758525(−81)
$\lvert x_3 - x_2 \rvert$	8.591040(−97)	6.139043(−81)	2.001202(−84)	2.162151(−77)	1.588247(−81)
$\lvert f(x_3) \rvert$	5.604505(−773)	1.805114(−644)	1.1671510(−672)	1.032863(−615)	3.278426(−649)
η	1.705849(−4)	8.081072(−3)	4.097965(−3)	1.953099(−2)	7.312887(−3)
ρ	7.999999	7.999999	7.999999	7.999999	7.999999

Example 4 (Eigenvalue Problem). *One of the challenging task of linear algebra is to calculate the eigenvalues of a large square matrix, especially when the required eigenvalues are the zeros of the characteristic polynomial obtained from the determinant of a square matrix of order greater than 4. Let us consider the following 9×9 matrix:*

$$A = \frac{1}{8} \begin{bmatrix} -12 & 0 & 0 & 19 & -19 & 76 & -19 & 18 & 437 \\ -64 & 24 & 0 & -24 & 24 & 64 & -8 & 32 & 376 \\ -16 & 0 & 24 & 4 & -4 & 16 & -4 & 8 & 92 \\ -40 & 0 & 0 & -10 & 50 & 40 & 2 & 20 & 242 \\ -4 & 0 & 0 & -1 & 41 & 4 & 1 & 2 & 25 \\ -40 & 0 & 0 & 18 & -18 & 104 & -18 & 20 & 462 \\ -84 & 0 & 0 & -29 & 29 & 84 & 21 & 42 & 501 \\ 16 & 0 & 0 & -4 & 4 & -16 & 4 & 16 & -92 \\ 0 & 0 & 0 & 0 & 0 & 0 & 0 & 0 & 24 \end{bmatrix}.$$

The corresponding characteristic polynomial of matrix A is given as follows:

$$f_4(x) = x^9 - 29x^8 + 349x^7 - 2261x^6 + 8455x^5 - 17663x^4 + 15927x^3 + 6993x^2 - 24732x + 12960. \tag{41}$$

The above function has one multiple zero at $\xi = 3$ of multiplicity 4 with initial approximation $x_0 = 3.1$.

Table 4. Comparison of different multiple root finding methods for $f_4(x)$.

	BM1	BM2	NS1	NS2	NS3
$\|x_1 - x_0\|$	1.577283(−1)	9.9275251(−2)	1283418(−1)	1.283182(−1)	1.283180(−1)
$\|f(x_1)\|$	9.361198(−4)	2.205656(−11)	5.299339(−5)	5.281568(−5)	5.281425(−5)
$\|x_2 - x_1\|$	5.772837(−2)	7.247474(−4)	2.834188(−2)	2.831824(−2)	2.831805(−2)
$\|f(x_2)\|$	9.059481(−49)	7.0590148(−38)	2.755794(−55)	8.779457(−55)	5.772523(−55)
$\|x_3 - x_2\|$	3.262145(−13)	2.278878(−10)	7.661066(−15)	1.023515(−14)	9.216561(−15)
$\|f(x_3)\|$	4.543117(−408)	2.278878(−117)	4.807225(−457)	1.869778(−452)	4.077620(−454)
η	2.644775(−3)	2.264227(15)	1.840177(−2)	2.474935(−2)	2.228752(−2)
ρ	7.981915	3.000250	7.989789	7.988696	7.989189

In Tables 1–4, we show the numerical results obtained by applying the different methods for approximating the multiple roots of $f_1(x) - f_4(x)$. The obtained values confirm the theoretical results. From the tables, it can be observed that our proposed schemes $NS1$, $NS2$, and $NS3$ exhibit a better performance in approximating the multiple root of f_1, f_2 and f_4 among other similar methods. Only in the case of the example for simple zeros Behl's scheme $BM1$ is performing slightly better than the other methods.

Dynamical Planes

The dynamical behavior of the test functions is presented in Figures 2–9. The dynamical planes have been generated using the routines published in Reference [28]. We used a mesh of 400×400 points in the region of the complex plane $[-100, 100] \times [-100, 100]$. We painted in orange the points whose orbit converged to the multiple root and in black those points whose orbit either diverged or converged to a strange fixed point or a cycle. We worked out with a tolerance of 10^{-3} and a maximum number of 80 iterations. The multiple root is represented in the different figures by a white star.

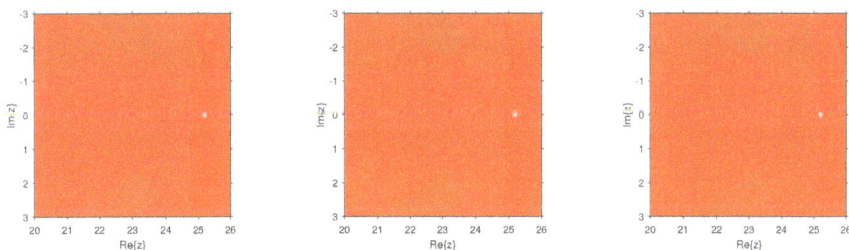

Figure 2. Dynamical planes of the methods $NS1$ (**Left**), $NS2$ (**Center**), and $NS3$ (**Right**) for $f_1(x)$.

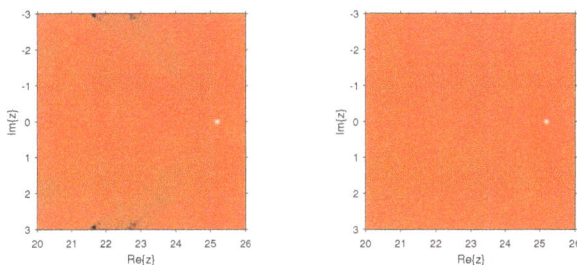

Figure 3. Dynamical planes of the methods $BM1$ (**Left**) and $BM2$ (**Right**) for $f_1(x)$.

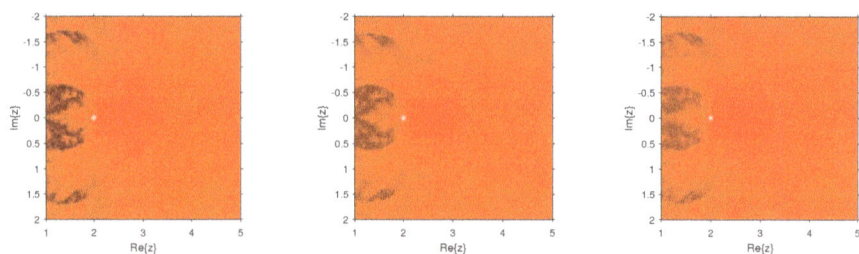

Figure 4. Dynamical planes of the methods *NS1* (**Left**), *NS2* (**Center**), and *NS3* (**Right**) for $f_2(x)$.

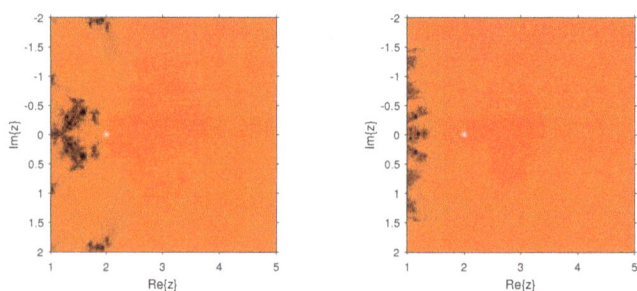

Figure 5. Dynamical planes of the methods *BM1* (**Left**) and *BM2* (**Right**) on $f_2(x)$.

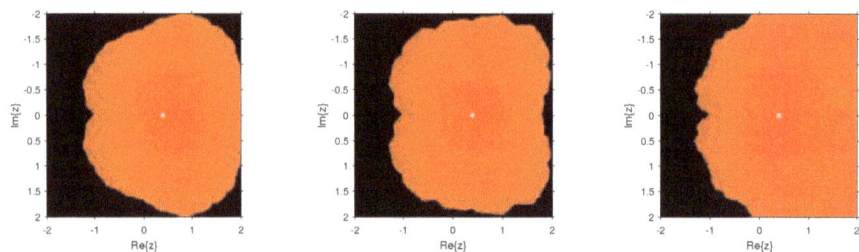

Figure 6. Dynamical planes of the methods *NS1* (**Left**), *NS2* (**Center**), and *NS3* (**Right**) for $f_3(x)$.

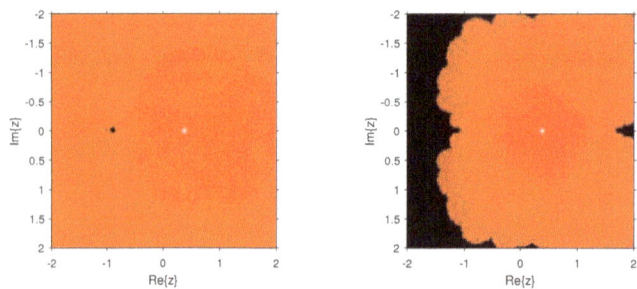

Figure 7. Dynamical planes of the methods *BM1* (**Left**) and *BM2* (**Right**) for $f_3(x)$.

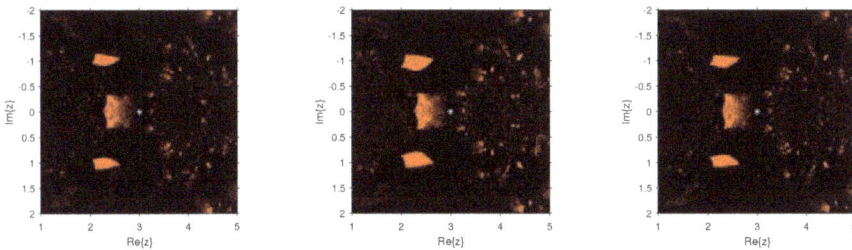

Figure 8. Dynamical planes of the methods $NS1$ (**Left**), $NS2$ (**Center**), and $NS3$ (**Right**) on $f_4(x)$.

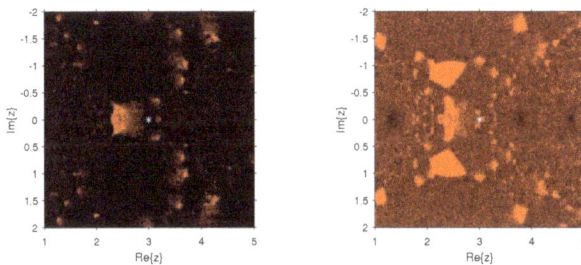

Figure 9. Dynamical planes of the methods $BM1$ (**Left**) and $BM2$ (**Right**) for $f_4(x)$.

Figures 2–9 study the convergence and divergence regions of the new schemes $NS1$, $NS2$, and $NS3$ in comparison with the other schemes of the same order. In the case of $f_1(x)$ and $f_2(x)$, we observed that the new schemes are more stable than $BM1$ and $BM2$ as they are almost divergence-free and also converge faster than $BM1$ and $BM2$ in their common regions of convergence. In the case of $f_3(x)$, $BM1$ performs better; however, $NS1$, $NS2$, and $NS3$ have an edge over $BM2$ for the region in spite of the analogous behavior to $BM2$, as the new schemes show more robustness. Similarly, in the case of $f_4(x)$, it can be clearly observed that the divergence region for $BM1$ is bigger than that for $NS1$, $NS2$, and $NS3$. Additionally, these schemes perform better than $BM2$ where they are convergent. The same behavior can be observed through the numerical comparison of these methods in Tables 1–4. As a future extension, we shall be trying to construct a new optimal eighth-order method whose stability analysis can allow to choose the optimal weight function for the best possible results.

5. Conclusions

In this manuscript, a new general class of optimal eighth-order methods for solving nonlinear equations with multiple roots was presented. This family was obtained using the procedure of weight functions with two functions: One univariate and another depending on three variables. To reach this optimal order, some conditions on the functions and their derivatives must be imposed. Several special cases were selected and applied to different real problems, comparing their performance with that of other known methods of the same order of convergence. Finally, their dependence on initial estimations was analyzed from their basins of attraction.

Author Contributions: methodology, F.Z.; writing original draft preparation, F.Z.; writing review and editing, J.R.T.; visualization, A.C.; supervision, J.R.T.

Funding: This research was partially supported byMinisterio de Economía y Competitividad MTM2014-52016-C2-2-P, by Generalitat Valenciana PROMETEO/2016/089 and Schlumberger Foundation-Faculty for Future Program.

Conflicts of Interest: The authors declare no conflict of interest.

References

1. Schroder, E. Uber unendlich viele Algorithmen zur Auflosung der Gleichungen. *Math. Ann.* **1870**, *2*, 317–365. [CrossRef]
2. Kung, H.T.; Traub, J.F. Optimal order of one-point and multipoint iteration. *J. Assoc. Comput. Mach.* **1974**, *21*, 643–651. [CrossRef]
3. Li, S.; Liao, X.; Cheng, L. A new fourth-order iterative method for finding multiple roots of nonlinear equations. *Appl. Math. Comput.* **2009**, *215*, 1288–1292.
4. Sharma, J.R.; Sharma, R. Modified Jarratt method for computing multiple roots. *Appl. Math. Comput.* **2010**, *217*, 878–881. [CrossRef]
5. Li, S.G.; Cheng, L.Z.; Neta, B. Some fourth-order nonlinear solvers with closed formulae for multiple roots. *Comput. Math. Appl.* **2010**, *59*, 126–135. [CrossRef]
6. Zhou, X.; Chen, X.; Song, Y. Constructing higher-order methods for obtaining the muliplte roots of nonlinear equations. *J. Comput. Math. Appl.* **2011**, *235*, 4199–4206. [CrossRef]
7. Sharifi, M.; Babajee, D.K.R.; Soleymani, F. Finding the solution of nonlinear equations by a class of optimal methods. *Comput. Math. Appl.* **2012**, *63*, 764–774. [CrossRef]
8. Soleymani, F.; Babajee, D.K.R.; Lofti, T. On a numerical technique forfinding multiple zeros and its dynamic. *J. Egypt. Math. Soc.* **2013**, *21*, 346–353. [CrossRef]
9. Soleymani, F.; Babajee, D.K.R. Computing multiple zeros using a class of quartically convergent methods. *Alex. Eng. J.* **2013**, *52*, 531–541. [CrossRef]
10. Liu, B.; Zhou, X. A new family of fourth-order methods for multiple roots of nonlinear equations. *Non. Anal. Model. Cont.* **2013**, *18*, 143–152.
11. Zhou, X.; Chen, X.; Song, Y. Families of third and fourth order methods for multiple roots of nonlinear equations. *Appl. Math. Comput.* **2013**, *219*, 6030–6038. [CrossRef]
12. Thukral, R. A new family of fourth-order iterative methods for solving nonlinear equations with multiple roots. *J. Numer. Math. Stoch.* **2014**, *6*, 37–44.
13. Behl, R.; Cordero, A.; Motsa, S.S.; Torregrosa, J.R. On developing fourth-order optimal families of methods for multiple roots and their dynamics. *Appl. Math. Comput.* **2015**, *265*, 520–532. [CrossRef]
14. Hueso, J.L.; Martínez, E.; Teruel, C. Determination of multiple roots of nonlinear equations and applications. *J. Math. Chem.* **2015**, *53*, 880–892. [CrossRef]
15. Behl, R.; Cordero, A.; Motsa, S.S.; Torregrosa, J.R.; Kanwar, V. An optimal fourth-order family of methods for multiple roots and its dynamics. *Numer. Algor.* **2016**, *71*, 775–796. [CrossRef]
16. Neta, B. Extension of Murakami's high-order non-linear solver to multiple roots. *Int. J. Comput. Math.* **2010**, *87*, 1023–1031. [CrossRef]
17. Geum, Y.H.; Kim, Y.I.; Neta, B. A class of two-point sixth-order multiple-zero finders of modified double-Newton type and their dynamics. *Appl. Math. Comput.* **2015**, *270*, 387–400 . [CrossRef]
18. Geum, Y.H.; Kim, Y.I.; Neta, B. A sixth-order family of three-point modified Newton-like multiple-root finders and the dynamics behind their extraneous fixed points. *Appl. Math. Comput.* **2016**, *283*, 120–140. [CrossRef]
19. Behl, R.; Cordero, A.; Motsa, S.S.; Torregrosa, J.R. An eighth-order family of optimal multiple root finders and its dynamics. *Numer. Algor.* **2017**. [CrossRef]
20. Zafar, F.; Cordero, A.; Rana, Q.; Torregrosa, J.R. Optimal iterative methods for finding multiple roots of nonlinear equations using free parameters. *J. Math. Chem.* **2017**. [CrossRef]
21. Geum, Y.H.; Kim, Y.I.; Neta, B. Constructing a family of optimal eighth-order modified Newton-type multiple-zero finders along with the dynamics behind their purely imaginary extraneous fixed points. *J. Comput. Appl. Math.* **2018**, *333*, 131–156. [CrossRef]
22. Behl, R.; Zafar, F.; Alshomrani, A.S.; Junjua, M.; Yasmin, N. An optimal eighth-order scheme for multiple zeros of univariate function. *Int. J. Comput. Math.* **2018**, *15*. [CrossRef]
23. Behl, R.; Alshomrani, A.S.; Motsa, S.S. An optimal scheme for multiple roots of nonlinear equations with eighth-order convergence. *J. Math. Chem.* **2018**. [CrossRef]
24. Jay, L.O. A note on Q-order of convergence. *BIT Numer. Math.* **2001**, *41*, 422–429. [CrossRef]
25. Edelstein-Keshet, L. *Differential Calculus for the Life Sciences*; Univeristy of British Columbia: Vancouver, BC, Canada, 2017.

26. Zachary, J.L. *Introduction to Scientific Programming: Computational Problem Solving Using Maple and C*; Springer: New York, NY, USA, 2012.

27. Khoury, R. *Douglas Wilhelm Harder, Numerical Methods and Modelling for Engineering*; Springer International Publishing: Berlin, Germany, 2017.

28. Chicharro, F.I.; Cordero, A.; Torregrosa, J.R. Drawing dynamical and parameters planes of iterative families and methods. *Sci. World J.* **2013**, *2013*, 780153. [CrossRef] [PubMed]

mathematics

MDPI

Article

A Higher Order Chebyshev-Halley-Type Family of Iterative Methods for Multiple Roots

Ramandeep Behl [1], Eulalia Martínez [2,*], Fabricio Cevallos [3] and Diego Alarcón [4]

[1] Department of Mathematics, King Abdualziz University, Jeddah 21589, Saudi Arabia; ramanbehl87@yahoo.in

[2] Instituto Universitario de Matemática Multidisciplinar, Universitat Politècnica de València, Camino de Vera s/n, 46022 Valencia, Spain

[3] Fac. de Ciencias Económicas, Universidad Laica "Eloy Alfaro de Manabí", Manabí 130214, Ecuador; alfa2205@gmail.com

[4] Departamento de Matemática Aplicada, Universitat Politècnica de València, Camino de Vera s/n, 46022 Valencia, Spain; diealcor@doctor.upv.es

* Correspondence: eumarti@mat.upv.es

Received: 15 January 2019; Accepted: 22 March 2019; Published: 9 April 2019

Abstract: The aim of this paper is to introduce new high order iterative methods for multiple roots of the nonlinear scalar equation; this is a demanding task in the area of computational mathematics and numerical analysis. Specifically, we present a new Chebyshev–Halley-type iteration function having at least sixth-order convergence and eighth-order convergence for a particular value in the case of multiple roots. With regard to computational cost, each member of our scheme needs four functional evaluations each step. Therefore, the maximum efficiency index of our scheme is 1.6818 for $\alpha = 2$, which corresponds to an optimal method in the sense of Kung and Traub's conjecture. We obtain the theoretical convergence order by using Taylor developments. Finally, we consider some real-life situations for establishing some numerical experiments to corroborate the theoretical results.

Keywords: nonlinear equations; multiple roots; Chebyshev–Halley-type; optimal iterative methods; efficiency index

1. Introduction

One important field in the area of computational methods and numerical analysis is to find approximations to the solutions of nonlinear equations of the form:

$$f(x) = 0, \tag{1}$$

where $f : \mathbb{D} \subset \mathbb{C} \to \mathbb{C}$ is the analytic function in the enclosed region $\mathbb{D}$, enclosing the required solution. It is almost impossible to obtain the exact solution in an analytic way for such problems. Therefore, we concentrate on obtaining approximations of the solution up to any specific degree of accuracy by means of an iterative procedure, of course doing it also with the maximum efficiency. In [1], Kung and Traub conjectured that a method without memory that uses $n + 1$ functional evaluations per iteration can have at most convergence order $p = 2^n$. If this bound is reached, the method is said to be optimal.

For solving nonlinear Equation (1) by means of iterations, we have the well-known cubically-convergent family of Chebyshev–Halley methods [2], which is given by:

$$x_{n+1} = x_n - \left[1 + \frac{1}{2} \frac{L_f(x_n)}{1 - \alpha L_f(x_n)} \right] \frac{f(x_n)}{f'(x_n)}, \quad \alpha \in \mathbb{R}, \tag{2}$$

where $L_f(x_n) = \frac{f''(x_n)f(x_n)}{\{f'(x_n)\}^2}$. A great variety of iterative methods can be reported in particular cases. For example, the classical Chebyshev's method [1,3], Halley's method [1,3], and the super-Halley method [1,3] can be obtained if $\alpha = 0$, $\alpha = \frac{1}{2}$, and $\alpha = 1$, respectively. Despite the third-order convergence, the scheme (2) is considered less practical from a computational point of view because of the computation of the second-order derivative.

For this reason, several variants of Chebyshev–Halley's methods free from the second-order derivative have been presented in [4–7]. It has been shown that these methods are comparable to the classical third-order methods of the Chebyshev–Halley-type in their performance and can also compete with Newton's method. One family of these methods is given as follows:

$$
\begin{aligned}
y_n &= x_n - \frac{f(x_n)}{f'(x_n)}, \\
x_{n+1} &= x_n - \left(1 + \frac{f(y_n)}{f(x_n) - \alpha f(y_n)}\right)\frac{f(x_n)}{f'(x_n)}, \quad \alpha \in \mathbb{R}
\end{aligned}
\tag{3}
$$

We can easily obtain some well-known third-order methods proposed by Potra and Pták [4] and Sharma [5] (the Newton-secant method (NSM)) for $\alpha = 0$ and $\alpha = 1$. In addition, we have Ostrowski's method [8] having optimal fourth-order convergence, which is also a special case for $\alpha = 2$. This family is important and interesting not only because of not using a second- or higher order derivative. However, this scheme also converges at least cubically and has better results in comparison to the existing ones. Moreover, we have several higher order modifications of the Chebyshev–Halley methods available in the literature, and some of them can be seen in [9–12].

In this study, we focus on the case of the multiple roots of nonlinear equations. We have some fourth-order optimal and non-optimal modifications or improvements of Newton's iteration function for multiple roots in the research articles [13–17]. Furthermore, we can find some higher order methods for this case, but some of them do not reach maximum efficiency [18–23]; so, this topic is of interest in the current literature.

We propose a new Chebyshev–Halley-type iteration function for multiple roots, which reaches a high order of convergence. Specifically, we get a family of iterative methods with a free parameter α, with sixth-order convergence. Therefore, the efficiency index is $6^{1/4}$, and for $\alpha = 2$, this index is $8^{1/4}$, which is the maximum value that one can get with four functional evaluations, reaching optimality in the sense of Kung and Traub's conjecture. Additionally, an extensive analysis of the convergence order is presented in the main theorem.

We recall that $\xi \in \mathbb{C}$ is a multiple root of the equation $f(x) = 0$, if it is verified that:

$$
f(\xi) = 0, f'(\xi) = 0, \cdots, f^{(m-1)}(\xi) = 0 \quad and \quad f^{(m)}(\xi) \neq 0,
$$

the positive integer ($m \geq 1$) being the multiplicity of the root.

We deal with iterative methods in which the multiplicity must be known in advance, because this value, m, is used in the iterative expression. However, we point out that these methods also work when one uses an estimation of the multiplicity, as was proposed in the classical study carried out in [24].

Finally, we consider some real-life situations that start from some given conditions to investigate and some standard academic test problems for numerical experiments. Our iteration functions here are found to be more comparable and effective than the existing methods for multiple roots in terms of residual errors and errors among two consecutive iterations, and also, we obtain a more stable computational order of convergence. That is, the proposed methods are competitive.

2. Construction of the Higher Order Scheme

In this section, we present the new Chebyshev–Halley-type methods for multiple roots of nonlinear equations, for the first time. In order to construct the new scheme, we consider the following scheme:

$$y_n = x_n - m \frac{f(x_n)}{f'(x_n)},$$

$$z_n = x_n - m \left(1 + \frac{\eta}{1 - \alpha\eta}\right) \frac{f(x_n)}{f'(x_n)}, \tag{4}$$

$$x_{n+1} = z_n - H(\eta, \tau) \frac{f(x_n)}{f'(x_n)},$$

where the function:

$$H(\eta, \tau) = \frac{\eta\tau \left(\beta - (\alpha - 2)^2\eta^2(\eta + 1) + \tau^3 + \tau^2\right)}{(\eta + 1)(\tau + 1)}$$

with:

$$\eta = \left(\frac{f(y_n)}{f(x_n)}\right)^{\frac{1}{m}},$$

$$\tau = \left(\frac{f(z_n)}{f(y_n)}\right)^{\frac{1}{m}},$$

$$\beta = m\left((\alpha(\alpha + 2) + 9)\eta^3 + \eta^2(\alpha(\alpha + 3) - 6\tau - 3) + \eta(\alpha + 8\tau + 1) + 2\tau + 1\right),$$

where $\alpha \in \mathbb{R}$ is a free disposable variable. For $m = 1$, we can easily obtain the scheme (3) from the first two steps of the scheme (4).

In Theorem 1, we illustrate that the constructed scheme attains at least sixth-order convergence and for $\alpha = 2$, it goes to eighth-order without using any extra functional evaluation. It is interesting to observe that $H(\eta, \tau)$ plays a significant role in the construction of the presented scheme (for details, please see Theorem 1).

Theorem 1. *Let us consider $x = \xi$ to be a multiple zero with multiplicity $m \geq 1$ of an analytic function $f : \mathbb{C} \to \mathbb{C}$ in the region containing the multiple zero ξ of $f(x)$. Then, the present scheme (4) attains at least sixth-order convergence for each α, but for a particular value of $\alpha = 2$, it reaches the optimal eighth-order convergence.*

Proof. We expand the functions $f(x_n)$ and $f'(x_n)$ about $x = \xi$ with the help of a Taylor's series expansion, which leads us to:

$$f(x_n) = \frac{f^{(m)}(\xi)}{m!} e_n^m \left(1 + c_1 e_n + c_2 e_n^2 + c_3 e_n^3 + c_4 e_n^4 + c_5 e_n^5 + c_6 e_n^6 + c_7 e_n^7 + c_8 e_n^8 + O(e_n^9)\right), \tag{5}$$

and:

$$f'(x_n) = \frac{f^m(\xi)}{m!} e_n^{m-1} \left(m + (m+1)c_1 e_n + (m+2)c_2 e_n^2 + (m+3)c_3 e_n^3 + (m+4)c_4 e_n^4 + (m+5)c_5 e_n^5 \right.$$
$$\left. + (m+6)c_6 e_n^6 + (m+7)c_7 e_n^7 + (m+8)c_8 e_n^8 + O(e_n^9)\right), \tag{6}$$

respectively, where $c_k = \frac{m!}{(m-1+k)!} \frac{f^{m-1+k}(\xi)}{f^m(\xi)}$, $k = 2, 3, 4 \ldots, 8$ and $e_n = x_n - \xi$ is the error in the nth iteration.

Inserting the above expressions (5) and (6) into the first substep of scheme (4) yields:

$$y_n - \xi = \frac{c_1}{m} e_n^2 + \sum_{i=0}^{5} \phi_i e_n^{i+3} + O(e_n^9), \tag{7}$$

where $\phi_i = \phi_i(m, c_1, c_2, \ldots, c_8)$ are given in terms of $m, c_2, c_3, \ldots, c_8$, for example $\phi_0 = \frac{1}{m^2}(2mc_2 - (m + 1)c_1^2)$ and $\phi_1 = \frac{1}{m^3}[3m^2c_3 + (m+1)^2c_1^3 - m(3m+4)c_1c_2]$, etc.

Using the Taylor series expansion and the expression (7), we have:

$$f(y_n) = f^{(m)}(\xi)e_n^{2m}\left[\frac{\left(\frac{c_1}{m}\right)^m}{m!} + \frac{(2mc_2 - (m+1)c_1^2)\left(\frac{c_1}{m}\right)^m e_n}{m!c_1} + \left(\frac{c_1}{m}\right)^{1+m}\frac{1}{2m!c_1^3}\{(3 + 3m + 3m^2 + m^3)c_1^4 \right.$$
$$\left. - 2m(2 + 3m + 2m^2)c_1^2c_2 + 4(m-1)m^2c_2^2 + 6m^2c_1c_3\}e_n^2 + \sum_{i=0}^{5}\bar{\phi}_ie_n^{i+3} + O(e_n^9)\right]. \tag{8}$$

We obtain the following expression by using (5) and (8):

$$\eta = \frac{c_1e_n}{m} + \frac{2mc_2 - (m+2)c_1^2}{m^2}e_n^2 + \theta_0e_n^3 + \theta_1e_n^4 + \theta_2e_n^5 + O(e_n^6), \tag{9}$$

where $\theta_0 = \frac{(2m^2 + 7m + 7)c_1^3 + 6m^2c_3 - 2m(3m+7)c_1c_2}{2m^3}$, $\theta_1 = -\frac{1}{6m^4}\left[12m^2(2m + 5)c_1c_3 + 12m^2((m+3)c_2^2 - 2mc_4) - 6m(4m^2 + 16m + 17)c_1^2c_2 + (6m^3 + 29m^2 + 51m + 34)c_1^4\right]$ and $\theta_2 = \frac{1}{24m^5}\left[12m^2(10m^2 + 43m + 49)c_1^2c_3 - 24m^3((5m+17)c_2c_3 - 5mc_5) + 12m^2\left((10m^2 + 47m + 53)c_2^2 - 2m(5m+13)c_4\right)c_1 - 4m(30m^3 + 163m^2 + 306m + 209)c_1^3c_2 + (24m^4 + 146m^3 + 355m^2 + 418m + 209)c_1^5\right]$.

With the help of Expressions (5)–(9), we obtain:

$$z_n - \xi = -\frac{(\alpha - 2)c_1^2}{m^2}e_n^3 + \sum_{i=0}^{4}\psi_ie_n^{i+4} + O(e_n^9), \tag{10}$$

where $\psi_i = \psi_i(\alpha, m, c_1, c_2, \ldots, c_8)$ are given in terms of $\alpha, m, c_2, c_3, \ldots, c_8$ with the first two coefficients explicitly written as $\psi_0 = -\frac{1}{2m^3}\left[(2\alpha^2 - 10\alpha + (7 - 4\alpha)m + 11)c_1^3 + 2m(4\alpha - 7)c_1c_2\right]$ and $\psi_1 = \frac{1}{6m^4}\left[(-6\alpha^3 + 42\alpha^2 - 96\alpha + (29 - 18\alpha)m^2 + 6(3\alpha^2 - 14\alpha + 14)m + 67)c_1^4 + 12m^2(5 - 3\alpha)c_1c_3 + 12m^2(3 - 2\alpha)c_2^2 + 12m(-3\alpha^2 + 14\alpha + (5\alpha - 8)m - 14)c_1^2c_2\right]$.

By using the Taylor series expansion and (10), we have:

$$f(z_n) = f^{(m)}(\xi)e_n^{3m}\left[\frac{\left(-\frac{(\alpha-2)c_1^2}{m^2}\right)^m}{m!} + \sum_{i=1}^{5}\psi_ie_n^i + O(e_n^6)\right]. \tag{11}$$

From Expressions (8) and (11), we further have:

$$\tau = -\frac{(\alpha - 2)c_1}{m}e_n + \frac{((-2\alpha^2 + 8\alpha + (2\alpha - 3)m - 7)c_1^2 + 2m(3 - 2\alpha)c_2)}{2m^2}e_n^2 + \gamma_1e_n^3 + \gamma_2e_n^4 + O(e_n^5), \tag{12}$$

where $\gamma_1 = \frac{1}{3m^3}\left[(-3\alpha^3 + 18\alpha^2 - 30\alpha + (4 - 3\alpha)m^2 + 3(2\alpha^2 - 7\alpha + 5)m + 11)c_1^3 + 3m^2(4 - 3\alpha)c_3 + 3m(-4\alpha^2 + 14\alpha + 3\alpha m - 4m - 10)c_1c_2\right]$ and $\gamma_2 = \frac{1}{24m^4}\left[24m^2(-6\alpha^2 + 20\alpha + (4\alpha - 5)m - 14)c_1c_3 + 12m^2((-8\alpha^2 + 24\alpha + 4\alpha m - 5m - 13)c_2^2 + 2m(5 - 4\alpha)c_4) - 12m(12\alpha^3 - 66\alpha^2 + 100\alpha + 2(4\alpha - 5)m^2 + (-20\alpha^2 + 64\alpha - 41)m - 33)c_1^2c_2 + (-24\alpha^4 + 192\alpha^3 - 492\alpha^2 + 392\alpha + 6(4\alpha - 5)m^3 + (-72\alpha^2 + 232\alpha - 151)m^2 + 6(12\alpha^3 - 66\alpha^2 + 100\alpha - 33)m + 19)c_1^4\right]$.

By using Expressions (9) and (12), we obtain:

$$H(\eta, \tau) = -\frac{(\alpha - 2)c_1^2}{m^2}e_n^2 + \lambda_1e_n^3 + \lambda_2e_n^4 + O(e_n^5) \tag{13}$$

where $\lambda_1 = \frac{c_1}{2m^3}\left[c_1^2\left(-2\alpha^2 + 8\alpha + (4\alpha - 7)m - 7\right) + 2(7 - 4\alpha)c_2 m\right]$ and $\lambda_2 = \frac{1}{6m^3}\left[c_1^4\left(-6\alpha^3 + 36\alpha^2 - 66\alpha + (29 - 18\alpha)m^2 + 3(6\alpha^2 - 22\alpha + 17)m + 34\right) + 12(5 - 3\alpha)c_3 c_1 m^2 + 12(3 - 2\alpha)c_2^2 m^2 + 6c_2 c_1^2 m\left(-6\alpha^2 + 22\alpha + 2(5\alpha - 8)m - 17\right)\right]$.

Now, we use the expressions (5)–(13) in the last substep of Scheme (4), and we get:

$$e_{n+1} = \sum_{i=1}^{3} L_i e_n^{i+5} + O(e_n^9), \tag{14}$$

where $L_1 = \frac{(\alpha - 2)^2 c_1^3}{m^6}\left[c_1^2\left(\alpha^2 - \alpha + m^2 - (\alpha^2 + 4\alpha - 17)m - 3\right) - 2c_2(m - 1)m\right]$, $L_2 = (\alpha - 2)c_1^2\big[-12c_2c_1^2 m\{10\alpha^3 - 24\alpha^2 - 39\alpha + (16\alpha - 27)m^2 - (10\alpha^3 + 27\alpha^2 - 262\alpha + 301)m + 91\} + 12c_3 c_1 m^2(-4\alpha + (4\alpha - 7)m + 8) + 12c_2^2 m^2(-12\alpha + 4(3\alpha - 5)m + 21) + c_1^4\{-24\alpha^4 + 168\alpha^3 - 156\alpha^2 - 662\alpha + (52\alpha - 88)m^3 - (60\alpha^3 + 162\alpha^2 - 1616\alpha + 1885)m^2 + 2(18\alpha^4 - 12\alpha^3 - 711\alpha^2 + 2539\alpha - 2089)m + 979\}\big]$ and

$L_3 = \frac{c_1}{24m^8}\Big[-24c_2c_3c_1 m^3\big((42\alpha^2 - 146\alpha + 125)m - 6(7\alpha^2 - 26\alpha + 24)\big) - 24c_2^3 m^3\big(-24\alpha^2 + 84\alpha + (24\alpha^2 - 80\alpha + 66)m - 73\big) + 12c_3 c_1^3 m^2\{2(15\alpha^4 - 63\alpha^3 - 5\alpha^2 + 290\alpha - 296) + (54\alpha^2 - 190\alpha + 165)m^2 + (-30\alpha^4 - 28\alpha^3 + 968\alpha^2 - 2432\alpha + 1697)m\} + 12c_1^2 m^2\Big\{c_2^2\big(80\alpha^4 - 304\alpha^3 - 226\alpha^2 + 1920\alpha + 2(81\alpha^2 - 277\alpha + 234)m^2 + (-80\alpha^4 - 112\alpha^3 + 2712\alpha^2 - 6410\alpha + 4209)m - 1787\big) - 4(\alpha - 2)c_4 m(-3\alpha + (3\alpha - 5)m + 6)\Big\} - 2c_2 c_1^4 m\Big\{-3(96\alpha^5 - 804\alpha^4 + 1504\alpha^3 + 2676\alpha^2 - 10612\alpha + 8283) + 4(177\alpha^2 - 611\alpha + 521)m^3 - 3(220\alpha^4 + 280\alpha^3 - 7556\alpha^2 + 18400\alpha - 12463)m^2 + 4(108\alpha^5 - 234\alpha^4 - 4302\alpha^3 + 22902\alpha^2 - 38593\alpha + 20488)m\Big\} + c_1^6\big\{48\alpha^6 - 480\alpha^5 + 996\alpha^4 + 5472\alpha^3 - 29810\alpha^2 + 50792\alpha + (276\alpha^2 - 956\alpha + 818)m^4 + (-360\alpha^4 - 448\alpha^3 + 12434\alpha^2 - 30518\alpha + 20837)m^3 + (432\alpha^5 - 1236\alpha^4 - 16044\alpha^3 + 92306\alpha^2 - 161292\alpha + 88497)m^2 + (-168\alpha^6 + 888\alpha^5 + 5352\alpha^4 - 55580\alpha^3 + 173290\alpha^2 - 224554\alpha + 97939)m - 29771\big\}\Big]$.

It is noteworthy that we reached at least sixth-order convergence for all α. In addition, we can easily obtain $L_1 = L_2 = 0$ by using $\alpha = 2$.

Now, by adopting $\alpha = 2$ in Expression (14), we obtain:

$$e_{n+1} = \frac{A_0\left(12c_3c_1 m^3 - 12c_2 c_1^2 m(3m^2 + 30m - 1) + 12c_2^2 m^2(2m - 1) + c_1^4(10m^3 + 183m^2 + 650m - 3)\right)}{24m^8} e_n^8 + O(e_n^9), \tag{15}$$

where $A_0 = (c_1^3(m + 1) - 2c_1 c_2 m)$. The above Expression (15) demonstrates that our proposed Scheme (4) reaches eighth-order convergence for $\alpha = 2$ by using only four functional evaluations per full iteration. Hence, it is an optimal scheme for a particular value of $\alpha = 2$ according to the Kung–Traub conjecture, completing the proof. $\square$

3. Numerical Experiments

In this section, we illustrate the efficiency and convergence behavior of our iteration functions for particular values $\alpha = 0$, $\alpha = 1$, $\alpha = 1.9$, and $\alpha = 2$ in Expression (4), called $OM1, OM2, OM3$, and $OM4$, respectively. In this regards, we choose five real problems having multiple and simple zeros. The details are outlined in the examples (1)–(3).

For better comparison of our iterative methods, we consider several existing methods of order six and the optimal order eight. Firstly, we compare our methods with the two-point family of sixth-order methods proposed by Geum et al. in [18], and out of them, we pick Case 4c, which is mentioned as follows:

$$y_n = x_n - m\frac{f(x_n)}{f'(x_n)}, \quad m > 1,$$

$$x_{n+1} = y_n - \left[\frac{m + a_1 u_n}{1 + b_1 u_n + b_2 u_n^2} \times \frac{1}{1 + c_1 s_n}\right]\frac{f(y_n)}{f'(y_n)}, \tag{16}$$

where:

$$a_1 = \frac{2m\left(4m^4 - 16m^3 + 31m^2 - 30m + 13\right)}{(m-1)\left(4m^2 - 8m + 7\right)}, \quad b_1 = \frac{4\left(2m^2 - 4m + 3\right)}{(m-1)\left(4m^2 - 8m + 7\right)},$$

$$b_2 = -\frac{4m^2 - 8m + 3}{4m^2 - 8m + 7}, \quad c_1 = 2(m-1),$$

$$u_n = \left(\frac{f(y_n)}{f(x_n)}\right)^{\frac{1}{m}}, \quad s_n = \left(\frac{f'(y_n)}{f'(x_n)}\right)^{\frac{1}{m-1}},$$

called *GM1*.

In addition, we also compare them with one more non-optimal family of sixth-order iteration functions given by the same authors of [19], and out of them, we choose Case 5YD, which is given by:

$$y_n = x_n - m\frac{f(x_n)}{f'(x_n)}, \quad m \geq 1,$$

$$w_n = x_n - m\left[\frac{(u_n - 2)(2u_n - 1)}{(u_n - 1)(5u_n - 2)}\right]\frac{f(x_n)}{f'(x_n)}, \tag{17}$$

$$x_{n+1} = x_n - m\left[\frac{(u_n - 2)(2u_n - 1)}{(5u_n - 2)(u_n + v_n - 1)}\right]\frac{f(x_n)}{f'(x_n)},$$

where $u_n = \left(\frac{f(y_n)}{f(x_n)}\right)^{\frac{1}{m}}$ and $v_n = \left(\frac{f(w_n)}{f(x_n)}\right)^{\frac{1}{m}}$, and this method is denoted as *GM2*.

Moreover, we compare our methods with the optimal eighth-order iterative methods proposed by Zafar et al. [21]. We choose the following two schemes out of them:

$$y_n = x_n - m\frac{f(x_n)}{f'(x_n)},$$

$$w_n = y_n - mu_n\left(6u_n^3 - u_n^2 + 2u_n + 1\right)\frac{f(x_n)}{f'(x_n)}, \tag{18}$$

$$x_{n+1} = w_n - mu_nv_n(1 + 2u_n)(1 + v_n)\left(\frac{2w_n + 1}{A_2P_0}\right)\frac{f(x_n)}{f'(x_n)}$$

and:

$$y_n = x_n - m\frac{f(x_n)}{f'(x_n)},$$

$$w_n = y_n - mu_n\left(\frac{1 - 5u_n^2 + 8u_n^3}{1 - 2u_n}\right)\frac{f(x_n)}{f'(x_n)}, \tag{19}$$

$$x_{n+1} = w_n - mu_nv_n(1 + 2u_n)(1 + v_n)\left(\frac{3w_n + 1}{A_2P_0(1 + w_n)}\right)\frac{f(x_n)}{f'(x_n)},$$

where $u_n = \left(\frac{f(y_n)}{f(x_n)}\right)^{\frac{1}{m}}$, $v_n = \left(\frac{f(w_n)}{f(y_n)}\right)^{\frac{1}{m}}$, $w_n = \left(\frac{f(w_n)}{f(x_n)}\right)^{\frac{1}{m}}$, and these iterative methods are denoted in our tables as *ZM1* and *ZM2*, respectively.

Finally, we demonstrate their comparison with another optimal eighth-order iteration function given by Behl et al. [22]. However, we consider the following the best schemes (which was claimed by them):

$$y_n = x_n - m\frac{f(x_n)}{f'(x_n)},$$

$$z_n = y_n - m\frac{f(x_n)}{f'(x_n)}h_n(1 + 2h_n), \tag{20}$$

$$x_{n+1} = z_n + m\frac{f(x_n)}{f'(x_n)}\frac{t_nh_n}{1 - t_n}\left[-1 - 2h_n - h_n^2 + 4h_n^3 - 2k_n\right]$$

and:

$$y_n = x_n - m \frac{f(x_n)}{f'(x_n)},$$

$$z_n = y_n - m \frac{f(x_n)}{f'(x_n)} h_n (1 + 2h_n),$$

$$x_{n+1} = z_n - m \frac{f(x_n)}{f'(x_n)} \frac{t_n h_n}{1 - t_n} \left[\frac{1 + 9h_n^2 + 2k_n + h_n(6 + 8k_n)}{1 + 4h_n} \right],$$

$$(21)$$

with $h_n = \left(\frac{f(y_n)}{f(x_n)} \right)^{\frac{1}{m}}$, $k_n = \left(\frac{f(z_n)}{f(x_n)} \right)^{\frac{1}{m}}$ $t_n = \left(\frac{f(z_n)}{f(y_n)} \right)^{\frac{1}{m}}$, which are denoted *BM1* and *BM2*, respectively.

In order to compare these schemes, we perform a numerical experience, and in Tables 1 and 2, we display the difference between two consecutive iterations $|x_{n+1} - x_n|$, the residual error in the corresponding function $|f(x_n)|$, and the computational order of convergence (ρ) (we used the formula given by Cordero and Torregrosa [25]:

$$\rho \approx \frac{\ln(|x_{k+1} - x_k| / |x_k - x_{k-1}|)}{\ln(|x_k - x_{k-1}| / |x_{k-1} - x_{k-2}|)}$$

$$(22)$$

We make our calculations with several significant digits (a minimum of 3000 significant digits) to minimize the round-off error. Moreover, the computational order of convergence is provided up to five significant digits. Finally, we display the initial guess and approximated zeros up to 25 significant digits in the corresponding example where an exact solution is not available.

All computations have been performed using the programming package Mathematica 11 with multiple precision arithmetic. Further, the meaning of $a(\pm b)$ is shorthand for $a \times 10^{(\pm b)}$ in the numerical results.

Example 1. *Population growth problem:*
The law of population growth is defined as follows:

$$\frac{dN(t)}{dt} = \gamma N(t) + \eta,$$

where N(t) = the population at time t, η = the fixed/constant immigration rate, and γ = the fixed/constant birth rate of the population. We can easily obtain the following solution of the above differential equation:

$$N(t) = N_0 e^{\gamma t} + \frac{\eta}{\gamma} (e^{\gamma t - 1}),$$

where N_0 is the initial population.

For a particular case study, the problem is given as follows: Suppose a certain population contains 1,000,000 individuals initially, that 300,000 individuals immigrate into the community in the first year, and that 1,365,000 individuals are present at the end of one year. Find the birth rate (γ) of this population.

To determine the birth rate, we must solve the equation:

$$f_1(x) = 1365 - 1000e^x - \frac{300}{x}(e^x - 1).$$

$$(23)$$

wherein $x = \gamma$ and our desired zero of the above function f_1 is 0.05504622451335177827483421. The reason for considering the simple zero problem is to confirm that our methods also work for simple zeros. We choose the starting point as $x_0 = 0.5$.

Example 2. *The van der Waals equation of state:*

$$\left(P + \frac{a_1 n^2}{V^2}\right)(V - na_2) = nRT,$$

explains the behavior of a real gas by introducing in the ideal gas equations two parameters, a_1 and a_2, specific for each gas. The determination of the volume V of the gas in terms of the remaining parameters requires the solution of a nonlinear equation in V,

$$PV^3 - (na_2 P + nRT)V^2 + a_1 n^2 V - a_1 a_2 n^2 = 0.$$

Given the constants a_1 and a_2 of a particular gas, one can find values for $n, P,$ and T, such that this equation has three simple roots. By using the particular values, we obtain the following nonlinear function:

$$f_2(x) = x^3 - 5.22x^2 + 9.0825x - 5.2675. \tag{24}$$

which has three zeros; out of them, one is the multiple zero $\alpha = 1.75$ of multiplicity two, and the other is the simple zero $\alpha = 1.72$. Our desired root is $\alpha = 1.75$, and we chose $x_0 = 1.8$ as the initial guess.

Example 3. *Eigenvalue problem:*
For this, we choose the following 8×8 matrix:

$$A = \begin{bmatrix} -12 & -12 & 36 & -12 & 0 & 0 & 12 & 8 \\ 148 & 129 & -397 & 147 & -12 & 6 & -109 & -74 \\ 72 & 62 & -186 & 66 & -8 & 4 & -54 & -36 \\ -32 & -24 & 88 & -36 & 0 & 0 & 24 & 16 \\ 20 & 13 & -45 & 19 & 8 & 6 & -13 & -10 \\ 120 & 98 & -330 & 134 & -8 & 24 & -90 & -60 \\ -132 & -109 & 333 & -115 & 12 & -6 & 105 & 66 \\ 0 & 0 & 0 & 0 & 0 & 0 & 0 & 4 \end{bmatrix}.$$

The corresponding characteristic polynomial of this matrix is as follows:

$$f_3(x) = (x-4)^3(x+4)(x-8)(x-20)(x-12)(x+12).$$

The above function has one multiple zero at $\alpha = 4$ of multiplicity three. In addition, we consider $x_0 = 2.7$ as the starting point.

Example 4. *Let us consider the following polynomial equation:*

$$f_4(z) = \left((x-1)^3 - 1\right)^{50}. \tag{25}$$

The desired zero of the above function f_4 is $\alpha = 2$ with multiplicity of order 50, and we choose initial guess $x_0 = 2.1$ for this problem.

Table 1. Comparison on the basis of the difference between two consecutive iterations $|x_{n+1} - x_n|$ for the functions f_1–f_4.

f	n	OM1	OM2	OM3	OM4	GM1	GM2	ZM1	ZM2	BM1	BM2
f_1	1	2.3 (−3)	8.4 (−4)	9.3 (−5)	3.5 (−5)	*	3.6 (−5)	1.6 (−4)	2.3 (−4)	7.6 (−5)	3.7 (−5)
	2	2.0 (−16)	9.0 (−20)	8.8 (−28)	2.0 (−37)	*	1.4 (−29)	4.2 (−31)	8.9 (−30)	2.6 (−34)	5.0 (−37)
	3	9.7 (−95)	1.3 (−115)	6.4 (−166)	2.5 (−295)	*	5.4 (−173)	1.0 (−243)	5.5 (−233)	5.4 (−270)	5.7 (−292)
	ρ	5.9997	6.0000	6.0001	8.0000	*	6.0000	8.0000	8.0000	8.0000	8.0000
f_2	1	1.3 (−3)	8.2 (−4)	4.0 (−3)	3.5 (−4)	9.5 (−4)	3.9 (−4)	3.9 (−4)	4.1 (−3)	2.7 (−4)	2.6 (−4)
	2	2.5 (−10)	4.2 (−12)	6.4 (−16)	8.7 (−18)	2.7 (−11)	1.0 (−14)	5.2 (−17)	9.8 (−17)	1.1 (−18)	1.4 (−19)
	3	2.0 (−50)	8.7 (−62)	6.5 (−87)	1.5 (−126)	2.0 (−56)	3.9 (−78)	5.9 (−120)	1.2 (−117)	6.3 (−134)	1.0 (−141)
	ρ	5.9757	5.9928	6.0214	7.9963	5.9836	5.9975	7.9945	7.9941	7.9971	8.0026
f_3	1	9.1 (−5)	3.6 (−5)	8.0 (−6)	6.0 (−6)	8.5 (−5)	4.8 (−5)	4.9 (−6)	5.2 (−6)	2.0 (−6)	1.8 (−6)
	2	1.8 (−28)	1.4 (−31)	9.8 (−38)	2.0 (−47)	1.0 (−28)	5.0 (−31)	6.0 (−48)	1.0 (−47)	1.5 (−51)	2.8 (−52)
	3	1.2 (−170)	4.4 (−190)	3.3 (−229)	2.5 (−379)	3.1 (−172)	5.8 (−187)	2.7 (−383)	2.3 (−381)	1.4 (−412)	1.3 (−418)
	ρ	6.0000	6.0000	6.0000	8.0000	6.0000	6.0000	8.0000	8.0000	8.0000	8.0000
f_4	1	2.4 (−5)	7.1 (−6)	4.2 (−7)	1.4 (−7)	1.8 (−5)	2.0 (−7)	4.8 (−7)	6.5 (−7)	1.9 (−7)	6.3 (−8)
	2	1.5 (−26)	1.7 (−30)	3.9 (−40)	6.7 (−54)	1.1 (−26)	1.8 (−41)	5.7 (−49)	8.4 (−48)	8.0 (−53)	4.2 (−57)
	3	7.5 (−154)	3.2 (−178)	2.6 (−438)	1.7 (−424)	6.6 (−154)	1.0 (−245)	2.2 (−384)	6.6 (−375)	9.6 (−416)	5.9 (−169)
	ρ	6.0000	6.0000	6.0000	8.0000	6.0000	6.0000	8.0000	8.0000	8.0000	2.2745

* means that the corresponding method does not work.

Table 2. Comparison on the basis of residual errors $|f(x_n)|$ for the functions f_1–f_4.

f	n	OM1	OM2	OM3	OM4	GM1	GM2	ZM1	ZM2	BM1	BM2
f_1	1	2.7	1.0	1.1 (−1)	4.2 (−2)	*	4.4 (−2)	1.9 (−1)	2.7 (−1)	9.2 (−2)	4.4 (−2)
	2	2.4 (−13)	1.1 (−16)	1.1 (−24)	2.4 (−34)	*	1.7 (−26)	5.1 (−28)	1.1 (−26)	3.2 (−31)	6.0 (−34)
	3	1.2 (−91)	1.6 (−112)	7.8 (−163)	3.0 (−292)	*	5.4 (−173)	1.2 (−240)	6.7 (−230)	6.5 (−267)	7.0 (−289)
f_2	1	5.0 (−8)	2.1 (−8)	4.8 (−9)	3.6 (−9)	2.8 (−8)	4.6 (−9)	4.6 (−9)	5.1 (−9)	2.3 (−9)	2.0 (−9)
	2	1.8 (−21)	5.3 (−25)	1.2 (−32)	2.3 (−36)	2.2 (−23)	3.2 (−30)	8.0 (−35)	2.9 (−34)	3.4 (−38)	5.9 (−40)
	3	1.2 (−101)	2.2 (−124)	1.3 (−174)	6.9 (−254)	1.2 (−113)	4.6 (−157)	1.1 (−240)	4.3 (−236)	1.2 (−268)	3.1 (−284)
f_3	1	4.9 (−8)	3.1 (−9)	3.1 (−11)	1.4 (−11)	4.1 (−8)	7.4 (−9)	7.8 (−12)	9.1 (−12)	5.2 (−13)	3.6 (−13)
	2	3.9 (−79)	1.8 (−88)	6.1 (−107)	4.9 (−136)	7.1 (−80)	8.0 (−87)	1.4 (−137)	6.9 (−137)	2.1 (−148)	1.5 (−150)
	3	1.0 (−505)	5.6 (−564)	2.4 (−681)	1.1 (−1131)	1.9 (−510)	1.2 (−554)	1.3 (−1143)	7.5 (−1138)	1.9 (−1231)	1.3 (−1249)
f_4	1	1.2 (−207)	2.7 (−234)	1.1 (−295)	3.3 (−319)	3.5 (−214)	1.0 (−311)	6.6 (−293)	2.3 (−286)	1.8 (−313)	6.2 (−337)
	2	1.9 (−1268)	2.6 (−1465)	3.8 (−1947)	1.6 (−2635)	1.9 (−1274)	9.8 (−2014)	3.4 (−2389)	9.4 (−2331)	9.8 (−2582)	1.1 (−2795)
	3	4.2 (−7633)	2.3 (−8851)	6.1 (−11856)	6.1 (−21166)	6.0 (−7636)	7.3 (−12226)	1.6 (−19159)	7.1 (−18686)	8.8 (−20728)	3.4 (−8388)

4. Conclusions

We presented an eighth-order modification of the Chebyshev–Halley-type iteration scheme having optimal convergence to obtain the multiple solutions of the scalar equation. The proposed scheme is optimal in the sense of the classical Kung–Traub conjecture. Thus, the efficiency index of the present methods is $E = \sqrt[4]{8} \approx 1.682$, which is better than the classical Newton's method $E = \sqrt[2]{2} \approx 1.414$. Finally, the numerical experience corroborates the theoretical results about the convergence order, and moreover, it can be concluded that our proposed methods are highly efficient and competitive.

Author Contributions: First two authors have contribute to the theoretical results and the other two authors have carried out the numerical experience.

Funding: This research was partially supported by Ministerio de Economía y Competitividad under Grant MTM2014-52016-C2-1-2-P and by the project of Generalitat Valenciana Prometeo/2016/089.

Conflicts of Interest: The authors declare no conflict of interest.

References

1. Traub, J.F. *Iterative Methods for the Solution of Equations*; Prentice-Hall: Englewood Cliffs, NJ, USA, 1964.
2. Gutiérrez, J.M.; Hernández, M.A. A family of Chebyshev–Halley type methods in Banach spaces. *Bull. Aust. Math. Soc.* **1997**, *55*, 113–130. [CrossRef]
3. Kanwar, V.; Singh, S.; Bakshi, S. Simple geometric constructions of quadratically and cubically convergent iterative functions to solve nonlinear equations. *Numer. Algorithms* **2008**, *47*, 95–107. [CrossRef]
4. Potra, F.A.; Pták, V. *Nondiscrete Induction and Iterative Processes*; Research Notes in Mathematics; Pitman: Boston, MA, USA, 1984; Volume 103.
5. Sharma, J.R. A composite third order Newton-Steffensen method for solving nonlinear equations. *Appl. Math. Comput.* **2005**, *169*, 242–246.
6. Argyros, I.K.; Ezquerro, J.A.; Gutiérrez, J.M.; Hernández, M.A.; Hilout, S. On the semilocal convergence of efficient Chebyshev-Secant-type methods. *J. Comput. Appl. Math.* **2011**, *235*, 3195–3206. [CrossRef]
7. Xiaojian, Z. Modified Chebyshev–Halley methods free from second derivative. *Appl. Math. Comput.* **2008**, *203*, 824–827. [CrossRef]
8. Ostrowski, A.M. *Solutions of Equations and System of Equations*; Academic Press: New York, NY, USA, 1960.
9. Amat, S.; Hernández, M.A.; Romero, N. A modified Chebyshev's iterative method with at least sixth order of convergence. *Appl. Math. Comput.* **2008**, *206*, 164–174. [CrossRef]
10. Kou, J.; Li, Y. Modified Chebyshev–Halley method with sixth-order convergence. *Appl. Math. Comput.* **2007**, *188*, 681–685. [CrossRef]
11. Li, D.; Liu, P.; Kou, J. An improvement of Chebyshev–Halley methods free from second derivative. *Appl. Math. Comput.* **2014**, *235*, 221–225. [CrossRef]
12. Sharma, J.R. Improved Chebyshev–Halley methods with sixth and eighth order convergence. *Appl. Math. Comput.* **2015**, *256*, 119–124. [CrossRef]
13. Neta, B. Extension of Murakami's high-order non-linear solver to multiple roots. *Int. J. Comput. Math.* **2010**, *87*, 1023–1031. [CrossRef]
14. Zhou, X.; Chen, X.; Song, Y. Constructing higher-order methods for obtaining the multiple roots of nonlinear equations. *J. Comput. Math. Appl.* **2011**, *235*, 4199–4206. [CrossRef]
15. Hueso, J.L.; Martínez, E.; Teruel, C. Determination of multiple roots of nonlinear equations and applications. *J. Math. Chem.* **2015**, *53*, 880-892. [CrossRef]
16. Behl, R.; Cordero, A.; Motsa, S.S.; Torregrosa, J.R. On developing fourth-order optimal families of methods for multiple roots and their dynamics. *Appl. Math. Comput.* **2015**, *265*, 520-532. [CrossRef]
17. Behl, R.; Cordero, A.; Motsa, S.S.; Torregrosa, J.R.; Kanwar, V. An optimal fourth-order family of methods for multiple roots and its dynamics. *Numer. Algorithms* **2016**, *71*, 775–796. [CrossRef]
18. Geum, Y.H.; Kim, Y.I.; Neta, B. A class of two-point sixth-order multiple-zero finders of modified double-Newton type and their dynamics. *Appl. Math. Comput.* **2015**, *270*, 387–400. [CrossRef]
19. Geum, Y.H.; Kim, Y.I.; Neta, B. A sixth-order family of three-point modified Newton-like multiple-root finders and the dynamics behind their extraneous fixed points. *Appl. Math. Comput.* **2016**, *283*, 120–140. [CrossRef]

20. Behl, R.; Cordero, A.; Motsa, S.S.; Torregrosa, J.R. An eighth-order family of optimal multiple root finders and its dynamics. *Numer. Algorithms* **2018**, *77*, 1249–1272. [CrossRef]
21. Zafar, F.; Cordero, A.; Motsa, S.S.; Torregrosa, J.R. Optimal iterative methods for finding multiple roots of nonlinear equations using free parameters. *J. Math. Chem.* **2018**, *56*, 1884–1901. [CrossRef]
22. Behl, R.; Alshomrani, A.S.; Motsa, S.S. An optimal scheme for multiple roots of nonlinear equations with eighth-order convergence. *J. Math. Chem.* **2018**, *56*, 2069–2084. [CrossRef]
23. Behl, R.; Zafar, F.; Alshomrani, A.S.; Junjuaz, M.; Yasmin, N. An optimal eighth-order scheme for multiple zeros of univariate functions. *Int. J. Comput. Methods* **2018**, 1843002. [CrossRef]
24. McNamee, J.M. A comparison of methods for accelerating convergence of Newton's method for multiple polynomial roots. *ACM Signum Newsl.* **1998**, *33*, 17–22. [CrossRef]
25. Cordero, A.; Torregrosa, J.R. Variants of Newton's method using fifth-order quadrature formulas. *Appl. Math. Comput.* **2007**, *190*, 686–698. [CrossRef]

![mathematics logo] *mathematics*

MDPI

Article

An Optimal Eighth-Order Family of Iterative Methods for Multiple Roots

Saima Akram *[iD], Fiza Zafar [iD] and Nusrat Yasmin

Centre for Advanced Studies in Pure and Applied Mathematics, Bahauddin Zakariya University,
Multan 60800, Pakistan
* Correspondence: saimaakram@bzu.edu.pk

Received: 5 June 2019; Accepted: 25 July 2019; Published: 27 July 2019

Abstract: In this paper, we introduce a new family of efficient and optimal iterative methods for finding multiple roots of nonlinear equations with known multiplicity ($m \geq 1$). We use the weight function approach involving one and two parameters to develop the new family. A comprehensive convergence analysis is studied to demonstrate the optimal eighth-order convergence of the suggested scheme. Finally, numerical and dynamical tests are presented, which validates the theoretical results formulated in this paper and illustrates that the suggested family is efficient among the domain of multiple root finding methods.

Keywords: nonlinear equations; optimal iterative methods; multiple roots; efficiency index

1. Introduction

The problem of solving nonlinear equation is recognized to be very old in history as many practical problems which arise are nonlinear in nature . Various one-point and multi-point methods are presented to solve nonlinear equations or systems of nonlinear equations [1–3]. The above-cited methods are designed for the simple root of nonlinear equations but the behavior of these methods is not similar when dealing with multiple roots of nonlinear equations. The well known Newton's method with quadratic convergence for simple roots of nonlinear equations decays to first order when dealing with multiple roots of nonlinear equations. These problems lead to minor troubles such as greater computational cost and severe troubles such as no convergence at all. The prior knowledge of multiplicity of roots make it simpler to manage these troubles. The strange behavior of the iterative methods while dealing with multiple roots has been well known since 19th century in the least when Schröder [4] developed a modification of classical Newton's method to conserve its second order of convergence for multiple roots. The nonlinear equations with multiple roots commonly arise from different topics such as complex variables, fractional diffusion or image processing, applications to economics and statistics (Lévy distributions), etc. By knowing the practical nature of multiple root finders, various one-point and multi-point root solvers have been developed in recent past [5–18] but most of them are not optimal as defined by Kung and Traub [19], who stated that an optimal without memory method can achieve its convergence order at the most 2^n requiring $n+1$ evaluations of functions or derivatives. As stated by Ostrowski [1], if an iterative method possess order of convergence as O and total number of functional evaluations is n per iterative step, then the index defined by $E = O^{1/n}$ is recognized as efficiency index of an iterative method.

Sharma and Sharma [17] proposed the following optimal fourth-order multiple root finder with known multiplicity m as follows:

$$y_n = x_n - \frac{2m}{m+2} \cdot \frac{f(x_n)}{f'(x_n)}, m > 1$$

$$x_{n+1} = x_n - \frac{m}{8} \Phi(x_n) \frac{f(x_n)}{f'(x_n)},$$

where $\Phi(x_n) = \left\{ (m^3 - 4m + 8) - (m+2)^2 \left(\frac{m}{m+2} \right)^m \frac{f'(x_n)}{f'(y_n)} \times 2(m-1)(m+2) \left(\frac{m}{m+2} \right)^m \frac{f'(x_n)}{f'(y_n)} \right\}$.

A two-step sixth-order non-optimal family for multiple roots presented by Geum et al. [9] is given by:

$$
\begin{aligned}
y_n &= x_n - m \cdot \frac{f(x_n)}{f'(x_n)}, m > 1, \\
x_{n+1} &= y_n - Q(r_n, s_n) \cdot \frac{f(y_n)}{f'(y_n)},
\end{aligned}
\tag{1}
$$

where, $r_n = \sqrt[m]{\frac{f(y_n)}{f(x_n)}}$, $s_n = \sqrt[m-1]{\frac{f'(y_n)}{f'(x_n)}}$ and $Q : \mathbb{C}^2 \to \mathbb{C}$ is holomorphic in a neighborhood of $(0,0)$. The following is a special case of their family:

$$
\begin{aligned}
y_n &= x_n - m \cdot \frac{f(x_n)}{f'(x_n)}, n \geq 0, m > 1, \\
x_{n+1} &= y_n - m \left[1 + 2(m-1)(r_n - s_n) - 4r_n s_n + s_n^2 \right] \cdot \frac{f(y_n)}{f'(y_n)}.
\end{aligned}
\tag{2}
$$

Another non-optimal family of three-point sixth-order methods for multiple roots by Geum et al. [10] is given as follows:

$$
\begin{aligned}
y_n &= x_n - m \cdot \frac{f(x_n)}{f'(x_n)}, m \geq 1, \\
w_n &= y_n - m \cdot G(p_n) \cdot \frac{f(x_n)}{f'(x_n)}, \\
x_{n+1} &= w_n - m \cdot K(p_n, v_n,) \cdot \frac{f(x_n)}{f'(x_n)},
\end{aligned}
\tag{3}
$$

where $p_n = \sqrt[m]{\frac{f(y_n)}{f(x_n)}}$ and $v_n = \sqrt[m]{\frac{f(w_n)}{f(x_n)}}$. The weight functions $G : \mathbb{C} \to \mathbb{C}$ is analytic in a neighborhood of 0 and $K : \mathbb{C}^2 \to \mathbb{C}$ is holomorphic in a neighborhood of $(0,0)$. The following is a special case of the family in Equation (3):

$$
\begin{aligned}
y_n &= x_n - m \cdot \frac{f(x_n)}{f'(x_n)}, m \geq 1, \\
w_n &= y_n - m \cdot \left[1 + p_n + 2p_n^2 \right] \cdot \frac{f(x_n)}{f'(x_n)}, \\
x_{n+1} &= w_n - m \cdot \left[1 + p_n + 2p_n^2 + (1 + 2p_n)v_n \right] \cdot \frac{f(x_n)}{f'(x_n)}.
\end{aligned}
\tag{4}
$$

The families in Equations (1) and (3) require four evaluations of function to produce convergence of order six having efficiency index $6^{\frac{1}{4}} = 1.5650$ and therefore are not optimal in the sense of the Kung–Traub conjecture [19].

Recently, Behl et al. [20] presented a multiple root finding family of iterative methods possessing convergence order eight given as:

$$
\begin{aligned}
y_n &= x_n - m \frac{f(x_n)}{f'(x_n)}, m \geq 1, \\
z_n &= y_n - u_n Q(h_n) \frac{f(x_n)}{f'(x_n)}, \\
x_{n+1} &= z_n - u_n t_n G(h_n, t_n) \frac{f(x_n)}{f'(x_n)},
\end{aligned}
\tag{5}
$$

where the functions $Q : \mathbb{C} \to \mathbb{C}$ and $G : \mathbb{C}^2 \to \mathbb{C}$ are restricted to be analytic functions in the regions nearby (0) and $(0,0)$, respectively, with $u_n = \left(\frac{f(y_n)}{f(x_n)}\right)^{\frac{1}{m}}$, $h_n = \frac{u_n}{a_1 + a_2 u_n}$ and $t_n = \left(\frac{f(z_n)}{f(y_n)}\right)^{\frac{1}{m}}$, being a_1 and a_2 complex non-zero free parameters.

We take Case (27) for $(a_1 = 1, a_2 = 1, G_{02} = 0)$ from the family of Behl et al. [20] and represent it by BM given by:

$$
\begin{aligned}
y_n &= x_n - m\frac{f(x_n)}{f'(x_n)}, \\
z_n &= y_n - \left(m + 2h_n m + \frac{1}{2}h_n^2(4m + 2m)\right)\frac{f(x_n)}{f'(x_n)}u_n \\
x_{n+1} &= z_n - \left(m + mt_n + 3mh_n^2 + mh_n(2 + 4t_n + h_n)\right)\frac{f(x_n)}{f'(x_n)}u_n t_n.
\end{aligned}
\tag{6}
$$

Most recently, another optimal eighth-order scheme presented by Zafar et al. [21] is given as:

$$
\begin{aligned}
y_n &= x_n - m\frac{f(x_n)}{f'(x_n)}, m \geq 1, \\
z_n &= y_n - m u_n H(u_n)\frac{f(x_n)}{f'(x_n)}, \\
x_{n+1} &= z_n - u_n t_n (B_1 + B_2 u_n) P(t_n) G(w_n)\frac{f(x_n)}{f'(x_n)},
\end{aligned}
\tag{7}
$$

where $B_1, B_2 \in \mathbb{R}$ are suppose to be free parameters and weight functions $H : \mathbb{C} \to \mathbb{C}$, $P : \mathbb{C} \to \mathbb{C}$ and $G : \mathbb{C} \to \mathbb{C}$ are restricted to be analytic in the regions nearby 0 with $u_n = \left(\frac{f(y_n)}{f(x_n)}\right)^{\frac{1}{m}}$, $t_n = \left(\frac{f(z_n)}{f(y_n)}\right)^{\frac{1}{m}}$ and $w_n = \left(\frac{f(z_n)}{f(x_n)}\right)^{\frac{1}{m}}$.

From the eighth-order family of Zafar et al. [21], we consider the following special case denoted by ZM:

$$
\begin{aligned}
y_n &= x_n - m\frac{f(x_n)}{f'(x_n)}, \\
z_n &= y_n - m u_n \left(6u_n^3 - u_n^2 + 2u_n + 1\right)\frac{f(x_n)}{f'(x_n)}, \\
x_{n+1} &= z_n - m u_n t_n (1 + 2u_n)(1 + t_n)(1 + 2w_n)\frac{f(x_n)}{f'(x_n)}.
\end{aligned}
\tag{8}
$$

The class of iterative methods referred as optimal is significant as compared to non-optimal methods due to their speed of convergence and efficiency index. Therefore, there was a need to develop optimal eighth-order schemes for finding multiple zeros $(m > 1)$ and simple zeros $(m = 1)$ due to their competitive efficiencies and order of convergence [1]; in addition, fewer iterations are needed to get desired accuracy as compared to iterative methods having order four and six given by Sharma and Geum [9,10,17], respectively. In this paper, our main concern is to find the optimal iterative methods for multiple root μ with known multiplicity $m \in \mathbb{N}$ of an adequately differentiable nonlinear function $f : I \subseteq \mathbb{R} \to \mathbb{R}$, where I represents an open interval. We develop an optimal eighth-order zero finder for multiple roots with known multiplicity $m \geq 1$. The beauty of the method lies in the fact that developed scheme is simple to implement with minimum possible number of functional evaluations. Four evaluations of the function are needed to obtain a family of convergence order eighth having efficiency index $8^{\frac{1}{4}} = 1.6817$.

The rest of the paper is organized as follows. In Section 2, we present the newly developed optimal iterative family of order eight for multiple roots of nonlinear equations. The discussion of analysis of convergence is also given in this section. In Section 3, some special cases of newly developed eighth-order schemes are presented. In Section 4, numerical results and comparison of the presented schemes with existing schemes of its domain is discussed. Concluding remarks are given in Section 5.

2. Development of the Scheme and Convergence Analysis

In this section, we suggest a new family of eighth-order method with known multiplicity $m \geq 1$ of the required multiple root as follows:

$$
\begin{aligned}
y_n &= x_n - m \cdot \frac{f(x_n)}{f'(x_n)}, n \geq 0, \\
z_n &= y_n - m \cdot t \cdot H(t) \cdot \frac{f(x_n)}{f'(x_n)}, \\
x_{n+1} &= z_n - m \cdot t \cdot L(s,u) \cdot \frac{f(x_n)}{f'(x_n)},
\end{aligned}
\tag{9}
$$

$$
\text{where } t = \sqrt[m]{\frac{f(y_n)}{f(x_n)}}, s = \sqrt[m]{\frac{f(z_n)}{f(y_n)}}, u = \sqrt[m]{\frac{f(z_n)}{f(x_n)}}.
$$

where the function $H : \mathbb{C} \to \mathbb{C}$ is restricted to be analytic function in the regions nearby 0 and weight function $L : \mathbb{C}^2 \to \mathbb{C}$ is holomorphic in the regions nearby $(0,0)$ and t, s and u are one-to-m multiple-valued functions.

In the next theorem, it is demonstrated that the proposed scheme in Equation (9) achieves the optimal eighth order of convergence without increasing the number of functional evaluations.

Theorem 1. *Suppose* $x = \mu$ *(say) is a multiple root having multiplicity* $m \geq 1$ *of an analytic function* $f : \mathbb{C} \to \mathbb{C}$ *in the region enclosing a multiple zero* μ *of* $f(x)$. *Which implies that the family of iterative methods defined by Equation (9) has convergence of order eighth when the following conditions are fulfilled:*

$$
H_0 = 1, H_1 = 2, H_2 = -2, H_3 = 36, L_{00} = 0, L_{10} = 1, L_{01} = 2, L_{11} = 4, L_{20} = 2.
\tag{10}
$$

Then, the proposed scheme in Equation (9) satisfies the following error equation:

$$
\begin{aligned}
e_{n+1} &= \frac{1}{24m^7} \{ c_1(c_1^2(11+m) - 2mc_2)((677 + 108m + 7m^2)c_1^4 \\
&\quad - 24m(9+m)c_1^2c_2 + 12m^2c_2^2 + 12m^2c_1c_3)e_n^8 \} + O(e_n^9),
\end{aligned}
\tag{11}
$$

where $e_n = x_n - \mu$ *and* $c_k = \frac{m!}{(m+k)!} \frac{f^{(m+k)}(\mu)}{f^{(m)}(\mu)}, k = 1, 2, 3, \cdots$.

Proof. Suppose $x = \mu$ is a multiple root of $f(x)$. We expand $f(x_n)$ and $f'(x_n)$ by Taylor's series expansion about $x = \mu$ using Mathematica (Computer based algebra software), to get

$$
f(x_n) = \frac{f^{(m)}(\mu)}{m!} e_n^m \left(1 + c_1 e_n + c_2 e_n^2 + c_3 e_n^3 + c_4 e_n^4 + c_5 e_n^5 + c_6 e_n^6 + c_7 e_n^7 + c_8 e_n^8 + O(e_n^9) \right),
\tag{12}
$$

and

$$
\begin{aligned}
f'(x_n) &= \frac{f^{(m)}(\mu)}{m!} e_n^{m-1} m + c_1(m+1)e_n + c_2(m+2)e_n^2 + c_3(m+3)e_n^3 + c_4(m+4)e_n^4 \\
&\quad + c_5(m+5)e_n^5 + c_6(m+6)e_n^6 + c_7(m+7)e_n^7 + c_8(m+8)e_n^8 + O(e_n^9),
\end{aligned}
$$

respectively. By utilizing the above Equations (11) and (12) in the first substep of Equation (9), we obtain

$$
y_n - \mu = \frac{c_1 e_n^2}{m} + \frac{(2c_2 m - c_1^2(m+1))e_n^3}{m^2} + \sum_{k=0}^{4} \lim G_k e_n^{k+4} + O(e_n^9),
\tag{13}
$$

where $G_k = G_k(m, c_1, c_2, \ldots, c_8)$ are expressed in terms of $m, c_1, c_2, c_3, \ldots, c_8$ and the two coefficients G_0 and G_1 can be explicitly written as $G_0 = \frac{1}{m^3}\{3c_3m^2 + c_1^3(m+1)^2 - c_1c_2m(3m+4)\}$ and $G_1 = -\frac{1}{m^4}\{c_1^4(m+1)^3 - 2c_2c_1^2m(2m^2 + 5m + 3) + 2c_3c_1m^2(2m+3) + 2m^2(c_2^2(m+2) - 2c_4m)\}$. By Taylor's expansion, we get

$$f(y_n) = f^{(m)}(\mu)e_n^{2m}\left[\frac{(\frac{c_1}{m})^m}{m!} + \frac{(2mc_2 - (m+1)c_1^2)(\frac{c_1}{m})^m e_n}{c_1 m!} + \sum_{k=0}^{6}\overline{G}_k e_n^{k+2} + O(e_n^9)\right]. \tag{14}$$

By using Equations (12) and (14), we get

$$u = \frac{c_1 e_n}{m} + \frac{(2mc_2 - (m+2)c_1^2)e_n^2}{m^2} + \psi_1 e_n^3 + \psi_2 e_n^4 + \psi_3 e_n^5 + O(e_n^6), \tag{15}$$

where $\psi_1 = \frac{1}{2m^3}[c_1^3(2m^2 + 7m + 7) + 6c_3m^2 - 2c_1c_2m(3m+7)]$, $\psi_2 = -\frac{1}{6m^4}[c_1^4(6m^3 + 29m^2 + 51m + 34) - 6c_2c_1^2m(4m^2 + 16m + 17) + 12c_1c_3m^2(2m+5) + 12m^2(c_2^2(m+3) - 2c_4m)]$, $\psi_3 = \frac{1}{24m^5}[-24m^3(c_2c_3(5m+17) - 5c_5m) + 12c_3c_1^2m^2(10m^2 + 43m + 49) + 12c_1m^2\{c_2^2(10m^2 + 47m + 53) - 2c_4m(5m+13)\} - 4c_2c_1^3m(30m^3 + 163m^2 + 306m + 209) + c_1^5(24m^4 + 146m^3 + 355m^2 + 418m + 209)]$.

Taylor series of $H(t)$ about 0 is given by:

$$H(t) = H_0 + H_1 t + \frac{H_2}{2!}t^2 + \frac{H_3}{3!}t^3 + O(e_n^4) \tag{16}$$

where $H_j = H^j(0)$ for $0 \leq j \leq 3$. Inserting Equations (13)–(16) in the second substep of the scheme in Equation (9), we get

$$\begin{aligned}
z_n = {} & \mu + \frac{-(1+H_0)c_1 e_n^2}{m} - \frac{(1+H_1+m-H_0(3+m)c_1^2) + 2(-1+H_0)mc_2)e_n^3}{m^2} \\
& + \frac{1}{2m^3}\left[(2+10H_1 - H_2 + 4m + 4H_1m + 2m^2 - H_0(13+11m+2m^2))c_1^3 \right. \\
& \left. + 2m(-4 - 4H_1 - 3m + H_0(11+3m)c_1c_2 - 6(-1+H_0)m^2c_3)e_n^4\right] + z_5 e_n^5 \\
& + z_6 e_n^6 + z_7 e_n^7 + O(e_n^8).
\end{aligned}$$

By selecting $H_0 = 1$ and $H_1 = 2$, we obtain

$$z_n = \mu + \frac{(c_1^3(9 - H_2 + m) - 2mc_1c_2)}{2m^3}e_n^4 + z_5 e_n^5 + z_6 e_n^6 + z_7 e_n^7 + O(e_n^8), \tag{17}$$

where $z_5 = -\frac{1}{6m^4}\{c_1^4(125 + H_3 + 84m + 7m^2 - 3H_2(7+3m) + 6m(-3H_2 + 4(7+m))c_1^2c_2 + 12c_2^2m^2 + 12c_2c_1m)\}$, $z_6 = \frac{1}{24m^5}\{1507 + 1850m + 677m^2 + 46m^3 + 4H_3(9+4m) - 6H2(59 + 53m + 12m^2))c_1^5 - 4m(925 + 8H_3 + 594m + 53m^2 - 3H_2(53 + 21m)c_1^3c_2 + 12m^2(83 - 9H_2 + 13m)c_1^2c_3 - 168m^3c_2c_3 + 12m^2c_1(115 - 12H2 + 17m)c_2^2 - 6mc_4)\}$ and $z_7 = -\{12c_1^3c_2m^2(36\beta + 13m + 11) + (37 - 168c_2c_3m^3 + 4c_1^3c_2m(96\beta^2 + 252\beta + 53m^2 + 18(14\beta + 5)m) + 12c_1m^2(c_2^2(48\beta + 17m + 19) - 6c_4m)\}$.

Again, we use the Taylor's expansion for Equation (17) to get:

$$\begin{aligned}
f(z_n) = {} & f^{(m)}(\mu)e_n^{4m}\frac{2^{-m}\left(\frac{c_1^3(9-H_2+m)-2mc_1c_2}{m^3}\right)^m}{m!} - \frac{2^{-m}\left(\frac{c_1^3(9-H_2+m)-2mc_1c_2}{m^3}\right)^{m-1}\rho_0}{3(m^3m!)}e_n \\
& + \sum \lim_{j=0}^{7}\overline{H}_j e_n^{j+1} + O(e_n^9),
\end{aligned}$$

where $\rho_0 = c_1^4(125 + H_3 + 84m + 7m^2 - 3H_2(7 + 3m))c_1^4 - 6m(-3H2 + 4(7 + m))c_1^2 c_2 + 12m^2 c_2^2 + 12c_3 c_1 m^2)$. With the help of Equations (12) and (18), we have

$$s = \frac{c_1^2(9 - H_2 + m) - 2mc_2}{2m^2}e_n^2 + \rho_1 e_n^3 + \rho_2 e_n^4 + \rho_3 e_n^5 + O(e_n^6), \tag{18}$$

where

$$\rho_1 = -\frac{1}{6m^3}\{c_1^3(98 + H_3 + 4m^2 + 54m - 6H_2(3 + m) - 12m(9 - H_2 + m)c_1 c_2 + 12m^2 c_3\},$$

$$\rho_2 = \frac{1}{24m^4}899 + 1002m + 313m^2 + 18m^3 + 4H_3(8 + 3m) - 6H_2(43 + 33m + 6m^2))c_1^4 -$$

$$12m(167 + 2H_3 + 87m + 6m^2 - H_2(33 + 10m)c_1^2 c_2 + 24m^2(26 - 3H_2 + 3m)c_1 c_3 +$$

$$12m^2(c_2^2(35 - 4H_2 + 3m) - 6mc_4)$$

and $\rho_3 = -\frac{1}{60m^5}[-4257 - 7270m - 4455m^2 - 101m^3 - 48m^4 - 10H_3(37 + 30m + 6m^2) + 30H_2(60 + 75m + 31m^2 + 4m^3)c_1^5 + 10m(1454 + 60H_3 + 1548m + 21H_3m + 454m^2 + 24m^3 - 18H_2(25 + 18m + 3m^2)c_1^3 c_2 - 30m^2(234 + 3H_3 + 118m + 8m^2 - 2H2(24 + 7m)c_1^2 c_3 - 60m^2 c_1(141 + 2H_3 + 67m + 4m^2 - 2H_2(15 + 4m)c_2^2 + 2(-17 + 2H_2 - 2m)mc_4) - 120m^3(-25 + 3H_2 - 2m)c_2 c_3 + 2mc_5\} + (\frac{1}{720m^6})((102047 + 180H_2^2 + 204435m + 187055m^2 + 81525m^3 + 14738m^4 + 600m^5 + 40H_3(389 + 498m + 214m^2 + 30m^3) - 45H_2(1223 + 2030m + 1353m^2 + 394m^3 + 40m^4)) - 30m(13629 + 22190m + 12915m^2 + 2746m^3 + 120m^4 + 16H_3(83 + 64m + 12m^2) - 6H_2(1015 + 1209m + 470m^2 + 56m^3)) + 120m^2(2063 + 2088m + 589m^2 + 30m^3 + H_3(88 + 30m) - 18H_2 + (36 + 25m + 4m^2)) + 80m^2(2323 + 2348m + 635m^2 + 30m^3 + 4H_3(28 + 9m) - 3H_2(259 + 173m + 26m^2)) - 2m(303 + 4H_3 + 149m + 10m^2 - 9H_2(7 + 2m)) - 720m^3((393 + 6H_3 + 178m + 10m^2 - H_2(87 + 22m))] + (-42 + 5H_2 - 5m)mc_5) + 20m^3((-473 - 8H_3 - 195m - 10m^2 + 12H_2(9 + 2m))c_2 c_3 + 6m(65 - 8H2 + 5m)c_2 + 3m(71 - 9H2 + 5m)c_{10}mc_6$.

Since it is obvious from Equation (15) that u possess order e_n, the expansion of weight function $L_f(s, u)$ by Taylor's series is possible in the regions nearby origin given as follows:

$$L(s, u) = L_{00} + sL_{10} + uL_{01} + suL_{11} + \frac{s^2}{2!}L_{20} \tag{19}$$

where $L_{i,j} = \frac{1}{i!j!}\frac{\partial^{i+j}}{\partial s^i \partial u^j}L(s, u)\Big|_{(0,0)}$. By using Equations (12)–(19) in the proposed scheme in Equation (9), we have

$$e_{n+1} = M_2 e_n^2 + M_3 e_n^3 + M_4 e_n^4 + M_5 e_n^5 + M_6 e_n^6 + M_7 e_n^7 + O(e_n^8), \tag{20}$$

where the coefficients $M_i(2 \leq i \leq 7)$ depend generally on m and the parameters $L_{i,j}$. To obtain at least fifth-order convergence, we have to choose $L_{00} = 0, L_{10} = 1$ and get

$$e_{n+1} = \frac{((-2 + L_{01})c_1^2((-9 + H_2 - m)c_1^2 + 2mc_2)}{2m^4}e_n^5 + \bar{M}_6 e_n^6 + \bar{M}_7 e_n^7 + O(e_n^8).$$

where the coefficients $\bar{M}_i(6 \leq i \leq 7)$ depend generally on m and the parameters $L_{i,j}$. To obtain eighth-order convergence, we are restricted to choosing the values of parameters given by:

$$H_2 = -2, H_3 = 36, L_{00} = 0, L_{10} = 1, L_{01} = 2, L_{20} = 2, L_{11} = 4. \tag{21}$$

This leads us to the following error equation:

$$\begin{aligned}
e_{n+1} &= \frac{1}{24m^7}[c_1(c_1^2(11 + m) - 2mc_2)((677 + 108m + 7m^2)c_1^4 - 24m(9 + m)c_1^2 c_2 \\
&\quad + 12m^2 c_2^2 + 12m^2 c_1 c_3)]e_n^8 + O(e_n^9).
\end{aligned} \tag{22}$$

The above error equation (Equation (22)) confirms that the presented scheme in Equation (9) achieves optimal order of convergence eight by utilizing only four functional evaluations (using $f(x_n), f'(x_n), f(y_n)$ and $f(z_n)$) per iteration. $\square$

3. Special Cases of Weight Functions

From Theorem 1, several choices of weight functions can be obtained. We have considered the following:

Case 1: The polynomial form of the weight function satisfying the conditions in Equation (10) can be represented as:

$$
\begin{aligned}
H(t) &= 1 + 2t - t^2 + 6t^3 \\
L(s,u) &= s + 2u + 4su + s^2
\end{aligned}
\tag{23}
$$

The particular iterative method related to Equation (23) is given by:
SM-1:

$$
\begin{aligned}
y_n &= x_n - m \cdot \frac{f(x_n)}{f'(x_n)}, n \geq 0, \\
z_n &= y_n - m \cdot t \cdot (1 + 2t - t^2 + 6t^3)\frac{f(x_n)}{f'(x_n)}, \\
x_{n+1} &= z_n - m \cdot t \cdot (s + s^2 + 2u + 4su) \cdot \frac{f(x_n)}{f'(x_n)}
\end{aligned}
$$

$$
\text{where } t = \sqrt[m]{\frac{f(y_n)}{f(x_n)}}, s = \sqrt[m]{\frac{f(z_n)}{f(y_n)}}, u = \sqrt[m]{\frac{f(z_n)}{f(x_n)}}
\tag{24}
$$

Case 2: The second suggested form of the weight functions in which $H(t)$ is constructed using rational weight function satisfying the conditions in Equation (10) is given by:

$$
\begin{aligned}
H(t) &= \frac{1 + 8t + 11t^2}{1 + 6t} \\
L(s,u) &= s + 2u + 4su + s^2
\end{aligned}
\tag{25}
$$

The corresponding iterative method in Equation (25) can be presented as:
SM-2:

$$
\begin{aligned}
y_n &= x_n - m \cdot \frac{f(x_n)}{f'(x_n)}, n \geq 0, \\
z_n &= y_n - m \cdot t \cdot \left(\frac{1 + 8t + 11t^2}{1 + 6t}\right)\frac{f(x_n)}{f'(x_n)}, \\
x_{n+1} &= z_n - m \cdot t \cdot (s + s^2 + 2u + 4su) \cdot \frac{f(x_n)}{f'(x_n)}
\end{aligned}
$$

$$
\text{where } t = \sqrt[m]{\frac{f(y_n)}{f(x_n)}}, s = \sqrt[m]{\frac{f(z_n)}{f(y_n)}}, u = \sqrt[m]{\frac{f(z_n)}{f(x_n)}}
\tag{26}
$$

Case 3: The third suggested form of the weight function in which $H(t)$ is constructed using trigonometric weight satisfying the conditions in Equation (10) is given by:

$$
\begin{aligned}
H(t) &= \frac{5 + 18t}{5 + 8t - 11t^2} \\
L(s,u) &= s + 2u + 4su + s^2
\end{aligned}
\tag{27}
$$

The corresponding iterative method obtained using Equation (27) is given by:

SM-3:

$$y_n = x_n - m \cdot \frac{f(x_n)}{f'(x_n)}, n \geq 0,$$

$$z_n = y_n - m \cdot t \cdot \left(\frac{5 + 18t}{5 + 8t - 11t^2}\right)\frac{f(x_n)}{f'(x_n)},$$

$$x_{n+1} = z_n - m \cdot t \cdot (s + s^2 + 2u + 4su) \cdot \frac{f(x_n)}{f'(x_n)}$$

$$\text{where } t = \sqrt[m]{\frac{f(y_n)}{f(x_n)}}, s = \sqrt[m]{\frac{f(z_n)}{f(y_n)}}, u = \sqrt[m]{\frac{f(z_n)}{f(x_n)}}. \tag{28}$$

4. Numerical Tests

In this section, we show the performance of the presented iterative family in Equation (9) by carrying out some numerical tests and comparing the results with existing method for multiple roots. All numerical computations were performed in Maple 16 programming package using 1000 significant digits of precision. When μ was not exact, we preferred to take the accurate value which has larger number of significant digits rather than the assigned precision. The test functions along with their roots μ and multiplicity m are listed in Table 1 [22]. The proposed methods SM-1 (Equation (24)), SM-2 (Equation (26)) and SM-3 (Equation (28)) are compared with the methods of Geum et al. given in Equations (2) and (4) denoted by GKM-1 and GKM-2 and with method of Bhel given in Equation (6) denoted by BM and Zafar et al. method given in Equation (8) denoted by ZM. In Tables 1–8, the error in first three iterations with reference to the sought zeros ($|x_n - \mu|$) is considered for different methods. The notation $E(-i)$ can be considered as $E \times 10^{-i}$. The test function along with their initial estimates x_0 and computational order of convergence (COC) is also included in these tables, which is computed by the following expression [23]:

$$COC \approx \frac{\log|(x_{k+1} - \mu)/(x_k - \mu)|}{\log|(x_k - \mu)/(x_{k-1} - \mu)|}.$$

Table 1. Test functions.

Test Functions	Exact Root μ	Multiplicity m
$f_1(x) = (cos(\frac{\pi x}{2}) + x^2 - \pi)^5$	2.034724896...	5
$f_2(x) = (e^x + x - 20)^2$	2.842438953...	2
$f_3(x) = (\ln x + \sqrt{(x^4 + 1)} - 2)^9$	1.222813963...	9
$f_4(x) = (\cos x - x)^3$	0.7390851332...	3
$f_5(x) = ((x - 1)^3 - 1)^{50}$	2.0	50
$f_6(x) = (x^3 + 4x^2 - 10)^6$	1.365230013...	6
$f_7(x) = (8xe^{-x^2} - 2x - 3)^8$	-1.7903531791...	8

Table 2. Comparison of different methods for multiple roots.

			$f_1(x), x_0 = 2.5$				
	GKM-1	GKM-2	SM-1	SM-2	SM-3	ZM	BM
$\|x_1 - \mu\|$	6.83(−4)	1.11(−3)	2.15(−4)	1.87(−4)	2.03(−4)	1.52(−4)	1.84(−4)
$\|x_2 - \mu\|$	3.42(−14)	2.53(−18)	2.37(−29)	3.53(−30)	1.25(−29)	9.69(−31)	2.89(−30)
$\|x_3 - \mu\|$	2.13(−55)	3.58(−106)	5.28(−229)	5.71(−236)	2.53(−231)	2.56(−240)	1.05(−236)
COC	4.00	6.00	8.00	8.00	8.00	8.00	8.00

Table 3. Comparison of different methods for multiple roots.

			$f_2(x), x_0 = 3.0$				
	GKM-1	GKM-2	SM-1	SM-2	SM-3	ZM	BM
$|x_1 - \mu|$	1.18(−7)	5.27(−6)	2.33(−7)	1.21(−7)	1.90(−7)	1.40(−7)	1.16(−7)
$|x_2 - \mu|$	2.62(−37)	1.15(−32)	1.30(−53)	2.21(−56)	1.99(−54)	1.30(−55)	1.57(−56)
$|x_3 - \mu|$	3.07(−221)	1.25(−192)	1.19(−423)	2.67(−446)	2.87(−430)	7.37(−440)	1.73(−447)
COC	6.00	6.00	8.00	8.00	8.00	8.00	8.00

Table 4. Comparison of different methods for multiple roots.

			$f_3(x), x_0 = 3.0$				
	GKM-1	GKM-2	SM-1	SM-2	SM-3	ZM	BM
$|x_1 - \mu|$	5.50(−1)	4.29(−2)	1.81(−2)	1.75(−2)	1.79(−2)	D*	D
$|x_2 - \mu|$	3.99(−7)	8.77(−10)	2.82(−15)	9.58(−16)	2.04(−15)	D	D
$|x_3 - \mu|$	1.13(−27)	7.51(−56)	2.06(−117)	8.21(−122)	6.49(−119)	D	D
COC	4.00	6.00	8.00	8.00	8.00	D	D

* D stands for divergence.

Table 5. Comparison of different methods for multiple roots.

			$f_4(x), x_0 = 1.0$				
	GKM-1	GKM-2	SM-1	SM-2	SM-3	ZM	BM
$|x_1 - \mu|$	2.77(−4)	2.55(−5)	6.78(−8)	5.45(−8)	6.29(−8)	4.90(−8)	5.15(−8)
$|x_2 - \mu|$	3.28(−14)	6.83(−36)	7.95(−60)	8.55(−61)	3.83(−60)	4.06(−61)	4.91(−61)
$|x_3 - \mu|$	5.86(−49)	2.51(−213)	2.82(−475)	3.11(−483)	7.18(−478)	8.99(−486)	3.36(−485)
COC	3.50	6.00	8.00	8.00	8.00	7.99	7.99

Table 6. Comparison of different methods for multiple roots.

			$f_5(x), x_0 = 2.1$				
	GKM-1	GKM-2	SM-1	SM-2	SM-3	ZM	BM
$|x_1 - \mu|$	7.68(−5)	1.12(−5)	7.58(−7)	4.85(−7)	6.52(−7)	4.77(−7)	4.65(−7)
$|x_2 - \mu|$	3.49(−17)	5.33(−29)	3.70(−47)	4.10(−49)	8.82(−48)	5.66(−49)	2.72(−49)
$|x_3 - \mu|$	1.46(−66)	6.11(−169)	1.19(−369)	1.06(−385)	9.93(−375)	2.22(−384)	3.79(−387)
COC	3.99	6.00	8.00	8.00	8.00	7.99	7.99

Table 7. Comparison of different methods for multiple roots.

			$f_6(x), x_0 = 3.0$				
	GKM-1	GKM-2	SM-1	SM-2	SM-3	ZM	BM
$|x_1 - \mu|$	5.44(−2)	1.01(−1)	5.40(−2)	5.30(−2)	5.36(−2)	4.36(−2)	5.39(−2)
$|x_2 - \mu|$	7.40(−7)	5.37(−7)	1.10(−10)	4.72(−11)	8.60(−11)	1.36(−11)	4.92(−11)
$|x_3 - \mu|$	3.54(−26)	1.86(−38)	5.28(−80)	2.43(−83)	5.76(−81)	1.80(−87)	3.14(−83)
COC	3.97	5.96	8.00	7.98	7.97	7.97	7.97

Table 8. Comparison of different methods for multiple roots.

			$f_7(x), x_0 = -1.2$				
	GKM-1	GKM-2	SM-1	SM-2	SM-3	ZM	BM
$|x_1 - \mu|$	2.65(−3)	2.15(−3)	4.38(−4)	4.24(−4)	4.32(−4)	3.41(−4)	4.26(−4)
$|x_2 - \mu|$	7.24(−12)	9.63(−17)	4.44(−27)	1.11(−27)	3.11(−27)	3.58(−28)	1.14(−27)
$|x_3 - \mu|$	4.05(−46)	7.81(−97)	4.97(−211)	2.55(−216)	2.28(−212)	5.27(−220)	3.06(−216)
COC	4.00	6.00	8.00	8.00	8.00	7.99	7.99

It is observed that the performance of the new method SM-2 is the same as BM for the function f_1 and better than ZM for the function f_2. The newly developed schemes SM-1, SM-2 and SM-3 are not only convergent but also their speed of convergence is better than GKM-1 and GKM-2 while ZM and BM show divergence for the function f_3. For functions f_4, f_5, f_6 and f_7, the newly developed schemes SM-1, SM-2 and SM-3 are comparable with ZM and BM. Hence, we conclude that the proposed family is comparable and robust among existing methods for multiple roots.

5. Dynamical Analysis

For the sake of stability comparison, we plot the dynamical planes corresponding to each scheme (SM-1, SM-2, SM-3, BM and ZM) for the nonlinear functions f_1, f_2, f_3, f_4, f_5, f_6, f_7 by using the procedure described in [24]. We draw a mesh of 400×400 points such that each point of the mesh is an initial-approximation of the required root of corresponding nonlinear function. The point is orange if the sequence of iteration method converges to the multiple root (with tolerance 10^{-3}) in fewer than 80 iterations and the point is black if the sequence does not converges to the multiple root. The multiple zero is represented by a white star in the figures. Figures 1–14 show that the basin of attraction drawn in orange is of the multiple zero only (i.e., a set of initial guesses converging to the multiple roots fills all the plotted regions of the complex plane). In general, convergence to other zeros or divergence can appear (referred to as strange stationary points). SM-1 has wider regions of convergence for f_1 as compared to ZM and BM in Figures 1 and 2; SM-1 and SM-3 have wider regions of convergence for f_2 as compared to ZM and BM in Figures 3 and 4. The convergence region of SM-2 for functions f_3, f_4 and f_6 is comparable with ZM and BM, as shown in Figures 5–8, 11 and 12. For function f_5 in Figures 9 and 10, the convergence region of SM-3 is better than ZM and BM. For function f_7, SM-1 and SM-3 have better convergence regions than ZM and BM, as shown in Figures 13 and 14. Figures 1–14 show that the region in orange is comparable or bigger for the presented methods SM-1, SM-2 and SM-3 than the regions obtained by schemes BM and ZM, which confirms the fast convergence and stability of the proposed schemes.

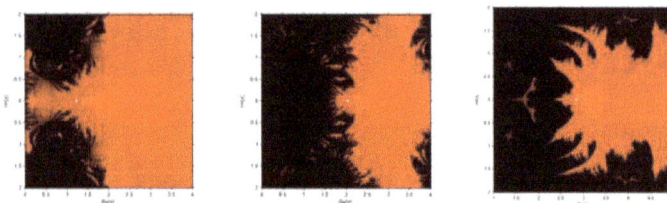

Figure 1. Basins of attraction of SM1 (**Left**), SM2 (**Middle**) and SM3 (**Right**) for $f_1(x)$.

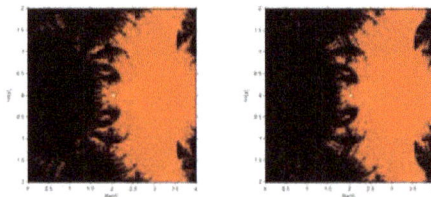

Figure 2. Basins of attraction of BM (**Left**) and ZM (**Right**) for $f_1(x)$.

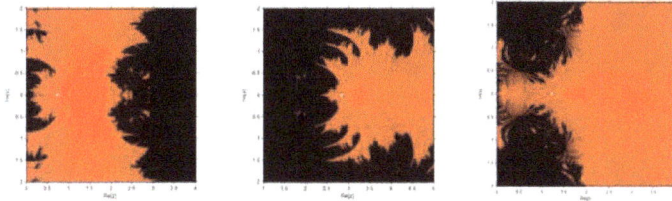

Figure 3. Basins of attraction of SM1 (**Left**), SM2 (**Middle**) and SM3 (**Right**) for $f_2(x)$.

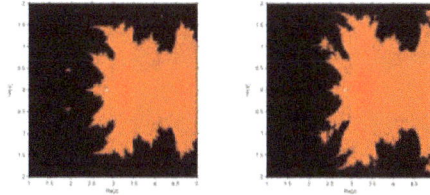

Figure 4. Basins of attraction of BM (**Left**) and ZM (**Right**) for $f_2(x)$.

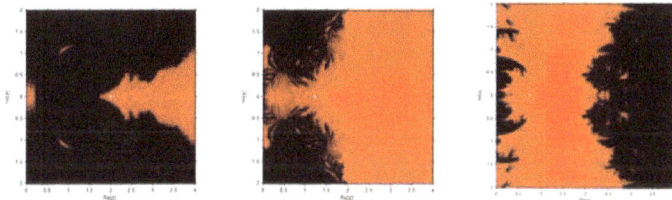

Figure 5. Basins of attraction of SM1 (**Left**), SM2 (**Middle**) and SM3 (**Right**) for $f_3(x)$.

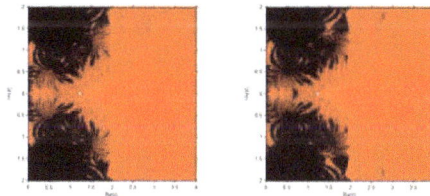

Figure 6. Basins of attraction of BM (**Left**) and ZM (**Right**) for $f_3(x)$.

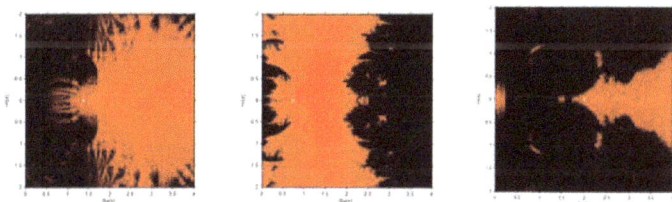

Figure 7. Basins of attraction of SM1 (**Left**), SM2 (**Middle**) and SM3 (**Right**) for $f_4(x)$.

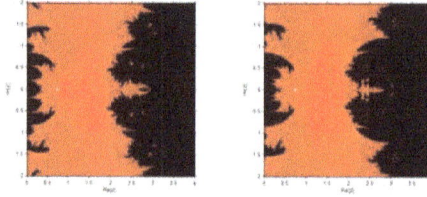

Figure 8. Basins of attraction of BM (**Left**) and ZM (**Right**) for $f_4(x)$.

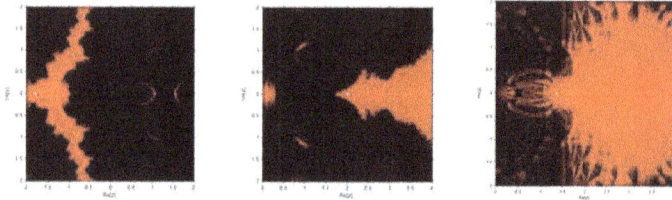

Figure 9. Basins of attraction of SM1 (**Left**), SM2 (**Middle**) and SM3 (**Right**) for $f_5(x)$.

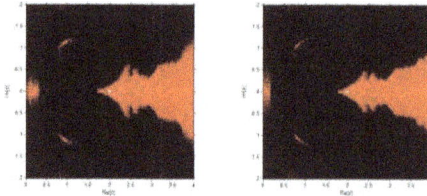

Figure 10. Basins of attraction of BM (**Left**) and ZM (**Right**) for $f_5(x)$.

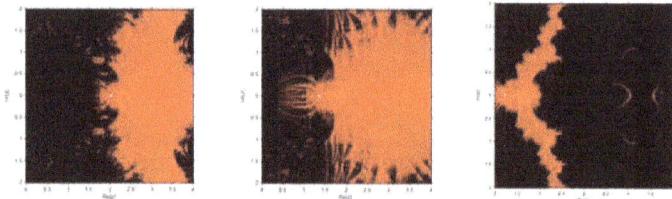

Figure 11. Basins of attraction of SM1 (**Left**), SM2 (**Middle**) and SM3 (**Right**) for $f_6(x)$.

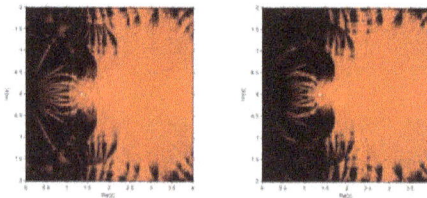

Figure 12. Basins of attraction of BM (**Left**) and ZM (**Right**) for $f_6(x)$.

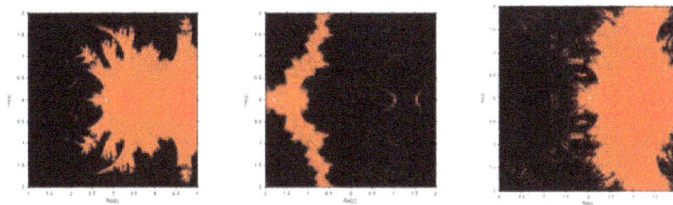

Figure 13. Basins of attraction of SM1 (**Left**), SM2 (**Middle**) and SM3 (**Right**) for $f_7(x)$.

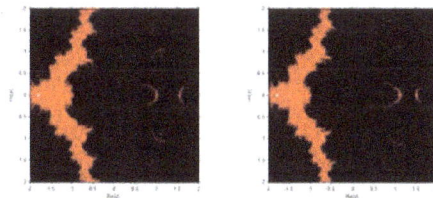

Figure 14. Basins of attraction of BM (**Left**) and ZM (**Right**) for $f_7(x)$.

6. Conclusions

In this paper, we present a new family of optimal eighth-order schemes to find multiple roots of nonlinear equations. An extensive convergence analysis is done, which verifies that the new family is optimal eighth-order convergent. The presented family required four functional evaluations to get optimal eighth-order convergence, having efficiency index $8^{\frac{1}{4}} = 1.6817$, which is higher than the efficiency index of the methods for multiple roots and of the families of Geum et al. [9,10]. Finally, numerical and dynamical tests confirmed the theoretical results and showed that the three members SM-1, SM-2 and SM-3 of the new family are better than existing methods for multiple roots. Hence, the proposed family is efficient among the domain of multiple root finding methods.

Author Contributions: Methodology, S.A.; writing—original draft preparation, S.A.; investigation, S.A.; writing—review and editing, F.Z. and N.Y.; and supervision, F.Z. and N.Y.

Funding: This research received no external funding.

Conflicts of Interest: The authors declare no conflict of interest.

References

1. Ostrowski, A.M. *Solution of Equations and Systems of Equations*; Academic Press: New York, NY, USA, 1960.
2. Petkovic, M.S.; Neta, B.; Petkovic, L.D.; Dzunic, J. *Multipoint Methods for Solving Nonlinear Equations*; Academic Press: New York, NY, USA, 2013.
3. Traub, J.F. *Iterative Methods for the Solution of Equations*; Prentice-Hall: Englewood Cliffs, NJ, USA, 1964.
4. Schröder, E. Über unendlich viele algorithmen zur auflösung der gleichungen. *Math. Annal.* **1870**, *2*, 317–365. [CrossRef]
5. Singh, A.; Jaiswal, P. An efficient family of optimal fourth-order iterative methods for finding multiple roots of nonlinear equations. *Proc. Natl. Acad. Sci. India Sec. A* **2015**, *85*, 439–450. [CrossRef]
6. Behl, R.; Cordero, A.; Motsa, S.S.; Torregrosa, J.R.; Kanwar, V. An optimal fourth-order family of methods for multiple roots and its dynamics. *Numer. Algorithms* **2016**, *71*, 775–796. [CrossRef]
7. Biazar, J.; Ghanbari, B. A new third-order family of nonlinear solvers for multiple roots. *Comput. Math. Appl.* **2010**, *59*, 3315–3319. [CrossRef]
8. Chun, C.; Bae, H.J.; Neta, B. New families of nonlinear third-order solvers for finding multiple roots. *Comput. Math. Appl.* **2009**, *57*, 1574–1582. [CrossRef]
9. Geum, Y.H.; Kim, Y.I.; Neta, B. A class of two-point sixth-order multiple-zero finders of modified double-Newton type and their dynamics. *Appl. Math. Comput.* **2015**, *270*, 387–400. [CrossRef]

10. Geum, Y.H.; Kim, Y.I.; Neta, B. A sixth-order family of three-point modified Newton-like multiple-root finders and the dynamics behind their extraneous fixed points. *Appl. Math. Comput.* **2016**, *283*, 120–140. [CrossRef]
11. Hueso, J.L.; Martinez, E.; Treuel, C. Determination of multiple roots of nonlinear equation and applications. *J. Math. Chem.* **2015**, *53*, 880–892. [CrossRef]
12. Lee, S.; Choe, H. On fourth-order iterative methods for multiple roots of nonlinear equation with high efficiency. *J. Comput. Anal. Appl.* **2015**, *18*, 109–120.
13. Lin, R.I.; Ren, H.M.; Šmarda, Z.; Wu, Q.B.; Khan, Y.; Hu, J.L. New families of third-order iterative methods for finding multiple roots. *J. Appl. Math.* **2014**, *2014*. [CrossRef]
14. Li, S.; Cheng, L.; Neta, B. Some fourth-order nonlinear solvers with closed formulae for multiple roots. *Comput. Math. Appl.* **2010**, *59*, 126–135. [CrossRef]
15. Ahmad, N.; Singh, V.P. Some New Three step Iterative methods for solving nonlinear equation using Steffensen's and Halley method. *Br. J. Math. Comp. Sci.* **2016**, *19*, 1–9. [CrossRef] [PubMed]
16. Neta, B. Extension of Murakami's high-order non-linear solver to multiple roots. *Int. J. Comput. Math.* **2010**, *87*, 1023–1031. [CrossRef]
17. Sharma, J.R.; Sharma, R. Modified Jarratt method for computing multiple roots. *Appl. Math. Comput.* **2010**, *217*, 878–881. [CrossRef]
18. Zhou, X.; Chen, X.; Song, Y. Constructing higher-order methods for obtaining the multiple roots of nonlinear equations. *J. Comput. Appl. Math.* **2011**, *235*, 4199–4206. [CrossRef]
19. Kung, H.T.; Traub, J.F. Optimal order of one-point and multipoint iteration. *J. Assoc. Comput. Mach.* **1974**, *21*, 643–651. [CrossRef]
20. Behl, R.; Cordero, A.; Motsa, S.S.; Torregrosa, J.R. An eighth-order family of optimal multiple root finders and its dynamics. *Numer. Algorithms* **2018**, *77*, 1249–1272. [CrossRef]
21. Zafar, F.; Cordero, A.; Quratulain, R.; Torregrosa, J.R. Optimal iterative methods for finding multiple roots of nonlinear equations using free parameters. *J. Math. Chem.* **2017**, *56*, 1884–1091. [CrossRef]
22. Neta, B.; Chun, C.; Scott, M. On the development of iterative methods for multiple roots. *Appl. Math. Comput.* **2013**, *224*, 358–361. [CrossRef]
23. Weerakoon, S.; Fernando, T.G.I. A variant of Newton's method with accelerated third-order convergence. *Appl. Math. Lett.* **2000**, *13*, 87–93. [CrossRef]
24. Chicharro, F.I.; Cordero, A.; Torregrosa, J.R. Drawing dynamical and pa- rameters planes of iterative families and methods. *Sci. World J.* **2013**, *2013*. [CrossRef] [PubMed]

mathematics

MDPI

Article

An Efficient Conjugate Gradient Method for Convex Constrained Monotone Nonlinear Equations with Applications [†]

Auwal Bala Abubakar [1,2], Poom Kumam [1,3,4,*], Hassan Mohammad [2] and Aliyu Muhammed Awwal [1,5]

[1] KMUTTFixed Point Research Laboratory, SCL 802 Fixed Point Laboratory, Science Laboratory Building, Department of Mathematics, Faculty of Science, King Mongkut's University of Technology Thonburi (KMUTT), 126 Pracha-Uthit Road, Bang Mod, Thrung Khru, Bangkok 10140, Thailand
[2] Department of Mathematical Sciences, Faculty of Physical Sciences, Bayero University, Kano 700241, Nigeria
[3] Center of Excellence in Theoretical and Computational Science (TaCS-CoE), Science Laboratory Building, King Mongkut's University of Technology Thonburi (KMUTT), 126 Pracha-Uthit Road, Bang Mod, Thrung Khru, Bangkok 10140, Thailand
[4] Department of Medical Research, China Medical University Hospital, China Medical University, Taichung 40402, Taiwan
[5] Department of Mathematics, Faculty of Science, Gombe State University, Gombe 760214, Nigeria
*   Correspondence: poom.kum@kmutt.ac.th
[†] This project was supported by Petchra Pra Jom Klao Doctoral Academic Scholarship for Ph.D. Program at KMUTT. Moreover, this project was partially supported by the Thailand Research Fund (TRF) and the King Mongkut's University of Technology Thonburi (KMUTT) under the TRF Research Scholar Award (Grant No. RSA6080047).

Received: 29 June 2019; Accepted: 6 August 2019; Published: 21 August 2019

Abstract: This research paper proposes a derivative-free method for solving systems of nonlinear equations with closed and convex constraints, where the functions under consideration are continuous and monotone. Given an initial iterate, the process first generates a specific direction and then employs a line search strategy along the direction to calculate a new iterate. If the new iterate solves the problem, the process will stop. Otherwise, the projection of the new iterate onto the closed convex set (constraint set) determines the next iterate. In addition, the direction satisfies the sufficient descent condition and the global convergence of the method is established under suitable assumptions. Finally, some numerical experiments were presented to show the performance of the proposed method in solving nonlinear equations and its application in image recovery problems.

Keywords: nonlinear monotone equations; conjugate gradient method; projection method; signal processing

MSC: 65K05; 90C52; 90C56; 92C55

1. Introduction

In this paper, we consider the following constrained nonlinear equation

$$F(x) = 0, \quad \text{subject to} \quad x \in \Psi, \tag{1}$$

where $F : \mathbb{R}^n \to \mathbb{R}^n$ is continuous and monotone. The constraint set $\Psi \subset \mathbb{R}^n$ is nonempty, closed and convex.

Monotone equations appear in many applications [1–3], for example, the subproblems in the generalized proximal algorithms with Bregman distance [4], reformulation of some ℓ_1-norm regularized problems arising in compressive sensing [5] and variational inequality problems are also converted into nonlinear monotone equations via fixed point maps or normal maps [6], (see References [7–9] for more examples). Among earliest methods for the case $\Psi = \mathbb{R}^n$ is the hyperplane projection Newton method proposed by Solodov and Svaiter in Reference [10]. Subsequently, many methods were proposed by different authors. Among the popular methods are spectral gradient methods [11,12], quasi-Newton methods [13–15] and conjugate gradient methods (CG) [16,17].

To solve the constrained case (1), the work of Solodov and Svaiter was extended by Wang et al. [18] which also involves solving a linear system in each iteration but it was shown later by some authors that the computation of the linear system is not necessary. For examples, Xiao and Zhu [19] presented a CG method, which is a combination the well known CG-DESCENT method in Reference [20] with the projection strategy by Solodov and Svaiter. Liu et al. [21] presented two CG method with sufficiently descent directions. In Reference [22], a modified version of the method in Reference [19] was presented by Liu and Li. The modification improves the numerical performance of the method in Reference [19]. Another extension of the Dai and Kou (DK) CG method combined with the projection method to solve (1) was proposed by Ding et al. in Reference [23]. Just recently, to popularize the Dai-Yuan (DY) CG method, Liu and Feng [24] modified the DY such that the direction will be sufficiently descent. A new hybrid spectral gradient projection method for solving convex constraints nonlinear monotone equations was proposed by Awwal et al. in Reference [25]. The method is a convex combination of two different positive spectral parameters together with the projection strategy. In addition, Abubakar et al. extended the method in Reference [17] to solve (1) and also solve some sparse signal recovery problems.

Inspired by some the above methods, we propose a descent conjugate gradient method to solve problem (1). Under appropriate assumptions, the global convergence is established. Preliminary numerical experiments were given to compare the proposed method with existing methods to solve nonlinear monotone equations and some signal and image reconstruction problems arising from compressive sensing.

The remaining part of this paper is organized as follows. In Section 2, we state the proposed algorithm as well as its convergence analysis. Finally, Section 3 reports some numerical results to show the performance of the proposed method in solving Equation (1), signal recovery problems and image restoration problems.

2. Algorithm: Motivation and Convergence Result

This section starts by defining the projection map together with some of its properties.

Definition 1. *Let $\Psi \subset \mathbb{R}^n$ be a nonempty closed convex set. Then for any $x \in \mathbb{R}^n$, its projection onto Ψ, denoted by $P_\Psi(x)$, is defined by*

$$P_\Psi(x) = \arg\min\{\|x - y\| : y \in \Psi\}.$$

Moreover, P_Ψ is nonexpansive, That is,

$$\|P_\Psi(x) - P_\Psi(y)\| \leq \|x - y\|, \quad \forall x, y \in \mathbb{R}^n. \tag{2}$$

All through this article, we assume the followings

(G_1) The mapping F is monotone, that is,

$$(F(x) - F(y))^T (x - y) \geq 0, \quad \forall x, y \in \mathbb{R}^n.$$

(G_2) The mapping F is Lipschitz continuous, that is there exists a positive constant L such that

$$\|F(x) - F(y)\| \le L\|x - y\|, \ \forall x, y \in \mathbb{R}^n.$$

(G_3) The solution set of (1), denoted by Ψ', is nonempty.

An important property that methods for solving Equation (1) must possess is that the direction d_k satisfy

$$F(x_k)^T d_k \le -c\|F(x_k)\|^2, \tag{3}$$

where $c > 0$ is a constant. The inequality (3) is called sufficient descent property if $F(x)$ is the gradient vector of a real valued function $f : \mathbb{R}^n \to \mathbb{R}$.

In this paper, we propose the following search direction

$$d_k = \begin{cases} -F(x_k), & \text{if } k = 0, \\ -F(x_k) + \beta_k d_{k-1} - \theta_k F(x_k), & \text{if } k \ge 1, \end{cases} \tag{4}$$

where

$$\beta_k = \frac{\|F(x_k)\|}{\|d_{k-1}\|} \tag{5}$$

and θ_k is determined such that Equation (3) is satisfied. It is easy to see that for $k = 0$, the equation holds with $c = 1$. Now for $k \ge 1$,

$$F(x_k)^T d_k = -F(x_k)^T F(x_k) + F(x_k)^T \frac{\|F(x_k)\|}{\|d_{k-1}\|} d_{k-1} - \theta_k F(x_k)^T F(x_k)$$

$$= -\|F(x_k)\|^2 + \frac{\|F(x_k)\|}{\|d_{k-1}\|} F(x_k)^T d_{k-1} - \theta_k \|F(x_k)\|^2 \tag{6}$$

$$= \frac{-\|F(x_k)\|^2 \|d_{k-1}\|^2 + \|F(x_k)\| \|d_{k-1}\| F(x_k)^T d_{k-1} - \theta_k \|F(x_k)\|^2 \|d_{k-1}\|^2}{\|d_{k-1}\|^2}.$$

Taking $\theta_k = 1$ we have

$$F(x_k)^T d_k \le -\|F(x_k)\|^2. \tag{7}$$

Thus, the direction defined by (4) satisfy condition (3) $\forall k$ where $c = 1$.

To prove the global convergence of Algorithm 1, the following lemmas are needed.

Algorithm 1: (DCG)

Step 0. Given an arbitrary initial point $x_0 \in \mathbb{R}^n$, parameters $\sigma > 0, 0 < \beta < 1, Tol > 0$ and set $k := 0$.

Step 1. If $\|F(x_k)\| \le Tol$, stop, otherwise go to **Step 2**.

Step 2. Compute d_k using Equation (4).

Step 3. Compute the step size $\alpha_k = \max\{\beta^i : i = 0, 1, 2, \cdots\}$ such that

$$-F(x_k + \alpha_k d_k)^T d_k \ge \sigma \alpha_k \|F(x_k + \alpha_k d_k)\| \|d_k\|^2. \tag{8}$$

Step 4. Set $z_k = x_k + \alpha_k d_k$. If $z_k \in \Psi$ and $\|F(z_k)\| \le Tol$, stop. Else compute

$$x_{k+1} = P_\Psi[x_k - \zeta_k F(z_k)]$$

where

$$\zeta_k = \frac{F(z_k)^T (x_k - z_k)}{\|F(z_k)\|^2}.$$

Step 5. Let $k = k + 1$ and go to **Step 1**.

Lemma 1. *The direction defined by Equation (4) satisfies the sufficient descent property, that is, there exist constants $c > 0$ such that (3) holds.*

Lemma 2. *Suppose that assumptions (G_1)–(G_3) holds, then the sequences $\{x_k\}$ and $\{z_k\}$ generated by Algorithm 1 (CGD) are bounded. Moreover, we have*

$$\lim_{k \to \infty} \|x_k - z_k\| = 0 \tag{9}$$

and

$$\lim_{k \to \infty} \|x_{k+1} - x_k\| = 0. \tag{10}$$

Proof. We will start by showing that the sequences $\{x_k\}$ and $\{z_k\}$ are bounded. Suppose $\bar{x} \in \Psi'$, then by monotonicity of F, we get

$$F(z_k)^T(x_k - \bar{x}) \geq F(z_k)^T(x_k - z_k). \tag{11}$$

Also by definition of z_k and the line search (8), we have

$$F(z_k)^T(x_k - z_k) \geq \sigma \alpha_k^2 \|F(z_k)\| \|d_k\|^2 \geq 0. \tag{12}$$

So, we have

$$\|x_{k+1} - \bar{x}\|^2 = \|P_\Psi[x_k - \zeta_k F(z_k)] - \bar{x}\|^2 \leq \|x_k - \zeta_k F(z_k) - \bar{x}\|^2$$

$$= \|x_k - \bar{x}\|^2 - 2\zeta_k F(z_k)^T(x_k - \bar{x}) + \|\zeta F(z_k)\|^2$$

$$\leq \|x_k - \bar{x}\|^2 - 2\zeta_k F(z_k)^T(x_k - z_k) + \|\zeta F(z_k)\|^2 \tag{13}$$

$$= \|x_k - \bar{x}\|^2 - \left(\frac{F(z_k)^T(x_k - z_k)}{\|F(z_k)\|}\right)^2$$

$$\leq \|x_k - \bar{x}\|^2$$

Thus the sequence $\{\|x_k - \bar{x}\|\}$ is non increasing and convergent and hence $\{x_k\}$ is bounded. Furthermore, from Equation (13), we have

$$\|x_{k+1} - \bar{x}\|^2 \leq \|x_k - \bar{x}\|^2, \tag{14}$$

and we can deduce recursively that

$$\|x_k - \bar{x}\|^2 \leq \|x_0 - \bar{x}\|^2, \quad \forall k \geq 0.$$

Then from Assumption (G_2), we obtain

$$\|F(x_k)\| = \|F(x_k) - F(\bar{x})\| \leq L\|x_k - \bar{x}\| \leq L\|x_0 - \bar{x}\|.$$

If we let $L\|x_0 - \bar{x}\| = \kappa$, then the sequence $\{F(x_k)\}$ is bounded, that is,

$$\|F(x_k)\| \leq \kappa, \quad \forall k \geq 0. \tag{15}$$

By the definition of z_k, Equation (12), monotonicity of F and the Cauchy-Schwatz inequality, we get

$$\sigma\|x_k - z_k\| = \frac{\sigma\|\alpha_k d_k\|^2}{\|x_k - z_k\|} \leq \frac{F(z_k)^T(x_k - z_k)}{\|x_k - z_k\|} \leq \frac{F(z_k)^T(x_k - z_k)}{\|x_k - z_k\|} \leq \|F(x_k)\|. \tag{16}$$

The boundedness of the sequence $\{x_k\}$ together with Equations (15) and (16), implies the sequence $\{z_k\}$ is bounded.

Since $\{z_k\}$ is bounded, then for any $\bar{x} \in \Psi$, the sequence $\{z_k - \bar{x}\}$ is also bounded, that is, there exists a positive constant $\nu > 0$ such that

$$\|z_k - \bar{x}\| \leq \nu.$$

This together with Assumption (G_2) yields

$$\|F(z_k)\| = \|F(z_k) - F(\bar{x})\| \leq L\|z_k - \bar{x}\| \leq L\nu.$$

Therefore, using Equation (13), we have

$$\frac{\sigma^2}{(L\nu)^2}\|x_k - z_k\|^4 \leq \|x_k - \bar{x}\|^2 - \|x_{k+1} - \bar{x}\|^2,$$

which implies

$$\frac{\sigma^2}{(L\nu)^2}\sum_{k=0}^{\infty}\|x_k - z_k\|^4 \leq \sum_{k=0}^{\infty}(\|x_k - \bar{x}\|^2 - \|x_{k+1} - \bar{x}\|^2) \leq \|x_0 - \bar{x}\| < \infty. \tag{17}$$

Equation (17) implies

$$\lim_{k\to\infty}\|x_k - z_k\| = 0.$$

However, using Equation (2), the definition of ζ_k and the Cauchy-Schwartz inequality, we have

$$\|x_{k+1} - x_k\| = \|P_\Psi[x_k - \zeta_k F(z_k)] - x_k\|$$

$$\leq \|x_k - \zeta_k F(z_k) - x_k\|$$

$$= \|\zeta_k F(z_k)\| \tag{18}$$

$$= \|x_k - z_k\|,$$

which yields

$$\lim_{k\to\infty}\|x_{k+1} - x_k\| = 0.$$

□

Equation (9) and definition of z_k implies that

$$\lim_{k\to\infty}\alpha_k\|d_k\| = 0. \tag{19}$$

Lemma 3. *Suppose d_k is generated by Algorithm 1 (CGD), then there exist $M > 0$ such the $\|d_k\| \leq M$*

Proof. By definition of d_k and Equation (15)

$$\|d_k\| = \| - 2F(x_k) + \frac{\|F(x_k)\|}{\|d_{k-1}\|} d_{k-1}\|$$

$$\leq 2\|F(x_k)\| + \frac{\|F(x_k)\|}{\|d_{k-1}\|} \|d_{k-1}\| \tag{20}$$

$$\leq 3\|F(x_k)\|$$

$$\leq 3\kappa.$$

Letting $M = 3\kappa$, we have the desired result. □

Theorem 1. *Suppose that assumptions (G_1)–(G_3) hold and let the sequence $\{x_k\}$ be generated by Algorithm 1, then*

$$\liminf_{k \to \infty} \|F(x_k)\| = 0, \tag{21}$$

Proof. To prove the Theorem, we consider two cases;

Case 1

Suppose $\liminf\limits_{k \to \infty} \|d_k\| = 0$, we have $\liminf\limits_{k \to \infty} \|F(x_k)\| = 0$. Then by continuity of F, the sequence $\{x_k\}$ has some accumulation point $\bar{x}$ such that $F(\bar{x}) = 0$. Because $\{\|x_k - \bar{x}\|\}$ converges and $\bar{x}$ is an accumulation point of $\{x_k\}$, therefore $\{x_k\}$ converges to $\bar{x}$.

Case 2

Suppose $\liminf\limits_{k \to \infty} \|d_k\| > 0$, we have $\liminf\limits_{k \to \infty} \|F(x_k)\| > 0$. Then by (19), it holds that $\lim\limits_{k \to \infty} \alpha_k = 0$. Also from Equation (8),

$$-F(x_k + \beta^{i-1} d_k)^T d_k < \sigma \beta^{i-1} \|F(x_k + \beta^{i-1} d_k)\| \|d_k\|^2$$

and the boundedness of $\{x_k\}, \{d_k\}$, we can choose a sub-sequence such that allowing k to go to infinity in the above inequality results

$$F(\bar{x})^T \bar{d} > 0. \tag{22}$$

On the other hand, allowing k to approach ∞ in (7), implies

$$F(\bar{x})^T \bar{d} \leq 0. \tag{23}$$

(22) and (23) imply contradiction. Hence, $\liminf\limits_{k \to \infty} \|F(x_k)\| > 0$ is not true and the proof is complete. □

3. Numerical Examples

This section gives the performance of the proposed method with existing methods such as PCG and PDY proposed in References [22,24], respectively, to solve monotone nonlinear equations using 9 benchmark test problems. Furthermore Algorithm 1 is applied to restore a blurred image. All codes were written in MATLAB R2018b and run on a PC with intel COREi5 processor with 4 GB of RAM and CPU 2.3 GHZ. All runs were stopped whenever $\|F(x_k)\| < 10^{-5}$.

The parameters chosen for the existing algorithm are as follows:

PCG method: All parameters are chosen as in Reference [22].

PDY method: All parameters are chosen as in Reference [24].

Algorithm 1: We have tested several values of $\beta \in (0, 1)$ and found that $\beta = 0.7$ gives the best result. In addition, to implement most of the optimization algorithms, the parameter σ is chosen as a very small number. Therefore, we chose $\beta = 0.7$ and $\sigma = 0.0001$ for the implementation of the proposed algorithm.

We test 9 different problems with dimensions ranging from $n = 1000, 5000, 10,000, 50,000, 100,000$ and 6 initial points: $x_1 = (0.1, 0.1, \cdots, 1)^T$, $x_2 = (0.2, 0.2, \cdots, 0.2)^T$, $x_3 = (0.5, 0.5, \cdots, 0.5)^T$, $x_4 = (1.2, 1.2, \cdots, 1.2)^T$, $x_5 = (1.5, 1.5, \cdots, 1.5)^T$, $x_6 = (2, 2, \cdots, 2)^T$. In Tables 1–9, the number of iterations (ITER), number of function evaluations (FVAL), CPU time in seconds (TIME) and the norm at the approximate solution (NORM) were reported. The symbol '$-$' is used when the number of iterations exceeds 1000 and/or the number of function evaluations exceeds 2000.

The test problems are listed below, where the function F is taken as $F(x) = (f_1(x), f_2(x), \ldots, f_n(x))^T$.

Problem 1 ([26]). *Exponential Function.*

$$f_1(x) = e^{x_1} - 1,$$
$$f_i(x) = e^{x_i} + x_i - 1, \text{ for } i = 2, 3, ..., n,$$
$$\text{and } \Psi = \mathbb{R}^n_+.$$

Problem 2 ([26]). *Modified Logarithmic Function.*

$$f_i(x) = \ln(x_i + 1) - \frac{x_i}{n}, \text{ for } i = 2, 3, ..., n,$$
$$\text{and } \Psi = \{x \in \mathbb{R}^n : \sum_{i=1}^{n} x_i \le n, x_i > -1, i = 1, 2, \ldots, n\}.$$

Problem 3 ([13]). *Nonsmooth Function.*

$$f_i(x) = 2x_i - \sin |x_i|, \ i = 1, 2, 3, ..., n,$$
$$\text{and } \Psi = \{x \in \mathbb{R}^n : \sum_{i=1}^{n} x_i \le n, x_i \ge 0, i = 1, 2, \ldots, n\}.$$

It is clear that Problem 3 is nonsmooth at $x = 0$.

Problem 4 ([26]). *Strictly Convex Function I.*

$$f_i(x) = e^{x_i} - 1, \text{ for } i = 1, 2, ..., n,$$
$$\text{and } \Psi = \mathbb{R}^n_+.$$

Problem 5 ([26]). *Strictly Convex Function II.*

$$f_i(x) = \frac{i}{n} e^{x_i} - 1, \text{ for } i = 1, 2, ..., n,$$
$$\text{and } \Psi = \mathbb{R}^n_+.$$

Problem 6 ([27]). *Tridiagonal Exponential Function*

$$f_1(x) = x_1 - e^{\cos(h(x_1 + x_2))},$$
$$f_i(x) = x_i - e^{\cos(h(x_{i-1} + x_i + x_{i+1}))}, \text{ for } i = 2, ..., n-1,$$
$$f_n(x) = x_n - e^{\cos(h(x_{n-1} + x_n))},$$
$$h = \frac{1}{n+1} \text{ and } \Psi = \mathbb{R}^n_+.$$

Problem 7 ([28]). *Nonsmooth Function*

$$f_i(x) = x_i - \sin|x_i - 1|, \ i = 1, 2, 3, ..., n.$$

$$\text{and } \Psi = \{x \in \mathbb{R}^n : \sum_{i=1}^{n} x_i \le n, x_i \ge -1, i = 1, 2, ..., n\}.$$

Problem 8 ([23]). *Penalty 1*

$$t_i = \sum_{i=1}^{n} x_i^2, \ c = 10^{-5}$$

$$f_i(x) = 2c(x_i - 1) + 4(t_i - 0.25)x_i, \ i = 1, 2, 3, ..., n.$$

$$\text{and } \Psi = \mathbb{R}_+^n.$$

Problem 9 ([29]). *Semismooth Function*

$$f_1(x) = x_1 + x_1^3 - 10,$$
$$f_2(x) = x_2 - x_3 + x_2^3 + 1,$$
$$f_3(x) = x_2 + x_3 + 2x_3^3 - 3,$$
$$f_4(x) = 2x_4^3,$$

$$\text{and } \Psi = \{x \in \mathbb{R}^4 : \sum_{i=1}^{4} x_i \le 3, x_i \ge 0, i = 1, 2, 3, 4\}.$$

In addition, we employ the performance profile developed in Reference [30] to obtain Figures 1–3, which is a helpful process of standardizing the comparison of methods. The measure of the performance profile considered are; number of iterations, CPU time (in seconds) and number of function evaluations. Figure 1 reveals that Algorithm 1 most performs better in terms of number of iterations, as it solves and wins 90 percent of the problems with less number of iterations, while PCG and PDY solves and wins less than 10 percent. In Figure 2, Algorithm 1 performed a little less by solving and winning over 80 percent of the problems with less CPU time as against PCG and PDY with similar performance of less than 10 percent of the problems considered. The translation of Figure 3 is identical to Figure 1. Figure 4 is the plot of the decrease in residual norm against number of iterations on problem 9 with x_4 as initial point. It shows the speed of the convergence of each algorithm using the convergence tolerance 10^{-5}, it can be observed that Algorithm 1 converges faster than PCG and PDY.

Table 1. Numerical Results for **Algorithm** 1 (DCG), **PCG** and **PDY** for Problem 1 with given initial points and dimensions.

DIMENSION	INITIAL POINT	Algorithm 1				PCG				PDY			
		ITER	FVAL	TME	NORM	ITER	FVAL	TIME	NORM	ITER	FVAL	TIME	NORM
1000	x_1	11	49	0.025557	8.88×10^{-6}	18	73	0.019295	5.72×10^{-6}	12	49	0.16248	9.18×10^{-6}
	x_2	12	53	0.014164	4.78×10^{-6}	18	73	0.011648	9.82×10^{-6}	13	53	0.03780	6.35×10^{-6}
	x_3	12	53	0.008524	8.75×10^{-6}	19	77	0.011197	7.1×10^{-6}	14	57	0.01550	5.59×10^{-6}
	x_4	13	57	0.011333	6.68×10^{-6}	18	73	0.022197	8.27×10^{-6}	15	61	0.01746	4.07×10^{-6}
	x_5	13	57	0.014202	6.09×10^{-6}	63	254	0.046072	9.58×10^{-6}	14	57	0.02193	9.91×10^{-6}
	x_6	13	57	0.011045	8.14×10^{-6}	61	246	0.031608	9.15×10^{-6}	40	162	0.03472	9.70×10^{-6}
5000	x_1	12	53	0.024311	5.82×10^{-6}	18	73	0.11431	7.42×10^{-6}	13	53	0.03158	6.87×10^{-6}
	x_2	13	57	0.027361	3.13×10^{-6}	19	77	0.03997	6.53×10^{-6}	14	57	0.04270	4.62×10^{-6}
	x_3	13	57	0.02541	5.73×10^{-6}	20	81	0.056159	5.2×10^{-6}	15	61	0.05433	4.18×10^{-6}
	x_4	14	61	0.032038	4.38×10^{-6}	19	77	0.038381	8.1×10^{-6}	15	61	0.04357	9.08×10^{-6}
	x_5	14	61	0.039044	3.98×10^{-6}	62	250	0.15836	9.53×10^{-6}	15	61	0.08960	7.30×10^{-6}
	x_6	14	61	0.027231	5.33×10^{-6}	60	242	0.13276	9.1×10^{-6}	39	158	0.11284	9.86×10^{-6}
10,000	x_1	12	53	0.05434	8.23×10^{-6}	18	73	0.073207	9.5×10^{-6}	13	53	0.06371	9.70×10^{-6}
	x_2	13	57	0.045664	4.43×10^{-6}	19	77	0.090771	8.15×10^{-6}	14	57	0.06336	6.53×10^{-6}
	x_3	13	57	0.041922	8.09×10^{-6}	20	81	0.070859	6.74×10^{-6}	15	61	0.06414	5.90×10^{-6}
	x_4	14	61	0.047641	6.2×10^{-6}	20	81	0.087357	5.11×10^{-6}	16	65	0.07920	4.28×10^{-6}
	x_5	14	61	0.05-5734	5.62×10^{-6}	62	250	0.24646	8.87×10^{-6}	39	158	0.22101	7.97×10^{-6}
	x_6	14	61	0.057104	7.54×10^{-6}	59	238	0.19949	9.96×10^{-6}	87	351	0.36237	9.93×10^{-6}
50,000	x_1	13	57	0.15384	5.41×10^{-6}	19	77	0.25487	8.8×10^{-6}	14	57	0.27607	7.12×10^{-6}
	x_2	13	57	0.15633	9.9×10^{-6}	20	81	0.32689	7.39×10^{-6}	15	61	0.26220	4.91×10^{-6}
	x_3	14	61	0.20801	5.32×10^{-6}	21	85	0.33649	6.31×10^{-6}	16	65	0.28260	4.37×10^{-6}
	x_4	15	65	0.1946	4.08×10^{-6}	21	85	0.32779	5.1×10^{-6}	38	154	0.60650	7.54×10^{-6}
	x_5	15	65	0.19799	3.69×10^{-6}	61	246	0.82615	8.85×10^{-6}	177	712	2.52330	9.44×10^{-6}
	x_6	15	65	0.22418	4.95×10^{-6}	59	238	0.79992	8.5×10^{-6}	361	1449	5.97950	9.74×10^{-6}
100,000	x_1	13	57	0.32291	7.65×10^{-6}	20	81	0.53846	5.52×10^{-6}	15	61	0.39342	3.39×10^{-6}
	x_2	14	61	0.33329	4.12×10^{-6}	21	85	0.61533	4.62×10^{-6}	15	61	0.42154	6.94×10^{-6}
	x_3	14	61	0.37048	7.52×10^{-6}	21	85	0.53638	8.78×10^{-6}	16	65	0.45851	6.18×10^{-6}
	x_4	15	65	0.36058	5.76×10^{-6}	21	85	0.62002	7.21×10^{-6}	175	704	4.36100	9.47×10^{-6}
	x_5	15	65	0.34975	5.22×10^{-6}	60	242	1.4564	9.73×10^{-6}	176	708	4.29180	9.91×10^{-6}
	x_6	15	65	0.3621	7.01×10^{-6}	58	234	1.4155	9.42×10^{-6}	360	1445	9.71190	9.99×10^{-6}

Table 2. Numerical Results for Algorithm 1 (DCG), PCG and PDY for Problem 2 with given initial points and dimensions.

DIMENSION	INITIAL POINT	Algorithm 1				PCG				PDY			
		ITER	FVAL	TIME	NORM	ITER	FVAL	TIME	NORM	ITER	FVAL	TIME	NORM
1000	x_1	9	38	3.1744	5.84×10^{-6}	15	59	0.049899	8.59×10^{-6}	10	39	0.01053	6.96×10^{-6}
	x_2	10	42	0.014633	6.25×10^{-6}	11	42	0.015089	9.07×10^{-6}	11	43	0.00937	9.23×10^{-6}
	x_3	9	38	0.017067	7.4×10^{-6}	17	66	0.016935	6.44×10^{-6}	13	51	0.01111	6.26×10^{-6}
	x_4	7	30	0.006392	6.53×10^{-6}	18	69	0.01436	6×10^{-6}	14	55	0.02154	9.46×10^{-6}
	x_5	11	46	0.011954	3.47×10^{-6}	13	48	0.00907	7.58×10^{-6}	15	59	0.01850	4.60×10^{-6}
	x_6	12	50	0.68666	6.74×10^{-6}	18	68	0.01352	5.4×10^{-6}	15	59	0.01938	7.71×10^{-6}
5000	x_1	10	42	0.11241	3.53×10^{-6}	16	63	0.041151	9.35×10^{-6}	11	43	0.03528	4.86×10^{-6}
	x_2	11	46	0.028723	3.81×10^{-6}	12	46	0.028706	8.8×10^{-6}	12	47	0.04032	6.89×10^{-6}
	x_3	10	42	0.029367	4.3×10^{-6}	18	70	0.047532	6.98×10^{-6}	14	55	0.04889	4.61×10^{-6}
	x_4	13	54	0.036231	3.67×10^{-6}	19	73	0.052164	6.45×10^{-6}	15	59	0.04826	6.96×10^{-6}
	x_5	11	46	0.04963	7.21×10^{-6}	14	52	0.040529	6.71×10^{-6}	16	63	0.05969	3.37×10^{-6}
	x_6	13	54	0.054971	4.05×10^{-6}	19	72	0.12303	5.71×10^{-6}	16	63	0.06253	5.64×10^{-6}
10,000	x_1	10	42	0.049614	4.98×10^{-6}	17	67	0.074779	6.6×10^{-6}	11	43	0.06732	6.85×10^{-6}
	x_2	11	46	0.061595	5.36×10^{-6}	13	50	0.08308	6.11×10^{-6}	12	47	0.12232	9.72×10^{-6}
	x_3	10	42	0.054587	6.02×10^{-6}	18	70	0.085554	9.83×10^{-6}	14	55	0.08288	6.51×10^{-6}
	x_4	13	54	0.073333	5.16×10^{-6}	19	73	0.10579	9.07×10^{-6}	15	59	0.08413	9.82×10^{-6}
	x_5	12	50	0.06306	2.83×10^{-6}	14	52	0.074982	9.18×10^{-6}	16	63	0.09589	4.75×10^{-6}
	x_6	13	54	0.062259	5.69×10^{-6}	19	72	0.099167	8.02×10^{-6}	16	64	0.11499	8.55×10^{-6}
50,000	x_1	11	46	0.20703	3.1×10^{-6}	18	71	0.39473	7.37×10^{-6}	12	47	0.27826	5.23×10^{-6}
	x_2	12	50	0.23251	3.35×10^{-6}	14	54	0.27346	6.74×10^{-6}	13	51	0.29642	7.11×10^{-6}
	x_3	11	46	0.21338	3.73×10^{-6}	20	78	0.37249	5.5×10^{-6}	15	59	0.35602	4.82×10^{-6}
	x_4	14	58	0.3232	3.22×10^{-6}	21	81	0.37591	5.07×10^{-6}	35	141	0.69470	6.69×10^{-6}
	x_5	12	50	0.22703	6.27×10^{-6}	16	60	0.26339	5.02×10^{-6}	35	141	0.68488	9.12×10^{-6}
	x_6	14	58	0.25979	3.54×10^{-6}	20	76	0.33814	8.93×10^{-6}	35	141	0.70973	9.91×10^{-6}
100,000	x_1	11	46	0.55511	4.38×10^{-6}	19	75	0.65494	5.22×10^{-6}	12	47	0.44541	7.39×10^{-6}
	x_2	12	50	0.54694	4.73×10^{-6}	14	54	0.4944	9.52×10^{-6}	14	55	0.53299	3.39×10^{-6}
	x_3	11	46	0.40922	5.27×10^{-6}	20	78	0.78319	7.78×10^{-6}	15	60	0.58603	8.71×10^{-6}
	x_4	14	58	0.62049	4.55×10^{-6}	21	81	0.76051	7.17×10^{-6}	72	290	2.70630	8.31×10^{-6}
	x_5	12	50	0.47039	8.86×10^{-6}	16	60	0.58545	7.07×10^{-6}	72	290	2.72220	8.68×10^{-6}
	x_6	14	58	0.71174	5.01×10^{-6}	21	80	0.77051	6.32×10^{-6}	72	290	2.75850	8.96×10^{-6}

Table 3. Numerical Results for **Algorithm** 1 (DCG), **PCG** and **PDY** for Problem 3 with given initial points and dimensions.

DIMENSION	INITIAL POINT	Algorithm 1 (DCG)				PCG				PDY			
		ITER	FVAL	TIME	NORM	ITER	FVAL	TIME	NORM	ITER	FVAL	TIME	NORM
1000	x_1	10	43	0.75322	9.9×10^{-6}	19	76	0.55752	5.62×10^{-6}	12	48	0.01255	4.45×10^{-6}
	x_2	11	47	0.06933	5.46×10^{-6}	20	80	0.010936	5.58×10^{-6}	12	48	0.01311	9.02×10^{-6}
	x_3	12	51	0.00676	3.48×10^{-6}	21	84	0.011048	6.58×10^{-6}	13	52	0.01486	8.34×10^{-6}
	x_4	12	51	0.09664	4.41×10^{-6}	22	88	0.011058	5.67×10^{-6}	14	56	0.01698	8.04×10^{-6}
	x_5	11	47	0.010487	9.06×10^{-6}	22	88	0.012198	5.64×10^{-6}	14	56	0.01551	9.72×10^{-6}
	x_6	13	55	0.012702	3.15×10^{-6}	21	84	0.018231	8.36×10^{-6}	14	56	0.01534	9.42×10^{-6}
5000	x_1	11	47	0.019458	6.19×10^{-6}	20	80	0.040808	6.29×10^{-6}	12	48	0.03660	9.94×10^{-6}
	x_2	12	51	0.021562	3.42×10^{-6}	21	84	0.06688	6.25×10^{-6}	13	52	0.03616	6.85×10^{-6}
	x_3	12	51	0.024274	7.79×10^{-6}	22	88	0.04144	7.37×10^{-6}	14	56	0.04594	6.14×10^{-6}
	x_4	12	51	0.026771	9.86×10^{-6}	23	92	0.052214	6.35×10^{-6}	15	60	0.04342	6.01×10^{-6}
	x_5	12	51	0.026814	5.67×10^{-6}	23	92	0.041444	6.31×10^{-6}	15	60	0.04296	7.25×10^{-6}
	x_6	13	55	0.023903	7.03×10^{-6}	22	88	0.040135	9.37×10^{-6}	32	129	0.10081	8.85×10^{-6}
10,000	x_1	11	47	0.044134	8.75×10^{-6}	20	80	0.064312	8.9×10^{-6}	13	52	0.06192	4.77×10^{-6}
	x_2	12	51	0.051947	4.83×10^{-6}	21	84	0.088102	8.84×10^{-6}	13	52	0.06442	9.68×10^{-6}
	x_3	13	55	0.057291	3.08×10^{-6}	23	92	0.07296	5.22×10^{-6}	14	56	0.09499	8.69×10^{-6}
	x_4	13	55	0.055134	3.9×10^{-6}	23	92	0.075265	8.99×10^{-6}	15	60	0.07696	8.5×10^{-6}
	x_5	12	51	0.047551	8.02×10^{-6}	23	92	0.073937	8.93×10^{-6}	33	133	0.18625	6.45×10^{-6}
	x_6	13	55	0.055069	9.95×10^{-6}	23	92	0.099888	6.64×10^{-6}	33	133	0.15548	7.51×10^{-6}
50,000	x_1	12	51	0.19938	5.47×10^{-6}	21	84	0.27031	9.97×10^{-6}	14	56	0.23642	3.51×10^{-6}
	x_2	13	55	0.22499	3.02×10^{-6}	22	88	0.2657	9.9×10^{-6}	14	56	0.24813	7.12×10^{-6}
	x_3	13	55	0.19396	6.89×10^{-6}	24	96	0.3246	5.85×10^{-6}	15	60	0.27049	6.53×10^{-6}
	x_4	13	55	0.20259	8.72×10^{-6}	25	100	0.32373	5.04×10^{-6}	34	137	0.54545	7.13×10^{-6}
	x_5	13	55	0.19452	5.01×10^{-6}	25	100	0.33764	5.01×10^{-6}	68	274	1.02330	9.99×10^{-6}
	x_6	14	59	0.22015	6.22×10^{-6}	24	96	0.33687	7.44×10^{-6}	69	278	1.03810	8.05×10^{-6}
100,000	x_1	12	51	0.39983	7.74×10^{-6}	22	88	0.63809	7.06×10^{-6}	14	56	0.45475	4.96×10^{-6}
	x_2	13	55	0.32765	4.28×10^{-6}	23	92	0.63458	7.02×10^{-6}	15	60	0.49018	3.39×10^{-6}
	x_3	13	55	0.30133	9.75×10^{-6}	24	96	0.71422	8.27×10^{-6}	15	60	0.49016	9.24×10^{-6}
	x_4	14	59	0.42865	3.45×10^{-6}	25	100	0.73524	7.13×10^{-6}	139	559	4.03110	9.01×10^{-6}
	x_5	13	55	0.34512	7.09×10^{-6}	25	100	0.70625	7.09×10^{-6}	70	282	2.07100	8.54×10^{-6}
	x_6	14	59	0.40387	8.8×10^{-6}	25	100	0.76777	5.27×10^{-6}	139	559	4.02440	9.38×10^{-6}

Table 4. Numerical Results for **Algorithm** 1 (DCG), **PCG** and **PDY** for Problem 4 with given initial points and dimensions.

		Algorithm 1				PCG				PDY			
DIMENSION	INITIAL POINT	ITER	FVAL	TIME	NORM	ITER	FVAL	TIME	NORM	ITER	FVAL	TIME	NORM
1000	x_1	10	43	0.15461	8.33×10^{-6}	18	72	0.11853	9.93×10^{-6}	12	48	0.00989	4.60×10^{-6}
	x_2	11	47	0.006276	3.84×10^{-6}	19	76	0.014318	8.75×10^{-6}	12	48	0.00966	9.57×10^{-6}
	x_3	11	47	0.009859	3.91×10^{-6}	20	80	0.0093776	7.15×10^{-6}	13	52	0.00887	8.49×10^{-6}
	x_4	11	47	0.007976	5.21×10^{-6}	47	189	0.023321	7.83×10^{-6}	12	48	0.01207	5.83×10^{-6}
	x_5	12	51	0.008382	4.09×10^{-6}	46	185	0.047105	9.76×10^{-6}	29	117	0.05371	9.43×10^{-6}
	x_6	12	51	0.008645	3.32×10^{-6}	41	165	0.027719	8.77×10^{-6}	29	117	0.02396	6.65×10^{-6}
5000	x_1	11	47	0.022024	5.21×10^{-6}	20	80	0.029445	5.57×10^{-6}	13	52	0.02503	3.49×10^{-6}
	x_2	11	47	0.020587	8.59×10^{-6}	20	80	0.033115	9.8×10^{-6}	13	52	0.02626	7.24×10^{-6}
	x_3	11	47	0.023714	8.75×10^{-6}	21	84	0.033318	8.01×10^{-6}	14	56	0.03349	6.29×10^{-6}
	x_4	12	51	0.024728	3.26×10^{-6}	49	197	0.071715	9.46×10^{-6}	13	52	0.02258	4.25×10^{-6}
	x_5	12	51	0.031015	9.14×10^{-6}	49	197	0.068565	8.68×10^{-6}	31	125	0.05471	7.59×10^{-6}
	x_6	12	51	0.030012	7.43×10^{-6}	44	177	0.070862	7.79×10^{-6}	63	254	0.10064	8.54×10^{-6}
10,000	x_1	11	47	0.041476	7.37×10^{-6}	20	80	0.043013	7.88×10^{-6}	13	52	0.03761	4.93×10^{-6}
	x_2	12	51	0.047866	3.4×10^{-6}	21	84	0.051685	6.94×10^{-6}	14	56	0.04100	3.37×10^{-6}
	x_3	12	51	0.042607	3.46×10^{-6}	22	88	0.050422	5.67×10^{-6}	14	56	0.03919	8.90×10^{-6}
	x_4	12	51	0.036406	4.61×10^{-6}	50	201	0.17563	9.84×10^{-6}	32	129	0.09613	6.02×10^{-6}
	x_5	13	55	0.041374	3.61×10^{-6}	50	201	0.20035	9.03×10^{-6}	32	129	0.09177	6.44×10^{-6}
	x_6	13	55	0.039847	2.94×10^{-6}	45	181	0.12214	8.11×10^{-6}	64	258	0.20791	9.39×10^{-6}
50,000	x_1	12	51	0.13928	4.61×10^{-6}	21	84	0.27145	8.83×10^{-6}	14	56	0.17193	3.63×10^{-6}
	x_2	12	51	0.18031	7.6×10^{-6}	22	88	0.23149	7.78×10^{-6}	14	56	0.15237	7.54×10^{-6}
	x_3	12	51	0.12526	7.74×10^{-6}	23	92	0.28789	6.36×10^{-6}	15	60	0.16549	6.66×10^{-6}
	x_4	13	55	0.14322	2.88×10^{-6}	53	213	0.61624	8.75×10^{-6}	67	270	0.76283	7.81×10^{-6}
	x_5	13	55	0.17904	8.08×10^{-6}	53	213	0.7119	8.02×10^{-6}	67	270	0.76157	8.80×10^{-6}
	x_6	13	55	0.13635	6.57×10^{-6}	47	189	0.48192	9.8×10^{-6}	269	1080	2.92510	9.41×10^{-6}
100,000	x_1	12	51	0.24293	6.52×10^{-6}	22	88	0.60822	6.25×10^{-6}	14	56	0.30229	5.13×10^{-6}
	x_2	13	55	0.27433	3.01×10^{-6}	23	92	0.52965	5.51×10^{-6}	15	60	0.31648	3.59×10^{-6}
	x_3	13	55	0.2714	3.06×10^{-6}	23	92	0.57064	8.99×10^{-6}	32	129	0.72838	9.99×10^{-6}
	x_4	13	55	0.26819	4.08×10^{-6}	54	217	1.1805	9.1×10^{-6}	135	543	2.86780	9.73×10^{-6}
	x_5	14	59	0.31696	3.2×10^{-6}	54	217	1.107	8.34×10^{-6}	272	1092	5.74140	9.91×10^{-6}
	x_6	13	55	0.2698	9.29×10^{-6}	49	197	1.0617	7.49×10^{-6}	548	2197	11.44130	9.87×10^{-6}

Table 5. Numerical Results for **Algorithm** 1 (DCG), PCG and PDY for Problem 5 with given initial points and dimensions.

DIMENSION	INITIAL POINT	Algorithm 1				PCG				PDY			
		ITER	FVAL	TIME	NORM	ITER	FVAL	TIME	NORM	ITER	FVAL	TIME	NORM
1000	x_1	19	78	0.71709	8.63×10^{-6}	22	83	0.099338	7.48×10^{-6}	16	63	0.07575	6.03×10^{-6}
	x_2	21	86	0.017127	7.65×10^{-6}	23	88	0.016014	7.31×10^{-6}	16	63	0.01470	5.42×10^{-6}
	x_3	23	95	0.013909	7.23×10^{-6}	23	90	0.016328	9.31×10^{-6}	33	132	0.02208	6.75×10^{-6}
	x_4	22	92	0.3165	8.64×10^{-6}	49	197	0.030124	8.45×10^{-6}	30	121	0.01835	8.39×10^{-6}
	x_5	35	145	0.024702	8.26×10^{-6}	53	213	0.039321	8.38×10^{-6}	32	129	0.02700	8.47×10^{-6}
	x_6	43	182	0.027471	8.7×10^{-6}	46	185	0.033627	8.8×10^{-6}	30	121	0.01712	6.95×10^{-6}
5000	x_1	146	592	0.23803	9.45×10^{-6}	24	91	0.060158	6.36×10^{-6}	17	67	0.04394	5.64×10^{-6}
	x_2	21	86	0.04337	9.46×10^{-6}	25	95	0.060385	6.24×10^{-6}	17	67	0.04635	5.07×10^{-6}
	x_3	24	99	0.054619	8.27×10^{-6}	25	98	0.040015	5.86×10^{-6}	35	140	0.08311	9.74×10^{-6}
	x_4	24	100	0.056424	6.66×10^{-6}	53	213	0.098097	9.11×10^{-6}	33	133	0.08075	6.02×10^{-6}
	x_5	38	157	0.071222	9.28×10^{-6}	58	233	0.10958	8.56×10^{-6}	35	141	0.10091	7.51×10^{-6}
	x_6	45	190	0.090276	7.14×10^{-6}	50	201	0.21521	7.65×10^{-6}	32	129	0.08054	8.55×10^{-6}
10,000	x_1	211	853	0.60357	9.65×10^{-6}	25	95	0.076427	5.4×10^{-6}	17	67	0.06816	8.81×10^{-6}
	x_2	22	90	0.08012	4.98×10^{-6}	25	95	0.098461	8.9×10^{-6}	17	67	0.08833	7.80×10^{-6}
	x_3	25	103	0.039269	5.89×10^{-6}	25	98	0.07495	8.64×10^{-6}	37	148	0.14732	6.36×10^{-6}
	x_4	25	104	0.11781	5.54×10^{-6}	55	221	0.19048	9.11×10^{-6}	37	149	0.14293	8.25×10^{-6}
	x_5	40	165	0.15859	7.43×10^{-6}	60	241	0.19751	9.01×10^{-6}	36	145	0.14719	8.23×10^{-6}
	x_6	46	194	0.1728	8.62×10^{-6}	51	205	0.28882	9.62×10^{-6}	74	298	0.26456	7.79×10^{-6}
50,000	x_1	225	909	2.1373	9.93×10^{-6}	26	99	0.34575	6.75×10^{-6}	42	169	0.58113	7.78×10^{-6}
	x_2	23	94	0.31098	4.48×10^{-6}	27	103	0.43806	5.16×10^{-6}	42	169	0.58456	7.13×10^{-6}
	x_3	26	107	0.36293	6.83×10^{-6}	27	106	0.4815	5.28×10^{-6}	41	165	0.58717	8.87×10^{-6}
	x_4	26	108	0.32427	9.72×10^{-6}	60	241	0.90868	8.66×10^{-6}	40	161	0.56431	7.17×10^{-6}
	x_5	43	177	0.48938	9.47×10^{-6}	65	261	0.7924	9.05×10^{-6}	82	330	1.08920	8.44×10^{-6}
	x_6	50	210	0.69117	8.12×10^{-6}	56	225	0.72334	8.19×10^{-6}	80	322	1.06670	7.82×10^{-6}
100,000	x_1	231	933	4.2588	9.85×10^{-6}	26	99	0.71242	9.73×10^{-6}	43	173	1.09620	8.47×10^{-6}
	x_2	139	564	2.7266	9.96×10^{-6}	27	103	0.62746	7.39×10^{-6}	43	173	1.10040	7.77×10^{-6}
	x_3	26	107	0.57505	9.92×10^{-6}	27	106	0.82989	7.77×10^{-6}	42	169	1.08330	9.66×10^{-6}
	x_4	27	112	0.62227	8.52×10^{-6}	62	249	1.5474	9×10^{-6}	85	342	2.11880	9.22×10^{-6}
	x_5	45	185	0.8992	7.79×10^{-6}	67	269	1.6692	9.5×10^{-6}	84	338	2.10640	9.78×10^{-6}
	x_6	52	218	1.4318	7.37×10^{-6}	58	233	1.4333	8.32×10^{-6}	167	671	4.06200	9.90×10^{-6}

Mathematics 2019, 7, 767

Table 6. Numerical Results for **Algorithm 1** (DCG), PCG and PDY for Problem 6 with given initial points and dimensions.

DIMENSION	INITIAL POINT	Algorithm 1				PCG				PDY			
		ITER	FVAL	TIME	NORM	ITER	FVAL	TIME	NORM	ITER	FVAL	TIME	NORM
1000	x_1	13	55	1.38	5.68×10^{-6}	23	92	0.4038	9.28×10^{-6}	15	60	0.01671	4.35×10^{-6}
	x_2	13	55	0.013339	5.47×10^{-6}	23	92	0.016325	8.92×10^{-6}	15	60	0.01346	4.18×10^{-6}
	x_3	13	55	0.066142	4.81×10^{-6}	23	92	0.023045	7.86×10^{-6}	15	60	0.01630	3.68×10^{-6}
	x_4	13	55	0.026838	3.3×10^{-6}	23	92	0.016172	5.38×10^{-6}	14	56	0.01339	7.48×10^{-6}
	x_5	12	51	0.009864	9.45×10^{-6}	22	88	0.03785	8.62×10^{-6}	14	56	0.01267	6.01×10^{-6}
	x_1	12	51	0.009881	5.57×10^{-6}	22	88	0.015013	5.08×10^{-6}	14	56	0.01685	3.54×10^{-6}
5000	x_1	14	59	0.042533	3.56×10^{-6}	25	100	0.061642	5.22×10^{-6}	15	60	0.05038	9.73×10^{-6}
	x_2	14	59	0.036648	3.43×10^{-6}	25	100	0.092952	5.02×10^{-6}	15	60	0.04775	9.36×10^{-6}
	x_3	14	59	0.043452	3.02×10^{-6}	24	96	0.068141	8.82×10^{-6}	15	60	0.04923	8.25×10^{-6}
	x_4	13	55	0.032579	7.38×10^{-6}	24	96	0.084625	6.04×10^{-6}	15	60	0.05793	5.64×10^{-6}
	x_5	13	55	0.03295	5.92×10^{-6}	23	92	0.086122	9.67×10^{-6}	15	60	0.04597	4.53×10^{-6}
	x_6	13	55	0.033062	3.49×10^{-6}	23	92	0.093318	5.7×10^{-6}	14	56	0.05070	7.93×10^{-6}
10,000	x_1	14	59	0.064917	5.04×10^{-6}	25	100	0.21424	7.38×10^{-6}	68	274	0.40724	9.06×10^{-6}
	x_2	14	59	0.069913	4.84×10^{-6}	25	100	0.13978	7.09×10^{-6}	68	274	0.41818	8.72×10^{-6}
	x_3	14	59	0.08473	4.27×10^{-6}	25	100	0.1731	6.25×10^{-6}	34	137	0.21905	6.22×10^{-6}
	x_4	14	59	0.075847	2.92×10^{-6}	24	96	0.14744	8.54×10^{-6}	15	60	0.10076	7.98×10^{-6}
	x_5	13	55	0.07974	8.38×10^{-6}	24	96	0.14169	6.85×10^{-6}	15	60	0.12680	6.40×10^{-6}
	x_6	13	55	0.063129	4.94×10^{-6}	23	92	0.15294	8.06×10^{-6}	15	60	0.11984	3.78×10^{-6}
50,000	x_1	15	63	0.25329	3.15×10^{-6}	26	104	0.64669	8.26×10^{-6}	143	575	3.09120	9.42×10^{-6}
	x_2	15	63	0.36394	3.03×10^{-6}	26	104	0.67717	7.95×10^{-6}	143	575	3.06200	9.06×10^{-6}
	x_3	14	59	0.2413	9.54×10^{-6}	26	104	0.5562	7×10^{-6}	142	571	3.04950	9.04×10^{-6}
	x_4	14	59	0.27502	6.53×10^{-6}	25	100	0.56171	9.56×10^{-6}	69	278	1.53920	9.14×10^{-6}
	x_5	14	59	0.36404	5.24×10^{-6}	25	100	0.57982	7.67×10^{-6}	68	274	1.49490	9.43×10^{-6}
	x_6	14	59	0.2506	3.09×10^{-6}	24	96	0.58645	9.03×10^{-6}	15	60	0.38177	8.44×10^{-6}
100,000	x_1	15	63	0.84781	4.45×10^{-6}	27	108	1.3215	5.86×10^{-6}	292	1172	13.59530	9.53×10^{-6}
	x_2	15	63	0.66663	4.28×10^{-6}	27	108	1.5062	5.63×10^{-6}	290	1164	13.30930	9.75×10^{-6}
	x_3	15	63	0.66683	3.77×10^{-6}	26	104	1.166	9.9×10^{-6}	144	579	6.68150	9.96×10^{-6}
	x_4	14	59	0.62697	9.24×10^{-6}	26	104	1.3961	6.78×10^{-6}	141	567	6.50800	9.92×10^{-6}
	x_5	14	59	0.62891	7.41×10^{-6}	26	104	1.2711	5.44×10^{-6}	70	282	3.30510	8.07×10^{-6}
	x_6	14	59	0.62422	4.37×10^{-6}	25	100	1.1685	6.4×10^{-6}	34	137	1.64510	6.37×10^{-6}

Table 7. Numerical Results for **Algorithm 1** (DCG), **PCG** and **PDY** for Problem 7 with given initial points and dimensions.

DIMENSION	INITIAL POINT	Algorithm 1				PCG				PDY			
		ITER	FVAL	TIME	NORM	ITER	FVAL	TIME	NORM	ITER	FVAL	TIME	NORM
1000	x_1	6	28	0.25689	2×10^{-6}	17	69	1.2275	6.98×10^{-6}	14	57	0.00953	5.28×10^{-6}
	x_2	6	28	0.008469	1.26×10^{-6}	15	61	0.23396	9.89×10^{-6}	13	53	0.00896	9.05×10^{-6}
	x_3	4	20	0.003619	9.25×10^{-6}	16	65	0.008095	5.79×10^{-6}	3	12	0.00426	8.47×10^{-6}
	x_4	5	24	0.004345	5.7×10^{-6}	16	65	0.010077	5.21×10^{-6}	15	61	0.01169	6.73×10^{-6}
	x_5	6	28	0.007146	4.42×10^{-6}	19	77	0.05354	4.95×10^{-6}	31	126	0.03646	9.03×10^{-6}
	x_6	6	27	0.004299	4.43×10^{-6}	18	72	0.025677	8.93×10^{-6}	15	60	0.01082	3.99×10^{-6}
5000	x_1	6	28	0.012915	4.47×10^{-6}	18	73	0.17722	7.6×10^{-6}	15	61	0.03215	4.25×10^{-6}
	x_2	6	28	0.012272	2.81×10^{-6}	17	69	0.027729	5.25×10^{-6}	14	57	0.02942	7.40×10^{-6}
	x_3	5	24	0.014669	1.16×10^{-6}	17	69	0.02985	6.31×10^{-6}	4	16	0.01107	1.01×10^{-7}
	x_4	6	28	0.012765	7.14×10^{-7}	17	69	0.028176	5.68×10^{-6}	16	65	0.04331	5.43×10^{-6}
	x_5	6	28	0.C1331	9.89×10^{-6}	20	81	0.032213	5.39×10^{-6}	33	134	0.09379	7.78×10^{-6}
	x_6	6	27	0.015828	9.91×10^{-6}	19	76	0.044328	9.73×10^{-6}	15	60	0.04077	8.92×10^{-6}
10,000	x_1	6	28	0.022346	6.32×10^{-6}	19	77	0.17863	5.23×10^{-6}	15	61	0.06484	6.01×10^{-6}
	x_2	6	28	0.022669	3.97×10^{-6}	17	69	0.049242	7.42×10^{-6}	15	61	0.07734	3.77×10^{-6}
	x_3	5	24	0.039342	1.64×10^{-6}	17	69	0.048238	8.92×10^{-6}	4	16	0.02707	1.42×10^{-7}
	x_4	6	28	0.021017	1.01×10^{-6}	17	69	0.04807	8.03×10^{-6}	16	65	0.07941	7.69×10^{-6}
	x_5	7	32	0.031654	7.83×10^{-7}	20	81	0.063156	7.62×10^{-6}	34	138	0.14942	6.83×10^{-6}
	x_6	7	31	0.023456	7.85×10^{-7}	20	80	0.059438	6.7×10^{-6}	34	138	0.15224	8.81×10^{-6}
50,000	x_1	7	32	0.092452	7.91×10^{-7}	20	81	1.0808	5.7×10^{-6}	16	65	0.25995	4.89×10^{-6}
	x_2	6	28	0.1068	8.88×10^{-6}	18	73	0.32804	8.08×10^{-6}	15	61	0.24674	8.42×10^{-6}
	x_3	5	24	0.055684	3.66×10^{-6}	18	73	0.2189	9.71×10^{-6}	4	16	0.09405	3.18×10^{-7}
	x_4	6	28	0.10193	2.26×10^{-6}	18	73	0.3497	8.75×10^{-6}	36	146	0.55207	6.39×10^{-6}
	x_5	7	32	0.095676	1.75×10^{-6}	21	85	0.22595	8.3×10^{-6}	35	142	0.54679	9.05×10^{-6}
	x_6	7	31	0.092855	1.76×10^{-6}	21	84	0.22374	7.3×10^{-6}	36	146	0.55764	7.59×10^{-6}
100,000	x_1	7	32	0.17597	1.12×10^{-6}	20	81	2.1675	8.06×10^{-6}	17	69	0.52595	5.68×10^{-6}
	x_2	7	32	0.1741	7.03×10^{-7}	19	77	0.45553	5.57×10^{-6}	16	65	0.52102	4.34×10^{-6}
	x_3	5	24	0.17522	5.18×10^{-6}	19	77	0.43219	6.69×10^{-6}	4	16	0.14864	4.50×10^{-7}
	x_4	6	28	0.20785	3.19×10^{-6}	19	77	0.52259	6.03×10^{-6}	36	146	1.05360	9.04×10^{-6}
	x_5	7	32	0.23979	2.48×10^{-6}	22	89	0.6171	5.72×10^{-6}	74	299	2.10730	8.55×10^{-6}
	x_6	7	31	0.23128	2.48×10^{-6}	22	88	0.57384	5.03×10^{-6}	37	150	1.08240	6.66×10^{-6}

Table 8. Numerical Results for **Algorithm 1** (DCG), **PCG** and **PDY** for Problem 8 with given initial points and dimensions.

DIMENSION	INITIAL POINT	Algorithm 1				PCG				PDY			
		ITER	FVAL	TIME	NORM	ITER	FVAL	TIME	NORM	ITER	FVAL	TIME	NORM
1000	x_1	7	28	0.11495	3.03×10^{-6}	9	32	0.85797	7.6×10^{-6}	69	279	0.05538	8.95×10^{-6}
	x_2	7	28	0.005034	3.03×10^{-6}	9	32	0.034675	7.6×10^{-6}	270	1085	0.18798	9.72×10^{-6}
	x_3	7	28	0.006743	3.03×10^{-6}	9	32	0.005985	7.6×10^{-6}	24	52	0.02439	6.57×10^{-6}
	x_4	7	28	0.005856	3.03×10^{-6}	9	32	0.004808	7.6×10^{-6}	27	58	0.01520	7.59×10^{-6}
	x_5	7	28	0.004635	3.03×10^{-6}	9	32	0.015026	7.6×10^{-6}	28	61	0.04330	9.21×10^{-6}
	x_6	7	28	0.006487	3.03×10^{-6}	9	32	0.15778	7.6×10^{-6}	40	85	0.02116	8.45×10^{-6}
5000	x_1	5	22	0.009068	4.52×10^{-6}	7	26	0.67239	1.3×10^{-6}	658	2639	1.13030	9.98×10^{-6}
	x_2	5	22	0.009369	4.52×10^{-6}	7	26	0.010651	1.3×10^{-6}	27	58	0.05101	7.59×10^{-6}
	x_3	5	22	0.010895	4.52×10^{-6}	7	26	0.015758	1.3×10^{-6}	49	104	0.08035	8.11×10^{-6}
	x_4	5	22	0.014958	4.52×10^{-6}	7	26	0.014935	1.3×10^{-6}	40	85	0.07979	8.45×10^{-6}
	x_5	5	22	0.01507	4.52×10^{-6}	7	26	0.01524	1.3×10^{-6}	18	40	0.09128	9.14×10^{-6}
	x_6	5	22	0.008716	4.52×10^{-6}	7	26	0.1999	1.3×10^{-6}	17	38	0.18528	8.98×10^{-6}
10,000	x_1	6	27	0.031198	3.81×10^{-6}	5	19	0.044387	5.06×10^{-6}	49	104	0.20443	7.62×10^{-6}
	x_2	6	27	0.02098	3.81×10^{-6}	5	19	0.0223	5.06×10^{-6}	40	85	0.15801	8.45×10^{-6}
	x_3	6	27	0.01991	3.81×10^{-6}	5	19	0.018209	5.06×10^{-6}	19	42	0.37880	7.66×10^{-6}
	x_4	6	27	0.025402	3.81×10^{-6}	5	19	0.021654	5.06×10^{-6}	90	187	1.25802	9.7×10^{-6}
	x_5	6	27	0.025816	3.81×10^{-6}	5	19	0.017353	5.06×10^{-6}	988	1988	12.68259	9.93×10^{-6}
	x_6	6	27	0.025065	3.81×10^{-6}	5	19	0.019763	5.06×10^{-6}	27	58	0.32859	7.59×10^{-6}
50,000	x_1	4	21	0.083641	2.34×10^{-7}	8	33	0.42902	5.15×10^{-6}	19	42	0.52291	6.42×10^{-6}
	x_2	4	21	0.074156	2.34×10^{-7}	8	33	0.11525	5.15×10^{-6}	148	304	3.93063	9.92×10^{-6}
	x_3	4	21	0.078596	2.34×10^{-7}	8	33	0.14432	5.15×10^{-6}	937	1886	22.97097	9.87×10^{-6}
	x_4	4	21	0.078289	2.34×10^{-7}	8	33	0.11562	5.15×10^{-6}	27	58	0.68467	7.59×10^{-6}
	x_5	4	21	0.073535	2.34×10^{-7}	8	33	0.11674	5.15×10^{-6}	346	702	8.45043	9.79×10^{-6}
	x_6	4	21	0.081909	2.34×10^{-7}	8	33	0.10486	5.15×10^{-6}	40	85	0.99230	8.45×10^{-6}
100,000	x_1	4	22	0.1663	6.25×10^{-6}	6	25	1.2922	6.81×10^{-7}	-	-	-	-
	x_2	4	22	0.15147	6.25×10^{-6}	6	25	0.18839	6.81×10^{-7}	-	-	-	-
	x_3	4	22	0.15582	6.25×10^{-6}	6	25	0.16153	6.81×10^{-7}	-	-	-	-
	x_4	4	22	0.15465	6.25×10^{-6}	6	25	0.17397	6.81×10^{-7}	-	-	-	-
	x_5	4	22	0.16744	6.25×10^{-6}	6	25	0.18586	6.81×10^{-7}	-	-	-	-
	x_6	4	22	0.1687	6.25×10^{-6}	6	25	0.17938	6.81×10^{-7}	-	-	-	-

Table 9. Numerical Results for **Algorithm 1** (DCG), **PCG** and **PDY** for Problem 9 with given initial points and dimensions.

DIMENSION	INITIAL POINT	Algorithm 1				PCG				PDY			
		ITER	FVAL	TIME	NORM	ITER	FVAL	TIME	NORM	ITER	FVAL	TIME	NORM
	x_1	51	215	0.23665	9.01×10^{-6}	79	321	0.5978	9.76×10^{-6}	59	241	0.71268	9.36×10^{-6}
	x_2	51	215	0.04968	9.99×10^{-6}	77	313	0.016326	9.85×10^{-6}	58	237	0.045441	9.73×10^{-6}
4	x_3	53	223	0.017211	9.46×10^{-6}	80	325	0.16529	9.38×10^{-6}	59	241	0.019552	9.9×10^{-6}
	x_4	53	223	0.019004	9.68×10^{-6}	83	337	0.041713	9.57×10^{-6}	62	253	0.022007	8.07×10^{-6}
	x_5	57	239	0.023447	8.87×10^{-6}	81	329	0.11972	9.04×10^{-6}	61	249	0.040117	8.36×10^{-6}
	x_6	54	227	0.020832	9.31×10^{-6}	82	333	0.016127	9.3×10^{-6}	61	249	0.017374	9.18×10^{-6}

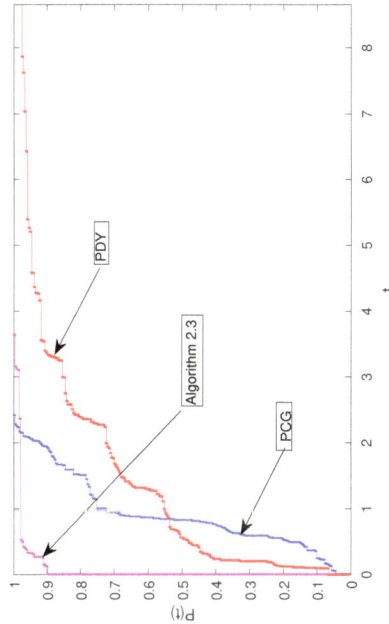

Figure 1. Performance profiles for the number of iterations.

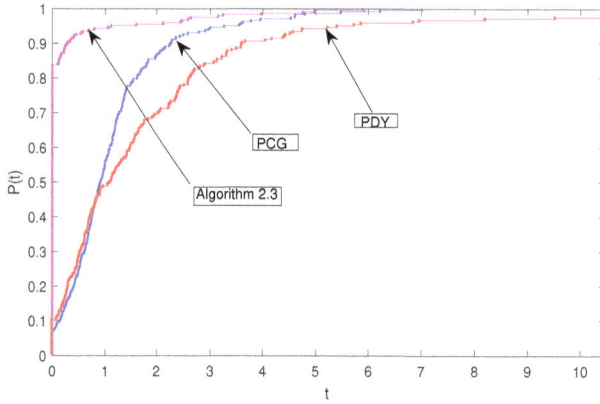

Figure 2. Performance profiles for the CPU time (in seconds).

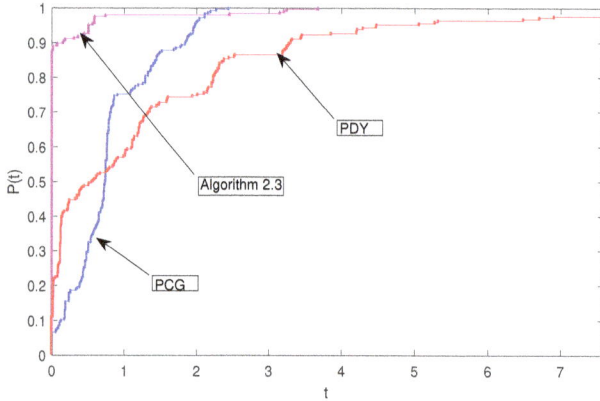

Figure 3. Performance profiles for the number of function evaluations.

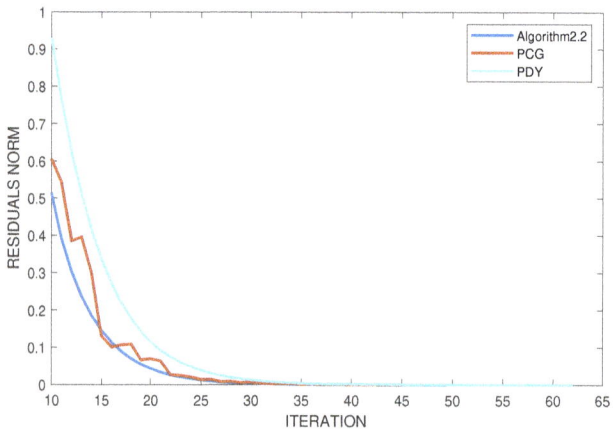

Figure 4. Convergence histories of **Algorithm** 1, **PCG** and **PDY** on Problem 9.

Applications in Compressive Sensing

There are many problems in signal processing and statistical inference involving finding sparse solutions to ill-conditioned linear systems of equations. Among popular approach is minimizing an objective function which contains quadratic (ℓ_2) error term and a sparse ℓ_1−regularization term, that is,

$$\min_x \frac{1}{2}\|y - Bx\|_2^2 + \eta\|x\|_1, \tag{24}$$

where $x \in R^n$, $y \in R^k$ is an observation, $B \in R^{k \times n}$ ($k << n$) is a linear operator, η is a non-negative parameter, $\|x\|_2$ denotes the Euclidean norm of x and $\|x\|_1 = \sum_{i=1}^n |x_i|$ is the ℓ_1−norm of x. It is easy to see that problem (24) is a convex unconstrained minimization problem. Due to the fact that if the original signal is sparse or approximately sparse in some orthogonal basis, problem (24) frequently appears in compressive sensing and hence an exact restoration can be produced by solving (24).

Iterative methods for solving (24) have been presented in many papers (see References [5,31–35]). The most popular method among these methods is the gradient based method and the earliest gradient projection method for sparse reconstruction (GPRS) was proposed by Figueiredo et al. [5]. The first step of the GPRS method is to express (24) as a quadratic problem using the following process. Let $x \in \mathbb{R}^n$ and splitting it into its positive and negative parts. Then x can be formulated as

$$x = u - v, \qquad u \geq 0, \quad v \geq 0,$$

where $u_i = (x_i)_+$, $v_i = (-x_i)_+$ for all $i = 1, 2, ..., n$ and $(.)_+ = \max\{0, .\}$. By definition of ℓ_1-norm, we have $\|x\|_1 = e_n^T u + e_n^T v$, where $e_n = (1, 1, ..., 1)^T \in R^n$. Now (24) can be written as

$$\min_{u,v} \frac{1}{2}\|y - B(u - v)\|_2^2 + \eta e_n^T u + \eta e_n^T v, \qquad u \geq 0, \quad v \geq 0, \tag{25}$$

which is a bound-constrained quadratic program. However, from Reference [5], Equation (25) can be written in standard form as

$$\min_z \frac{1}{2}z^T Dz + c^T z, \qquad \text{such that} \quad z \geq 0, \tag{26}$$

where $z = \begin{pmatrix} u \\ v \end{pmatrix}$, $c = we_{2n} + \begin{pmatrix} -b \\ b \end{pmatrix}$, $b = B^T y$, $D = \begin{pmatrix} B^T B & -B^T B \\ -B^T B & B^T B \end{pmatrix}$.

Clearly, D is a positive semi-definite matrix, which implies that Equation (26) is a convex quadratic problem.

Xiao et al. [19] translated (26) into a linear variable inequality problem which is equivalent to a linear complementarity problem. Furthermore, it was noted that z is a solution of the linear complementarity problem if and only if it is a solution of the nonlinear equation:

$$F(z) = \min\{z, Dz + c\} = 0. \tag{27}$$

The function F is a vector-valued function and the "min" is interpreted as component-wise minimum. It was proved in References [36,37] that $F(z)$ is continuous and monotone. Therefore problem (24) can be translated into problem (1) and thus Algorithm 1 (DCG) can be applied to solve it.

In this experiment, we consider a simple compressive sensing possible situation, where our goal is to restore a blurred image. We use the following well-known gray test images; (P1) Cameraman, (P2) Lena, (P3) House and (P4) Peppers for the experiments. We use 4 different Gaussian blur kernals with standard deviation σ to compare the robustness of DCG method with CGD method proposed in Reference [19]. CGD method is an extension of the well-known conjugate gradient method for unconstrained optimization CG-DESCENT [20] to solve the ℓ_1-norm regularized problems.

To access the performance of each algorithm tested with respect to metrics that indicate a better quality of restoration, in Table 10 we reported the number of iterations, the objective function (ObjFun) value at the approximate solution, the mean of squared error (MSE) to the original image $\tilde{x}$,

$$MSE = \frac{1}{n}\|\tilde{x} - x_*\|^2,$$

where x_* is the reconstructed image and the signal-to-noise-ratio (SNR) which is defined as

$$SNR = 20 \times \log_{10}\left(\frac{\|\tilde{x}\|}{\|x - \tilde{x}\|}\right).$$

We also reported the structural similarity (SSIM) index that measure the similarity between the original image and the restored image [38]. The MATLAB implementation of the SSIM index can be obtained at http://www.cns.nyu.edu/~lcv/ssim/.

Table 10. Efficiency comparison based on the value of the number of iterations (Iter), objective function (ObjFun) value, mean-square-error (MSE) and signal-to-noise-ratio (SNR) under different Pi (σ).

Image	Iter		ObjFun		MSE		SNR	
	DCG	CGD	DCG	CGD	DCG	CGD	DCG	CGD
P1(1E-8)	8	9	4.397×10^3	4.398×10^3	3.136×10^{-2}	3.157×10^{-2}	9.42	9.39
P1(1E-1)	8	9	4.399×10^3	4.401×10^3	3.147×10^{-2}	3.163×10^{-2}	9.40	9.38
P1(0.11)	11	8	4.428×10^3	4.432×10^3	3.229×10^{-2}	3.232×10^{-2}	9.29	9.29
P1(0.25)	12	8	4.468×10^3	4.473×10^3	3.365×10^{-2}	3.289×10^{-2}	9.11	9.21
P1(1E-8)	9	9	4.555×10^3	4.556×10^3	3.287×10^{-2}	3.3412×10^{-2}	9.14	9.07
P1(1E-1)	9	9	4.558×10^3	4.559×10^3	3.298×10^{-2}	3.348×10^{-2}	9.12	9.06
P1(0.11)	12	12	4.588×10^3	4.591×10^3	3.416×10^{-2}	3.446×10^{-2}	8.97	8.93
P1(0.25)	7	8	4.628×10^3	4.630×10^3	3.621×10^{-2}	3.500×10^{-2}	8.72	8.86
P1(1E-8)	9	9	5.179×10^3	5.179×10^3	3.209×10^{-2}	3.3259×10^{-2}	10.03	9.96
P1(1E-1)	9	9	5.182×10^3	5.182×10^3	3.231×10^{-2}	3.267×10^{-2}	10.00	9.95
P1(0.11)	7	9	5.209×10^3	5.209×10^3	3.436×10^{-2}	3.344×10^{-2}	9.73	9.85
P1(0.25)	10	8	5.250×10^3	5.254×10^3	3.557×10^{-2}	3.438×10^{-2}	9.58	9.73
P1(1E-8)	9	9	4.388×10^3	4.389×10^3	3.299×10^{-2}	3.335×10^{-2}	9.03	8.99
P1(1E-1)	9	9	4.391×10^3	4.393×10^3	3.308×10^{-2}	3.340×10^{-2}	9.02	8.98
P1(0.11)	12	8	4.421×10^3	4.424×10^3	3.425×10^{-2}	3.411×10^{-2}	8.87	8.89
P1(0.25)	7	8	4.461×10^3	4.463×10^3	3.621×10^{-2}	3.483×10^{-2}	8.63	8.80

The original, blurred and restored images by each of the algorithm are given in Figures 5–8. The figures demonstrate that both the two tested algorithm can restored the blurred images. It can be observed from Table 10 and Figures 5–8 that Algorithm 1 (DCG) compete with the CGD algorithm, therefore, it can be used as an alternative to CGD for restoring blurred image.

Figure 5. The original image (**top left**), the blurred image (**top right**), the restored image by CGD (**bottom left**) with SNR = 20.05, SSIM = 0.83 and by DCG (**bottom right**) with SNR = 20.12, SSIM = 0.83.

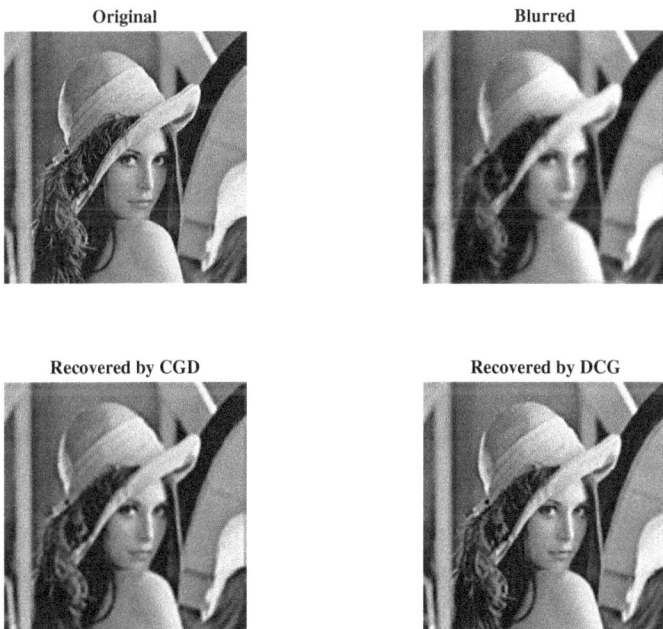

Figure 6. The original image (**top left**), the blurred image (**top right**), the restored image by CGD (**bottom left**) with SNR = 22.93, SSIM = 0.87 and by DCG (**bottom right**) with SNR = 24.36, SSIM = 0.90.

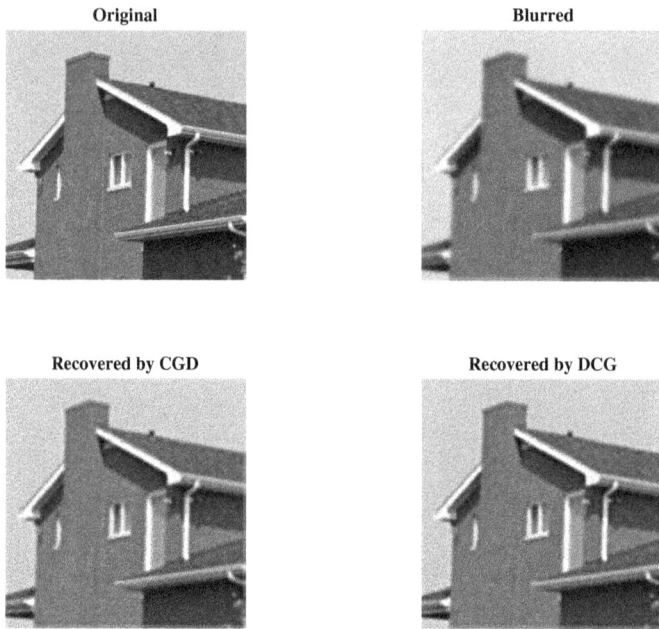

Figure 7. The original image (**top left**), the blurred image (**top right**), the restored image by CGD (**bottom left**) with SNR = 25.65, SSIM = 0.86 and by DCG (**bottom right**) with SNR = 26.37, SSIM = 0.87.

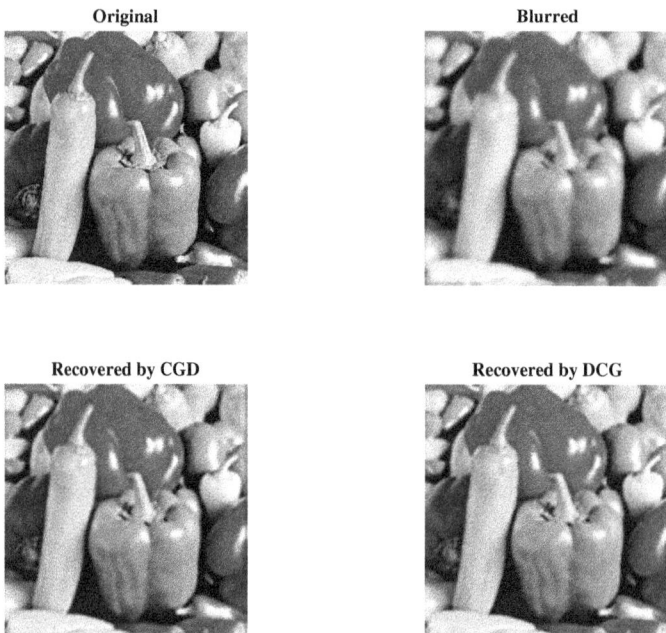

Figure 8. The original image (**top left**), the blurred image (**top right**), the restored image by CGD (**bottom left**) with SNR = 21.50, SSIM = 0.84 and by DCG (**bottom right**) with SNR = 21.81, SSIM = 0.85.

4. Conclusions

In this research article, we present a CG method which possesses the sufficient descent property for solving constrained nonlinear monotone equations. The proposed method has the ability to solve non-smooth equations as it does not require matrix storage and Jacobian information of the nonlinear equation under consideration. The sequence of iterates generated converge the solution under appropriate assumptions. Finally, we give some numerical examples to display the efficiency of the proposed method in terms of number of iterations, CPU time and number of function evaluations compared with some related methods for solving convex constrained nonlinear monotone equations and its application in image restoration problems.

Author Contributions: conceptualization, A.B.A.; methodology, A.B.A.; software, H.M.; validation, P.K. and A.M.A.; formal analysis, P.K. and H.M.; investigation, P.K. and A.M.A.; resources, P.K.; data curation, A.B.A. and H.M.; writing—original draft preparation, A.B.A.; writing—review and editing, H.M.; visualization, A.M.A.; supervision, P.K.; project administration, P.K.; funding acquisition, P.K.

Funding: Petchra Pra Jom Klao Doctoral Scholarship for Ph.D. program of King Mongkut's University of Technology Thonburi (KMUTT) and Theoretical and Computational Science (TaCS) Center. Moreover, this project was partially supported by the Thailand Research Fund (TRF) and the King Mongkut's University of Technology Thonburi (KMUTT) under the TRF Research Scholar Award (Grant No. RSA6080047).

Acknowledgments: We thank Associate Professor Jin Kiu Liu for providing us with the access of the CGD-CS MATLAB codes. The authors acknowledge the financial support provided by King Mongkut's University of Technology Thonburi through the "KMUTT 55th Anniversary Commemorative Fund". This project is supported by the theoretical and computational science (TaCS) center under computational and applied science for smart research innovation (CLASSIC), Faculty of Science, KMUTT. The first author was supported by the "Petchra Pra Jom Klao Ph.D. Research Scholarship from King Mongkut's University of Technology Thonburi".

Conflicts of Interest: The authors declare no conflict of interest.

References

1. Gu, B.; Sheng, V.S.; Tay, K.Y.; Romano, W.; Li, S. Incremental support vector learning for ordinal regression. *IEEE Trans. Neural Netw. Learn. Syst.* **2015**, *26*, 1403–1416. [CrossRef] [PubMed]
2. Li, J.; Li, X.; Yang, B.; Sun, X. Segmentation-based image copy-move forgery detection scheme. *IEEE Trans. Inf. Forensics Secur.* **2015**, *10*, 507–518.
3. Wen, X.; Shao, L.; Xue, Y.; Fang, W. A rapid learning algorithm for vehicle classification. *Inf. Sci.* **2015**, *295*, 395–406. [CrossRef]
4. Michael, S.V.; Alfredo, I.N. Newton-type methods with generalized distances for constrained optimization. *Optimization* **1997**, *41*, 257–278.
5. Figueiredo, M.A.T.; Nowak, R.D.; Wright, S.J. Gradient projection for sparse reconstruction: Application to compressed sensing and other inverse problems. *IEEE J. Sel. Top. Signal Process.* **2007**, *1*, 586–597. [CrossRef]
6. Magnanti, T.L.; Perakis, G. Solving variational inequality and fixed point problems by line searches and potential optimization. *Math. Program.* **2004**, *101*, 435–461. [CrossRef]
7. Pan, Z.; Zhang, Y.; Kwong, S. Efficient motion and disparity estimation optimization for low complexity multiview video coding. *IEEE Trans. Broadcast.* **2015**, *61*, 166–176.
8. Xia, Z.; Wang, X.; Sun, X.; Wang, Q. A secure and dynamic multi-keyword ranked search scheme over encrypted cloud data. *IEEE Trans. Parallel Distrib. Syst.* **2016**, *27*, 340–352. [CrossRef]
9. Zheng, Y.; Jeon, B.; Xu, D.; Wu, Q.M.; Zhang, H. Image segmentation by generalized hierarchical fuzzy c-means algorithm. *J. Intell. Fuzzy Syst.* **2015**, *28*, 961–973.
10. Solodov, M.V.; Svaiter, B.F. A globally convergent inexact newton method for systems of monotone equations. In *Reformulation: Nonsmooth, Piecewise Smooth, Semismooth and Smoothing Methods*; Springer: Dordrecht, The Netherlands, 1998; pp. 355–369.
11. Mohammad, H.; Abubakar, A.B. A positive spectral gradient-like method for nonlinear monotone equations. *Bull. Comput. Appl. Math.* **2017**, *5*, 99–115.
12. Zhang, L.; Zhou, W. Spectral gradient projection method for solving nonlinear monotone equations. *J. Comput. Appl. Math.* **2006**, *196*, 478–484. [CrossRef]

13. Zhou, W.J.; Li, D.H. A globally convergent BFGS method for nonlinear monotone equations without any merit functions. *Math. Comput.* **2008**, *77*, 2231–2240. [CrossRef]
14. Zhou, W.; Li, D. Limited memory BFGS method for nonlinear monotone equations. *J. Comput. Math.* **2007**, *25*, 89–96.
15. Abubakar, A.B.; Waziria, M.Y. A matrix-free approach for solving systems of nonlinear equations. *J. Mod. Methods Numer. Math.* **2016**, *7*, 1–9. [CrossRef]
16. Abubakar, A.B.; Kumam, P. An improved three-term derivative-free method for solving nonlinear equations. *Comput. Appl. Math.* **2018**, *37*, 6760–6773. [CrossRef]
17. Abubakar, A.B.; Kumam, P.; Awwal, A.M. A descent dai-liao projection method for convex constrained nonlinear monotone equations with applications. *Thai J. Math.* **2018**, *17*, 128–152.
18. Wang, C.; Wang, Y.; Xu, C. A projection method for a system of nonlinear monotone equations with convex constraints. *Math. Methods Oper. Res.* **2007**, *66*, 33–46. [CrossRef]
19. Xiao, Y.; Zhu, H. A conjugate gradient method to solve convex constrained monotone equations with applications in compressive sensing. *J. Math. Anal. Appl.* **2013**, *405*, 310–319. [CrossRef]
20. Hager, W.; Zhang, H. A new conjugate gradient method with guaranteed descent and an efficient line search. *SIAM J. Optim.* **2005**, *16*, 170–192. [CrossRef]
21. Liu, S.-Y.; Huang, Y.-Y.; Jiao, H.-W. Sufficient descent conjugate gradient methods for solving convex constrained nonlinear monotone equations. *Abstr. Appl. Anal.* **2014**, *2014*, 305643. [CrossRef]
22. Liu, J.K.; Li, S.J. A projection method for convex constrained monotone nonlinear equations with applications. *Comput. Math. Appl.* **2015**, *70*, 2442–2453. [CrossRef]
23. Ding, Y.; Xiao, Y.; Li, J. A class of conjugate gradient methods for convex constrained monotone equations. *Optimization* **2017**, *66*, 2309–2328. [CrossRef]
24. Liu, J.; Feng, Y. A derivative-free iterative method for nonlinear monotone equations with convex constraints. *Numer. Algorithms* **2018**, 1–18. [CrossRef]
25. Muhammed, A.A.; Kumam, P.; Abubakar, A.B.; Wakili, A.; Pakkaranang, N. A new hybrid spectral gradient projection method for monotone system of nonlinear equations with convex constraints. *Thai J. Math.* **2018**, *16*, 125–147.
26. La Cruz, W.; Martínez, J.; Raydan, M. Spectral residual method without gradient information for solving large-scale nonlinear systems of equations. *Math. Comput.* **2006**, *75*, 1429–1448. [CrossRef]
27. Bing, Y.; Lin, G. An efficient implementation of Merrill's method for sparse or partially separable systems of nonlinear equations. *SIAM J. Optim.* **1991**, *1*, 206–221. [CrossRef]
28. Yu, Z.; Lin, J.; Sun, J.; Xiao, Y.H.; Liu, L.Y.; Li, Z.H. Spectral gradient projection method for monotone nonlinear equations with convex constraints. *Appl. Numer. Math.* **2009**, *59*, 2416–2423. [CrossRef]
29. Yamashita, N.; Fukushima, M. Modified Newton methods for solving a semismooth reformulation of monotone complementarity problems. *Math. Program.* **1997**, *76*, 469–491. [CrossRef]
30. Dolan, E.D.; Moré, J.J. Benchmarking optimization software with performance profiles. *Math. Program.* **2002**, *91*, 201–213. [CrossRef]
31. Figueiredo, M.A.T.; Nowak, R.D. An EM algorithm for wavelet-based image restoration. *IEEE Trans. Image Process.* **2003**, *12*, 906–916. [CrossRef]
32. Hale, E.T.; Yin, W.; Zhang, Y. *A Fixed-Point Continuation Method for ℓ_1-Regularized Minimization with Applications to Compressed Sensing*; CAAM TR07-07; Rice University: Houston, TX, USA, 2007; pp. 43–44.
33. Beck, A.; Teboulle, M. A fast iterative shrinkage-thresholding algorithm for linear inverse problems. *SIAM J. Imaging Sci.* **2009**, *2*, 183–202. [CrossRef]
34. Van Den Berg, E.; Friedlander, M.P. Probing the pareto frontier for basis pursuit solutions. *SIAM J. Sci. Comput.* **2008**, *31*, 890–912. [CrossRef]
35. Birgin, E.G.; Martínez, J.M.; Raydan, M. Nonmonotone spectral projected gradient methods on convex sets. *SIAM J. Optim.* **2000**, *10*, 1196–1211. [CrossRef]
36. Xiao, Y.; Wang, Q.; Hu, Q. Non-smooth equations based method for ℓ_1-norm problems with applications to compressed sensing. *Nonlinear Anal. Theory Methods Appl.* **2011**, *74*, 3570–3577. [CrossRef]

37. Pang, J.-S. Inexact Newton methods for the nonlinear complementarity problem. *Math. Program.* **1986**, *36*, 54–71. [CrossRef]

38. Wang, Z.; Bovik, A.C.; Sheikh, H.R.; Simoncelli, E.P. Image quality assessment: From error visibility to structural similarity. *IEEE Trans. Image Process.* **2004**, *13*, 600–612. [CrossRef]

mathematics

MDPI

Article

Advances in the Semilocal Convergence of Newton's Method with Real-World Applications

Ioannis K. Argyros [1], Á. Alberto Magreñán [2,*], Lara Orcos [3] and Íñigo Sarría [4]

[1] Department of Mathematics Sciences Lawton, Cameron University, Lawton, OK 73505, USA; iargyros@cameron.edu

[2] Departamento de Matemáticas y Computación, Universidad de La Rioja, 26006 Logroño, Spain

[3] Departamento de Matemática Aplicada, Universidad Politècnica de València, 46022 València, Spain; laorpal@doctor.upv.es

[4] Escuela Superior de Ingeniería y Tecnología, Universidad Internacional de La Rioja, 26006 Logroño; Spain; inigo.sarria@unir.net

* Correspondence: angel-alberto.magrenan@unirioja.es

Received: 3 March 2019; Accepted: 21 March 2019; Published: 24 March 2019

Abstract: The aim of this paper is to present a new semi-local convergence analysis for Newton's method in a Banach space setting. The novelty of this paper is that by using more precise Lipschitz constants than in earlier studies and our new idea of restricted convergence domains, we extend the applicability of Newton's method as follows: The convergence domain is extended; the error estimates are tighter and the information on the location of the solution is at least as precise as before. These advantages are obtained using the same information as before, since new Lipschitz constant are tighter and special cases of the ones used before. Numerical examples and applications are used to test favorable the theoretical results to earlier ones.

Keywords: Banach space; Newton's method; semi-local convergence; Kantorovich hypothesis

1. Introduction

In this study we are concerned with the problem of approximating a locally unique solution z^* of equation

$$G(x) = 0, \tag{1}$$

where G is a Fréchet-differentiable operator defined on a nonempty, open convex subset D of a Banach space E_1 with values in a Banach space E_2.

Many problems in Computational disciplines such us Applied Mathematics, Optimization, Mathematical Biology, Chemistry, Economics, Medicine, Physics, Engineering and other disciplines can be solved by means of finding the solutions of equations in a form like Equation (1) using Mathematical Modelling [1–7]. The solutions of this kind of equations are rarely found in closed form. That is why most solutions of these equations are given using iterative methods. A very important problem in the study of iterative procedures is the convergence region. In general this convergence region is small. Therefore, it is important to enlarge the convergence region without additional hypotheses.

The study of convergence of iterative algorithms is usually centered into two categories: Semi-local and local convergence analysis. The semi-local convergence is based on the information around an initial point, to obtain conditions ensuring the convergence of theses algorithms while the local convergence is based on the information around a solution to find estimates of the computed radii of the convergence balls.

Newton's method defined for all $n = 0, 1, 2, \ldots$ by

$$z_{n+1} = z_n - G'(z_n)^{-1} G(z_n), \tag{2}$$

is undoubtedly the most popular method for generating a sequence $\{z_n\}$ approximating z^*, where z_0 is an initial point. There is a plethora of convergence results for Newton's method [1–4,6,8–14]. We shall increase the convergence region by finding a more precise domain where the iterates $\{z_n\}$ lie leading to smaller Lipschitz constants which in turn lead to a tighter convergence analysis for Newton's method than before. This technique can apply to improve the convergence domain of other iterative methods in an analogous way.

Let us consider the conditions:

There exist $z_0 \in \Omega$ and $\eta \geq 0$ such that

$$G'(z_0)^{-1} \in \mathbb{L}(E_2, E_1) \text{ and } \|G'(z_0)^{-1}G(z_0)\| \leq \eta;$$

There exists $T \geq 0$ such that the Lipschitz condition

$$\|G'(z_0)^{-1}(G'(x) - G'(y))\| \leq T\|x - y\|$$

holds for all $x, y \in \Omega$.

Then, the sufficient convergence condition for Newton's method is given by the famous for its simplicity and clarity Kantorovich sufficient convergence criterion for Newton's method

$$h_K = 2T\eta \leq 1. \tag{3}$$

Let us consider a motivational and academic example to show that this condition is not satisfied. Choose $E_1 = E_2 = \mathbb{R}$, $z_0 = 1$, $p \in [0, 0.5)$, $D = S(z_0, 1 - p)$ and define function G on D by

$$G(x) = z^3 - p.$$

Then, we have $T = 2(2 - p)$. Then, the Kantorovich condition is not satisfied, since $h_K > 1$ for all $p \in (0, 0.5)$. We set $I_K = \varnothing$ to be the set of point satisfying Equation (3). Hence, there is no guarantee that Newton's sequence starting at z_0 converges to $z^* = \sqrt[3]{p}$.

The rest of the paper is structured as follows: In Section 2 we present the semi-local convergence analysis of Newton's method Equation (2). The numerical examples and applications are presented in Section 3 and the concluding Section 4.

2. Semi-Local Convergence Analysis

We need an auxiliary result on majorizing sequences for Newton's method.

Lemma 1. *Let $H > 0$, $K > 0$, $L > 0$, $L_0 > 0$ and $\eta > 0$ be parameters. Suppose that:*

$$h_4 = L_4\eta \leq 1, \tag{4}$$

where

$$L_4^{-1} = \begin{cases} \dfrac{1}{L_0 + H}, & \text{if } b = LK + 2\delta L_0(K - 2H) = 0 \\[3mm] 2\dfrac{-\delta(L_0 + H) + \sqrt{\delta^2(L_0 + H)^2 + \delta(LK + 2\delta L_0(K - 2H))}}{LK + 2\delta L_0(K - 2H)}, & \text{if } b > 0 \\[3mm] -2\dfrac{\delta(L_0 + H) + \sqrt{\delta^2(L_0 + H)^2 + \delta(LK + 2\delta L_0(K - 2H))}}{LK + 2\delta L_0(K - 2H)}, & \text{if } b < 0 \end{cases}$$

and

$$\delta = \frac{2L}{L + \sqrt{L^2 + 8L_0L}}.$$

holds. Then, scalar sequence $\{t_n\}$ given by

$$t_0 = 0, \quad t_1 = \eta, \quad t_2 = t_1 + \frac{K\,(t_1 - t_0)^2}{2\,(1 - H\,t_1)},$$

$$t_{n+2} = t_{n+1} + \frac{L\,(t_{n+1} - t_n)^2}{2\,(1 - L_0\,t_{n+1})} \quad \text{for all} \quad n = 1, 2, \cdots, \tag{5}$$

is well defined, increasing, bounded from above by

$$t^{**} = \eta + \left(1 + \frac{\delta_0}{1 - \delta}\right) \frac{K\,\eta^2}{2\,(1 - H\,\eta)} \tag{6}$$

and converges to its unique least upper bound t^* which satisfies

$$t_2 \leq t^* \leq t^{**}, \tag{7}$$

where $\delta_0 = \dfrac{L(t_2 - t_1)}{2(1 - L_0 t_2)}$. Moreover, the following estimates hold:

$$0 < t_{n+2} - t_{n+1} \leq \delta_0\,\delta^{n-1} \frac{K\,\eta^2}{2\,(1 - H\,\eta)} \quad \text{for all} \quad n = 1, 2, \cdots \tag{8}$$

and

$$t^* - t_n \leq \frac{\delta_0\,(t_2 - t_1)}{1 - \delta}\,\delta^{n-2} \quad \text{for all} \quad n = 2, 3, \cdots. \tag{9}$$

Proof. By induction, we show that

$$0 < \frac{L\,(t_{k+1} - t_k)}{2\,(1 - L_0\,t_{k+1})} \leq \delta \tag{10}$$

holds for all $k = 1, 2, \cdots$. Estimate Equation (10) is true for $k = 1$ by Equation (4). Then, we have by Equation (5)

$$0 < t_3 - t_2 \leq \delta_0\,(t_2 - t_1) \implies t_3 \leq t_2 + \delta_0\,(t_2 - t_1)$$

$$\implies t_3 \leq t_2 + (1 + \delta_0)\,(t_2 - t_1) - (t_2 - t_1)$$

$$\implies t_3 \leq t_1 + \frac{1 - \delta_0^2}{1 - \delta_0}\,(t_2 - t_1) < t^{**}$$

and for $m = 2, 3, \cdots$

$$t_{m+2} \quad \leq \quad t_{m+1} + \delta_0\,\delta^{m-1}\,(t_2 - t_1)$$

$$\leq \quad t_m + \delta_0\,\delta^{m-2}\,(t_2 - t_1) + \delta_0\,\delta^{m-1}\,(t_2 - t_1)$$

$$\leq \quad t_1 + (1 + \delta_0\,(1 + \delta + \cdots + \delta^{m-1}))\,(t_2 - t_1)$$

$$= \quad t_1 + (1 + \delta_0\,\tfrac{1 - \delta^m}{1 - \delta})\,(t_2 - t_1) \leq t^{**}.$$

Assume that Equation (10) holds for all natural integers $n \leq m$. Then, we get by Equations (5) and (10) that

$$0 < t_{m+2} - t_{m+1} \leq \delta_0 \, \delta^{m-1} \, (t_2 - t_1) \leq \delta^m \, (t_2 - t_1)$$

and

$$t_{m+2} \leq t_1 + (1 + \delta_0 \frac{1 - \delta^m}{1 - \delta}) \, (t_2 - t_1) \leq t_1 + \frac{1 - \delta^{m+1}}{1 - \delta} \, (t_2 - t_1) < t^{**}.$$

Evidently estimate Equation (10) is true, if m is replaced by $m + 1$ provided that

$$\frac{L}{2} \, (t_{m+2} - t_{m+1}) \leq \delta \, (1 - L_0 \, t_{m+2})$$

or

$$\frac{L}{2} \, (t_{m+2} - t_{m+1}) + \delta \, L_0 \, t_{m+2} - \delta \leq 0$$

or

$$\frac{L}{2} \, \delta^m \, (t_2 - t_1) + \delta \, L_0 \left(t_1 + \frac{1 - \delta^{m+1}}{1 - \delta} \, (t_2 - t_1) \right) - \delta \leq 0. \tag{11}$$

Estimate Equation (11) motivates us to define recurrent functions $\{\psi_k\}$ on $[0, 1)$ by

$$\psi_m(s) = \frac{L}{2} \, (t_2 - t_1) \, t^{m+1} + s \, L_0 \, (1 + s + t^2 + \cdots + t^m) \, (t_2 - t_1) - (1 - L_0 \, t_1) \, s.$$

We need a relationship between two consecutive functions ψ_k. We get that

$$\begin{aligned}
\psi_{m+1}(s) &= \tfrac{L}{2} \, (t_2 - t_1) \, t^{m+2} + s \, L_0 \, (1 + s + t^2 + \cdots + t^{m+1}) \, (t_2 - t_1) \\
&\quad -(1 - L_0 \, t_1) \, s \\
&= \tfrac{L}{2} \, (t_2 - t_1) \, t^{m+2} + s \, L_0 \, (1 + s + t^2 + \cdots + t^{m+1}) \, (t_2 - t_1) \\
&\quad -(1 - L_0 \, t_1) \, s - \tfrac{L}{2} \, (t_2 - t_1) \, t^m \\
&\quad -s \, L_0 \, (1 + s + t^2 + \cdots + t^m) \, (t_2 - t_1) + (1 - L_0 \, t_1) \, s + \psi_k(s).
\end{aligned}$$

Therefore, we deduce that

$$\psi_{m+1}(s) = \psi_m(s) + \frac{1}{2} \, (2 \, L_0 \, t^2 + L \, s - L) \, t^m \, (t_2 - t_1). \tag{12}$$

Estimate Equation (11) is satisfied, if

$$\psi_m(\delta) \leq 0 \quad \text{holds for all} \quad m = 1, 2, \cdots. \tag{13}$$

Using Equation (12) we obtain that

$$\psi_{m+1}(\delta) = \psi_m(\delta) \quad \text{for all} \quad m = 1, 2, \cdots.$$

Let us now define function ψ_∞ on $[0, 1)$ by

$$\psi_\infty(s) = \lim_{m \to \infty} \psi_m(s). \tag{14}$$

Then, we have by Equation (14) and the choice of δ that

$$\psi_\infty(\delta) = \psi_m(\delta) \quad \text{for all} \quad m = 1, 2, \cdots .$$

Hence, Equation (13) is satisfied, if

$$\psi_\infty(\delta) \leq 0. \tag{15}$$

Using Equation (11) we get that

$$\psi_\infty(\delta) = \left(\frac{L_0}{1-\delta} (t_2 - t_1) + L_0 t_1 - 1 \right) \delta. \tag{16}$$

It then, follows from Equations (2.1) and (2.13) that Equation (15) is satisfied. The induction is now completed. Hence, sequence $\{t_n\}$ is increasing, bounded from above by t^{**} given by Equation (6), and as such it converges to its unique least upper bound t^* which satisfies Equation (7). $\square$

Let $S(z, \varrho)$, $\bar{S}(z, \varrho)$ stand, respectively for the open and closed ball in E_1 with center $z \in E_1$ and of radius $\varrho > 0$.

The conditions (A) for the semi-local convergence are:

(A_1) $G : D \subset E_1 \to E_2$ is Fréchet differentiable and there exist $z_0 \in D$, $\eta \geq 0$ such that $G'(z_0)^{-1} \in \mathcal{L}(E_2, E_1)$ and

$$\|G'(z_0)^{-1} G(z_0)\| \leq \eta.$$

(A_2) There exists $L_0 > 0$ such that for all $x \in D$

$$\|G'(z_0)^{-1}(G'(x) - G'(z_0))\| \leq L_0 \|x - z_0\|.$$

(A_3) $L_0 \eta < 1$ and there exists $L > 0$ such that

$$\|G'(z_0)^{-1}(G'(x) - G'(y))\| \leq L \|x - y\|.$$

for all $x, y \in D_0 := S(z_1, \frac{1}{L_0} - \|G'(z_0)^{-1} G(z_0)\|) \cap D.$

(A_4) There exists $H > 0$ such that

$$\|G'(z_0)^{-1}(G'(z_1) - G'(z_0))\| \leq H \|z_1 - z_0\|,$$

where $z_1 = z_0 - G'(z_0)^{-1} G(z_0).$

(A_5) There exists $K > 0$ such that for all $\theta \in [0, 1]$

$$\|G'(z_0)^{-1}(G'(z_0 + \theta(z_1 - z_0)) - G'(z_0))\| \leq K\theta \|z_1 - z_0\|.$$

Notice that $(A_2) \implies (A_3) \implies (A_5) \implies (A_4)$. Clearly, we have that

$$H \leq K \leq L_0 \tag{17}$$

and $\frac{L}{L_0}$ can be arbitrarily large [9]. It is worth noticing that (A_3)–(A_5) are not additional to (A_2) hypotheses, since in practice the computation of Lipschitz constant T requires the computation of the other constants as special cases.

Next, first we present a semi-local convergence result relating majorizing sequence $\{t_n\}$ with Newton's method and hypotheses (A).

Theorem 1. *Suppose that hypotheses* (A), *hypotheses of Lemma 1 and* $\overline{S}(z_0, t^*) \subseteq D$ *hold, where* t^* *is given in Lemma 1. Then, sequence* $\{z_n\}$ *generated by Newton's method is well defined, remains in* $\overline{S}(z_0, t^*)$ *and converges to a solution* $z^* \in \overline{S}(z_0, t^*)$ *of equation* $G(x) = 0$. *Moreover, the following estimates hold*

$$\|z_{n+1} - z_n\| \leq t_{n+1} - t_n \tag{18}$$

and

$$\|z_n - z^*\| \leq t^* - t_n \quad \text{for all} \quad n = 0, 1, 2, \cdots, \tag{19}$$

where sequence $\{t_n\}$ *is given in Lemma 1. Furthermore, if there exists* $R \geq t^*$ *such that*

$$\overline{S}(z_0, R) \subseteq D \quad \text{and} \quad L_0 (t^* + R) < 2,$$

then, the solution z^* *of equation* $G(x) = 0$ *is unique in* $\overline{S}(z_0, R)$.

Proof. We use mathematical induction to prove that

$$\|z_{k+1} - x_k\| \leq t_{k+1} - t_k \tag{20}$$

and

$$\overline{S}(z_{k+1}, t^* - t_{k+1}) \subseteq \overline{S}(z_k, t^* - t_k) \quad \text{for all} \quad k = 1, 2, \cdots. \tag{21}$$

Let $z \in \overline{S}(z_1, t^* - t_1)$.

Then, we obtain that

$$\|z - z_0\| \leq \|z - z_1\| + \|z_1 - z_0\| \leq t^* - t_1 + t_1 - t_0 = t^* - t_0,$$

which implies $z \in \overline{S}(z_0, t^* - t_0)$. Note also that

$$\|z_1 - z_0\| = \|G'(z_0)^{-1} G(z_0)\| \leq \eta = t_1 - t_0.$$

Hence, estimates Equations (20) and (21) hold for $k = 0$. Suppose these estimates hold for $n \leq k$. Then, we have that

$$\|z_{k+1} - z_0\| \leq \sum_{i=1}^{k+1} \|z_i - z_{i-1}\| \leq \sum_{i=1}^{k+1} (t_i - t_{i-1}) = t_{k+1} - t_0 = t_{k+1}$$

and

$$\|z_k + \theta (z_{k+1} - z_k) - z_0\| \leq t_k + \theta (t_{k+1} - t_k) \leq t^*$$

for all $\theta \in (0, 1)$. Using Lemma 1 and the induction hypotheses, we get in turn that

$$\|G'(z_0)^{-1}(G'(z_{k+1}) - G'(z_0))\| \leq M\|x_{k+1} - z_0\| \leq M(t_{k+1} - t_0) \leq M t_{k+1} < 1, \tag{22}$$

where

$$M = \begin{cases} H & \text{if} \quad k = 0 \\ L_0 & \text{if} \quad k = 1, 2, \cdots. \end{cases}$$

It follows from Equation (22) and the Banach lemma on invertible operators that $G'(z_{m+1})^{-1}$ exists and

$$\|G'(z_{k+1})^{-1} G'(z_0)\| \leq (1 - M\|z_{k+1} - z_0\|)^{-1} \leq (1 - M t_{k+1})^{-1}. \tag{23}$$

Using iteration of Newton's method, we obtain the approximation

$$
\begin{aligned}
G(z_{k+1}) \;&=\; G(z_{k+1}) - G(z_k) - G'(z_k)\,(z_{k+1} - z_k) \\
&=\; \int_0^1 \bigl(G'(z_k + \theta\,(z_{k+1} - z_k)) - G'(z_m)\bigr)\,(z_{k+1} - z_k)\,d\theta .
\end{aligned} \tag{24}
$$

Then, by Equation (24) we get in turn

$$
\begin{aligned}
&\|G'(z_0)^{-1}\,G(z_{k+1})\| \\
&\le \int_0^1 \|G'(z_0)^{-1}\,(G'(z_k + \theta\,(z_{k+1} - z_k)) - G'(z_k))\|\,\|z_{k+1} - z_k\|\,d\theta \\
&\le M_1 \int_0^1 \|\theta\,(z_{k+1} - z_k)\|\,\|z_{k+1} - z_k\|\,d\theta \le \frac{M_1}{2}\,(t_{k+1} - t_k)^2 ,
\end{aligned} \tag{25}
$$

where

$$
M_1 = \begin{cases} K & \text{if } k = 0 \\ L & \text{if } k = 1, 2, \cdots . \end{cases}
$$

Moreover, by iteration of Newton's method, Equations (23) and (25) and the induction hypotheses we get that

$$
\begin{aligned}
\|z_{k+2} - z_{k+1}\| \;&=\; \|(G'(z_{k+1})^{-1} G'(z_0))\,(G'(z_0)^{-1} G(z_{k+1}))\| \\[4pt]
&\le\; \|G'(z_{k+1})^{-1} G'(z_0)\|\,\|G'(z_0)^{-1} G(z_{k+1})\| \\[4pt]
&\le\; \frac{\frac{M_1}{2}\,(t_{k+1} - t_k)^2}{1 - M\,t_{k+1}} = t_{k+2} - t_{k+1} .
\end{aligned}
$$

That is, we showed Equation (20) holds for all $k \ge 0$. Furthermore, let $z \in \overline{S}(z_{k+2}, t^* - t_{k+2})$. Then, we have that

$$
\begin{aligned}
\|z - x_{k+1}\| \;&\le\; \|z - z_{k+2}\| + \|z_{k+2} - z_{k+1}\| \\[4pt]
&\le\; t^* - t_{k+2} + t_{k+2} - t_{k+1} = t^* - t_{k+1} .
\end{aligned}
$$

That is, $z \in \overline{S}(z_{k+1}, t^* - t_{k+1})$. The induction for Equations (20) and (21) is now completed. Lemma 1 implies that sequence $\{s_n\}$ is a complete sequence. It follows from Equations (20) and (21) that $\{z_n\}$ is also a complete sequence in a Banach space E_1 and as such it converges to some $z^* \in \overline{S}(z_0, t^*)$ (since $\overline{S}(z_0, t^*)$ is a closed set). By letting $k \longrightarrow \infty$ in Equation (25) we get $G(*) = 0$. Estimate Equation (19) is obtained from Equation (18) (cf. [4,6,12]) by using standard majorization techniques. The proof for the uniqueness part has been given in [9]. □

The sufficient convergence criteria for Newton's method using the conditions (A), constants L, L_0 and η given in affine invariant form are:

- Kantorovich [6]

$$
h_K = 2T\eta \le 1. \tag{26}
$$

- Argyros [9]

$$
h_1 = (L_0 + T)\eta \le 1. \tag{27}
$$

- Argyros [3]

$$
h_2 = \frac{1}{4}\left(T + 4L_0 + \sqrt{T^2 + 8L_0 T}\right)\eta \le 1 \tag{28}
$$

- Argyros [11]

$$
h_3 = \frac{1}{4}\left(4L_0 + \sqrt{L_0 T + 8L_0^2} + \sqrt{L_0 T}\right)\eta \le 1 \tag{29}
$$

- Argyros [12]

$$h_4 = \tilde{L}_4 \eta \leq 1,$$
$$\tilde{L}_4 = L_4(T), \quad \delta = \delta(T). \tag{30}$$

If $H = K = L_0 = L$, then Equations (27)–(30) coincide with Equations (26). If $L_0 < T$, then $L < T$

$$h_K \leq 1 \Rightarrow h_1 \leq 1 \Rightarrow h_2 \leq 1 \Rightarrow h_3 \leq 1 \Rightarrow h_4 \leq 1 \Rightarrow h_5 \leq 1,$$

but not vice versa. We also have that for $\dfrac{L_0}{T} \to 0$:

$$\frac{h_1}{h_K} \to \frac{1}{2}, \quad \frac{h_2}{h_K} \to \frac{1}{4}, \quad \frac{h_2}{h_1} \to \frac{1}{2}$$

$$\frac{h_3}{h_K} \to 0, \quad \frac{h_3}{h_1} \to 0, \quad \frac{h_3}{h_2} \to 0 \tag{31}$$

Conditions Equations (31) show by how many times (at most) the better condition improves the less better condition.

Remark 1. (a) *The majorizing sequence* $\{t_n\}$, t^*, t^{**} *given in [12] under conditions* (A) *and Equation (29) is defined by*

$$t_0 = 0, \quad t_1 = \eta, \quad t_2 = t_1 + \frac{L_0(t_1 - t_0)^2}{2(1 - L_0 t_1)}$$

$$t_{n+2} = t_{n+1} + \frac{T(t_{n+1} - t_n)^2}{2(1 - L_0 t_{n+1})}, \quad n = 1, 2, \ldots \tag{32}$$

$$t^* = \lim_{n \to \infty} t_n \leq t^{**} = \eta + \frac{L_0 \eta^2}{2(1 - \delta)(1 - L_0 \eta)}.$$

Using a simple inductive argument and Equation (32) we get for $L_1 < L$ *that*

$$t_n < t_{n-1}, n = 3, 4, \ldots, \tag{33}$$

$$t_{n+1} - t_n < t_n - t_{n-1}, n = 2, 3, \ldots, \tag{34}$$

and

$$t^* \leq t^{**} \tag{35}$$

Estimates for Equations (5)–(7) show the new error bounds are more precise than the old ones and the information on the location of the solution z^* *is at least as precise as already claimed in the abstract of this study (see also the numerical examples). Clearly the new majorizing sequence* $\{t_n\}$ *is more precise than the corresponding ones associated with other conditions.*

(b) *Condition* $\bar{S}(z_0, t^*) \subseteq D$ *can be replaced by* $S(z_0, \frac{1}{L_0})$ *(or D_0). In this case condition* $(A_2)'$ *holds for all* $x, y \in S(z_0, \frac{1}{L_0})$ *(or D_0).*

(c) *If* $L_0 \eta \leq 1$, *then, we have that* $z_0 \in \bar{S}(z_1, \frac{1}{L_0} - \|G'(z_0)^{-1} G(z_0)\|)$, *since* $\bar{S}(z_1, \frac{1}{L_0} - \|G'(z_0)^{-1} G(z_0)\|) \subseteq S(z_0, \frac{1}{L_0})$.

3. Numerical Examples

Example 1. *Returning back to the motivational example, we have* $L_0 = 3 - p$. *Conditions Equations (27)–(29) are satisfied, respectively for*

$$p \in I_1 := [0.494816242, 0.5),$$

$$p \in I_2 := [0.450339002, 0.5)$$

and

$$p \in I_3 := [0.4271907643, 0.5).$$

We are now going to consider such an initial point which previous conditions cannot be satisfied but our new criteria are satisfied. That is, the improvement that we get with our new weaker criteria.

We get that

$$H = \frac{5 + p}{3},$$

$$K = 2,$$

$$L = \frac{2}{3(3 - p)}(-2p^2 + 5p + 6).$$

Using this values we obtain that condition Equation (4) is satisfied for $p \in [0.0984119, 0.5)$, However, must also have that

$$L_0 \eta < 1$$

which is satisfied for $p \in I_4 := (0, 0.5]$. That is, we must have $p \in I_4$, so there exist numerous values of p for which the previous conditions cannot guarantee the convergence but our new ones can. Notice that we have

$$I_K \subseteq I_1 \subseteq I_2 \subseteq I_3 \subseteq I_4$$

Hence, the interval of convergence cannot be improved further under these conditions. Notice that the convergence criterion is even weaker than the corresponding one for the modified Newton's method given in [11] by $L_0(\eta) < 0.5$.

For example, we choose different values of p and we see in Table 1.

Table 1. Convergence of Newton's method choosing $z_0 = 1$, for different values of p.

p	0.41	0.43	0.45
z_1	0.803333	0.810000	0.816667
z_2	0.747329	0.758463	0.769351
z_3	0.742922	0.754802	0.766321
z_4	0.742896	0.754784	0.766309
z_5	0.742896	0.754784	0.766309

Example 2. Consider $E_1 = E_2 = \mathcal{A}[0,1]$. Let $D^* = \{x \in \mathcal{A}[0,1]; \|x\| \le R\}$, such that $R > 0$ and G defined on D^* as

$$G(x)(u_1) = x(u_1) - f(u_1) - \lambda \int_0^1 \mu(u_1, u_2)x(u_2)^3 \, du_2, \quad x \in C[0,1], \ u_1 \in [0,1],$$

where $f \in \mathcal{A}[0,1]$ is a given function, λ is a real constant and the kernel μ is the Green function. In this case, for all $x \in D^*$, $G'(x)$ is a linear operator defined on D^* by the following expression:

$$[G'(x)(v)](u_1) = v(u_2) - 3\lambda \int_0^1 \mu(u_1, u_2)x(u_2)^2 v(u_2) \, du_2, \quad v \in C[0,1], \ u_1 \in [0,1].$$

If we choose $z_0(u_1) = f(u_1) = 1$, it follows

$$\|I - G'(z_0)\| \le 3|\lambda|/8.$$

Hence, if
$$|\lambda| < 8/3,$$

$G'(z_0)^{-1}$ is defined and
$$\|G'(z_0)^{-1}\| \le \frac{8}{8 - 3|\lambda|},$$

$$\|G(z_0)\| \le \frac{|\lambda|}{8},$$

$$\eta = \|G'(z_0)^{-1}G(z_0)\| \le \frac{|\lambda|}{8 - 3|\lambda|}.$$

Consider $\lambda = 1.00$, we get
$$\eta = 0.2,$$
$$T = 3.8,$$
$$L_0 = 2.6,$$
$$K = 2.28,$$
$$H = 1.28$$

and
$$L = 1.38154\ldots.$$

By these values we conclude that conditions (26)–(29) are not satisfied, since

$$h_K = 1.52 > 1,$$

$$h_1 = 1.28 > 1,$$

$$h_2 = 1.19343\ldots > 1,$$

$$h_3 > 1.07704\ldots > 1,$$

but condition (2.27) and condition (4) are satisfied, since

$$h_4 = 0.985779\ldots < 1$$

and
$$h_5 = 0.97017\ldots < 1.$$

Hence, Newton's method converges by Theorem 1.

4. Application: Planck's Radiation Law Problem

We consider the following problem [15] :

$$\varphi(\lambda) = \frac{8\pi c P \lambda^{-5}}{e^{\frac{cP}{\lambda BT}} - 1} \tag{36}$$

which calculates the energy density within an isothermal blackbody. The maxima for φ occurs when density $\varphi(\lambda)$. From (36), we get

$$\varphi'(\lambda) = \left(\frac{8\pi c P \lambda^{-6}}{e^{\frac{cP}{\lambda BT}} - 1}\right)\left(\frac{(\frac{cP}{\lambda BT})e^{\frac{cP}{\lambda BT}} - 1}{e^{\frac{cP}{\lambda kT}} - 1} - 5\right) = 0, \tag{37}$$

that is when

$$\frac{\left(\frac{cP}{\lambda BT}\right)e^{\frac{cP}{\lambda BT}-1}}{e^{\frac{cP}{\lambda BT}-1}} = 5. \tag{38}$$

After using the change of variable $x = \frac{cP}{\lambda BT}$ and reordering terms, we obtain

$$f(x) = e^{-x} - 1 + \frac{x}{5}. \tag{39}$$

As a consequence, we need to find the roots of Equation (39).
We consider $\Omega = \overline{E(5,1)} \subset \mathbb{R}$ and we obtain

$$\eta = 0.0348643\ldots,$$

$$L_0 = 0.0599067\ldots,$$

$$K = 0.0354792\ldots,$$

$$H = 0.0354792\ldots$$

and

$$L = 0.094771\ldots.$$

So (A) are satisfied. Moreover, as $b = 0.000906015 > 0$, then

$$L_4 = 10.0672\ldots,$$

which satisfies

$$L_4\eta = 0.350988\ldots < 1$$

and that means that conditions of Lemmal 1 are also satisfied. Finally, we obtain that

$$t^* = 0.0348859\ldots.$$

Hence, Newton's method converges to the solution $x^* = 4.965114231744276\ldots$ by Theorem 1.

Author Contributions: All authors have contributed in a similar way.

Funding: This research was supported in part by Programa de Apoyo a la investigación de la fundación Séneca-Agencia de Ciencia y Tecnología de la Región de Murcia 19374/PI/14, by the project MTM2014-52016-C2-1-P of the Spanish Ministry of Science and Innovation.

Conflicts of Interest: The authors declare no conflict of interest.

References

1. Amat, S.; Busquier, S.; Gutiérrez, J.M. Geometric constructions of iterative functions to solve nonlinear equations. *J. Comput. Appl. Math.* **2003**, *157*, 197–205. [CrossRef]
2. Amat, S.; Busquier, S. Third-order iterative methods under Kantorovich conditions. *J. Math. Anal. Appl.* **2007**, *336*, 243–261. [CrossRef]
3. Argyros, I.K. A semi-local convergence analysis for directional Newton methods. *Math. Comput.* **2011**, *80*, 327–343. [CrossRef]
4. Argyros, I.K.; Magreñán, Á.A. *Iterative Methods and Their Dynamics with Applications: A Contemporary Study*; CRC Press: Bocaratón, FL, USA, 2017.
5. Farhane, N.; Boumhidi, I.; Boumhidi, J. Smart Algorithms to Control a Variable Speed Wind Turbine. *Int. J. Interact. Multimed. Artif. Intell.* **2017**, *4*, 88–95. [CrossRef]
6. Kantorovich, L.V.; Akilov, G.P. *Functional Analysis*; Pergamon Press: Oxford, UK, 1982.
7. Kaur, R.; Arora, S. Nature Inspired Range Based Wireless Sensor Node Localization Algorithms. *Int. J. Interact. Multimed. Artif. Intell.* **2017**, *4*, 7–17. [CrossRef]

Mathematics **2019**, *7*, 299

8. Amat, S.; Busquier, S.; Negra, M. Adaptive approximation of nonlinear operators. *Numer. Funct. Anal. Optim.* 2004, *25*, 397–405. [CrossRef]

9. Argyros, I.K. On the Newton-Kantorovich hypothesis for solving equations. *J. Comput. Math.* **2004**, *169*, 315–332. [CrossRef]

10. Argyros, I.K.; González, D. Extending the applicability of Newton's method for k-Fréchet differentiable operators in Banach spaces. *Appl. Math. Comput.* **2014**, *234*, 167–178. [CrossRef]

11. Argyros, I.K.; Hilout, S. Weaker conditions for the convergence of Newton's method. *J. Complex.* **2012**, *28*, 364–387. [CrossRef]

12. Argyros, I.K.; Hilout, S. On an improved convergence analysis of Newton's method. *Appl. Math. Comput.* **2013**, *225*, 372–386. [CrossRef]

13. Ezquerro, J.A.; Hernández, M.A. How to improve the domain of parameters for Newton's method. *Appl. Math. Lett.* **2015**, *48*, 91–101. [CrossRef]

14. Gutiérrez, J.M.; Magreñán, Á.A.; Romero, N. On the semi-local convergence of Newton-Kantorovich method under center-Lipschitz conditions. *Appl. Math. Comput.* **2013**, *221*, 79–88.

15. Divya, J. Families of Newton-like methods with fourth-order convergence. *Int. J. Comput. Math.* **2013**, *90*, 1072–1082.

mathematics

MDPI

Article

A Third Order Newton-Like Method and Its Applications

D. R. Sahu [1,*], Ravi P. Agarwal [2] and Vipin Kumar Singh [3]

[1] Department of Mathematics, Banaras Hindu University, Varanasi-221005, India
[2] Department of Mathematics, Texas A&M University-Kingsville, Kingsville, TX 78363-8202, USA;
   Ravi.Agarwal@tamuk.edu
[3] Department of Mathematics, Banaras Hindu University, Varanasi-221005, India;
   vipinkumarsingh666@gmail.com
* Correspondence: drsahudr@gmail.com

Received: 8 October 2018; Accepted: 7 December 2018; Published: 30 December 2018

check for
updates

Abstract: In this paper, we design a new third order Newton-like method and establish its convergence theory for finding the approximate solutions of nonlinear operator equations in the setting of Banach spaces. First, we discuss the convergence analysis of our third order Newton-like method under the ω-continuity condition. Then we apply our approach to solve nonlinear fixed point problems and Fredholm integral equations, where the first derivative of an involved operator does not necessarily satisfy the Hölder and Lipschitz continuity conditions. Several numerical examples are given, which compare the applicability of our convergence theory with the ones in the literature.

Keywords: nonlinear operator equation; Fréchet derivative; ω-continuity condition; Newton-like method; Frédholm integral equation

---

1. Introduction

Our purpose of this paper is to compute solution of nonlinear operator equation of the form

$$F(x) = 0, \tag{1}$$

where $F : D \subset X \to Y$ is a nonlinear operator defined on an open convex subset D of a Banach space X with values into a Banach space Y.

A lot of challenging problems in physics, numerical analysis, engineering, and applied mathematics are formulated in terms of finding roots of the equation of the form Equation (1). In order to solve such problems, we often use iterative methods. There are many iterative methods available in literature. One of the central method for solving such problems is the Newton method [1,2] defined by

$$x_{n+1} = x_n - (F'_{x_n})^{-1} F(x_n) \tag{2}$$

for each $n \geq 0$, where F'_x denotes the Fréchet derivative of F at point $x \in D$.

The Newton method and the Newton-like method are attractive because it converges rapidly from any sufficient initial guess. A number of researchers [3–20] have generalized and established local as well as semilocal convergence analysis of the Newton method Equation (2) under the following conditions:

(a) *Lipschitz condition:* $\|F'_x - F'_y\| \leq K\|x - y\|$ for all $x, y \in D$ and for some $K > 0$;
(b) *Hölder Lipschitz condition:* $\|F'_x - F'_y\| \leq K\|x - y\|^p$ for all $x, y \in D$ and for some $p \in (0, 1]$ and $K > 0$;

(c) ω-*continuity condition*: $\|F'_x - F'_y\| \leq \omega(\|x - y\|)$ for all $x, y \in D$,where $\omega : [0, \infty) \to [0, \infty)$ is a nondecreasing and continuous function.

One can observe that the condition (c) is more general than the conditions (a) and (b). One can find numerical examples where the Lipschitz condition (a) and the Hölder continuity condition (b) on the first Fréchet derivative do not hold, but the ω-continuity condition (c) on first Fréchet derivative holds (see Example 1, [21]).

On the other hand, many mathematical problems such as differential equations, integral equations, economics theory, game theory, variational inequalities, and optimization theory ([22,23]) can be formulated into the fixed point problem:

$$\text{Find } x \in C \text{ such that } x = G(x), \tag{3}$$

where $G : C \to X$ is an operator defined on a nonempty subset C of a metric space X. The easiest iterative method for constructing a sequence is Picard iterative method [24] which is given by

$$x_{n+1} = G(x_n) \tag{4}$$

for each $n \geq 0$. The Banach contraction principle (see [1,22,23,25]) provides sufficient conditions for the convergence of the iterative method Equation (4) to the fixed point of G. Banach spaces have more geometrical stricture with respect to metric spaces. For study fixed points of nonlinear smooth operators, Banach space structure is required. More details of Banach space theory and fixed point theory of nonlinear operators can be found in [1,22,23,26–28].

The Newton method and its variant [29,30] are also used to solve the fixed point problem of the form:

$$(I - G)(x) = 0, \tag{5}$$

where I is the identity operator defined on X and $G : D \subset X \to X$ is a nonlinear Fréchet differentiable operator defined on an open convex subset D of a Banach space X. For finding approximate solution of the Equation (5), Bartle [31] used the Newton-like iterative method of the form

$$x_{n+1} = x_n - (I - G'_{y_n})^{-1}(I - G(x_n)) \tag{6}$$

for each $n \geq 0$, where G'_x is Fréchet derivative of G at point $x \in D$ and $\{y_n\}$ is the sequence of arbitrary points in D which are sufficiently closed to the desired solution of the Equation (5). Bartle [31] has discussed the convergence analysis of the form Equation (6) under the assumption that G is Fréchet differentiable at least at desired points and a modulus of continuity is known for G' as a function of x. The Newton method Equation (2) and the modified Newton method are the special cases of the form Equation (6).

Following the idea of Bartle [31], Rall [32] introduced the following Stirling method for finding a solution of the fixed point problem Equation (5):

$$\begin{cases} y_n = G(x_n), \\ x_{n+1} = x_n - (I - G'_{y_n})^{-1}(x_n - G(x_n)) \end{cases} \tag{7}$$

for each $n \geq 0$. Many researchers [33–35] have studied the Stirling method Equation (7) and established local as well as semilocal convergence analysis of the Stirling-like method.

Recently, Parhi and Gupta [21,36] have discussed the semilocal convergence analysis of the following Stirling-like iterative method for computing a solution of operator Equation (5):

$$\begin{cases} z_n = G(x_n), \\ y_n = x_n - (I - G'_{z_n})^{-1}(x_n - G(x_n)), \\ x_{n+1} = y_n - (I - G'_{z_n})^{-1}(y_n - G(y_n)) \end{cases} \tag{8}$$

for each $n \geq 0$. More precisely, Parhi and Gupta [21] have studied the semilocal convergence analysis of Equation (8) for computing a solution of the operator Equation (5), where $G : D \subset X \to X$ is a nonlinear Fréchet differentiable operator defined on an open convex subset D under the condition:

(Ω) $\|G'_x\| \leq k$ for all $x \in D$ and for some $k \in (0, \frac{1}{3}]$.

There are some nonlinear Fréchet differentiable operators $G : D \subset X \to X$ defined on an open convex subset D which fail to satisfy the condition (Ω) (see Example 1). Therefore, ref. [21] (Theorem 1) is not applicable for such operators. So, there is the following natural question:

Problem 1. *Is it possible to develop the Stirling-like iterative method for computing a solution of the operator Equation (5), where the condition (Ω) does not hold?*

The main purpose of this paper is to design a new Newton-like method for solving the operator Equation (1) and provide an affirmative answer of the Problem 1. We prove our proposed Newton-like method has R-order of convergence at least $2p + 1$ under the ω-continuity condition and it covers a wide variety of iterative methods. We derive the Stirling-like iterative method for computing a solution of the fixed point problem Equation (5), where (Ω) does not hold and hence it gives an affirmative answer to Question 1 and generalizes the results of Parhi and Gupta [21,36] in the context of the condition (Ω).

In Section 2, we summarize some known concepts and results. In Section 3, we introduce a new Newton-like method for solving the operator Equation (1) and establish convergence theory of the proposed Newton-like method. In Section 4, we derive the Stirling-like iterative method from the proposed Newton-like method and establish a convergence theorem for computing a solution of the fixed point problem. Applications to Fredholm integral equations are also presented in Section 5, together with several numerical examples, which compare the applicability of our iterative technique with the ones in the literature.

2. Preliminary

In this section, we discuss some technical results. Throughout the paper, we denote $B(X, Y)$ a collection of bounded linear operators from a Banach space X into a Banach space Y and $B(X) = B(X, X)$. For some $r > 0$, $B_r[x]$ and $B_r(x)$ are the closed and open balls with center x and radius r, respectively, $\mathbb{N}_0 = \mathbb{N} \cup \{0\}$ and Φ denote the collection of nonnegative, nondecreasing, continuous real valued functions defined on $[0, \infty)$.

Lemma 1. *(Rall [37] (p. 50)) Let $\mathcal{L}$ be a bounded linear operator on a Banach space X. Then $\mathcal{L}^{-1}$ exists if and only if there is a bounded linear operator M in X such that M^{-1} exists and*

$$\|M - \mathcal{L}\| < \frac{1}{\|M^{-1}\|}.$$

If $\mathcal{L}^{-1}$ exists, then we have

$$\left\|\mathcal{L}^{-1}\right\| \leq \frac{\|M^{-1}\|}{1 - \|1 - M^{-1}\mathcal{L}\|} \leq \frac{\|M^{-1}\|}{1 - \|M^{-1}\| \|M - \mathcal{L}\|}.$$

Lemma 2. *Let $0 < k \leq \frac{1}{3}$ be a real number. Assume that $q = \frac{1}{p+1} + k^p$ for any $p \in (0, 1]$ and the scalar equation*

$$(1 - k^p(1 + qt)^p t)^{p+1} - \left(\frac{q^p t^p}{p+1} + k^p\right)^p q^p t^{2p} = 0$$

has a minimum positive root α. Then we have the following:

(1) $q > k$ for all $p \in (0, 1]$.
(2) $\alpha \in (0, 1)$.

Proof.   (1)   This part is obvious. Indeed, we have

$$\frac{1}{p+1} + k^p - \frac{1}{3} = \frac{2-p}{3(1+p)} + k^p > 0$$

for all $p \in (0, 1]$ and $0 < k \leq \frac{1}{3}$.
(2)   Set

$$g(t) = (1 - k^p(1 + qt)^p t)^{p+1} - \left(\frac{q^p t^p}{p+1} + k^p\right)^p q^p t^{2p}. \tag{9}$$

It is clear from the definition of $g(t)$ that $g(0) > 0, g(1) < 0$ and $g'(t) < 0$ in $(0, 1)$. Therefore, $g(t)$ is decreasing in $(0, 1)$ and hence the Equation (9) has a minimum positive root $\alpha \in (0, 1)$. This completes the proof.   □

Lemma 3.   *Let $b_0 \in (0, \alpha)$ be a number such that $k^p(1 + qb_0)^p b_0 < 1$, where k, p, α and q are same as in Lemma 2. Define the real sequences $\{b_n\}, \{\theta_n\}$ and $\{\gamma_n\}$ by*

$$b_{n+1} = \frac{\left(\frac{q^p b_n^p}{p+1} + k^p\right)^p q^p b_n^{2p}}{(1 - k^p(1 + qb_n)^p b_n)^{p+1}} b_n, \tag{10}$$

$$\theta_n = \frac{\left(\frac{q^p b_n^p}{p+1} + k^p\right) qb_n^2}{1 - k^p(1 + qb_n)^p b_n}, \quad \gamma_n = \frac{1}{1 - k^p(1 + qb_n)^p b_n} \tag{11}$$

for each $n \in \mathbb{N}_0$. Then we have the following:

(1) $\dfrac{\left(\frac{q^p b_0^p}{p+1} + k^p\right)^p q^p b_0^{2p}}{(1 - k^p(1 + qb_0)^p b_0)^{p+1}} < 1.$
(2)   The sequence $\{b_n\}$ is decreasing, that is $b_{n+1} \leq b_n$ for all $n \in \mathbb{N}_0$.
(3) $k^p(1 + qb_n)^p b_n < 1$ for all $n \in \mathbb{N}_0$.
(4) $b_{n+1} \leq \zeta^{(2p+1)^n} b_n$ for all $n \in \mathbb{N}_0$.
(5) $\theta_n \leq \zeta^{\frac{(2p+1)^n - 1}{p}} \theta$ for all $n \in \mathbb{N}_0$, where $\theta_0 = \theta$ and $\zeta = \gamma_0 \theta^p$.

Proof.   (1)   Since the scalar equation $g(t) = 0$ defined by Equation (9) has a minimum positive root $\alpha \in (0, 1)$ and $g(t)$ is decreasing in $(0, 1)$ with $g(0) > 0$ and $g(1) < 0$. Therefore, $g(t) > 0$ in the interval $(0, \alpha)$ and hence

$$\frac{\left(\frac{q^p b_0^p}{p+1} + k^p\right)^p q^p b_0^{2p}}{(1 - k^p(1 + qb_0)^p b_0)^{p+1}} < 1.$$

(2)   From (1) and Equation (10), we have $b_1 \leq b_0$. This shows that (2) is true for $n = 0$. Let $j \geq 0$ be a fixed positive integer. Assume that (2) is true for $n = 0, 1, 2, \cdots, j$. Now, using Equation (10), we have

$$b_{j+2} = \frac{\left(\frac{q^p b_{j+1}^p}{p+1} + k^p\right)^p q^p b_{j+1}^{2p}}{(1 - k^p(1 + qb_{j+1})^p b_{j+1})^{p+1}} b_{j+1} \leq \frac{\left(\frac{q^p b_j^p}{p+1} + k^p\right)^p q^p b_j^{2p}}{(1 - k^p(1 + qb_j)^p b_n)^{p+1}} b_j = b_{j+1}.$$

Thus (2) holds for $n = j + 1$. Therefore, by induction, (2) holds for all $n \in \mathbb{N}_0$.

(3) Since $b_n < b_{n-1}$ for each $n = 1, 2, 3, \cdots$ and $k^p(1 + qb_0)^p b_0 < 1$ for all $p \in (0, 1]$, it follows that

$$k^p(1 + qb_n)^p b_n < k^p(1 + qb_0)^p b_0 < 1.$$

(4) From (3), one can easily prove that the sequences $\{\gamma_n\}$ and $\{\theta_n\}$ are well defined. Using Equations (10) and (11), one can easily observe that

$$b_{n+1} = \gamma_n \theta_n^p b_n \tag{12}$$

for each $n \in \mathbb{N}_0$. Put $n = 0$ and $n = 1$ in Equation (12), we have

$$b_1 = \gamma_0 \theta^p b_0 = \zeta^{(2p+1)^0} b_0$$

and

$$
\begin{aligned}
b_2 &= \frac{\left(\frac{q^p b_1^p}{p+1} + k^p\right)^p q^p b_1^{2p}}{(1 - k^p(1 + qb_1)^p b_1)^{p+1}} b_1 \\
&\leq \frac{\left(\frac{q^p b_0^p}{p+1} + k^p\right)^p q^p (\zeta b_0)^{2p}}{(1 - k^p(1 + qb_0)^p b_0)^{p+1}} b_1 \\
&\leq \zeta^{2p} \frac{\left(\frac{q^p b_0^p}{p+1} + k^p\right)^p q^p b_0^{2p}}{(1 - k^p(1 + qb_0)^p b_0)^{p+1}} b_1 \\
&= \zeta^{2p} \gamma_0 \theta^p b_1 = \zeta^{2p+1} b_1.
\end{aligned}
$$

Hence (4) holds for $n = 0$ and $n = 1$. Let $j > 1$ be a fixed integer. Assume that (4) holds for each $n = 0, 1, 2 \cdots, j$. From Equations (11) and (12), we have

$$
\begin{aligned}
b_{j+2} &= \frac{\left(\frac{q^p b_{j+1}^p}{p+1} + k^p\right)^p q^p b_{j+1}^{2p}}{(1 - k^p(1 + qb_{j+1})^p b_{j+1})^{p+1}} b_{j+1} \\
&\leq \frac{\left(\frac{q^p b_{j+1}^p}{p+1} + k^p\right)^p q^p (\zeta^{(2p+1)^j} b_j)^{2p}}{(1 - k^p(1 + qb_{j+1})^p b_{j+1})^{p+1}} b_{j+1} \\
&\leq (\zeta^{2p(2p+1)^j}) \frac{\left(\frac{q^p b_j^p}{p+1} + k^p\right)^p q^p b_j^{2p}}{(1 - k^p(1 + qb_j)^p b_j)^{p+1}} b_{j+1} \\
&\leq \zeta^{2p(2p+1)^j} \zeta^{2p(2p+1)^{j-1}} \cdots \zeta^{2p(2p+1)} \zeta^{(2p+1)} b_{j+1} \\
&= \zeta^{(2p+1)^{j+1}} b_{j+1}.
\end{aligned}
$$

Thus (4) holds for $n = j + 1$. Therefore, by induction, (4) holds for all $n \in \mathbb{N}_0$.

(5) From Equation (11) and (4), one can easily observe that

$$\theta_1 = \frac{\left(\frac{q^p b_1^p}{p+1} + k^p\right) qb_1^2}{1 - k^p(1 + qb_1)^p b_1} \leq \frac{\left(\frac{q^p b_0^p}{p+1} + k^p\right) q(\zeta b_0)^2}{1 - k^p(1 + qb_0)^p b_0} \leq \zeta^{\frac{(2p+1)^1 - 1}{p}} \theta.$$

Hence (5) holds for $n = 1$. Let $j > 1$ be a fixed integer. Assume that (5) holds for each $n = 0, 1, 2 \cdots, j$. From Equation (11), we have

$$
\begin{aligned}
\theta_{j+1} &= \frac{\left(\frac{q^p b_{j+1}^p}{p+1} + k^p\right) q b_{j+1}^2}{1 - k^p \left(1 + q b_{j+1}\right)^p b_{j+1}} \\
&\leq \frac{\left(\frac{q^p b_0^p}{p+1} + k^p\right) q \left(\xi^{\frac{(2p+1)^{j+1}-1}{2p}} b_0\right)^2}{1 - k^p \left(1 + q b_0\right)^p b_0} \\
&= \xi^{\frac{(2p+1)^{j+1}-1}{p}} \theta.
\end{aligned}
$$

Thus (5) holds for $n = j + 1$. Therefore, by induction, (v) holds for all $n \in \mathbb{N}_0$. This completes the proof. $\square$

3. Computation of a Solution of the Operator Equation (1)

Let X and Y be Banach spaces and D be a nonempty open convex subset of X. Let $F : D \subset X \to Y$ be a nonlinear operator such that F is Fréchet differentiable at each point of D and let $L \in B(Y, X)$ such that $(I - LF)(D) \subseteq D$. To solve the operator Equation (1), we introduce the Newton-like algorithm as follows:

Starting with $x_0 \in D$ and after $x_n \in D$ is defined, we define the next iterate x_{n+1} as follows:

$$
\begin{cases}
z_n = (I - LF)(x_n), \\
y_n = (I - F_{z_n}'^{-1} F)(x_n), \\
x_{n+1} = (I - F_{z_n}'^{-1} F)(y_n)
\end{cases}
\tag{13}
$$

for each $n \in \mathbb{N}_0$.

If we take $X = Y, F = I - G$ and $L = I \in B(X)$ in Equation (13), then the iteration process Equation (13) reduces to the Stirling-like iteration process Equation (8).

Before proving the main result of the paper, we establish the following:

Proposition 1. *Let D be a nonempty open convex subset of a Banach space X, $F : D \subset X \to Y$ be a Fréchet differentiable at each point of D with values in a Banach space Y and $L \in B(Y, X)$ such that $(I - LF)(D) \subseteq D$. Let $\omega : [0, \infty) \to [0, \infty)$ be a nondecreasing and continuous real-valued function. Assume that F satisfies the following conditions:*

(1) $\|F_x' - F_y'\| \leq \omega(\|x - y\|)$ *for all $x, y \in D$;*
(2) $\|I - LF_x'\| \leq c$ *for all $x \in D$ and for some $c \in (0, \infty)$.*

Define a mapping $T : D \to D$ by

$$
T(x) = (I - LF)(x)
\tag{14}
$$

for all $x \in D$. Then we have

$$
\|I - F_{Tx}'^{-1} F_{Ty}'\| \leq \|F_{Tx}'^{-1}\| \omega(c\|x - y\|)
$$

for all $x, y \in D$.

Proof. For any $x, y \in D$, we have

$$
\begin{aligned}
\|I - F_{Tx}'^{-1}F_{Ty}'\| &\leq \|F_{Tx}'^{-1}\|\|F_{Tx}' - F_{Ty}'\| \\
&\leq \|F_{Tx}'^{-1}\|\omega(\|Tx - Ty\|) \\
&= \|F_{Tx}'^{-1}\|\omega(\|x - y - L(F(x) - F(y))\|) \\
&= \|F_{Tx}'^{-1}\|\omega\left(\left\|x - y - L\int_0^1 F_{y+t(x-y)}'(x-y)dt\right\|\right) \\
&\leq \|F_{Tx}'^{-1}\|\omega\left(\int_0^1 \|I - LF_{y+t(x-y)}'\|dt\|x - y\|\right) \\
&\leq \|F_{Tx}'^{-1}\|\omega(c\|x - y\|).
\end{aligned}
$$

This completes the proof. □

Now, we are ready to prove our main results for solving the problem Equation (1) in the framework of Banach spaces.

Theorem 1. *Let D be a nonempty open convex subset of a Banach space X, $F : D \subset X \to Y$ a Fréchet differentiable at each point of D with values in a Banach space Y and $L \in B(Y, X)$ such that $(I - LF)(D) \subseteq D$. Let $x_0 \in D$ be such that $z_0 = x_0 - LF(x_0)$ and $F_{z_0}'^{-1} \in B(Y, X)$ exist. Let $\omega \in \Phi$ and let α be the solution of the Equation (9). Assume that the following conditions hold:*

(C1) $\|F_x' - F_y'\| \leq \omega(\|x - y\|)$ for all $x, y \in D$;

(C2) $\|I - LF_x'\| \leq k$ for all $x \in D$ and for some $k \in (0, \frac{1}{3}]$;

(C3) $\|F_{z_0}'^{-1}\| \leq \beta$ for some $\beta > 0$;

(C4) $\|F_{z_0}'^{-1}F(x_0)\| \leq \eta$ for some $\eta > 0$;

(C5) $\omega(ts) \leq t^p\omega(s), s \in [0, \infty), t \in [0, 1]$ and $p \in (0, 1]$;

(C6) $b_0 = \beta\omega(\eta) < \alpha, q = \frac{1}{p+1} + k^p, \theta = \dfrac{\left(\frac{q^p b_0^p}{p+1} + k^p\right)qb_0^2}{1 - k^p(1 + qb_0)^p b_0}$ and $B_r[x_0] \subset D$, where $r = \frac{1+q}{1-\theta}\eta$.

Then we have the following:

(1) *The sequence $\{x_n\}$ generated by Equation (13) is well defined, remains in $B_r[x_0]$ and satisfies the following estimates:*

$$
\begin{cases}
\|y_{n-1} - z_{n-1}\| \leq k\|y_{n-1} - x_{n-1}\|, \\
\|x_n - y_{n-1}\| \leq qb_{n-1}\|y_{n-1} - x_{n-1}\|, \\
\|x_n - x_{n-1}\| \leq (1 + qb_{n-1})\|y_{n-1} - x_{n-1}\|, \\
F_{z_n}'^{-1} \text{ exists and } \|F_{z_n}'^{-1}\| \leq \gamma_{n-1}\|F_{z_{n-1}}'^{-1}\|, \\
\|y_n - x_n\| \leq \theta_{n-1}\|y_{n-1} - x_{n-1}\| \leq \theta^n\|y_0 - x_0\|, \\
\|F_{z_n}'^{-1}\|\omega(\|y_n - x_n\|) \leq b_n
\end{cases}
\tag{15}
$$

for all $n \in \mathbb{N}$, where $z_n, y_n \in B_r[x_0]$, the sequences $\{b_n\}, \{\theta_n\},$ and $\{\gamma_n\}$ are defined by Equations (10) and (11), respectively.

(2) *The sequence $\{x_n\}$ converges to the solution $x^* \in B_r[x_0]$ of the Equation (1).*

(3) *The priory error bounds on x^* is given by:*

$$
\|x_n - x^*\| \leq \frac{(1 + qb_0)\eta}{\xi^{1/2p^2}\left(1 - \xi^{-\frac{(2p+1)^n}{p}}\gamma_0^{-\frac{1}{p}}\right)\gamma_0^{n/p}}\left(\xi^{1/2p^2}\right)^{(2p+1)^n}
$$

for each $n \in \mathbb{N}_0$.

(4) *The sequence $\{x_n\}$ has R-order of convergence at least $2p + 1$.*

Proof. (1) First, we show that Equation (15) is true for $n = 1$. Since $x_0 \in D$, $y_0 = x_0 - F_{z_0}'^{-1}F(x_0)$ is well defined. Note that

$$\|y_0 - x_0\| = \|F_{z_0}'^{-1}F(x_0)\| \leq \eta < r.$$

Hence $y_0 \in B_r[x_0]$. Using Equation (13), we have

$$
\begin{aligned}
\|y_0 - z_0\| &= \| - F_{z_0}'^{-1}F(x_0) + LF(x_0)\| \\
&= \|y_0 - x_0 - LF_{z_0}'(y_0 - x_0)\| \\
&\leq \|I - LF_{z_0}'\|\|y_0 - x_0\| \\
&\leq k\|y_0 - x_0\|.
\end{aligned}
$$

By Proposition 1 and (C2), we have

$$
\begin{aligned}
\|x_1 - y_0\| &= \|F_{z_0}'^{-1}(F(y_0) - F(x_0) - F_{z_0}'(y_0 - x_0))\| \\
&\leq \int_0^1 \|F_{z_0}'^{-1}(F_{x_0+t(y_0-x_0)}' - F_{y_0}' + F_{y_0}' - F_{z_0}')\|\|y_0 - x_0\|dt \\
&\leq \beta \left[\int_0^1 \|(F_{x_0+t(y_0-x_0)}' - F_{y_0}')\|dt + \|F_{y_0}' - F_{z_0}'\|\right]\|y_0 - x_0\| \\
&= \beta \left[\int_0^1 \omega((1-t)\|y_0 - x_0\|)dt + \omega(\|y_0 - z_0\|)\right]\|y_0 - x_0\| \\
&= \beta \left[\int_0^1 (1-t)^p\omega(\|y_0 - x_0\|)dt + \omega(k\|y_0 - x_0\|)\right]\|y_0 - x_0\| \\
&= \beta \left[\int_0^1 (1-t)^p\omega(\|y_0 - x_0\|)dt + k^p\omega(\|y_0 - x_0\|)\right]\|y_0 - x_0\| \\
&\leq \beta \left[\frac{1}{p+1} + k^p\right]\omega(\|y_0 - x_0\|)\|y_0 - x_0\| \\
&\leq q\beta\omega(\eta)\|y_0 - x_0\| \leq qb_0\|y_0 - x_0\|.
\end{aligned}
$$

Thus we have

$$
\begin{aligned}
\|x_1 - x_0\| &\leq \|x_1 - y_0\| + \|y_0 - x_0\| \leq qb_0\|y_0 - x_0\| + \|y_0 - x_0\| \\
&\leq (1 + qb_0)\|y_0 - x_0\| < r,
\end{aligned}
\tag{16}
$$

which shows that $x_1 \in B_r[x_0]$. Note that $z_1 = (I - LF)(x_1) \in D$. Using Proposition 1 and (C3)–(C5), we have

$$
\begin{aligned}
\|I - F_{z_0}'^{-1}F_{z_1}'\| &\leq \|F_{z_0}'^{-1}\|\omega(k\|x_1 - x_0\|) \\
&\leq \beta\omega(k(1 + qb_0)\|y_0 - x_0\|) \\
&\leq \beta k^p(1 + qb_0)^p\omega(\|y_0 - x_0\|) \\
&\leq k^p(1 + qb_0)^p\beta\omega(\eta) \\
&\leq (k(1 + qb_0))^p b_0 < 1.
\end{aligned}
$$

Therefore, by Lemma 1, $F_{z_1}'^{-1}$ exists and

$$\|F_{z_1}'^{-1}\| \leq \frac{\|F_{z_0}'^{-1}\|}{1 - (k(1 + qb_0))^p b_0} = \gamma_0\|F_{z_0}'^{-1}\|.
\tag{17}$$

Subsequently, we have

$$
\begin{aligned}
\|y_1 - x_1\| &= \|F_{z_1}'^{-1} F(x_1)\| \\
&= \|F_{z_1}'^{-1} (F(x_1) - F(y_0) - F_{z_0}'(x_1 - y_0))\| \\
&\leq \|F_{z_1}'^{-1}\| \left[\int_0^1 \|(F_{y_0 + t(x_1 - y_0)}' - F_{y_0}')\| dt + \|F_{y_0}' - F_{z_0}'\| \right] \|x_1 - y_0\| \\
&\leq \|F_{z_1}'^{-1}\| \left[\tfrac{1}{p+1} \omega(\|x_1 - y_0\|) + \omega(k\|y_0 - x_0\|) \right] \|x_1 - y_0\| \\
&\leq \|F_{z_1}'^{-1}\| \left[\tfrac{1}{p+1} \omega(q b_0 \|x_0 - y_0\|) + k^p \omega(\|y_0 - x_0\|) \right] q b_0 \|x_0 - y_0\| \\
&\leq \|F_{z_1}'^{-1}\| \left[\tfrac{q^p b_0^p}{p+1} \omega(\|y_0 - x_0\|) + k^p \omega(\|y_0 - x_0\|) \right] q b_0 \|y_0 - x_0\| \\
&\leq \gamma_0 \left[\tfrac{q^p b_0^p}{p+1} + k^p \right] \beta \omega(\eta) q b_0 \|y_0 - x_0\| \\
&\leq \tfrac{\left(\tfrac{q^p b_0^p}{p+1} + k^p \right) q b_0^2}{1 - (k(1 + q b_0))^p b_0} \|y_0 - x_0\| \\
&\leq \theta \|y_0 - x_0\|.
\end{aligned}
\tag{18}
$$

From Equations (16) and (18), we have

$$
\begin{aligned}
\|y_1 - x_0\| &\leq \|y_1 - x_1\| + \|x_1 - x_0\| \\
&\leq \theta \|y_0 - x_0\| + (1 + q b_0) \|y_0 - x_0\| \\
&\leq (1 + q b_0) \theta \|y_0 - x_0\| + (1 + q b_0) \|y_0 - x_0\| \\
&\leq (1 + q b_0)(1 + \theta) \eta < r
\end{aligned}
$$

and

$$
\begin{aligned}
\|z_1 - x_0\| &\leq \|z_1 - y_1\| + \|y_1 - x_1\| + \|x_1 - x_0\| \\
&\leq (1 + k)\|y_1 - x_1\| + (1 + q b_0)\|y_0 - x_0\| \\
&\leq (1 + q)\theta \eta + (1 + q)\eta \\
&= (1 + q)(1 + \theta)\eta < r.
\end{aligned}
$$

This shows that $z_1, y_1 \in B_r[x_0]$. From Equations (17) and (18), we get

$$
\begin{aligned}
\|F_{z_1}'^{-1}\| \omega(\|y_1 - x_1\|) &\leq \gamma_0 \|F_{z_0}'^{-1}\| \omega(\theta \|y_0 - x_0\|) \\
&\leq \gamma_0 \theta^p \beta \omega(\eta) \\
&\leq \gamma_0 \theta^p b_0 = b_1.
\end{aligned}
$$

Thus we see that Equation (15) holds for $n = 1$.

Let $j > 1$ be a fixed integer. Assume that Equation (15) is true for $n = 1, 2, \cdots, j$. Since $x_j \in B_r[x_0]$, it follows $z_j = (I - LF)(x_j) \in D$. Using (C3), (C4), Equations (13) and (15), we have

$$
\begin{aligned}
\|y_j - z_j\| &= \|LF(x_j) - F_{z_j}'^{-1} F(x_j)\| = \|(L - F_{z_j}'^{-1}) F(x_j)\| \\
&= \|(L - F_{z_j}'^{-1}) F_{z_j}'(x_j - y_j)\| \\
&\leq \|I - L F_{z_j}'\| \|y_j - x_j\| \\
&\leq k \|y_j - x_j\|.
\end{aligned}
\tag{19}
$$

Using Equations (13) and (19), we have

$$
\begin{aligned}
\|x_{j+1} - y_j\| &= \|F_{z_j}'^{-1} F(y_j)\| \\
&\leq \|F_{z_j}'^{-1}\| \|F(y_j) - F(x_j) - F_{z_j}'(y_j - x_j)\| \\
&\leq \|F_{z_j}'^{-1}\| \left[\int_0^1 \|F_{x_j + t(y_j - x_j)}' - F_{z_j}'\| dt \right] \|y_j - x_j\| \\
&\leq \|F_{z_j}'^{-1}\| \left[\int_0^1 \|F_{x_j + t(y_j - x_j)}' - F_{y_j}'\| dt + \|F_{y_j}' - F_{z_j}'\| \right] \|y_j - x_j\| \\
&= \|F_{z_j}'^{-1}\| \left[\int_0^1 \omega(x_j + t(y_j - x_j) - y_j) dt + \omega(\|y_j - z_j\|) \right] \|y_j - x_j\| \\
&\leq \|F_{z_j}'^{-1}\| \left[\int_0^1 \omega((1 - t)\|y_j - x_j\|) dt + \omega(k\|y_j - x_j\|) \right] \|y_j - x_j\| \\
&\leq \|F_{z_j}'^{-1}\| \left[\int_0^1 ((1 - t)^p + k^p)\omega(\|y_j - x_j\|) dt \right] \|y_j - x_j\| \\
&\leq \|F_{z_j}'^{-1}\| \left[\frac{1}{p+1} + k^p \right] \omega(\|y_j - x_j\|) \|y_j - x_j\| \\
&= q\|F_{z_j}'^{-1}\| \omega(\|y_j - x_j\|) \|y_j - x_j\| \\
&= q b_j \|y_j - x_j\|.
\end{aligned}
\tag{20}
$$

From Equation (20), we have

$$
\begin{aligned}
\|x_{j+1} - x_j\| &\leq \|x_{j+1} - y_j\| + \|y_j - x_j\| \\
&\leq q b_j \|y_j - x_j\| + \|y_j - x_j\| \\
&\leq (1 + q b_j)\|y_j - x_j\|.
\end{aligned}
\tag{21}
$$

Using Equations (20) and (21) and the triangular inequality, we have

$$
\begin{aligned}
\|x_{j+1} - x_0\| &\leq \sum_{s=0}^{j} \|x_{s+1} - x_s\| \\
&\leq \sum_{s=0}^{j} (1 + q b_s)\|y_s - x_s\| \\
&\leq \sum_{s=0}^{j} (1 + q b_0)\theta^s \|y_0 - x_0\| \\
&\leq (1 + q b_0)\frac{1 - \theta^{j+1}}{1 - \theta}\eta \\
&\leq \frac{(1 + q)\eta}{1 - \theta} = r,
\end{aligned}
$$

which implies that $x_{k+1} \in B_r[x_0]$. Again, by using Proposition 1, (C2), (C5), and Equation (21), we have

$$
\begin{aligned}
\|I - F_{z_j}'^{-1} F_{z_{j+1}}'\| &\leq \|F_{z_j}'^{-1}\| \omega(k\|x_{j+1} - x_j\|) \\
&\leq \|F_{z_j}'^{-1}\| k^p (1 + q b_j)^p \omega(\|y_j - x_j\|) \\
&\leq k^p (1 + q b_j)^p b_j < 1.
\end{aligned}
$$

Therefore, by Lemma 1, $F_{z_{j+1}}'^{-1}$ exists and

$$
\|F_{z_{j+1}}'^{-1}\| \leq \frac{\|F_{z_j}'^{-1}\|}{1 - k^p (1 + q b_j)^p b_j} = \gamma_j \|F_{z_j}'^{-1}\|.
$$

Using Equations (13), (C2), and (21), we have

$$
\begin{aligned}
\|y_{j+1} - x_{j+1}\| &= \|F_{z_{j+1}}'^{-1} F(x_{j+1})\| \\
&= \|F_{z_{j+1}}'^{-1} (F(x_{j+1}) - F(y_j) - F_{z_j}'(x_{j+1} - y_j))\| \\
&\leq \|F_{z_{j+1}}'^{-1}\| \left[\int_0^1 \|F_{y_j+t(x_{j+1}-y_j)}' - F_{y_j}'\| dt + \|F_{y_j}' - F_{z_j}'\| \right] \|x_{j+1} - y_j\| \\
&\leq \|F_{z_{j+1}}'^{-1}\| \left[\int_0^1 \omega(t\|x_{j+1} - y_j\|) dt + \omega(\|y_j - z_j\|) \right] \|x_{j+1} - y_j\| \\
&\leq \|F_{z_{j+1}}'^{-1}\| \left[\int_0^1 \omega(tqb_j\|y_j - x_j\|) dt + \omega(k\|y_j - x_j\|) \right] qb_j\|y_j - x_j\| \\
&\leq \gamma_j \|F_{z_j}'^{-1}\| \left[\frac{q^p b_j^p}{p+1} \omega(\|y_j - x_j\|) + k^p \omega(\|y_j - x_j\|) \right] qb_j\|y_j - x_j\| \\
&= \gamma_j \left[\frac{q^p b_j^p}{p+1} + k^p \right] \|F_{z_j}'^{-1}\| \omega(\|y_j - x_j\|) qb_j\|y_j - x_j\| \\
&\leq \gamma_j \left[\frac{q^p b_j^p}{p+1} + k^p \right] qb_j^2 \|y_j - x_j\| \\
&\leq \theta_j \|y_j - x_j\| \leq \theta^{j+1} \|y_0 - x_0\|,
\end{aligned}
$$

$$
\begin{aligned}
\|y_{j+1} - x_0\| &\leq \|y_{j+1} - x_{j+1}\| + \|x_{j+1} - x_0\| \\
&\leq \theta^{j+1} \|y_0 - x_0\| + \sum_{s=0}^{j} \|x_{s+1} - x_s\| \\
&\leq \theta^{j+1} \|y_0 - x_0\| + \sum_{s=0}^{j} (1 + qb_0)\theta^s \|y_0 - x_0\| \\
&\leq (1 + qb_0) \sum_{s=0}^{j+1} \theta^s \eta \\
&\leq \frac{(1+q)\eta}{1-\theta} = r
\end{aligned}
$$

and

$$
\begin{aligned}
\|z_{j+1} - x_0\| &\leq \|z_{j+1} - y_{j+1}\| + \|y_{j+1} - x_{j+1}\| + \|x_{j+1} - x_0\| \\
&\leq (1+k)\|y_{j+1} - x_{j+1}\| + \sum_{s=0}^{j} (1 + qb_0)\theta^s \eta \\
&\leq (1+q)\theta^{j+1}\eta + \sum_{s=0}^{j} (1+q)\theta^s \eta \\
&\leq \sum_{s=0}^{j+1} (1+q)\theta^s \eta < r
\end{aligned}
$$

which implies that $z_{j+1}, y_{j+1} \in B_r(x_0)$. Also, we have

$$
\begin{aligned}
\|F_{z_{j+1}}'^{-1}\| \omega(\|y_{j+1} - x_{j+1}\|) &\leq \gamma_j \|F_{z_j}'^{-1}\| \omega(\theta_j \|y_j - x_j\|) \\
&\leq \gamma_j \theta_j^p \|F_{z_j}'^{-1}\| \omega(\|y_j - x_j\|) \\
&\leq \gamma_j \theta_j^p b_j = b_{j+1}.
\end{aligned}
$$

Hence we conclude that Equation (15) is true for $n = j + 1$. Therefore, by induction, Equation (15) is true for all $n \in \mathbb{N}_0$.

(2) First, we show that the sequence $\{x_n\}$ is a Cauchy sequence. For this, letting $m, n \in \mathbb{N}_0$ and using Lemma 3, we have

$$
\begin{aligned}
\|x_{m+n} - x_n\| & \leq \sum_{j=n}^{m+n-1} \|x_{j+1} - x_j\| \\
& \leq \sum_{j=n}^{m+n-1} (1 + qb_j) \|y_j - x_j\| \\
& \leq (1 + qb_0) \sum_{j=n}^{m+n-1} \prod_{i=0}^{j-1} \theta_i \|y_0 - x_0\| \\
& \leq (1 + qb_0) \sum_{j=n}^{m+n-1} \prod_{i=0}^{j-1} \zeta^{\frac{(2p+1)^i - 1}{p}} \theta \|y_0 - x_0\| \\
& \leq (1 + qb_0) \sum_{j=n}^{m+n-1} \prod_{i=0}^{j-1} \zeta^{\frac{(2p+1)^i}{p}} \gamma_0^{-\frac{1}{p}} \|y_0 - x_0\| \\
& = (1 + qb_0) \sum_{j=n}^{m+n-1} \zeta^{\sum_{i=0}^{j-1} \frac{(2p+1)^i}{p}} \gamma_0^{-\frac{1}{p}} \|y_0 - x_0\| \\
& \leq (1 + qb_0) \left(\sum_{j=n}^{m+n-1} \zeta^{\frac{(2p+1)^j - 1}{2p^2}} \gamma_0^{-\frac{j}{p}} \right) \|y_0 - x_0\|.
\end{aligned}
$$

By Bernoulli's inequality, for each $j \geq 0$ and $y > -1$, we have $(1 + y)^j \geq 1 + jy$. Hence we have

$$
\begin{aligned}
& \|x_{m+n} - x_n\| \\
& \leq (1 + qb_0) \zeta^{-\frac{1}{2p^2}} \gamma_0^{-\frac{n}{p}} \left(\zeta^{\frac{(2p+1)^n}{2p^2}} + \zeta^{\frac{(2p+1)^n(2p+1)}{2p^2}} \gamma_0^{-\frac{1}{p}} + \cdots + \zeta^{\frac{(2p+1)^n(2p+1)^{m-1}}{2p^2}} \gamma_0^{-\frac{(m-1)}{p}} \right) \eta \\
& \leq (1 + qb_0) \zeta^{-\frac{1}{2p^2}} \gamma_0^{-\frac{n}{p}} \left(\zeta^{\frac{(2p+1)^n}{2p^2}} + \zeta^{\frac{(2p+1)^n(1+2p)}{2p^2}} \gamma_0^{-\frac{1}{p}} + \cdots + \zeta^{\frac{(2p+1)^n(1+2(m-1)p)}{2p^2}} \gamma_0^{-\frac{(m-1)}{p}} \right) \eta \\
& = (1 + qb_0) \zeta^{-\frac{1}{2p^2}} \gamma_0^{-\frac{n}{p}} \left(\zeta^{\frac{(2p+1)^n}{2p^2}} + \zeta^{(2p+1)^n \left(\frac{1}{2p^2} + \frac{1}{p} \right)} \gamma_0^{-\frac{1}{p}} + \cdots + \zeta^{(2p+1)^n \left(\frac{1}{2p^2} + \frac{m-1}{p} \right)} \gamma_0^{-\frac{(m-1)}{p}} \right) \eta \qquad (22) \\
& = (1 + qb_0) \zeta^{\frac{(2p+1)^n - 1}{2p^2}} \gamma_0^{-\frac{n}{p}} \left(1 + \left(\zeta^{(2p+1)^n} \gamma_0^{-1} \right)^{\frac{1}{p}} + \cdots + \left(\zeta^{(2p+1)^n} \gamma_0^{-1} \right)^{\frac{m-1}{p}} \right) \eta \\
& = (1 + qb_0) \zeta^{\frac{(2p+1)^n - 1}{2p^2}} \gamma_0^{-\frac{n}{p}} \left(\frac{1 - \left(\zeta^{(2p+1)^n} \gamma_0^{-1} \right)^{\frac{m}{p}}}{1 - \left(\zeta^{(2p+1)^n} \gamma_0^{-1} \right)^{\frac{1}{p}}} \right) \eta.
\end{aligned}
$$

Since the sequence $\{x_n\}$ is a Cauchy sequence and hence it converges to some point $x^* \in B_r[x_0]$. From Equations (13), (C2), and (15), we have

$$
\begin{aligned}
\|LF(x_n)\| & \leq \|z_n - y_n\| + \|y_n - x_n\| \\
& \leq k\|y_n - x_n\| + \|y_n - x_n\| \\
& \leq (1 + k)\theta^n \eta.
\end{aligned}
$$

Taking the limit as $n \to \infty$ and using the continuity of F and the linearity of L, we have

$$F(x^*) = 0.$$

(3) Taking the limit as $m \to \infty$ in Equation (22), we have

$$\|x^* - x_n\| \leq \frac{(1 + qb_0)\eta}{\xi^{1/2p^2}\left(1 - \xi^{\frac{(2p+1)^n}{p}} \gamma_0^{-\frac{1}{p}}\right)\gamma_0^{n/p}} \left(\xi^{1/2p^2}\right)^{(2p+1)^n} \tag{23}$$

for each $n \in \mathbb{N}_0$.

(4) Here we prove

$$\frac{\|x_{n+1} - x^*\|}{\|x_n - x^*\|^{2p+1}} \leq K$$

for all $n \in \mathbb{N}_0$ and for some $K > 0$. One can easily observe that there exists $n_0 > 0$ such that

$$\|x_n - x^*\| < 1 \tag{24}$$

whenever $n \geq n_0$. Using Equations (13) and (24), we have

$$
\begin{aligned}
\|z_n - x^*\| &= \|x_n - x^* - LF(x_n)\| \\
&= \|x_n - x^* - L(F(x_n) - F(x^*))\| \\
&= \left\|x_n - x^* - L\int_0^1 F'_{x^*+t(x_n-x^*)}(x_n - x^*)dt\right\| \\
&\leq \int_0^1 \|I - LF'_{x^*+t(x_n-x^*)}\|\|x_n - x^*\|dt \\
&\leq k\|x_n - x^*\|
\end{aligned}
$$

and

$$
\begin{aligned}
\|y_n - x^*\| &= \|x_n - x^* - F'^{-1}_{z_n}F(x_n)\| \\
&= \|F'^{-1}_{z_n}[F'_{z_n}(x_n - x^*) - F(x_n)]\| \\
&\leq \|F'^{-1}_{z_n}\|\|F(x_n) - x^* - F'_{z_n}(x_n - x^*)\| \\
&= \|F'^{-1}_{z_n}\|\|\int_0^1 (F'_{x^*+t(x_n-x^*)} - F'_{z_n})(x_n - x^*)\|dt \\
&\leq \|F'^{-1}_{z_n}\|\int_0^1 \|F'_{x^*+t(x_n-x^*)} - F'_{z_n}\|\|x_n - x^*\|dt \\
&\leq \|F'^{-1}_{z_n}\|\int_0^1 (\|F'_{x^*+t(x_n-x^*)} - F'_{x^*}\| + \|F'_{x^*} - F'_{z_n}\|)\|x_n - x^*\|dt \\
&\leq \|F'^{-1}_{z_n}\|\int_0^1 (\omega(t\|x_n - x^*\|) + \omega(\|z_n - x^*\|))\|x_n - x^*\|dt \\
&\leq \|F'^{-1}_{z_n}\|\int_0^1 (t^p\|x_n - x^*\|^p\omega(1) + \omega(k\|x_n - x^*\|))\|x_n - x^*\|dt \\
&\leq \|F'^{-1}_{z_n}\|\left(\frac{1}{p+1} + k^p\right)\omega(1)\|x_n - x^*\|^{p+1} \\
&= \|F'^{-1}_{z_n}\|q\omega(1)\|x_n - x^*\|^{p+1}.
\end{aligned}
\tag{25}
$$

Using Equations (13), (24) and (25), we have

$$\|x_{n+1} - x^*\|$$
$$= \|y_n - F_{z_n}'^{-1} F(y_n) - x^*\|$$
$$\leq \|F_{z_n}'^{-1}\| \|F(y_n) - F(x^*) - F_{z_n}'(y_n - x^*)\|$$
$$= \|F_{z_n}'^{-1}\| \left\| \int_0^1 (F_{x^* + t(y_n - x^*)}' - F_{z_n}')(y_n - x^*)\,dt \right\|$$
$$\leq \|F_{z_n}'^{-1}\| \int_0^1 \|F_{x^* + t(y_n - x^*)}' - F_{z_n}'\| \|y_n - x^*\|\,dt$$
$$= \|F_{z_n}'^{-1}\| \int_0^1 (\|F_{x^* + t(y_n - x^*)}' - F_{x^*}'\| + \|F_{x^*}' - F_{z_n}'\|) \|y_n - x^*\|\,dt$$
$$\leq \|F_{z_n}'^{-1}\| \int_0^1 (\omega(t\|y_n - x^*\|) + \omega(k\|x_n - x^*\|)) \|y_n - x^*\|\,dt$$
$$= \|F_{z_n}'^{-1}\| \int_0^1 \left(t^p \omega \left(\|F_{z_n}'^{-1}\| q \omega(1) \|x_n - x^*\|^{p+1} \right) + \omega(k\|x_n - x^*\|) \right) dt$$
$$\times \|F_{z_n}'^{-1}\| q \omega(1) \|x_n - x^*\|^{p+1}$$
$$\leq \|F_{z_n}'^{-1}\|^2 \int_0^1 \left(t^p \|x_n - x^*\|^{p(p+1)} \omega \left(\|F_{z_n}'^{-1}\| q \omega(1) \right) + k^p \|x_n - x^*\|^p \omega(1) \right) dt$$
$$\times q \omega(1) \|x_n - x^*\|^{p+1}$$
$$= \|F_{z_n}'^{-1}\|^2 \left(\frac{\|x_n - x^*\|^{p^2} \omega \left(\|F_{z_n}'^{-1}\| q \omega(1) \right)}{p+1} + k^p \right) q \omega(1) \|x_n - x^*\|^{2p+1}$$
$$= K_n \|x_n - x^*\|^{2p+1},$$

where

$$K_n = \|F_{z_n}'^{-1}\|^2 \left(\frac{\|x_n - x^*\|^{p^2} \omega \left(\|F_{z_n}'^{-1}\| q \omega(1) \right)}{p+1} + k^p \right) q \omega(1).$$

Let $\|F_{x^*}'^{-1}\| \leq d$ and $0 < d < \omega(\sigma)^{-1}$, where $\sigma > 0$. Then, for all $x \in B_\sigma(x^*)$, we have

$$\|I - F_{x^*}'^{-1} F_x'\| \leq \|F_{x^*}'^{-1}\| \|F_{x^*}' - F_x'\| \leq d\omega(\sigma) < 1$$

and so, by Lemma 1, we have

$$\|F_x'^{-1}\| \leq \frac{d}{1 - d\omega(\sigma)} := \lambda.$$

Since $x_n \to x^*$ and $z_n \to x^*$ as $n \to \infty$, there exists a positive integer N_0 such that

$$\|F_{z_n}'^{-1}\| \leq \frac{d}{1 - d\omega(\sigma)}$$

for all $n \geq N_0$. Thus, for all $n \geq N_0$, one can easily observe that

$$K_n \leq \lambda^2 \left(\frac{\sigma^{p^2} \omega (\lambda q \omega(1))}{p+1} + k^p \right) q \omega(1) = K.$$

This shows that the R-order of convergence at least $(2p + 1)$. This completes the proof. □

4. Applications

4.1. Fixed Points of Smooth Operators

Let X be a Banach spaces and D be a nonempty open convex subset of X. Let $G : D \subset X \to X$ be a nonlinear operator such that D is Fréchet differentiable at each point of D and let $L \in B(X, X)$ such that $(I - L(I - G))(D) \subseteq D$. For $F = I - G$, the Newton-like algorithm Equation (13) reduces to the following Stirling-like method for computing fixed point of the operator G:

Starting with $x_0 \in D$ and after $x_n \in D$ is defined, we define the next iterate x_{n+1} as follows:

$$\begin{cases} z_n = (I - L(I - G))(x_n), \\ y_n = (I - (I - G'_{z_n})^{-1}(I - G))(x_n), \\ x_{n+1} = (I - (I - G'_{z_n})^{-1}(I - G))(y_n) \end{cases} \tag{26}$$

for each $n \in \mathbb{N}_0$.

For the choice of $X = Y$ and $F = I - G$, Theorem 1 reduces to the following:

Theorem 2. *Let D be a nonempty open convex subset of a Banach space X, $G : D \to X$ be a Fréchet differentiable at each point of D with values into itself. Let $L \in B(X)$ be such that $(I - L(I - G))(D) \subseteq D$. Let $x_0 \in D$ be such that $z_0 = x_0 - L(x_0 - G(x_0))$ and let $(I - G'_{z_0})^{-1} \in B(X)$ exist. Let $\omega \in \Phi$ and α be a solution of the Equation* (9). *Assume that the conditions* (C5)–(C6) *and the following conditions hold:*

(C7) $\|(I - G'_{z_0})^{-1}\| \leq \beta$ *for some* $\beta > 0$;
(C8) $\|(I - G'_{z_0})^{-1}(x_0 - G(x_0))\| \leq \eta$ *for some* $\eta > 0$;
(C9) $\|G'_x - G'_y\| \leq \omega(\|x - y\|)$ *for all* $x, y \in D$;
(C10) $\|I - L(I - G'_x)\| \leq k$ *for all* $x \in D$ *and for some* $k \in (0, \frac{1}{3}]$.

Then the sequence $\{x_n\}$ generated by Equation (26) *is well defined, remains in $B_r[x_0]$ and converges to the fixed point $x^* \in B_r[x_0]$ of the operator G and the sequence $\{x_n\}$ has R-order of convergence at least $2p + 1$.*

We give an example to illustrate Theorem 2.

Example 1. *Let $X = Y = \mathbb{R}$ and $D = (-1, 1) \subset X$. Define a mapping $G : D \to \mathbb{R}$ by*

$$G(x) = \frac{1.1x^3 - x}{6} \tag{27}$$

for all $x \in D$. Define $L : \mathbb{R} \to \mathbb{R}$ by $L(x) = \frac{7.9}{7}x$ for all $x \in \mathbb{R}$. One can easily observe that

$$(I - L(I - G))(x) \in D$$

for all $x \in D$. Clearly, G is differentiable on D and its derivative at $x \in D$ is $G'_x = \frac{3.3x^2 - 1}{6}$ and G'_x is bounded with $\|G'_x\| \leq 0.3833$ for all $x \in D$ and G' satisfies the Lipschitz condition

$$\|G'_x - G'_y\| \leq K\|x - y\|$$

for all $x, y \in D$, where $K = 1.1$. For $x_0 = 0.3$, we have

$$z_0 = (I - L(I - G))(x_0) = -0.0894135714, \quad \|(I - G'_{z_0})^{-1}\| \leq 0.860385626 = \beta,$$

$$\|(I - G'_{z_0})^{-1}(x_0 - G(x_0))\| \leq 0.29687606 = \eta.$$

For $p = 1, q = \frac{5}{6}$ and $\omega(t) = Kt$ for all $t \geq 0$, we have

$$b_0 = \beta\omega(\eta) = 0.280970684 < 1,$$

$$\theta = \frac{\left(\frac{q^p b_0^p}{p+1} + k^p\right)qb_0^2}{1 - k(1 + qb_0)b_0} = 0.0335033167 < 1$$

and $r = 0.563139828$. Hence all the conditions of Theorem 2 are satisfied. Therefore, the sequence $\{x_n\}$ generated by Equation (26) *is in $B_r[x_0]$ and it converges to the fixed point $x^* = 0 \in B_r[x_0]$ of G.*

Remark 1. *In Example 1, $\|G'_x\| \le 0.38333 > \frac{1}{3}$. Thus the condition (Ω) does not hold and so we can not apply Parhi and Gupta [21] (Theorem 1) for finding fixed points of the operators like G defined by (27). Thus the Stirling-like method defined by Equation (26) provides an affirmative answer of the Problem 1.*

If the condition (Ω) holds, then Theorem 2 with $L = I$ reduces to the main result of Parhi and Gupta [21] as follows:

Corollary 1. *[21] (Theorem 1) Let D be a nonempty open convex subset of a Banach space X and $G : D \rightarrow D$ be a Fréchet differentiable operator and let $x_0 \in D$ with $z_0 = G(x_0)$. Let $(I - G'_{z_0})^{-1} \in B(X)$ exists and $\omega \in \Phi$. Assume that the conditions (C5)–(C9) and the following condition holds:*

(C11) $\|G'_x\| \le k$ for all $x \in D$ and for some $k \in (0, \frac{1}{3}]$.

Then the sequence $\{x_n\}$ generated by Equation (8) is well defined, remains in $B_r[x_0]$ and converges to the fixed point $x^* \in B_r[x_0]$ of the operator G with $R-$order of the convergence at least $2p + 1$.

Example 2. *Let $X = Y = \mathbb{R}$ and $D = (-6, 6) \subset X$. Define a mapping $G : D \rightarrow \mathbb{R}$ by*

$$G(x) = 2 + e^{\frac{\sin x}{5}}$$

for all $x \in D$. It is obvious that G is Fréchet differentiable on D and its Fréchet derivative at $x \in D$ is $G'_x = \frac{\cos x}{5} e^{\frac{\sin x}{5}}$. Clearly, G'_x is bounded with $\|G'_x\| \le 0.22 < \frac{1}{3} = k$ and

$$\|G'_x - G'_y\| \le K\|x - y\|$$

for all $x, y \in D$, where $K = 0.245$. For $x_0 = 0$, we have

$$z_0 = G(x_0) = 3, \quad \|(I - G'_{z_0})^{-1}\| \le 0.834725586524139 = \beta$$

and

$$\|(I - G'_{z_0})^{-1}(x_0 - G(x_0))\| \le 2.504176759572418 = \eta.$$

For $p = 1, q = \frac{5}{6}$ and $\omega(t) = Kt$ for all $t \ge 0$, we have

$$b_0 = \beta K \eta = 0.512123601526580 < 1,$$

$$\theta = \frac{\left(\frac{q^p b_0^p}{p+1} + k^p\right) q b_0^2}{1 - k(1 + q b_0)b_0} = 0.073280601270728 < 1$$

and $r = 5.147038576039456$.

Hence all the conditions of Theorem 2 with $L = I$ are satisfied. Therefore, the sequence $\{x_n\}$ generated by Equation (26) is in $B_r[x_0]$ and it converges to the fixed point $x^* = 3.023785446275295 \in B_r[x_0]$ of G (Table 1).

Table 1. A priori error bounds.

n	$\|x_n - x^*\|$
0	3.0237854462752
1	$1.7795738211156 \times 10^{-2}$
2	$6.216484249588206 \times 10^{-6}$
3	$2.335501569916687 \times 10^{-9}$
4	$8.775202786637237 \times 10^{-13}$
5	$4.440892098500626 \times 10^{-16}$

4.2. Fredholm Integral Equations

Let X be a Banach space over the field $\mathbb{F}$ ($\mathbb{R}$ or $\mathbb{C}$) with the norm $\|\cdot\|$ and D be an open convex subset of X. Further, let $B(X)$ be the Banach space of bounded linear operators from X into itself. Let $S \in B(X)$, $u \in X$ and $\lambda \in \mathbb{F}$. We investigate a solution $x \in X$ of the nonlinear Fredholm-type operator equation:

$$x - \lambda S Q(x) = u, \tag{28}$$

where $Q : D \to X$ is continuously Fréchet differentiable on D. The operator Equation (28) has been discussed in [10,38,39]. Define an operator $F : D \to X$ by

$$F(x) = x - \lambda S Q(x) - u \tag{29}$$

for all $x \in D$. Then solving the operator Equation (29) is equivalent to solving the operator Equation (1). From Equation (29), we have

$$F'_x(h) = h - \lambda S Q'_x(h) \tag{30}$$

for all $h \in X$. Now, we apply Theorem 1 to solve the operator Equation (28).

Theorem 3. *Let X be a Banach space and D an open convex subset of X. Let $Q : D \to X$ be a continuously Fréchet differentiable mapping at each point of D. Let $L, S \in B(X)$ and $u \in X$. Assume that, for any $x_0 \in D$, $z_0 = x_0 - L(x_0 - \lambda S Q(x_0) - u)$ and $(I - \lambda S Q'_{z_0})^{-1}$ exist. Assume that the condition (C6) and the following conditions hold:*

(C12) $(I - L(I - \lambda S Q))(x) - u \in D$ *for all $x \in D$;*
(C13) $\|(I - \lambda S Q'_{z_0})^{-1}\| \leq \beta$ *for some $\beta > 0$;*
(C14) $\|(I - \lambda S Q'_{z_0})^{-1}(x_0 - \lambda S Q(x_0) - u)\| \leq \eta$ *for some $\eta > 0$;*
(C15) $\|Q'_x - Q'_y\| \leq \omega_0(\|x - y\|)$ *for all $x, y \in D$, where $\omega_0 \in \Phi$;*
(C16) $\omega_0(st) \leq s^p \omega_0(t), s \in [0,1]$ *and $t \in [0, \infty)$;*
(C17) $\|I - L(I - \lambda S Q'_x)\| \leq k, k \leq \frac{1}{3}$ *for all $x \in D$.*

Then we have the following:
(1) The sequence $\{x_n\}$ generated by

$$\begin{cases} z_n = x_n - L(x_n - \lambda S Q(x_n) - u), \\ y_n = x_n - (I - \lambda S Q'_{z_n})^{-1}(x_n - \lambda S Q(x_n) - u), \\ x_{n+1} = y_n - (I - \lambda S Q'_{z_n})^{-1}(y_n - \lambda S Q(y_n) - u) \end{cases} \tag{31}$$

for each $n \in \mathbb{N}_0$ is well defined, remains in $B_r[x_0]$ and converges to a solution x^* of the Equation (28).
(2) The R-order convergence of sequence $\{x_n\}$ is at least $2p + 1$.

Proof. Let $F : D \to X$ be an operator defined by Equation (29). Clearly, F is Fréchet differentiable at each point of D and its Fréchet derivative at $x \in D$ is given by Equation (30). Now, from (C13) and Equation (30), we have $\|F'^{-1}_{z_0}\| \leq \beta$ and so it follows that (C3) holds. From (C14), Equations (29) and (30), we have $\|F'^{-1}_{z_0}(F(x_0))\| \leq \eta$. Hence (C4) is satisfied. For all $x, y \in D$, using (C15), we have

$$\begin{aligned} \|F'_x - F'_y\| &= \sup\{\|(F'_x - F'_y)z\| : z \in X, \|z\| = 1\} \\ &\leq |\lambda|\|S\| \sup\{\|Q'_x - Q'_y\|\|z\| : z \in X, \|z\| = 1\} \\ &\leq |\lambda|\|S\|\omega_0(\|x - y\|) \\ &= \omega(\|x - y\|), \end{aligned}$$

where $\omega(t) = |\lambda|\|S\|\omega_0(t)$. Clearly, $\omega \in \Phi$ and, from (C16), we have

$$\omega(st) \le s^p \omega(t)$$

for all $s \in [0,1]$ and $t \in (0,\infty]$. Thus (C1) and (C5) hold. (C2) follows from (C17) for $c = k \in (0, \frac{1}{3}]$. Hence all the conditions of Theorem 1 are satisfied. Therefore, Theorem 3 follows from Theorem 1. This completes the proof. $\square$

Let $D = X = Y = C[a,b]$ be the space of all continuous real valued functions defined on $[a,b] \subset \mathbb{R}$ with the norm $\|x\| = \sup\limits_{t\in[a,b]} |x(t)|$. Consider, the following nonlinear integral equation:

$$x(s) = g(s) + \lambda \int_a^b K(s,t)(\mu(x(t))^{1+p} + \nu(x(t))^2)dt \tag{32}$$

for all $s \in [a,b]$ and $p \in (0,1]$, where $g, x \in C[a,b]$ with $g(s) \ge 0$ for all $s \in [a,b]$, $K : [a,b] \times [a,b] \to \mathbb{R}$ is a continuous nonnegative real-valued function and $\mu, \nu, \lambda \in \mathbb{R}$. Define two mappings $S, Q : D \to X$ by

$$Sx(s) = \int_a^b K(s,t)x(t)dt \tag{33}$$

for all $s \in [a,b]$ and

$$Qx(s) = \mu(x(s))^{1+p} + \nu(x(s))^2 \tag{34}$$

for all $\mu, \nu \in \mathbb{R}$ and $s \in [a,b]$.

One can easily observe that K is bounded on $[a,b] \times [a,b]$, that is, there exists a number $M \ge 0$ such that $|K(s,t)| \le M$ for all $s,t \in [a,b]$. Clearly, S is bounded linear operator with $\|S\| \le M(b-a)$ and Q is Fréchet differentiable and its Fréchet derivative at $x \in D$ is given by

$$Q'_x h(s) = (\mu(1+p)x^p + 2\nu x)h(s) \tag{35}$$

for all $h \in C[a,b]$. For all $x, y \in D$, we have

$$
\begin{aligned}
\|Q'_x - Q'_y\| &= \sup\{\|(Q'_x - Q'_y)h\| : h \in C[a,b], \|h\| = 1\} \\
&\le \sup\{\|(\mu(1+p)(x^p - y^p) + 2\nu(x-y))h\| : h \in C[a,b], \|h\| = 1\} \\
&\le \sup\{(|\mu|(1+p)\|x^p - y^p\| + 2|\nu|\|x-y\|)\|h\| : h \in C[a,b], \|h\| = 1\} \\
&\le |\mu|(1+p)\|x-y\|^p + 2|\nu|\|x-y\| \\
&= \omega_0(\|x-y\|),
\end{aligned}
\tag{36}
$$

where $\omega_0(t) = |\mu|(1+p)t^p + 2|\nu|t, t \ge 0$ with

$$\omega_0(st) \le s^p \omega_0(t) \tag{37}$$

for all $s \in [0,1]$ and $t \in [0,\infty)$. For any $x \in D$, using Equations (33) and (35), we have

$$
\begin{aligned}
&\|SQ'_x\| \\
&= \sup\{\|SQ'_x h\| : h \in X, \|h\| = 1\} \\
&= \sup\left\{\sup_{s\in[a,b]} \left|\int_a^b K(s,t)(\mu(1+p)(x(t))^p + 2\nu x(t))h(t)dt\right| : h \in X, \|h\| = 1\right\} \\
&\le \sup\left\{\int_a^b |K(s,t)|(|\mu|(1+p)|x(t)|^p + 2|\nu||x(t)|)|h(t)|dt : h \in X, \|h\| = 1\right\} \\
&\le (|\mu|(1+p)\|x\|^p + 2|\nu|\|x\|)M(b-a) < 1.
\end{aligned}
\tag{38}
$$

We now apply Theorem 3 to solve the Fredholm integral Equation (32).

Theorem 4. *Let $D = X = Y = C[a,b]$ and $\mu, \nu, \lambda, M \in \mathbb{R}$. Let $S, Q : D \to X$ be operators defined by Equations* (33) *and* (34), *respectively. Let $L \in B(X)$ and $x_0 \in D$ be such that $z_0 = x_0 - L(x_0 - \lambda SQ(x_0) - g) \in D$. Assume that the condition* (C6) *and the following conditions hold:*

(C18) $\quad \frac{1}{1 - |\lambda|(|\mu|(1+p)\|z_0\|^p + 2|\nu|\|z_0\|)M(b-a)} = \beta$ *for some $\beta > 0$;*

(C19) $\quad \frac{\|x_0 - g\| + |\lambda|(|\mu|\|x_0\|^{p+1} + 2|\nu|\|x_0\|^2)M(b-a)}{1 - |\lambda|(|\mu|(1+p)\|z_0\|^p + 2|\nu|\|z_0\|)M(b-a)} = \eta$ *for some $\eta > 0$;*

(C20) $\quad \|I - L\| + |\lambda|\|L\|(|\mu|(1+p)\|x\|^p + 2|\nu|\|x\|)M(b-a) \leq \frac{1}{3}$ *for all $x \in D$.*

Then the sequence generated by Equation (31) *with $u = g \in X$ is well defined, remains in $B_r[x_0]$ and converges to the solution $x^* \in B_r[x_0]$ of the Equation* (32) *with R-order convergence at least $(2p + 1)$.*

Proof. Note that $D = X = Y = C[a,b]$. Obviously, (C12) holds. Using Equations (C20), (33), (35) and (38), we have

$$\|I - (I - \lambda SQ'_{z_0})\| \leq |\lambda|(|\mu|(1+p)\|z_0\|^p + 2|\nu|\|z_0\|)M(b-a) < 1.$$

Therefore, by Lemma 1, $(I - \lambda SQ'_{z_0})^{-1}$ exists and

$$\|(I - \lambda SQ'_{z_0})^{-1}\| \leq \frac{1}{1 - |\lambda|(|\mu|(1+p)\|z_0\|^p + 2|\nu|\|z_0\|)M(b-a)}. \tag{39}$$

Hence Equations (C18) and (39) implies (C13) holds. Using Equations (C19), (38) and (39), we have

$$
\begin{aligned}
&\|(I - \lambda SQ'_{z_0})^{-1}(x_0 - \lambda SQ(x_0) - g)\| \\
\leq\ & \|(I - \lambda SQ'_{z_0})^{-1}\|(\|x_0 - g\| + \|\lambda SQ(x_0)\|) \\
\leq\ & \frac{\|x_0 - g\| + |\lambda|(|\mu|\|x_0\|^{p+1} + 2|\nu|\|x_0\|^2)M(b-a)}{1 - |\lambda|(|\mu|(1+p)\|z_0\|^p + 2|\nu|\|z_0\|)M(b-a)} \\
\leq\ & \eta.
\end{aligned}
$$

Thus the condition (C14) is satisfied. The conditions (C15) and (C16) follow from Equations (36) and (37), respectively. Now, from Equation (C20) and (38), we have

$$
\begin{aligned}
\|I - L(I - \lambda SQ'_x)\| \ &\leq\ \|I - L\| + \|L\|\|\lambda SQ'_x\| \\
&\leq\ \|I - L\| + \|L\||\lambda|(|\mu|(1+p)\|x\|^p + 2|\nu|\|x\|)M(b-a) \\
&\leq\ \frac{1}{3}.
\end{aligned}
$$

This implies that (C17) holds. Hence all the conditions of Theorem 3 are satisfied. Therefore, Theorem 4 follows from Theorem 3. This completes the proof. $\quad\square$

Now, we give one example to illustrate Theorem 3.

Example 3. *Let $X = Y = C[0,1]$ be the space of all continuous real valued functions defined on $[0,1]$. Let $D = \{x : x \in C[0,1], \|x\| < \frac{3}{2}\} \subset C[0,1]$. Consider the following nonlinear integral equation:*

$$x(s) = \sin(\pi s) + \frac{1}{10} \int_0^1 \cos(\pi s) \sin(\pi t)(x(t))^{p+1} dt, p \in (0,1]. \tag{40}$$

Define two mappings $S : X \to X$ and $Q : D \to Y$ by

$$S(x)(s) = \int_0^1 K(s,t)x(t)dt, \quad Q(x)(s) = (x(s))^{p+1},$$

where $K(s,t) = \cos(\pi s)\sin(\pi t)$. For $u = \sin(\pi s)$, the problem Equation (40) is equivalent to the problem Equation (28). Here, one can easily observe that S is bounded linear operator with $\|S\| \leq 1$ and Q is Fréchet differentiable with $Q'_x h(s) = (p+1)(x(s))^p h(s)$ for all $h \in X$ and $s \in [0,1]$. For all $x,y \in D$, we have

$$\|Q'_x - Q'_y\| \leq (p+1)\|x-y\|^p = \omega_0(\|x-y\|),$$

where $\omega_0(t) = (p+1)t^p$ for any $t \geq 0$. Clearly, $\omega_0 \in \Phi$. Define a mapping $F : D \to X$ by

$$F(x)(s) = x(s) - \frac{1}{10}SQ(x)(s) - \sin(\pi s).$$

Clearly, F is Fréchet differentiable on D. We now show that (C12) holds for $L = I \in B(X)$. Note that

$$\|(I - L(I - \lambda SQ))(x) - u\| = \left\|\frac{1}{10}SQ(x)(s) + \sin(\pi s)\right\| \leq \frac{1}{10}\left(\frac{3}{2}\right)^{p+1} + 1 < \frac{3}{2}$$

for all $x \in D$. Thus $(I - L(I - \lambda SQ))(x) - u \in D$ for all $x \in D$. For all $x \in D$, we have

$$\|I - F'_x\| \leq \frac{p+1}{10}\|x\|^p \leq \frac{p+1}{10}\left(\frac{3}{2}\right)^p = k.$$

Therefore, by Lemma 1, F'^{-1}_x exists and

$$F'^{-1}_x U(s) = U(s) + \frac{(p+1)\cos(\pi s)\int_0^1 \sin(\pi t)(x(t))^p U(t)dt}{10 - (p+1)\int_0^1 \sin(\pi t)\cos(\pi t)(x(t))^p dt} \tag{41}$$

for all $U \in Y$.

Let $x_0(s) = \sin(\pi s), \omega(t) = \frac{\omega_0(t)}{10} = \frac{p+1}{10}t^p$. Then we have the following:

(a) $x_0 \in X, F(x_0(s)) = -\frac{\cos(\pi s)}{10}\int_0^1 (\sin(\pi t))^{p+2}dt;$

(b) $z_0(s) = x_0(s) - F(x_0(s)) = \sin(\pi s) + \frac{\cos(\pi s)}{10}\int_0^1 (\sin(\pi t))^{p+2}dt;$

(c) $\|F'^{-1}_{z_0}\| \leq \frac{10^{p+1}}{10^{p+1}-(p+1)11^p} = \beta;$

(d) $\|F'^{-1}_{z_0}F(x_0)\| \leq \frac{10^p}{10^{p+1}-(p+1)11^p} = \eta;$

(e) $b_0 = \beta\omega(\eta) = \frac{(p+1)10^{p(p+1)}}{(10^{p+1}-(p+1)11^p)^{p+1}}$ and $q = \frac{1}{p+1} + \left(\frac{p+1}{10}\right)^p \left(\frac{3}{2}\right)^{p^2}.$

One can easily observe that $\theta = \frac{\left(\frac{qb_0}{2}+k\right)qb_0^2}{1-k(1+q)b_0} < 1$ for all $p \in (0,1]$ and $r = \frac{(1+qb_0)\eta}{1-\theta}$. Hence all the conditions of Theorem 3 are satisfied. Therefore, the sequence $\{x_n\}$ generated by Equation (31) is well defined, remains in $B_r[x_0]$ and converges to a solution of the integral Equation (40).

For $p = 1$, the convergence behavior of Newton-like method Equation (31) is given in Table 2.

Table 2. Iterates of Newton-like method Equation (31).

n	$x_n(s)$	$z_n(s)$	$y_n(s)$
0	$\sin(\pi s)$	$\sin(\pi s) + 0.0424413182\cos(\pi s)$	$\sin(\pi s) + 0.0425947671\cos(\pi s)$
1	$\sin(\pi s) + 0.0424794035\cos(\pi s)$	$\sin(\pi s) + 0.0424791962\cos(\pi s)$	$\sin(\pi s) + 0.0424796116\cos(\pi s)$
2	$\sin(\pi s) + 0.0424795616\cos(\pi s)$	$\sin(\pi s) + 0.042479611\cos(\pi s)$	$\sin(\pi s) + 0.0424796112\cos(\pi s)$
3	$\sin(\pi s) + 0.0424796111\cos(\pi s)$	$\sin(\pi s) + 0.0424796109\cos(\pi s)$	$\sin(\pi s) + 0.0424796113\cos(\pi s)$

5. Conclusions

The semilocal convergence of the third order Newton-like method for finding zeros of an operator from a Banach space to another Banach space and the corresponding Stirling-like method for finding fixed points of an operator on a Banach space are established under the ω-continuity condition. Our iterative technique is applied to nonlinear Fredholm-type operator equations. The R-order of our methods are clearly shown to be equal to at least $2p + 1$ for any $p \in (0, 1]$. Some numerical examples are given in support of our work, where earlier work cannot apply. In future, our iterative techniques can be applied in optimization problems.

Author Contributions: The authors have contributed equally to this paper.

Funding: This research has received no exteranl funding.

Acknowledgments: We would like to thank the reviewers for their valuable suggestions for improvement of this paper.

Conflicts of Interest: The authors declare no conflict of interest.

Reference

1. Kantorovich, L.V.; Akilov, G.P. *Functional Analysis*; Aregamon Press: Oxford, UK, 1982.
2. Kantorovich, L.V. On Newton's method for functional equations. *Dokl. Akad. Nauk. SSSR* **1948**, *59*, 1237–1240. (In Russian)
3. Rheinbolt, W.C. A unified Convergence theory for a class of iterative processes. *SIAM J. Numer. Anal.* **1968**, *5*, 42–63. [CrossRef]
4. Argyros, I.K.; Cho, Y.J.; Hilout, S. *Numerical Methods for Equations and its Applications*; CRC Press/Taylor & Francis Group Publ. Comp.: New York, NY, USA, 2012.
5. Argyros, I.K.; Hilout, S. Improved generaliged differentiability conditions for Newton-like methods. *J. Complex.* **2010**, *26*, 316–333. [CrossRef]
6. Argyros, I.K.; Hilout, S. Majorizing sequences for iterative methods. *J. Comput. Appl. Math.* **2012**, *236*, 1947–1960. [CrossRef]
7. Argyros, I.K. An improved error analysis for Newton-like methods under generalized conditions. *J. Comput. Appl. Math.* **2003**, *157*, 169–185. [CrossRef]
8. Argyros, I.K.; Hilout, S. On the convergence of Newton-type methods under mild differentiability conditions. *Number. Algorithms* **2009**, *52*, 701–726. [CrossRef]
9. Sahu, D.R.; Singh, K.K.; Singh, V.K. Some Newton-like methods with sharper error estimates for solving operator equations in Banach spaces. *Fixed Point Theory Appl.* **2012**, *78*, 1–20. [CrossRef]
10. Sahu, D.R.; Singh, K.K.; Singh, V.K. A Newton-like method for generalized operator equations in Banach spaces. *Numer. Algorithms* **2014**, *67*, 289–303. [CrossRef]
11. Sahu, D.R.; Cho, Y.J.; Agarwal, R.P.; Argyros, I.K. Accessibility of solutions of operator equations by Newton-like Methods. *J. Comlex.* **2015**, *31*, 637–657. [CrossRef]
12. Argyros, I.K. A unifying local-semilocal convergence analysis and applications for two-point Newton-like methods in Banach space. *J. Math. Anal. Appl.* **2004**, *298*, 374–397. [CrossRef]
13. Argyros, I.K.; Cho, Y.J.; George, S. On the "Terra incognita" for the Newton-Kantrovich method. *J. Korean Math. Soc.* **2014**, *51*, 251–266. [CrossRef]
14. Argyros, I.K.; Cho, Y.J.; George, S. Local convergence for some third-order iterative methods under weak conditions. *J. Korean Math. Soc.* **2016**, *53*, 781–793. [CrossRef]
15. Ren, H.; Argyros, I.K.; Cho, Y.J. Semi-local convergence of Steffensen-type algorithms for solving nonlinear equations. *Numer. Funct. Anal. Optim.* **2014**, *35*, 1476–1499. [CrossRef]
16. Ezquerro, J.A.; Hernández, M.A.; Salanova, M.A. A discretization scheme for some conservative problems. *J. Comput. Appl. Math.* **2000**, *115*, 181–192. [CrossRef]
17. Ezquerro, J.A.; Hernández, M.A.; Salanova, M.A. A newton like method for solving some boundery value problems. *Numer. Funct. Anal. Optim.* **2002**, *23*, 791–805. [CrossRef]
18. Ezquerro, J.A.; Hernández, M.A. Generalized differentiability conditions for Newton's method. *IMA J. Numer. Anal.* **2002**, *22*, 187–205. [CrossRef]
19. Proinov, P.D. New general convergence theory for iterative process and its applications to Newton-Kantorovich type theores. *J. Complex.* **2010**, *26*, 3–42. [CrossRef]
20. Sahu, D.R.; Yao, J.C.; Singh, V.K.; Kumar, S. Semilocal Convergence Analysis of S-iteration Process of Newton-Kantorovich Like in Banach Spaces. *J. Optim. Theory Appl.* **2017**, *172*, 102–127. [CrossRef]

21. Parhi, S.K.; Gupta, D.K. Convergence of a third order method for fixed points in Banach spaces. *Numer. Algorithms* **2012**, *60*, 419–434. [CrossRef]
22. Agarwal, R.P.; O'Regan, D.; Sahu, D.R. *Fixed Point Theory for Lipschitzian-Type Mappings with Applications*; Topological Fixed Point Theory and its Applications; Springer: New York, NY, USA, 2009; p. 6.
23. Agarwal, R.P.; Meehan, M.; O'Regan, D. *Fixed Point Theory and Applications*; Cambridge University Press: Cambridge, UK, 2004.
24. Picard, E. Memorire sur la theorie des equations aux derivees partielles et la methode aes approximations successive. *J. Math. Pures Appl.* **1980**, *6*, 145–210.
25. Cho, Y.J. Survey on metric fixed point theory and applications. In *Advances on Real and Complex Analysis with Applications*; Trends in Mathematics; Ruzhahsky, M., Cho, Y.J., Agarwal, P., Area, I., Eds.; Birkhäuser, Springer: Basel, Switzerland, 2017.
26. Granas, A.; Dugundji, J. *Fixed Point Theory*; Springer: New York, NY, USA, 2003.
27. Zeidler, E. *Nonlinear Functional Analysis and Its Applications I: Fixed-Point Theorems*; Springer: New York, NY, USA, 1986.
28. Zeidler, E. *Nonlinear Functional Analysis and Its Applications III: Variational Methods and Applications*; Springer: New York, NY, USA, 1985.
29. Argyros, I.K. On Newton's method under mild differentiability conditions and applications. *Appl. Math. Comput.* **1999**, *102*, 177–183. [CrossRef]
30. Ortega, J.M.; Rheinboldt, W.C. *Iterative Solution of Nonlinear Equations in Several Variables*; Academic Press: New York, NY, USA; London, UK, 1970.
31. Bartle, R.G. Newton's method in Banach spaces. *Proc. Am. Math. Soc.* **1955**, *6*, 827–831.
32. Rall, L.B. Convergence of Stirling's method in Banaeh spaces. *Aequa. Math.* **1975**, *12*, 12–20. [CrossRef]
33. Parhi, S.K.; Gupta, D.K. Semilocal convergence of a Stirling-like method in Banach spaces. *Int. J. Comput. Methods* **2010**, *7*, 215–228. [CrossRef]
34. Argyros, I.K.; Muruster, S.; George, S. On the Convergence of Stirling's Method for Fixed Points Under Not Necessarily Contractive Hypotheses. *Int. J. Appl. Comput. Math.* **2017**, *3*, 1071–1081. [CrossRef]
35. Alshomrani, A.S.; Maroju, P.; Behl, R. Convergence of a Stirling-like method for fixed points in Banach spaces. *J. Comput. Appl. Math.* **2018**. [CrossRef]
36. Parhi, S.K.; Gupta, D.K. A third order method for fixed points in Banach spaces. *J. Math. Anal. Appl.* **2009**, *359*, 642–652. [CrossRef]
37. Rall, L.B. *Computational Solution of Nonlinear Operator Equations*; John Wiley and Sons: New York, NY, USA, 1969.
38. Hernández, M.A.; Salanova, M.A. A Newton-like iterative process for the numerical solution of Fredholm nonlinear integral equations. *J. Integr. Equ. Appl.* **2005**, *17*, 1–17. [CrossRef]
39. Kohaupt, L. A Newton-like method for the numerical solution of nonlinear Fredholm-type operator equations. *Appl. Math. Comput.* **2012**, *218*, 10129–10148. [CrossRef]

mathematics

MDPI

Article

Ball Comparison for Some Efficient Fourth Order Iterative Methods Under Weak Conditions

Ioannis K. Argyros [1,†] **and Ramandeep Behl** [2,*,†]

1   Department of Mathematics Sciences, Cameron University, Lawton, OK 73505, USA; iargyros@cameron.edu
2   Department of Mathematics, Faculty of Science, King Abdulaziz University, Jeddah 21589, Saudi Arabia
*   Correspondence: ramanbehl87@yahoo.in; Tel.: +96-570-811-650
†   These authors contributed equally to this work.

Received: 17 December 2018; Accepted: 9 January 2019; Published: 16 January 2019

Abstract: We provide a ball comparison between some 4-order methods to solve nonlinear equations involving Banach space valued operators. We only use hypotheses on the first derivative, as compared to the earlier works where they considered conditions reaching up to 5-order derivative, although these derivatives do not appear in the methods. Hence, we expand the applicability of them. Numerical experiments are used to compare the radii of convergence of these methods.

Keywords: fourth order iterative methods; local convergence; banach space; radius of convergence

MSC: 65G99; 65H10; 47H17; 49M15

1. Introduction

Let $\mathbb{E}_1$, $\mathbb{E}_2$ be Banach spaces and $\mathbb{D} \subset \mathbb{E}_1$ be a nonempty and open set. Set $\mathbb{LB}(\mathbb{E}_1, \mathbb{E}_2) = \{M : \mathbb{E}_1 \to \mathbb{E}_2\}$, bounded and linear operators. A plethora of works from numerous disciplines can be phrased in the following way:

$$\lambda(x) = 0, \tag{1}$$

using mathematical modelling, where $\lambda : \mathbb{D} \to \mathbb{E}_2$ is a continuously differentiable operator in the Fréchet sense. Introducing better iterative methods for approximating a solution s_* of expression (1) is a very challenging and difficult task in general. Notice that this task is extremely important, since exact solutions of Equation (1) are available in some occasions.

We are motivated by four iterative methods given as

$$\begin{cases} y_j = x_j - \dfrac{2}{3}\lambda'(x_j)^{-1}\lambda(x_j) \\ x_{n+1} = x_j - \dfrac{1}{2}\left[\left(3\lambda'(y_j) - \lambda'(x_j)\right)^{-1}\left(3\lambda'(y_j) + \lambda'(x_j)\right)\right]\lambda'(x_j)^{-1}\lambda(x_j), \end{cases} \tag{2}$$

$$\begin{cases} y_j = x_j - \dfrac{2}{3}\lambda'(x_j)^{-1}\lambda(x_j) \\ x_{n+1} = x_j - \left[-\dfrac{1}{2}I + \dfrac{9}{8}B_j + \dfrac{3}{8}A_j\right]\lambda'(x_j)^{-1}\lambda(x_j), \end{cases} \tag{3}$$

$$\begin{cases} y_j = x_j - \dfrac{2}{3}\lambda'(x_j)^{-1}\lambda(x_j) \\ x_{n+1} = x_j - \left[I + \dfrac{1}{4}(A_j - I) + \dfrac{3}{8}(A_j - I)^2\right]\lambda'(y_j)^{-1}\lambda(x_j), \end{cases} \tag{4}$$

and

$$\begin{cases} y_j = x_j - H_j \lambda'(x_j)^{-1} \lambda(x_j) \\ x_{n+1} = z_j - \left[3I - H_j \lambda'(x_j)^{-1} [x_j, z_j; \lambda] \right] \lambda'(x_j)^{-1} \lambda(z_j), \end{cases} \tag{5}$$

where $H_j^0 = H^0(x_j)$, $x_0, y_0 \in \mathbb{D}$ are initial points, $H(x) = 2I + H^0(x)$, $H_j = H(X_j) \in \mathcal{LB}(\mathbb{E}_1, \mathbb{E}_1)$, $A_j = \lambda'(x_j)^{-1} \lambda'(y_j)$, $z_j = \frac{x_j + y_j}{2}$, $B_j = \lambda'(y_j)^{-1} \lambda'(x_j)$, and $[\cdot, \cdot; \lambda] : \mathbb{D} \times \mathbb{D} \to \mathcal{LB}(\mathbb{E}_1, \mathbb{E}_1)$ is a first order divided difference. These methods specialize to the corresponding ones (when $\mathbb{E}_1 = \mathbb{E}_2 = \mathbb{R}^i$, i is a natural number) studied by Nedzhibov [1], Hueso et al. [2], Junjua et al. [3], and Behl et al. [4], respectively. The 4-order convergence of them was established by Taylor series and conditions on the derivatives up to order five. Even though these derivatives of higher-order do not appear in the methods (2)–(5). Hence, the usage of methods (2)–(5) is very restricted. Let us start with a simple problem. Set $\mathbb{E}_1 = \mathbb{E}_2 = \mathbb{R}$ and $\mathbb{D} = [-\frac{5}{2}, \frac{3}{2}]$. We suggest a function $\lambda : A \to \mathbb{R}$ as

$$\lambda(t) = \begin{cases} 0, & t = 0 \\ t^5 + t^3 \ln t^2 - t^4, & t \neq 0 \end{cases}.$$

Then, $s_* = 1$ is a zero of the above function and we have

$$\lambda'(t) = 5t^4 + 3t^2 \ln t^2 - 4t^3 + 2t^2,$$

$$\lambda''(t) = 20t^3 + 6t \ln t^2 - 12t^2 + 10t,$$

and

$$\lambda'''(t) = 60t^2 + 6 \ln t^2 - 24t + 22.$$

Then, the third-order derivative of function $\lambda'''(x)$ is not bounded on $\mathbb{D}$. The methods (2)–(5) cannot be applicable to such problems or their special cases that require the hypotheses on the third or higher-order derivatives of λ. Moreover, these works do not give a radius of convergence, estimations on $\|x_j - s_*\|$, or knowledge about the location of s_*. The novelty of our work is that we provide this information, but requiring only the derivative of order one, for these methods. This expands the scope of utilization of them and similar methods. It is vital to note that the local convergence results are very fruitful, since they give insight into the difficult operational task for choosing the starting points/guesses.

Otherwise with the earlier approaches: (i) We use the Taylor series and high order derivative, (ii) we do not have any clue for the choice of the starting point x_0, (iii) we have no estimate in advance about the number of iterations needed to obtain a predetermined accuracy, and (iv) we have no knowledge of the uniqueness of the solution.

The work lays out as follows: We give the convergence of these iterative schemes (2)–(5) with some main theorems in Section 2. Some numerical problems are discussed in the Section 3. The final conclusions are summarized in Section 4.

2. Local Convergence Analysis

Let us consider that $I = [0, \infty)$ and $\varphi_0 : I \to I$ be a non-decreasing and continuous function with $\varphi_0(0) = 0$.

Assume that the following equation

$$\varphi_0(t) = 1 \tag{6}$$

has a minimal positive solution ρ_0. Let $I_0 = [0, \rho_0)$. Let $\varphi : I_0 \to I$ and $\varphi_1 : I_0 \to I$ be continuous and non-decreasing functions with $\varphi(0) = 0$. We consider functions on the interval I_0 as

$$\psi_1(t) = \frac{\int_0^1 \varphi((1 - \tau)t) d\tau + \frac{1}{3} \int_0^1 \varphi_1(\tau t) d\tau}{1 - \varphi_0(t)}$$

and
$$\bar{\psi}_1(t) = \psi_1(t) - 1.$$

Suppose that
$$\varphi_0(t) < 3. \tag{7}$$

Then, by (7), $\bar{\psi}_1(0) < 0$ and $\bar{\psi}_1(t) \to \infty$, as $t \to \rho_0^-$. On the basis of the classical intermediate value theorem, the function $\bar{\psi}_1(t)$ has a minimal solution R_1 in $(0, \rho_0)$. In addition, we assume
$$q(t) = 1 \tag{8}$$

has a minimal positive solution ρ_q, where
$$q(t) = \frac{1}{2}\Big(3\varphi_0(\psi_1(t)t) + \varphi_0(t)\Big).$$

Set $\rho = \min\{\rho_0, \rho_q\}$.

Moreover, we consider two functions ψ_2 and $\bar{\psi}_2$ on $I_1 = [0, \rho)$ as
$$\psi_2(t) = \frac{\int_0^1 \varphi((1-\tau)t)d\tau}{1 - \varphi_0(t)} + \frac{3}{2}\frac{\Big(\varphi_0(\psi_1(t)t) + \varphi_0(t)\Big)\int_0^1 \varphi_1(\tau t)d\tau}{(1 - q(t))(1 - \varphi_0(t))}$$

and
$$\bar{\psi}_2(t) = \psi_2(t) - 1.$$

Then, $\bar{\psi}_2(0) = -1$, and $\bar{\psi}_2(t) \to \infty$, with $t \to \rho^-$. We recall R_2 as the minimal solution of $\bar{\psi}_2(t) = 0$. Set
$$R = \min\{R_1, R_2\}. \tag{9}$$

It follows from (9) that for every $t \in [0, R)$
$$0 \le \varphi_0(t) < 1, \tag{10}$$

$$0 \le \psi_1(t) < 1, \tag{11}$$

$$0 \le q(t) < 1 \tag{12}$$

and
$$0 \le \psi_2(t) < 1 \tag{13}$$

Define by $S(s_*, r) = \{y \in \mathbb{E}_1 : \|s_* - y\| < r,\}$ and denote by $\bar{S}(s_*, r)$ the closure of $S(s_*, r)$. The local convergence of method (2) uses the conditions (A):

(a_1) $\lambda : \mathbb{D} \to \mathbb{E}_2$ is a continuously differentiable operator in the Fréchet sense, and there exists $s_* \in \mathbb{D}$.
(a_2) There exists a function $\varphi_0 : I \to I$ non-decreasing and continuous with $\varphi_0(0) = 0$ for all $x \in \mathbb{D}$
$$\left\|\lambda'(s_*)^{-1}\Big(\lambda'(x) - \lambda'(s_*)\Big)\right\| \le \varphi_0(\|x - s_*\|).$$

Set $\mathbb{D}_0 = \mathbb{D} \cap S(s_*, \rho_0)$, where ρ_0 is given in (6).
(a_3) There exist functions $\varphi : I_0 \to I$, $\varphi_1 : I_0 \to I$ non-decreasing and continuous with $\varphi(0) = 0$ so that for all $x, y \in \mathbb{D}_0$
$$\left\|\lambda'(s_*)^{-1}\Big(\lambda'(y) - \lambda'(x)\Big)\right\| \le \varphi(\|y - x\|)$$

and
$$\left\|\lambda'(s_*)^{-1}\lambda'(x)\right\| \le \varphi_1(\|y - x\|)$$

(a_4) $S(s_*, R) \subset \mathbb{D}$, radii ρ_0, ρ_q as given, respectively by (6), (8) exist; the condition (7) holds, where R is defined in (9).

(a_5)

$$\int_0^1 \varphi_0(\tau R^*) d\tau < 1, \text{ for some } R^* \geq R.$$

Set $\mathbb{D}_1 = D \cap S(s_*, R^*)$.

We can now proceed with the local convergence study of Equation (2) adopting the preceding notations and the conditions (A).

Theorem 1. *Under the conditions* (A) *sequence* $\{x_j\}$ *starting at* $x_0 \in S(s_*, R) - \{s_*\}$ *converges to* s_*, $\{x_j\} \subset S(x, R)$ *so that*

$$\|y_j - s_*\| \leq \psi_1(\|x_j - s_*\|)\|x_j - s_*\| \leq \|x_j - s_*\| < R \tag{14}$$

and

$$\|x_{n+1} - s_*\| \leq \psi_2(\|x_j - s_*\|)\|x_j - s_*\| \leq \|x_j - s_*\|, \tag{15}$$

with ψ_1 *and* ψ_2 *functions considered previously and* R *is given in* (9). *Moreover,* s_* *is a unique solution in the set* $\mathbb{D}_1$.

Proof. We proof the estimates (14) and (15) by adopting mathematical induction. Therefore, we consider $x \in S(s_*, R) - \{s_*\}$. By ($a_1$), ($a_2$), (9), and (10), we have

$$\|\lambda'(s_*)^{-1}(\lambda'(s_*) - \lambda'(x))\| \leq \varphi_0(\|s_* - x_0\|) < \varphi_0(R) < 1, \tag{16}$$

hence $\lambda'(x)^{-1} \in \mathcal{LB}(\mathbb{E}_2, \mathbb{E}_1)$ and

$$\|\lambda'(x)^{-1}\lambda'(s_*)\| \leq \frac{1}{1 - \varphi_0(\|s_* - x_0\|)}. \tag{17}$$

The point y_0 is also exists by (17) for $n = 0$. Now, by using (a_1), we have

$$\lambda(x) = \lambda(x) - \lambda(s_*) = \int_0^1 \lambda'(s_* + \tau(x - s_*)) d\tau (x - s_*). \tag{18}$$

From (a_3) and (18), we yield

$$\left\|\lambda'(s_*)^{-1}\lambda(x)\right\| \leq \int_0^1 \varphi_1(\tau\|x - s_*\|) d\tau \|x - s_*\|. \tag{19}$$

We can also write by method (2) for $n = 0$

$$y_0 - s_* = \left(x_0 - s_* - \lambda'(x_0)^{-1}\lambda(x_0)\right) + \frac{1}{3}\lambda'(x_0)^{-1}\lambda(x_0). \tag{20}$$

By expressions (9), (11), (17), (19), and (20), we obtain in turn that

$$
\begin{aligned}
\|y_0 - s_*\| &\leq \left\|\lambda'(x_0)^{-1}\lambda'(s_*)\right\| \left\|\int_0^1 \lambda'(s_*)^{-1}\left(\lambda'(s_* + \tau(x_0 - s_*)) - \lambda'(x_0)\right)(x_0 - s_*) d\tau\right\| \\
&\quad + \frac{1}{3}\left\|\lambda'(x_0)^{-1}\lambda'(s_*)\right\| \left\|\lambda'(s_*)^{-1}\lambda(x_0)\right\| \\
&\leq \frac{\int_0^1 \varphi((1-\tau)\|x_0 - s_*\|) d\tau + \frac{1}{3}\int_0^1 \varphi(\tau\|x_0 - s_*\|) d\tau}{1 - \varphi_0(\|x_0 - s_*\|)} \\
&= \psi_1(\|x_0 - s_*\|)\|x_0 - s_*\| \leq \|x_0 - s_*\| < R,
\end{aligned}
\tag{21}
$$

which confirms $y_0 \in S(s_*, R)$ and (14) for $n = 0$. We need to show that $\left(3\lambda'(y_0) - 3\lambda'(x_0)\right)^{-1} \in \mathcal{LB}(\mathbb{E}_2, \mathbb{E}_1)$.

In view of (a_2), (12), and (21), we have

$$
\begin{aligned}
&\left\| (2\lambda'(s_*))^{-1} \left[3\lambda'(y_0) - \lambda'(x_0) - 3\lambda'(s_*) + \lambda'(s_*) \right] \right\| \\
&\leq \frac{1}{2} \left[3 \left\| \lambda'(s_*)^{-1}(\lambda'(y_0) - \lambda'(s_*)) \right\| + \left\| \lambda'(s_*)^{-1}(\lambda'(x_0) - \lambda'(s_*)) \right\| \right] \\
&\leq \frac{1}{2} \left[\varphi_0(\|y_0 - s_*\|) + \varphi_0(\|x_0 - s_*\|) \right] \\
&\leq \frac{1}{2} \left[\varphi_0(\psi_1(\|x_0 - s_*\|)\|x_0 - s_*\|) + \varphi_0(\|x_0 - s_*\|) \right] \\
&= q(\|x_0 - s_*\|) < 1,
\end{aligned}
\tag{22}
$$

so

$$
\left\| \left(3\lambda'(y_0) - \lambda'(x_0) \right)^{-1} \lambda'(s_*) \right\| \leq \frac{1}{1 - q(\|x_0 - s_*\|)}.
\tag{23}
$$

Using (9), (13), (17), (a_3), (21), (23), and the second substep of method (2) (since x_1 exists by (23)), we can first write

$$
\begin{aligned}
x_1 - s_* = {} & x_0 - s_* - \lambda'(x_0)^{-1}\lambda(x_0) \\
& + \left[I - \frac{1}{2}\left(3\lambda'(y_0) - \lambda'(x_0) \right)^{-1} \left(3\lambda'(y_0) + \lambda'(x_0) \right) \right] \lambda'(x_0)^{-1}\lambda(x_0)
\end{aligned}
\tag{24}
$$

so

$$
\begin{aligned}
\|x_1 - s_*\| \leq {} & \|x_0 - s_* - \lambda'(x_0)^{-1}\lambda(x_0)\| + \frac{3}{2}\|(3\lambda'(y_0) - \lambda'(x_0))^{-1}\lambda'(s_*)\| \\
& \times \left[\|\lambda'(s_*)^{-1}(\lambda'(y_0) - \lambda'(x_0))\| + \|\lambda'(s_*)^{-1}(\lambda'(x_0) - \lambda'(s_*))^{-1}\| \right] \|\lambda'(x_0)^{-1}\lambda(s_*)\| \|\lambda'(x_0)^{-1}\lambda(x_0)\| \\
\leq {} & \left[\frac{\int_0^1 \varphi((1-\tau)t)t\,d\tau}{1 - \varphi_0(t)} + \frac{3}{2} \frac{\left(\varphi_0(\|y_0 - s_*\|) + \varphi_0(\|x_0 - s_*\|) \right)\int_0^1 \varphi_1(\tau\|x_0 - s_*\|)d\tau}{(1 - q(\|x_0 - s_*\|))(1 - \varphi_0(\|x_0 - s_*\|))} \right] \|x_0 - s_*\| \\
\leq {} & \psi_2(\|x_0 - s_*\|)\|x_0 - s_*\| \leq \|x_0 - s_*\|.
\end{aligned}
\tag{25}
$$

So, (15) holds and $x_1 \in S(s_*, R)$.

To obtain estimate (25), we also used the estimate

$$
\begin{aligned}
& I - \frac{1}{2}\left(3\lambda'(y_0) - \lambda'(x_0) \right)^{-1} \left(3\lambda'(y_0) + \lambda'(x_0) \right) \\
& = \frac{1}{2}\left(3\lambda'(y_0) - \lambda'(x_0) \right)^{-1} \left[2(3\lambda'(y_0) - \lambda'(x_0)) - (3\lambda'(y_0) + \lambda'(x_0)) \right] \\
& = \frac{3}{2}\left(3\lambda'(y_0) - \lambda'(x_0) \right)^{-1} \left[(\lambda'(y_0) - \lambda'(s_*)) + (\lambda'(s_*) - \lambda'(x_0)) \right]
\end{aligned}
\tag{26}
$$

The induction for (14) and (15) can be finished, if x_m, y_m, x_{m+1} replace x_0, y_0, x_1 in the preceding estimations. Then, from the estimate

$$
\|x_{m+1} - s_*\| \leq \mu \|x_m - s_*\| < R, \quad \mu = \varphi_2(\|x_0 - s_*\|) \in [0, 1),
\tag{27}
$$

we arrive at $\lim_{m \to \infty} x_m = s_*$ and $x_{m+1} \in S(s_*, R)$. Let us consider that $K = \int_0^1 \lambda'(y_* + \tau(s_* - y_*))d\tau$ for $y^* \in \mathbb{D}_1$ with $K(y_*) = 0$. From (a_1) and (a_5), we obtain

$$
\begin{aligned}
\|\lambda'(s_*)^{-1}(\lambda'(s_*) - K)\| & \leq \int_0^1 \varphi_0(\tau\|s_* - y_*\|)d\tau \\
& \leq \int_0^1 \varphi_0(\tau R)d\tau < 1.
\end{aligned}
\tag{28}
$$

So, $K^{-1} \in \mathcal{LB}(\mathbb{E}_1, \mathbb{E}_2)$, and $s_* = y_*$ by the identity

$$0 = K(s_*) - K(y_*) = K(s_* - y_*).$$ (29)

□

Proof. Next, we deal with method (3) in an analogous way. We shall use the same notation as previously. Let $\varphi_0, \varphi, \varphi_1, \rho_0, \psi_1, R_1$, and $\bar{\psi}_1$, be as previously.

We assume

$$\varphi_0(\psi_1(t)t) = 1$$ (30)

has a minimal solution ρ_1. Set $\rho = \min\{\rho_0, \rho_1\}$. Define functions ψ_2 and $\bar{\psi}_2$ on interval $I_2 = [0, \rho)$ by

$$\psi_2(t) = \frac{\int_0^1 \varphi((1-\tau)t)d\tau}{1 - \varphi_0(t)} + \left[2 + \frac{3\big(\varphi_0(\psi_1(t)t) + \varphi_0(t)\big)}{8(1 - \varphi_0(t))} + \frac{9\big(\varphi_0(\psi_1(t)t) + \varphi_0(t)\big)}{8(1 - \varphi_0(\psi_1(t)t))} \right] \frac{\int_0^1 \varphi_1(\tau t)d\tau}{1 - \varphi_0(t)}$$

and

$$\bar{\psi}_2(t) = \psi_2(t) - 1.$$

Then, $\bar{\psi}_2(0) = -1$ and $\bar{\psi}_2(t) \to \infty$, with $t \to \rho^-$. R_2 is known as the minimal solution of equation $\bar{\psi}_2(t) = 0$ in $(0, \rho)$, and set

$$R = \min\{R_1, R_2\}.$$ (31)

Replace ρ_q by ρ_1 in the conditions (A) and call the resulting conditions $(A)'$.

Moreover, we use the estimate obtained for the second substep of method (3)

$$\begin{aligned}
x_1 - s_* &= x_0 - s_* - \lambda'(x_0)^{-1}\lambda(x_0) + \left[\frac{3}{2}I - \frac{9}{8}B_0 - \frac{9}{16}A_0\right]\lambda'(x_0)^{-1}\lambda(x_0) \\
&= x_0 - s_* - \lambda'(x_0)^{-1}\lambda(x_0) + \left[-2I + \frac{3}{8}(I - A_0) + \frac{9}{8}(I - B_0)\right]\lambda'(x_0)^{-1}\lambda(x_0) \\
&= x_0 - s_* - \lambda'(x_0)^{-1}\lambda(x_0) + \left[-2I + \frac{3}{8}\lambda'(x_0)^{-1}\big(\lambda'(x_0) - \lambda'(y_0)\big)\right. \\
&\quad \left. + \frac{9}{8}\lambda'(y_0)^{-1}\big(\lambda'(y_0) - \lambda'(x_0)\big)\right]\lambda'(x_0)^{-1}\lambda(x_0).
\end{aligned}$$ (32)

Then, by replacing (24) by (32) in the proof of Theorem 1, we have instead of (25)

$$\begin{aligned}
\|x_1 - s_*\| &= \left[\frac{\int_0^1 \varphi((1-\tau)\|s_* - x_0\|)d\tau}{1 - \varphi(\|s_* - x_0\|)} + \left\{2 + \frac{3\big(\varphi(\|y_0 - s_*\|) + \varphi_0(\|s_* - x_0\|)\big)}{8(1 - \varphi_0(\|s_* - x_0\|))}\right.\right. \\
&\quad \left.\left. + \frac{9\big(\varphi_0(\|y_0 - s_*\|) + \varphi_0(\|s_* - x_0\|)\big)}{8(1 - \varphi_0(\|y_0 - s_*\|))}\right\} \frac{\int_0^1 \varphi_1(\|s_* - x_0\|)d\tau}{1 - \varphi_0(\|s_* - x_0\|)}\right] \|s_* - x_0\| \\
&\leq \psi_2(\|s_* - x_0\|)\|s_* - x_0\| \leq \|s_* - x_0\|.
\end{aligned}$$ (33)

The rest follows as in Theorem 1. □

Hence, we arrived at the next Theorem.

Theorem 2. *Under the conditions $(A)'$, the conclusions of Theorem 1 hold for method (3).*

Proof. Next, we deal with method (4) in the similar way. Let $\varphi_0, \varphi, \varphi_1, \rho_0, \rho_1, \rho, \psi_1, R_1$, and $\bar{\psi}_1$, be as in the case of method (3). We consider functions ψ_2 and $\bar{\psi}_2$ on I_1 as

$$\psi_2(t) = \frac{\int_0^1 \varphi((1-\tau)t)d\tau}{1 - \varphi_0(t)} + \frac{\varphi_0(\psi_1(t)t) + \varphi_0(t)}{(1 - \varphi_0(t))(1 - \varphi_0(\psi_1(t)t))} + \frac{1}{4}\frac{(\varphi_0(\psi_1(t)t) + \varphi_0(t))}{(1 - \varphi_0(t))}$$
$$+ \frac{3}{8}\left(\frac{\varphi_0(\psi_1(t)t) + \varphi_0(t)}{1 - \varphi_0(t)}\right)^2$$

and

$$\bar{\psi}_2(t) = \psi_2(t) - 1.$$

The minimal zero of $\bar{\psi}_2(t) = 0$ is denoted by R_2 in $(0, \rho)$, and set

$$R = \min\{R_1, R_2\}. \tag{34}$$

Notice again that from the second substep of method (4), we have

$$x_1 - s_* = x_0 - s_* - \lambda'(x_0)^{-1}\lambda(x_0) + \left[\lambda'(x_0)^{-1} - \lambda'(y_0)^{-1} - \frac{1}{4}(A_0 - I) - \frac{3}{8}(I - A_0)^2\right]\lambda(x_0)$$

$$= x_0 - s_* - \lambda'(x_0)^{-1}\lambda(x_0) + \left\{\lambda'(x_0)^{-1}\left[(\lambda'(y_0) - \lambda'(s_*)) + (\lambda'(s_*) - \lambda'(x_0))\right]\right.$$

$$- \frac{1}{4}\lambda'(x_0)^{-1}\left[(\lambda'(y_0) - \lambda'(s_*)) + (\lambda'(s_*) - \lambda'(x_0))\right]$$

$$- \frac{3}{8}\lambda'(x_0)^{-1}\left[(\lambda'(y_0) - \lambda'(s_*)) + (\lambda'(s_*) - \lambda'(x_0))\right]^2\left.\right\}\lambda(x_0),$$

$\tag{35}$

so

$$\|x_1 - s_*\| \leq \left[\frac{\int_0^1 \varphi((1-\tau)\|s_* - x_0\|)d\tau}{1 - \varphi(\|s_* - x_0\|)} + \frac{\varphi_0(\varphi(\|s_* - x_0\|)\|s_* - x_0\|) + \varphi_0(\|s_* - x_0\|)}{(1 - \varphi_0(\|s_* - x_0\|))(1 - \varphi_0(\varphi(\|s_* - x_0\|)\|s_* - x_0\|))}\right.$$

$$+ \frac{1}{4}\frac{(\varphi(\psi_1(\|s_* - x_0\|)\|s_* - x_0\|) + \varphi_0(\|s_* - x_0\|))}{(1 - \varphi_0(\|s_* - x_0\|))}$$

$$+ \frac{3}{8}\left(\frac{\varphi(\psi_1(\|s_* - x_0\|)\|s_* - x_0\|) + \varphi_0(\|s_* - x_0\|)}{(1 - \varphi_0(\|s_* - x_0\|))}\right)^2\left.\right]\|s_* - x_0\|$$

$$\leq \psi_2(\|s_* - x_0\|)\|s_* - x_0\| \leq \|s_* - x_0\|.$$

$\tag{36}$

The rest follows as in Theorem 1. $\square$

Hence, we arrived at the next following Theorem.

Theorem 3. *Under the conditions* $(A)'$, *conclusions of Theorem 1 hold for scheme* (4).

Proof. Finally, we deal with method (5). Let $\varphi_0, \varphi, \varphi_1, \rho_0, I_0$ be as in method (2). Let also $\varphi_2 : I_0 \to I$, $\varphi_3 : I_0 \to I$, $\varphi_4 : I_0 \to I$ and $\varphi_5 : I_0 \times I_0 \to I$ be continuous and increasing functions with $\varphi_3(0) = 0$. We consider functions ψ_1 and $\bar{\psi}_1$ on I_0 as

$$\psi_1(t) = \frac{\int_0^1 \varphi((1-\tau)t)d\tau + \varphi_2(t)\int_0^1 \varphi_1(\tau t)d\tau}{1 - \varphi_0(t)}$$

and

$$\bar{\psi}_1(t) = \psi_1(t) - 1.$$

Suppose that

$$\varphi_1(0)\varphi_2(0) < 1. \tag{37}$$

Then, by (6) and (37), we yield $\bar{\psi}_1(0) < 0$ and $\bar{\psi}_1(t) \to \infty$ with $t \to \rho_0^-$. R_1 is known as the minimal zero of $\bar{\psi}_1(t) = 0$ in $(0, \rho_0)$. We assume

$$\varphi_0\big(g(t)t\big) = 1, \tag{38}$$

where $g(t) = \frac{1}{2}(1 + \psi_1(t))$, has a minimal positive solution ρ_1. Set $I_1 = [0, \rho)$, where $\rho = \min\{\rho_0, \rho_1\}$. We suggest functions ψ_2 and $\bar{\psi}_2$ on I_1 as

$$\psi_2(t) = \left[\frac{\int_0^1 \varphi((1-\tau)g(t)t)d\tau}{1 - \varphi_0(g(t)t)} + \frac{\big(\varphi_0(g(t)t) + \varphi_0(t)\big)\int_0^1 \varphi_1(\tau g(t)t)d\tau}{(1 - \varphi_0(g(t)))(1 - \varphi_0(t))} \right.$$
$$\left. + 2\frac{\varphi_3\left(\frac{t}{2}(1 + \psi_1(t))\right)\int_0^1 \varphi_1(\tau g(t)t)d\tau}{(1 - \varphi_0(t))^2} + \frac{\varphi_4(t)\varphi_5(t, \psi_1(t))\int_0^1 \varphi_1(\tau g(t)t)d\tau}{(1 - \varphi_0(t))^2} \right]g(t)$$

and

$$\bar{\psi}_2(t) = \psi_2(t) - 1.$$

Suppose that

$$\Big(2\varphi_3(0) + \varphi_4(0)\varphi_5(0,0)\Big)\varphi_1(0) < 1. \tag{39}$$

By (39) and the definition of I_1, we have $\bar{\psi}_2(0) < 0$, $\bar{\psi}_2(t) \to \infty$ with $t \to \rho^-$. We assume R_2 as the minimal solution of $\bar{\psi}_2(t) = 0$. Set

$$R = \min\{R_1, R_2\}. \tag{40}$$

The study of local convergence of scheme (5) is depend on the conditions (C):

$(c_1) = (a_1)$.
$(c_2) = (a_2)$.
(c_3) There exist functions $\varphi : I_1 \to I$, $\varphi_1 : I_0 \to I$, $\varphi_2 : I_0 \to I$, $\varphi_3 : I_0 \to I$, $\varphi_4 : I_0 \to I$, and $\varphi_5 : I_0 \times I_0 \to I$, increasing and continuous functions with $\varphi(0) = \varphi_3(0) = 0$ so all $x, y \in D_0$

$$\|\lambda'(s_*)^{-1}\big(\lambda'(y) - \lambda'(x)\big)\| \le \varphi(\|y - x\|),$$
$$\|\lambda'(s_*)^{-1}\lambda'(x)\| \le \varphi_1(\|x - s_*\|),$$
$$\|I - H(x)\| \le \varphi_2(\|x - s_*\|),$$
$$\|\lambda'(s_*)^{-1}\big([x, y; \lambda] - \lambda'(x)\big)\| \le \varphi_3(\|y - x\|),$$
$$\|H^0(x)\| \le \varphi_4(\|x - s_*\|),$$

and

$$\|\lambda'(s_*)^{-1}[x, y; \lambda]\| \le \varphi_5(\|x - s_*\|, \|y - s_*\|),$$

(c_4) $S(s_*, R) \subseteq D$, ρ_0, ρ_1 given, respectively by (6), (38) exist, (37) and (38) hold, and R is defined in (40).
$(c_5) = (a_5)$.

Then, using the estimates

$$
\begin{aligned}
\|y_0 - s_*\| &= \|x_0 - s_* - \lambda'(x_0)^{-1}\lambda(x_0) + (I - H_0)\lambda'(x_0)^{-1}\lambda(x_0)\| \\
&\leq \frac{\int_0^1 \varphi((1-\tau)\|x_0 - s_*\|)d\tau \|x_0 - s_*\|}{1 - \varphi_0(\|x_0 - s_*\|)} + \|I - H_0\|\|\lambda'(x_0)^{-1}\lambda'(s_*)\|\|\lambda'(s_*)^{-1}\lambda(x_0)\| \\
&\leq \left[\frac{\int_0^1 \varphi((1-\tau)\|x_0 - s_*\|)d\tau + \varphi_2(\|x_0 - s_*\|)\int_0^1 \varphi_1(\tau\|x_0 - s_*\|)d\tau}{1 - \varphi_0(\|x_0 - s_*\|)} \right] \|x_0 - s_*\| \\
&\leq \psi_1(\|x_0 - s_*\|)\|x_0 - s_*\| \leq \|x_0 - s_*\| < R,
\end{aligned}
\tag{41}
$$

and

$$
\begin{aligned}
\|x_1 - s_*\| &= \|z_0 - s_* - \lambda'(z_0)^{-1}\lambda(z_0) + \lambda'(z_0)^{-1}(\lambda'(x_0) - \lambda'(z_0))\lambda'(x_0)^{-1}\lambda(z_0)\| \\
&\quad + 2\lambda'(x_0)^{-1}([x_0, z_0; \lambda] - \lambda'(x_0))\lambda'(x_0)^{-1}\lambda(z_0) + H_j^0 \lambda'(x_0)^{-1}[x_0, z_0; \lambda]\lambda'(x_0)^{-1}\lambda(z_0)\| \\
&\leq \left[\frac{\int_0^1 \varphi((1-\tau)g(\|x_0 - s_*\|)\|x_0 - s_*\|)d\tau}{1 - \varphi_0(g(\|x_0 - s_*\|)\|x_0 - s_*\|)} \right. \\
&\quad + \frac{(\varphi_0(\|x_0 - s_*\|) + \varphi_0(g(\|x_0 - s_*\|)\|x_0 - s_*\|))\int_0^1 \varphi_1(\tau g(\|x_0 - s_*\|)\|x_0 - s_*\|)d\tau}{(1 - \varphi_0(g(\|x_0 - s_*\|)\|x_0 - s_*\|))(1 - \varphi_0(\|x_0 - s_*\|))} \\
&\quad + 2\frac{\varphi_3\left(\frac{(1+\psi_1(\|x_0 - s_*\|))\|x_0 - s_*\|}{2}\right)\int_0^1 \varphi_1(\tau g(\|x_0 - s_*\|)\|x_0 - s_*\|)d\tau}{(1 - \varphi_0(\|x_0 - s_*\|))^2} \\
&\quad \left. + \frac{\varphi_4(\|x_0 - s_*\|)\varphi_5(\|x_0 - s_*\|, \|y_0 - s_*\|)\int_0^1 \varphi_1(\tau g(\|x_0 - s_*\|)\|x_0 - s_*\|)d\tau}{(1 - \varphi_0(\|x_0 - s_*\|))^2} \right] \|z_0 - s_*\| \\
&\leq \psi_2(\|x_0 - s_*\|)\|x_0 - s_*\| \leq \|x_0 - s_*\|.
\end{aligned}
\tag{42}
$$

Here, recalling that $z_0 = \frac{x_0 + y_0}{2}$, we also used the estimates

$$
\begin{aligned}
\|z_0 - s_*\| &= \left\| \frac{x_0 + y_0}{2} - s_* \right\| \leq \frac{1}{2}(\|x_0 - s_*\| + \|y_0 - s_*\|) \\
&\leq \frac{1}{2}(1 + \psi_1(\|x_0 - s_*\|))\|x_0 - s_*\|,
\end{aligned}
\tag{43}
$$

$$
\alpha = \lambda'(z_0)^{-1} - \lambda'(x_0)^{-1} = \lambda'(z_0)^{-1}\left[(\lambda'(x_0) - \lambda'(s_*)) + (\lambda'(s_*) - \lambda'(z_0))\right]\lambda'(x_0)^{-1},
$$
$$
\beta = (-2I + H_0\lambda'(x_0)^{-1}[x_0, z_0; \lambda])\lambda'(x_0)^{-1},
$$

and

$$
\begin{aligned}
\gamma &= -2I + (2I + H_0^0)\lambda'(x_0)^{-1}[x_0, z_0; \lambda] \\
&= -2I + 2I\lambda'(x_0)^{-1}[x_0, z_0; \lambda] + 2H_0^0\lambda'(x_0)^{-1}[x_0, z_0; \lambda] \\
&= 2\lambda'(x_0)^{-1}([x_0, z_0; \lambda] - \lambda'(x_0)) + H_0^0\lambda'(x_0)^{-1}[x_0, z_0; \lambda]
\end{aligned}
$$

to obtain (41) and (42). $\square$

Hence, we arrived at the next following Theorem.

Theorem 4. *Under the conditions* (C), *the conclusions of Theorem* 1 *hold for method* (5).

3. Numerical Applications

We test the theoretical results on many examples. In addition, we use five examples and out of them: The first one is a counter example where the earlier results are applicable; the next three are real life problems, e.g., a chemical engineering problem, an electron trajectory in the air gap among two parallel surfaces problem, and integral equation of Hammerstein problem, which are displayed in

Examples 1–5. The last one compares favorably (5) to the other three methods. Moreover, the solution to corresponding problem are also listed in the corresponding example which is correct up to 20 significant digits. However, the desired roots are available up to several number of significant digits (minimum one thousand), but due to the page restriction only 30 significant digits are displayed.

We compare the four methods namely (2)–(5), denoted by NM, HM, JM, and BM, respectively on the basis of radii of convergence ball and the approximated computational order of convergence

$$\rho = \frac{\log\left[\|x_{(j+1)} - x_{(j)}\| / \|x_{(j)} - x_{(j-1)}\|\right]}{\log\left[\|x_{(j)} - x_{(j-1)}\| / \|x_{(j-1)} - x_{(j-2)}\|\right]}, j = 2, 3, 4, \ldots \text{ (for the details please see Cordero and Torregrosa [5])}$$

($ACOC$). We have included the radii of ball convergence in the following Tables 1–6 except, the Table 4 that belongs to the values of abscissas t_j and weights w_j. We use the *Mathematica* 9 programming package with multiple precision arithmetic for computing work.

We choose in all examples $H^0(x) = 0$ and $H(x) = 2I$, so $\varphi_2(t) = 1$ and $\varphi_4(t) = 0$. The divided difference is $[x, y; \lambda] = \int_0^1 \lambda'(y + \theta(x - y))d\theta$. In addition, we choose the following stopping criteria (i) $\|x_{j+1} - x_j\| < \epsilon$ and (ii) $\|\lambda(x_j)\| < \epsilon$, where $\epsilon = 10^{-250}$.

Example 1. *Set* $\mathbb{X} = \mathbb{Y} = \mathbb{R}$*. We suggest a function* λ *on* $\mathbb{D} = [-\frac{1}{\pi}, \frac{2}{\pi}]$ *as*

$$\lambda(x) = \begin{cases} 0, & x = 0 \\ x^5 \sin(1/x) + x^3 \log(\pi^2 x^2), & x \neq 0 \end{cases}.$$

But, $\lambda'''(x)$ *is unbounded on* Ω *at* $x = 0$*. The solution of this problem is* $s_* = \frac{1}{\pi}$*. The results in Nedzhibov [1], Hueso et al. [2], Junjua et al. [3], and Behl et al. [4] cannot be utilized. In particular, conditions on the 5th derivative of* λ *or may be even higher are considered there to obtain the convergence of these methods. But, we need conditions on* λ' *according to our results. In additon, we can choose*

$$H = \frac{80 + 16\pi + (\pi + 12\log 2)\pi^2}{2\pi + 1}, \quad \varphi_1(t) = 1 + Ht, \quad \varphi_0(t) = \varphi(t) = Ht,$$

$$\varphi_5(s, t) = \frac{1}{2}(\varphi_1(s) + \varphi_1(t)) \quad \text{and} \quad \varphi_3(t) = \frac{1}{2}\varphi_2(t).$$

The distinct radius of convergence, number of iterations n, and COC (ρ) *are mentioned in Table 1.*

Table 1. Comparison on the basis of different radius of convergence for Example 1.

Schemes	R_1	R_2	R	x_0	n	ρ
NM	0.011971	0.010253	0.010253	0.30831	4	4.0000
HM	0.011971	0.01329	0.011971	0.32321	4	4.0000
JM	0.011971	0.025483	0.011971	0.32521	4	4.0000
BM	0	0	0	-	-	-

Equation (39) is violated with these choices of φ_i. This is the reason that R is zero in the method BM. Therefore, our results hold only, if $x_0 = s_*$.

Example 2. *The function*

$$\lambda_2(x) = x^4 - 1.674 - 7.79075x^3 + 14.7445x^2 + 2.511x. \tag{44}$$

appears in the conversion to ammonia of hydrogen-nitrogen [6,7]. The function λ_2 *has 4 zeros, but we choose* $s_* \approx 3.9485424455620457727 + 0.3161235708970163733i$*. Moreover, we have*

$$\varphi_0(t) = \varphi(t) = 40.6469t, \quad \varphi_1(t) = 1 + 40.6469t, \quad \varphi_3(t) = \frac{1}{2}\varphi_2(t), \quad \text{and} \quad \varphi_5(s, t) = \frac{1}{2}(\varphi_1(s) + \varphi_1(t)).$$

The distinct radius of convergence, number of iterations n, and COC (ρ) *are mentioned in Table 2.*

Table 2. Comparison on the basis of different radius of convergence for Example 2.

Schemes	R_1	R_2	R	x_0	n	ρ
NM	0.0098841	0.0048774	0.0048774	$3.953 + 0.3197i$	4	4.0000
HM	0.0098841	0.016473	0.016473	$3.9524 + 0.32i$	4	4.0000
JM	0.0098841	0.0059094	0.0059094	$3.9436 + 0.3112i$	4	4.0000
BM	0	0	0	-	-	-

Equation (39) is violated with these choices of φ_i. This is the reason that R is zero in the method *BM*. Therefore, our results hold only, if $x_0 = s_*$.

Example 3. *An electron trajectory in the air gap among two parallel surfaces is formulated given as*

$$x(t) = x_0 + \left(v_0 + e\frac{E_0}{m\omega}\sin(\omega t_0 + \alpha)\right)(t - t_0) + e\frac{E_0}{m\omega^2}\left(\cos(\omega t + \alpha) + \sin(\omega + \alpha)\right), \tag{45}$$

where e, m, x_0, v_0, and $E_0 \sin(\omega t + \alpha)$ are the charge, the mass of the electron at rest, the position, velocity of the electron at time t_0, and the RF electric field among two surfaces, respectively. For particular values of these parameters, the following simpler expression is provided:

$$f_3(x) = x + \frac{\pi}{4} - \frac{1}{2}\cos(x). \tag{46}$$

The solution of function f_3 is $s_* \approx -0.3090932715417949452741986808924$. Moreover, we have

$$\varphi(t) = \varphi_0(t) = 0.5869t, \quad \varphi_1(t) = 1 + 0.5869t, \quad \varphi_3(t) = \frac{1}{2}\varphi_2(t) \quad and \quad \varphi_5(s,t) = \frac{1}{2}\left(\varphi_1(s) + \varphi_1(t)\right).$$

The distinct radius of convergence, number of iterations n, and COC (ρ) are mentioned in Table 3.

Table 3. Comparison on the basis of different radius of convergence for Example 3.

Schemes	R_1	R_2	R	x_0	n	ρ
NM	0.678323	0.33473	0.33473	0.001	4	4.0000
HM	0.678323	1.13054	0.678323	−0.579	4	4.0000
JM	0.678323	0.40555	0.40555	0.091	5	4.0000
BM	0	7.60065×10^{-18}	0	-	-	-

Equation (39) is violated with these choices of φ_i. This is the reason that R is zero in the method *BM*. Therefore, our results hold only, if $x_0 = s_*$.

Example 4. *Considering mixed Hammerstein integral equation Ortega and Rheinbolt [8], as*

$$x(s) = 1 + \frac{1}{5}\int_0^1 U(s,t)x(t)^3 dt, \quad x \in C[0,1], \ s,t \in [0,1], \tag{47}$$

where the kernel U is

$$U(s,t) = \begin{cases} s(1-t), s \leq t, \\ (1-s)t, t \leq s. \end{cases}$$

We phrase (47) by using the Gauss-Legendre quadrature formula with $\int_0^1 \phi(t)dt \simeq \sum_{k=1}^{10} w_k\phi(t_k)$, where t_k and w_k are the abscissas and weights respectively. Denoting the approximations of $x(t_i)$ with x_i ($i = 1,2,3,...,10$), then we yield the following 8×8 system of nonlinear equations

$$5x_i - 5 - \sum_{k=1}^{10} a_{ik}x_k^3 = 0, \ i = 1,2,3...,10,$$

$$a_{ik} = \begin{cases} w_k t_k (1 - t_i), & k \le i, \\ w_k t_i (1 - t_k), & i < k. \end{cases}$$

The values of t_k and w_k can be easily obtained from Gauss-Legendre quadrature formula when $k = 8$ mentioned in Table 4.

Table 4. Values of abscissas t_j and weights w_j.

j	t_j	w_j
1	0.013046735741414139961101799	0.033335672154344068796784440
2	0.067468316665550774463395165	0.074725674575290296572888816
3	0.160295215850487796882883632	0.109543181257991021997776746
4	0.283302302935375407640460036703	0.134633359654998177545611346
5	0.425562830509184394557587000	0.147762112357376435086946499
6	0.574437169490815605442413000	0.147762112357376435086946499
7	0.716697697064623595399632297	0.134633359654998177545611346
8	0.839704784149512203117163368	0.109543181257991021997776746
9	0.932531683344492255366604834	0.074725674575290296572888816
10	0.986953264258585860038982011	0.033335672154344068796784440

The required approximate root is $s_* \approx (1.001377, \dots, 1.006756, \dots, 1.014515, \dots, 1.021982, \dots,$
$1.026530, \dots, 1.026530, \dots, 1.021982, \dots, 1.014515, \dots, 1.006756, \dots, 1.001377, \dots)^T$. Moreover, we have

$$\varphi_0(t) = \varphi(t) = \frac{3}{20}t, \quad \varphi_1(t) = 1 + \frac{3}{20}t, \quad \varphi_3(t) = \frac{1}{2}\varphi_2(t) \quad \text{and} \quad \varphi_5(s, t) = \frac{1}{2}(\varphi_1(s) + \varphi_1(t)).$$

The distinct radius of convergence, number of iterations n, and COC (ρ) are mentioned in Table 5.

Table 5. Comparison on the basis of different radius of convergence for Example 4.

Schemes	R_1	R_2	R	x_0	n	ρ
NM	2.6667	1.3159	1.3159	$(1,1,\dots,1)$	4	4.0000
HM	2.6667	4.4444	2.6667	$(1.9,1.9,\dots,1.9)$	5	4.0000
JM	2.6667	1.5943	1.5943	$(2.1,2.1,\dots,2.1)$	5	4.0000
BM	0	0	0	-	-	-

Equation (39) is violated with these choices of φ_i. This is the reason that R is zero in the method BM. Therefore, our results hold only, if $x_0 = s_*$.

Example 5. We consider a boundary value problem from [8], which is defined as follows:

$$t'' = \frac{1}{2}t^3 + 3t' - \frac{3}{2-x} + \frac{1}{2}, \quad t(0) = 0, \; t(1) = 1. \tag{48}$$

We assume the following partition on $[0, 1]$

$$x_0 = 0 < x_1 < x_2 < \cdots < x_j, \quad \text{where} \; x_{j+1} = x_j + h, \; h = \frac{1}{j}.$$

We discretize this BVP (48) by

$$t'_i \approx \frac{t_{i+1} - t_{i-1}}{2h}, \quad t''_i \approx \frac{t_{i-1} - 2t_i + t_{i+1}}{h^2}, \quad i = 1, 2, \dots, j - 1.$$

Then, we obtain a $(k-1) \times (k-1)$ order nonlinear system, given by

$$t_{i+1} - 2t_i + t_{i-1} - \frac{h^2}{2}t_i^3 - \frac{3}{2-x_i}h^2 - 3\frac{t_{i+1} - t_{i-1}}{2}h - \frac{1}{h^2} = 0, \; i = 1, 2, \dots, j - 1,$$

where $t_0 = t(x_0) = 0$, $t_1 = t(x_1)$, ..., $t_{j-1} = t(x_{j-1})$, $t_j = t(x_j) = 1$ *and initial approximation* $t_0^{(0)} = \left(\frac{1}{2}, \frac{1}{2}, \frac{1}{2}, \frac{1}{2}, \frac{1}{2}, \frac{1}{2}\right)^T$. *In particular, we choose* $k = 6$ *so that we can obtain a* 5×5 *nonlinear system. The required solution of this problem is*

$$\bar{x} \approx \left(0.09029825\ldots, 0.1987214\ldots, 0.3314239\ldots, 0.4977132\ldots, 0.7123306\ldots\right)^T.$$

The distinct radius of convergence, number of iterations n, and COC (ρ) *are mentioned in Table 6.*

Table 6. Convergence behavior of distinct fourth-order methods for Example 5.

Methods	j	$\|F(x^{(j)})\|$	$\|x^{(j+1)} - x^{(j)}\|$	ρ
MM	1	8.1 (−6)	2.0 (−4)	
	2	1.0 (−23)	3.1 (−23)	
	3	9.1 (−95)	2.4 (−94)	
	4	3.7 (−379)	9.0 (−379)	3.9996
HM	1	7.8 (−6)	1.9 (−5)	
	2	7.6 (−24)	2.4 (−23)	
	3	2.7 (−95)	7.2 (−95)	
	4	2.6 (−381)	6.3 (−381)	3.9997
JM	1	7.8 (−6)	1.9 (−5)	
	2	7.6 (−24)	2.4 (−23)	
	3	2.7 (−95)	7.2 (−95)	
	4	2.6 (−381)	6.3 (−381)	3.9997
BM	1	7.2 (−6)	1.7 (−5)	
	2	4.2 (−24)	1.3 (−23)	
	3	1.9 (−96)	5.2 (−96)	
	4	5.6 (−386)	1.4 (−385)	3.9997

4. Conclusions

The convergence order of iterative methods involves Taylor series, and the existence of high order derivatives. Consequently, upper error bounds on $\|x_j - s_*\|$ and uniqueness results are not reported with this technique. Hence, the applicability of these methods is limited to functions with high order derivatives. To address these problems, we present local convergence results based on the first derivative. Moreover, we compare methods (2)–(5). Notice that our convergence criteria are sufficient but not necessary. Therefore, if e.g., the radius of convergence for the method (5) is zero, that does not necessarily imply that the method does not converge for a particular numerical example. Our method can be adopted in order to expand the applicability of other methods in an analogous way.

Author Contributions: Both the authors have equal contribution for this paper.

Funding: This research received no external funding.

Conflicts of Interest: The authors declare no conflict of interest.

References

1. Nedzhibov, G.H. A family of multi-point iterative methods for solving systems of nonlinear equations. *Comput. Appl. Math.* **2008**, *222*, 244–250. [CrossRef]
2. Hueso, J.L.; Martínez, E.; Teruel, C. Convergence, Efficiency and Dynamics of new fourth and sixth order families of iterative methods for nonlinear systems. *Comp. Appl. Math.* **2015**, *275*, 412–420. [CrossRef]
3. Junjua, M.; Akram, S.; Yasmin, N.; Zafar, F. A New Jarratt-Type Fourth-Order Method for Solving System of Nonlinear Equations and Applications. *Appl. Math.* **2015**, *2015*, 805278. [CrossRef]
4. Behl, R.; Cordero, A.; Torregrosa, J.R.; Alshomrani, A.S. New iterative methodsfor solving nonlinear problems with one and several unknowns. *Mathematics* **2018**, *6*, 296. [CrossRef]

5. Cordero, A.; Torregrosa, J.R. Variants of Newton's method using fifth-order quadrature formulas. *Appl. Math. Comput.* **2007**, *190*, 686–698. [CrossRef]

6. Balaji, G.V.; Seader, J.D. Application of interval Newton's method to chemical engineering problems. *Reliab. Comput.* **1995**, *1*, 215–223. [CrossRef]

7. Shacham, M. An improved memory method for the solution of a nonlinear equation. *Chem. Eng. Sci.* **1989**, *44*, 1495–1501. [CrossRef]

8. Ortega, J.M.; Rheinbolt, W.C. *Iterative Solutions of Nonlinears Equations in Several Variables*; Academic Press: Cambridge, MA, USA, 1970.

mathematics

MDPI

Article

Extended Local Convergence for the Combined Newton-Kurchatov Method Under the Generalized Lipschitz Conditions

Ioannis K. Argyros [1] and Stepan Shakhno [2,*]

[1] Department of Mathematical Sciences, Cameron University, Lawton, OK 73505, USA; iargyros@cameron.edu
[2] Department of Theory of Optimal Processes, Ivan Franko National University of Lviv, 79000 Lviv, Ukraine
[*] Correspondence: stepan.shakhno@lnu.edu.ua

Received: 4 February 2019; Accepted: 20 February 2019; Published: 23 February 2019

Abstract: We present a local convergence of the combined Newton-Kurchatov method for solving Banach space valued equations. The convergence criteria involve derivatives until the second and Lipschitz-type conditions are satisfied, as well as a new center-Lipschitz-type condition and the notion of the restricted convergence region. These modifications of earlier conditions result in a tighter convergence analysis and more precise information on the location of the solution. These advantages are obtained under the same computational effort. Using illuminating examples, we further justify the superiority of our new results over earlier ones.

Keywords: nonlinear equation; iterative process; non-differentiable operator; Lipschitz condition

MSC: 65H10; 65J15; 47H17

1. Introduction

Consider the nonlinear equation

$$F(x) + Q(x) = 0, \tag{1}$$

where F is a Fréchet-differentiable nonlinear operator on an open convex subset D of a Banach space E_1 with values in a Banach space E_2, and $Q : D \to E_2$ is a continuous nonlinear operator.

Let x, y be two points of D. A linear operator from E_1 into E_2, denoted $Q(x, y)$, which satisfies the condition

$$Q(x, y)(x - y) = Q(x) - Q(y) \tag{2}$$

is called a divided difference of Q at points x and y.

Let x, y, z be three points of D. A operator $Q(x, y, z)$ will be called a divided difference of the second order of the operator Q at the points x, y and z, if it satisfies the condition

$$Q(x, y, z)(y - z) = Q(x, y) - Q(x, z). \tag{3}$$

A well-known simple difference method for solving nonlinear equations $F(x) = 0$ is the Secant method

$$x_{n+1} = x_n - (F(x_{n-1}, x_n))^{-1}F(x_n), \ n = 0, 1, 2, \ldots, \tag{4}$$

where $F(x_{n-1}, x_n)$ is a divided difference of the first order of $F(x)$ and x_0, x_{-1} are given.

Secant method for solving nonlinear operator equations in a Banach space was explored by the authors [1–6] under the condition that the divided differences of a nonlinear operator F satisfy the Lipschitz (Hölder) condition with constant L of type

$$\|F(x,y) - F(u,v)\| \le L(\|x - u\| + \|y - v\|).$$

In [7] a one-point iterative Secant-type method with memory was psoposed.

In [8,9] the Kurchatov method under the classical Lipschitz conditions for the divided differences of the first and second order was explored and its quadratic convergence of it was determined. The iterative formula of Kurchatov method has the form [1,8–11]

$$x_{n+1} = x_n - (F(2x_n - x_{n-1}, x_{n-1}))^{-1}F(x_n),\ n = 0,1,2,\ldots. \tag{5}$$

Related articles but with stronger convergence criteria exist; see works of Argyros, Ezquerro, Hernandez, Rubio, Gutierrez, Wang, Li [1,12–15] and references therein.

In [14] which dealt with the study of the Newton method, it was proposed that there are generalized Lipschitz conditions for the nonlinear operator, in which instead of constant L, some positive integrable function is used.

In our work [16], we introduced, for the first time, a similar generalized Lipschitz condition for the operator of the first order divided difference, and under this condition, the convergence of the Secant method was studied and it was found that its convergence order is $(1 + \sqrt{5})/2$.

In [17], we introduced a generalized Lipschitz condition for the divided differences of the second order, and we have studied the local convergence of the Kurchatov method (5).

Note that in many papers, such as [3,18–21], the authors investigated the Secant and Secant-type methods under the generalized conditions for the first divided differences of the form

$$\|(F(x,y) - F(u,v)))\| \le \omega(\|x - y\|, \|u - v\|) \quad \forall x,y,u,v \in D, \tag{6}$$

where $\omega : \mathbf{R}_+ \times \mathbf{R}_+ \longrightarrow \mathbf{R}_+$ is continuous nondecreasing function in their two arguments. Under these same conditions, in the work of Argyros [10], it was proven that there is a semi-local convergence of the Kurchatov method and in [22] of Ren and Argyros the semi-local convergence of a combined Kurchatov method and Secant method was demonstrated. In both cases, only the linear convergence of the methods is received.

We also refer the reader to the intersting paper by Donchev et al. [23], where several other relaxed Lipschitz conditions are used in the setting of fixed points for these conditions. Clearly, our results can be written in this setting too in an analogous way.

In [24], we first proposed and studied the local convergence of the combined Newton-Kurchatov method

$$x_{n+1} = x_n - (F'(x_n) + Q(2x_n - x_{n-1}, x_{n-1}))^{-1}(F(x_n) + Q(x_n)),\ n = 0,1,2,\ldots, \tag{7}$$

where $F'(u)$ is a Fréchet derivative, $Q(u,v)$ is a divided difference of the first order, x_0, x_{-1} are given, which is built on the basis of the mentioned Newton and Kurchatov methods. Semi-local convergence of the method (7) under the classical Lipschitz conditions is studied in the mentioned article, but the convergence only with the order $(1 + \sqrt{5})/2$ has been determined.

In [25], we studied the method (7) under relatively weak, generalized Lipschitz conditions for the derivatives and divided differences of nonlinear operators. Setting $Q(x) \equiv 0$, we receive the results for the Newton method [14], and when $F(x) \equiv 0$ we got the known results for Kurchatov method [9,17]. We proved the quadratic order of convergence of the method (7), which is higher than the convergence order $(1 + \sqrt{5})/2$ for the Newton–Secant method [1,26–28]

$$x_{n+1} = x_n - (F'(x_n) + Q(x_{n-1}, x_n))^{-1}(F(x_n) + Q(x_n)),\ n = 0,1,2,\ldots, . \tag{8}$$

The results of the numerical study of the method (7) and other combined methods on the test problems are provided in our works [24,28].

In this work, we continue to study a combined method (7) for solving nonlinear Equation (1), but with optimization considerations resulting in a tighter analysis than in [25].

The rest of the article is structured as follows: In Section 2, we present the local convergence analysis of the method (7) and the uniqueness ball for solution of the equation. Section 3 contains the Corollaries of Theorems from Section 2. In Section 4, we provide the numerical example. The article ends with some conclusions.

2. Local Convergence of Newton-Kurchatov Method (7)

Let us denote $B(x_0, r) = \{x : \|x - x_0\| < r\}$ an open ball of radius $r > 0$ with center at point $x_0 \in D$, $B(x_0, r) \subset D$.

Condition on the divided difference operator $Q(x, y)$

$$\|Q(x, y) - Q(u, v)\| \le L(\|x - u\| + \|y - v\|) \quad \forall x, y, u, v \in D \tag{9}$$

is called Lipschitz condition in domain D with constant $L > 0$. If the condition is being fulfilled

$$\|Q(x, y) - Q'(x_0)\| \le L(\|x - x_0\| + \|y - x_0\|) \quad \forall x, y \in B(x_0, r), \tag{10}$$

then we call it the center Lipschitz condition in the ball $B(x_0, r)$ with constant L.

However, L in Lipschitz conditions can be not a constant, and can be a positive integrable function. In this case, if for $x_* \in D$ inverse operator $[F'(x_*)]^{-1}$ exists, then the conditions (9) and (10) for $x_0 = x_*$ can be replaced respectively for

$$\|Q'(x_*)^{-1}(Q(x, y) - Q(u, v)))\| \le \int_0^{\|x-y\|+\|u-v\|} L(t)dt \quad \forall x, y, u, v \in D \tag{11}$$

and

$$\|Q'(x_*)^{-1}(Q(x, y) - Q'(x_*))\| \le \int_0^{\|x-x_*\|+\|y-x_*\|} L(t)dt \quad \forall x, y \in B(x_*, r). \tag{12}$$

Simultaneously

Lipschitz conditions (11) and (12) are called generalized Lipschitz conditions or Lipschitz conditions with the L average.

Similarly, we introduce the generalized Lipschitz condition for the divided difference of the second order

$$\|Q'(x_*)^{-1}(Q(u, x, y) - Q(v, x, y))\| \le \int_0^{\|u-v\|} N(t)dt \quad \forall x, y, u, v \in B(x_*, r), \tag{13}$$

where N is a positive integrable function.

Remark 1. *Note than the operator F is Fréchet differentiable on D when the Lipschitz conditions (9) or (11) are fulfilled $\forall x, y, u, v \in D$ (the divided differences $F(x, y)$ are Lipschitz continuous on D) and $F(x, x) = F'(x)$ $\forall x \in D$ [29].*

Suppose that equation

$$\int_0^r L_1^0(u)du + \int_0^{2r} L_2^0(u)du + 2r \int_0^{2r} N_0(u)du = 1.$$

has at least one positive solution. Denote by r_0 the smallest such solution. Set $D_0 = D \cap B(x_*, r_0)$

The radius of the convergence ball and the convergence order of the combined Newton–Kurchatov method (7) are determined in next theorem.

Theorem 1. *Let F and Q be continuous nonlinear operators defined in open convex domain D of a Banach space E_1 with values in the Banach space E_2. Let us suppose, that: (1) $H(x) \equiv F(x) + Q(x) = 0$ has a solution $x_* \in D$, for which there exists a Fréchet derivative $H'(x_*)$ and it is invertible; (2) F has the Fréchet derivative of the first order, and Q has divided differences of the first and second order on $B(x_*, 3r) \subset D$, so that for each $x, y, u, v \in D$*

$$\|H'(x_*)^{-1}(F'(x) - F'(x_*))\| \leq \int_0^{\rho(x)} L_1^0(u)du, \tag{14}$$

$$\|H'(x_*)^{-1}(Q(x,y) - Q(x_*, x_*))\| \leq \int_0^{\|x-x_*\|+\|y-x_*\|} L_2^0(t)dt, \tag{15}$$

$$\|H'(x_*)^{-1}(Q(u,x,y) - Q(v,x,y))\| \leq \int_0^{\|u-v\|} N_0(t)dt, \tag{16}$$

and for each $x, y, u, v \in D_0$

$$\|H'(x_*)^{-1}(F'(x) - F'(x^\theta))\| \leq \int_{\theta\rho(x)}^{\rho(x)} L_1(u)du, \ 0 \leq \tau \leq 1, \tag{17}$$

$$\|H'(x_*)^{-1}(Q(x,y) - Q(u,v))\| \leq \int_0^{\|x-u\|+\|y-v\|} L_2(t)dt, \tag{18}$$

$$\|H'(x_*)^{-1}(Q(u,x,y) - Q(v,x,y))\| \leq \int_0^{\|u-v\|} N(t)dt, \tag{19}$$

where $x^\theta = x_* + \theta(x - x_*)$, $\varrho(x) = \|x - x_*\|$, $L_1^0, L_2^0, N_0 L_1, L_2$ and N are positive nondecreasing integrable functions and $r > 0$ satisfies the equation

$$\frac{\frac{1}{r}\int_0^r L_1(u)udu + \int_0^r L_2(u)du + 2r\int_0^{2r} N(u)du}{1 - \left(\int_0^r L_1^0(u)du + \int_0^{2r} L_2^0(u)du + 2r\int_0^{2r} N_0(u)du\right)} = 1. \tag{20}$$

Then for all $x_0, x_{-1} \in B(x_*, r)$ the iterative method (7) is well defined and the generated by it sequence $\{x_n\}_{n\geq 0}$, which belongs to $B(x_*, r)$, converges to x_* and satisfies the inequality

$$\|x_{n+1} - x_*\| \leq e_n :=$$
$$\frac{\frac{1}{\rho(x_n)}\int_0^{\rho(x_n)} L_1(u)udu + \int_0^{\rho(x_n)} L_2(u)du + \int_0^{\|x_n - x_{n-1}\|} N(u)du\|x_n - x_{n-1}\|}{1 - \left(\int_0^{\rho(x_n)} L_1^0(u)du + \int_0^{2\rho(x_n)} L_2^0(u)du + \int_0^{\|x_n - x_{n-1}\|} N_0(u)du\|x_n - x_{n-1}\|\right)}\|x_n - x_*\|. \tag{21}$$

Proof. First we show that $f(t) = \frac{1}{t^2}\int_0^t L_1(u)udu$, $g(t) = \frac{1}{t}\int_0^t L_2(u)du$, $h(t) = \frac{1}{t}\int_0^t N(u)du$, $f_0(t) = \frac{1}{t^2}\int_0^t L_1^0(u)udu$, $g_0(t) = \frac{1}{t}\int_0^t L_2^0(u)du$, $h_0(t) = \frac{1}{t}\int_0^t N_0(u)du$ monotonically nondecreasing with respect to t. Indeed, under the monotony of L_1, L_2, N we have

$$\left(\frac{1}{t_2^2}\int_0^{t_2} - \frac{1}{t_1^2}\int_0^{t_1}\right)L_1(u)udu = \left(\frac{1}{t_2^2}\int_{t_1}^{t_2} + \left(\frac{1}{t_2^2} - \frac{1}{t_1^2}\right)\int_0^{t_1}\right)L_1(u)udu \geq$$

$$\geq L(t_1)\left(\frac{1}{t_2^2}\int_{t_1}^{t_2} + \left(\frac{1}{t_2^2} - \frac{1}{t_1^2}\right)\int_0^{t_1}\right)udu = L_1(t_1)\left(\frac{1}{t_2^2}\int_0^{t_2} - \frac{1}{t_1^2}\int_0^{t_1}\right)udu = 0,$$

$$\left(\frac{1}{t_2}\int_0^{t_2} - \frac{1}{t_1}\int_0^{t_1}\right)L_2(u)du = \left(\frac{1}{t_2}\int_{t_1}^{t_2} + \left(\frac{1}{t_2} - \frac{1}{t_1}\right)\int_0^{t_1}\right)L_2(u)du \geq$$

$$\geq L_2(t_1)\left(\frac{1}{t_2}\int_{t_1}^{t_2} + \left(\frac{1}{t_2} - \frac{1}{t_1}\right)\int_0^{t_1}\right)du = L_2(t_1)\left(\frac{t_2 - t_1}{t_2} + t_1\left(\frac{1}{t_2} - \frac{1}{t_1}\right)\right) = 0$$

for $0 < t_1 < t_2$. So, $f(t), g(t)$ are nondecreasing with respect to t. Similarly we get for $h(t), f_0(t), g_0(t)$ and $h_0(t)$.

We denote by A_n linear operator $A_n = F'(x_n) + Q(2x_n - x_{n-1}, x_{n-1})$. Easy to see that if $x_n, x_{n-1} \in B(x_*, r)$, then $2x_n - x_{n-1}, x_{n-1} \in B(x_*, 3r)$. Then A_n is invertible and the inequality holds

$$\|A_n^{-1}H'(x_*)\| = \|[I - (I - H'(x_*)^{-1}A_n)]^{-1}\| \leq$$

$$\leq \left(1 - \left(\int_0^{\rho(x_n)} L_1^0(u)du + \int_0^{2\rho(x_n)} L_2^0(u)du + \int_0^{\|x_n - x_{n-1}\|} N_0(u)du\|x_n - x_{n-1}\|\right)\right)^{-1}. \tag{22}$$

Indeed from the formulas (14)–(16) we get

$$\|I - H'(x_*)^{-1}A_n\| = \|H'(x_*)^{-1}(F'(x_*) - F'(x_n) + Q(x_*, x_*) - Q(x_n, x_n)+$$

$$+ Q(x_n, x_n) - Q(2x_n - x_{n-1}, x_{n-1}))\| \leq \int_0^{\rho(x_n)} L_1^0(u)du + \|H'(x_*)^{-1}(Q(x_*, x_*)-$$

$$- Q(x_n, x_n) + Q(x_n, x_n) - Q(x_n, x_{n-1}) + Q(x_n, x_{n-1}) - Q(2x_n - x_{n-1}, x_{n-1}))\| \leq$$

$$\leq \int_0^{\rho(x_n)} L_1^0(u)du + \int_0^{2\rho(x_n)} L_2^0(u)du+$$

$$+ \|H'(x_*)^{-1}(Q(x_n, x_{n-1}, x_n) - Q(2x_n - x_{n-1}, x_{n-1}, x_n))(x_n - x_{n-1})\| \leq$$

$$\leq \int_0^{\rho(x_n)} L_1^0(u)du + \int_0^{2\rho(x_n)} L_2^0(u)du + \int_0^{\|x_n - x_{n-1}\|} N_0(u)du\|x_n - x_{n-1}\|.$$

From the definition r_0 (20), we get

$$\int_0^{r_0} L_1(u)du + \int_0^{2r_0} L_2(u)du + 2r\int_0^{2r_0} N(u)du < 1, \tag{23}$$

since $r < r_0$.

Using the Banach theorem on inverse operator [30], we get formula (22). Then we can write

$$\|x_{n+1} - x_*\| = \|x_n - x_* - A_n^{-1}(F(x_n) - F(x_*) + Q(x_n) - Q(x_*))\| =$$

$$= \| - A_n^{-1}\left(\int_0^1 (F'(x_n^\tau) - F'(x_n))d\tau + Q(x_n, x_*) - Q(2x_n - x_{n-1}, x_{n-1}))(x_n - x_*)\| \leq$$

$$\leq \|A_n^{-1}H'(x_*)\|(\|H'(x_*)^{-1}\int_0^1 \int_{\tau\rho(x_n)}^{\rho(x_n)} L_1(u)dud\tau + \|H'(x_*)^{-1}(+Q(x_n, x_*)- \tag{24}$$

$$- Q(2x_n - x_{n-1}, x_{n-1}))\|)\|x_n - x_*\|.$$

According to the condition (17)–(19) of the theorem we get

$$\|H'(x_*)^{-1}\left(\int_0^1 \int_{\tau\rho(x_n)}^{\rho(x_n)} L_1(u)\,du\,d\tau + Q(x_n,x_*) - A_n\right)\| =$$

$$= \frac{1}{\rho(x_n)}\int_0^{\rho(x_n)} L_1(u)u\,du + \|H'(x_*)^{-1}(Q(x_n,x_*) - Q(x_n,x_n)+$$

$$+Q(x_n,x_n) - Q(x_n,x_{n-1}) + Q(x_n,x_{n-1}) - Q(2x_n - x_{n-1},x_{n-1}))\| \le$$

$$\le \frac{1}{\rho(x_n)}\int_0^{\rho(x_n)} L_1(u)u\,du + \|H'(x_*)^{-1}(Q(x_n,x_*) - Q(x_n,x_n))\|+$$

$$+\|H'(x_*)^{-1}(Q(x_n,x_{n-1},x_n) - Q(2x_n - x_{n-1},x_{n-1},x_n))(x_n - x_{n-1})\| \le$$

$$\le \frac{1}{\rho(x_n)}\int_0^{\rho(x_n)} L_1(u)u\,du + \int_0^{\rho(x_n)} L_2(u)\,du + \int_0^{\|x_n-x_{n-1}\|} N(u)\,du\|x_n - x_{n-1}\|.$$

From (22) and (24) shows that fulfills (21). Then from (21) and (20) we get

$$\|x_{n+1} - x_*\| < \|x_n - x_*\| < \ldots < \max\{\|x_0 - x_*\|, \|x_{-1} - x_*\|\} < r.$$

Therefore, the iterative process (5) is correctly defined and the sequence that it generates belongs to $B(x_*, r)$. From the last inequality and estimates (21) we get $\lim_{n\to\infty} \|x_n - x_*\| = 0$. Since the sequence $\{x_n\}_{n\ge 0}$ converges to x_*, then

$$\|x_n - x_{n-1}\| \le \|x_n - x_*\| + \|x_{n-1} - x_*\| \le 2\|x_{n-1} - x_*\|$$

and $\lim_{n\to\infty} \|x_n - x_{n-1}\| = 0$. $\square$

Corollary 1. *The order of convergence of the iterative procedure (7) is quadratic.*

Proof. Let us denote $\rho_{\max} = \max\{\rho(x_0), \rho(x_{-1})\}$. Since $g(t)$ and $h(t)$ are monotonically nondecreasing, then with taking into account the expressions

$$\frac{1}{\rho(x_n)}\int_0^{\rho(x_n)} L_1(u)u\,du = \frac{\int_0^{\rho(x_n)} L_1(u)u\,du\rho(x_n))}{(\rho(x_n))^2} \le \frac{\int_0^{\rho_{\max}} L_1(u)u\,du\rho(x_n)}{(\rho_{\max})^2} =: A_1\rho(x_n),$$

$$\int_0^{\rho(x_n)} L_2(u)\,du = \frac{\int_0^{\rho(x_n)} L_2(u)\,du\rho(x_n)}{\rho(x_n)} \le \frac{\int_0^{\rho_{\max}} L_2(u)\,du\rho(x_n)}{\rho_{\max}} =: A_2\rho(x_n),$$

$$\int_0^{\|x_n-x_{n-1}\|} N(u)\,du = \frac{\int_0^{\|x_n-x_{n-1}\|} N(u)\,du\|x_n - x_{n-1}\|}{\|x_n - x_{n-1}\|} <$$

$$< \frac{\int_0^{\|x_0-x_{-1}\|} N(u)\,du\|x_n - x_{n-1}\|}{\|x_0 - x_{-1}\|} =: A_3\|x_n - x_{n-1}\|$$

and

$$\left(1 - \left(\int_0^{\rho(x_n)} L_1^0(u)\,du + 2\int_0^{\rho(x_n)} L_2^0(u)\,du + \int_0^{\|x_n-x_{n-1}\|} N_0(u)\,du\|x_n - x_{n-1}\|\right)\right)^{-1} <$$

$$< \left(1 - \left(\int_0^{\rho_{\max}} L_1^0(u)\,du + 2\int_0^{\rho_{\max}} L_2^0(u)\,du + \int_0^{\|x_0-x_{-1}\|} N_0(u)\,du\|x_0 - x_{-1}\|\right)\right)^{-1} =: A_4,$$

from the inequality (21) follows

$$\|x_{n+1} - x_*\| \le A_4(A_1\rho(x_n) + A_2\rho(x_n) + A_3\|x_n - x_{n-1}\|^2)\|x_n - x_*\|.$$

or

$$\|x_{n+1} - x_*\| \le C_3\|x_n - x_*\|^2 + C_4\|x_n - x_{n-1}\|^2\|x_n - x_*\|. \tag{25}$$

Here $A_k, k = 1, ..., 4, C_3, C_4$ are some positive constants.

Assume that the order of convergence of the iterative process (7) is not lower 2, therefore there exist $C_5 \ge 0$ and $N > 0$, that for all $n \ge N$ the inequality holds

$$\|x_n - x_*\| \ge C_5\|x_{n-1} - x_*\|^2.$$

Since

$$\|x_n - x_{n-1}\|^2 \le (\|x_n - x_*\| + \|x_{n-1} - x_*\|)^2 \le 4\|x_{n-1} - x_*\|^2,$$

then from (44) we get

$$\|x_{n+1} - x_*\| \le C_3\|x_n - x_*\|^2 + 4C_4\|x_{n-1} - x_*\|^2\|x_n - x_*\|$$

$$\le (C_3 + 4C_4/C_5)\|x_n - x_*\|^2 = C_6\|x_n - x_*\|^2. \tag{26}$$

inequality (26) means that the order of convergence is not lower than 2. Thus, the convergence rate of sequence $\{x_n\}_{n\ge0}$ to x_* is quadratic. $\quad\square$

3. Uniqueness Ball of the Solution

The next theorem determines the ball of uniqueness of the solution x_* of (1) in $B(x_*, r)$.

Theorem 2. *Let us assume that: (1) $H(x) \equiv F(x) + Q(x) = 0$ has a solution $x_* \in D$, in which there exists a Fréchet derivative $H'(x_*)$ and it is invertible; (2) F has a continuous Fréchet derivative in $B(x_*, r)$, F' satisfies the generalized Lipschitz condition*

$$\|H'(x_*)^{-1}(F'(x) - F'(x_*))\| \le \int_0^{\rho(x)} L_1^0(u)du \quad \forall x \in B(x_*, r),$$

the divided difference $Q(x, y)$ satisfies the generalized Lipschitz condition

$$\|H'(x_*)^{-1}(Q(x, x_*) - G'(x_*))\| \le \int_0^{\rho(x)} L_2^0(u)du \quad \forall x \in B(x_*, r),$$

where L_1 and L_2 are positive integrable functions. Let $r > 0$ satisfy

$$\frac{1}{r}\int_0^r (r - u)L_1^0(u)du + \int_0^r L_2^0(u)du \le 1.$$

Then the equation $H(x) = 0$ has a unique solution x_* in $B(x_*, r)$.

Proof analogous to [27,31].

4. Corollaries

In the study of iterative methods, the traditional assumption is that the derivatives and/or the divided differences satisfy the classical Lipschitz conditions. Assuming that L_1, L_2 and N are constants, we get from Theorems 1 and 2 important corollaries, which are of interest.

Corollary 2. *Let us assume that: (1) $H(x) \equiv F(x) + Q(x) = 0$ has a solution $x_* \in D$, in which there exists Fréchet derivative $H'(x_*)$ and it is invertible; (2) F has a continuous Fréchet derivative and Q has divided differences of the first and second order $Q(x,y)$ and $Q(x,y,z)$ in $B(x_*, 3r) \subset D$, which satisfy the Lipschitz conditions for each $x, y, u, v \in D$*

$$\|H'(x_*)^{-1}(F'(x) - F'(x_*))\| \leq L_1^0 \|x - x_*\|,$$

$$\|H'(x_*)^{-1}(Q(x,y) - Q(u,v))\| \leq L_2^0(\|x - u\| + \|y - v\|),$$

for $x, y, u, v \in D_0$

$$\|H'(x_*)^{-1}(Q(u,x,y) - Q(v,x,y))\| \leq N_0 \|u - v\|,$$

$$\|H'(x_*)^{-1}(F'(x) - F'(x_* + \tau(x - x_*)))\| \leq (1 - \tau)L_1 \|x - x_*\|,$$

$$\|H'(x_*)^{-1}(Q(x,y) - Q(u,v))\| \leq L_2(\|x - u\| + \|y - v\|),$$

$$\|H'(x_*)^{-1}(Q(u,x,y) - Q(v,x,y))\| \leq N \|u - v\|,$$

where L_1^0, L_2^0, N_0, L_1, L_2 and N are positive numbers,

$$r_0 = \frac{2}{L_1^0 + 2L_2^0 + \sqrt{(L_1^0 + 2L_2^0)^2 + 16N_0}},$$

and r is the positive root of the equation

$$\frac{L_1 r/2 + L_2 r + 4Nr^2}{1 - L_1^0 r - 2L_2^0 r - 4N_0 r^2} = 1.$$

Then Newton-Kurchatov method (5) converges for all $x_{-1}, x_0 \in B(x_*, r)$ and there fulfills

$$\|x_{n+1} - x_*\| \leq \frac{(L_1/2 + L_2)\|x_n - x_*\| + N\|x_n - x_{n-1}\|^2}{1 - \left(L_1^0 + 2L_2^0 \|x_n - x_*\| + N_0\|x_n - x_{n-1}\|^2\right)}.$$

Moreover, r is the best of all possible.

Note that value of $r = \dfrac{2}{3L}$ improves $\bar{r} = \dfrac{2}{3L_1^1}$ for Newton method for solving equation $F(x) - 0$ [14,32,33], and with $r = 2/(3L_2 + \sqrt{9L_2^2 + 32N})$ improves $\bar{r} = 2/(3L_2^1 + \sqrt{9(L_2^1)^2 + 32N_1})$ for Kurchatov method for solving the equation $Q(x) = 0$, as derived in [8].

Corollary 3. *Suppose that: (1) $H(x) \equiv F(x) + Q(x) = 0$ has a solution $x_* \in D$, in which there exists the Fréchet derivative $H'(x_*)$ and it is invertible; (2) F has continuous derivative and Q has divided difference $Q(x, x_*)$ in $B(x_*, r) \subset D$, which satisfy the Lipschitz conditions*

$$\|H'(x_*)^{-1}(F'(x) - F'(x_*))\| \leq L_1^0 \|x - x_*\|,$$

$$\|H'(x_*)^{-1}(Q(x, x_*) - G'(x_*))\| \leq L_2^0 \|x - x_*\|$$

for all $x \in B(x_*, r)$, where L_1^0 and L_2^0 are positive numbers and $r = \dfrac{2}{L_1^0 + 2L_2^0}$. Then x_* is the only solution in $B(x_*, r)$ of $H(x) = 0$, r does not depend on F and Q and is the best choice.

Note that the resulting radius of the uniqueness ball of the solution $r = \dfrac{2}{L_1}$ improves $\bar{r} = \dfrac{2}{L_1^1}$ for Newton method for solving the equation $F(x) = 0$ [14] and $r = \dfrac{1}{L_2}$ improves $\bar{r} = \dfrac{1}{L_2^1}$ for Kurchatov method for solving the equation $Q(x) = 0$ [8]. (See also the numerical examples).

Remark 2. *We compare the results in [25] with the new results in this article. In order to do this, let us consider the conditions given in [25] corresponding to our conditions (15)–(17):*
For each $x, y, u, v \in D$

$$\|H'(x_*)^{-1}(F'(x) - F'(x^\theta))\| \leq \int_{\theta\rho(x)}^{\rho(x)} L_1^1(u)du, \ 0 \leq \theta \leq 1, \tag{27}$$

$$\|H'(x_*)^{-1}(Q(x,y) - Q(u,v))\| \leq \int_0^{\|x-u\|+\|y-v\|} L_2^1(t)dt, \tag{28}$$

$$\|H'(x_*)^{-1}(Q(u,x,y) - Q(v,x,y))\| \leq \int_0^{\|u-v\|} N^1(t)dt, \tag{29}$$

$$\frac{\frac{1}{\bar{r}}\int_0^{\bar{r}} L_1^1(u)udu + \int_0^{\bar{r}} L_2^1(u)du + 2\bar{r}\int_0^{2\bar{r}} N^1(u)du}{1 - \left(\int_0^{\bar{r}} L_1^1(u)du + \int_0^{2\bar{r}} L_2^1(u)du + 2\bar{r}\int_0^{2\bar{r}} N^1 1(u)du\right)} = 1, \tag{30}$$

$$\|x_{n+1} - x_n\| \leq \bar{e}_n. \tag{31}$$

It follows from (14)–(16), (17)–(19), (27)–(29), that

$$L_1^0(t) \leq L_1^1(t), \tag{32}$$

$$L_1(t) \leq L_1^1(t), \tag{33}$$

$$L_2^0(t) \leq L_2^1(t), \tag{34}$$

$$L_2(t) \leq L_2^1(t), \tag{35}$$

$$N_0(t) \leq N^1(t), \tag{36}$$

$$N(t) \leq N^1(t), \tag{37}$$

leading to

$$\bar{r} \leq r, \tag{38}$$

$$e_n \leq \bar{e}_n, \tag{39}$$

$$A_l \leq \bar{A}_l, \ l = 1, 2, 3, 4, \tag{40}$$

$$C_l \leq \bar{C}_l, \ l = 1, 2, 3, 4, 6 \tag{41}$$

and

$$C_5 \geq \bar{C}_5, \tag{42}$$

$$\bar{e}_n :=$$

$$\frac{\frac{1}{\rho(x_n)}\int_0^{\rho(x_n)} L_1^1(u)udu + \int_0^{\rho(x_n)} L_2^1(u)du + \int_0^{\|x_n - x_{n-1}\|} N^1(u)du\|x_n - x_{n-1}\|}{1 - \left(\int_0^{\rho(x_n)} L_1^1(u)du + \int_0^{2\rho(x_n)} L_2^1(u)du + \int_0^{\|x_n - x_{n-1}\|} N^1(u)du\|x_n - x_{n-1}\|\right)}\|x_n - x_*\|, \tag{43}$$

$$\|x_{n+1} - x_*\| \leq \bar{A}_4(\bar{A}_1\rho(x_n) + \bar{A}_2\rho(x_n) + \bar{A}_3\|x_n - x_{n-1}\|^2)\|x_n - x_*\|,$$

or

$$\|x_{n+1} - x_*\| \leq \bar{C}_3 \|x_n - x_*\|^2 + \bar{C}_4 \|x_n - x_{n-1}\|^2 \|x_n - x_*\|, \tag{44}$$

or

$$\|x_{n+1} - x_*\| \leq \bar{C}_6 \|x_n - x_*\|^2$$

with

$$\bar{C}_6 = (\bar{C}_3 + 4\bar{C}_4 / \bar{C}_5)$$

for some

$$\|x_n - x_*\| \geq \bar{C}_5 \|x_{n-1} - x_*\|^2.$$

Hence, we obtain the impovements:

(1) At least as many initial choices x_{-1}, x_0 as before.
(2) At least as few iterations than before to obtain a predetermined error accuracy.
(3) At least as precice information on the location of the solution as before.

Moreover, if any of (32)–(37) holds as a strict inequality, then so do (38)–(42). Furthermore, we notice that these improvements are found using the same information, since the functions L_1^0, L_2^0, N_0, L_1, L_2, N are special cases of functions L_1^1, L_2^1, N^1 used in [25]. Finally, if $G = 0$ or $F = 0$, we obtain the results for Newton's method or the Kurchatov method as special cases. Clearly, the results for these methods are also improved. Our technique can also be used to improve the results of other iterative methods in an analogous way.

5. Numerical Examples

Example 1. *Let $E_1 = E_2 = R^3$ and $\Omega = S(x_*, 1)$. Define functions F and Q for $v = (v_1, v_2, v_3)^T$ on Ω by*

$$F(v) = \left(e^{v_1} - 1, \frac{e-1}{2}v_2^2 + v_2, v_3\right)^T, \tag{45}$$
$$Q(v) = \left(|v_1|, |v_2|, |v_2|, |\sin(v_3)|\right)^T$$

$$F'(v) = \text{diag}\left(e^{v_1}, (e-1)v_2 + 1, 1\right), \tag{46}$$
$$Q(v, \bar{v}) = \text{diag}\left(\frac{|\bar{v}_1| - |v_1|}{\bar{v}_1 - v_1}, \frac{|\bar{v}_2| - |v_2|}{\bar{v}_2 - v_2}, \frac{|\sin(\bar{v}_3)| - |\sin(v_3)|}{\bar{v}_3 - v_3}\right)$$

Choose:

$$H(x) = F(x) + Q(x),$$

$$\|H'(x_*)^{-1}\| = 1, L_1^0 = e - 1, L_2^0 = 1, N_0 = \frac{1}{2},$$

$$L_1 = e^{\frac{1}{e-1}}, L_2 = 1, N = \frac{1}{2},$$

$$L_1^1 = e, L_2^1 = 1, N^1 = \frac{1}{2}.$$

Then compute:
r using (20), $r = 0.1599$;
$\bar{r}$ using (30), $\bar{r} = 0.1315$.
Also, $\bar{r} < r$.
Notice that $L_1^0 < L_1 < L_1^1$, so the improvements stated in Remark 1 hold.

6. Conclusions

In [1,8,34], we studied the local convergence of Secant and Kurchatov methods in the case of fulfilment of Lipschitz conditions for the divided differences, which hold for some Lipschitz constants. In [14], the convergence of the Newton method is shown for the generalized Lipschitz conditions for the Fréchet derivative of the first order. We explored the local convergence of the Newton-Kurchatov method under the generalized Lipschitz conditions for Fréchet derivative of a differentiable part of the operator and the divided differences of the nondifferentiable part. Our results contain known parts as partial cases.

By using our idea of restricted convergence regions, we find tighter Lipschitz constants leading to a finer local convergence analysis of method (7) and its special cases compared to in [25].

Author Contributions: All authors contributed equally and significantly to writing this article. All authors read and approved the final manuscript.

Funding: This research received no external funding.

Acknowledgments: The authors would like to express their sincere gratitude to the referees for their valuable comments which have significantly improved the presentation of this paper.

Conflicts of Interest: The authors declare no conflict of interest.

References

1. Argyros, I.K. *Convergence and Applications of Newton-Type Iterations*; Springer: New York, NY, USA, 2008.
2. Hernandez, M.A.; Rubio, M.J. The Secant method and divided differences Hölder continuous. *Appl. Math. Comput.* **2001**, *124*, 139–149. [CrossRef]
3. Hernandez, M.A.; Rubio, M.J. The Secant method for nondifferentiable operators. *Appl. Math. Lett.* **2002**, *15*, 395–399. [CrossRef]
4. Ortega, J.M.; Rheinboldt, W.C. *Iterative Solution of Nonlinear Equations in Several Variables*; Academic Press: New York, NY, USA, 1970.
5. Traub, J.F. *Iterative Methods for the Solution of Equations*; Prentice-Hall, Inc.: Englewood Cliffs, NJ, USA, 1964.
6. Shakhno, S.M. Application of nonlinear majorants for intestigation of the secant method for solving nonlinear equations. *Matematychni Studii* **2004**, *22*, 79–86.
7. Ezquerro, J.A.; Grau-Sánchez, M.; Hernández, M.A. Solving non-differentiable equations by a new one-point iterative method with memory. *J. Complex.* **2012**, *28*, 48–58. [CrossRef]
8. Shakhno, S.M. On a Kurchatov's method of linear interpolation for solving nonlinear equations. *Proc. Appl. Math. Mech.* **2004**, *4*, 650–651. [CrossRef]
9. Shakhno, S.M. About the difference method with quadratic convergence for solving nonlinear operator equations. *Matematychni Studii* **2006**, *26*, 105–110. (In Ukrainian)
10. Argyros, I.K. A Kantorovich-type analysis for a fast iterative method for solving nonlinear equations. *J. Math. Anal. Appl.* **2007**, *332*, 97–108. [CrossRef]
11. Kurchatov, V.A. On a method of linear interpolation for the solution of functional equations. *Dokl. Akad. Nauk SSSR* **1971**, *198*, 524–526. (In Russian); translation in *Soviet Math. Dokl.* **1971**, *12*, 835–838.
12. Ezquerro, J.A.; Hernández, M. Generalized differentiability conditions for Newton's method. *IMA J. Numer. Anal.* **2002**, *22*, 187–205. [CrossRef]
13. Gutiérrez, J.M.; Hernández, M.A. Newton's method under weak Kantorovich conditions. *IMA J. Numer. Anal.* **2000**, *20*, 521–532. [CrossRef]
14. Wang, X.H. Convergence of Newton's method and uniqueness of the solution of equations in Banach space. *IMA J. Numer. Anal.* **2000**, *20*, 123–134. [CrossRef]
15. Wang, X.H.; Li, C. Local and global behavior for algorithms of solving equations. *Chin. Sci. Bull.* **2001**, *46*, 444–451. [CrossRef]
16. Shakhno, S.M. On the secant method under generalized Lipschitz conditions for the divided difference operator. *Proc. Appl. Math. Mech.* **2007**, *7*, 2060083–2060084. [CrossRef]
17. Shakhno, S.M. Method of linear interpolation of Kurchatov under generalized Lipschitz conditions for divided differences of first and second order. *Visnyk Lviv. Univ. Ser. Mech. Math.* **2012**, *77*, 235–242. (In Ukrainian)

18. Amat, S. On the local convergence of Secant-type methods. *Intern. J. Comput. Math.* **2004**, *81*, 1153–1161. [CrossRef]
19. Amat, S.; Busquier, S. On a higher order Secant method. *Appl. Math. Comput.* **2003**, *141*, 321–329. [CrossRef]
20. Argyros, I.K.; Ezquerro, J.A.; Gutiérrez, J.M.; Hernández, M.A.; Hilout, S. Chebyshev-Secant type methods for non-differentiable operator. *Milan J. Math.* **2013**, *81*, 25–35.
21. Ren, H. New sufficient convergence conditions of the Secant method nondifferentiable operators. *Appl. Math. Comput.* **2006**, *182*, 1255–1259. [CrossRef]
22. Ren, H.; Argyros, I.K. A new semilocal convergence theorem with nondifferentiable operators. *J. Appl. Math. Comput.* **2010**, *34*, 39–46. [CrossRef]
23. Donchev, T.; Farkhi, E.; Reich, S. Fixed set iterations for relaxed Lipschitz multimaps. *Nonlinear Anal.* **2003**, *53*, 997–1015. [CrossRef]
24. Shakhno, S.M.; Yarmola, H.P. Two-point method for solving nonlinear equation with nondifferentiable operator. *Matematychni Studii.* **2011**, *36*, 213–220. (In Ukrainian)
25. Shakhno, S.M. Combined Newton-Kurchatov method under the generalized Lipschitz conditions for the derivatives and divided differences. *J. Numer. Appl. Math.* **2015**, *2*, 78–89.
26. Catinas, E. On some iterative methods for solving nonlinear equations. *Revue d'Analyse Numérique et de Théorie de l'Approximation* **1994**, *23*, 47–53.
27. Shakhno, S. Convergence of combined Newton-Secant method and uniqueness of the solution of nonlinear equations. *Visnyk Ternopil Nat. Tech. Univ.* **2013**, *69*, 242–252. (In Ukrainian)
28. Shakhno, S.M.; Mel'nyk, I.V.; Yarmola, H.P. Analysis of convergence of a combined method for the solution of nonlinear equations. *J. Math. Sci.* **2014**, *201*, 32–43. [CrossRef]
29. Argyros, I. K. On the secant method. *Publ. Math. Debr.* **1993**, *43*, 233–238.
30. Kantorovich, L.V.; Akilov, G.P. *Functional Analysis*; Pergamon Press: Oxford, UK, 1982.
31. Shakhno, S.M. Convergence of the two-step combined method and uniqueness of the solution of nonlinear operator equations. *J. Comput. Appl. Math.* **2014**, *261*, 378–386. [CrossRef]
32. Potra, F.A. On an iterative algorithm of order 1.839... for solving nonlinear operator equations. *Numer. Funct. Anal. Optim.* **1985**, *7*, 75–106. [CrossRef]
33. Traub, J.F.; Woźniakowski, H. Convergence and complexity of Newton iteration for operator equations. *J. Assoc. Comput. Mach.* **1979**, *26*, 250–258. [CrossRef]
34. Hernandez, M.A.; Rubio, M.J. A uniparametric family of iterative processes for solving nondifferentiable equations. *J. Math. Anal. Appl.* **2002**, *275*, 821–834. [CrossRef]

![Σ mathematics logo] *mathematics*

MDPI

Article

Study of a High Order Family: Local Convergence and Dynamics

Cristina Amorós [1], Ioannis K. Argyros [2], Ruben González [1], Á. Alberto Magreñán [3], Lara Orcos [4] and Íñigo Sarría [1,*]

[1]   Escuela Superior de Ingeniería y Tecnología, Universidad Internacional de La Rioja, 26006 Logroño, Spain; cristina.amoros@unir.net (C.A.); ruben.gonzalez@unir.net (R.G.)
[2]   Department of Mathematics Sciences, Cameron University, Lawton, OK 73505, USA; iargyros@cameron.edu
[3]   Departamento de Matemáticas y Computación, Universidad de La Rioja, 26004 Logroño, Spain; angel-alberto.magrenan@unirioja.es
[4]   Facultad de Educación, Universidad Internacional de La Rioja, 26006 Logroño, Spain; lara.orcos@unir.net
*   Correspondence: inigo.sarria@unir.net

Received: 10 December 2018; Accepted: 25 February 2019; Published: 28 February 2019

Abstract: The study of the dynamics and the analysis of local convergence of an iterative method, when approximating a locally unique solution of a nonlinear equation, is presented in this article. We obtain convergence using a center-Lipschitz condition where the ball radii are greater than previous studies. We investigate the dynamics of the method. To validate the theoretical results obtained, a real-world application related to chemistry is provided.

Keywords: high order; sixteenth order convergence method; local convergence; dynamics

---

1. Introduction

A well known problem is that of approximating a locally unique solution x^* of equation

$$F(x) = 0, \tag{1}$$

where F is a differentiable function defined on a nonempty convex subset D of S with values in Ω, where Ω can be $\mathbb{R}$ or $\mathbb{C}$. In this article, we are going to deal with it.

Mathematics is always changing and the way we teach it also changes as it is presented in [1,2]. In the literature [3–8], we can find many problems in engineering and applied sciences that can be solved by finding solutions of equations in a way such as (1). Finding exact solutions for this type of equation is not easy. Only in a few special cases can we find the solutions of these equations in closed form. We must look for other ways to find solutions to these equations. Normally we resort to iterative methods to be able to find solutions. Once we propose to find the solution iteratively, it is mandatory to study the convergence of the method. This convergence is usually seen in two different ways, which gives rise to two different categories, the semilocal convergence analysis and the local convergence analysis. The first of these, the semilocal convergence analysis, is based on information around an initial point, which will provide us with criteria that will ensure the convergence of an iteration procedure. On the other hand, the local convergence analysis is generally based on information about a solution to find values of the calculated radii of the convergence balls. The local results obtained are fundamental since they provide the degree of difficulty to choose the initial points.

We must also deal with the domain of convergence in the study of iterative methods. Normally, the convergence domain is very small and it is necessary to be able to extend this convergence domain without adding any additional hypothesis. Another important problem is finding more accurate estimates of error in distances. $\|x_{n+1} - x_n\|$, $\|x_n - x^*\|$. Therefore, to extend the domain without the need for additional hypotheses and to find more precise estimates of the error committed, in addition to the study of dynamic behavior, will be our objectives in this work.

The iterative methods can be applied to polynomials, and the dynamic properties related to this method will give us important information about its stability and reliability. Recently in some studies, authors such as Amat et al. [9–11], Chun et al. [12], Gutiérrez et al. [13], Magreñán [14–16], and many others [8,13,17–30] have studied interesting dynamic planes, including periodic behavior and other anomalies detected. For all the above, in this article, we are going to study the parameter spaces associated with a family of iterative methods, which will allow us to distinguish between bad and good methods, always speaking in terms of their numerical properties.

We present the dynamics and the local convergence of the four step method defined for each $n = 0, 1, 2, \ldots$ by

$$
\begin{aligned}
y_n &= x_n - \alpha F'(x_n)^{-1} F(x_n) \\
z_n &= y_n - C_1(x_n) F'(x_n)^{-1} F(y_n) \\
v_n &= z_n - C_2(x_n) F'(x_n)^{-1} F(z_n) \\
x_{n+1} &= z_n - C_3(x_n) F'(x_n)^{-1} F(v_n),
\end{aligned}
\tag{2}
$$

where $\alpha \in \mathbb{R}$ is a parameter, x_0 is an initial point and $C_i : \mathbb{R} \to \mathbb{R}$, $i = 1, 2, 3$ are continuous functions given. Numerous methods of more than one step are particular cases of the previous method (2). For example, for certain values of the parameters this family can be reduced to:

- Artidiello et al. method [31]
- Petković et al. method [32]
- Kung-Traub method [29]
- Fourth order King family
- Fourth order method given by Zhao et al. in [33]
- Eighth order method studied by Dzunic et al. [34].

It should be noted that to demonstrate the convergence of all methods after the method (2), in all cases Taylor expansions have been used as well as hypotheses involving derivatives of order greater than one, usually the third derivative or greater. However, in these methods only the first derivative appears. In this article we will perform the analysis of local convergence of the method (2) using hypotheses that involve only the first derivative of the function F. In this way we save the tedious calculation of the successive derivatives (in this case the second and third derivatives) in each step. The order of convergence (COC) is found using and an approximation of the COC (ACOC) using that do not require the usage of derivatives of order higher than one (see Remark 1). Our objective will also be able to provide a computable radius of convergence and error estimates based on the Lipschitz constants.

We must also realize that there are a lot of iterative methods to approximate solutions of nonlinear equations defined in $\mathbb{R}$ or $\mathbb{C}$ [32,35–38]. These studies show that if the initial point x_0 is close enough to the solution x^*, the sequence $\{x_n\}$ converges to x^*. However, from the initial estimate, how close to the solution x^* should it be? In these cases, the local results do not provide us with information about the radius of the convergence ball for the corresponding method. We will approach this question for the method (2) in Section 2. Similarly, we can use the same technique with other different methods.

2. Method's Local Convergence

Let us define, respectively, $U(v, \rho)$ and $\bar{U}(v, \rho)$ as open and closed balls in S, of radius $\rho > 0$ and with center $v \in \Omega$.

To study the analysis of local convergence of the method (2), we are going to define a series of conditions that we will name (C):

(C_1) $F : D \subset \Omega \to \Omega$ is a differentiable function.

We know that exist a constant $x^* \in D$, $L_0 > 0$, such that for each $x \in D$ is fulfilled

(C$_2$) $F(x^*) = 0$, $F'(x^*) \neq 0$.

(C$_3$) $\|F'(x^*)^{-1}(F'(x) - F'(x^*))\| \leq L_0 \|x - x^*\|$

Let $D_0 := D \cap U(x^*, \frac{1}{L_0})$. There exist constants $L > 0$, $M \geq 1$ such that for each $x, y \in D_0$

(C$_4$) $\|F'(x^*)^{-1}(F'(x) - F'(y))\| \leq L\|x - y\|$

(C$_5$) $\|F'(x^*)^{-1}F'(x)\| \leq M$.

There exist parameters γ_i and continuous nondecreasing functions $\psi_i : [0, \gamma_i) \to \mathbb{R}$ such that $i = 0, 1, 2, 3$:

(C$_6$) $\gamma_{i+1} \leq \gamma_i \leq \frac{1}{L_0}$

and

(C$_7$) $\psi_i(t) \to a +\infty$ or a number greater than 0 as $t \to \gamma_i^{-1}$. For $\alpha \in \mathbb{R}$, consider the functions

$$q_j : [0, \gamma_j) \to \mathbb{R} \quad j = 0, 1, 2, 3 \text{ by}$$

$$q_j(t) = \begin{cases} M|1 - \alpha|, & j = 0 \\ \\ M^{i+j}|1 - \alpha| \prod_{i=0}^{j} \psi_1(t) \cdots \psi_j(t), & j = 1, 2, 3 \end{cases}$$

(C$_8$) $p_j := q_j(0) < 1, \quad j = 0, 1, 2, 3,$

(C$_9$) $C_i : \Omega \to \Omega$ are continuous functions such that for each $x \in D_0$, $\|C_i(x)\| \leq \psi_i(\|x - x^*\|)$ and

(C$_{10}$) $\bar{U}(x^*, r) \subset D$ for some $r > 0$ to be appointed subsequently.

We are going to introduce some parameters and some functions for the local convergence analysis of the method (2). We define the function g_0 on the interval $[0, \frac{1}{L_0})$ by

$$g_0(t) = \frac{1}{2(1 - L_0 t)}(Lt + 2M|1 - \alpha|)$$

and parameters r_0, ϱ_A by

$$r_0 = \frac{2(1 - M|1 - \alpha|)}{2L_0 + L}, \quad \varrho_A = \frac{2}{2L_0 + L}.$$

Then, since $p_0 = M|1 - \alpha| < 1$ by (C$_8$), we have that $0 < r_0 < \varrho_A$, $g_0(r_1) = 1$ and for each $t \in [0, r_1)$ $0 \leq g_0(t) < 1$. Define functions g_i, h_i on the interval $[0, \gamma_i)$ by

$$g_i(t) = (1 + \frac{M\psi_i(t)}{1 - L_0 t})g_{i-1}(t)$$

and

$$h_i(t) = g_i(t) - 1$$

for $i = 1, 2, 3$. We have by (C$_8$) that $h_i(0) = p_i - 1 < 0$ and by (C$_6$) and (C$_7$) $h_i(t) \to$ a positive number or $+\infty$. Applying the intermediate value theorem, we know that functions h_i have zeros in the interval $[0, \gamma_i)$. Denote by r_i the smallest such zero. Set

$$r = \min\{r_j\}, \quad j = 0, 1, 2, 3. \tag{3}$$

Therefore, we can write that

$$0 \leq r < r_A \tag{4}$$

moreover for each $j = 0, 1, 2, 3, t \in [0, r)$

$$0 \leq g_j(t) < 1. \tag{5}$$

Now, making use of the conditions (C) and the previous notation, we will show the results of local convergence for the method (2).

Theorem 1. *Let us assume that (C) conditions hold, if we take the radius r in (C_{10}) that has been defined previously. Then, the sequence $\{x_n\}$ generated by our method (2) and considering $x_0 \in U(x^*, r) \setminus \{x^*\}$ is well defined, remains in the ball $U(x^*, r)$ for each $n \geq 0$ and converges to the solution x^*. On the other hand, we see that the estimates are true:*

$$\|y_n - x^*\| \leq g_0(\|x_n - x^*\|)\|x_n - x^*\| < \|x_n - x^*\| < r, \tag{6}$$

$$\|z_n - x^*\| \leq g_1(\|x_n - x^*\|)\|x_n - x^*\| < \|x_n - x^*\|, \tag{7}$$

$$\|v_n - x^*\| \leq g_2(\|x_n - x^*\|)\|x_n - x^*\| < \|x_n - x^*\| \tag{8}$$

and

$$\|x_{n+1} - x^*\| \leq g_3(\|x_n - x^*\|)\|x_n - x^*\| < \|x_n - x^*\|, \tag{9}$$

where the "g" functions are defined previously. Furthermore, for

$$T \in [r, \frac{2}{L_0}) \tag{10}$$

the unique solution of equation $F(x) = 0$ in $\bar{U}(x^*, T) \cap D$ is the bound point x^*.

Proof. Using mathematical induction we shall prove estimates (6) and (10). By hypothesis $x_0 \in U(x, r) \setminus \{x^*\}$, the conditions (C_1), (C_3) and (3), we have that

$$\|F'(x^*)^{-1}(F'(x_0) - F'(x^*))\| \leq L_0\|x_0 - x^*\| < L_0 r < 1. \tag{11}$$

Taking into account the Banach lemma on invertible functions [5,7,39] we can write that $F'(x_0)^{-1} \in L(S, S)$ and

$$\|F'(x_0)^{-1}F'(x^*)\| \leq \frac{1}{1 - L_0\|x_0 - x^*\|}. \tag{12}$$

consequently, y_0 is well defined by the first substep of the method (2) for $n = 0$. We can set using the conditions (C_1) and (C_2) that

$$F(x_0) = F(x_0) - F(x^*) = \int_0^1 F'(x^* + \theta(x_0 - x^*))(x_0 - x^*)d\theta. \tag{13}$$

Remark that $\|x^* + \theta(x_0 - x^*) - x^*\| = \theta\|x_0 - x^*\| < r$, so $x^* + \theta(x_0 - x^*) \in U(x^*, r)$. Then, using (13) and condition (C_5), we have that

$$\|F'(x^*)^{-1}F(x_0)\| \leq \|\int_0^1 F'(x^*)^{-1}F'(x^* + \theta(x_0 - x^*))(x_0 - x^*)d\theta\| \leq M\|x_0 - x^*\|. \tag{14}$$

In view of conditions (C_2), (C_4), (3) and (5) (for $j = 0$) and (12) and (14), we obtain that

$$
\begin{aligned}
\|y_0 - x^*\| &= \|x_0 - x^* - F'(x_0)^{-1}F(x_0) + (1-\alpha)F'(x_0)^{-1}F(x_0)\| \\
&\leq \|x_0 - x^* - F'(x_0)^{-1}F(x_0)\| + |1-\alpha|\|F(x_0)^{-1}F'(x^*)\|\|F'(x^*)^{-1}F(x_0)\| \\
&\leq \|F(x_0)^{-1}F'(x^*)\|\|\int_0^1 F'(x^*)^{-1}(F'(x^* + \theta(x_0 - x^*)) - F'(x_0))(x_0 - x^*)d\theta \\
&\quad + \frac{|1-\alpha|M\|x_0 - x^*\|}{1 - L_0\|x_0 - x^*\|} \\
&\leq \frac{L\|x_0 - x^*\|^2}{2(1 - L_0\|x_0 - x^*\|)} + \frac{|1-\alpha|M\|x_0 - x^*\|}{1 - L_0\|x_0 - x^*\|} \\
&= g_0(\|x_0 - x^*\|)\|x_0 - x^*\| < \|x_0 - x^*\| < r,
\end{aligned}
$$

(15)

which evidences (6) for $n = 0$ and $y_0 \in U(x^*, r)$. Then, applying (C_9) condition, (3) and (5) (for $j = 1$), (12) and (14) (for $y_0 = x_0$) and (15), we achieve that

$$
\|z_0 - x^*\| \leq g_1(\|x_0 - x^*\|)\|x_0 - x^*\| \leq \|x_0 - x^*\|,
$$

(16)

which displays (7) for $n = 0$ and $z_0 \in U(x^*, r)$. In the same way, we show estimates (8) and (9) for $n = 0$ and $v_0, x_1 \in U(x^*, r)$. Just substituting x_0, y_0, z_0, v_0, x_1 by $x_k, y_k, z_k, v_k, x_{k+1}$ in the preceding estimates, we deduct that (6)–(9). Using the estimates $\|x_{k+1} - x^*\| \leq c\|x_k - x^*\| < r, c = g_3(\|x_0 - x^*\|) \in [0,1)$, we arrive at $\lim_{k \to \infty} x_k = x^*$ and $x_{k+1} \in U(x^*, r)$. We have yet to see the uniqueness, let $y^* \in \bar{U}(x^*, T)$ be such that $F(y^*) = 0$. Define $B = \int_0^1 F'(y^* + \theta(x^* - y^*))d\theta$. Taking into account the condition (C_2), we obtain that

$$
\|F'(x^*)^{-1}(B - F'(x^*))\| \leq \frac{L_0}{2}\|y^* - x^*\| \leq \frac{L_0}{2}T < 1.
$$

(17)

Hence, $B \neq 0$. Using the identity $0 = F(y^*) - F(x^*) = B(y^* - x^*)$, we can deduct that $x^* = y^*$. $\square$

Remark 1.

1. *Considering* (10) *and the next value*

$$
\begin{aligned}
\|F'(x^*)^{-1}F'(x)\| &= \|F'(x^*)^{-1}(I + F'(x) - F'(x^*))\| \\
&\leq \|F'(x^*)^{-1}(F'(x) - F'(x^*))\| + 1 \\
&\leq L_0\|x_0 - x^*\| + 1
\end{aligned}
$$

    we can clearly eliminate the condition (10) *and M can be turned into*

$$
M(t) = 1 + L_0 t \text{ or what is the same } M(t) = M = 2, \text{ because } t \in [0, \frac{1}{L_0}).
$$

2. *The results that we have seen, can also be applied for F operators that satisfy the autonomous differential equation* [5,7] *of the form*

$$
F'(x) = P(F(x)),
$$

    where P is a known continuous operator. As $F'(x^*) = P(F(x^*)) = P(0)$, *we are able to use the previous results without needing to know the solution* x^*. *Take for example* $F(x) = e^x - 1$. *Now, we can take* $P(x) = x + 1$. *However, we do not know the solution.*

3. In the articles [5,7] was shown that the radius ϱ_A has to be the convergence radius for Newton's method using (10) and (11) conditions. If we apply the definition of r_1 and the estimates (8), the convergence radius r of the method (2) it can no be bigger than the convergence radius ϱ_A of the second order Newton's method. The convergence ball given by Rheinboldt [8] is

$$\varrho_R = \tfrac{2}{3L_1}. \tag{18}$$

In particular, for $L_0 < L_1$ or $L < L_1$ we have that

$$\varrho_R < \varrho_A$$

and

$$\frac{\varrho_R}{\varrho_A} \to \frac{1}{3} \quad as \quad \frac{L_0}{L_1} \to 0.$$

That is our convergence ball r_1 which is maximum three times bigger than Rheinboldt's. The precise amount given by Traub in [28] for ϱ_R.

4. We should note that family (3) stays the same if we use the conditions of Theorem 1 instead of the stronger conditions given in [15,36]. Concerning, for the error bounds in practice we can use the approximate computational order of convergence (ACOC) [36]

$$\zeta = \frac{\ln\frac{\|x_{n+2}-x_{n+1}\|}{\|x_{n+1}-x_n\|}}{\ln\frac{\|x_{n+1}-x_n\|}{\|x_n-x_{n-1}\|}}, \quad for\ each\ n = 1, 2, \ldots$$

or the computational order of convergence (COC) [40]

$$\zeta^* = \frac{\ln\frac{\|x_{n+2}-x^*\|}{\|x_{n+1}-x^*\|}}{\ln\frac{\|x_{n+1}-x^*\|}{\|x_n-x^*\|}}, \quad for\ each\ n = 0, 1, 2, \ldots$$

And these order of convergence do not require higher estimates than the first Fréchet derivative used in [19,23,32,33,41].

Remark 2. *Let's see how we can choose the functions in the case of the method* (2). *In this case we have that*

$$\overline{C_1}\left(\frac{F(y_n)}{F(x_n)}\right) = C_1(x_n), \quad \overline{C_2}\left(\frac{F(y_n)}{F(x_n)}, \frac{F(z_n)}{F(y_n)}\right) = C_2(x_n), \quad \overline{C_3}\left(\frac{F(y_n)}{F(x_n)}, \frac{F(z_n)}{F(y_n)}, \frac{F(v_n)}{F(z_n)}\right) = C_3(x_n)$$

To begin, the condition (C_3) can be eliminated because in this case we have $\alpha = 1$. Then, if $x_n \neq x^*$, the following inequality holds

$$\|(F'(x^*)(x_n - x^*))^{-1}[F(x_n) - F(x^*) - F'(x^*)(x_n - x^*)]\|$$

$$\leq \|x_n - x^*\|^{-1}\frac{L_0}{2}\|x_n - x^*\| = \frac{L_0}{2}\|x_n - x^*\| < \frac{L_0}{2}r < 1.$$

Hence, we have that

$$\|F'(x_n)^{-1}F(x^*)\| \leq \frac{1}{\|x_n - x^*\|(1 - \frac{L_0}{2}\|x_n - x^*\|)}.$$

Consequently, we get that

$$\left\|\frac{F(y_n)}{F(x_n)}\right\| = \|F'(x_n)^{-1}F'(x^*)\|\|F'(x^*)^{-1}F(y_n)\|$$

$$\leq \frac{M\|y_n - x^*\|}{\|x_n - x^*\|(1 - \frac{L_0}{2}\|x_0 - x^*\|)} \tag{19}$$

$$\leq \frac{Mg_0(\|x_n - x^*\|)^2}{1 - L_0\|x_n - x^*\|}.$$

Similarly, we obtain

$$\|F(y_n)^{-1}F'(x^*)\| \leq \frac{1}{\|y_n - x^*\|(1 - \frac{L_0}{2}\|y_n - x^*\|)},$$

$$\left\|\frac{F(z_n)}{F(y_n)}\right\| \leq \frac{M(1 + \frac{M\psi_1(\|x_n - x^*\|)}{1 - L_0\|x_n - x^*\|})}{1 - \frac{L_0}{2}g_0(\|x_n - x^*\|)\|x_n - x^*\|}, \tag{20}$$

$$\|F(z_n)^{-1}F'(x^*)\| \leq \frac{1}{\|z_n - x^*\|(1 - \frac{L_0}{2}\|y_n - x^*\|)},$$

and

$$\left\|\frac{F(z_n)}{F(y_n)}\right\| \leq \frac{M(1 + \frac{M\psi_2(\|x_n - x^*\|)}{1 - L_0\|x_n - x^*\|})}{1 - \frac{L_0}{2}g_0(\|x_n - x^*\|)\|x_n - x^*\|}, \tag{21}$$

Let us choose C_i, $i = 1, 2, 3, 4$ as in [31]:

$$C_1(a) = 1 + 2a + 4a^3 - 3a^4 \tag{22}$$

$$C_2(a, b) = 1 + 2a + b + a^2 + 4ab + 3a^2b + 4ab^2 + 4a^3b - 4a^2b^2 \tag{23}$$

and

$$C_3(a, b, c) = 1 + 2a + b + c + a^2 + 4ab + 2ac + 4a^2b + a^2c + 6ab^2 + 8abc - b^3 + 2bc. \tag{24}$$

As these functions, they fulfill the terms imposed in Theorem 1 in [31], So, we have that the order of convergence of the method (2) has to reach at least order 16.

Set

$$a = a(t) = \frac{Mg_0(t)}{1 - L_0t}, \tag{25}$$

$$b = b(t) = \frac{M(1 + \frac{M\psi_1(t)}{1 - L_0t})}{1 - \frac{L_0}{2}t}, \tag{26}$$

$$c = c(t) = \frac{M(1 + \frac{M\psi_2(t)}{1 - L_0t})}{1 - \frac{L_0}{2}t}, \tag{27}$$

and

$$\gamma_i = \frac{1}{L_0}, \quad i = 0, 1, 2, 3.$$

Then it follows from (19)–(24) that functions ψ_i can be defined by

$$\psi_1(t) = 1 + 2a + 4a^3 + 3a^4 \tag{28}$$

$$\psi_2(t) = 1 + 2a + b + a^2 + 4ab + 3a^2b + 4ab^2 + 4a^3b + 4a^2b^2 \tag{29}$$

and

$$\psi_3(t) = 1 + 2a + b + c + a^2 + 4ab + 2ac + 4a^2b + a^2c + 6ab^2 + 8abc + b^3 + 2bc. \tag{30}$$

3. Dynamical Study of a Special Case of the Family (2)

In this article, the concepts of critical point, fixed point, strange fixed point, attraction basins, parameter planes and convergence planes are going to be assumed. We refer the reader to see [5,7,16,38] to recall the basic dynamical concepts.

In this third section we will study the complex dynamics of a particular case of the method (2), which consists in select:

$$C_1(x_n) = F'(y_n)^{-1}F'(x_n),$$

$$C_2(x_n) = F'(z_n)^{-1}F'(x_n)$$

and

$$C_3(x_n) = F'(y_n)^{-1}F'(x_n).$$

Let be a polynomial of degree two with two roots, that they are not the same. If we apply this operator on the previous polynomial and using the Möebius map $h(z) = \frac{z-A}{z-B}$, we obtain

$$G(z, \alpha) = \frac{z^8(1 - \alpha + z)^8}{(-1 - z + \alpha z)^8}. \tag{31}$$

The fixed points of this operator are:

- 0
- ∞
- And 15 more, which are:

  – 1 (related to original ∞).
  – The roots of a 14 degree polynomial.

In Figure 1 the bifurcation diagram of all fixed points, extraneous or not, is presented.

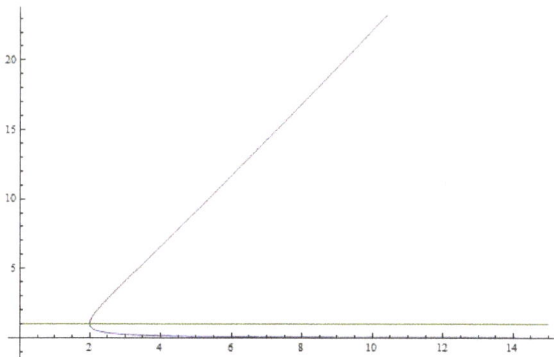

Figure 1. Fixed points's bifurcation diagram.

Now, we are going to compute the critical points, i.e., the roots of

$$G'(z, \alpha) = -\frac{8(-1+\alpha-z)^7 z^7 \left(-1+\alpha-2z-z^2+\alpha z^2\right)}{(-1-z+\alpha z)^9}$$

The free critical points are: $cp_1(\alpha) = -1 + \alpha$, $cp_2(\alpha) = \frac{1-\sqrt{-(-2+\alpha)\alpha}}{-1+\alpha}$ and $cp_3(\alpha) = \frac{1+\sqrt{-(-2+\alpha)\alpha}}{-1+\alpha}$.
We also have the following results.

Lemma 1.

(a) *If $\alpha = 0$*

   (i) $cp_1(\alpha) = cp_2(\alpha) = cp_3(\alpha) = -1$.

(b) *If $\alpha = 2$*

   (i) $cp_1(\alpha) = cp_2(\alpha) = cp_3(\alpha) = 1$.

You can easily verify that for every value of α we have to $cp_2(\alpha) = \frac{1}{cp_3(\alpha)}$

It is easy to see that there is only one independent critical point. So, we assume that $cp_2(\alpha)$ is the only free critical point without loss of generality. Taking $cp_2(\alpha)$, we perform the study of the parameter space associated with the free critical point. This will allow us to find the some members of the family, and we want to stay with the best members.

We are going to show different planes of parameters. In Figure 2 we show the parameter spaces associated to critical point $cp_2(\alpha)$. Now let us paint a point of cyan if the iteration of the method starting in $z_0 = cp_1(\alpha)$ converges to the fixed point 0 (related to root A) or if it converges to ∞ (allied to root B). That is, the points relative to the roots of the quadratic polynomial will be painted cyan and a point is painted in yellow if the iteration converges to 1 (related to ∞). Therefore, all convergence will be painted cyan. On the other hand, convergence to strange fixed points or cycles appears in other colors. As an immediate consequence, all points of the plane that are not cyan are not a good choice of α in terms of numerical behavior.

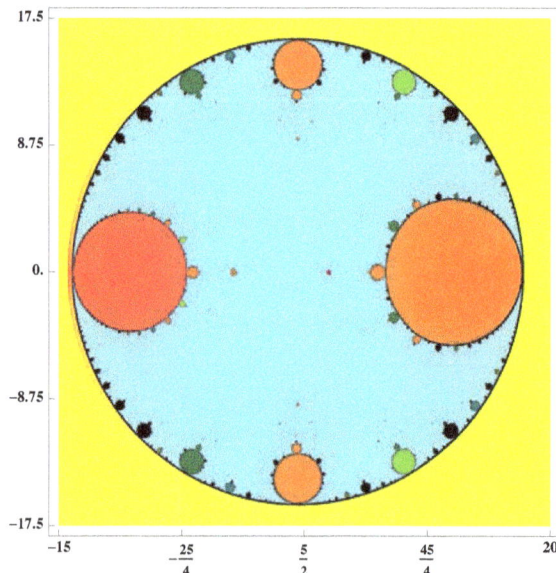

Figure 2. Parameter space of the free critical point $cp_2(\alpha)$.

Once we have detected the anomalies, we can go on to describe the dynamic planes. To understand the colors we have used in these dynamic planes, we have to indicate that if after a maximum of 1000 iterations and with a tolerance of 10^{-6} convergence has not been achieved to the roots, we have painted in black. Conversely, we colored in magenta the convergence to 0 and colored in cyan the convergence to ∞. Then, the cyan or magenta regions identify the convergence.

If we focus our attention on the region shown in Figure 2, it is clear that there are family members with complicated behaviors. We will also show dynamic planes in Figures 3 and 4, of a family member with convergence regions to any of the strange fixed points.

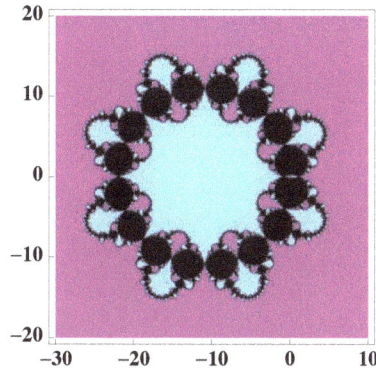

Figure 3. Attraction basins associated to $\alpha = -10$.

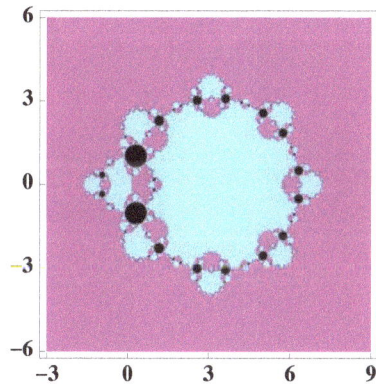

Figure 4. Attraction basins associated to $\alpha = 4.25$.

In the following figures, we will show the dynamic planes of family members with convergence to different attracting n-cycles. For example, in the Figures 5 and 6, we see the dynamic planes to an attracting 2-cycle and in the Figure 7 the dynamic plane of family members with convergence to an attracting 3-cycle that was painted in green in the parameter planes.

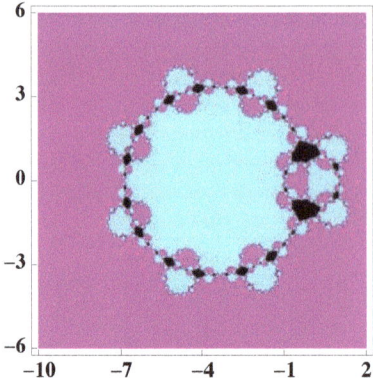

Figure 5. Attraction basins associated to $\alpha = -2.5$.

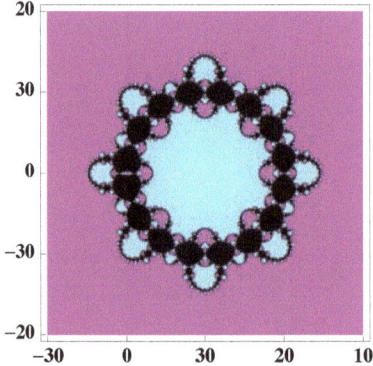

Figure 6. Attraction basins associated to $\alpha = 11$.

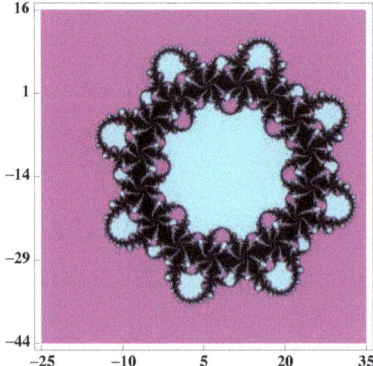

Figure 7. Attraction basins associated to $\alpha = 10 - 13i$.

Other particular cases are shown in Figures 8 and 9. The basins of attraction for different α values in which we see the convergence to the roots of the method can be seen.

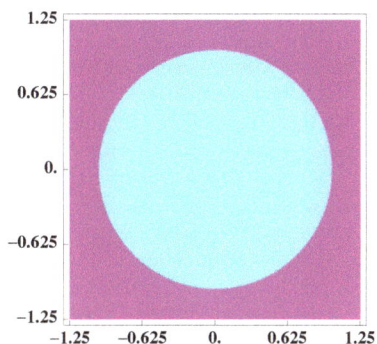

Figure 8. Attraction basins associated to $\alpha = 0.5$.

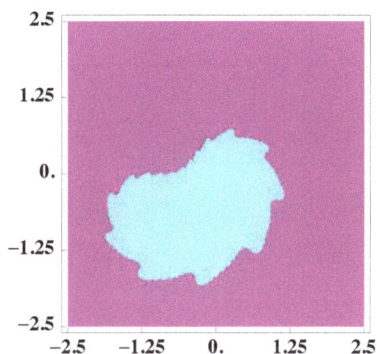

Figure 9. Attraction basins associated to $\alpha = -0.5i$.

4. Example Applied

Next, we want to show the applicability of the theoretical part previously seen in a real problem. Chemistry is a discipline in which many equations are handled. In this concrete case, let us consider the quartic equation that can describe the fraction or amount of the nitrogen-hydrogen feed that is turned into ammonia, which is known as fractional conversion and is shown in [42,43].

If the pressure is 250 atm. and the temperature reaches a value of 500 °C, the previous equation reduces to: $g(x) = x^4 - 7.79075x^3 + 14.7445x^2 + 2.511x - 1.674$. We define S as all real line, D as the interval $[0, 1]$ and $\xi = 0$. We consider the function F defined on D. If we now take the functions $\psi_i(t)$ with $i = 1, 2, 3$ and choosing the value of as $\alpha = 1.025$, we obtain: $L_0 = 2.594\ldots$, $L = 3.282\ldots$. It is clear that in this case $L_0 < L$, so we improve the results. Now, we compute $M = 1.441\ldots$. Additionally, computing the zeros of the functions previously defined, we get: $r_0 = 0.227\ldots$, $\varrho_A = 0.236\ldots$, $r_1 = 0.082\ldots$, $r_2 = 0.155\ldots$, $r_3 = 0.245\ldots$, and as a result of it we get $r = r_1 = 0.082\ldots$. Then we can guarantee that the method (2) converges for $\alpha = 1.025$ due to Theorem 1. The applicability of our family of methods is thus proven.

Author Contributions: All authors have contributed equally in writing this article. All authors read and approved the final manuscript

Funding: This research was funded by Programa de Apoyo a la investigaciń de la fundaciń Séneca-Agencia de Ciencia y Tecnología de la Región de Murcia19374/PI/14' and by the project MTM2014-52016-C2-1-P of the Spanish Ministry of Science and Innovation.

Conflicts of Interest: The authors declare no conflict of interest

References

1. Tello, J.I.C.; Orcos, L.; Granados, J.J.R. Virtual forums as a learning method in Industrial Engineering Organization. *IEEE Latin Am. Trans.* **2016**, *14*, 3023–3028. [CrossRef]
2. LeTendre, G.; McGinnis, E.; Mitra, D.; Montgomery, R.; Pendola, A. The American Journal of Education: Challenges and opportunities in translational science and the grey area of academic. *Rev. Esp. Pedag.* **2018**, *76*, 413–435. [CrossRef]
3. Argyros, I.K.; González, D. Local convergence for an improved Jarratt–type method in Banach space. *Int. J. Interact. Multimed. Artif. Intell.* **2015**, *3*, 20–25. [CrossRef]
4. Argyros, I.K.; George, S. Ball convergence for Steffensen–type fourth-order methods. *Int. J. Interact. Multimed. Artif. Intell.* **2015**, *3*, 27–42. [CrossRef]
5. Argyros, I.K.; Magreñán, Á.A. *Iterative Methods and Their Dynamics with Applications: A Contemporary Study*; CRC-Press: Boca Raton, FL, USA, 2017.
6. Behl, R.; Sarría, Í.; González-Crespo, R.; Magreñán, Á.A. Highly efficient family of iterative methods for solving nonlinear models. *J. Comput. Appl. Math.* **2019**, *346*, 110–132. [CrossRef]
7. Magreñán, Á.A.; Argyros, I.K. *A Contemporary Study of Iterative Methods: Convergence, Dynamics and Applications*; Elsevier: Amsterdam, The Netherlands, 2018.
8. Rheinboldt, W.C. An adaptive continuation process for solving systems of nonlinear equations. *Pol. Acad. Sci.* **1978**, *3*, 129–142. [CrossRef]
9. Amat, S.; Busquier, S.; Plaza, S. Dynamics of the King and Jarratt iterations. *Aequ. Math.* **2005**, *69*, 212–223. [CrossRef]
10. Amat, S.; Busquier, S.; Plaza, S. Chaotic dynamics of a third-order Newton-type method. *J. Math. Anal. Appl.* **2010**, *366*, 24–32. [CrossRef]
11. Amat, S.; Hernández, M.A.; Romero, N. A modified Chebyshev's iterative method with at least sixth order of convergence. *Appl. Math. Comput.* **2008**, *206*, 164–174. [CrossRef]
12. Chicharro, F.; Cordero, A.; Torregrosa, J.R. Drawing dynamical and parameters planes of iterative families and methods. *Sci. World J.* **2013**, *2013*, 780153. [CrossRef] [PubMed]
13. Gutiérrez, J.M.; Hernández, M.A. Recurrence relations for the super-Halley method. *Comput. Math. Appl.* **1998**, *36*, 1–8. [CrossRef]
14. Kou, J.; Li, Y. An improvement of the Jarratt method. *Appl. Math. Comput.* **2007**, *189*, 1816–1821. [CrossRef]
15. Li, D.; Liu, P.; Kou, J. An improvement of the Chebyshev-Halley methods free from second derivative. *Appl. Math. Comput.* **2014**, *235*, 221–225. [CrossRef]
16. Magreñán, Á.A. Different anomalies in a Jarratt family of iterative root-finding methods. *Appl. Math. Comput.* **2014**, *233*, 29–38.
17. Budzko, D.; Cordero, A.; Torregrosa, J.R. A new family of iterative methods widening areas of convergence. *Appl. Math. Comput.* **2015**, *252*, 405–417. [CrossRef]
18. Bruns, D.D.; Bailey, J.E. Nonlinear feedback control for operating a nonisothermal CSTR near an unstable steady state. *Chem. Eng. Sci.* **1977**, *32*, 257–264. [CrossRef]
19. Candela, V.; Marquina, A. Recurrence relations for rational cubic methods I: The Halley method. *Computing* **1990**, *44*, 169–184. [CrossRef]
20. Candela, V.; Marquina, A. Recurrence relations for rational cubic methods II: The Chebyshev method. *Computing* **1990**, *45*, 355–367. [CrossRef]
21. Ezquerro, J.A.; Hernández, M.A. New iterations of R-order four with reduced computational cost. *BIT Numer. Math.* **2009**, *49*, 325–342. [CrossRef]
22. Ezquerro, J.A.; Hernández, M.A. On the R-order of the Halley method. *J. Math. Anal. Appl.* **2005**, *303*, 591–601. [CrossRef]
23. Ganesh, M.; Joshi, M.C. Numerical solvability of Hammerstein integral equations of mixed type. *IMA J. Numer. Anal.* **1991**, *11*, 21–31. [CrossRef]
24. Hernández, M.A. Chebyshev's approximation algorithms and applications. *Comput. Math. Appl.* **2001**, *41*, 433–455. [CrossRef]
25. Hernández, M.A.; Salanova, M.A. Sufficient conditions for semilocal convergence of a fourth order multipoint iterative method for solving equations in Banach spaces. *Southwest J. Pure Appl. Math.* **1999**, *1*, 29–40.

26. Jarratt, P. Some fourth order multipoint methods for solving equations. *Math. Comput.* **1966**, *20*, 434–437. [CrossRef]
27. Ren, H.; Wu, Q.; Bi, W. New variants of Jarratt method with sixth-order convergence. *Numer. Algorithms* **2009**, *52*, 585–603. [CrossRef]
28. Traub, J.F. *Iterative Methods for the Solution of Equations*; Prentice-Hall Series in Automatic Computation: Englewood Cliffs, NJ, USA, 1964.
29. Wang, X.; Kou, J.; Gu, C. Semilocal convergence of a sixth-order Jarratt method in Banach spaces. *Numer. Algorithms* **2011**, *57*, 441–456. [CrossRef]
30. Cordero, A.; Torregrosa, J.R.; Vindel, P. Dynamics of a family of Chebyshev-Halley type methods. *Appl. Math. Comput.* **2013**, *219*, 8568–8583. [CrossRef]
31. Artidiello, S.; Cordero, A.; Torregrosa, J.R.; Vassileva, M.P. Optimal high order methods for solving nonlinear equations. *J. Appl. Math.* **2014**, *2014*, 591638. [CrossRef]
32. Petković, M.; Neta, B.; Petković, L.; Džunić, J. *Multipoint Methods for Solving Nonlinear Equations*; Elsevier: Amsterdam, The Netherlands, 2013.
33. Zhao, L.; Wang, X.; Guo, W. New families of eighth-order methods with high efficiency index for solving nonlinear equations. *Wseas Trans. Math.* **2012**, *11*, 283–293.
34. Džunić, J.; Petković, M. A family of Three-Point methods of Ostrowski's Type for Solving Nonlinear Equations. *J. Appl. Math.* **2012**, *2012*, 425867. [CrossRef]
35. Chun, C. Some improvements of Jarratt's method with sixth-order convergence. *Appl. Math. Comput.* **1990**, *190*, 1432–1437. [CrossRef]
36. Cordero, A.; Torregrosa, J.R. Variants of Newton's method using fifth-order quadrature formulas. *Appl. Math. Comput.* **2007**, *190*, 686–698. [CrossRef]
37. Ezquerro, J.A.; Hernández, M.A. Recurrence relations for Chebyshev-type methods. *Appl. Math. Optim.* **2000**, *41*, 227–236. [CrossRef]
38. Magreñán, Á.A. A new tool to study real dynamics: The convergence plane. *Appl. Math. Comput.* **2014**, *248*, 215–224. [CrossRef]
39. Rall, L.B. *Computational Solution of Nonlinear Operator Equations*; Robert E. Krieger: New York, NY, USA, 1979.
40. Weerakon, S.; Fernando, T.G.I. A variant of Newton's method with accelerated third-order convergence. *Appl. Math. Lett.* **2000**, *13*, 87–93. [CrossRef]
41. Cordero, A.; García-Maimó, J.; Torregrosa, J.R.; Vassileva, M.P.; Vindel, P. Chaos in King's iterative family. *Appl. Math. Lett.* **2013**, *26*, 842–848. [CrossRef]
42. Gopalan, V.B.; Seader, J.D. Application of interval Newton's method to chemical engineering problems. *Reliab. Comput.* **1995**, *1*, 215–223.
43. Shacham, M. An improved memory method for the solution of a nonlinear equation. *Chem. Eng. Sci.* **1989**, *44*, 1495–1501. [CrossRef]

mathematics

MDPI

Article

On the Semilocal Convergence of the Multi–Point Variant of Jarratt Method: Unbounded Third Derivative Case

Zhang Yong [1,*], **Neha Gupta** [2], **J. P. Jaiswal** [2,*] **and Kalyanasundaram Madhu** [3]

[1] School of Mathematics and Physics, Changzhou University, Changzhou 213164, China
[2] Department of Mathematics, Maulana Azad National Institute of Technology, Bhopal 462003, India; neha.gupta.mh@gmail.com
[3] Department of Mathematics, Saveetha Engineering College, Chennai 602105, India; kalyan742@pec.edu
[*] Correspondence: yzhangcczu@aliyun.com (Z.Y.); asstprofjpmanit@gmail.com (J.P.J.)

Received: 10 May 2019; Accepted: 10 June 2019; Published: 13 June 2019

Abstract: In this paper, we study the semilocal convergence of the multi-point variant of Jarratt method under two different mild situations. The first one is the assumption that just a second-order Fréchet derivative is bounded instead of third-order. In addition, in the next one, the bound of the norm of the third order Fréchet derivative is assumed at initial iterate rather than supposing it on the domain of the nonlinear operator and it also satisfies the local ω-continuity condition in order to prove the convergence, existence-uniqueness followed by a priori error bound. During the study, it is noted that some norms and functions have to recalculate and its significance can be also seen in the numerical section.

Keywords: Banach space; semilocal convergence; ω-continuity condition; Jarratt method; error bound

MSC: 65J15; 65H10; 65G99; 47J25

1. Introduction

The problem of finding a solution of the nonlinear equation affects a large area of various fields. For instance, kinetic theory of gases, elasticity, applied mathematics and also engineering dynamic systems are mathematically modeled by difference or differential equations. Likewise, there are numerous problems in the field of medical, science, applied mathematics and engineering that can be reduced in the form of a nonlinear equation. Many of those problems cannot be solved directly through any of the methods. For this, we opt for numerical procedure and are able to find at least an approximate solution of the problem using various iterative methods. In this concern, Newton's method [1] is one of the best and most renowned quadratically convergent iterative methods in Banach spaces, which is frequently used by the authors as it is an efficient method and has a smooth execution. Now, consider a nonlinear equation having the form

$$L(m) = 0, \tag{1}$$

where L is a nonlinear operator defined as $L : B \subseteq \nabla_1 \to \nabla_2$, where B is a non-empty open convex domain of a Banach space ∇_1 with values in a Banach space ∇_2 which is usually known as the Newton–Kantorovich method that can be defined as

$$\begin{cases} m_0 \ given \ in \ B, \\ m_n = m_{n-1} - [L'(m_{n-1})]^{-1}L(m_{n-1}), \ n \in \mathbb{N}, \end{cases}$$

where $L'(m_{n-1})$ is the Fréchet derivative of L at m_{n-1}. The results on semilocal convergence have been originally studied by L.V. Kantorovich in [2]. In the early stages, he gave the method of recurrence relations and afterwards described the method of majorant principle. Subsequently, Rall in [3] and many researchers have studied the improvements of the results based on recurrence relations. A large number of researchers studied iterative methods of various order to solve the nonlinear equations extensively. The convergence of iterative methods generally relies on two types: semilocal and local convergence analysis. In the former type, the convergence of iterative methods depends upon the information available around the starting point, whereas, in the latter one, it depends on the information around the given solution.

In the literature, researchers have developed various higher order schemes in order to get better efficiency and also discussed their convergence. Various types of convergence analysis using different types of continuity conditions viz. Lipschitz continuity condition has been studied by Wang et al. in [4,5], Singh et al. in [6], and Jaiswal in [7], to name a few. Subsequently, many authors have studied the weaker continuity condition than Lipschitz namely Hölder by Hernández in [8], Parida and Gupta in [9,10], Wang and Kou in [11] are some of them. Usually, there are some nonlinear equations that neither satisfy Lipschitz nor Hölder continuity conditions; then, we need a generalized form of continuity condition such as ω-continuity, which has been studied by Ezquerro and Hernández in [12,13], Parida and Gupta in [14,15], Prashanth and Gupta in [16,17], Wang and Kou in [18–20], etc.

The algorithms having higher order of convergence plays an important role where the quick convergence is required like in the stiff system of equations. Thus, it is quite interesting to study higher order methods. In this article, we target our study on the semilocal convergence analysis using recurrence relations technique on the multi-point variant of Jarratt method when the third order Fréchet derivative becomes unbounded in the given domain.

2. The Method and Some Preliminary Results

Throughout the paper, we use the below mentioned notations:

$B \equiv$ non-empty open subset of ∇_1; $B_0 \subseteq B$ is a non-empty convex subset; ∇_1, $\nabla_2 \equiv$ Banach spaces, $U(m,b) = \{n \in \nabla_1 : \|n - m\| < b\}$, $\overline{U(m,b)} = \{n \in \nabla_1 : \|n - m\| \le b\}$.

Here, we consider the multi-point variant of the Jarratt method suggested in [21]

$$
\begin{aligned}
n_n &= m_n + \tfrac{2}{3}(p_n - m_n), \\
o_n &= m_n - Y_L(m_n)\wp_n L(m_n), \\
m_{n+1} &= o_n - \left[\tfrac{3}{2}L'(n_n)^{-1}Y_L(m_n) + \wp_n\left(I - \tfrac{3}{2}Y_L(m_n)\right)\right]L(o_n),
\end{aligned}
\tag{2}
$$

where $Y_L(m_n) = [6L'(n_n) - 2L'(m_n)]^{-1}[3L'(n_n) + L'(m_n)]$, $\wp_n = [L'(m_n)]^{-1}$, $p_n = m_n - \wp_n L(m_n)$ and I is the identity operator. In the same article for deriving semilocal convergence results, the researchers have assumed the following hypotheses:

$(A1) \|\wp_0 L(m_0)\| \le \kappa,$
$(A2) \|\wp_0\| \le \lambda,$
$(A3) \|L''(m)\| \le P, m \in B,$
$(A4) \|L'''(m)\| \le Q, m \in B,$
$(A5) \|L'''(m) - L'''(n)\| \le \omega(\|m - n\|), \ \forall m, n \in B,$

where $\omega : R_+ \to R_+$, is a continuous and non-decreasing function for $m > 0$ such that $\omega(m) \ge 0$ and satisfying $\omega(\epsilon z) \le \phi(\epsilon)\omega(z), \epsilon \in [0,1]$ and $z \in [0, +\infty)$ with $\phi : [0,1] \to R_+$, is also continuous and non-decreasing. One can realize that, if $\omega(m) = Lm$, then this condition is reduced into Lipschitz and when $\omega(m) = Lm^q$, $q \in (0,1]$ to the Hölder. Furthermore, we found some nonlinear functions which are unbounded in a given domain but seem to be bounded on a particular point of the domain.

For a motivational example, consider a function h on $(-2, 2)$. We can verify the above fact by considering the following example [22]

$$h(m) = \begin{cases} m^3 \ln(m^2) - 6m^2 - 3m + 8 & m \in (-2, 0) \cup (0, 2), \\ 0, & m = 0. \end{cases} \tag{3}$$

Clearly, we can see this fact that $h'''(m)$ is unbounded in $(-2, 2)$. Hence, for avoiding the unboundedness of the function, we replace the condition $(A4)$ by the milder condition since the given example is bounded at $m = 1$. Thus, here we can assume that the norm of the third order Fréchet derivative is bounded on the initial iterate as:

$(B1) \|L'''(m_0)\| \leq \overline{A}$, $m_0 \in B_0$,
where m_0 be an initial approximation. Moreover, we also assume
$(B2) \|L'''(m) - L'''(n)\| \leq \omega(\|m - n\|) \ \forall \ m, n \in B(m_0, \varepsilon)$,
where $\varepsilon > 0$. For now, we choose $\varepsilon = \frac{\kappa}{\tilde{\tau}_0}$, where $\tilde{\tau}_0$ will be defined later and the rationality of this choice of such ε will be proved. Moreover, some authors have considered partial convergence conditions. The following nonlinear integral equation of mixed Hammerstein type [23]

$$m(s) = 1 + \int_0^1 G(s, t) \left(\frac{1}{2} m(t)^{\frac{5}{2}} + \frac{7}{16} m(t)^3 \right) dt, s \in [0, 1], \tag{4}$$

where $m \in [0, 1], t \in [0, 1], G(s, t)$ is the Green function defined by

$$G(s, t) = \begin{cases} (1 - s)t & t \leq s, \\ s(1 - t) & s \leq t, \end{cases}$$

is an example that justified this idea which will be proved later in the numerical application section. In this study, on using recurrence relations, we first discuss the semilocal convergence of the above-mentioned algorithm by just assuming that the second-order Fréchet derivative is bounded. In addition, next, we restrict the domain of the nonlinear operator and consider the bound of the norm of the third-order Fréchet derivative on an initial iterate only rather than supposing it on the given domain of the nonlinear operator.

We start with a nonlinear operator $L : B \subseteq \nabla_1 \to \nabla_2$ and let the Hypotheses $(A1)$–$(A3)$ be fulfilled. Consider the following auxiliary scalar functions out of which Δ and Λ function are taken from the reference [21] and Γ and Θ have been recalculated:

$$\begin{aligned} \Gamma(\theta) \ &= 1 + \frac{1}{2} \frac{\theta}{1-\theta} + \left[1 + \frac{\theta}{1-\frac{2}{3}\theta} \left(1 + \frac{1}{2} \frac{\theta}{1-\theta} \right) \right] \\ &\times \frac{\theta}{2} \left[\frac{1}{1-\theta} + \left(1 + \frac{1}{2} \frac{\theta}{1-\theta} \right)^2 \right], \end{aligned} \tag{5}$$

$$\Delta(\theta) \ = \ \frac{1}{1 - \theta \Gamma(\theta)}, \tag{6}$$

$$\begin{aligned} \Theta(\theta) = \ & \left[\frac{\theta}{1-\frac{2}{3}\theta} \left(1 + \frac{1}{2} \frac{\theta}{1-\theta} \right) + \theta \left[1 + \frac{\theta}{1-\frac{2}{3}\theta} \left(1 + \frac{1}{2} \frac{\theta}{1-\theta} \right) \right] \right. \\ & + \frac{\theta^2}{2(1-\theta)} \left[1 + \frac{\theta}{1-\frac{2}{3}\theta} \left(1 + \frac{1}{2} \frac{\theta}{1-\theta} \right) \right] \\ & \left. + \frac{\theta}{2} \left[1 + \frac{\theta}{1-\frac{2}{3}\theta} \left(1 + \frac{1}{2} \frac{\theta}{1-\theta} \right) \right]^2 \Lambda(\theta) \right] \Lambda(\theta), \end{aligned} \tag{7}$$

where

$$\Lambda(\theta) \; = \; \frac{\theta}{2}\left[\frac{1}{1-\theta} + \left(1 + \frac{1}{2}\frac{\theta}{1-\theta}\right)^2\right]. \tag{8}$$

Next, we study some of the properties of the above-stated functions. Let $k(\theta) = \Gamma(\theta)\theta - 1$. Since $k(0) = -1 < 0$ and $k(\frac{1}{2}) \approx 1.379 > 0$, then the function $k(t)$ has at least one real root in $(0, \frac{1}{2})$. Suppose γ is the smallest positive root, then clearly $\gamma < \frac{1}{2}$. Now, we begin with the following lemmas that will be used later in the main theorem(s).

Lemma 1. *Let the functions Γ, Δ and Θ be given in Equations* (5)–(7), *respectively, and γ be the smallest positive real root of $\Gamma(\theta)\theta - 1$. Then,*

(a) $\Gamma(\theta)$ *and $\Delta(\theta)$ are increasing and $\Gamma(\theta) > 1$, $\Delta(\theta) > 1$ for $\theta \in (0, \gamma)$,*
(b)   *for $\theta \in (0, \gamma)$, $\Theta(\theta)$ is an increasing function.*

Proof. The proof is straightforward from the expressions of Γ, Δ and Θ given in Relations (5)–(7), respectively.   □

Define $\kappa_0 = \kappa$, $\lambda_0 = \lambda$, $\tau_0 = P\lambda\kappa$ and $\zeta_0 = \Delta(\tau_0)\Theta(\tau_0)$. Furthermore, we designate the following sequences as:

$$\kappa_{n+1} \; = \; \zeta_n \kappa_n, \tag{9}$$
$$\lambda_{n+1} \; = \; \Delta(\tau_n)\lambda_n, \tag{10}$$
$$\tau_{n+1} \; = \; P\lambda_{n+1}\kappa_{n+1} = \Delta(\tau_n)\zeta_n\tau_n, \tag{11}$$
$$\zeta_{n+1} \; = \; \Delta(\tau_{n+1})\Theta(\tau_{n+1}), \tag{12}$$

where $n \geq 0$. Some important properties of the immediate sequences are given by the following lemma.

Lemma 2. *If $\tau_0 < \gamma$ and $\Delta(\tau_0)\zeta_0 < 1$, where γ is the smallest positive root of $\Gamma(\theta)\theta - 1 = 0$, then we have*

(a) $\Delta(\tau_n) > 1$ *and $\zeta_n < 1$ for $n \geq 0$,*
(b)   *the sequences $\{\kappa_n\}$, $\{\tau_n\}$ and $\{\zeta_n\}$ are decreasing,*
(c) $\Gamma(\tau_n)\tau_n < 1$ *and $\Delta(\tau_n)\zeta_n < 1$ for $n \geq 0$.*

Proof. The proof can be done readily using mathematical induction.   □

Lemma 3. *Let the functions Γ, Δ and Θ be given in the Relations* (5)–(7), *respectively. Assume that $\alpha \in (0, 1)$, then $\Gamma(\alpha\theta) < \Gamma(\theta)$, $\Delta(\alpha\theta) < \Delta(\theta)$ and $\Theta(\alpha\theta) < \alpha^2\Theta(\theta)$, for $\theta \in (0, \gamma)$.*

Proof. For $\alpha \in (0, 1)$, $\theta \in (0, \gamma)$ and by using the Equations (5)–(7), this lemma can be proved.   □

3. Recurrence Relations for the Method

Here, we characterized some norms which are already derived in the reference [21] for the Method (2) and some are recalculated here.

For $n = 0$, the existence of $\wp_0$ implies the existence of p_0, n_0 and further, we have

$$\|p_0 - m_0\| \leq \kappa_0, \ \|n_0 - m_0\| \leq \frac{2}{3}\kappa_0, \tag{13}$$

i.e., p_0 and $n_0 \in U(m_0, \rho\kappa)$, where $\rho = \frac{\Gamma(\tau_0)}{1-\zeta_0}$. Let $R(m_0) = \wp_0[L'(n_0) - L'(m_0)]$ also; since $\tau_0 < 1$, we have

$$\|R(m_0)\| \leq \frac{2}{3}\tau_0 , \quad \left\|\left[I + \frac{3}{2}R(m_0)\right]^{-1}\right\| \leq \frac{1}{1-\tau_0}. \tag{14}$$

Moreover,

$$\begin{aligned}
\|Y_L(m_0)\| &= \left\|I - \frac{3}{4}\left[I + \frac{3}{2}R(m_0)\right]^{-1} R(m_0)\right\| \\
&\leq 1 + \left\|\frac{3}{4}\left[I + \frac{3}{2}R(m_0)\right]^{-1}\right\| \|R(m_0)\| \\
&\leq 1 + \frac{1}{2}\frac{\tau_0}{1-\tau_0}.
\end{aligned} \tag{15}$$

From the second sub-step of the considered scheme, it is obvious that

$$\|o_0 - m_0\| \leq \left[1 + \frac{1}{2}\frac{\tau_0}{1-\tau_0}\right]\kappa_0. \tag{16}$$

It is similar to obtain

$$\|o_0 - p_0\| \leq \left[\frac{1}{2}\frac{\tau_0}{1-\tau_0}\right]\kappa_0. \tag{17}$$

Using the Banach Lemma, we realize that $L'(n_0)^{-1}$ exists and can be bounded as

$$\|L'(n_0)^{-1}\| \leq \frac{\lambda_0}{1 - \frac{2}{3}\tau_0}. \tag{18}$$

From Taylor's formula, we have

$$\begin{aligned}
L(o_0) &= L(m_0) + L'(m_0)(o_0 - m_0) \\
&+ \int_0^1 [L'(m_0 + \theta(o_0 - m_0)) - L'(m_0)]d\theta(o_0 - m_0).
\end{aligned} \tag{19}$$

From the above relation, it follows that

$$\|L(o_0)\| \leq \Lambda(\tau_0)\frac{\kappa}{\lambda}. \tag{20}$$

Though in the considered reference [21] the norm $\|m_1 - o_0\|$ has already been calculated, here we are recalculating it in a more precise way such that the recalculated norm becomes finer than the given in the reference [21] and its significance can be seen in the numerical section. The motivation for recalculating this norm has been also discussed later. From the last sub-step of the Equation (2),

$$\begin{aligned}
m_1 - o_0 &= -\left[\frac{3}{2}L'(n_0)^{-1}Y_L(m_0) + \wp_n\left(I - \frac{3}{2}Y_L(m_0)\right)\right]L(o_0) \\
&= -\left[\wp_0 + \frac{3}{2}[L'(n_0)^{-1} + L'(n_0)^{-1}]\right]Y_L(m_0)L(o_0).
\end{aligned}$$

On taking the norm, we have

$$\begin{aligned}
\|m_1 - o_0\| &\leq \frac{\kappa_0}{2}\left[1 + \frac{\tau_0}{1-\frac{2}{3}\tau_0}\left(1 + \frac{1}{2}\frac{\tau_0}{1-\tau_0}\right)\right] \\
&\times \left[\frac{\tau_0}{1-\tau_0} + \tau_0\left(1 + \frac{1}{2}\frac{\tau_0}{1-\tau_0}\right)^2\right],
\end{aligned} \tag{21}$$

and thus we obtain

$$\|m_1 - m_0\| \le \|m_1 - o_0\| + \|o_0 - m_0\| \le \Gamma(\tau_0)\kappa_0. \tag{22}$$

Hence, $m_1 \in U(m_0, \rho\kappa)$. Now, since the assumption $\zeta_0 < \frac{1}{\Delta(\tau_0)} < 1$, notice that $\tau_0 < \gamma$ hence $\Gamma(\tau_0) < \Gamma(\gamma)$ and it can be written as

$$\|I - \wp_0 L'(m_1)\| \le \tau_0 \Lambda(\tau_0) < 1. \tag{23}$$

Thus, $\wp_1 = [L'(m_1)]^{-1}$ exists and, by virtue of Banach lemma, it may be written as

$$\|\wp_1\| \le \frac{\lambda_0}{1 - \tau_0 \Gamma(\tau_0)} = \lambda_1.$$

Again by Taylor's expansion along o_n, we can write

$$L(m_{n+1}) = L(o_n) + L'(p_n)(m_{n+1} - o_n) \\ + \int_0^1 [L'(o_n + \theta(m_{n+1} - o_n)) - L'(p_n)]d\theta(m_{n+1} - o_n), \tag{24}$$

and

$$L'(p_n) = L'(m_n) + \int_0^1 L''(m_n + \theta(p_n - m_n))d\theta(p_n - m_n). \tag{25}$$

On using the above relation and, for $n = 0$, Equation (24) assumes the form

$$L(m_1) = L(o_0) + L'(m_0)(m_1 - o_0) \\ + \int_0^1 L''(m_0 + \theta(p_0 - m_0))d\theta(p_0 - m_0)(m_1 - o_0) \\ + \int_0^1 [L'(o_0 + \theta(m_1 - o_0)) - L'(p_0)]d\theta(m_1 - o_0).$$

Using the last sub-step of the Scheme given in the Equation (2), the above expression can be rewritten as

$$L(m_1) = \frac{3}{2}[L'(n_0) - L'(m_0)]L'(n_0)^{-1}Y_L(m_0)L(o_0) \\ + \int_0^1 L''(m_n + \theta(p_n - m_n))d\theta(p_n - m_n)(m_{n+1} - o_n) \\ + \int_0^1 [L'(o_n + \theta(m_{n+1} - o_n)) - L'(p_n)]d\theta(m_{n+1} - o_n).$$

In addition, thus,

$$\|L(m_1)\| \le \Theta(\tau_0)\frac{\kappa}{\lambda}. \tag{26}$$

Hence,

$$\|p_1 - m_1\| \le \Delta(\tau_0)\Theta(\tau_0)\kappa_0 = \kappa_1.$$

In addition, because $\Gamma(\tau_0) > 1$ and by triangle inequality, we find

$$\|p_1 - m_0\| \le \rho\kappa,$$

and

$$\|n_1 - m_0\| \le \|m_1 - m_0\| + \left\|\frac{2}{3}(p_1 - m_1)\right\| \le (\Gamma(\tau_0) + \zeta_0)\kappa_0 < \rho\kappa,$$

which implies $p_1, n_1 \in U(m_0, \rho\kappa)$. Furthermore, we have

$$P\|\wp_1\|\|\wp_1 L(m_1)\| \leq \Delta^2(\tau_0)\Theta(\tau_0)\tau_0 = \tau_1. \tag{27}$$

Moreover, we can state the following lemmas.

Lemma 4. *Under the hypotheses of Lemma* 2, *let* $\sigma = \Delta(\tau_0)\zeta_0$ *and* $\varsigma = \frac{1}{\Delta(\tau_0)}$, *we have*

$$\zeta_i \leq \varsigma\sigma^{3^n}, \tag{28}$$

$$\prod_{i=0}^{n} \zeta_i \leq \varsigma^{n+1}\sigma^{\frac{3^{n+1}-1}{2}}, \tag{29}$$

$$\kappa_n \leq \kappa\varsigma^n\sigma^{\frac{3^n-1}{2}}, \tag{30}$$

$$\sum_{i=n}^{n+m} \kappa_i \leq \kappa\varsigma^n\sigma^{\frac{3^n-1}{2}}\left(\frac{1-\varsigma^{m+1}\sigma^{\frac{3^n(3^m+1)}{2}}}{1-\varsigma\sigma^{3^n}}\right), \tag{31}$$

where $n \geq 0$ *and* $m \geq 1$.

Proof. In order to prove this lemma, first, we need to derive

$$\zeta_n \leq \varsigma\sigma^{3^n}.$$

We will prove it by executing the induction. By Lemma 3 and since $\tau_1 = \sigma\tau_0$, hence for $n = 1$,

$$\zeta_1 = \Delta(\sigma\tau_0)\Theta(\sigma\tau_0) < \sigma^2\zeta_0 < \varsigma\sigma^{3^1}.$$

Let it be true for $n = k$, then

$$\zeta_k \leq \varsigma\sigma^{3^k}, k \geq 1.$$

Now, we will prove it for $n = k + 1$. Thus,

$$\zeta_{k+1} < \Delta(\sigma\tau_k)\Theta(\sigma\tau_k) < \varsigma\sigma^{3^{k+1}}.$$

Therefore, $\zeta_n \leq \varsigma\sigma^{3^n}$ is true for $n \geq 0$. Making use of this inequality, we have

$$\prod_{i=0}^{k} \zeta_i \leq \prod_{i=0}^{k} \varsigma\sigma^{3^i} = \varsigma^{k+1}\prod_{i=0}^{k}\sigma^{3^i} = \varsigma^{k+1}\sigma^{\frac{3^{k+1}-1}{2}}, k \geq 0.$$

By making use of the above-derived inequality in the Relation (9), we have

$$\kappa_n = \zeta_{n-1}\kappa_{n-1} = \zeta_{n-1}\zeta_{n-2}\kappa_{n-2} = \cdots = \kappa_0\prod_{i=0}^{n-1} \zeta_i \leq \kappa\varsigma^n\sigma^{\frac{3^n-1}{2}}, n \geq 0.$$

With the evidence that $0 < \varsigma < 1$ and $0 < \sigma < 1$, we can say that $\kappa_n \to 0$ as $n \to \infty$. Let us denote

$$\omega = \sum_{i=k}^{k+m} \varsigma^i\sigma^{\frac{3^i}{2}}, k \geq 0, m \geq 1.$$

The above equation may also be rewritten in the following form

$$
\begin{aligned}
\omega &\leq \varsigma^k \sigma^{\frac{3k}{2}} + \varsigma\sigma^{3k} \sum_{i=k}^{k+m-1} \varsigma^i \sigma^{\frac{3i}{2}} \\
&= \varsigma^k \sigma^{\frac{3k}{2}} + \varsigma\sigma^{3k} \left(\omega - \varsigma^{k+m} \sigma^{\frac{3k+m}{2}} \right),
\end{aligned}
$$

and then it becomes

$$
\omega < \varsigma^k \sigma^{\frac{3k}{2}} \left(\frac{1 - \varsigma^{m+1} \sigma^{\frac{3^k(3^m+1)}{2}}}{1 - \varsigma\sigma^{3k}} \right).
$$

Moreover,

$$
\sum_{i=k}^{k+m} \kappa_i \leq \sum_{i=k}^{k+m} \kappa\varsigma^i \sigma^{\frac{3i-1}{2}} \leq \kappa\varsigma^k \sigma^{\frac{3k-1}{2}} \left(\frac{1 - \varsigma^{m+1} \sigma^{\frac{3^k(3^m+1)}{2}}}{1 - \varsigma\sigma^{3k}} \right).
$$

□

Lemma 5. *Let the hypotheses of Lemma 2 and the conditions (A1)–(A3) hold; then, the following conditions are true for all $n \geq 0$:*

$$
\begin{aligned}
&(i)\, \wp_n = [L'(m_n)]^{-1} \text{exists and } \|\wp_n\| \leq \lambda_n, \\
&(ii)\, \|\wp_n L(m_n)\| \leq \kappa_n, \\
&(iii)\, P\|\wp_n\|\|\wp_n L(m_n)\| \leq \tau_n, \\
&(iv)\, \|p_n - m_n\| \leq \kappa_n, \\
&(v)\, \|m_{n+1} - m_n\| \leq \Gamma(\tau_n)\kappa_n, \\
&(vi)\, \|m_{n+1} - m_0\| \leq \rho\kappa, \text{ where } \rho = \frac{\Gamma(\tau_0)}{1 - \zeta_0}.
\end{aligned}
\tag{32}
$$

Proof. By using the mathematical induction of Lemma 4, we can prove $(i) - (v)$ for $n \geq 0$. Now, for $n \geq 1$, by making use of Relation (31) and the above results, we get

$$
\|m_{n+1} - m_0\| \leq \sum_{i=0}^{n} \|m_{i+1} - m_i\| < \rho\kappa.
$$

□

Lastly, the following lemma can be proved in a similar way of the article by Wang and Kou [22].

Lemma 6. *Let $\rho = \frac{\Gamma(\tau_0)}{1 - \zeta_0}$ and $\Delta(\tau_0)\zeta_0 < 1$ and $\tau_0 < \gamma$, where γ is the smallest positive root of $\Gamma(\theta)\theta - 1 = 0$; then, $\rho < \frac{1}{\tau_0}$.*

4. Semilocal Convergence When L''' Condition Is Omitted

In the ensuing section, our objective is to prove the convergence of the Algorithm mentioned in the Equation(2) by assuming the Hypotheses $(A1)$–$(A3)$ only. Furthermore, we will find a ball with center m_0 and of radius $\rho\kappa$ in which the solution exists and will be unique as well together with which we will define its error bound.

Theorem 1. *Suppose $L : B \subseteq \nabla_1 \to \nabla_2$ is a continuously second-order Fréchet differentiable on B. Suppose the hypotheses (A1)–(A3) are true and $m_0 \in B$. Assume that $\tau_0 = P\lambda\kappa$ and $\zeta_0 = \Delta(\tau_0)\Theta(\tau_0)$ satisfy $\tau_0 < \gamma$ and $\Delta(\tau_0)\zeta_0 < 1$, where γ is the smallest root of $\Gamma(\theta)\theta - 1 = 0$ and Γ, Δ and Θ are defined by Equations (5)–(7), respectively. In addition, suppose $\overline{U(m_0, \rho\kappa)} \subseteq B$, where $\rho = \frac{\Gamma(\tau_0)}{1 - \zeta_0}$. Then, initiating* with m_0, the iterative sequence $\{m_n\}$ creating from the Scheme given in the Equation (2) converges to a zero m^* of $L(m) = 0$ with $m_n, m^* \in \overline{U(m_0, \rho\kappa)}$ and m^* is an exclusive zero of $L(m) = 0$ in $U(m_0, \frac{2}{P\lambda} - \rho\kappa) \cap B$. Furthermore, its error bound is given by

$$\|m_n - m^*\| \leq \Gamma(\tau_0)\kappa\varsigma^n\sigma^{\frac{3^n-1}{2}}\left(\frac{1}{1-\varsigma\sigma^{3^n}}\right),\tag{33}$$

where $\sigma = \Delta(\tau_0)\zeta_0$ and $\varsigma = \frac{1}{\Delta(\tau_0)}$.

Proof. Clearly, the sequence $\{m_n\}$ is well established in $\overline{U(m_0, \rho\kappa)}$. Now,

$$\begin{aligned}\|m_{k+l} - m_k\| &\leq \sum_{i=k}^{k+l-1}\|m_{i+1} - m_i\| \\ &\leq \Gamma(\tau_0)\kappa\varsigma^k\sigma^{\frac{3^k-1}{2}}\left(\frac{1-\varsigma^l\sigma^{\frac{3^k(3^l-1}{2}+1)}}{1-\varsigma\sigma^{3^k}}\right),\end{aligned}\tag{34}$$

which shows that $\{m_k\}$ is a Cauchy sequence. Hence, there exists m^* satisfying

$$\lim_{k\to\infty} m_k = m^*.$$

Letting $k = 0, l \to \infty$ in Equation (34), we obtain

$$\|m^* - m_0\| \leq \rho\kappa,$$

which implies that $m^* \in \overline{U(m_0, \rho\kappa)}$. Next, we will show that m^* is a zero of $L(m) = 0$. Because

$$\|\wp_0\|\,\|L(m_n)\| \leq \|\wp_n\|\,\|L(m_n)\|,$$

and in the above inequality by tending $n \to \infty$ and using the continuity of L in B, we find that $L(m^*) = 0$. Finally, for unicity of m^* in $U(m_0, \frac{2}{P\lambda} - \rho\kappa) \cap B$, let m^{**} be another solution of $L(m)$ in $U(m_0, \frac{2}{P\lambda} - \rho\kappa) \cap B$. Using Taylor's theorem, we get

$$0 = L(m^{**}) - L(m^*) = \int_0^1 L'((1 - t\theta)m^* + \theta m^{**})d\theta(m^{**} - m^*).$$

In addition,

$$\begin{aligned}\|\wp_0\|\left\|\int_0^1 [L'((1 - \theta)m^* + \theta m^{**}) - L'(m_0)]d\theta\right\| \\ \leq\quad P\lambda \int_0^1 [(1 - \theta)\|m^* - m_0\| + \theta\|m^{**} - m_0\|]d\theta \\ \leq\quad \frac{P\lambda}{2}\left[\rho\kappa + \frac{2}{P\lambda} - \rho\kappa\right] = 1,\end{aligned}$$

which implies $\int_0^1 L'((1 - \theta)m^* + \theta m^{**})d\theta$ is invertible and hence $m^{**} = m^*$. □

5. Semilocal Convergence When L''' Is Bounded on Initial Iterate

In the current section, we establish the existence and uniqueness theorem of the solution based on the weaker conditions $(A1)$–$(A3)$, $(B1)$ and $(B2)$. Define the sequences as

$$\tilde{\kappa}_{n+1} = \tilde{\zeta}_n\tilde{\kappa}_n,\tag{35}$$

$$\tilde{\lambda}_{n+1} = \Delta(\tilde{\tau}_n)\tilde{\lambda}_n,\tag{36}$$

$$\tilde{\tau}_{n+1} = P\tilde{\lambda}_{n+1}\tilde{\kappa}_{n+1} = \Delta(\tilde{\tau}_n)\tilde{\zeta}_n\tilde{\tau}_n, \tag{37}$$

$$\tilde{\mu}_{n+1} = Q\tilde{\lambda}_{n+1}\tilde{\kappa}^2_{n+1} = \Delta(\tilde{\tau}_n)\tilde{\zeta}^2_n\tilde{\mu}_n, \tag{38}$$

$$\tilde{\nu}_{n+1} = \tilde{\lambda}_{n+1}\tilde{\kappa}^2_{n+1}\omega(\tilde{\kappa}_{n+1}) \leq \Delta(\tilde{\tau}_n)\phi(\tilde{\zeta}_n)\tilde{\zeta}^2_n\tilde{\nu}_n, \tag{39}$$

$$\tilde{\zeta}_{n+1} = \Delta(\tilde{\tau}_{n+1})\Theta'(\tilde{\tau}_{n+1}, \tilde{\mu}_{n+1}, \tilde{\nu}_{n+1}), \tag{40}$$

where $n \geq 0$ and $Q = \overline{A} + \omega\left(\frac{\kappa}{\tilde{\tau}_0}\right)$. Here, we assign $\tilde{\kappa}_0 = \kappa$, $\tilde{\lambda}_0 = \lambda$, $\tilde{\tau}_0 = P\lambda\kappa$, $\tilde{\mu}_0 = Q\lambda\kappa^2$, $\tilde{\nu}_0 = \lambda\kappa^2\omega(\kappa)$ and $\tilde{\zeta}_0 = \Delta(\tilde{\tau}_0)\Theta'(\tilde{\tau}_0, \tilde{\mu}_0, \tilde{\nu}_0)$. From Lemma (5), it is known that

$$\|m_n - m_0\| < \rho\kappa < \frac{\kappa}{\tilde{\tau}_0}.$$

Therefore, $m_n \in U(m_0, \frac{\kappa}{\tilde{\tau}_0})$. Similarly, for $t \in [0,1]$ and $n \geq 1$ and using Lemma (6), we get

$$
\begin{aligned}
\|m_n + st(p_n - m_n) - m_0\| &\leq \|m_n - m_0\| + \|p_n - m_n\| \\
&\leq \sum_{i=0}^{n-1} \|m_{i+1} - m_i\| + \tilde{\kappa}_n \\
&\leq \Gamma(\tilde{\tau}_0)\sum_{i=0}^{n}\tilde{\kappa}_i \leq \rho\kappa < \frac{\kappa}{\tilde{\tau}_0}.
\end{aligned}
$$

Therefore, $\{m_n + st(p_n - m_n)\} \in U(m_0, \frac{\kappa}{\tilde{\tau}_0})$. This shows that the choice for $\varepsilon = \frac{\kappa}{\tilde{\tau}_0}$ is relevant. Assume that there exists a root $\tilde{\tau}_0 \in (0, \gamma)$ of the equation

$$m = \left[\overline{A} + \omega\left(\frac{\kappa}{m}\right)\right]\lambda\kappa^2.$$

It is obvious that $\tilde{\mu}_0 = Q\lambda\kappa^2$, where $Q = \overline{A} + \omega\left(\frac{\kappa}{\tilde{\tau}_0}\right)$. Notice that here we don't define $\tilde{\tau}_0$ as the root of the following equation:

$$m = \left[\overline{A} + \omega\left(\frac{\Gamma(m)\kappa}{1 - \Delta(m)\Theta'(m, \tilde{\mu}_0, \tilde{\nu}_0)}\right)\right]\lambda\kappa^2.$$

It would be remembered that, for all $m \in U(m_0, \frac{\kappa}{\tilde{\tau}_0})$, we have

$$
\begin{aligned}
\|L'''(m)\| &= \|L'''(m_0)\| + \|L'''(m) - L'''(m_0)\| \\
&\leq \overline{A} + \omega(\|m - m_0\|) \\
&\leq \overline{A} + \omega\left(\frac{\kappa}{\tilde{\tau}_0}\right) = Q.
\end{aligned}
$$

Here, we include two auxiliary scalar functions taken from the reference [21]

$$
\begin{aligned}
\Theta'(\theta, \eta, \xi) = &\left[\frac{5}{6}\eta + \frac{(3\theta+\eta)(6\theta+2\eta)}{27-18\theta} + \frac{(2\theta+\eta)(3\theta+\eta)}{6-4\theta}\right. \\
&\left. + \frac{(2+2\theta+\eta)(3\theta+\eta)\theta}{(12-8\theta)(1-\theta)}\right]\tilde{\Lambda}(\theta, \eta, \xi) \\
&+ \frac{1}{2}\frac{\theta^2}{1-\theta}\left[\frac{9}{6-4\theta}\left(1 + \frac{1}{2}\frac{\theta}{1-\theta}\right) + \frac{3\theta}{4(1-\theta)} + \frac{1}{2}\right]\tilde{\Lambda}(\theta, \eta, \xi) \\
&+ \frac{\theta}{2}\left[\frac{9}{6-4\theta}\left(1 + \frac{1}{2}\frac{\theta}{1-\theta}\right) + \frac{3\theta}{4(1-\theta)} + \frac{1}{2}\right]^2\tilde{\Lambda}(\theta, \eta, \xi)^2,
\end{aligned}
\tag{41}
$$

where

$$\tilde{\Lambda}(\theta, \eta, \xi) = \frac{1}{8}\frac{\theta^3}{(1-\theta)^2} + \frac{1}{12}\frac{\theta\eta}{1-\theta} + \left(D_1 + \frac{1}{3}D_2\right)\xi. \tag{42}$$

$D_1 = \int_0^1 \int_0^1 \phi(s\theta)\theta(1-\theta)ds d\theta$ and $D_2 = \int_0^1 \int_0^1 \phi\left(\frac{2}{3}s\theta\right)\theta ds d\theta$.

Using the property of the induction and from the conditions $(A1)$–$(A3)$, $(B1)$ and $(B2)$, the following relations are true for all $n \geq 0$:

$$\begin{aligned}
&(i)\,\wp_n = [L'(m_n)]^{-1}exists\ and\ \|\wp_n\| \leq \tilde{\lambda}_n,\\
&(ii)\,\|\wp_n L(m_n)\| \leq \tilde{\kappa}_n,\\
&(iii)\,P\|\wp_n\|\|\wp_n L(m_n)\| \leq \tilde{\tau}_n,\\
&(iv)\,Q\|\wp_n\|\|\wp_n L(m_n)\| \leq \tilde{\mu}_n,\\
&(v)\,\|\wp_n\|\|\wp_n L(m_n)\|^2 \omega(\|\wp_n L(m_n)\|) \leq \tilde{v}_n,\\
&(vi)\,\|m_{n+1} - m_n\| \leq \Gamma(\tilde{\tau}_n)\tilde{\kappa}_n,\\
&(vii)\,\|m_{n+1} - m_0\| \leq \tilde{\rho}\kappa,\ where\ \tilde{\rho} = \frac{\Gamma(\tilde{\tau}_0)}{1-\tilde{\zeta}_0}.
\end{aligned} \tag{43}$$

The second theorem of this article is based on the weaker assumptions, which is stated as:

Theorem 2. *Suppose $L : B \subseteq \nabla_1 \to \nabla_2$ is a continuously third-order Fréchet differentiable on a non-empty open convex subset $B_0 \subseteq B$. Suppose the hypotheses $(A1)$–$(A3)$, $(B1)$ and $(B2)$ are true and $m_0 \in B_0$. Assume that $\tilde{\tau}_0 = P\lambda\kappa, \tilde{\mu}_0 = Q\lambda\kappa^2, \tilde{v}_0 = \lambda\kappa^2\omega(\kappa)$ and $\tilde{\zeta}_0 = \Delta(\tilde{\tau}_0)\Theta'(\tilde{\tau}_0, \tilde{\mu}_0, \tilde{v}_0)$ satisfy $\tilde{\tau}_0 < \gamma$ and $\Delta(\tilde{\tau}_0)\tilde{\zeta}_0 < 1$, where γ is the smallest root of $\Gamma(\theta)\theta - 1 = 0$ and Γ, Δ and Θ' are defined by Equations (5), (6) and (41). In addition, suppose $\overline{U(m_0, \tilde{\rho}\kappa)} \subseteq B_0$, where $\tilde{\rho} = \frac{\Gamma(\tilde{\tau}_0)}{1-\tilde{\zeta}_0}$. Then, initiating with m_0, the iterative sequence $\{m_n\}$ created from the Scheme given in the Equation (2) converges to a zero m^* of $L(m) = 0$ with $m_n, m^* \in \overline{U(m_0, \tilde{\rho}\kappa)}$ and m^* is an exclusive zero of $L(m) = 0$ in $U(m_0, \frac{2}{P\lambda} - \tilde{\rho}\kappa) \cap B$. Furthermore, its error bound is given by*

$$\|m_n - m^*\| \leq \Gamma(\tilde{\tau}_0)\kappa\tilde{\zeta}^n\tilde{\sigma}^{\frac{5^n-1}{4}}\left(\frac{1}{1-\tilde{\zeta}\tilde{\sigma}^{5^n}}\right), \tag{44}$$

where $\tilde{\sigma} = \Delta(\tilde{\tau}_0)\tilde{\zeta}_0$ and $\tilde{\zeta} = \frac{1}{\Delta(\tilde{\tau}_0)}$.

Proof. Analogous to the proof of Theorem 1. $\square$

6. Numerical Example

Example 1. *Consider nonlinear integral equation from the reference [23] already mentioned in the introduction is given as*

$$m(s) = 1 + \int_0^1 G(s,t)\left(\frac{1}{2}m(t)^{\frac{5}{2}} + \frac{7}{16}m(t)^3\right)dt, s \in [0,1], \tag{45}$$

where $m \in [0,1], t \in [0,1]$ and G is the Green's function defined by

$$G(s,t) = \begin{cases} (1-s)t & t \leq s, \\ s(1-t) & s \leq t. \end{cases}$$

Proof. Solving Equation (45) is equivalent to find the solution for $L(m) = 0$, where $L : B \subseteq C[0,1] \to C[0,1]$:

$$[L(m)](s) = m(s) - 1 - \int_0^1 G(s,t) \left(\frac{1}{2} m(t)^{\frac{5}{2}} + \frac{7}{16} m(t)^3 \right) dt, s \in [0,1].$$

The Fréchet derivatives of L are given by

$$L'(m)n(s) = n(s) - \int_0^1 G(s,t) \left(\frac{5}{4} m(t)^{\frac{3}{2}} + \frac{21}{16} m(t)^2 \right) n(t) dt, n \in B,$$

$$L''(m)no(s) = - \int_0^1 G(s,t) \left(\frac{15}{8} m(t)^{\frac{1}{2}} + \frac{21}{8} m(t) \right) n(t)o(t) dt, \quad n, o \in B.$$

Using the max-norm and taking into account that a solution m^* of Equation (45) in $C[0,1]$ must satisfy

$$\|m^*\| - \frac{1}{16} \|m^*\|^{\frac{5}{2}} - \frac{7}{128} \|m^*\|^3 - 1 \leq 0,$$

i.e., $\|m^*\| \leq s_1 = 1.18771$ and $\|m^*\| \geq s_2 = 2.54173$, where s_1 and s_2 are the positive roots of the real equation $t - \frac{t^{\frac{5}{2}}}{16} - \frac{7}{128} t^3 - 1 = 0$. Consequently, if we look for a solution m^* such that $\|m^*\| \leq s_1$, we can consider $U(0,s) \subseteq C[0,1]$, where $s \in (s_1, s_2)$, as a non-empty open convex domain. We choose, for example, $s = 2$ and therefore $B = U(0,2)$. If $m_0 = 1$, then

$$\|\wp_0\| = \frac{128}{87} = \lambda, \|\wp_0 L(m_0)\| \leq \frac{15}{87} = \kappa, \|L''(m)\| \leq \frac{15\sqrt{2}}{64} + \frac{21}{32} = P.$$

Thus, $\tau_0 \approx 0.2505$. Hence, $\tau_0 \Gamma(\tau_0) = 0.4068 < 1$ and $\Delta(\tau_0)\zeta_0 = 0.790 < 1$ (It is noticeable that, if we choose the function $\Gamma(m)$ from the reference [21], then we get $\Delta(\tau_0)\zeta_0 = 1.280 > 1$ which violates one of the assumed hypotheses considered in Theorem 1 and hence this motivates us to recalculate the function $\Gamma(m)$). In addition, $U(m_0, \rho\kappa) = U(1, 0.5270) \subseteq U(0,2) = B$. Thus, the conditions of Theorem 1 of Section 4 are satisfied and the nonlinear Equation (45) has the solution m^* in the region $\{u \in C[0,1] : \|u - 1\| \leq 0.5270\}$, which is unique in $\{u \in C[0,1] : \|u - 1\| < 0.8492\} \cap B$. Hence, we can deduce that the existence ball of solution based on our result is superior to that of Wang and Kou in [23], but our uniqueness ball is inferior. $\square$

Example 2. *Now, consider another example discussed in [22] and also mentioned in the introduction, is given by*

$$h(m) = \begin{cases} m^3 \ln(m^2) - 6m^2 - 3m + 8, & m \in (-2,0) \cup (0,2), \\ 0, & m = 0. \end{cases} \tag{46}$$

Proof. Taking $U(0,2) = B$. Let $m_0 = 1$ be an initial approximation. The derivatives of h are given by

$$\begin{aligned} h'(m) &= 3m^2 \ln(m^2) + 2m^2 - 12m - 3, \\ h''(m) &= 6m \ln(m^2) + 10m - 12, \\ h'''(m) &= 6\ln(m^2) + 22. \end{aligned}$$

Clearly, h''' is unbounded in B and does not satisfy the condition $(A4)$ but satisfies assumption $(B1)$, and we have

$$\|\wp_0\| = \frac{1}{13} = \lambda, \|\wp_0 h(m_0)\| = \frac{1}{13} = \kappa, \|h''(m)\| \leq 12\ln(4) + 32 = P.$$

$\|h'''(m_0)\| = 22, \|h'''(m) - h'''(n)\| \leq \dfrac{12}{1 - \frac{13}{32+12log(4)}}|m - n|$, for all $m, n \in U\left(1, \dfrac{13}{32+12ln(4)}\right)$. Here, $\omega(z) = \dfrac{12}{1 - \frac{13}{32+12log(4)}}z$ and $\phi(\epsilon) = 1$. Here, $\tau_0 \approx 0.2878$ and since $\tau_0\Gamma(\tau_0) = 0.51440 < 1$, $\Delta(\tau_0)\zeta_0 = 0.01742 < 1$. Thus, the assumptions of Theorem 2 of Section 5 are satisfied. In addition, thus, the solution lies in the ball $m \in U(1, 0.13867)$, which is unique in $U(1, 0.39592) \cap B$. Table 1 shows the comparison of error bounds for the considered Algorithm mentioned in the Equation 2 but with two different values of function $\Gamma(m)$ (One is given in the reference [21] and the other is recalculated here). This table also confirms that the value of the recalculated function is prominent.

Table 1. Comparison of the error bounds for Method 2.

n	With Recalculated $\Gamma(m)$	With $\Gamma(m)$ Calculated in [21]
1	0.00085294	0.0019139
2	1.4091×10^{-13}	2.7117×10^{-11}
3	3.1182×10^{-61}	2.0135×10^{-48}
4	2.9759×10^{-298}	5.9108×10^{-232}

□

7. Conclusions

In this contribution, we have analyzed the semilocal convergence of a well defined multi-point variant of the Jarratt method in Banach spaces. This iterative method can be used to solve various kinds of nonlinear equations that satisfy the assumed set of hypotheses. The analysis of this method has been examined using recurrence relations by relaxing the assumptions in two different approaches. In the first approach, we have softened the classical convergence conditions to the prove convergence, existence and uniqueness results together with a priori error bounds. In another way, we have assumed the norm of the third order Fréchet derivative on an initial iterate, so that it never gets unbounded on the given domain and, in addition, it satisfies the local ω-continuity condition as well. Two numerical applications are mentioned that sustain our theoretical consideration.

Author Contributions: For research All the authors have similar contribution.

Funding: This paper is supported by two project funds: the Natural Science Foundation of the Jiangsu Higher Education Institutions of China (Grant No: 16KJB110002.) and the National Science Foundation for Young Scientists of China (Grant No: 11701048).

Acknowledgments: The authors are grateful to the reviewers for their significant criticism which made the paper more elegant and readable.

Conflicts of Interest: The authors declare no conflict of interest.

References

1. Ortega, J.M.; Rheinboldt, W.C. *Iterative Solution of Nonlinear Equation in Several Variables*; Academic Press: New York, NY, USA; London, UK, 1970.
2. Kantorovich, L.V.; Akilov, G.P. *Functional Analysis*; Pergamon Press: Oxford, UK, 1982.
3. Rall, L.B. *Computational Solution of Nonlinear Operator Equations*; Robert E Krieger: New York, NY, USA, 1979.
4. Wang, X.; Gu, C.; Kou, J. Semilocal convergence of a multipoint fourth-order Super–Halley method in Banach spaces. *Numer. Algor.* **2011**, *56*, 497–516. [CrossRef]
5. Wang, X.; Kou, J.; Gu, C. Semilocal convergence of a sixth-order Jarratt method in Banach spaces. *Numer. Algor.* **2011**, *57*, 441–456. [CrossRef]
6. Singh, S.; Gupta, D.K.; Martínez, E.; Hueso, J.L. Semilocal convergence analysis of an iteration of order five using recurrence relations in Banach spaces. *Mediterr. J. Math.* **2016**, *13*, 4219–4235. [CrossRef]
7. Jaiswal, J.P. Semilocal convergence of an eighth-order method in Banach spaces and its computational efficiency. *Numer. Algor.* **2016**, *71*, 933–951. [CrossRef]

8. Hernández, M.A. Chebyshev's approximation algorithms and applications. *Comput. Math. Appl.* **2001**, *41*, 433–445. [CrossRef]
9. Parida, P.K.; Gupta, D.K. Recurrence relations for semilocal convergence of a Newton-like method in Banach spaces. *J. Math. Anal. Appl.* **2008**, *345*, 350–361. [CrossRef]
10. Parida, P.K.; Gupta, D.K. Semilocal convergence of a family of third-order methods in Banach spaces under Hölder continuous second derivative. *Non. Anal. Theo. Meth. Appl.* **2008**, *69*, 4163–4173. [CrossRef]
11. Wang, X.; Kou, J. Convergence for modified Halley-like methods with less computation of inversion. *J. Diff. Eqn. Appl.* **2013**, *19*, 1483–1500. [CrossRef]
12. Ezquerro, J.A.; Hernández, M.A. On the *R*-order of the Halley method. *J. Math. Anal. Appl.* **2005**, *303*, 591–601. [CrossRef]
13. Ezquerro, J.A.; Hernández, M.A. A generalization of the Kantorovich type assumptions for Halley's method. *Int. J. Comput. Math.* **2007**, *84*, 1771–1779. [CrossRef]
14. Parida, P.K.; Gupta, D.K. Semilocal convergence of a family of third-order Chebyshev-type methods under a mild differentiability condition. *Int. J. Comput. Math.* **2010**, *87*, 3405–3419. [CrossRef]
15. Parida, P.K.; Gupta, D.K.; Parhi, S.K. On Semilocal convergence of a multipoint third order method with *R*-order $(2 + p)$ under a mild differentiability condition. *J. Appl. Math. Inf.* **2013**, *31*, 399–416. [CrossRef]
16. Prashanth, M.; Gupta, D.K. Convergence of a parametric continuation method. *Kodai Math. J.* **2014**, *37*, 212–234. [CrossRef]
17. Prashanth, M.; Gupta, D.K. Semilocal convergence for Super-Halley's method under ω-differentiability condition. *Jpn. J. Ind. Appl. Math.* **2015**, *32*, 77–94. [CrossRef]
18. Wang, X.; Kou, J. Semilocal convergence of multi-point improved Super-Halley-type methods without the second derivative under generalized weak condition. *Numer. Algor.* **2016**, *71*, 567–584. [CrossRef]
19. Wang, X., Kou, J. Semilocal convergence analysis on the modifications for Chebyshev–Halley methods under generalized condition. *Appl. Math. Comput.* **2016**, *281*, 243–251. [CrossRef]
20. Wang, X.; Kou, J. Semilocal convergence on a family of root-finding multi-point methods in Banach spaces under relaxed continuity condition. *Numer. Algor.* **2017**, *74*, 643–657. [CrossRef]
21. Wang, X.; Kou, J. Semilocal convergence of a modified multi-point Jarratt method in Banach spaces under general continuity condition. *Numer. Algor.* **2012**, *60*, 369–390. [CrossRef]
22. Wang, X.; Kou, J. Convergence for a class of improved sixth-order Chebyshev–Halley type method. *Appl. Math. Comput.* **2016**, *273*, 513–524. [CrossRef]
23. Wang, X.; Kou, J. Convergence for a family of modified Chebyshev methods under weak condition. *Numer. Algor.* **2014**, *66*, 33–48. [CrossRef]

mathematics

MDPI

Article

The Modified Inertial Iterative Algorithm for Solving Split Variational Inclusion Problem for Multi-Valued Quasi Nonexpansive Mappings with Some Applications

Pawicha Phairatchatniyom [1], Poom Kumam [1,2,*], Yeol Je Cho [3,4], Wachirapong Jirakitpuwapat [1] and Kanokwan Sitthithakerngkiet [5]

1 KMUTTFixed Point Research Laboratory, Room SCL 802 Fixed Point Laboratory, Science Laboratory Building, Department of Mathematics, Faculty of Science, King Mongkut's University of Technology Thonburi, 126 Pracha Uthit Rd., Bang Mod, Thung Khru, Bangkok 10140, Thailand; pairatchat@gmail.com (P.P.); wachirapong.jira@hotmail.com (W.J.)

2 Department of Medical Research, China Medical University Hospital, China Medical University, Taichung 40402, Taiwan

3 Department of Mathematics Education, Gyeongsang National University, Jinju 52828, Korea; yjchomath@gmail.com

4 School of Mathematical Sciences, University of Electronic Science and Technology of China, Chengdu 611731, China

5 Intelligent and Nonlinear Dynamic Innovations Research Center, Department of Mathematics, Faculty of Applied Science, King Mongkut's University of Technology North Bangkok (KMUTNB), Wongsawang, Bangsue, Bangkok 10800, Thailand; kanokwan.s@sci.kmutnb.ac.th

* Correspondence: poom.kum@kmutt.ac.th; Tel.: +66-24708994

Received: 28 April 2019; Accepted: 28 May 2019; Published: 19 June 2019

Abstract: Based on the very recent work by Shehu and Agbebaku in Comput. Appl. Math. 2017, we introduce an extension of their iterative algorithm by combining it with inertial extrapolation for solving split inclusion problems and fixed point problems. Under suitable conditions, we prove that the proposed algorithm converges strongly to common elements of the solution set of the split inclusion problems and fixed point problems.

Keywords: variational inequality problem; split variational inclusion problem; multi-valued quasi-nonexpasive mappings; Hilbert space

MSC: 47H06; 47H09; 47J05; 47J25

1. Introduction

The *split monotone variational inclusion problem* (**SMVIP**) was introduced by Moudafi [1]. This problem is as follows:

$$\text{Find a point } x^* \in H_1 \text{ such that } 0 \in \hat{f}(x^*) + B_1(x^*) \tag{1}$$

and such that

$$y^* = Ax^* \in H_2 \text{ solves } 0 \in \hat{g}(y^*) + B_2(y^*), \tag{2}$$

where 0 is the zero vector, H_1 and H_2 are real Hilbert spaces, $\hat{f}$ and $\hat{g}$ are given single-valued operators defined on H_1 and H_2, respectively, B_1 and B_2 are multi-valued maximal monotone mappings defined on H_1 and H_2, respectively, and A is a bounded linear operator defined on H_1 to H_2.

It is well known (see [1]) that

$$0 \in \hat{f}(x^*) + B_1(x^*) \iff x^* = J_\lambda^{B_1}(x^* - \lambda \hat{f}(x^*)),$$

and that

$$0 \in \hat{g}(y^*) + B_2(y^*) \iff y^* = J_\lambda^{B_2}(y^* - \lambda \hat{g}(y^*)), \quad y^* = Ax^*,$$

where $J_\lambda^{B_1} := (I + \lambda B_1)^{-1}$ and $J_\lambda^{B_2} := (I + \lambda B_2)^{-1}$ are the resolvent operators of B_1 and B_2, respectively, with $\lambda > 0$. Note that $J_\lambda^{B_1}$ and $J_\lambda^{B_2}$ are nonexpansive and firmly nonexpansive.

Recently, Shehu and Agbebaku [2] proposed an algorithm involving a step-size selected and proved strong convergence theorem for split inclusion problem and fixed point problem for multi-valued quasi-nonexpansive mappings. In [1], Moudafi pointed out that the problem **(SMVIP)** [3–5] includes, as special cases, the split variational inequality problem [6], the split zero problem, the split common fixed point problem [7–9] and the split feasibility problem [10,11], which have already been studied and used in image processing and recovery [12], sensor networks in computerized tomography and data compression for models of inverse problems [13].

If $\hat{f} \equiv 0$ and $\hat{g} \equiv 0$ in the problem **(SMVIP)**, then the problem reduces to the *split variational inclusion problem* **(SVIP)** as follows:

$$\text{Find a point } x^* \in H_1 \text{ such that } 0 \in B_1(x^*) \tag{3}$$

and such that

$$y^* = Ax^* \in H_2 \text{ solves } 0 \in B_2(y^*). \tag{4}$$

Note that the problem **(SVIP)** is equivalent to the following problem:

$$\text{Find a point } x^* \in H_1 \text{ such that } x^* = J_\lambda^{B_1}(x^*) \text{ and } y^* = J_\lambda^{B_2}(y^*), \quad y^* = Ax^*$$

for some $\lambda > 0$.

We denote the solution set of the problem **(SVIP)** by Ω, i.e.,

$$\Omega = \{x^* \in H_1 : 0 \in B_1(x^*) \text{ and } 0 \in B_2(y^*), \ y^* = Ax^*\}.$$

Many works have been developed to solve the split variational inclusion problem **(SVIP)**. In 2002, Byrne et al. [7] introduced the iterative method $\{x_n\}$ as follows: For any $x_0 \in H_1$,

$$x_{n+1} = J_\lambda^{B_1}(x_n + \gamma A^*(J_\lambda^{B_2} - I)Ax_n) \tag{5}$$

for each $n \geq 0$, where A^* is the adjoint of the bounded linear operator A, $\gamma \in (0, 2/L)$, $L = \|A^*A\|$ and $\lambda > 0$. They have shown the weak and strong convergence of the above iterative method for solving the problem **(SVIP)**.

Later, inspired by the above iterative algorithm, many authors have extended the algorithm $\{x_n\}$ generated by (5). In particular, Kazmi and Rizvi [4] proposed an algorithm $\{x_n\}$ for approximating a solution of the problem **(SVIP)** as follows:

$$\begin{cases} u_n = J_\lambda^{B_1}(x_n + \gamma_n A^*(J_\lambda^{B_2} - I)Ax_n), \\ x_{n+1} = \alpha_n f_n(x_n) + (1 - \alpha_n)Su_n \end{cases} \tag{6}$$

for each $n \geq 0$, where $\{\alpha_n\}$ is a sequence in $(0,1)$, $\lambda > 0$, $\gamma \in (0, 1/L)$, L is the spectral radius of the operator A^*A, $f : H_1 \to H_1$ is a contraction and $S : H_1 \to H_1$ is a nonexpansive mapping. In 2015,

Sitthithakerngkiet et al. [5] proposed an algorithm $\{x_n\}$ for solving the problem **(SVIP)** and the fixed point problem **(FPP)** of a countable family of nonexpansive mappings as follows:

$$\begin{cases} y_n = J_\lambda^{B_1}(x_n + \gamma_n A^*(J_\lambda^{B_2} - I)Ax_n), \\ x_{n+1} = \alpha_n f(x_n) + (1 - \alpha_n D)S_n y_n \end{cases} \tag{7}$$

for each $n \geq 0$, where $\{\alpha_n\}$ is a sequence in $(0,1)$, $\lambda > 0$, $\gamma \in (0,1/L)$, L is the spectral radius of the operator A^*A, $f : H_1 \to H_1$ is a contraction, $D : H_1 \to H_2$ is strongly positive bounded linear operator and, for each $n \geq 1$, $S_n : H_1 \to H_1$ is a nonexpansive mapping.

In both their works, they obtained some strong convergence results by using their proposed iterative methods (for some more results on algorithms, see [14,15]).

Recall that a point $x^* \in H_1$ is called a fixed point of a given multi-valued mapping $S : H_1 \to 2^{H_1}$ if

$$x^* \in Sx^* \tag{8}$$

and the *fixed point problem* **(FPP)** for a multi-valued mapping $S : H_1 \to 2^{H_1}$ is as follows:

$$\text{Find a point } x^* \in H_1 \text{ such that } x^* \in Sx^*.$$

The set of fixed points of the multi-valued mapping S is denoted by $F(S)$.

As applications, the fixed point theory for multi-valued mappings was applied to various fields, especially mathematical economics and game theory (see [16–18]).

Recently, motivated by the results of Byrne et al. [7], Kazmi and Rizvi [4] and Sitthithakerngkiet [5], Shehu and Agbebaku [2] introduced the *split fixed point inclusion problem* **(SFPIP)** from the problems **(SVIP)** and **(FPP)** for a multi-valued quasi-nonexpansive mapping $S : H_1 \to 2^{H_1}$ as follows:

$$\text{Find a point } x^* \in H_1 \text{ such that } 0 \in B_1(x^*), \ x^* \in Sx^* \tag{9}$$

and such that

$$y^* = Ax^* \in H_2 \text{ solves } 0 \in B_2(y^*), \tag{10}$$

where H_1 and H_2 are real Hilbert spaces, B_1 and B_2 are multi-valued maximal monotone mappings defined on H_1 and H_2, respectively, and A is a bounded linear operator defined on H_1 to H_2.

Note that the problem **(SFPIP)** is equivalent to the following problem: for some $\lambda > 0$,

$$\text{Find a point } x^* \in H_1 \text{ such that } x^* = J_\lambda^{B_1}(x^*), \ x^* \in Sx^* \text{ and } Ax^* = J_\lambda^{B_2}(Ax^*).$$

The solution set of the problem **(SFPIP)** is denoted by $F(S) \bigcap \Omega$, i.e.,

$$F(S) \bigcap \Omega = \{x^* \in H_1 : 0 \in B_1(x^*), \ x^* \in Sx^* \text{ and } 0 \in B_2(Ax^*)\}.$$

Notice that, if S is the identity operator, then the problem **(SFPIP)** reduces to the problem **(SVIP)**. Moreover, if $J_\lambda^{B_1} = J_\lambda^{B_2} = A = I$, then the problem **(SFPIP)** reduces to the problem **(FPP)** for a multi-valued quasi-nonexpansive mapping.

Furthermore, Shehu and Agbebaku [2] introduced an algorithm $\{x_n\}$ for solving the problem **(SFPIP)** for a multi-valued quai-nonexpasive mapping S as follows: For any $x_1 \in H_1$,

$$\begin{cases} u_n = J_\lambda^{B_1}(x_n + \gamma_n A^*(J_\lambda^{B_2} - 1)Ax_n), \\ x_{n+1} = \alpha_n f_n(x_n) + \beta_n x_n + \delta_n(\sigma w_n + (1 - \sigma)u_n), \ w_n \in Sx_n, \end{cases} \tag{11}$$

for each $n \geq 1$, where $\{\alpha_n\}$, $\{\beta_n\}$ and $\{\delta_n\}$ are the real sequences in $(0, 1)$ such that

$$\alpha_n + \beta_n + \delta_n = 1, \quad \sigma \in (0, 1), \quad \gamma_n := \tau_n \frac{\|(J_\lambda^{B_2} - I) A x_n\|^2}{\|A^*(J_\lambda^{B_2} - I)\|^2},$$

where $0 < a \leq \tau_n \leq b < 1$, and $\{f_n(x)\}$ is the uniform convergence sequence for any x in a bounded subset D of H_1, and proved that the sequences $\{u_n\}$ and $\{x_n\}$ generated by (11) both converge strongly to $p \in F(S) \cap \Omega$, where $p = P_{F(S) \cap \Omega} f(p)$.

In optimization theory, the second-order dynamical system, which is called the heavy ball method, is used to accelerate the convergence rate of algorithms. This method is a two-step iterative method for minimizing a smooth convex function which was firstly introduced by Polyak [19].

The following is a modified heavy ball method for the improvement of the convergence rate, which was introduced by Nesterov [20]:

$$\begin{cases} y_n = x_n + \theta_n(x_n - x_{n-1}), \\ x_{n+1} = y_n - \lambda_n \nabla f(y_n) \end{cases}$$

for each $n \geq 1$, where $\lambda_n > 0$, $\theta_n \in [0, 1)$ is an extrapolation factor. Here, the term $\theta_n(x_n - x_{n-1})$ is the inertia (for more recent results on the inertial algorithms, see [21,22]).

The following method is called the *inertial proximal point algorithm*, which was introduced by Alvarez and Attouch [23]. This method combined the proximal point algorithm [24] with the inertial extrapolation [25,26]:

$$\begin{cases} y_n = x_n + \theta_n(x_n - x_{n-1}), \\ x_{n+1} = (I + \lambda_n \widehat{T})^{-1}(y_n) \end{cases} \tag{12}$$

for each $n \geq 1$, where I is identity operator and $\widehat{T}$ is a maximal monotone operator. It was proven that, if a positive sequence λ_n is non-decreasing, $\theta_n \in [0, 1)$ and the following summability condition holds:

$$\sum_{n=1}^{\infty} \theta_n \|x_n - x_{n-1}\|^2 < \infty, \tag{13}$$

then $\{x_n\}$ generated by (12) converges to a zero point of T.

In fact, recently, some authors have pointed out some problems in this summability condition (13) given in [27], that is, to satisfy this summability condition (13) of the sequence $\{x_n\}$, one needs to calculate $\{\theta_n\}$ at each step. Recently, Bot et al. [28] improved this condition, that is, they got rid of the summability condition (13) and replaced the other conditions.

In this paper, inspired by the results of Shehu and Agbebaku [2], Nesterov [20] and Alvarez and Attouch [23], we proposed a new algorithm by combining the iterative algorithm (11) with the inertial extrapolation for solving the problem **(SFPIP)** and prove some strong convergence theorems of the proposed algorithm to show the existence of a solution of the problem **(SFPIP)**. Furthermore, as applications, we consider our proposed algorithm for solving the variational inequality problem and give some applications in game theory.

2. Preliminaries

In this section, we recall some definitions and results which will be used in the proof of the main results.

Let H_1 and H_2 be two real Hilbert spaces with the inner product $\langle \cdot, \cdot \rangle$ and the norm $\| \cdot \|$. Let C be a nonempty closed and convex subset of H_1 and D be a nonempty bounded subset of H_1. Let $A : H_1 \to H_2$ be a bounded linear operator and $A^* : H_2 \to H_1$ be the adjoint of A.

Let $\{x_n\}$ be a sequence in H, we denote the strong and weak convergence of a sequence $\{x_n\}$ by $x_n \to x$ and $x_n \rightharpoonup x$, respectively.

Recall that a mapping $T : C \to C$ is said to be:

(1)   *Lipschitz* if there exists a positive constant α such that, for all $x, y \in C$,

$$\|Tx - Ty\| \leq \alpha \|x - y\|.$$

If $\alpha \in (0, 1)$ and $\alpha = 1$, then the mapping T is contractive and nonexpansive, respectively.

(2)   *firmly nonexpansive* if

$$\|Tx - Ty\|^2 \leq \langle Tx - Ty, x - y \rangle$$

for all $x, y \in C$.

A mapping P_C is said to be the *metric projection* of H_1 onto C if, for all point $x \in H_1$, there exists a unique nearest point in C, denoted by $P_C x$, such that

$$\|x - P_C x\| \leq \|x - y\|$$

for all $y \in C$.

It is well known that P_C is nonexpansive mapping and satisfies

$$\langle x - y, P_C x - P_C y \rangle \leq \|P_C x - P_C y\|^2$$

for all $x, y \in H_1$. Moreover, $P_C x$ is characterized by the fact $P_C x \in C$ and

$$\langle x - P_C x, y - P_C x \rangle \leq 0$$

for all $y \in C$ and $x \in H_1$ (see [6,22]).

A multi-valued mapping $B_1 : H_1 \to 2^{H_1}$ is said to be *monotone* if, for all $x, y \in H_1$, $u \in B_1(x)$ and $v \in B_1(y)$,

$$\langle x - y, u - v \rangle \geq 0.$$

A monotone mapping $B_1 : H_1 \to 2^{H_1}$ is said to be *maximal* if the graph $G(B_1)$ of B_1 is not properly contained in the graph of any other monotone mapping. It is known that a monotone mapping B_1 is maximal if and only if, for all $(x, u) \in H_1 \times H_1$,

$$\langle x - y, u - v \rangle \geq 0$$

for all $(y, v) \in G(B_1)$ implies that $u \in B_1(x)$.

Let $B_1 : H_1 \to 2^{H_1}$ be a multi-valued maximal monotone mapping. Then the *resolvent mapping* $J_\lambda^{B_1} : H_1 \to H_1$ associated with B_1 is defined by

$$J_\lambda^{B_1}(x) := (I + \lambda B_1)^{-1}(x)$$

for all $x \in H_1$ and for some $\lambda > 0$, where I is the identity operator on H_1. It is well known that, for any $\lambda > 0$, the resolvent operator $J_\lambda^{B_1}$ is single-valued firmly nonexpansive (see [2,5,6,14]).

Definition 1. *Suppose that* $\{f_n(x)\}$ *is a sequence of functions defined on a bounded set D. Then* $f_n(x)$ *converges uniformly to the function* $f(x)$ *on D if, for all* $x \in D$,

$$f_n(x) \to f(x) \ as \ n \to \infty.$$

Let $f_n : D \to H_1$ be a uniformly convergent sequence of contraction mappings on D, i.e., there exists $\mu_n \in (0, 1)$ such that

$$f_n(x) - f_n(y)\| \le \mu_n \|x - y\|$$

for all $x, y \in D$.

Let $CB(H_1)$ denote the family of nonempty closed and bounded subsets of H_1. The *Hausdorff metric* on $CB(H_1)$ is defined by

$$\widehat{H}(x, y) = \max \left\{ \sup_{x \in A} \inf_{y \in B} \|x - y\|, \sup_{y \in B} \inf_{x \in A} \|x - y\| \right\}$$

for all $A, B \in CB(H_1)$ (see [18]).

Definition 2. [2] *Let* $S : H_1 \to CB(H_1)$ *be a multi-valued mapping. Assume that* $p \in H_1$ *is a fixed point of S, that is,* $p \in Sp$. *The mapping S is said to be:*

(1)   *nonexpansive if, for all* $x, y \in H_1$,

$$\widehat{H}(Sx, Sy) \le \|x - y\|.$$

(2)   *quasi-nonexpansive if* $F(S) \ne \emptyset$ *and, for all* $x \in H_1$ *and* $p \in F(S)$,

$$\widehat{H}(Sx, Sp) \le \|x - p\|$$

Definition 3. [2] *A single-valued mapping* $S : H \to H$ *is said to be demiclosed at the origin if, for any sequence* $\{x_n\} \subset H$ *with* $x_n \rightharpoonup x$ *and* $Sx_n \to 0$, *we have* $Sx = 0$.

Definition 4. [2] *A multi-valued mapping* $S : H_1 \to CB(H_1)$ *is said to be demiclosed at the origin if, for any sequence* $\{x_n\} \subset H$ *with* $x_n \rightharpoonup x$ *and* $d(x_n, Sx_n) \to 0$, *we have* $x \in Sx$.

Lemma 1. [29,30] *Let H be a Hilbert space. Then, for any* $x, y, z \in X$ *and* $\alpha, \beta, \gamma \in [0, 1]$ *with* $\alpha + \beta + \gamma = 1$, *we have*

$$\|\alpha x + \beta y + \gamma z\|^2 = \alpha \|x\|^2 + \beta \|y\|^2 + \gamma \|z\|^2 - \alpha\beta \|x - y\|^2 - \alpha\gamma \|x - z\|^2 - \beta\gamma \|y - z\|^2.$$

Lemma 2. [2,31] *Let H be a real Hilbert space. Then the following results hold:*

(1) $\|x - y\|^2 = \|x\|^2 - 2\langle x, y \rangle + \|y\|^2$.
(2) $\|x + y\|^2 = \|x\|^2 + 2\langle x, y \rangle + \|y\|^2$.
(3) $\|x + y\|^2 \le \|x\|^2 + 2\langle y, x + y \rangle$ *for all* $x, y \in H$.

Lemma 3. [2,32,33] *Let* $\{a_n\}$, $\{c_n\} \subset \mathbb{R}_+$, $\{\sigma_n\} \subset (0, 1)$ *and* $\{b_n\} \subset \mathbb{R}$ *be sequences such that*

$$a_{n+1} \le (1 - \sigma_n)a_n + b_n + c_n \ \ for \ all \ \ n \ge 0.$$

Assume $\sum_{n=0}^{\infty} |c_n| < \infty$. *Then the following results hold:*

(1)   *If* $b_n \le \beta\sigma_n$ *for some* $\beta \ge 0$, *then* $\{a_n\}$ *is a bounded sequence.*

(2)   If we have

$$\sum_{n=0}^{\infty} \sigma_n = \infty \quad and \quad \limsup_{n\to\infty} \frac{b_n}{\sigma_n} \le 0,$$

then $\lim_{n\to\infty} a_n = 0$.

Lemma 4. [32,33] *Let $\{s_n\}$ be a sequence of non-negative real numbers such that*

$$s_{n+1} \le (1 - \lambda_n)s_n + \lambda_n t_n + r_n$$

for each $n \ge 1$, where

(a) $\{\lambda_n\} \subset [0,1]$ *and* $\sum_{n=1}^{\infty} \lambda_n = \infty$;
(b) $\limsup t_n \le 0$;
(c) $r_n \ge 0$ *and* $\sum_{n=1}^{\infty} r_n < \infty$.

Then $s_n \to 0$ as $n \to \infty$.

3. The Main Results

In this section, we prove some strong convergence theorems of the proposed algorithm for solving the problem **(SFPIP)**.

Theorem 1. *Let H_1, H_2 be two real Hilbert spaces, $A : H_1 \to H_2$ be bounded operator with adjoint operator A^* and $B_1 : H_1 \to 2^{H_1}$, $B_2 : H_2 \to 2^{H_2}$ be maximal monotone mappings. Let $S : H_1 \to CB(H_1)$ be a multi-valued quasi-nonexpansive mapping and S be demiclosed at the origin. Let $\{f_n\}$ be a sequence of μ_n-contractions $f_n : H_1 \to H_1$ with $0 < \mu_* \le \mu_n \le \mu^* < 1$ and $\{f_n(x)\}$ be uniformly convergent for any x in a bounded subset D of H_1. Suppose that $F(S) \cap \Omega \ne \emptyset$. For any $x_0, x_1 \in H_1$, let the sequences $\{y_n\}$, $\{u_n\}$, $\{z_n\}$ and $\{x_n\}$ be generated by*

$$\begin{cases} y_n = x_n + \theta_n(x_n - x_{n-1}), \\ u_n = J_\lambda^{B_1}(y_n + \gamma_n A^*(J_\lambda^{B_2} - I)Ay_n), \\ z_n = \zeta v_n + (1 - \zeta)u_n, \ v_n \in Sx_n, \\ x_{n+1} = \alpha_n f_n(x_n) + \beta_n x_n + \delta_n z_n \end{cases} \quad (14)$$

for each $n \ge 1$, where $\zeta \in (0,1)$, $\gamma_n := \tau_n \frac{\|(J_\lambda^{B_2} - I)Ay_n\|^2}{\|A^*(J_\lambda^{B_2} - I)Ay_n\|^2}$ with $0 < \tau_* \le \tau_n \le \tau^* < 1$, $\{\theta_n\} \subset [0, \bar{\omega})$ for some $\bar{\omega} > 0$ and $\{\alpha_n\}, \{\beta_n\}, \{\delta_n\} \in (0,1)$ with $\alpha_n + \beta_n + \delta_n = 1$ satisfying the following conditions:

(C1) $\lim_{n\to\infty} \alpha_n = 0$;
(C2) $\sum_{n=1}^{\infty} \alpha_n = \infty$;
(C3) $0 < \epsilon_1 \le \beta_n$ *and* $0 < \epsilon_2 \le \delta_n$;
(C4) $\lim_{n\to\infty} \frac{\theta_n}{\alpha_n} \|x_n - x_{n-1}\| = 0$.

Then $\{x_n\}$ generated by (14) *converges strongly to $p \in F(S) \cap \Omega$, where $p = P_{F(S)\cap\Omega} f(p)$.*

Proof. First, we show that $\{x_n\}$ is bounded. Let $p = P_{F(S)\cap\Omega} f(p)$. Then $p \in F(S) \cap \Omega$ and so $J_\lambda^{B_1} p = p$ and $J_\lambda^{B_2} Ap = Ap$. By the triangle inequality, we get

$$\begin{aligned} \|y_n - p\| &= \|x_n + \theta_n(x_n - x_{n-1}) - p\| \\ &\le \|x_n - p\| + \theta_n \|x_n - x_{n-1}\|. \end{aligned} \quad (15)$$

By the Cauchy-Schwarz inequality and Lemma 2 (1) and (2), we get

$$
\begin{aligned}
\|y_n - p\|^2 &= \|x_n + \theta_n(x_n - x_{n-1}) - p\|^2 \\
&= \|x_n - p\|^2 + \theta_n^2 \|x_n - x_{n-1}\|^2 + 2\theta_n\langle x_n - p, x_n - x_{n-1}\rangle \\
&\le \|x_n - p\|^2 + \theta_n^2 \|x_n - x_{n-1}\|^2 + 2\theta_n\|x_n - x_{n-1}\|\|x_n - p\|.
\end{aligned}
\tag{16}
$$

By using (15) and the fact that S is quasi-nonexpansive S, we get

$$
\begin{aligned}
\|z_n - p\| &= \|\xi v_n + (1 - \xi)u_n - p\| \\
&= \|\xi(v_n - p) + (1 - \xi)(u_n - p)\| \\
&\le \xi\|v_n - p\| + (1 - \xi)\|u_n - p\| \\
&\le \xi d(v_n, Sp) + (1 - \xi)\|y_n - p\| \\
&\le \xi \widehat{H}(Sx_n, Sp) + (1 - \xi)[\|x_n - p\| + \theta_n\|x_n - x_{n-1}\|] \\
&\le \xi\|x_n - p\| + (1 - \xi)\|x_n - p\| + (1 - \xi)\theta_n\|x_n - x_{n-1}\| \\
&\le \|x_n - p\| + \theta_n\|x_n - x_{n-1}\|,
\end{aligned}
\tag{17}
$$

which implies that

$$
\begin{aligned}
\|z_n - p\|^2 &\le (\|x_n - p\| + \theta_n\|x_n - x_{n-1}\|)^2 \\
&= \|x_n - p\|^2 + 2\theta_n\|x_n - x_{n-1}\|\|x_n - p\| + \theta_n^2\|x_n - x_{n-1}\|^2.
\end{aligned}
\tag{18}
$$

Since $J_\lambda^{B_1}$ is nonexpansive, by Lemma 2 (2), we get

$$
\begin{aligned}
\|u_n - p\|^2 &= \|J_\lambda^{B_1}(y_n + \gamma_n A^*(J_\lambda^{B_2} - I)Ay_n) - p\|^2 \\
&= \|J_\lambda^{B_1}(y_n + \gamma_n A^*(J_\lambda^{B_2} - I)Ay_n) - J_\lambda^{B_1}p\|^2 \\
&\le \|y_n + \gamma_n A^*(J_\lambda^{B_2} - I)Ay_n - p\|^2 \\
&= \|y_n - p\|^2 + \gamma_n^2\|A^*(J_\lambda^{B_2} - I)Ay_n\|^2 + 2\gamma_n\langle y_n - p, A^*(J_\lambda^{B_2} - I)Ay_n\rangle.
\end{aligned}
\tag{19}
$$

Again, by Lemma 2 (2), we get

$$
\begin{aligned}
&\langle y_n - p, A^*(J_\lambda^{B_2} - I)Ay_n\rangle \\
&= \langle A(y_n - p), (J_\lambda^{B_2} - I)Ay_n\rangle \\
&= \langle J_\lambda^{B_2}Ay_n - Ap - (J_\lambda^{B_2} - I)Ay_n, (J_\lambda^{B_2} - I)Ay_n\rangle \\
&= \langle J_\lambda^{B_2}Ay_n - Ap, (J_\lambda^{B_2} - I)Ay_n\rangle - \langle(J_\lambda^{B_2} - I)Ay_n, (J_\lambda^{B_2} - I)Ay_n\rangle \\
&= \langle J_\lambda^{B_2}Ay_n - Ap, (J_\lambda^{B_2} - I)Ay_n\rangle - \|(J_\lambda^{B_2} - I)Ay_n\|^2 \\
&= \frac{1}{2}\left(\|J_\lambda^{B_2}Ay_n - Ap\|^2 + \|(J_\lambda^{B_2} - I)Ay_n\|^2\right. \\
&\quad \left. - \|J_\lambda^{B_2}Ay_n - Ap - (J_\lambda^{B_2} - I)Ay_n\|^2\right) - \|(J_\lambda^{B_2} - I)Ay_n\|^2 \\
&= \frac{1}{2}\left(\|J_\lambda^{B_2}Ay_n - Ap\|^2 + \|(J_\lambda^{B_2} - I)Ay_n\|^2 - \|J_\lambda^{B_2}Ay_n - Ap - J_\lambda^{B_2}Ay_n + Ay_n\|^2\right) \\
&\quad - \|(J_\lambda^{B_2} - I)Ay_n\|^2 \\
&= \frac{1}{2}\left(\|J_\lambda^{B_2}Ay_n - Ap\|^2 + \|(J_\lambda^{B_2} - I)Ay_n\|^2 - \|Ay_n - Ap\|^2\right) - \|(J_\lambda^{B_2} - I)Ay_n\|^2 \\
&= \frac{1}{2}\left(\|J_\lambda^{B_2}Ay_n - Ap\|^2 - \|Ay_n - Ap\|^2 - \|(J_\lambda^{B_2} - I)Ay_n\|^2\right) \\
&\le \frac{1}{2}\left(\|Ay_n - Ap\|^2 - \|Ay_n - Ap\|^2 - \|(J_\lambda^{B_2} - I)Ay_n\|^2\right) \\
&= -\frac{1}{2}\|(J_\lambda^{B_2} - I)Ay_n\|^2.
\end{aligned}
\tag{20}
$$

Using (20) into (19), we get

$$\|u_n - p\|^2 \leq \|y_n - p\|^2 + \gamma_n^2 \|A^*(J_\lambda^{B_2} - I)Ay_n\|^2 - \gamma_n \|(J_\lambda^{B_2} - I)Ay_n\|^2$$
$$= \|y_n - p\|^2 - \gamma_n (\|(J_\lambda^{B_2} - I)Ay_n\|^2 - \gamma_n \|A^*(J_\lambda^{B_2} - I)Ay_n\|^2). \tag{21}$$

By the definition of γ_n, (21) can then be written as follows:

$$\|u_n - p\|^2 \leq \|y_n - p\|^2 - \gamma_n(1 - \tau_n)\|(J_\lambda^{B_2} - I)Ay_n\|^2 \leq \|y_n - p\|^2.$$

Thus we have

$$\|u_n - p\| \leq \|y_n - p\|. \tag{22}$$

Using the condition (C3) and (17), we get

$$\begin{aligned}
\|x_{n+1} - p\| &= \|\alpha_n f_n(x_n) + \beta_n x_n + \delta_n z_n - p\| \\
&= \|\alpha_n(f_n(x_n) - f_n(p)) + \alpha_n(f_n(p) - p) + \beta_n(x_n - p) + \delta_n(z_n - p)\| \\
&\leq \alpha_n \|f_n(x_n) - f_n(p)\| + \alpha_n \|f_n(p) - p\| + \beta_n \|x_n - p\| + \delta_n \|z_n - p\| \\
&\leq \alpha_n \mu_n \|x_n - p\| + \alpha_n \|f_n(p) - p\| + \beta_n \|x_n - p\| + \delta_n(\|x_n - p\| \\
&\quad + (1 - \xi)\theta_n \|x_n - x_{n-1}\|) \\
&\leq (\alpha_n \mu^* + (\beta_n + \delta_n))\|x_n - p\| + (1 - \xi)\delta_n \theta_n \|x_n - x_{n-1}\| + \alpha_n \|f_n(p) - p\| \\
&= (1 - \alpha_n(1 - \mu^*))\|x_n - p\| + (1 - \xi)\delta_n \alpha_n \frac{\theta_n}{\alpha_n} \|x_n - x_{n-1}\| + \alpha_n \|f_n(p) - p\|.
\end{aligned}$$

Since $\{f_n\}$ is the uniform convergence on D, there exists a constant $M > 0$ such that

$$\|f_n(p) - p\| \leq M$$

for each $n \geq 1$. So we can choose $\beta := \dfrac{M}{1 - \mu^*}$ and set

$$a_n := \|x_n - p\|, \quad b_n := \alpha_n \|f_n(p) - p\|,$$

$$c_n := (1 - \xi)\delta_n \alpha_n \frac{\theta_n}{\alpha_n} \|x_n - x_{n-1}\|, \quad \sigma_n := \alpha_n(1 - \mu^*).$$

By Lemma 3 (1) and our assumptions, it follows that $\{x_n\}$ is bounded. Moreover, $\{u_n\}$ and $\{y_n\}$ are also bounded.

Now, by Lemma 2, we get

$$\|x_{n+1} - p\|^2$$
$$= \|\alpha_n(f_n(x_n) - f_n(p)) + \alpha_n(f_n(p) - p) + \beta_n(x_n - p) + \delta_n(z_n - p)\|^2$$
$$\leq \|\alpha_n(f_n(x_n) - f_n(p)) + \beta_n(x_n - p) + \delta_n(z_n - p)\|^2 + 2\alpha_n\langle f_n(p) - p, x_{n+1} - p\rangle$$
$$= \|\beta_n(x_n - p) + \delta_n(z_n - p)\|^2 + \alpha_n^2\|f_n(x_n) - f_n(p)\|^2$$
$$+ 2\alpha_n\langle f_n(x_n) - f_n(p), \beta_n(x_n - p) + \delta_n(z_n - p)\rangle + 2\alpha_n\langle f_n(p) - p, x_{n+1} - p\rangle$$
$$\leq \beta_n^2\|x_n - p\|^2 + \delta_n^2\|z_n - p\|^2 + 2\beta_n\delta_n\langle x_n - p, z_n - p\rangle + \alpha_n^2\mu_n^2\|x_n - p\|^2$$
$$+ 2\alpha_n\langle f_n(p) - p, x_{n+1} - p\rangle + 2\alpha_n\|f_n(x_n) - f_n(p)\|\|\beta_n(x_n - p) + \delta_n(z_n - p)\|$$
$$\leq \beta_n^2\|x_n - p\|^2 + \delta_n^2\|z_n - p\|^2 + \beta_n\delta_n\left(\|x_n - p\|^2 + \|z_n - p\|^2 - \|x_n - z_n\|^2\right)$$
$$+ \alpha_n^2\mu^{*2}\|x_n - p\|^2 + 2\alpha_n\mu_n\|x_n - p\|(\beta_n\|x_n - p\| + \delta_n\|z_n - p\|)$$
$$+ 2\alpha_n\langle f_n(p) - p, x_{n+1} - p\rangle$$
$$\leq \beta_n(\beta_n + \delta_n)\|x_n - p\|^2 + \delta_n(\beta_n + \delta_n)\|z_n - p\|^2 - \beta_n\delta_n\|x_n - z_n\|^2 + \alpha_n^2\mu^{*2}\|x_n - p\|^2$$
$$+ 2\mu^*\alpha_n(\beta_n + \delta_n)\|x_n - p\|^2 + 2\mu^*\alpha_n(1 - \xi)\delta_n\theta_n\|x_n - x_{n-1}\|\|x_n - p\|$$
$$+ 2\alpha_n\langle f_n(p) - p, x_{n+1} - p\rangle$$
$$\leq \beta_n(\beta_n + \delta_n)\|x_n - p\|^2 + \delta_n(\beta_n + \delta_n)\left(\|x_n - p\|^2 + \theta_n^2\|x_n - x_{n-1}\|^2\right)$$
$$+ 2\theta_n\|x_n - x_{n-1}\|\|x_n - p\|) - \beta_n\delta_n\|x_n - z_n\|^2 + \alpha_n^2\mu^{*2}\|x_n - p\|^2$$
$$+ 2\mu^*\alpha_n(\beta_n + \delta_n)\|x_n - p\|^2 + 2\mu^*\alpha_n(1 - \xi)\delta_n\theta_n\|x_n - x_{n-1}\|\|x_n - p\|$$
$$+ 2\alpha_n\langle f_n(p) - p, x_{n+1} - p\rangle$$
$$= \left((1 - \alpha_n)^2 + \alpha_n^2\mu^{*2} + 2\mu^*\alpha_n(1 - \alpha_n)\right)\|x_n - p\|^2 - \beta_n\delta_n\|x_n - z_n\|^2$$
$$+ 2\left(1 - \alpha_n(1 - \mu^*(1 - \xi))\right)\delta_n\theta_n\|x_n - x_{n-1}\|\|x_n - p\| + (1 - \alpha_n)\delta_n\theta_n^2\|x_n - x_{n-1}\|^2$$
$$+ 2\alpha_n\langle f_n(p) - p, x_{n+1} - p\rangle. \tag{23}$$

Now, we consider two steps for the proof as follows:

Case 1. Suppose that there exists $n_0 \in \mathbb{N}$ such that $\{\|x_n - p\|\}_{n=n_0}^\infty$ is non-increasing and then $\{\|x_n - p\|\}$ converges. By Lemma 1, we get

$$\|x_{n+1} - p\|^2 = \|\alpha_n f_n(x_n) + \beta_n x_n + \delta_n z_n - p\|^2$$
$$= \alpha_n\|f_n(x_n) - p\|^2 + \beta_n\|x_n - p\|^2 + \delta_n\|z_n - p\|^2 - \alpha_n\beta_n\|f_n(x_n) - x_n\|^2$$
$$- \alpha_n\gamma_n\|f_n(x_n) - z_n\|^2 - \beta_n\gamma_n\|x_n - z_n\|^2$$
$$\leq \alpha_n\|f_n(x_n) - p\|^2 + \beta_n\|x_n - p\|^2 + \delta_n\|z_n - p\|^2$$
$$\leq \alpha_n\|f_n(x_n) - p\|^2 + \beta_n\|x_n - p\|^2 + \delta_n(\xi\|x_n - p\|^2 + (1 - \xi)\|u_n - p\|^2)$$
$$\leq \alpha_n\|f_n(x_n) - p\|^2 + (\beta_n + \xi\delta_n)\|x_n - p\|^2 + (1 - \xi)\delta_n\|u_n - p\|^2,$$

which implies that

$$-\|u_n - p\|^2 \leq \frac{1}{(1 - \xi)\delta_n}\left(\alpha_n\|f_n(x_n) - p\|^2 + (\beta_n + \xi\delta_n)\|x_n - p\|^2 - \|x_{n+1} - p\|^2\right). \tag{24}$$

Applying (16) and (24) to (21), we get

$$\gamma_n(\|(J_\lambda^{B_2} - I)Ay_n\|^2 - \gamma_n\|A^*(J_\lambda^{B_2} - I)Ay_n\|^2)$$
$$\leq \|y_n - p\|^2 - \|u_n - p\|^2$$
$$\leq \|x_n - p\|^2 + 2\theta_n\|x_{n-1} - p\|\|x_n - p\| + \theta_n^2\|x_n - x_{n-1}\|^2$$
$$+ \frac{1}{(1-\xi)\delta_n}(\alpha_n\|f_n(x_n) - p\|^2 + (\beta_n + \xi\delta_n)\|x_n - p\|^2 - \|x_{n+1} - p\|^2)$$
$$= \frac{\beta_n + \delta_n}{(1-\xi)\delta_n}\|x_n - p\|^2 + \frac{\alpha_n}{(1-\xi)\delta_n}\|f_n(x_n) - p\|^2 - \frac{1}{(1-\xi)\delta_n}\|x_{n+1} - p\|^2$$
$$+ \theta_n\|x_n - x_{n-1}\|(2\|x_n - p\| + \theta_n\|x_n - x_{n-1}\|)$$
$$\leq \frac{1}{(1-\xi)\epsilon_2}(\|x_n - p\|^2 - \|x_{n+1} - p\|^2) + \frac{\alpha_n}{(1-\xi)\epsilon_2}\left(\|f_n(x_n) - p\|^2 - \|x_n - p\|^2\right.$$
$$+ \frac{\theta_n}{\alpha_n}\|x_n - x_{n-1}\|\left(2\|x_n - p\| + \alpha_n\frac{\theta_n}{\alpha_n}\|x_n - x_{n-1}\|\right)\right).$$

Since $\{\|x_n - p\|\}$ is convergent, we have $\|x_n - p\| - \|x_{n+1} - p\| \to 0$ as $n \to \infty$. By the conditions (C2) and (C4), we get

$$\gamma_n(\|(J_\lambda^{B_2} - I)Ay_n\|^2 - \gamma_n\|A^*(J_\lambda^{B_2} - I)Ay_n\|^2) \to 0 \text{ as } n \to \infty.$$

From the definition of γ_n, we get

$$\frac{\tau_n(1-\tau_n)\|(J_\lambda^{B_2} - I)Ay_n\|^4}{\|A^*(J_\lambda^{B_2} - I)Ay_n\|^2} \to 0 \text{ as } n \to \infty$$

or

$$\frac{\|(J_\lambda^{B_2} - I)Ay_n\|^2}{\|A^*(J_\lambda^{B_2} - I)Ay_n\|} \to 0 \text{ as } n \to \infty.$$

Since

$$\|A^*(J_\lambda^{B_2} - I)Ay_n\| \leq \|A^*\|\|(J_\lambda^{B_2} - I)Ay_n\| = \|A\|\|(J_\lambda^{B_2} - I)Ay_n\|,$$

it is easy to see that

$$\|(J_\lambda^{B_2} - I)Ay_n\| \leq \|A\|\frac{\|(J_\lambda^{B_2} - I)Ay_n\|^2}{\|A^*(J_\lambda^{B_2} - I)Ay_n\|}.$$

Consequently, we get

$$\|(J_\lambda^{B_2} - I)Ay_n\| \to 0 \text{ as } n \to \infty \tag{25}$$

and also

$$\|A^*(J_\lambda^{B_2} - I)Ay_n\| \to 0 \text{ as } n \to \infty. \tag{26}$$

Similarly, from (23) and our assumptions, we get

$$\|x_n - z_n\|^2$$
$$= \frac{1}{\beta_n \delta_n}\{\|x_n - p\|^2 - \|x_{n+1} - p\|^2 + (1 - \alpha_n)\delta_n \theta_n^2 \|x_n - x_{n-1}\|^2$$
$$+ 2(1 - \alpha_n(1 - \mu^*(1 - \xi)))\delta_n \theta_n \|x_n - x_{n-1}\|\|x_n - p\|$$
$$+ \alpha_n [(\alpha_n(1 + \mu^{*2}) - 2(1 - \mu^*(1 - \alpha_n)))\|x_n - p\|^2 + 2\langle f_n(p) - p, x_{n+1} - p\rangle]\}$$
$$\leq \frac{1}{\epsilon_1 \epsilon_2}\{\|x_n - p\|^2 - \|x_{n+1} - p\|^2 + \frac{\theta_n}{\alpha_n}\|x_n - x_{n-1}\|[\delta_n(1 - \alpha_n)\alpha_n^2 \frac{\theta_n}{\alpha_n}\|x_n - x_{n-1}\|$$
$$+ 2\delta_n(1 - \alpha_n(1 - \mu^*(1 - \xi)))\theta_n \|x_n - p\|] + \alpha_n [2\langle f_n(p) - p, x_{n+1} - p\rangle$$
$$+ (\alpha_n(1 + \mu^{*2}) - 2(1 - \mu^*(1 - \alpha_n)))\|x_n - p\|^2]\} \to 0 \text{ as } n \to \infty.$$

Therefore, we have

$$\|x_n - z_n\| \to 0 \text{ as } n \to \infty. \tag{27}$$

By the condition (C2) and (27), we get

$$\|x_{n+1} - x_n\| = \|\alpha_n f_n(x_n) + \beta_n x_n + \delta_n z_n - x_n\|$$
$$\leq \alpha_n \|f_n(x_n) - x_n\| + \delta_n \|x_n - z_n\| \to 0 \text{ as } n \to \infty.$$

Thus we have

$$\|x_{n+1} - z_n\| \leq \|x_{n+1} - x_n\| + \|x_n - z_n\| \to 0 \text{ as } n \to \infty.$$

Since $J_\lambda^{B_1}$ is firmly nonexpansive, we have

$$\|u_n - p\|^2$$
$$= \|J_\lambda^{B_1}(y_n + \gamma_n A^*(J_\lambda^{B_2} - I)Ay_n) - J_\lambda^{B_1}p\|^2$$
$$\leq \langle u_n - p, y_n + \gamma_n A^*(J_\lambda^{B_2} - I)Ay_n - p\rangle$$
$$= \frac{1}{2}(\|u_n - p\|^2 + \|y_n + \gamma_n A^*(J_\lambda^{B_2} - I)Ay_n - p\|^2 - \|u_n - y_n - \gamma_n A^*(J_\lambda^{B_2} - I)Ay_n\|^2)$$
$$= \frac{1}{2}(\|u_n - p\|^2 + \|y_n - p\|^2 + \gamma_n^2 \|A^*(J_\lambda^{B_2} - I)Ay_n\|^2 + 2\langle y_n - p, \gamma_n A^*(J_\lambda^{B_2} - I)Ay_n\rangle$$
$$- \|u_n - y_n\|^2 - \gamma_n^2 \|A^*(J_\lambda^{B_2} - I)Ay_n\|^2 + 2\langle u_n - y_n, \gamma_n A^*(J_\lambda^{B_2} - I)Ay_n\rangle)$$
$$\leq \frac{1}{2}(\|y_n - p\|^2 + \|y_n - p\|^2 - \|u_n - y_n\|^2 + 2\langle u_n - p, \gamma_n A^*(J_\lambda^{B_2} - I)Ay_n\rangle)$$
$$\leq \frac{1}{2}(2\|y_n - p\|^2 - \|u_n - y_n\|^2 + 2\gamma_n \|u_n - p\|\|A^*(J_\lambda^{B_2} - I)Ay_n\|)$$
$$\leq \|y_n - p\|^2 - \frac{1}{2}\|u_n - y_n\|^2 + \gamma_n \|u_n - p\|\|A^*(J_\lambda^{B_2} - I)Ay_n\|$$

or

$$\|u_n - y_n\|^2 \leq 2(\|y_n - p\|^2 - \|u_n - p\|^2 + \gamma_n \|u_n - p\|\|A^*(J_\lambda^{B_2} - 1)Ay_n\|). \tag{28}$$

From (28), (16), (24) and (26) and our assumptions, it follows that

$$
\begin{aligned}
\|u_n - y_n\|^2 &\leq 2\big[\|x_n - p\|^2 + 2\theta_n\|x_n - x_{n-1}\|\|x_n - p\| + \theta_n^2\|x_n - x_{n-1}\|^2 \\
&\quad + \frac{1}{(1-\xi)\delta_n}\big(\alpha_n\|f_n(x_n) - p\|^2 + (\beta_n + \xi\delta_n)\|x_n - p\|^2 - \|x_{n+1} - p\|^2\big) \\
&\quad + \gamma_n\|u_n - p\|\|A^*(J_\lambda^{B_2} - 1)Ay_n\|\big] \\
&= 2\big[\frac{1}{(1-\xi)\epsilon_2}\big(\|x_n - p\|^2 - \|x_{n+1} - p\|^2\big) + \gamma_n\|u_n - p\|\|A^*(J_\lambda^{B_2} - 1)Ay_n\| \\
&\quad + \frac{\alpha_n}{(1-\xi)\epsilon_2}\big(\|f_n(x_n) - p\|^2 - \|x_n - p\|^2 \\
&\quad + \frac{\theta_n}{\alpha_n}\|x_n - x_{n-1}\|\big(2\|x_n - p\| + \alpha_n\frac{\theta_n}{\alpha_n}\|x_n - x_{n-1}\|\big)\big)\big] \to 0 \text{ as } n \to \infty,
\end{aligned}
$$

that is, we have

$$\|u_n - y_n\| \to 0 \text{ as } n \to \infty. \tag{29}$$

From $y_n := x_n + \theta_n(x_n - x_{n-1})$, we get

$$\|y_n - x_n\| = \|x_n + \theta_n(x_n - x_{n-1}) - x_n\| = \alpha_n\frac{\theta_n}{\alpha_n}\|x_n - x_{n-1}\|,$$

which, with the condition (C4), implies that

$$\|y_n - x_n\| \to 0 \text{ as } n \to \infty. \tag{30}$$

In addition, using (27), (29) and (30), we obtain

$$
\begin{aligned}
\|z_n - u_n\| &\leq \|u_n - y_n\| + \|y_n - z_n\| \\
&\leq \|u_n - y_n\| + \|y_n - x_n\| + \|x_n - z_n\| \to 0 \text{ as } n \to \infty.
\end{aligned}
$$

From $z_n := \xi v_n + (1-\xi)u_n$, we get

$$\|v_n - u_n\| = \frac{1}{\xi}\|z_n - u_n\| \to 0 \text{ as } n \to \infty. \tag{31}$$

Thus, by (29)–(31), we also get

$$
\begin{aligned}
\|x_n - v_n\| &\leq \|x_n - u_n\| + \|u_n - v_n\| \\
&\leq \|x_n - y_n\| + \|y_n - u_n\| + \|u_n - v_n\| \to 0 \text{ as } n \to \infty.
\end{aligned}
$$

Therefore, we have

$$d(x_n, Sx_n) \leq \|x_n - v_n\| \to 0 \text{ as } n \to \infty. \tag{32}$$

Since $\{x_n\}$ is bounded, there exists a subsequence $\{x_{n_k}\}$ of $\{x_n\}$ such that $x_{n_k} \rightharpoonup x^* \in H_1$ and, consequently, $\{u_{n_k}\}$ and $\{y_{n_k}\}$ converge weakly to the point x^*.

From (32), Lemma 4 and the demiclosedness principle for a multi-valued mapping S at the origin, we get $x^* \in Sx^*$, which implies that

$$x^* \in F(S).$$

Next, we show that $x^* \in \Omega$. Let $(v, z) \in G(B_1)$, that is, $z \in B_1(v)$. On the other hand, $u_{n_k} = J_\lambda^{B_1}(y_{n_k} + \gamma_{n_k}A^*(J_\lambda^{B_2} - I)Ay_{n_k})$ can be written as

$$y_{n_k} + \gamma_{n_k}A^*(J_\lambda^{B_1} - I)Ay_{n_k} \in u_{n_k} + \lambda B_1(u_{n_k}),$$

or, equivalently,

$$\frac{(y_{n_k} - u_{n_k}) + \gamma_{n_k} A^* (J_\lambda^{B_1} - I) A y_{n_k}}{\lambda} \in B_1(u_{n_k}).$$

Since B_1 is maximal monotone, we get

$$\left\langle v - u_{n_k}, z - \frac{(y_{n_k} - u_{n_k}) + \gamma_{n_k} A^* (J_\lambda^{B_2} - I) A y_{n_k}}{\lambda} \right\rangle \geq 0.$$

Therefore, we have

$$\langle v - u_{n_k}, z \rangle \geq \left\langle v - u_{n_k}, \frac{(y_{n_k} - u_{n_k}) + \gamma_{n_k} A^* (J_\lambda^{B_2} - I) A y_{n_k}}{\lambda} \right\rangle$$

$$= \left\langle v - u_{n_k}, \frac{y_{n_k} - u_{n_k}}{\lambda} \right\rangle + \left\langle v - u_{n_k}, \frac{\gamma_{n_k} A^* (J_\lambda^{B_2} - I) A y_{n_k}}{\lambda} \right\rangle. \tag{33}$$

Since $u_{n_k} \rightharpoonup x^*$, we have

$$\lim_{k \to \infty} \langle v - u_{n_k}, z \rangle = \langle v - x^*, z \rangle.$$

By (26) and (29), it follows that (33) becomes $\langle v - x^*, z \rangle \geq 0$, which implies that

$$0 \in B_1(x^*).$$

Moreover, from (29), we know that $\{ A y_{n_k} \}$ converges weakly to Ax^* and, by (25), the fact that $J_\lambda^{B_2}$ is nonexpansive and the demiclosedness principle for a multi-valued mapping, we have

$$0 \in B_2(Ax^*),$$

which implies that $x^* \in \Omega$. Thus $x^* \in F(S) \cap \Omega$. Since $\{ f_n(x) \}$ is uniformly convergent on D, we get

$$\limsup_{n \to \infty} \langle f_n(p) - p, x_{n+1} - p \rangle = \limsup_{j \to \infty} \langle f_{n_j}(p) - p, x_{n_j+1} - p \rangle$$

$$= \langle f(p) - p, x^* - p \rangle \leq 0.$$

From (23), we get

$$\|x_{n+1} - p\|^2 \leq \left(1 - 2\alpha_n(1 - \mu^*(1 - \alpha_n)) + \alpha_n^2(1 + \mu^{*2})\right) \|x_n - p\|^2 - \beta_n \delta_n \|x_n - z_n\|^2$$

$$+ 2\left(1 - \alpha_n(1 - \mu^*(1 - \xi))\right) \delta_n \theta_n \|x_n - x_{n-1}\| \|x_n - p\|$$

$$+ (1 - \alpha_n) \delta_n \theta_n^2 \|x_n - x_{n-1}\|^2 + 2\alpha_n \langle f_n(p) - p, x_{n+1} - p \rangle$$

$$\leq \left(1 - 2\alpha_n(1 - \mu^*)\right) \|x_n - p\|^2 + 2\alpha_n(1 - \mu^*) \frac{\langle f_n(p) - p, x_{n+1} - p \rangle}{1 - \mu^*}$$

$$+ \alpha_n \Big[\delta_n \frac{\theta_n}{\alpha_n} \|x_n - x_{n-1}\| \left(2\left(1 - \alpha_n(1 - \mu^*(1 - \xi))\right)\right) \|x_n - p\|$$

$$+ \left((1 - \alpha_n) \alpha_n \frac{\theta_n}{\alpha_n} \|x_n - x_{n-1}\|\right) + \alpha_n(1 + \mu^{*2}) \|x_n - p\|^2 \Big].$$

By Lemma 4, we obtain

$$\lim_{n \to \infty} x_n = p.$$

Case 2. Suppose that $\{ \|x_n - p\| \}_{n=n_0}^\infty$ is not a monotonically decreasing sequence for some n_0 large enough. Set $\Gamma_n = \|x_n - p\|^2$ and let $\tau : \mathbb{B} \to \mathbb{N}$ be a mapping defined by

$$\tau(n) := \max\{ k \in \mathbb{N} : k \leq n, \Gamma_k \leq \Gamma_{k+1} \}$$

for all $n \geq n_0$. Obviously, τ is a non-decreasing sequence. Thus we have

$$0 \leq \Gamma_{\tau(n)} \leq \Gamma_{\tau(n)+1}$$

for all $n \geq n_0$. That is, $\|x_{\tau(n)} - p\| \leq \|x_{\tau(n)+1} - p\|$ for all $n \geq n_0$. Thus $\lim_{n\to\infty} \|x_{\tau(n)} - p\|$ exists. As in Case 1, we can show that

$$\lim_{n\to\infty} \|(J_\lambda^{B_2} - I)Ay_{\tau(n)}\| = 0, \quad \lim_{n\to\infty} \|A^*(J_\lambda^{B_2} - I)Ay_{\tau(n)}\| = 0, \tag{34}$$

$$\lim_{n\to\infty} \|x_{\tau(n)+1} - x_{\tau(n)}\| = 0, \quad \lim_{n\to\infty} \|u_{\tau(n)} - x_{\tau(n)}\| = 0, \tag{35}$$

$$\lim_{n\to\infty} \|v_{\tau(n)} - u_{\tau(n)}\| = 0, \quad \lim_{n\to\infty} \|x_{\tau(n)} - v_{\tau(n)}\| = 0. \tag{36}$$

Therefore, we have

$$d(x_{\tau(n)}, Sx_{\tau(n)}) \leq \|x_{\tau(n)} - v_{\tau(n)}\| \to 0 \text{ as } n \to \infty. \tag{37}$$

Since $\{x_{\tau(n)}\}$ is bounded, there exists a subsequence $\{u_{\tau(n)}\}$ of $\{x_{\tau(n)}\}$ that converges weakly to a point $x^* \in H_1$. From $\|u_{\tau(n)} - x_{\tau(n)}\| \to 0$, it follows that $u_{\tau(n)} \rightharpoonup x^* \in H_1$.

Moreover, as in Case 1, we show that $x^* \in F(S) \cap \Omega$. Furthermore, since $\{f_n(x)\}$ is uniformly convergent on $D \subset H_1$, we obtain that

$$\limsup_{n\to\infty} \langle f_{\tau(n)}(p) - p, x_{\tau(n)+1} - p \rangle \leq 0.$$

From (23), we get

$$\begin{aligned}
\|x_{\tau(n)+1} - p\|^2 &\leq \left(1 - 2\alpha_{\tau(n)}(1 - \mu^*(1 - \alpha_{\tau(n)})) + \alpha_{\tau(n)}^2(1 + \mu^{*2})\right)\|x_{\tau(n)} - p\|^2 \\
&\quad - \beta_{\tau(n)}\delta_{\tau(n)}\|x_{\tau(n)} - z_{\tau(n)}\|^2 + 2\alpha_{\tau(n)}\langle f_{\tau(n)}(p) - p, x_{\tau(n)+1} - p \rangle \\
&\quad + 2\left(1 - \alpha_{\tau(n)}(1 - \mu^*(1 - \xi))\right)\delta_{\tau(n)}\theta_{\tau(n)}\|x_{\tau(n)} - x_{\tau(n)-1}\|\|x_{\tau(n)} - p\| \\
&\quad + (1 - \alpha_{\tau(n)})\delta_{\tau(n)}\theta_{\tau(n)}^2\|x_{\tau(n)} - x_{\tau(n)-1}\|^2 \\
&\leq \left(1 - 2\alpha_{\tau(n)}(1 - \mu^*)\right)\|x_{\tau(n)} - p\|^2 + \alpha_{\tau(n)}^2(1 + \mu^{*2})\|x_{\tau(n)} - p\|^2 \\
&\quad + \delta_{\tau(n)}\theta_n\|x_{\tau(n)} - x_{\tau(n)-1}\|\left(2(1 - \alpha_{\tau(n)}(1 - \mu^*))\|x_{\tau(n)} - p\| \right. \\
&\quad \left. + (1 - \alpha_{\tau(n)})\theta_{\tau(n)}\|x_{\tau(n)} - x_{\tau(n)-1}\|\right) + 2\alpha_{\tau(n)}\langle f_{\tau(n)}(p) - p, x_{\tau(n)+1} - p \rangle,
\end{aligned}$$

which implies that

$$\begin{aligned}
2\alpha_{\tau(n)}(1 - \mu^*)\|x_{\tau(n)} - p\|^2 &\leq \|x_{\tau(n)} - p\|^2 - \|x_{\tau(n)+1} - p\|^2 + \alpha_{\tau(n)}^2(1 + \mu^{*2})\|x_{\tau(n)} - p\|^2 \\
&\quad + \delta_{\tau(n)}\theta_n\|x_{\tau(n)} - x_{\tau(n)-1}\|\left(2(1 - \alpha_{\tau(n)}(1 - \mu^*))\|x_{\tau(n)} - p\| \right. \\
&\quad \left. + (1 - \alpha_{\tau(n)})\theta_{\tau(n)}\|x_{\tau(n)} - x_{\tau(n)-1}\|\right) \\
&\quad + 2\alpha_{\tau(n)}\langle f_{\tau(n)}(p) - p, x_{\tau(n)+1} - p \rangle,
\end{aligned}$$

or

$$\begin{aligned}
2(1 - \mu^*)\|x_{\tau(n)} - p\|^2 &\leq \alpha_{\tau(n)}(1 + \mu^{*2})\|x_{\tau(n)} - p\|^2 + 2\langle f_{\tau(n)}(p) - p, x_{\tau(n)+1} - p \rangle \\
&\quad + \delta_{\tau(n)}\frac{\theta_{\tau(n)}}{\alpha_{\tau(n)}}\|x_{\tau(n)} - x_{\tau(n)-1}\|\left(2(1 - \alpha_{\tau(n)}(1 - \mu^*))\|x_{\tau(n)} - p\| \right. \\
&\quad \left. + (1 - \alpha_{\tau(n)})\alpha_{\tau(n)}\frac{\theta_{\tau(n)}}{\alpha_{\tau(n)}}\|x_{\tau(n)} - x_{\tau(n)-1}\|\right).
\end{aligned}$$

Thus we have

$$\limsup_{n\to\infty} \|x_{\tau(n)} - p\| \le 0$$

and so

$$\lim_{n\to\infty} \|x_{\tau(n)} - p\| = 0. \tag{38}$$

By (35) and (38), we get

$$\|x_{\tau(n)+1} - p\| \le \|x_{\tau(n)+1} - x_{\tau(n)}\| + \|x_{\tau(n)} - p\| \to 0, \ n \to \infty.$$

Furthermore, for all $n \ge n_0$, it is easy to see that $\Gamma_{\tau(n)} \le \Gamma_{\tau(n)+1}$ if $n \ne \tau(n)$ (that is, $\tau(n) < n$) because of $\Gamma_j \ge \Gamma_{j+1}$ for $\tau(n) + 1 \le j \le n$. Consequently, it follows that, for all $n \ge n_0$,

$$0 \le \Gamma_n \le \max\{\Gamma_{\tau(n)}, \Gamma_{\tau(n)+1}\} = \Gamma_{\tau(n)+1}.$$

Therefore, $\lim \Gamma_n = 0$, that is, $\{x_n\}$ converges strongly to the point x^*. This completes the proof. $\square$

Remark 1. [22] *The condition (C4) is easily implemented in numerical results because the value of $\|x_n - x_{n-1}\|$ is known before choosing θ_n. Indeed, we can choose the parameter θ_n such as*

$$\theta_n = \begin{cases} \min\left\{\bar{\omega}, \frac{\omega_n}{\|x_n - x_{n-1}\|}\right\}, & \text{if } \|x_n - x_{n-1}\| \ne 0, \\ \bar{\omega}, & \text{otherwise,} \end{cases}$$

where $\{\omega_n\}$ is a positive sequence such that $\omega_n = o(\alpha_n)$. Moreover, in the condition (C4), we can take $\alpha_n = \dfrac{1}{n+1}, \bar{\omega} = \dfrac{4}{5}$ and

$$\theta_n = \begin{cases} \min\left\{\bar{\omega}, \frac{\alpha_n^2}{\|x_n - x_{n-1}\|}\right\}, & \text{if } \|x_n - x_{n-1}\| \ne 0, \\ \bar{\omega}, & \text{otherwise,} \end{cases}$$

or

$$\theta_n = \begin{cases} \min\left\{\frac{4}{5}, \frac{1}{(n+1)^2 \|x_n - x_{n-1}\|}\right\}, & \text{if } \|x_n - x_{n-1}\| \ne 0, \\ \frac{4}{5}, & \text{otherwise.} \end{cases}$$

If the multi-valued quasi-nonexpansive mapping S in Theorem 1 is a single-valued quasi-nonexpansive mapping, then we obtain the following:

Corollary 1. *Let H_1 and H_2 be two real Hilbert spaces. Suppose that $A : H_1 \to H_2$ is a bounded linear operator with adjoint operator A^*. Let $\{f_n\}$ be a sequence of μ_n-contractions $f_n : H_1 \to H_1$ with $0 < \mu_* \le \mu_n \le \mu^* < 1$ and $\{f_n(x)\}$ be uniformly convergent for any x in a bounded subset D of H_1. Suppose that $S : H_1 \to H_1$ is a single-valued quasi-nonexpansive mapping, $I - S$ is demiclosed at the origin and $F(S) \cap \Omega \ne \emptyset$. For any $x_0, x_1 \in H_1$, let the sequences $\{y_n\}, \{u_n\}, \{z_n\}$ and $\{x_n\}$ be generated by*

$$\begin{cases} y_n = x_n + \theta_n(x_n - x_{n-1}), \\ u_n = J_\lambda^{B_1}(y_n + \gamma_n A^*(J_\lambda^{B_2} - I)Ay_n), \\ z_n = \xi S x_n + (1 - \xi)u_n, \\ x_{n+1} = \alpha_n f_n(x_n) + \beta_n x_n + \delta_n z_n \end{cases} \tag{39}$$

for each $n \geq 1$, where $\zeta \in (0,1)$, $\gamma_n := \tau_n \frac{\|(J_\lambda^{B_2}-I)Ay_n\|^2}{\|A^*(J_\lambda^{B_2}-I)Ay_n\|^2}$ with $0 < \tau_* \leq \tau_n \leq \tau^* < 1$, $\{\theta_n\} \subset [0,\bar{\omega})$ for some $\bar{\omega} > 0$ and $\{\alpha_n\}, \{\beta_n\}, \{\delta_n\} \in (0,1)$ with $\alpha_n + \beta_n + \delta_n = 1$ satisfying the following conditions:

(C1) $\lim_{n \to \infty} \alpha_n = 0$;

(C2) $\sum_{n=1}^\infty \alpha_n = \infty$;

(C3) $0 < \epsilon_1 \leq \beta_n$ and $0 < \epsilon_2 \leq \delta_n$;

(C4) $\lim_{n \to \infty} \frac{\theta_n}{\alpha_n} \|x_n - x_{n-1}\| = 0$.

Then the sequence $\{x_n\}$ generated by (39) converges strongly to a point $p \in F(S) \cap \Omega$, where $p = P_{F(S) \cap \Omega} f(p)$.

Remark 2. *If $\theta_n = 0$, then the iterative scheme (14) in Theorem 1 reduces to the iterative (11).*

4. Applications

In this section, we give some applications of the problem **(SFPIP)** in the variational inequality problem and game theory. First, we introduce variational inequality problem in [34] and game theory (see [35]).

4.1. The Variational Inequality Problem

Let C be a nonempty closed and convex subset of a real Hilbert space H_1. Suppose that an operator $F : H_1 \to H_1$ is monotone.

Now, we consider the following variational inequality problem **(VIP)**:

$$\text{Find a point } x^* \in C \text{ such that } \langle Fx^*, y - x^* \rangle \geq 0 \text{ for all } y \in C. \tag{40}$$

The solution set of the problem **(VIP)** is denoted by Γ.

Moreover, it is well-known that x^* is a solution of the problem **(VIP)** if and only if x^* is a solution of the problem **(FPP)** [34], that is, for any $\gamma > 0$,

$$x^* = P_C(x^* - \gamma F x^*).$$

The following lemma is extracted from [2,36]. This lemma is used for finding a solution of the split inclusion problem and the variational inequality problem:

Lemma 5. *Let H_1 be a real Hilbert space, $F : H_1 \to H_1$ be a monotone and L-Lipschitz operator on a nonempty closed and convex subset C of H_1. For any $\gamma > 0$, let $T = P_C(I - \gamma F(P_C(I - \gamma F)))$. Then, for any $y \in \Gamma$ and $L\gamma < 1$, we have*

$$\|Tx - Ty\| \leq \|x - y\|,$$

$I - T$ is demiclosed at the origin and $F(T) = \Gamma$.

Now, we apply our Theorem 1, by combining with Lemma 5, to find a solution of the problem **(VIP)**, that is, a point in the set Γ.

let $B_1 : H_1 \to 2^{H_1}$ and $B_2 : H_2 \to 2^{H_2}$ be maximal monotone mappings defined on H_1 and H_2, respectively, and $A : H_1 \to H_2$ be a bounded linear operator with its adjoint A^*.

Now, we consider the *split fixed point variational inclusion problem* **(SFPVIP)** as follows:

$$\text{Find a point } x^* \in H_1 \text{ such that } 0 \in B_1(x^*), \ x^* \in \Gamma \tag{41}$$

and

$$y^* = Ax^* \in H_2 \text{ such that } 0 \in B_2(y^*). \tag{42}$$

Theorem 2. *Let H_1 and H_2 be two real Hilbert spaces, $A : H_1 \to H_2$ be a bounded linear operator with its adjoint A^*. Let $\{f_n\}$ be a sequence of μ_n-contractions $f_n : H_1 \to H_1$ with $0 < \mu_* \leq \mu_n \leq \mu^* < 1$ and $\{f_n(x)\}$ be uniformly convergent for any x in a bounded subset D of H_1. For any $\lambda > 0$, let $T = P_C(I - \gamma F(P_C(I - \gamma F)))$ with $L\gamma < 1$, where $F : H_1 \to H_1$ is a L-Lipschitz and monotone operator on $C \subset H_1$ and $F(T) \cap \Omega \neq \emptyset$. For any $x_0, x_1 \in H_1$, let the sequences $\{y_n\}$, $\{u_n\}$, $\{z_n\}$ and $\{x_n\}$ be generated by*

$$\begin{cases} y_n = x_n + \theta_n(x_n - x_{n-1}), \\ u_n = J_\lambda^{B_1}(y_n + \gamma_n A^*(J_\lambda^{B_2} - I)Ay_n), \\ z_n = \xi T x_n + (1 - \xi)u_n, \\ x_{n+1} = \alpha_n f_n(x_n) + \beta_n x_n + \delta_n z_n \end{cases} \tag{43}$$

for each $n \geq 1$, where $\xi \in (0, 1)$, $\gamma_n := \tau_n \dfrac{\|(J_\lambda^{B_2} - I)Ay_n\|^2}{\|A^*(J_\lambda^{B_2} - I)Ay_n\|^2}$ with $0 < \tau_* \leq \tau_n \leq \tau^* < 1$, $\{\theta_n\} \subset [0, \bar{\omega})$ for some $\bar{\omega} > 0$ and $\{\alpha_n\}, \{\beta_n\}, \{\delta_n\} \in (0, 1)$ with $\alpha_n + \beta_n + \delta_n = 1$ satisfying the following conditions:

(C1) $\lim_{n \to \infty} \alpha_n = 0$;

(C2) $\sum_{n=1}^{\infty} \alpha_n = \infty$;

(C3) $0 < \epsilon_1 \leq \beta_n, 0 < \epsilon_2 \leq \delta_n$;

(C4) $\lim_{n \to \infty} \dfrac{\theta_n}{\alpha_n} \|x_n - x_{n-1}\| = 0$.

Then the sequence $\{x_n\}$ generated by (43) *converges strongly to a point $p \in F(T) \cap \Omega = \Gamma \cap \Omega$, where $p = P_{\Gamma \cap \Omega} f(p)$.*

Proof. Since $I - T$ is demiclosed at the origin and $F(T) = \Gamma$, by using Lemma (5) and Corollary (1), the sequence $\{x_n\}$ converges strongly to a point $p \in F(T) \cap \Omega$, that is, the sequence $\{x_n\}$ converges strongly to a point $p \in \Gamma$. $\square$

4.2. Game Theory

Now, we consider a game of N players in strategic form

$$G = (p_i, S_i),$$

where $i = 1, \cdots, N$, $p_i : S = S_1 \times S_2 \times \cdots \times S_N \to \mathbb{R}$ is the pay-off function (continuous) of the ith player and $S_i \in \mathbb{R}^{M_i}$ is the set of strategy of the ith player such that $M_i = |S_i|$.

Let S_i be nonempty compact and convex set, $s_i \in S_i$ be the strategy of the ith player and $s = (s_1, s_2, \cdots, s_N)$ be the collective strategy of all players. For any $s \in S$ and $z_i \in S_i$ of the ith player for each i, the symbols S_{-i}, s_{-i} and (z_i, s_{-i}) are defined by

- $S_{-i} := (S_1 \times \cdots \times S_{i-1} \times S_{i+1} \times \cdots \times S_N)$ is the set of strategies of the remaining players when s_i was chosen by ith player,

- $s_{-i} := (s_1, \cdots, s_{i-1}, s_{i+1}, \cdots, s_N)$ is the strategies of the remaining players when ith player has s_i and

- $(z_i, s_{-i}) := (s_1, \cdots, s_{i-1}, z_i, s_{i+1}, \cdots, s_N)$ is the strategies of the situation that z_i was chosen by ith player when the rest of the remaining players have chosen s_{-i}.

Moreover, $\bar{s}_i$ is a special strategy of the ith player, supporting the player to maximize his pay-off, which equivalent to the following:

$$p_i(\bar{s}_i, s_{-i}) = \max_{z_i \in S_i} p_i(z_i, s_{-i}).$$

Definition 5. [37,38] *Given a game of N players in strategic form, the collective strategies $s^* \in S$ is said to be a Nash equilibrium point if*

$$p_i(s^*) = \max_{z_i \in S_i} p_i(z_i, s_i^*)$$

for all $i = 1, \cdots, N$ and $s_i^* \in S_{-i}$.

If no player can change his strategy to bring advantages, then the collective strategies $s^* = (s_i^*, s_{-i}^*)$ is a Nash equilibrium point. Furthermore, a Nash equilibrium point is the collective strategies of all players, i.e., s_i^* (for each $i \geq 1$) is the best response of ith player. There is a multi-valued mapping $T_i : S_{-i} \to 2^{S_i}$ such that

$$
\begin{aligned}
T_i(s_{-i}) &= \arg \max p_i(z_i, s_{-i}) \\
&= \{s_i \in S_i : p_i(s_i, s_{-i}) = \max_{z_i \in S_i} p_i(z_i, s_{-i})\}
\end{aligned}
$$

for all $s_{-i} \in S_{-i}$. Therefore, we can define the mapping $T : S \to 2^S$ by

$$T := T_1 \times T_2 \times \cdots \times T_N$$

such that the Nash equilibrium point is the collective strategies s^*, where $s^* \in F(T)$. Note that $s^* \in F(T)$ is equivalent to $s_i^* \in T(s_{-i}^*)$.

Let H_1 and H_2 be two real Hilbert spaces, $B_1 : H_1 \to 2^{H_1}$ and $B_2 : H_2 \to 2^{H_2}$ be multi-valued mappings. Suppose S is nonempty compact and convex subset of $H_1 = \mathbb{R}^{M_N}$, $H_2 = \mathbb{R}$ and the rest of the players have made their best responses s_{-i}^*. For each $s \in S$, define a mapping $A : S \to H_2$ by

$$As = p_i(s) - p_i(z_i, s_{-i}^*),$$

where p_i is linear, bounded and convex. Indeed, A is also linear, bounded and convex.

The *Nash equilibrium problem* **(NEP)** is the following:

$$\text{Find a point} \quad s^* \in S \quad \text{such that} \quad As^* > 0, \quad 0 \in H_2. \tag{44}$$

However, the solution to the problem **(NEP)** may not be single-valued. Then the problem **(NEP)** reduces to finding the fixed point problem **(FPP)** of a multi-valued mapping, i.e.,

$$\text{Find a point} \quad s^* \in S \quad \text{such that} \quad s^* \in Ts^*, \tag{45}$$

where T is multi-valued pay-off function.

Now, we apply our Theorem 1 to find a solution to the problem **(FPP)**.

Let $B_1 : H_1 \to 2^{H_1}$ and $B_2 : H_2 \to 2^{H_2}$ be maximal monotone mappings defined on H_1 and H_2, respectively, and $A : H_1 \to H_2$ be a bounded linear operator with its adjoint A^*.

Now, we consider the following problem:

$$\text{Find a point} \quad s^* \in H_1 \quad \text{such that} \quad 0 \in B_1(s^*), \quad s^* \in Ts^* \tag{46}$$

and

$$y^* = As^* \in H_2 \quad \text{such that} \quad 0 \in B_2(y^*). \tag{47}$$

Theorem 3. *Assume that B_1 and B_2 are maximal monotone mappings defined on Hilbert spaces H_1 and H_2, respectively. Let $T : S \to CB(S)$ be a multi-valued quasi-nonexpansive mapping such that T is demiclosed at the origin. Let $\{f_n\}$ be a sequence of μ_n-contractions $f_n : H_1 \to H_1$ with $0 < \mu_* \leq \mu_n \leq \mu^* < 1$ and $\{f_n(x)\}$ be uniformly convergent for any x in a bounded subset D of H_1. Suppose that the problem **(NEP)** has* *a nonempty solution and $F(T) \cap \Omega \neq \emptyset$. For arbitrarily chosen $x_0, x_1 \in H_1$, let the sequences $\{y_n\}, \{u_n\}, \{z_n\}$ and $\{x_n\}$ be generated by*

$$
\begin{cases}
y_n = x_n + \theta_n(x_n - x_{n-1}), \\
u_n = J_\lambda^{B_1}(y_n + \gamma_n A^*(J_\lambda^{B_2} - I)Ay_n), \\
z_n = \xi v_n + (1-\xi)u_n, \quad v_n \in Tx_n, \\
x_{n+1} = \alpha_n f_n(x_n) + \beta_n x_n + \delta_n z_n
\end{cases}
\tag{48}
$$

for each $n \geq 1$, where $\xi \in (0,1)$, $\gamma_n := \tau_n \dfrac{\|(J_\lambda^{B_2}-I)Ay_n\|^2}{\|A^*(J_\lambda^{B_2}-I)Ay_n\|^2}$ with $0 < \tau_* \leq \tau_n \leq \tau^* < 1$, $\{\theta_n\} \subset [0, \bar{\omega})$ for some $\bar{\omega} > 0$ and $\{\alpha_n\}, \{\beta_n\}, \{\delta_n\} \in (0,1)$ with $\alpha_n + \beta_n + \delta_n = 1$ satisfying the following conditions:

(C1) $\lim_{n\to\infty} \alpha_n = 0$;

(C2) $\sum_{n=1}^{\infty} \alpha_n = \infty$;

(C3) $0 < \epsilon_1 \leq \beta_n$ and $0 < \epsilon_2 \leq \delta_n$;

(C4) $\lim_{n\to\infty} \dfrac{\theta_n}{\alpha_n} \|x_n - x_{n-1}\| = 0$.

Then the sequence $\{x_n\}$ generated by Equation (48) converges strongly to Nash equilibrium point.

Proof. By Theorem 1, the sequence $\{x_n\}$ converges strongly to a point $p \in F(T) \cap \Omega$, then the sequence $\{x_n\}$ converges strongly to a Nash equilibrium point. $\square$

Author Contributions: All five authors contributed equally to work. All authors read and approved the final manuscript. P.K. and K.S. conceived and designed the experiments. P.P., W.J. and Y.J.C. analyzed the data. P.P. and W.J. wrote the paper. Authorship must be limited to those who have contributed substantially to the work reported.

Funding: The Royal Golden Jubilee PhD Program (Grant No. PHD/0167/2560). The Petchra Pra Jom Klao Ph.D. Research Scholarship (Grant No. 10/2560). The King Mongkut's University of Technology North Bangkok, Contract No. KMUTNB-KNOW-61-035.

Acknowledgments: The authors acknowledge the financial support provided by King Mongkut's University of Technology Thonburi through the "KMUTT 55th Anniversary Commemorative Fund". Pawicha Phairatchatniyom would like to thank the "*Science Graduate Scholarship*", Faculty of Science, King Mongkut's University of Technology Thonburi (KMUTT) (Grant No. 11/2560). Wachirapong Jirakitpuwapat would like to thank the Petchra Pra Jom Klao Ph.D. Research Scholarship and the King Mongkut's University of Technology Thonburi (KMUTT) for financial support. Moreover, Kanokwan Sitthithakerngkiet was funded by King Mongkut's University of Technology North Bangkok, Contract No. KMUTNB-KNOW-61-035.

Conflicts of Interest: The authors declare no conflict of interest.

References

1. Moudafi. Split monotone variational inclusions. *J. Opt. Theory Appl.* **2011**, *150*, 275–283. [CrossRef]

2. Shehu, Y.; Agbebaku, D. On split inclusion problem and fixed point problem for multi-valued mappings. *Comput. Appl. Math.* **2017**, *37*. [CrossRef]

3. Shehu, Y.; Ogbuisi, F.U. An iterative method for solving split monotone variational inclusion and fixed point problems. *Rev. Real Acad. Cienc. Exact. Fís. Nat. Serie A. Mat.* **2015**, *110*, 503–518. [CrossRef]

4. Kazmi, K.; Rizvi, S. An iterative method for split variational inclusion problem and fixed point problem for a nonexpansive mapping. *Opt. Lett.* **2013**, *8*. [CrossRef]

5. Sitthithakerngkiet, K.; Deepho, J.; Kumam, P. A hybrid viscosity algorithm via modify the hybrid steepest descent method for solving the split variational inclusion in image reconstruction and fixed point problems. *Appl. Math. Comput.* **2015**, *250*, 986–1001. [CrossRef]

6. Censor, Y.; Gibali, A.; Reich, S. Algorithms for the split variational inequality problem. *Num. Algorithms* **2012**, *59*, 301–323. [CrossRef]

7. Byrne, C.; Censor, Y.; Gibali, A.; Reich, S. Weak and strong convergence of algorithms for the split common null point problem. Technical Report. *arXiv* **2011**, arXiv:1108.5953.

8. Moudafi, A. The split common fixed-point problem for demicontractive mappings. *Inverse Prob.* **2010**, *26*, 055007. [CrossRef]

9. Yao, Y.; Liou, Y.C.; Postolache, M. Self-adaptive algorithms for the split problem of the demicontractive operators. *Optimization* **2017**, *67*, 1309–1319. [CrossRef]

10. Dang, Y.; Gao, Y. The strong convergence of a KM-CQ-like algorithm for a split feasibility problem. *Inverse Prob.* **2011**, *27*, 015007. [CrossRef]

11. Sahu, D.R.; Pitea, A.; Verma, M. A new iteration technique for nonlinear operators as concerns convex programming and feasibility problems. *Numer. Algorithms* **2019**. [CrossRef]

12. Censor, Y.; Elfving, T. A multiprojection algorithm using Bregman projections in a product space. *Numer. Algorithms* **1994**, *8*, 221–239. [CrossRef]

13. Combettes, P. The convex feasibility problem in image recovery. *Adv. Imag. Electron. Phys.* **1996**, *95*, 155–270. [CrossRef]

14. Kazmi, K.R.; Rizvi, S.H. Iterative approximation of a common solution of a split equilibrium problem, a variational inequality problem and a fixed point problem. *J. Egypt. Math. Soc.* **2013**, *21*, 44–51. [CrossRef]

15. Peng, J.W.; Wang, Y.; Shyu, D.S.; Yao, J.C. Common solutions of an iterative scheme for variational inclusions, equilibrium problems, and fixed point problems. *J. Inequal. Appl.* **2008**, *15*, 720371. [CrossRef]

16. Jung, J.S. Strong convergence theorems for multivalued nonexpansive nonself-mappings in Banach spaces. *Nonlinear Anal. Theory, Meth. Appl.* **2007**, *66*, 2345–2354. [CrossRef]

17. Panyanak, B. Mann and Ishikawa iterative processes for multivalued mappings in Banach spaces. *Comput. Math. Appl.* **2007**, *54*, 872–877. [CrossRef]

18. Shahzad, N.; Zegeye, H. On Mann and Ishikawa iteration schemes for multi-valued maps in Banach spaces. *Nonlinear Anal.* **2009**, *71*, 838–844. [CrossRef]

19. Polyak, B. Some methods of speeding up the convergence of iteration methods. *USSR Comput. Math. Math. Phys.* **1964**, *4*, 1–17. [CrossRef]

20. Nesterov, Y. A method of solving a convex programming problem with convergence rate O(1/sqr(k)). *Sov. Math. Dokl.* **1983**, *27*, 372–376.

21. Dang, Y.; Sun, J.; Xu, H. Inertial accelerated algorithms for solving a split feasibility problem. *J. Ind. Manag. Optim.* **2017**, *13*, 1383–1394. [CrossRef]

22. Suantai, S.; Pholasa, N.; Cholamjiak, P. The modified inertial relaxed CQ algorithm for solving the split feasibility problems. *J. Ind. Manag. Opt.* **2017**, *13*, 1–21. [CrossRef]

23. Alvarez, F.; Attouch, H. An inertial proximal method for maximal monotone operators via discretization of a nonlinear oscillator with damping. Wellposedness in optimization and related topics (Gargnano, 1999). *Set-Valued Anal.* **2001**, *9*, 3–11. [CrossRef]

24. Rockafellar, R.T. Monotone operators and the proximal point algorithm. *SIAM J. Control Opt.* **1976**, *14*, 877–898. [CrossRef]

25. Attouch, H.; Peypouquet, J.; Redont, P. A dynamical approach to an inertial forward-backward algorithm for convex minimization. *SIAM J. Opt.* **2014**, *24*, 232–256. [CrossRef]

26. Bot, R.I.; Csetnek, E.R. An inertial alternating direction method of multipliers. *Min. Theory Appl.* **2016**, *1*, 29–49.

27. Maingé, P.E. Convergence theorems for inertial KM-type algorithms. *J. Comput. Appl. Math.* **2008**, *219*, 223–236. [CrossRef]

28. Bot, R.I.; Csetnek, E.R.; Hendrich, C. Inertial Douglas-Rachford splitting for monotone inclusion problems. *Appl. Math. Comput.* **2015**, *256*, 472—487.

29. Chuang, C.S. Strong convergence theorems for the split variational inclusion problem in Hilbert spaces. *Fix. Point Theory Appl.* **2013**, *2013*. [CrossRef]

30. Che, H.; Li, M. Solving split variational inclusion problem and fixed point problem for nonexpansive semigroup without prior knowledge of operator norms. *Math. Prob. Eng.* **2015**, *2015*, 1–9. [CrossRef]

31. Ansari, Q.H.; Rehan, A.; Wen, C.F. Split hierarchical variational inequality problems and fixed point problems for nonexpansive mappings. *J. Inequal. Appl.* **2015**, *16*, 274. [CrossRef]

32. Xu, H.K. Iterative algorithms for nonlinear operators. *J. Lond. Math. Soc.* **2002**, *66*, 240–256. [CrossRef]

33. Maingé, P.E. Approximation methods for common fixed points of nonexpansive mappings in Hilbert spaces. *J. Math. Anal. Appl.* **2007**, *325*, 469–479. [CrossRef]

34. Glowinski, R.; Tallec, P. *Augmented Lagrangian and Operator-Splitting Methods in Nonlinear Mechanics*; SIAM Studies in Applied Mathematics; Society for Industrial and Applied Mathematics: Philadelphia, PA, USA, 1989.

35. Von Neumann, J.; Morgenstern, O. *Theory of Games and Economic Behavior*; Princeton University Press: Princeton, NJ, USA, 1947.

36. Kraikaew, R.; Saejung, S. Strong convergence of the Halpern subgradient extragradient method for solving variational inequalities in Hilbert spaces. *J. Opt. Theory Appl.* **2014**, *163*, 399–412. [CrossRef]

37. Nash, J.F., Jr. Equilibrium points in *n*-person games. *Proc. Nat. Acad. Sci. USA* **1950**, *36*, 48–49. [CrossRef] [PubMed]

38. Nash, J. Non-cooperative games. *Ann. Math.* **1951**, *54*, 286–295. [CrossRef]

mathematics

MDPI

Article

A Fast Derivative-Free Iteration Scheme for Nonlinear Systems and Integral Equations

Mozafar Rostami, Taher Lotfi * and Ali Brahmand

Department of Mathematics, Hamedan Branch, Islamic Azad University, Hamedan 15743-65181, Iran
* Correspondence: lotfi@iauh.ac.ir

Received: 23 June 2019; Accepted: 11 July 2019; Published: 18 July 2019

Abstract: Derivative-free schemes are a class of competitive methods since they are one remedy in cases at which the computation of the Jacobian or higher order derivatives of multi-dimensional functions is difficult. This article studies a variant of Steffensen's method with memory for tackling a nonlinear system of equations, to not only be independent of the Jacobian calculation but also to improve the computational efficiency. The analytical parts of the work are supported by several tests, including an application in mixed integral equations.

Keywords: integral equation; efficiency index; nonlinear models; iterative methods; higher order

MSC: 65H10; 45D05

1. Introductory Notes

1.1. Background

There exist many works handling the approximate solution of linear and nonlinear integral equations. However, tackling nonlinear integral equations would be more challenging due to the presence of nonlinearity which might be expensive for different solvers [1,2].

Some authors discussed the asymptotic error expansion of collocation-type and Nystrom-type methods for Volterra–Fredholm integral equations with nonlinearity, see [3] for a complete discussion on this issue. One class of nonlinear internal equations is the mixed Hammerstein integral equations with several application in engineering problems [2].

Since in the process of finding the solution of such integral equations, most of the time a system of algebraic equation would occur that must be solved quickly and accurately, thus we here bring the attention to develop and study a useful numerical solution scheme for solving nonlinear systems with application in tackling nonlinear integral equations.

Clearly, there are some other nonlinear problems in literature which could yield in tackling nonlinear system of equations, see e.g., [4,5].

1.2. Definition

Consider a nonlinear system of equations of algebraic type as follows [6]:

$$
\begin{cases}
a_1(x_1, x_2, \ldots, x_m) = 0, \\
a_2(x_1, x_2, \ldots, x_m) = 0, \\
\vdots \\
a_m(x_1, x_2, \ldots, x_m) = 0,
\end{cases}
\tag{1}
$$

which contains m equations with m unknowns and $A(x) = (a_1(x), a_2(x), \ldots, a_m(x))^T$ while $a_1(x), a_2(x),$ $\ldots, a_m(x)$ are the functions of coordinate. We can also write (1) using $x = (x_1, x_2, \ldots, x_m)$ in a more compact form as

$$
A(x) = 0.
\tag{2}
$$

The purpose of this work is to study finding the solution of system (1) via iteration process and discuss its application in solving nonlinear integral equations. As such, now let us briefly review some of the existing methods for finding its simple roots in the next subsection.

1.3. Existing Solvers

The Steffensen's scheme for solving nonlinear systems is written as follows [7]:

$$
\begin{cases}
w^{(n)} = x^{(n)} + A(x^{(n)}), \quad x^{(0)} \in \mathbb{R}^m, \\
x^{(n+1)} = x^{(n)} - [x^{(n)}, w^{(n)}; A]^{-1} A(x^{(n)}), \quad n = 0, 1, 2, \cdots,
\end{cases}
\tag{3}
$$

which is based upon the divided difference operator (DDO). The 1st order DDO of A for the multidimensional nodes x and y is expressed by a component-to-component procedure as follows [8]:

$$
[x, y; A]_{i,j} = \frac{A_i(x_1, \ldots, x_j, y_{j+1}, \ldots, y_m) - A_i(x_1, \ldots, x_{j-1}, y_j, \ldots, y_m)}{x_j - y_j}, \quad 1 \leq i, j \leq m.
\tag{4}
$$

Recall that first-order divided difference of A on $\mathbb{R}^m$ is a mapping as follows:

$$
[\cdot, \cdot; A] : D \subset \mathbb{R}^m \times \mathbb{R}^m \to \mathcal{L}(\mathbb{R}^m),
\tag{5}
$$

that reads

$$
[y, x; A](y - x) = A(y) - A(x), \quad \forall x, y \in D.
\tag{6}
$$

Here $\mathcal{L}(\cdot)$ shows the set of bounded linear functions. By considering $h = y - x$, one we can also express the first-order DDO as follows [8]:

$$
[x + h, x; A] = \int_0^1 A'(x + th)dt, \quad \forall (x, h) \in \mathbb{R}^m \times \mathbb{R}^m.
\tag{7}
$$

Traub in [9] investigated another way based on the function $J(x, H)$ for approximating the Jacobian matrix of the Newton's method and to obtain the Steffensen's scheme based on a point-wise definition.

An improvement of (3) was given in [10,11] as follows:

$$
\begin{cases}
z^{(n)} = x^{(n)} - [x^{(n)}, w^{(n)}; A]^{-1} A(x^{(n)}), \\
x^{(n+1)} = z^{(n)} - [x^{(n)}, w^{(n)}; A]^{-1} A(z^{(n)}),
\end{cases}
\tag{8}
$$

wherein

$$w^{(n)} = x^{(n)} + \beta A(x^{(n)}), \ \beta \in \mathbb{R}. \tag{9}$$

The point in (8) in contrast to (3) is that it applies two steps and of course two m-D functional evaluations to reach a higher rate than quadratic. Here, the idea is to freeze the DDO per cycle and then increase the sub steps so as to gain as much as possible of order improvement, as well as some improvements in the computational efficiency index of the scheme.

Let us also recall some of the iteration schemes having the requirement of Jacobian computation now. The Jarratt's iteration having fourth rate of convergence for solving (1) is given by [12]:

$$\begin{cases} z^{(n)} = x^{(n)} - \frac{2}{3}A'(x^{(n)})^{-1}A(x^{(n)}), \\ x^{(n+1)} = x^{(n)} - \frac{1}{2}(3A'(z^{(n)}) - A'(x^{(n)}))^{-1} \\ \qquad \times (3A'(z^{(n)}) + A'(x^{(n)}))A'(x^{(n)})^{-1}A(x^{(n)}). \end{cases} \tag{10}$$

This fourth-order iteration expression requires the computation of two matrix inverses (based on the resolution of linear systems) to achieve its rate, which manifest that getting higher rate of convergence in the form of a two-step method is costly.

1.4. Motivation

All methods discussed until now are without memory; some improvements over such schemes can be done by considering additional memory terms.

Our motivation of pursuing this aim is not only limited at tackling nonlinear systems, but a motivation is to apply such schemes for practical engineering problems such as the nonlinear mixed integral equations, see e.g., [13–16].

The goal in our development is to reach a higher computational efficiency using as low as possible number of linear systems of equations and the functional evaluations. This is directly interlinked with the concept of scientific computing and numerical analysis which gives a meaning to the investigation and proposing novel numerical procedures.

1.5. Achievement and Contribution

The objective of this work is to present a two-step higher order scheme to solve system of nonlinear equations. As such, we present an iteration method with memory for finding both real and complex zeros. Our scheme does not require computing the Fréchet derivatives of the function.

1.6. Organization

We unfold this article as follows. In Section 2, the derivation and contribution of an iteration expression is furnished. Section 3 provides an error analysis for its convergence rate. The computational efficiency of different solvers by including not only the number of functional evaluations, but also the number of system of involved linear equations, the number of LU (Lower-Upper) factorizations as well as the other similar operations will be discussed in Section 4 in detail. Section 5 discusses the application of the proposed scheme. Concluding remarks are given in Section 6.

2. A Derivative-Free Scheme

Here our attempt is to increase the computational efficiency index of (8) without imposing several more steps or further DDDs in each cycle. To complete such a task, we rely on the concept of methods with memory which state that the convergence speed and efficiency of iterative methods could be improved by saving and using the already computed values of the functions and nodes.

In fact, the error equation of the uni-parametric family of methods (8) includes a term of the form below:

$$I + \beta A'(\alpha) = 0. \tag{11}$$

The free nonzero parameter β in (11) can clearly affect not only on the domain of convergence (attraction basins of the iterative method) but also to the improvement of the convergence order. When tackling a nonlinear system of equations, and since α is not known, we can use an approximation for $A'(\alpha)$ to make the whole relation (11) approximately zero. Therefore, we may write

$$\beta \simeq -A'(\bar{\alpha})^{-1}, \tag{12}$$

wherein $\bar{\alpha}$ is an approximation of the solution (per cycle).

It is important to discuss how we approximate the matrix $\beta := B^{(n)}(n \geq 1)$ by employing some estimates to $-A'(\alpha)$ computed via the existing data.

To improve the performance of (8) using the notion of methods with memory, we consider the following iteration expression:

$$\begin{cases} w^{(n)} &= x^{(n)} + \beta A(x^{(n)}) \\ z^{(n)} &= x^{(n)} - [x^{(n)}, w^{(n)}; A]^{-1} A(x^{(n)}), \\ x^{(n+1)} &= z^{(n)} - [x^{(n)}, w^{(n)}; A]^{-1} A(z^{(n)}). \end{cases} \tag{13}$$

To ease up the implementation of the scheme with memory, let us first consider

$$\beta := B^{(n)} = -[w^{(n-1)}, x^{(n-1)}; A]^{-1} = -M^{(n-1)^{-1}} \approx -A'(\alpha)^{-1}. \tag{14}$$

and

$$\begin{cases} M^{(n-1)}\delta^{(n)} = A(x^{(n)}), \\ M^{(n-1)}\gamma^{(n)} = A(y^{(n)}). \end{cases} \tag{15}$$

Thus, now we contribute the following scheme:

$$\begin{cases} B^{(n)} = -[w^{(n-1)}, x^{(n-1)}; A]^{-1}, & n \geq 1, \\ w^{(n)} = x^{(n)} + B^{(n)} A(x^{(n)}), n \geq 1, \\ y^{(n)} = x^{(n)} + \delta^{(n)}, & n \geq 0, \\ x^{(n+1)} = y^{(n)} + \gamma^{(n)}. \end{cases} \tag{16}$$

Lemma 1. *Let $D \subset \mathbb{R}^m$ be a nonempty convex domain. Suppose that A is thrice Fréchet differentiable on D, and that $[u, v; A] \in \mathcal{L}(D, D)$, for any $u, v \in D (u \neq v)$ and the initial value $x^{(0)}$ and the solution α are close to each other. By considering $B^{(n)} = -[w^{(n-1)}, x^{(n-1)}; A]^{-1}$ and $d^{(n)} := I + B^{(n)} A'(\alpha)$, one obtains the error relation below*

$$d^{(n)} \sim e^{(n-1)}. \tag{17}$$

Proof. See [17] for more details. □

To implement (16), one needs to solve some linear systems of algebraic equations. This means that at each new step, a new LU factorization is needed, and no information can be exploited from the previous steps. However, there exists a body of literature about recycling this kind of information to obtain updated preconditioner for iterative solvers, [18–20]. We leave future discussion about constructing and imposing such a preconditioner for future works in this field.

As long as the coefficient matrices are sparse or large and sparse, a Krylov subspace method can be employed to speed up the process. However, the merit in (16) is that the two linear systems have one same coefficient matrix. Hence, only one LU factorization would be enough and by saving the decomposition, one can act it to two different right-hand-side vectors to get the solution vectors per sub cycles of (16).

A challenging part of the implementation using (16) is the incorporation of $B^{(n)}$. This is a not anymore a constant and it should be defined as a matrix. In this paper, whenever required the initial matrix $B^{(0)}$ is specified by:

$$B^{(0)^{-1}} = \text{diag}\left(-\frac{1}{1000}\right). \tag{18}$$

The choice of the initial matrix $B^{(0)}$ affects directly on the whole process in order to arrive in the convergence phase as quickly as possible. Here, (18) is in agreement with the dynamical studies of Steffensen-type methods with memory at which the basins of attractions are larger as long as the free parameter is close to zero.

Noting also that updating $B^{(n)}$ per cycle is again based on the already computed LU factorization while it should only act on the identity matrix to proceed.

3. Rate of Convergence

It is known that via the Taylor expansion of $A'(x+th)$ in the node x and integrating, one obtains:

$$\int_0^1 A'(x+th)dt = A'(x) + \frac{1}{2}A''(x)h + \frac{1}{6}A'''(x)h^2 + \frac{1}{24}A^{(iv)}(x)h^3 + \mathcal{O}(h^4). \tag{19}$$

It is here assumed that $A'(\alpha)$ is not singular and $e^{(n)} = x^{(n)} - \alpha$ is called the error at the n-th iterate and [6,21]:

$$e^{(n+1)} = He^{(n)^p} + \mathcal{O}(e^{(n)^{p+1}}), \tag{20}$$

(20) is the equation of error, whereas H is a p-linear function. This means that $H \in \mathcal{L}(\mathbb{R}^m, \mathbb{R}^m, \dots, \mathbb{R}^m)$. Moreover, we consider:

$$e^{(n)^p} = \underbrace{(e^{(n)}, e^{(n)}, \dots, e^{(n)})}_{p \text{ times}}, \tag{21}$$

which would be a matrix.

Before stating the main theorem, it is pointed out that if A be differentiable in terms of Fréchet concept in D sufficiently. As in [22], the l-th differentiation of A at $u \in \mathbb{R}^m$, $l \geq 1$, is the following l-linear function

$$A^{(l)}(u) : \mathbb{R}^m \times \cdots \times \mathbb{R}^m \longrightarrow \mathbb{R}^m, \tag{22}$$

so that $A^{(l)}(u)(v_1, v_2, \dots, v_l) \in \mathbb{R}^m$. It is also famous that, for $\alpha + h \in \mathbb{R}^m$ locating in a neighborhood of a root α of (1), the Taylor expansion could be written and we have [22]:

$$A(\alpha + h) = A'(\alpha)\left[h + \sum_{l=2}^{p-1} C_l h^l\right] + \mathcal{O}(h^p), \tag{23}$$

wherein

$$C_l = (1/l!)[A'(\alpha)]^{-1}A^{(l)}(\alpha), \ l \geq 2. \tag{24}$$

One finds $C_l h^l \in \mathbb{R}^m$, because $A^{(l)}(\alpha) \in \mathcal{L}(\mathbb{R}^m \times \cdots \times \mathbb{R}^m, \mathbb{R}^m)$ and $[A'(\alpha)]^{-1} \in \mathcal{L}(\mathbb{R}^m)$. Moreover, for A' we have:

$$A'(\alpha + h) = A'(\alpha) \left[I + \sum_{l=2}^{p-1} l C_l h^{l-1} \right] + \mathcal{O}(h^p), \tag{25}$$

where I is the unit matrix of appropriate size. Here, $l C_l h^{l-1} \in \mathcal{L}(\mathbb{R}^m)$.

Theorem 1. *Assume that in* (1), $A : D \subseteq \mathbb{R}^m \longrightarrow \mathbb{R}^m$ *is Fréchet differentiable sufficiently at any points of D at* $\alpha \in \mathbb{R}^m$. *Here we assume* $A(\alpha) = 0$ *and* $\det(A'(x)) \neq 0$. *Then,* (16) *with a choice of suitable initial vector has 3.30 R-order of convergence.*

Proof. For the iteration scheme (16) in the case without memory and using (23)–(25), we can obtain:

$$e^{(n+1)} = (\beta A'(\alpha) + I)(\beta A'(\alpha) + 2I) C_2^2 e^{(n)^3} + \mathcal{O}(e^{(n)^4}). \tag{26}$$

Let us now re-write (26) in the asymptotical form as comes next:

$$e^{(n+1)} \sim d_1^{(n)} e^{(n)^3}. \tag{27}$$

Several symbolic calculations by taking into account that the coefficient of the error terms in our *m*-D case are all matrices, Lemma 1, and their multiplications does not admit commutativity, one obtains that:

$$d_1^{(n)} \sim e^{(n-1)}, \quad \forall n \geq 1. \tag{28}$$

Therefore, one attains:

$$d_1^{(n)^2} \sim e^{(n-1)^2}, \quad \forall n \geq 0. \tag{29}$$

Combining (28) and (29) into (27), we attain:

$$e^{(n+1)} \sim e^{(n-1)^1} e^{(n)^3}. \tag{30}$$

It shows that

$$\frac{1}{p} + 3 = p, \tag{31}$$

wherein its convergence *r*-order is given by:

$$p = \frac{1}{2} \left(\sqrt{13} + 3 \right) \simeq 3.30278. \tag{32}$$

The proof is ended. $\quad\square$

4. Efficiency

Here we only need to compute one LU factorization per cycle and act it two times for different linear systems with two different right hand sides and one time on an identity matrix for the acceleration matrix $B^{(n)}$ to achieve a higher speed rate.

It is recalled that the classical index of efficiency is defined by [8]:

$$E = p^{\frac{1}{c}}, \tag{33}$$

wherein p is the convergence rate and $\mathcal{C}$ is the whole burden per cycle considering the number of functional evaluations.

When dealing with nonlinear system of equations, the cost of functional evaluations per cycle can be expressed as follows:

- To evaluate A, m evaluations of functions are required.
- To evaluate the associated Jacobian matrix A' needs m^2 evaluations of functions.
- To evaluate the first-order DDO, we need $m^2 - m$ evaluations of functions.
- In addition, the LU factorization cost is $\theta(2\frac{m^3}{3})$ plus $\theta(2m^2)$ in tackling the two involved triangular systems.

wherein θ is a weight that connects the cost of 1 evaluation of function with one flops. Here it is assumed that $\theta = 1$. No preconditioning is imposed in each cycle of these methods for solving the linear systems. This is done for all the compared methods.

To be more precise, we consider that the cost for computing each of the scalar functions is unit. The cost for computing other involved calculations are all also a factor of this unity cost. This is the way to give a flops-like efficiency index [23].

Considering only the consumed functional evaluations per cycle might not be a key element for reporting the indices of efficiency when solving nonlinear systems of equations. The number of matrix products, scalar products, decomposition of LU and the solution of the triangular systems of algebraic linear equations are significant in estimating the real cost and superiority of a scheme in comparison to the existing solvers in literature [23].

Hence, the results can be summarized as follows for large m:

$$2^{\frac{1}{\frac{2m^3}{3}+4m^2}} < 2^{\frac{1}{\frac{2m^3}{3}+3m^2+m}} < 3^{\frac{1}{\frac{2m^3}{3}+6m^2+m}} < 3.30^{\frac{1}{\frac{2m^3}{3}+6m^2+m}}. \tag{34}$$

In our comparisons, we applied the Newton's quadratically convergent iteration expression (NM) and also Steffensen's method (SM), the third-order expression of Amat et al. (8) denoted by AM, and the presented approach (16) showed by PM, for tackling our nonlinear systems of algebraic equations. This is also plotted in Figure 1 showing the competitiveness of the scheme with memory (16).

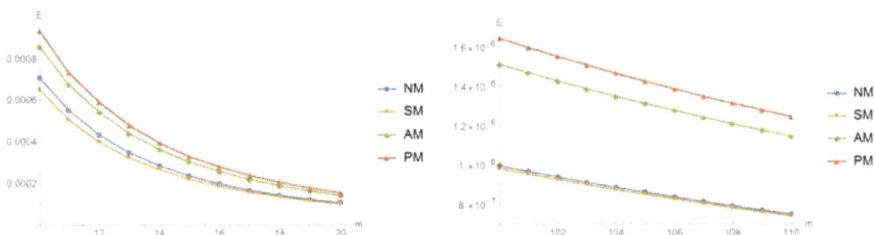

Figure 1. The comparison of flops-like efficiency indices for various schemes with and without memory by changing m.

5. Computational Tests

The aim of this section is to reveal the application of our proposed nonlinear solver for some practical problems. The software Mathematica 11.0, [24,25] was used for doing all calculations regarding the compared methods. We avoided computing any matrix inverse and the linear systems were solved applying the command LinearSolver[]. For the implementation of such schemes a possible stopping criterion can be defined based on the residual norm and imposed as follows:

$$||A(x^{(k)})||_2 \le \varepsilon, \tag{35}$$

wherein ε is the required accuracy. $||\cdot||_2$ is the l^2 norm.

To confirm the theoretical convergence speed in our numerical tests, we obtain the the numerical rate of convergence by employing the following definition:

$$\rho \approx \frac{\ln(||A(x^{(k+1)})||_2/||A(x^{(k)})||_2)}{\ln(||A(x^{(k)})||_2/||A(x^{(k-1)})||_2)}. \tag{36}$$

5.1. An Academical Test

Example 1. *Here a nonlinear system of equations $A(x) = 0$, having complex root is examined as comes next:*

$$A(x) = \begin{cases} 5\exp(x_1 - 2)x_2 + 2x_7{}^{x_{10}} + 8x_3{}^{x_4} - 5x_6{}^3 - x_9, \\ 5\tan(x_1 + 2) + \cos(x_9{}^{x_{10}}) + x_2{}^3 + 7x_3{}^4 - 2\sin^3(x_6), \\ x_1{}^2 - x_{10}x_5x_6x_7x_8x_9 + \tan(x_2) + 2x_3{}^{x_4} - 5x_6{}^3, \\ 2\tan(x_1{}^2) + 2^{x_2} + x_3{}^2 - 5x_5{}^3 - x_6 + x_8{}^{\cos(x_9)}, \\ 10x_1{}^2 - x_{10} + \cos(x_2) + x_3{}^2 - 5x_6{}^3 - 2x_8 - 4^{x_9}, \\ \cos^{-1}(x_1{}^2)\sin(x_2) - 2x_{10}x_5{}^4x_6x_9 + x_3{}^2, \\ x_1x_2{}^{x_7} - x_8{}^{x_{10}} + x_3{}^5 - 5x_5{}^3 + x_7, \\ \cos^{-1}(-10x_{10} + x_8 + x_9) + x_4\sin(x_2) + x_3 - 15x_5{}^2 + x_7, \\ 10x_1 + x_3{}^2 - 5x_5{}^2 + 10x_6{}^{x_8} - \sin(x_7) + 2x_9, \\ x_1\sin(x_2) - 2x_{10}{}^{x_8} + x_{10} - 5x_6 - 10x_9, \end{cases} \tag{37}$$

wherein the exact solution just shown up to 10 decimal places as follows:

$$\begin{aligned} \alpha \simeq \quad & (1.3273490437 + 0.3502924960i, 1.058599346 - 1.748724664i, \\ & 1.0276186794 - 0.0141308051i, 3.273950008 + 0.127828308i, \\ & 0.8318243937 + 0.0017551949i, -0.4853245912 + 0.6848776400i, \\ & 0.1693667630 + 0.1840917580i, 1.534419958 - 0.321214766i, \\ & 2.086379651 + 0.426342755i, -1.989592331 + 1.478395393i)^*. \end{aligned} \tag{38}$$

The numerical evidences and the computational order of convergence ρ for this experiment are reported forward in Table 1 using 1000 fixed floating point arithmetic and the starting value $x^{(0)} = (1.2 + 0.3I, 1.1 - 1.9I, 1.0 - 0.1I, 2.5 + 0.5I, 0.8 - 0.1I, -0.4 + 1.I, 0.1 + 0.1I, 1.3 - 0.7I, 2.0 + 0.5I, -1.9 + 1.4I)^*$. Here, the residual norm $||\cdot||_2$ is reported.

Table 1. Comparison evidences for Example 1.

Met.	$\|A(x^{(3)})\|$	$\|A(x^{(4)})\|$	$\|A(x^{(5)})\|$	$\|A(x^{(6)})\|$	$\|A(x^{(7)})\|$	$\|A(x^{(8)})\|$	$\|A(x^{(9)})\|$	ρ
NM	$8.19E-1$	$2.73E-2$	$1.79E-5$	$1.28E-11$	$2.52E-23$	$8.28E-47$	$2.50E-94$	2.02
SM	$7.68E-1$	$1.83E-2$	$7.33E-6$	$5.17E-12$	$1.51E-24$	$2.03E-49$	$3.91E-99$	1.99
AM	$8.50E-2$	$6.50E-6$	$4.98E-17$	$3.68E-50$	$1.26E-149$	$4.83E-448$		3.00
PM	$6.80E-2$	$1.12E-7$	$5.15E-26$	$1.76E-86$	$6.47E-287$			3.31

5.2. An Integral Equation Using a Collocation Approach

Example 2. *The purpose of this test was to examine the performance of the new derivative-free scheme with memory for the following mixed Hammerstein integral equation* [6]:

$$y(s) = 1 + \frac{1}{5} \int_0^1 G(s,t)y(t)^3 dt, \tag{39}$$

wherein $y \in C[0,1]$, $s,t \in [0,1]$ *and the kernel G is defined as follows:*

$$G(s,t) = \begin{cases} (1-s)t, & t \leq s, \\ s(1-t), & t > s. \end{cases} \tag{40}$$

By employing the well-resulted Gauss-Legendre quadrature formula in discretization of integral equations given in the following form, we will be able to tackle (39):

$$\int_0^1 y(t)dt \approx \sum_{j=1}^{\chi} w_j y(t_j), \tag{41}$$

where the abscissas t_j and the weights w_j were determined via the formula of Gauss–Legendre quadrature.

The lower limit of integration in standard Gauss–Legendre quadrature formula is -1. In order to approximate the integral (41) over $[0,1]$, we should map the roots of Legendre polynomials t_j on this segment and scale the weights w_j.

Showing the estimation $y(t_i)$ via x_i ($i = 1,2,\cdots,\chi$), one would be able to transfigure the process of solving nonlinear mixed integral equations into a set of nonlinear algebraic equations as comes next:

$$5x_i - 5 - \sum_{j=1}^{\chi} c_{ij} x_j^3 = 0, \quad i = 1,2,\cdots,\chi, \tag{42}$$

where

$$c_{ij} = \begin{cases} w_j t_j (1-t_i), & if \ j \leq i, \\ w_j t_i (1-t_j), & if \ i < j. \end{cases} \tag{43}$$

For this example, we employed 200 digits floating point in the computations with the stop termination as the following residual norm (35) with $\varepsilon = 10^{-100}$. The initial vector was selected as $x^{(0)} = (3,3,\ldots,3)^*$, while the results are shown in Figure 2 using $\chi = 40$ as a list log plot of the function values by performing the cycle. It reveals a stable and fast performance of the new scheme with memory in solving integral equations. Recalling that the Figure 2 can be interpreted only as an error of numerical solution of the system (42) but not the error of solution of the source integral Equation (39).

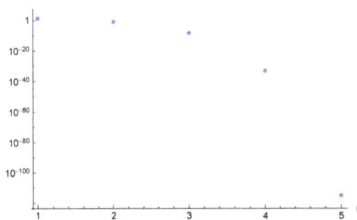

Figure 2. Error history for solving the integral equation in Example 2 using PM (Performed by Mathematica).

6. Summary

For derivative-involved iteration schemes in solving nonlinear systems, we use the $m \times m$ Jacobian matrix, i.e., $F'(x)$, with entries $F'(x)_{jk} = \partial_{x_k} f_j(x)$. Higher order schemes, such as Chebyshev methods, need higher multi dimensional derivatives which make them less practical. To be more precise, the first Fréchet derivative is a matrix with m^2 elements, while the 2nd order Fréchet differentiation has m^3 entries (ignoring the symmetric feature).

In this work, we have developed and introduced a variant of Steffensen's method with memory for tackling nonlinear problems. The scheme consists of two steps and requires the the computation of only one LU factorization which makes its computational efficiency index higher than some of the existing solvers in the literature.

The application of the iteration scheme for nonlinear integral equations via the collocation approach was discussed and its application for other types of nonlinear discretized set of equations obtained from practical problems such as the ones in [26,27] can be investigated similarly.

Author Contributions: All authors contributed equally in preparing and writing this work.

Funding: This work was supported by Hamedan Branch of Islamic Azad university.

Acknowledgments: We are grateful to three anonymous referees for several comments which improved the readability of this work.

Conflicts of Interest: The authors declare no conflict of interest.

References

1. Qasim, S.; Ali, Z.; Ahmad, F.; Serra-Capizzano, S.; Ullah, M.Z.; Mahmood, A. Solving systems of nonlinear equations when the nonlinearity is expensive. *Comput. Math. Appl.* **2016**, *71*, 1464–1478. [CrossRef]
2. Wazwaz, A.-M. *Linear and Nonlinear Integral Equations*; Higher Education Press: Beijing, China; Springer: Berlin/Heidelberg, Germany, 2011.
3. Mashayekhi, S.; Razzaghi, M.; Tripak, O. Solution of the nonlinear mixed Volterra-Fredholm integral equations by hybrid of block-pulse functions and Bernoulli polynomials. *Sci. World J.* **2014**, *2014*. [CrossRef] [PubMed]
4. Alzahrani, E.O.; Al-Aidarous, E.S.; Younas, A.M.M.; Ahmad, F.; Ahmad, S.; Ahmad, S. A higher order frozen Jacobian iterative method for solving Hamilton-Jacobi equations. *J. Nonlinear Sci. Appl.* **2016**, *9*, 6210–6227. [CrossRef]
5. Soleymani, F. Pricing multi-asset option problems: A Chebyshev pseudo–spectral method. *BIT Numer. Math.* **2019**, *59*, 243–270. [CrossRef]
6. Ortega, J.M.; Rheinboldt, W.C. *Iterative Solution of Nonlinear Equations in Several Variables*; Academic Press: New York, NY, USA, 1970.
7. Noda, T. The Steffensen iteration method for systems of nonlinear equations. *Proc. Jpn. Acad.* **1987**, *63*, 186–189. [CrossRef]
8. Grau-Sánchez, M.; Grau, À.; Noguera, M. On the computational efficiency index and some iterative methods for solving systems of nonlinear equations. *J. Comput. Appl. Math.* **2011**, *236*, 1259–1266. [CrossRef]
9. Traub, J.F. *Iterative Methods for the Solution of Equations*; Prentice Hall: New York, NY, USA, 1964.
10. Amat, S.; Busquier, S. Convergence and numerical analysis of a family of two-step Steffensen's methods. *Comput. Math. Appl.* **2005**, *49*, 13–22. [CrossRef]
11. Soleymani, F.; Sharifi, M.; Shateyi, S.; Haghani, F.K. A class of Steffensen-type iterative methods for nonlinear systems. *J. Appl. Math.* **2014**, *2014*. [CrossRef]
12. Babajee, D.K.R.; Dauhoo, M.Z.; Darvishi, M.T.; Barati, A. A note on the local convergence of iterative methods based on Adomian decomposition method and 3-node quadrature rule. *Appl. Math. Comput.* **2008**, *200*, 452–458. [CrossRef]

13. Alaidarous, E.S.; Ullah, M.Z.; Ahmad, F.; Al-Fhaid, A.S. An efficient higher-order quasilinearization method for solving nonlinear BVPs. *J. Appl. Math.* **2013**, *2013*. [CrossRef]
14. Hanaç, E. The phase plane analysis of nonlinear equation. *J. Math. Anal.* **2018**, *9*, 89–97.
15. Hasan, P.M.A.; Sulaiman, N.A. Numerical treatment of mixed Volterra-Fredholm integral equations using trigonometric functions and Laguerre polynomials. *ZANCO J. Pure Appl. Sci.* **2018**, *30*, 97–106.
16. Qasim, U.; Ali, Z.; Ahmad, F.; Serra-Capizzano, S.; Ullah, M.Z.; Asma, M. Constructing frozen Jacobian iterative methods for solving systems of nonlinear equations, associated with ODEs and PDEs using the homotopy method. *Algorithms* **2016**, *9*, 18. [CrossRef]
17. Ahmad, F.; Soleymani, F.; Khaksar Haghani, F.; Serra-Capizzano, S. Higher order derivative-free iterative methods with and without memory for systems of nonlinear equations. *Appl. Math. Comput.* **2017**, *314*, 199–211. [CrossRef]
18. Bellavia, S.; Bertaccini, D.; Morini, B. Nonsymmetric preconditioner updates in Newton-Krylov methods for nonlinear systems. *SIAM J. Sci. Comput.* **2011**, *33*, 2595–2619. [CrossRef]
19. Bellavia, S.; Morini, B.; Porcelli, M. New updates of incomplete LU factorizations and applications to large nonlinear systems. *Optim. Methods Softw.* **2014**, *29*, 321–340. [CrossRef]
20. Bertaccini, D.; Durastante, F. Interpolating preconditioners for the solution of sequence of linear systems. *Comput. Math. Appl.* **2016**, *72*, 1118–1130. [CrossRef]
21. Sharma, J.R.; Kumar, D.; Argyros, I.K.; Magreñán, Á.A. On a bi-parametric family of fourth order composite Newton-Jarratt methods for nonlinear systems. *Mathematics* **2019**, *7*, 492. [CrossRef]
22. Cordero, A.; Hueso, J.L.; Martínez, E.; Torregrosa, J.R. A modified Newton-Jarratt's composition. *Numer. Algorithms* **2010**, *55*, 87–99. [CrossRef]
23. Montazeri, H.; Soleymani, F.; Shateyi, S.; Motsa, S.S. On a new method for computing the numerical solution of systems of nonlinear equations. *J. Appl. Math.* **2012**, *2012*, 1–15. [CrossRef]
24. Sánchez León, J.G. *Mathematica Beyond Mathematics: The Wolfram Language in the Real World*; Taylor & Francis Group: Boca Raton, FL, USA, 2017.
25. Wagon, S. *Mathematica in Action*, 3rd ed.; Springer: Berlin, Germany, 2010.
26. Soheili, A.R.; Soleymani, F. Iterative methods for nonlinear systems associated with finite difference approach in stochastic differential equations. *Numer. Algorithms* **2016**, *71*, 89–102. [CrossRef]
27. Soleymani, F.; Barfeie, M. Pricing options under stochastic volatility jump model: A stable adaptive scheme. *Appl. Numer. Math.* **2019**, *145*, 69–89. [CrossRef]

![Sigma logo] *mathematics*

MDPI

Article

A Unified Convergence Analysis for Some Two-Point Type Methods for Nonsmooth Operators

Sergio Amat [1,*,†], Ioannis Argyros [2,†], Sonia Busquier [1,†], Miguel Ángel Hernández-Verón [3,†] and María Jesús Rubio [3,†]

1 Departamentode Matemática Aplicada y Estadística, Universidad Politécnica de Cartagena, 11003 Cádiz, Spain
2 Department of Mathematical Sciences, Cameron University, Lawton, OK 73505, USA
3 Departamento de Matemáticas y Computación, Universidad de La Rioja, Calle Madre de Dios, 53, 26006 Logrono, Spain
* Correspondence: sergio.amat@upct.es; Tel.: +34-968-325-651
† These authors contributed equally to this work.

Received: 18 June 2019; Accepted: 31 July 2019; Published: 3 August 2019

Abstract: The aim of this paper is the approximation of nonlinear equations using iterative methods. We present a unified convergence analysis for some two-point type methods. This way we compare specializations of our method using not necessarily the same convergence criteria. We consider both semilocal and local analysis. In the first one, the hypotheses are imposed on the initial guess and in the second on the solution. The results can be applied for smooth and nonsmooth operators.

Keywords: iterative methods; nonlinear equations; Newton-type methods; smooth and nonsmooth operators

1. Introduction

One of the most important techniques in order to approximate nonlinear equations are iterative methods [1–6]. In this paper, we present a unified approach for two-point Newton-type methods for smooth and nonsmooth operators [7–10]. We will consider two types of convergence. The semilocal convergence is where the hypotheses are imposed on the initial guess; and local convergence is where the hypotheses are imposed on the solution. Our family includes a great variety of methods. We are interested also in the application of these methods in practice (nonlinear systems, boundary problems and image processing).

For a greater generality, in this study, let X and Y be two Banach spaces and D a nonempty, open, and convex set; let $F_1 : D \subset X \to Y$ and $F_2 : D \subset X \to Y$ be continuous operators. Moreover, we assume that the operator F_1 has a continuous Fréchet derivative and F_2 is a continuous operator whose differentiability is not assumed. We consider the equation

$$F(x) = F_1(x) + F_2(x) = 0. \tag{1}$$

To solve this equation, we use the two-point Newton-type methods defined by

$$x_{k+1} = x_k - L_{k-1,k}^{-1}(F_1(x_k) + F_2(x_k)) \tag{2}$$

for each $k = 0, 1, 2, ...$, where $x_{-1}, x_0 \in D$ are the initial points, $L(.,.) : D \times D \to \mathfrak{L}(X, Y)$ and $\mathfrak{L}(X, Y)$ is the space of bounded linear operators from X into Y. We have denoted by $L_{k-1,k} = L(x_{k-1}, x_k)$.

If $F_2(x) \neq 0$, we have that the operator F is not Fréchet differentiable. In general, to approximate a solution of (1) in this situation, derivative-free iterative methods are used [11–14]. To obtain this type of iterative processes, it is common to approximate derivatives by difference divided. Remember that, given an operator $H : D \subset X \to Y$, we call $[x, y; H] \in \mathcal{L}(X, Y)$ a first order divided differences operator for H on the points x and y $(x \neq y)$ in D if

$$[x, y; H](x - y) = H(x) - H(y). \tag{3}$$

So, to solve (1) with iterative methods given from (2), we can consider at least two different procedures. Firstly, we have the Zincenko method [15], given by the following algorithm:

$$\begin{cases} \text{Given } x_{-1}, \ x_0 \in D, \\ x_{k+1} = x_k - [F_1'(x_k)]^{-1} F(x_k), \quad n \geq 0, \end{cases} \tag{4}$$

where we directly eliminate the nondifferentiable part of F, i.e., F_2. So, in this case, $L_{k-1,k} = F_1'(x_k)$ in (2). Secondly, we can consider an approximation of F' by divided differences, the secant-type methods [16,17]:

$$\begin{cases} \text{Given } x_{-1}, \ x_0 \in D, \\ y_k = \lambda x_k + (1 - \lambda) x_{k-1}, \quad \lambda \in [0, 1), \\ x_{k+1} = x_k - [y_k, x_k; F]^{-1} F(x_k), \quad n \geq 0, \end{cases} \tag{5}$$

where the secant method, for $\lambda = 0$, is obtained. So, in this case, $L_{k-1,k} = [y_k, x_k; F]$ in (2). But, if we consider a better approximation of the derivative of F, an approximation of second order, we have the Kurchatov method [18]:

$$\begin{cases} \text{Given } x_{-1}, \ x_0 \in D, \\ x_{k+1} = x_k - [x_{k-1}, 2x_k - x_{k-1}; F]^{-1} F(x_n), \quad n \geq 0, \end{cases} \tag{6}$$

in this case, $L_{k-1,k} = [x_{k-1}, 2x_k - x_{k-1}; F]$ in (2).

By using this procedure of decomposition for operator F, we see that we can also consider the application of iterative methods that use derivatives when F is nondifferentiable. So, if we consider decomposition of F given in (1), we can use the Newton-secant-type algorithm:

$$\begin{cases} \text{Given } x_{-1}, \ x_0 \in D, \\ y_k = \lambda x_k + (1 - \lambda) x_{k-1}, \quad \lambda \in [0, 1), \\ x_{k+1} = x_k - \left(F_1'(x_k) + [y_k, x_k; F] \right)^{-1} F(x_k), \quad n \geq 0, \end{cases} \tag{7}$$

where $L_{k-1,k} = F_1'(x_k) + [y_k, x_k; F]$ in (2). The other possibility, from the decomposition method, is to consider the Newton–Kurchatov [19] algorithm:

$$\begin{cases} \text{Given } x_{-1}, \ x_0 \in D, \\ x_{k+1} = x_k - \left(F_1'(x_k) + [x_{k-1}, 2x_k - x_{k-1}; F] \right)^{-1} F(x_k), \quad n \geq 0, \end{cases} \tag{8}$$

where $L_{k-1,k} = F_1'(x_k) + [x_{k-1}, 2x_k - x_{k-1}; F]$ in (2). Another possibility is to consider Steffensen-type methods, that is, the methods associated to divided differences like $[x_k, x_k + F(x_k); F]$.

As we can see, there are a lot of iterative methods that can be written as algorithms (2).

The main aim of this paper is to obtain a general study for the convergence, local and semilocal, for these Newton-type of iterative methods given in (2).

2. Convergence Analysis for Two-Point Newton-Type Methods

In this section, we present both semilocal and local convergence analysis. In the first one, the hypotheses are imposed on the initial guess; and in the second, on the solution. The results can be applied for smooth and nonsmooth operators.

2.1. Local Convergence Analysis

We start the local analysis of method (2). Let $v_0 : [0, +\infty) \times [0, +\infty) \to [0, +\infty)$ be a nondecreasing continuous function. Assume that the equation

$$v_0(t, t) = 1 \tag{9}$$

has at least one positive root r_0. Let also $v : [0, r_0) \times [0, r_0) \to [0, +\infty)$ be a nondecreasing continuous function. Define function $\bar{v}$ on the interval $[0, r_0)$ by

$$\bar{v}(t) = \frac{v(t, t)}{1 - v_0(t, t)} - 1. \tag{10}$$

Assume that the equation

$$\bar{v}(t) = 0 \tag{11}$$

has a minimal positive solution r. It follows that for each $t \in [0, r)$

$$0 \le v_0(t, t) < 1 \tag{12}$$

and

$$0 \le \bar{v}(t) < 1. \tag{13}$$

Our analysis of method (2) will use the conditions (A):

- (a_1) There exist a solution $x^* \in D$ of Equation (1), and $B \in \mathcal{L}(X, Y)$ such that $B^{-1} \in \mathcal{L}(Y, X)$.
- (a_2) Condition (9) holds and for each $x, u \in D$

$$\|B^{-1}(L(x, u) - B)\| \le v_0(\|x - x^*\|, \|u - y^*\|),$$

where v_0 is defined previously, and r_0 is given in (9).

Set $D_0 = D \cap \bar{U}(x^*, r_0)$.

- (a_3) For each $x, z \in D_0$, and any solution $y \in D$ of Equation (1)

$$\|B^{-1}(F_1(x) + F_2(x) - L(z, x)y)\| \le v(\|z - y\|, \|x - y\|)\|x - y\|,$$

where v is defined previously, and $L(\cdot, \cdot) : D_0 \times D_0 \to \mathcal{L}(X, Y)$.
- (a_4) $\bar{U}(x^*, r) \subset D$, where r is given in (10).
- (a_5)

$$\xi := \frac{v(r, r)}{1 - v_0(r, r)} \in [0, 1).$$

We are able to perform our local analysis of method (2) based on the aformentioned conditions (A).

Theorem 1. *Assume that the conditions (A) hold. Then, sequence x_k, defined by method (2) for $x_{-1}, x_0 \in U(x^*, r) - x^*$, is well defined in $U(x^*, r)$; remains in $U(x^*, r)$; and converges to x^*. Finally, the following estimates hold.*

$$\|x_{k+1} - x^*\| \le \frac{v(\|x_{k-1} - x^*\|, \|x_k - x^*\|)}{1 - v_0(\|x_{k-1} - x^*\|, \|x_k - x^*\|)} \|x_k - x^*\| \le \|x_k - x^*\| < r. \tag{14}$$

The vector x^* is the only solution of Equation (1) in $\bar{U}(x^*, r)$.

Proof. We will use mathematical induction on k.

Let $x, u \in U(x^*, r)$.

Using (2), (a_1) and (a_2), we obtain

$$\|B^{-1}(L(x, u) - B)\| \le v_0(\|x - x^*\|, \|u - x^*\|) \le v_0(r, r) < 1. \tag{15}$$

Using the Banach lemma on invertible operators [20] and (15), we deduce that $L(x, u)^{-1} \in \mathfrak{L}(Y, X)$, and

$$\|L(x, u)^{-1}B\| \le \frac{1}{1 - v_0(\|x - x^*\|, \|u - x^*\|)}. \tag{16}$$

In particular, estimate (16) holds for $x = x_0$, so x_1 is well defined by method (2) for $k = 0$. Using the definition of the method (2) (for $k = 0$); (a_1), (a_3), (13), and (16) (for $k = 0$) that

$$
\begin{aligned}
\|x_1 - x^*\| &= \|x_0 - x^* - L(x_{-1}, x_0)^{-1}(x_{-1}, x_0)(F_1(x_0) + F_2(x_0))\| \\
&= \|[L(x_{-1}, x_0)^{-1}B][B^{-1}(F_1(x_0) + F_2(x_0) - L(x_{-1}, x_0)(x_0 - x^*))]\| \\
&\le \|L(x_{-1}, x_0)^{-1}B\| \|B^{-1}(F_1(x_0) + F_2(x_0) - L(x_{-1}, x_0)(x_0 - x^*))\| \\
&\le \frac{v(\|x_{-1} - x^*\|, \|x_0 - x^*\|)}{1 - v_0(\|x_{-1} - x^*\|, \|x_0 - x^*\|)} \|x_0 - x^*\| \le \|x_0 - x^*\| < r,
\end{aligned} \tag{17}
$$

which shows estimate (14) for $k = 0$ and $x_1 \in U(x^*, r)$.

Replace x_0, x_1 by x_i, x_{i+1} in the preceding estimates to complete the induction for estimate (14). Then, from the estimate

$$\|x_{i+1} - x^*\| \le \mu \|x_i - x^*\| < r, \tag{18}$$

where

$$\mu = \frac{v(\|x_{-1} - x\|, \|x_0 - x^*\|)}{1 - v_0(\|x_{-1} - x^*\|, \|x_0 - x^*\|)} \in [0, 1),$$

thus, $\lim_{i \to +\infty} x_i = x^*$ and $x_{i+1} \in U(x^*, r)$. Moreover, for the uniqueness part, let $y^* \in \bar{U}(x^*, r)$ with $F_1(y^*) + F_2(y^*) = 0$. Using ($a3$), ($a5$), and estimate (17), we obtain in turn that

$$
\begin{aligned}
\|x_{i+1} - y^*\| &\le \|L(x_{i-1}, x_i)^{-1}B\| \|B^{-1}(F_1(x_i) + F_2(x_i) - L(x_{i-1}, x_i)(x_i - y^*))\| \\
&\le \frac{v(\|x_{i-1} - y^*\|, \|x_i - y^*\|)}{1 - v_0(\|x_{i-1} - y^*\|, \|x_i - y^*\|)} \|x_i - y^*\| \\
&\le \mu \|x_i - y^*\| < \mu^{i+1} \|x_0 - y^*\|,
\end{aligned} \tag{19}
$$

which shows $\lim_{i \to +\infty} x_i = y^*$—but, we showed $\lim_{i \to +\infty} x_i = x^*$. Hence, we conclude that $x^* = y^*$. $\square$

Remark 1. • *Condition (a_3) can be replaced by the stronger: for each $x, y \in D_0$*

$$\|B^{-1}(F_1(x) + F_2(x) - L(x)(x - y))\| \le v_1(\|x - y\|)\|x - y\|,$$

where function v_1 is as v. However, for each $t, t' \ge 0$

$$v(t, t') \le v_1(t, t').$$

• *Linear operator B does not necessarily depend on q, where $q = x^*$ or $q = x_0$. It is used to determine the invertibility of linear operator $L(\cdot, \cdot)$ appearing in the method. The invertibility of B can be assured by* an additional condition of the form $||I - B|| < 1$ or in some other way. A possible choice for B is $B = B(q)$ or $B = F_1'(q)$.

- It follows from the definition of r_0 and r that $r_0 \geq r$.

2.2. Semilocal Convergence Analysis

For the semilocal case, we also define some functions and parameters. Let $w_0 : [0, +\infty) \times [0, +\infty) \to [0, +\infty)$ be a continuous and nondecreasing function.

Assume that the equation

$$w_0(t, t) = 1, \tag{20}$$

has one smallest positive root that we denote by ρ_0. Let $w : [0, \rho_0) \times [0, \rho_0) \times [0, \rho_0) \to [0, +\infty)$ be a nondecreasing continuous function. Moreover, for $\eta, \bar{\eta} \geq 0$, define parameters C_1 and C_2 by

$$C_1 = \frac{w(\bar{\eta}, \eta, 0)}{1 - w_0(0, \eta)},$$

$$C_2 = \frac{w(0, \frac{\eta}{1-C_1}, \eta)}{1 - w_0(\eta, \frac{\eta}{1-C_1})}$$

and function $C : [0, \rho_0) \to [0, +\infty)$ by $C(t) = \frac{w(t,t,t)}{1-w_0(t,t)}$. Assume that the equation

$$(\frac{C_1 C_2}{1 - C(t)} + C_1 + 1)\eta - t = 0 \tag{21}$$

has one smallest positive root that we denote by ρ.

The semilocal convergence analysis of method (2) will be based on conditions (H):

- ($h1$) There exists $x_{-1}, x_0 \in D$, and $B \in \mathcal{L}(X, Y)$ such that $B^{-1} \in \mathcal{L}(Y, X)$.
- ($h2$) Condition (20) holds, and for each $x \in D$

$$||B^{-1}(L(z, x) - B)|| \leq w_0(||z - x_0||, ||x - x_0||),$$

where w_0 is defined previously and ρ_0 is given in (20).

Set $D_1 = D \cap \bar{U}(x_0, \rho_0)$.

- ($h3$) For $L(\cdot, \cdot) : D_1 \times D_1 \to \mathcal{L}(X, Y)$, and each $x, y, z \in D_1$

$$||B^{-1}(F_1(y) - F_1(x) + F_2(y) - F_2(x) - L(z, x)(y - x))||$$
$$\leq w(||z - x_0||, ||y - x_0||, ||x - x_0||)||y - x||,$$

where w is defined previously.

- ($h4$) $\bar{U}(x_0, \rho) \subseteq D$ and condition (21) holds for ρ, where $||x_1 - x_0|| \leq \eta$ and $||x_{-1} - x_0|| \leq \bar{\eta}$.

Then, using the hypotheses (H), we obtain the estimates:

$$
\begin{aligned}
\|x_2 - x_1\| &\leq \frac{w(\|x_{-1} - x_0\|, \|x_1 - x_0\|, \|x_0 - x_0\|)\|x_1 - x_0\|}{1 - w_0(\|x_0 - x_0\|, \|x_1 - x_0\|)} = C_1\|x_1 - x_0\|, \\
\|x_2 - x_0\| &\leq \|x_2 - x_1\| + \|x_1 - x_0\| \leq (1 + C_1)\|x_1 - x_0\| \\
&= \frac{1 - C_1^2}{1 - C_1}\|x_1 - x_0\| \\
&< \frac{\eta}{1 - C_1} < \rho, \\
\|x_3 - x_2\| &\leq \frac{w(\|x_0 - x_0\|, \|x_2 - x_0\|, \|x_1 - x_0\|)}{1 - w_0(\|x_1 - x_0\|, \|x_2 - x_0\|)}\|x_2 - x_1\| \\
&\leq \frac{w(0, \frac{\eta}{1-C_1}, \eta)}{1 - w_0(\eta, \frac{\eta}{1-C_1})}\|x_2 - x_1\| = C_2\|x_2 - x_1\|, \\
\|x_3 - x_0\| &\leq \|x_3 - x_2\| + \|x_2 - x_1\| + \|x_1 - x_0\| \\
&\leq C_2\|x_2 - x_1\| + C_1\|x_1 - x_0\| + \|x_1 - x_0\| \\
&\leq (C_2 C_1 + C_1 + 1)\|x_1 - x_0\|, \\
\|x_4 - x_3\| &\leq \frac{w(\|x_1 - x_0\|, \|x_3 - x_0\|, \|x_2 - x_0\|)}{1 - w_0(\|x_2 - x_0\|, \|x_3 - x_0\|)}\|x_3 - x_2\| \\
&\leq C(\rho)\|x_3 - x_2\| \leq C(\rho)C_2\|x_2 - x_1\| \\
&\leq C(\rho)C_2 C_1\|x_1 - x_0\|,
\end{aligned}
\tag{22}
$$

similarly for $i = 3, 4, \ldots$

$$
\begin{aligned}
\|x_{i+1} - x_i\| &\leq C(\rho)\|x_i - x_{i-1}\| \leq C(\rho)^{i-2}\|x_3 - x_2\|, \\
\|x_{i+1} - x_0\| &\leq \|x_{i+1} - x_i\| + \ldots + \|x_4 - x_3\| + \|x_3 - x_0\| \\
&\leq C(\rho)\|x_i - x_{i-1}\| + \ldots + C(\rho)\|x_3 - x_2\| \\
&\quad + (C_2 C_1 + C_1 + 1)\|x_1 - x_0\| \\
&\leq C(\rho)^{i-2}\|x_3 - x_2\| + \ldots + C(\rho)\|x_3 - x_2\| \\
&\quad + (C_2 C_1 + C_1 + 1)\|x_1 - x_0\| \\
&\leq \left(\frac{1 - C(\rho)^{i-1}}{1 - C(\rho)} C_2 C_1 + C_1 + 1\right)\|x_1 - x_0\| \\
&< \left(\frac{C_1 C_2}{1 - C(\rho)} + C_1 + 1\right)\eta \leq \rho,
\end{aligned}
\tag{23}
$$

$$
\begin{aligned}
\|x_{i+j} - x_i\| &\leq \|x_{i+j} - x_{i+j-1}\| + \|x_{i+j-1} - x_{i+j-2}\| + \ldots + \|x_{i+1} - x_i\| \\
&\leq (C(\rho)^{i+j-3} + \ldots + C(\rho)^{i-2})\|x_3 - x_2\| \\
&\leq C(\rho)^{i-2}\frac{1 - C(\rho)^{j-1}}{1 - C(\rho)}\|x_3 - x_2\| \\
&\leq C(\rho)^{i-2}\frac{1 - C(\rho)^{j-1}}{1 - C(\rho)}C_2 C_1\|x_1 - x_0\| \\
&\leq C(\rho)^{i-2}\frac{1 - C(\rho)^{j-1}}{1 - C(\rho)}C_2 C_1 \eta.
\end{aligned}
\tag{24}
$$

It follows from (23) that $x_i \in U(x_0, \rho)$; and from (24) that sequence x_i is complete in a Banach space X. In particular, it converges to some $x^* \in \bar{U}(x_0, \rho)$. By letting $i \to +\infty$ in the estimate

$$\|B^{-1}(F_1(x_i) + F_2(x_i))\| = \|B^{-1}(F_1(x_i) + F_2(x_i) - F_1(x_{i-1}) - F_2(x_{i-1}) - L_{i-2,i-1}(x_i - x_{i-1}))\|$$

$$\leq \frac{w(\|x_{i-2} - x_0\|, \|x_i - x_0\|, \|x_{i-1} - x_0\|)\|x_i - x_{i-1}\|}{1 - w_0(\|x_{i-1} - x_0\|, \|x_i - x_0\|)} \leq \frac{w(\rho, \rho, \rho)}{1 - w_0(\rho, \rho)}\|x_i - x_{i-1}\|,$$

we obtain $F_1(x^*) + F_2(x^*) = 0$. The uniqueness part is omitted as analogous to the one in the local convergence case.

Hence, we can present our semilocal convergence result associated to the method (2).

Theorem 2. *Assume that the conditions (H) hold. Then, sequence x_k, defined by the method (2) for $x_{-1}, x_0 \in D$, is well defined in $U(x_0, \rho)$; remains in $U(x_0, \rho)$; and converges to a solution $x^* \in \bar{U}(x_0, \rho)$ of Equation (1). On the other hand, the vector x^* is the only solution of Equation (1) in $\bar{U}(x_0, \rho)$.*

The same comments given in the previous remark hold.

3. Numerical Experiment

Consider the nondifferentiable system of equations

$$\begin{cases} 3x_1^2 x_2 + x_2^2 - 1 + |x_1 - 1|^{3/2} = 0, \\ x_1^4 + x_1 x_2^3 - 1 + |x_2|^{3/2} = 0. \end{cases} \tag{25}$$

We therefore have an operator $F : \mathbb{R}^2 \to \mathbb{R}^2$ such that $F = F_1 + F_2$, as in (1), with $F_1, F_2 : \mathbb{R}^2 \to \mathbb{R}^2$, $F_1 = (F_1^{\ 1}, F_1^{\ 2})$, $F_2 = (F_2^{\ 1}, F_2^{\ 2})$, being

$$F_1^1(x_1, x_2) = 3x_1^2 x_2 + x_2^2 - 1 \quad \text{and} \quad F_1^2(x_1, x_2) = x_1^4 + x_1 x_2^3 - 1,$$

$$F_2^1(x_1, x_2) = |x_1 - 1|^{3/2} \quad \text{and} \quad F_2^2(x_1, x_2) = |x_2|^{3/2},$$

where the operator F_1 is continuously Fréchet-differentiable and F_2 is continuous but is a Fréchet nondifferentiable operator.

For $\mathbf{u} = (u_1, u_2^T)$, $\mathbf{v} = (v_1, v_2)^T \in \mathbb{R}^2$, we consider the divided difference of first order defined by $[\mathbf{u}, \mathbf{v}; F] = ([\mathbf{u}, \mathbf{v}; F]_{ij})_{i,j=1}^2 \in \mathcal{L}(\mathbb{R}^2, \mathbb{R}^2)$, where

$$[\mathbf{u}, \mathbf{v}; F]_{i1} = \begin{cases} \dfrac{F_i(u_1, u_2, \ldots, v_m) - F_i(u_1, v_2, \ldots, v_m)}{u_1 - v_1}, & \text{if } u_2 \neq v_2, \\ 0, & \text{if } u_1 = v_1, \end{cases}$$

$$[\mathbf{u}, \mathbf{v}; F]_{i2} = \begin{cases} \dfrac{F_i(u_1, u_2, \ldots, v_m) - F_i(u_1, v_2, \ldots, v_m)}{u_2 - v_2}, & \text{if } u_2 \neq v_2, \\ 0, & \text{if } u_2 = v_2, \end{cases}$$

for $i = 1, 2$.

The iterative processes given by (1) allow us to consider direct iterative processes, such as (5) and (6); as well as iterative processes that use the decomposition method, such as (7) and (8). In this experiment, for the nondifferentiable system (25), we check that the application of the iterative processes that use the decomposition method have better behavior than the direct methods.

To carry out this study, we will consider as an approximate solution of system (25):

$$\mathbf{x}^* = (0.9383410452297656, 0.3312445136375143),$$

the starting points $\mathbf{x}_{-1} = (5,5)$ and $\mathbf{x}_0 = (1,0)$, and use a tolerance $\|\mathbf{x}_{n+1} - \mathbf{x}_n\| \le 10^{-16}$. In these conditions, in Tables 1 and 2 we can see the results of the application of the direct iterative processes, the secant-type, and Kurchatov methods. Whereas in Tables 3 and 4 we can see the results of the application of the iterative processes that use the decomposition method, Newton-secant-type and Newton–Kuchatov methods. Observing the results obtained, it is evident that the best behavior of the iterative processes is given by (2) using the decomposition method.

Table 1. $\|\mathbf{x}^* - \mathbf{x}_n\|$ for secant-type methods (5) and different values of the parameter λ.

n	$\lambda = 0$	$\lambda = 0.5$	$\lambda = 0.99$
1	3.18484×10^{-1}	2.965×10^{-1}	4.54388×10^{-2}
2	5.21264×10^{-2}	4.13083×10^{-2}	3.74494×10^{-3}
3	3.66108×10^{-3}	2.35344×10^{-3}	2.94716×10^{-5}
4	2.59348×10^{-4}	7.2935×10^{-5}	3.69966×10^{-9}
5	1.30031×10^{-6}	1.24012×10^{-7}	2.10942×10^{-15}
6	4.42187×10^{-10}	6.07747×10^{-12}	
7	1.11022×10^{-15}	1.11022×10^{-16}	

Table 2. Kurchatov method (6).

n	$\|\mathbf{x}^* - \mathbf{x}_n\|$
1	2.99754×10^{-1}
2	1.07269×10^{-1}
3	4.20963×10^{-2}
4	8.37098×10^{-3}
5	2.78931×10^{-4}
6	9.0784×10^{-8}
7	2.9826×10^{-11}

Table 3. $\|\mathbf{x}^* - \mathbf{x}_n\|$ for Newton-secant-type methods (7) and different values of the parameter λ.

n	$\lambda = 0$	$\lambda = 0.5$	$\lambda = 0.99$
1	2.3538×10^{-1}	1.00278×10^{-1}	4.29554×10^{-2}
2	3.48717×10^{-1}	2.88094×10^{-2}	2.53626×10^{-3}
3	1.47537×10^{-1}	1.90518×10^{-3}	9.06208×10^{-6}
4	3.4371×10^{-2}	8.39107×10^{-6}	1.9925×10^{-10}
5	3.08399×10^{-3}	4.78016×10^{-9}	
6	4.63665×10^{-5}	7.38298×10^{-15}	
7	4.05776×10^{-8}		
8	2.96929×10^{-13}		

Table 4. Newton–Kurchatov method (8).

n	$\|\overline{x^* - \overline{x}_n}\|$
1	6.1659×10^{-1}
2	4.75269×10^{-3}
3	6.29174×10^{-5}
4	3.37027×10^{-9}

Remark 2. *In the above example, we have selected the initial guess in a region where the operator is not smooth. The methods can be applied to systems where the operator is not smooth at the solution.*

For instance, for the system:

$$
\begin{cases}
|x_1^2 - 1| + x_2 - 1 = 0, \\
x_1 + x_2^2 - 1 = 0,
\end{cases}
\tag{26}
$$

the solution is $(1,1)$. If we take as initial guess $(0.5, 0.5)$, the Steffensen method gives as errors 3.12×10^{-2}, 5.48×10^{-4}, 1.92×10^{-7}, 2.15×10^{-14}, we observe its second order.

4. Boundary Value Problem: Discretization via the Multiple Shooting Method

We will use the multiple shooting method for the discretization of boundary problems of the type

$$
y''(t) = f(t, y(t), y'(t)), \qquad y(a) = \alpha, \qquad y(b) = \beta.
\tag{27}
$$

Thus, we should find the solution of the following nonlinear system of equations $F(s) = 0$, where $F : \mathbb{R}^N \longrightarrow \mathbb{R}^N$ and

$$
\begin{cases}
F_1(s_0, s_1, \ldots, s_{N-1}) &= s_1 - y'(t_1; s_0) \\
F_2(s_0, s_1, \ldots, s_{N-1}) &= s_2 - y'(t_2; s_0, s_1) \\
\quad \vdots \\
F_{N-1}(s_0, s_1, \ldots, s_{N-1}) &= s_{N-1} - y'(t_{N-1}; s_0, s_1, \ldots, s_{N-2}) \\
F_N(s_0, s_1, \ldots, s_{N-1}) &= \beta - y(t_N; s_0, s_1, s_{N-2}, s_{N-1}).
\end{cases}
$$

for a discretization of $[a, b]$ with N subintervals,

$$
TNj, \qquad T = b - a, \qquad j = 0, 1, \ldots, N.
$$

We consider the secant-type method

$$
\begin{cases}
\text{Given } y_{-1}, y_0 \in D, \\
z_n = \lambda_n y_n + (1 - \lambda_n) y_{n-1}, \quad \lambda_n \in [0, 1), \\
y_{n+1} = y_n - [z_n, y_n; F]^{-1} F(y_n), \quad n \geq 0,
\end{cases}
\tag{28}
$$

where λ_n is such that $||z_n - x_n|| \leq Tol$ for a given tolerance, and Newton's method

$$
\begin{cases}
\text{Given } y_0 \in D, \\
y_{n+1} = y_n - F(y_n)^{-1} F(y_n), \quad n \geq 0.
\end{cases}
\tag{29}
$$

We perform a numerical comparison between both methods. As we can see, in the multiple shooting method, the iterative schemes are used as black boxes.

For the initial slope $\vec{s}_0 = (s_0^0, s_1^0, \ldots, s_{N-1}^0)$, we propose

$$
\begin{cases}
s_0^0 &= \dfrac{\beta - \alpha}{b - a} = \dfrac{y(t_N) - y(t_0)}{t_N - t_0}, \\[2mm]
s_1^0 &= \dfrac{y(t_N) - y(t_1; s_0)}{t_N - t_1}, \\[2mm]
s_2^0 &= \dfrac{y(t_N) - y(t_2; s_0, s_1)}{t_N - t_2}, \\[2mm]
&\vdots \\[2mm]
s_{N-1}^0 &= \dfrac{y(t_N) - y(t_{N-1}; s_0, s_1, \ldots, s_{N-2})}{t_N - t_{N-1}}.
\end{cases}
$$

We analyze this particular example ([21], p. 554):

$$
\begin{aligned}
y''(t) &= \tau \cdot \sinh(\tau \cdot y(t)), \\
y(0) &= 0, \quad y(1) = 1.
\end{aligned}
$$

We take $\tau = 2.5$ and $N = 4$ subintervals.

This problem admits the solution:

$$
y(t) = \frac{2}{\tau} \arg\sinh\left(\frac{s}{2} \cdot \frac{\mathrm{sn}\,(\tau t, 1 - s^2/4)}{\mathrm{cn}\,(\tau t, 1 - s^2/4)} \right),
$$

where

$$
s = y'(0) = 0.3713363932677645
$$

and $\mathrm{sn}(\cdot, \cdot)$ and $\mathrm{cn}(\cdot, \cdot)$ are the Jacobi elliptic functions.

Newton's method (29),

n	$\|F(\vec{s}_n)\|_\infty$	$\|y(t) - y_n\|_\infty$	$\|y'(t) - y_n'\|_\infty$
0	10^0	10^{-1}	10^0
1	10^{-1}	10^{-1}	10^{-1}
2	10^{-2}	10^{-2}	10^{-2}
3	10^{-4}	10^{-4}	10^{-4}
4	10^{-7}	10^{-7}	10^{-7}
5	10^{-15}	10^{-15}	10^{-14}

Secant-type method (28),

n	$\|F(\vec{s}_n)\|_\infty$	$\|y(t) - y_n\|_\infty$	$\|y'(t) - y_n'\|_\infty$	$\|F'(y_n) - [y_n, x_n; F]\|_\infty$
0	10^0	10^{-1}	10^0	10^{-6}
1	10^{-1}	10^{-1}	10^{-1}	10^{-6}
2	10^{-2}	10^{-2}	10^{-2}	10^{-7}
3	10^{-4}	10^{-4}	10^{-4}	10^{-6}
4	10^{-7}	10^{-7}	10^{-7}	10^{-6}
5	10^{-15}	10^{-15}	10^{-14}	10^{-6}

The methods using Jacobians obtain their order of convergence. However, in this example, the computation of the derivatives involves the approximation of a more complicated problem. For this reason, the methods free of derivatives are preferred, see [21]. Of course, we need to compute a good approximation to the Jacobian, this is the motivation of our parameters λ_n. For more similar examples and conclusions, we refer [22].

Remark 3. *In many cases, when we manipulate an image, some random noise appears. This noise makes the later steps of processing the image difficult and inaccurate.*

Let $f : \Omega \to \mathbb{R}$ be a noise signal or image.

Introducing the variable w:

$$w = \frac{\nabla u}{\sqrt{|\nabla u|^2}},$$

the Total-Variation model is equivalent to the nonlinear and nondifferentiable system:

$$
\begin{aligned}
-\nabla \cdot w + \lambda(u - f) &= 0, \\
w\sqrt{|\nabla u|^2} - \nabla u &= 0.
\end{aligned}
$$

This system should be discretized using finite differences and the associated nonlinear system of equations can be approximated by our family (see [23] for more details).

5. Conclusions

This paper was devoted to the analysis of a general family of two-point Newton-type methods for smooth and nonsmooth operators. We have considered two types of convergence—semilocal and local. The family includes a great number of methods. We have applied the schemes to several interesting problems, in particular to nonsmooth nonlinear systems, boundary problems, and image denoising models.

Author Contributions: All the authors have been contributed similarly and they are participated in all the work.

Funding: This research received no external funding.

Acknowledgments: Research of the first and third authors supported in part by Programa de Apoyo a la investigación de la fundación Séneca-Agencia de Ciencia y Tecnología de la Región de Murcia 20928/PI/18 and by MTM2015-64382-P. Research of the fourth and fifth authors supported by Ministerio de Economía y Competitividad under grant MTM2014-52016-C2-1P.

Conflicts of Interest: The authors declare no conflict of interest.

References

1. Amat, S.; Busquier, S. A modified secant method for semismooth equations. *Appl. Math. Lett.* **2003**, *16*, 877–881. [CrossRef]
2. Amat, S.; Busquier, S. On a higher order secant method. *Appl. Math. Comput.* **2003**, *141*, 321–329. [CrossRef]
3. Amat, S.; Busquier, S. A two-step Steffensen's method under modified convergence conditions. *J. Math. Anal. Appl.* **2006**, *324*, 1084–1092. [CrossRef]
4. Argyros, I.K.; Magreñán, A.A. *Iterative Methods and Their Dynamics with Applications: A Contemporary Study*; CRC Press: Boca Raton, FL, USA, 2017.
5. Argyros, I.K.; Magreñán, A.A. *A Contemporary Study of Iterative Methods*; Academic Press: Cambridge, MA, USA, 2018.
6. Grau-Sánchez, M. Noguera, M.; Amat, S. On the approximation of derivatives using divided difference operators preserving the local convergence order of iterative methods. *J. Comput. Appl. Math.* **2013**, *237*, 363–372.
7. Amat, S.; Bermúdez, C.; Busquier, S.; Mestiri, D. A family of Halley-Chebyshev iterative schemes for non-Fréchet differentiable operators. *J. Comput. Appl. Math.* **2009**, *228*, 486-493. [CrossRef]

8. Bartoň, M. Solving polynomial systems using no-root elimination blending schemes. *Comput.-Aided Des.* **2011**, *43*, 1870–1878. [CrossRef]

9. Bartoň, M.; Elber, G.; Hanniel, I. Topologically guaranteed univariate solutions of underconstrained polynomial systems via no-loop and single-component tests. *Comput.-Aided Des.* **2011**, *43*, 1035–1044. [CrossRef]

10. Hanniel, I.; Elber, G. Subdivision termination criteria in subdivision multivariate solvers using dual hyperplanes representations. *Comput.-Aided Des.* **2007**, *39*, 369–378. [CrossRef]

11. Chen, J.; Shen, Z. Convergence analysis of the secant type methods. *Appl. Math. Comput.* **2007**, *188*, 514–524. [CrossRef]

12. Hongmin, R.; Qingiao, W. The convergence ball of the Secant method under Hölder continuous divided differences. *J. Comput. Appl. Math.* **2006**, *194*, 284–293. [CrossRef]

13. Kewei, L. Homocentric convergence ball of the Secant method. *Appl. Math. J. Chin. Univ. Ser. B* **2007**, *22*, 353–365.

14. Ren, H.; Argyros, I.K. Local convergence of efficient Secant-type methods for solving nonlinear equations. *Appl. Math. Comput.* **2012**, *218*, 7655–7664. [CrossRef]

15. Zincenko, A.I. *Some Approximate Methods of Solving Equations with Non-Differentiable Operators*; Dopovidi Akad Nauk: 1963; pp. 156–161.

16. Hernández, M.A.; Rubio, M.J. A uniparametric family of iterative processes for solving nondifferentiable equations. *J. Math. Anal. Appl.* **2002**, *275*, 821–834. [CrossRef]

17. Hernández, M.A.; Rubio, M.J.; Ezquerro, J.A. Secant-like methods for solving nonlinear integral equations of the Hammerstein type. *J. Comput. Appl. Math.* **2000**, *115*, 245–254. [CrossRef]

18. Kurchatov, V.A. On a method of linear interpolation for the solution of funcional equations. *Dolk. Akad. Nauk SSSR* **1971**, *198*, 524–526; translation in *Soviet Math. Dolk.* **1971**, *12*, 835–838. (In Russian)

19. Hernández, M.A.; Rubio, M.J. On a Newton-Kurchatov-type Iterative Process. *Numer. Funct. Anal.* **2016**, *37*, 65–79. [CrossRef]

20. Kantorovich, L.V.; Akilov, G.P. *Functional Analysis*; Pergamon Press: Oxford, UK, 1982.

21. Stoer, J.; Bulirsch, R. *Introduction to Numerical Analysis*, 2nd ed.; Springer: New York, NY, USA, 1993.

22. Alarcón, V.; Amat, S.; Busquier, S.; López, D.J. A Steffensen's type method in Banach spaces with applications on boundary value problems. *J. Comput. Appl. Math.* **2008**, *216*, 243–250. [CrossRef]

23. Amat, S.; Argyros, I.K.; Busquier, S.; Hernández-Verón, M.A.; Martínez, E. A unified convergence analysis for single step-type methods for non-smooth operators. 2019, submitted.

![Σ mathematics logo]

mathematics

MDPI

Article

A Modified Fletcher–Reeves Conjugate Gradient Method for Monotone Nonlinear Equations with Some Applications

Auwal Bala Abubakar [1,2] (ORCID), Poom Kumam [1,3,4,*] (ORCID), Hassan Mohammad [2] (ORCID),
Aliyu Muhammed Awwal [1,5] (ORCID) and Kanokwan Sitthithakerngkiet [6] (ORCID)

[1] KMUTTFixed Point Research Laboratory, SCL 802 Fixed Point Laboratory, Science Laboratory Building, Department of Mathematics, Faculty of Science, King Mongkut's University of Technology Thonburi (KMUTT), 126 Pracha-Uthit Road, Bang Mod, Thrung Khru, Bangkok 10140, Thailand
[2] Department of Mathematical Sciences, Faculty of Physical Sciences, Bayero University, Kano 700241, Nigeria
[3] Center of Excellence in Theoretical and Computational Science (TaCS-CoE), Science Laboratory Building, King Mongkut's University of Technology Thonburi (KMUTT), 126 Pracha-Uthit Road, Bang Mod, Thrung Khru, Bangkok 10140, Thailand
[4] Department of Medical Research, China Medical University Hospital, China Medical University, Taichung 40402, Taiwan
[5] Department of Mathematics, Faculty of Science, Gombe State University, Gombe 760214, Nigeria
[6] Department of Mathematics, Faculty of Applied Science, King Mongkut's University of Technology North Bangkok, 1518 Pracharat 1 Road, Wongsawang, Bangsue, Bangkok 10800, Thailand
* Correspondence: poom.kum@kmutt.ac.th

Received: 24 June 2019; Accepted: 5 August 2019; Published: 15 August 2019

Abstract: One of the fastest growing and efficient methods for solving the unconstrained minimization problem is the conjugate gradient method (CG). Recently, considerable efforts have been made to extend the CG method for solving monotone nonlinear equations. In this research article, we present a modification of the Fletcher–Reeves (FR) conjugate gradient projection method for constrained monotone nonlinear equations. The method possesses sufficient descent property and its global convergence was proved using some appropriate assumptions. Two sets of numerical experiments were carried out to show the good performance of the proposed method compared with some existing ones. The first experiment was for solving monotone constrained nonlinear equations using some benchmark test problem while the second experiment was applying the method in signal and image recovery problems arising from compressive sensing.

Keywords: nonlinear equations; conjugate gradient method; projection method; convex constraints; signal and image processing

MSC: 65K05; 90C52; 90C56; 94A08

1. Introduction

In this paper, we are considering a system of nonlinear monotone equations of the form

$$F(x) = 0, \quad \text{subject to} \quad x \in E, \tag{1}$$

where $E \subseteq R^n$ is closed and convex, $F : R^n \to R^m$, $(m \geq n)$ is continuous and monotone, which means

$$\langle F(x) - F(y), (x - y) \rangle \geq 0, \quad \forall x, y \in R^n.$$

A well-known fact is that under the above assumption, the solution set of (1) is convex unless is empty. It is important to mention that nonlinear monotone equations arise in many practical applications. These and other reasons motivate researchers to develop a large number of class of Iterative methods for solving such systems, for example, see [1–7] among others. In addition, convex constrained equations have application in many scientific fields, some of which are the economic equilibrium problems [8], the chemical equilibrium systems [9], etc. Several algorithms were developed to solve (1), among them, are the trust-region [10] and the Levenberg-Marquardt method [11]. Moreover, the requirement to compute and store the matrix in every iteration makes them ineffective for large-scale nonlinear equations.

Conjugate gradient (CG) methods are efficient for solving large-scale optimization and nonlinear systems because of their low memory requirements. This forms part of the reason several Iterative methods with CG-like directions are proposed in recent years [12,13]. Initially, CG methods and their modified versions are proposed for unconstrained optimization problems [14–19]. Inspired by them, in the last decade, many authors used the CG direction to solve nonlinear monotone equations for both constrained and unconstrained cases. Since in this article, we are interested in solving nonlinear monotone equations with convex constraints, we will only discuss existing methods with such properties.

Many methods for solving nonlinear monotone equations with convex constraints have been presented in the last decade. For examples, Xiao and Zhu [20] presented a CG method, which combines the well-known CG-DESCENT method in [17] and the projection method by Solodov and Svaiter [21]. Liu et al. [22] proposed two CG methods with projection strategy for solving (1). In [23], a modification of the method in [20] was presented by Liu and Li. One of the reasons for the modification was to improve the numerical performance of the method in [20]. Also, Sun and Liu [24] presented derivative-free projection methods for solving nonlinear equations with convex constraints. These methods are the combination of some existing CG methods and the well-known projection method. In addition, a hybrid CG projection method for convex constrained equations was developed in [25]. Ou and Li [26] proposed a combination of a scaled CG method and the projection strategy to solve (1). Furthermore, Ding et al. [27] extended the Dai and Kou (DK) CG method to solve (1) by also combining it with the projection method. Just recently, to popularize the Dai-Yuan (DY) method, Liu and Feng [28] proposed a modified DY method for solving convex constraints monotone equation. The global convergence was also obtained under certain assumptions and finally, some numerical results were reported to show its efficiency.

Inspired by some the above proposals, we present a simple modification of the Fletcher–Reeves (FR) conjugate gradient method [19] considered in [12] to solve nonlinear monotone equations with convex constraints. The modification ensures that the direction is automatically descent, improves its numerical performance and still inherits the nice convergence properties of the method. Under suitable assumptions, we establish the global convergence of the proposed algorithm. Numerical experiments presented show the good performance and competitiveness of the method. In addition, the proposed method has the advantages of the direct methods [29] such as boundary control method by Belishev and Kuryiev [30], the globally convergent method proposed by Beilina and Klibanov [31] and method based on the multidimensional analogs of Gelfand–Levitan–Krein equations [32,33]. The proposed method can be seen as a local method that looks for the closest root. However, there are several global nonlinear solvers that guarantee finding all roots inside a domain and within a very fine double-float accuracy. In some cases a combination of subdivision-based polynomial solver with a decomposition algorithm are employed in order to handle large and complex systems (see for examples [34–36] and references therein).

The remaining part of this article is organized as follows. In Section 2, we mention some preliminaries and present the proposed method. The global convergence of the method is established in Section 3. Finally, Section 4 reports some numerical results to show the performance of the method in solving monotone nonlinear equations with convex constraints, and also apply it to recover a noisy signal and a blurred image.

2. Algorithm

In this section, we define the projection map together with its well-known properties, give some useful assumptions and finally present the proposed algorithm. Throughout this article, $\|\cdot\|$ denotes the Euclidean norm.

Definition 1. *Let $E \subset R^n$ be nonempty closed and convex set. Then for any $x \in R^n$, its projection onto E is defined as*

$$P_E(x) = \arg\min\{\|x - y\| : y \in E.\}$$

The following lemma gives some properties of the projection map.

Lemma 1 ([37]). *Suppose $E \subset R^n$ is nonempty, closed and convex set. Then the following statements are true:*

1. $\langle x - P_E(x), P_E(x) - z \rangle \geq 0, \quad \forall x, z \in R^n.$
2. $\|P_E(x) - P_E(y)\| \leq \|x - y\|, \quad \forall x, y \in R^n.$
3. $\|P_E(x) - z\|^2 \leq \|x - z\|^2 - \|x - P_E(x)\|^2, \quad \forall x, z \in R^n.$

Throughout, we suppose the followings

(C_1) The solution set of (1), denoted by E', is nonempty.
(C_2) The mapping F is monotone.
(C_3) The mapping F is Lipschitz continuous, that is there exists a positive constant L such that $\|F(x) - F(y)\| \leq L\|x - y\|, \quad \forall x, y \in R^n.$

Our algorithm is motivated by the work of Papp and Rapajić in [12]. In the paper, they modified the well known Fletcher–Reeves conjugate gradient method to solve unconstrained nonlinear monotone equation. The modification was adding the term $-\theta_k F(x_k)$ to the direction of Fletcher–Reeves. The parameter θ_k was then determined in three different ways and three different directions were proposed, namely, M3TFR1, M3TFR2 and M3TFR3. The direction we are interested in is M3TFR1 and is defined as:

$$d_k = \begin{cases} -F(x_k), & \text{if } k = 0, \\ -F(x_k) + \beta_k^{FR} w_{k-1} + \theta_k F(x_k), & \text{if } k \geq 1, \end{cases} \tag{2}$$

where,

$$\beta_k^{FR} = \frac{\|F(x_k)\|^2}{\|F(x_{k-1})\|^2}, \quad \theta_k = -\frac{F(x_k)^T w_{k-1}}{\|F(x_{k-1})\|^2}, \quad w_{k-1} = z_{k-1} - x_{k-1}, \ z_{k-1} = x_{k-1} + \alpha_{k-1} d_{k-1}.$$

It follows that

$$F(x_k)^T d_k = -\|F(x_k)\|^2.$$

Using same modification proposed in [3], we modify the direction (2) as follows

$$d_k = \begin{cases} -F(x_k), & \text{if } k = 0, \\ -F(x_k) + \frac{\|F(x_k)\|^2 w_{k-1} - F(x_k)^T w_{k-1} F(x_k)}{\max\{\mu\|w_{k-1}\|\|F(x_k)\|, \|F(x_{k-1})\|^2\}}, & \text{if } k \geq 1, \end{cases} \tag{3}$$

where $\mu > 0$ is a positive constant. The difference between the M3TFR1 direction and the direction proposed in this paper is the scaling term appearing in the denominator of Equation (3) i.e., $\max\{\mu\|w_{k-1}\|\|F(x_k)\|, \|F(x_{k-1})\|^2\}$. This modification was shown to have a very good numerical performance in [3] and also helps in obtaining the boundedness of the direction easily.

Remark 1. *Note the the parameter μ is chosen to be strictly positive because if $\mu \leq 0$ then*

$$\max\{\mu\|w_{k-1}\|\|F(x_k)\|, \|F(x_{k-1})\|^2\} = \|F(x_{k-1})\|^2.$$

This means that the direction d_k will always be M3TFR1 given by (2).

3. Convergence Analysis

To prove the global convergence of Algorithm 1, the following results are needed.

Algorithm 1: A modified descent Fletcher–Reeves CG method (MFRM).

Step 0. Select the initial point $x_0 \in R^n$, parameters $\mu > 0, \sigma > 0, 0 < \rho < 1$, $Tol > 0$, and set $k := 0$.

Step 1. If $\|F(x_k)\| \leq Tol$, stop, otherwise go to **Step 2**.

Step 2. Find d_k using (3).

Step 3. Find the step length $\alpha_k = \gamma\rho^{m_k}$ where m_k is the smallest non-negative integer m such that

$$-\langle F(x_k + \alpha_k d_k), d_k\rangle \geq \sigma\alpha_k\|F(x_k + \alpha_k d_k)\|\|d_k\|^2. \tag{4}$$

Step 4. Set $z_k = x_k + \alpha_k d_k$. If $z_k \in E$ and $\|F(z_k)\| \leq Tol$, stop. Else compute

$$x_{k+1} = P_E[x_k - \zeta_k F(z_k)]$$

where

$$\zeta_k = \frac{F(z_k)^T(x_k - z_k)}{\|F(z_k)\|^2}.$$

Step 5. Let $k = k + 1$ and go to Step 1.

Lemma 2. *Let d_k be defined by Equation (3), then*

$$d_k^T F(x_k) = -\|F(x_k)\|^2 \tag{5}$$

and

$$\|F(x_k)\| \leq \|d_k\| \leq \left(1 + \frac{2}{\mu}\right)\|F(x_k)\|. \tag{6}$$

Proof. By Equation (3), suppose $k = 0$,

$$d_k^T F(x_k) = -F(x_k)^T F(x_k) = -\|F(x_k)\|^2.$$

Now suppose $k > 0$,

$$\begin{aligned}
d_k^T F(x_k) &= -F(x_k)^T F(x_k) + \frac{(\|F(x_k)\|^2 w_{k-1})^T F(x_k) - (F(x_k)^T w_{k-1} F(x_k))^T F(x_k)}{\max\{\mu\|w_{k-1}\|\|F(x_k)\|, \|F(x_{k-1})\|^2\}} \\
&= -\|F(x_k)\|^2 + \frac{\|F(x_k)\|^2 w_{k-1}^T F(x_k) - F(x_k)^T (w_{k-1}^T F(x_k)) F(x_k)}{\max\{\mu\|w_{k-1}\|\|F(x_k)\|, \|F(x_{k-1})\|^2\}} \\
&= -\|F(x_k)\|^2 + \frac{\|F(x_k)\|^2 w_{k-1}^T F(x_k) - \|F(x_k)\|^2 w_{k-1}^T F(x_k)}{\max\{\mu\|w_{k-1}\|\|F(x_k)\|, \|F(x_{k-1})\|^2\}} \\
&= -\|F(x_k)\|^2.
\end{aligned} \tag{7}$$

Using Cauchy–Schwartz inequality, we get

$$\|F(x_k)\| \le \|d_k\|. \tag{8}$$

Furthermore, since $\max\{\mu\|w_{k-1}\|\|F(x_k)\|, \|F(x_{k-1})\|^2\} \ge \mu\|w_{k-1}\|\|F(x_k)\|$, then,

$$
\begin{aligned}
\|d_k\| &= \left\| -F(x_k) + \frac{\|F(x_k)\|^2 w_{k-1} - (F(x_k)^T w_{k-1})F(x_k)}{\max\{\mu\|w_{k-1}\|\|F(x_k)\|, \|F(x_{k-1})\|^2\}} \right\| \\
&\le \| -F(x_k)\| + \frac{\|\|F(x_k)\|^2 w_{k-1} - (F(x_k)^T w_{k-1})F(x_k)\|}{\max\{\mu\|w_{k-1}\|\|F(x_k)\|, \|F(x_{k-1})\|^2\}} \\
&\le \|F(x_k)\| + \frac{\|F(x_k)\|^2\|w_{k-1}\|}{\mu\|w_{k-1}\|\|F(x_k)\|} + \frac{\|F(x_k)^T w_{k-1}F(x_k)\|}{\mu\|w_{k-1}\|\|F(x_k)\|} \\
&\le \|F(x_k)\| + \frac{\|F(x_k)\|^2\|w_{k-1}\|}{\mu\|w_{k-1}\|\|F(x_k)\|} + \frac{\|F(x_k)\|^2\|w_{k-1}\|}{\mu\|w_{k-1}\|\|F(x_k)\|} \\
&= \|F(x_k)\| + \frac{2\|F(x_k)\|}{\mu} \\
&= \left(1 + \frac{2}{\mu}\right)\|F(x_k)\|.
\end{aligned} \tag{9}
$$

Combining (8) and (9), we get the desired result. □

Lemma 3. *Suppose that assumptions (C_1)–(C_3) hold and the sequences $\{x_k\}$ and $\{z_k\}$ are generated by Algorithm 1. Then we have*

$$\alpha_k \ge \rho \min\left\{1, \frac{\|F(x_k)\|^2}{(L+\sigma)\|F(x_k + \frac{\alpha_k}{\rho}d_k)\|\|d_k\|^2}\right\}$$

Proof. Suppose $\alpha_k \ne \rho$, then $\frac{\alpha_k}{\rho}$ does not satisfy Equation (4), that is

$$-F\left(x_k + \frac{\alpha_k}{\rho}d_k\right) < \sigma\frac{\alpha_k}{\rho}\|F(x_k + \frac{\alpha_k}{\rho}d_k)\|\|d_k\|^2.$$

This combined with (7) and the fact that F is Lipschitz continuous yields

$$
\begin{aligned}
\|F(x_k)\|^2 &= -F(x_k)^T d_k \\
&= \left(F(x_k + \frac{\alpha_k}{\rho}d_k) - F(x_k)\right)^T d_k - F^T\left(x_k + \frac{\alpha_k}{\rho}d_k\right)d_k \\
&\le L\frac{\alpha_k}{\rho}\|F(x_k + \frac{\alpha_k}{\rho}d_k)\|\|d_k\|^2 + \sigma\frac{\alpha_k}{\rho}\|F(x_k + \frac{\alpha_k}{\rho}d_k)\|\|d_k\|^2 \\
&= \frac{L+\sigma}{\rho}\alpha_k\|F(x_k + \frac{\alpha_k}{\rho}d_k)\|\|d_k\|^2.
\end{aligned} \tag{10}
$$

The above equation implies

$$\alpha_k \ge \rho \min \frac{\|F(x_k)\|^2}{(L+\sigma)\|F(x_k + \frac{\alpha_k}{\rho}d_k)\|\|d_k\|^2},$$

which completes the proof. □

Lemma 4. *Suppose that assumptions* (C_1)–(C_3) *holds, then the sequences* $\{x_k\}$ *and* $\{z_k\}$ *generated by Algorithm 1 are bounded. Moreover, we have*

$$\lim_{k \to \infty} \|x_k - z_k\| = 0 \tag{11}$$

and

$$\lim_{k \to \infty} \|x_{k+1} - x_k\| = 0. \tag{12}$$

Proof. We will start by showing that the sequences $\{x_k\}$ and $\{z_k\}$ are bounded. Suppose $\bar{x} \in E'$, then by monotonicity of F, we get

$$\langle F(z_k), x_k - \bar{x} \rangle \geq \langle F(z_k), x_k - z_k \rangle. \tag{13}$$

Also by definition of z_k and the line search (4), we have

$$\langle F(z_k), x_k - z_k \rangle \geq \sigma \alpha_k^2 \|F(z_k)\| \|d_k\|^2 \geq 0. \tag{14}$$

So, we have

$$\|x_{k+1} - \bar{x}\|^2 = \|P_E[x_k - \zeta_k F(z_k)] - \bar{x}\|^2 \leq \|x_k - \zeta_k F(z_k) - \bar{x}\|^2$$

$$= \|x_k - \bar{x}\|^2 - 2\zeta \langle F(z_k), x_k - \bar{x} \rangle + \|\zeta F(z_k)\|^2$$

$$\leq \|x_k - \bar{x}\|^2 - 2\zeta_k \langle F(z_k), x_k - z_k \rangle + \|\zeta F(z_k)\|^2 \tag{15}$$

$$= \|x_k - \bar{x}\|^2 - \left(\frac{\langle F(z_k), x_k - z_k \rangle}{\|F(z_k)\|} \right)^2$$

$$\leq \|x_k - \bar{x}\|^2.$$

Thus the sequence $\{\|x_k - \bar{x}\|\}$ is non increasing and convergent, and hence $\{x_k\}$ is bounded. Furthermore, from Equation (15), we have

$$\|x_{k+1} - \bar{x}\|^2 \leq \|x_k - \bar{x}\|^2, \tag{16}$$

and we can deduce recursively that

$$\|x_k - \bar{x}\|^2 \leq \|x_0 - \bar{x}\|^2, \quad \forall k \geq 0.$$

Then from assumption (C_3), we obtain

$$\|F(x_k)\| = \|F(x_k) - F(\bar{x})\| \leq L\|x_k - \bar{x}\| \leq L\|x_0 - \bar{x}\|.$$

If we let $L\|x_0 - \bar{x}\| = \kappa$, then the sequence $\{F(x_k)\}$ is bounded, that is,

$$\|F(x_k)\| \leq \kappa, \quad \forall k \geq 0. \tag{17}$$

By the definition of z_k, Equation (14), monotonicity of F and the Cauchy–Schwatz inequality, we get

$$\sigma\|x_k - z_k\| = \frac{\sigma\|\alpha_k d_k\|^2}{\|x_k - z_k\|} \leq \frac{\langle F(z_k), x_k - z_k \rangle}{\|x_k - z_k\|} \leq \frac{\langle F(x_k), x_k - z_k \rangle}{\|x_k - z_k\|} \leq \|F(x_k)\|. \tag{18}$$

The boundedness of the sequence $\{x_k\}$ together with Equations (17) and (18), implies the sequence $\{z_k\}$ is bounded.

Now, as $\{z_k\}$ is bounded, then for any $\bar{x} \in E'$, the sequence $\{z_k - \bar{x}\}$ is also bounded, that is, there exists a positive constant $\nu > 0$ such that

$$\|z_k - \bar{x}\| \leq \nu.$$

This together with assumption (C_3), this yields

$$\|F(z_k)\| = \|F(z_k) - F(\bar{x})\| \leq L\|z_k - \bar{x}\| \leq L\nu.$$

Therefore, using Equation (15), we have

$$\frac{\sigma^2}{(L\nu)^2}\|x_k - z_k\|^4 \leq \|x_k - \bar{x}\|^2 - \|x_{k+1} - \bar{x}\|^2,$$

which implies

$$\frac{\sigma^2}{(L\nu)^2}\sum_{k=0}^{\infty}\|x_k - z_k\|^4 \leq \sum_{k=0}^{\infty}(\|x_k - \bar{x}\|^2 - \|x_{k+1} - \bar{x}\|^2) \leq \|x_0 - \bar{x}\| < \infty. \tag{19}$$

Equation (19) implies

$$\lim_{k\to\infty}\|x_k - z_k\| = 0.$$

However, using statement 2 of Lemma 1, the definition of ζ_k and the Cauchy-Schwartz inequality, we have

$$\|x_{k+1} - x_k\| = \|P_E[x_k - \zeta_k F(z_k)] - x_k\|$$

$$\leq \|x_k - \zeta_k F(z_k) - x_k\|$$

$$= \|\zeta_k F(z_k)\| \tag{20}$$

$$= \|x_k - z_k\|,$$

which yields

$$\lim_{k\to\infty}\|x_{k+1} - x_k\| = 0.$$

□

Remark 2. *By Equation (11) and definition of z_k, then*

$$\lim_{k\to\infty}\alpha_k\|d_k\| = 0. \tag{21}$$

Theorem 1. *Suppose that assumption (C_1)–(C_3) holds and let the sequence $\{x_k\}$ be generated by Algorithm 1, then*

$$\liminf_{k\to\infty}\|F(x_k)\| = 0. \tag{22}$$

Proof. Assume that Equation (22) is not true, then there exists a constant $\epsilon > 0$ such that

$$\|F(x_k)\| \geq \epsilon, \quad \forall k \geq 0. \tag{23}$$

Combining (8) and (23), we have

$$\|d_k\| \geq \|F(x_k)\| \geq \epsilon, \quad \forall k \geq 0.$$

As $w_k = x_k + \alpha_k d_k$ and $\lim_{k \to \infty} \|x_k - z_k\| = 0$, we get $\lim_{k \to \infty} \alpha_k \|d_k\| = 0$ and

$$\lim_{k \to \infty} \alpha_k = 0. \tag{24}$$

On the other side, if $M = \left(1 + \frac{2}{\mu}\right) \kappa$, Lemma 3 and Equation (9) implies $\alpha_k \|d_k\| \geq \rho \frac{\epsilon^2}{(L+\sigma)MLv}$, which contradicts with (24). Therefore, (22) must hold. $\square$

4. Numerical Experiments

To test the performance of the proposed method, we compare it with accelerated conjugate gradient descent (ACGD) and projected Dai-Yuan (PDY) methods in [27,28], respectively. In addition, MFRM method is applied to solve signal and image recovery problems arising in compressive sensing. All codes were written in MATLAB R2018b and run on a PC with intel COREi5 processor with 4GB of RAM and CPU 2.3GHZ. All runs were stopped whenever $\|F(x_k)\| < 10^{-5}$. The parameters chosen for each method are as follows:

MFRM method: $\gamma = 1, \rho = 0.9, \mu = 0.01, \sigma = 0.0001$.
ACGD method: all parameters are chosen as in [27].
PDY method: all parameters are chosen as in [28].

We tested eight problems with dimensions of $n = 1000, 5000, 10,000, 50,000, 100,000$ and 6 initial points: $x_1 = (0.1, 0.1, \cdots, 1)^T$, $x_2 = (0.2, 0.2, \cdots, 0.2)^T$, $x_3 = (0.5, 0.5, \cdots, 0.5)^T$, $x_4 = (1.2, 1.2, \cdots, 1.2)^T$, $x_5 = (1.5, 1.5, \cdots, 1.5)^T$, $x_6 = (2, 2, \cdots, 2)^T$. In Tables 1–8, the number of Iterations (Iter), number of function evaluations (Fval), CPU time in seconds (time) and the norm at the approximate solution (NORM) were reported. The symbol '$-$' is used when the number of Iterations exceeds 1000 and/or the number of function evaluations exceeds 2000.

The test problems are listed below, where the function F is taken as $F(x) = (f_1(x), f_2(x), \ldots, f_n(x))^T$.

Problem 1 [38] Exponential Function.

$$f_1(x) = e^{x_1} - 1,$$
$$f_i(x) = e^{x_i} + x_i - 1, \text{ for } i = 2, 3, ..., n,$$
$$\text{and } E = \mathbb{R}^n_+.$$

Problem 2 [38] Modified Logarithmic Function.

$$f_i(x) = \ln(x_i + 1) - \frac{x_i}{n}, \text{ for } i = 2, 3, ..., n,$$
$$\text{and } E = \{x \in \mathbb{R}^n : \sum_{i=1}^{n} x_i \leq n, x_i > -1, i = 1, 2, \ldots, n\}.$$

Problem 3 [6] Nonsmooth Function.

$$f_i(x) = 2x_i - \sin|x_i|, \ i = 1, 2, 3, ..., n,$$
$$\text{and } E = \{x \in \mathbb{R}^n : \sum_{i=1}^{n} x_i \leq n, x_i \geq 0, i = 1, 2, \ldots, n\}.$$

It is clear that problem 3 is nonsmooth at $x = 0$.

Problem 4 [38] Strictly Convex Function I.

$$f_i(x) = e^{x_i} - 1, \text{ for } i = 1, 2, ..., n,$$
$$\text{and } E = \mathbb{R}_+^n.$$

Problem 5 [38] Strictly Convex Function II.

$$f_i(x) = \frac{i}{n} e^{x_i} - 1, \text{ for } i = 1, 2, ..., n,$$
$$\text{and } E = \mathbb{R}_+^n.$$

Problem 6 [39] Tridiagonal Exponential Function

$$f_1(x) = x_1 - e^{\cos(h(x_1+x_2))},$$
$$f_i(x) = x_i - e^{\cos(h(x_{i-1}+x_i+x_{i+1}))}, \text{ for } i = 2, ..., n - 1,$$
$$f_n(x) = x_n - e^{\cos(h(x_{n-1}+x_n))},$$
$$h = \frac{1}{n+1} \text{ and } E = \mathbb{R}_+^n.$$

Problem 7 [40] Nonsmooth Function

$$f_i(x) = x_i - \sin|x_i - 1|, \ i = 1, 2, 3, ..., n.$$
$$\text{and } E = \{x \in \mathbb{R}^n : \sum_{i=1}^{n} x_i \leq n, x_i \geq -1, i = 1, 2, ..., n\}.$$

Problem 8 [27] Penalty 1

$$t_i = \sum_{i=1}^{n} x_i^2, \ c = 10^{-5}$$
$$f_i(x) = 2c(x_i - 1) + 4(t_i - 0.25)x_i, \ i = 1, 2, 3, ..., n.$$
$$\text{and } E = \mathbb{R}_+^n.$$

To show in detail the efficiency and robustness of all methods, we employ the performance profile developed in [41], which is a helpful process of standardizing the comparison of methods. Suppose that we have n_s solvers and n_l problems and we are interested in using either number of Iterations, CPU time or number of function evaluations as our measure of performance; so we let $k_{l,s}$ to be the number of iterations, CPU time or number of function evaluations required to solve problem by solver s. To compare the performance on problem l by a solver s with the best performance by any other solver on this problem, we use the performance ratio $r_{l,s}$ defined as

$$r_{l,s} = \frac{k_{l,s}}{\min\{k_{l,s} : s \in S\}},$$

where S is the set of solvers.

The overall performance of the solver is obtained using the (cumulative) distribution function for the performance ratio P. So if we let

$$P(t) = \frac{1}{n_l} size\{l \in L : r_{l,s} \leq t\},$$

315

then $P(t)$ is the probability for solver $s \in S$ that a performance ratio $r_{l,s}$ is within a factor $t \in R$ of the best possible ratio. If the set of problems L is large enough, then the solvers with the large probability $P(t)$ are considered as the best.

Table 1. Numerical results for modified Fletcher–Reeves (MFRM), accelerated conjugate gradient descent (ACGD) and projected Dai-Yuan (PDY) for problem 1 with given initial points and dimensions.

		MFRM				ACGD				PDY			
Dimension	Initial Point	Iter	Fval	Time	Norm	Iter	Fval	Time	Norm	Iter	Fval	Time	Norm
1000	x_1	23	98	0.42639	9.01×10^{-6}	8	34	0.21556	9.26×10^{-6}	12	49	0.19349	9.18×10^{-6}
	x_2	7	35	0.019885	8.82×10^{-6}	9	39	0.086582	3.01×10^{-6}	13	53	0.07318	6.35×10^{-6}
	x_3	8	40	0.011238	9.74×10^{-6}	9	38	0.034359	4.02×10^{-6}	14	57	0.01405	5.59×10^{-6}
	x_4	15	70	0.066659	6.01×10^{-6}	16	67	0.017188	9.22×10^{-6}	15	61	0.01421	4.07×10^{-6}
	x_5	5	31	0.16103	0	18	75	0.11646	4.46×10^{-6}	14	57	0.08690	9.91×10^{-6}
	x_6	31	134	0.03232	7.65×10^{-6}	25	104	0.042967	6.74×10^{-6}	40	162	0.04060	9.70×10^{-6}
5000	x_1	8	38	0.053865	5.63×10^{-6}	9	38	0.023729	3.89×10^{-6}	13	53	0.02775	6.87×10^{-6}
	x_2	8	40	0.036653	2.59×10^{-6}	9	38	0.021951	6.65×10^{-6}	14	57	0.02974	4.62×10^{-6}
	x_3	8	40	0.030089	6.41×10^{-6}	9	39	0.019317	8.01×10^{-6}	15	61	0.04353	4.18×10^{-6}
	x_4	16	74	0.081741	4.71×10^{-6}	17	71	0.05235	8.12×10^{-6}	15	61	0.03288	9.08×10^{-6}
	x_5	5	31	0.030748	0	18	75	0.038894	8.14×10^{-6}	15	61	0.03556	7.30×10^{-6}
	x_6	31	134	0.087531	8.1×10^{-6}	26	108	0.053473	7.96×10^{-6}	39	158	0.10419	9.86×10^{-6}
10,000	x_1	5	26	0.03829	3.7×10^{-6}	9	39	0.044961	5.5×10^{-6}	13	53	0.05544	9.70×10^{-6}
	x_2	8	40	0.055099	3.64×10^{-6}	9	39	0.0358	9.39×10^{-6}	14	57	0.06201	6.53×10^{-6}
	x_3	8	40	0.049974	5.44×10^{-6}	10	43	0.04176	2.12×10^{-6}	15	61	0.08704	5.90×10^{-6}
	x_4	16	74	0.125	6.61×10^{-6}	18	75	0.066316	4.58×10^{-6}	16	65	0.07797	4.28×10^{-6}
	x_5	5	31	0.048751	0	18	75	0.11807	7.86×10^{-6}	39	158	0.20751	7.97×10^{-6}
	x_6	28	122	0.13649	7.18×10^{-6}	27	112	0.10593	6.22×10^{-6}	87	351	0.36678	9.93×10^{-6}
50,000	x_1	5	26	0.1584	3.58×10^{-6}	10	43	0.15918	2.33×10^{-6}	14	57	0.23129	7.12×10^{-6}
	x_2	8	40	0.18044	8.1×10^{-6}	10	43	0.16252	3.97×10^{-6}	15	61	0.23975	4.91×10^{-6}
	x_3	8	40	0.186	4.54×10^{-6}	10	43	0.15707	4.67×10^{-6}	16	65	0.24735	4.37×10^{-6}
	x_4	17	78	0.31567	5.47×10^{-6}	19	79	0.27474	4.1×10^{-6}	38	154	0.55277	7.54×10^{-6}
	x_5	5	31	0.18586	0	18	75	0.27118	5.06×10^{-6}	177	712	2.29950	9.44×10^{-6}
	x_6	20	90	0.39237	6.44×10^{-6}	28	116	0.35197	7.69×10^{-6}	361	1449	4.63780	9.74×10^{-6}
100,000	x_1	5	26	0.26116	4.59×10^{-6}	10	42	0.28038	3.29×10^{-6}	15	61	0.50090	3.39×10^{-6}
	x_2	9	43	0.35288	1.59×10^{-6}	10	42	0.28999	5.62×10^{-6}	15	61	0.45876	6.94×10^{-6}
	x_3	8	40	0.35809	4.96×10^{-6}	10	42	0.29255	6.59×10^{-6}	16	65	0.51380	6.18×10^{-6}
	x_4	17	78	0.59347	7.73×10^{-6}	19	79	0.51261	5.79×10^{-6}	175	704	4.48920	9.47×10^{-6}
	x_5	32	138	0.98463	7.09×10^{-6}	18	75	0.46086	4.05×10^{-6}	176	708	4.49410	9.91×10^{-6}
	x_6	17	78	0.57701	9.31×10^{-6}	29	120	0.71678	6.05×10^{-6}	360	1445	9.10170	9.99×10^{-6}

Table 2. Numerical results for MFRM, ACGD and PDY for problem 2 with given initial points and dimensions.

		MFRM				ACGD				PDY			
Dimension	Initial Point	Iter	Fval	Time	Norm	Iter	Fval	Time	Norm	Iter	Fval	Time	Norm
1000	x_1	3	8	0.007092	5.17×10^{-7}	3	8	0.036061	5.17×10^{-7}	10	39	0.01053	6.96×10^{-6}
	x_2	3	8	0.012401	6.04×10^{-6}	3	8	0.006143	6.04×10^{-6}	11	43	0.00937	9.23×10^{-6}
	x_3	4	11	0.003993	4.37×10^{-7}	4	11	0.006476	4.37×10^{-7}	13	51	0.01111	6.26×10^{-6}
	x_4	5	14	0.010363	1.52×10^{-7}	5	14	0.005968	1.52×10^{-7}	14	55	0.02154	9.46×10^{-6}
	x_5	5	14	0.007234	1.1×10^{-6}	5	14	0.02349	1.1×10^{-6}	15	59	0.01850	4.60×10^{-6}
	x_6	6	17	0.006496	1.74×10^{-8}	6	17	0.00677	1.74×10^{-8}	15	59	0.01938	7.71×10^{-6}
5000	x_1	3	8	0.011561	1.75×10^{-7}	3	8	0.009794	1.75×10^{-7}	11	43	0.03528	4.86×10^{-6}
	x_2	3	8	0.010452	3.13×10^{-6}	3	8	0.009591	3.13×10^{-6}	12	47	0.04032	6.89×10^{-6}
	x_3	4	11	0.01516	1.42×10^{-7}	4	11	0.013767	1.42×10^{-7}	14	55	0.04889	4.61×10^{-6}
	x_4	5	14	0.019733	3.94×10^{-8}	5	14	0.014274	3.94×10^{-8}	15	59	0.04826	6.96×10^{-6}
	x_5	5	14	0.018462	4.05×10^{-7}	5	14	0.011728	4.05×10^{-7}	16	63	0.05969	3.37×10^{-6}
	x_6	6	17	0.028536	2.36×10^{-9}	6	17	0.016345	2.36×10^{-9}	16	63	0.06253	5.64×10^{-6}
10,000	x_1	3	8	0.019053	1.21×10^{-7}	3	8	0.0135	1.21×10^{-7}	11	43	0.06732	6.85×10^{-6}
	x_2	3	8	0.01791	2.79×10^{-6}	3	8	0.015807	2.79×10^{-6}	12	47	0.12232	9.72×10^{-6}
	x_3	4	11	0.033042	9.73×10^{-8}	4	11	0.020752	9.73×10^{-8}	14	55	0.08288	6.51×10^{-6}
	x_4	5	14	0.031576	2.56×10^{-8}	5	14	0.04483	2.56×10^{-8}	15	59	0.08413	9.82×10^{-6}
	x_5	5	14	0.032747	2.93×10^{-7}	5	14	0.026975	2.93×10^{-7}	16	63	0.09589	4.75×10^{-6}
	x_6	6	17	0.036002	1.24×10^{-9}	6	17	0.032445	1.24×10^{-9}	16	64	0.11499	8.55×10^{-6}
50,000	x_1	3	8	0.0737	6.32×10^{-8}	7	26	0.16925	2.94×10^{-6}	12	47	0.27826	5.23×10^{-6}
	x_2	3	8	0.06964	3.37×10^{-6}	9	34	0.18801	2.78×10^{-6}	13	51	0.29642	7.11×10^{-6}
	x_3	4	11	0.093027	4.87×10^{-8}	7	25	0.15375	9.11×10^{-6}	15	59	0.35602	4.82×10^{-6}
	x_4	5	14	0.11219	1.11×10^{-8}	7	24	0.15382	9.18×10^{-6}	35	141	0.69470	6.69×10^{-6}
	x_5	5	14	0.1173	1.84×10^{-7}	9	32	0.18164	6.21×10^{-6}	35	141	0.68488	9.12×10^{-6}
	x_6	6	17	0.13794	4.01×10^{-10}	6	19	0.11216	5.2×10^{-6}	35	141	0.70973	9.91×10^{-6}
100,000	x_1	3	8	0.13021	5.4×10^{-8}	7	26	0.2609	4.14×10^{-6}	12	47	0.44541	7.39×10^{-6}
	x_2	3	8	0.13267	4.27×10^{-6}	9	34	0.32666	3.93×10^{-6}	13	51	0.53299	3.39×10^{-6}
	x_3	4	11	0.17338	4.05×10^{-8}	8	29	0.3113	3.33×10^{-6}	15	60	0.58603	8.71×10^{-6}
	x_4	5	14	0.20036	8.15×10^{-9}	8	28	0.2997	3.34×10^{-6}	72	290	2.70630	8.31×10^{-6}
	x_5	5	14	0.25274	1.8×10^{-7}	9	32	0.32098	9.46×10^{-6}	72	290	2.72220	8.68×10^{-6}
	x_6	6	17	0.24952	2.71×10^{-10}	6	19	0.21972	7.01×10^{-6}	72	290	2.75850	8.96×10^{-6}

Table 3. Numerical results for MFRM, ACGD and PDY for problem 3 with given initial points and dimensions.

Dimension	Initial Point	MFRM				ACGD				PDY			
		Iter	Fval	Time	Norm	Iter	Fval	Time	Norm	Iter	Fval	Time	Norm
1000	x_1	6	24	0.024062	3.11×10^{-6}	6	40	0.02951	4.44×10^{-6}	12	48	0.01255	4.45×10^{-6}
	x_2	6	24	0.005345	5.94×10^{-6}	6	40	0.0077681	8.75×10^{-6}	12	48	0.01311	9.02×10^{-6}
	x_3	6	24	0.006109	9.94×10^{-6}	6	44	0.0067049	5.09×10^{-6}	13	52	0.01486	8.34×10^{-6}
	x_4	8	33	0.006127	3.1×10^{-6}	8	44	0.007142	5.04×10^{-6}	14	56	0.01698	8.04×10^{-6}
	x_5	11	46	0.010427	2.71×10^{-6}	11	40	0.010411	3.12×10^{-6}	14	56	0.01551	9.72×10^{-6}
	x_6	16	68	0.010682	8.38×10^{-6}	16	77	0.014759	5.98×10^{-6}	14	56	0.01534	9.42×10^{-6}
5000	x_1	6	24	0.020455	6.96×10^{-6}	6	40	0.020368	9.93×10^{-6}	12	48	0.03660	9.94×10^{-6}
	x_2	7	28	0.021552	1.33×10^{-6}	7	44	0.029622	5.09×10^{-6}	13	52	0.03616	6.85×10^{-6}
	x_3	7	28	0.023056	2.22×10^{-6}	7	48	0.030044	2.96×10^{-6}	14	56	0.04594	6.14×10^{-6}
	x_4	8	33	0.022984	6.92×10^{-6}	8	48	0.022777	2.93×10^{-6}	15	60	0.04342	6.01×10^{-6}
	x_5	11	46	0.031466	6.06×10^{-6}	11	40	0.019226	6.97×10^{-6}	15	60	0.04296	7.25×10^{-6}
	x_6	17	72	0.049308	7.67×10^{-6}	17	81	0.036095	6.05×10^{-6}	32	129	0.10081	8.85×10^{-6}
10,000	x_1	6	24	0.03064	9.85×10^{-6}	6	44	0.03997	3.65×10^{-6}	13	52	0.06192	4.77×10^{-6}
	x_2	7	28	0.035806	1.88×10^{-6}	7	44	0.037221	7.19×10^{-6}	13	52	0.06442	9.68×10^{-6}
	x_3	7	28	0.035795	3.14×10^{-6}	7	48	0.053226	4.18×10^{-6}	14	56	0.09499	8.69×10^{-6}
	x_4	8	33	0.041017	9.79×10^{-6}	8	48	0.057984	4.15×10^{-6}	15	60	0.07696	8.5×10^{-6}
	x_5	11	46	0.06448	8.58×10^{-6}	11	40	0.047413	9.85×10^{-6}	33	133	0.18625	6.45×10^{-6}
	x_6	18	76	0.09651	4.44×10^{-6}	18	81	0.085238	8.56×10^{-6}	33	133	0.15548	7.51×10^{-6}
50,000	x_1	7	28	0.14323	2.2×10^{-6}	7	44	0.17175	8.17×10^{-6}	14	56	0.23642	3.51×10^{-6}
	x_2	7	28	0.13625	4.2×10^{-6}	7	48	0.18484	4.18×10^{-6}	14	56	0.24813	7.12×10^{-6}
	x_3	7	28	0.13246	7.03×10^{-6}	7	48	0.1827	9.36×10^{-6}	15	60	0.27049	6.53×10^{-6}
	x_4	9	37	0.18261	4.16×10^{-6}	9	48	0.18993	9.27×10^{-6}	34	137	0.54545	7.13×10^{-6}
	x_5	12	50	0.21743	5.2×10^{-6}	12	44	0.17043	5.73×10^{-6}	68	274	1.02330	9.99×10^{-6}
	x_6	18	76	0.34645	9.93×10^{-6}	18	85	0.32938	8.66×10^{-6}	69	278	1.03810	8.05×10^{-6}
100,000	x_1	7	28	0.27078	3.11×10^{-6}	7	48	0.36144	3×10^{-6}	14	56	0.45475	4.96×10^{-6}
	x_2	7	28	0.26974	5.94×10^{-6}	7	48	0.37515	5.91×10^{-6}	15	60	0.49018	3.39×10^{-6}
	x_3	7	28	0.25475	9.94×10^{-6}	7	52	0.39071	3.44×10^{-6}	15	60	0.49016	9.24×10^{-6}
	x_4	9	37	0.3089	5.88×10^{-6}	9	52	0.35961	3.41×10^{-6}	139	559	4.03110	9.01×10^{-6}
	x_5	12	50	0.41839	7.35×10^{-6}	12	44	0.33105	8.1×10^{-6}	70	282	2.07100	8.54×10^{-6}
	x_6	19	80	0.64773	5.75×10^{-6}	19	89	0.61329	5.54×10^{-6}	139	559	4.02440	9.38×10^{-6}

Table 4. Numerical results for MFRM, ACGD and PDY for problem 4 with given initial points and dimensions.

Dimension	Initial Point	MFRM				ACGD				PDY			
		Iter	Fval	Time	Norm	Iter	Fval	Time	Norm	Iter	Fval	Time	Norm
1000	x_1	6	24	0.00855	1.65×10^{-6}	10	40	0.014662	3.65×10^{-6}	12	48	0.00989	4.60×10^{-6}
	x_2	5	20	0.004234	2.32×10^{-6}	10	40	0.0064115	5.79×10^{-6}	12	48	0.00966	9.57×10^{-6}
	x_3	10	42	0.007426	6.42×10^{-6}	10	40	0.0054818	3.29×10^{-6}	13	52	0.00887	8.49×10^{-6}
	x_4	21	90	0.011603	5.84×10^{-6}	27	110	0.012854	8.97×10^{-6}	12	48	0.01207	5.83×10^{-6}
	x_5	16	71	0.010735	8.48×10^{-6}	26	106	0.015603	5.97×10^{-6}	29	117	0.05371	9.43×10^{-6}
	x_6	1	15	0.005932	0	36	147	0.025039	9.56×10^{-6}	29	117	0.02396	6.65×10^{-6}
5000	x_1	6	24	0.019995	3.68×10^{-6}	10	40	0.018283	8.15×10^{-6}	13	52	0.02503	3.49×10^{-6}
	x_2	5	20	0.00934	5.2×10^{-6}	11	44	0.016733	3.36×10^{-6}	13	52	0.02626	7.24×10^{-6}
	x_3	11	46	0.02156	3.89×10^{-6}	10	40	0.017073	7.37×10^{-6}	14	56	0.03349	6.29×10^{-6}
	x_4	22	94	0.043325	6.81×10^{-6}	29	118	0.047436	7.09×10^{-6}	13	52	0.02258	4.25×10^{-6}
	x_5	18	79	0.096692	6.15×10^{-6}	27	110	0.058405	7.95×10^{-6}	31	125	0.05471	7.59×10^{-6}
	x_6	1	15	0.012199	0	39	159	0.059448	7.33×10^{-6}	63	254	0.10064	8.54×10^{-6}
10,000	x_1	6	24	0.019264	5.2×10^{-6}	11	44	0.026877	3×10^{-6}	13	52	0.03761	4.93×10^{-6}
	x_2	5	20	0.017891	7.35×10^{-6}	11	44	0.03118	4.76×10^{-6}	14	56	0.04100	3.37×10^{-6}
	x_3	11	46	0.036079	5.5×10^{-6}	11	44	0.034673	2.71×10^{-6}	14	56	0.03919	8.90×10^{-6}
	x_4	22	94	0.069778	9.63×10^{-6}	30	122	0.069971	5.97×10^{-6}	32	129	0.09613	6.02×10^{-6}
	x_5	18	79	0.062821	8.69×10^{-6}	28	114	0.066866	6.68×10^{-6}	32	129	0.09177	6.44×10^{-6}
	x_6	1	15	0.017237	0	40	163	0.093749	7.26×10^{-6}	64	258	0.20791	9.39×10^{-6}
50,000	x_1	7	28	0.093473	1.16×10^{-6}	11	44	0.16749	6.7×10^{-6}	14	56	0.17193	3.63×10^{-6}
	x_2	6	24	0.072206	1.64×10^{-6}	12	48	0.11391	2.77×10^{-6}	14	56	0.15237	7.54×10^{-6}
	x_3	12	50	0.14285	3.33×10^{-6}	11	44	0.11036	6.06×10^{-6}	15	60	0.16549	6.66×10^{-6}
	x_4	24	102	0.30313	5.86×10^{-6}	31	126	0.30903	7.94×10^{-6}	67	270	0.76283	7.81×10^{-6}
	x_5	20	87	0.28955	6.31×10^{-6}	29	118	0.30266	8.89×10^{-6}	67	270	0.76157	8.80×10^{-6}
	x_6	1	15	0.061327	0	42	171	0.41158	7.96×10^{-6}	269	1080	2.92510	9.41×10^{-6}
100,000	x_1	7	28	0.15038	1.65×10^{-6}	11	44	0.2434	9.48×10^{-6}	14	56	0.30229	5.13×10^{-6}
	x_2	6	24	0.13126	2.32×10^{-6}	12	48	0.2614	3.91×10^{-6}	15	60	0.31648	3.59×10^{-6}
	x_3	12	50	0.31585	4.71×10^{-6}	11	44	0.2161	8.57×10^{-6}	32	129	0.72838	9.99×10^{-6}
	x_4	24	102	0.58023	8.29×10^{-6}	32	130	0.65289	6.68×10^{-6}	135	543	2.86780	9.73×10^{-6}
	x_5	20	87	0.5122	8.92×10^{-6}	30	122	0.61637	7.48×10^{-6}	272	1092	5.74140	9.91×10^{-6}
	x_6	1	15	0.11696	0	43	175	0.82759	7.88×10^{-6}	548	2197	11.44130	9.87×10^{-6}

Table 5. Numerical results for MFRM, ACGD and PDY for problem 5 with given initial points and dimensions.

Dimension	Initial Point	MFRM Iter	Fval	Time	Norm	ACGD Iter	Fval	Time	Norm	PDY Iter	Fval	Time	Norm
	x_1	26	98	0.023555	3.51×10^{-6}	39	154	0.022285	9.7×10^{-6}	16	63	0.07575	6.03×10^{-6}
	x_2	40	154	0.024539	5.9×10^{-6}	22	85	0.015671	5.03×10^{-6}	16	63	0.01470	5.42×10^{-6}
1000	x_3	37	144	0.021659	7.11×10^{-6}	43	173	0.029569	7.96×10^{-6}	33	132	0.02208	6.75×10^{-6}
	x_4	49	206	0.030696	9.52×10^{-6}	30	122	0.014942	6.05×10^{-6}	30	121	0.01835	8.39×10^{-6}
	x_5	46	194	0.11589	7.06×10^{-6}	29	118	0.040406	6.5×10^{-6}	32	129	0.02700	8.47×10^{-6}
	x_6	43	182	0.027471	8.7×10^{-6}	40	163	0.0311	9.83×10^{-6}	30	121	0.01712	6.95×10^{-6}
	x_1	38	147	0.073315	4.96×10^{-6}	30	117	0.060877	9.56×10^{-6}	17	67	0.04394	5.64×10^{-6}
	x_2	20	77	0.056225	4.98×10^{-6}	16	60	0.027911	5.91×10^{-6}	17	67	0.04635	5.07×10^{-6}
5000	x_3	41	157	0.082151	8.92×10^{-6}	78	315	0.12774	9.7×10^{-6}	35	140	0.08311	9.74×10^{-6}
	x_4	48	202	0.10166	9.19×10^{-6}	31	126	0.067911	8.39×10^{-6}	33	133	0.08075	6.02×10^{-6}
	x_6	147	562	3.308158	8.44×10^{-7}	31	126	0.067856	7.81×10^{-6}	35	141	0.10091	7.51×10^{-6}
	x_7	45	190	0.090276	7.14×10^{-6}	44	179	0.09371	7.37×10^{-6}	32	129	0.08054	8.55×10^{-6}
	x_1	37	143	0.12665	9.28×10^{-6}	77	308	0.28678	9.85×10^{-6}	17	67	0.06816	8.81×10^{-6}
	x_2	22	84	0.077288	9.78×10^{-6}	16	60	0.071657	7.52×10^{-6}	17	67	0.08833	7.80×10^{-6}
10,000	x_3	39	149	0.1297	6.74×10^{-6}	105	424	0.34212	9.08×10^{-6}	37	148	0.14732	6.36×10^{-6}
	x_4	60	250	0.2175	7.56×10^{-6}	32	130	0.11937	7.17×10^{-6}	37	149	0.14293	8.25×10^{-6}
	x_5	44	186	0.1727	7.68×10^{-6}	32	130	0.11921	8.26×10^{-6}	36	145	0.14719	8.23×10^{-6}
	x_6	46	194	0.1728	8.62×10^{-6}	45	183	0.15634	9.01×10^{-6}	74	298	0.26456	7.79×10^{-6}
	x_1	44	170	0.62202	1×10^{-5}	90	539	31.75299	2.56×10^{-7}	42	169	0.58113	7.78×10^{-6}
	x_2	69	280	0.9662	6.87×10^{-6}	31	122	0.33817	7.09×10^{-6}	42	169	0.58456	7.13×10^{-6}
50,000	x_3	119	464	25.87657	9.34×10^{-7}	260	1047	2.8824	9.67×10^{-6}	41	165	0.58717	8.87×10^{-6}
	x_4	50	210	0.71599	8.38×10^{-6}	33	134	0.39039	9.98×10^{-6}	40	161	0.56431	7.17×10^{-6}
	x_5	46	194	0.65508	8.47×10^{-6}	35	142	0.40807	7.19×10^{-6}	82	330	1.08920	8.44×10^{-6}
	x_6	50	210	0.69117	8.12×10^{-6}	49	199	0.57702	8.97×10^{-6}	80	322	1.06670	7.82×10^{-6}
	x_1	31	121	0.84183	4.48×10^{-6}	88	530	61.97806	5.53×10^{-7}	43	173	1.09620	8.47×10^{-6}
	x_2	135	518	59.19294	8.37×10^{-7}	110	442	2.2661	9.55×10^{-6}	43	173	1.10040	7.77×10^{-6}
100,000	x_3	46	178	1.1322	6.99×10^{-6}	345	1388	7.1938	9.76×10^{-6}	42	169	1.08330	9.66×10^{-6}
	x_4	50	210	1.3737	8.85×10^{-6}	34	138	0.74362	8.65×10^{-6}	85	342	2.11880	9.22×10^{-6}
	x_5	47	198	1.3879	8.31×10^{-6}	36	146	0.79012	8.09×10^{-6}	84	338	2.10640	9.78×10^{-6}
	x_6	52	218	1.4318	7.37×10^{-6}	51	207	1.1601	8.42×10^{-6}	167	671	4.06200	9.90×10^{-6}

Table 6. Numerical Results for MFRM, ACGD and PDY for problem 6 with given initial points and dimensions.

Dimension	Initial Point	MFRM Iter	Fval	Time	Norm	ACGD Iter	Fval	Time	Norm	PDY Iter	Fval	Time	Norm
	x_1	11	44	0.011156	8.32×10^{-6}	12	48	0.02786	7.88×10^{-6}	15	60	0.01671	4.35×10^{-6}
	x_2	11	44	0.016092	7.32×10^{-6}	12	48	0.01042	7.58×10^{-6}	15	60	0.01346	4.18×10^{-6}
1000	x_3	11	44	0.010446	8.83×10^{-6}	12	48	0.0092	6.68×10^{-6}	15	60	0.01630	3.68×10^{-6}
	x_4	10	40	0.011233	7.38×10^{-6}	12	48	0.013617	4.57×10^{-6}	14	56	0.01339	7.48×10^{-6}
	x_5	9	36	0.011325	8.29×10^{-6}	12	48	0.011492	3.67×10^{-6}	14	56	0.01267	6.01×10^{-6}
	x_6	7	28	0.009452	8.25×10^{-6}	11	44	0.016351	8.32×10^{-6}	14	56	0.01685	3.54×10^{-6}
	x_1	8	32	0.026924	1.87×10^{-6}	13	52	0.036025	4.59×10^{-6}	15	60	0.05038	9.73×10^{-6}
	x_2	8	32	0.043488	1.8×10^{-6}	13	52	0.040897	4.42×10^{-6}	15	60	0.04775	9.36×10^{-6}
5000	x_3	8	32	0.02709	1.59×10^{-6}	13	52	0.039937	3.89×10^{-6}	15	60	0.04923	8.25×10^{-6}
	x_4	8	32	0.026351	1.1×10^{-6}	13	52	0.033013	2.66×10^{-6}	15	60	0.05793	5.64×10^{-6}
	x_5	7	28	0.023442	8.62×10^{-6}	12	48	0.030462	8.22×10^{-6}	15	60	0.04597	4.53×10^{-6}
	x_6	7	28	0.022952	5.08×10^{-6}	12	48	0.028786	4.85×10^{-6}	14	56	0.05070	7.93×10^{-6}
	x_1	8	32	0.061374	2.62×10^{-6}	13	52	0.092372	6.5×10^{-6}	68	274	0.40724	9.06×10^{-6}
	x_2	8	32	0.06285	2.52×10^{-6}	13	52	0.059778	6.25×10^{-6}	68	274	0.41818	8.72×10^{-6}
10,000	x_3	8	32	0.059913	2.22×10^{-6}	13	52	0.077326	5.5×10^{-6}	34	137	0.21905	6.22×10^{-6}
	x_4	8	32	0.057003	1.52×10^{-6}	13	52	0.087745	3.77×10^{-6}	15	60	0.10076	7.98×10^{-6}
	x_5	8	32	0.070307	1.22×10^{-6}	13	52	0.077217	3.02×10^{-6}	15	60	0.12680	6.40×10^{-6}
	x_6	7	28	0.052718	7.18×10^{-6}	12	48	0.067375	6.85×10^{-6}	15	60	0.11984	3.78×10^{-6}
	x_1	8	32	0.21258	5.85×10^{-6}	14	56	0.32965	3.78×10^{-6}	143	575	3.09120	9.42×10^{-6}
	x_2	8	32	0.21203	5.63×10^{-6}	14	56	0.31297	3.63×10^{-6}	143	575	3.06200	9.06×10^{-6}
50,000	x_3	8	32	0.20885	4.96×10^{-6}	14	56	0.30089	3.2×10^{-6}	142	571	3.04950	9.04×10^{-6}
	x_4	8	32	0.20483	3.4×10^{-6}	13	52	0.26855	8.42×10^{-6}	69	278	1.53920	9.14×10^{-6}
	x_5	8	32	0.21467	2.72×10^{-6}	13	52	0.26304	6.76×10^{-6}	68	274	1.49490	9.43×10^{-6}
	x_6	8	32	0.20933	1.61×10^{-6}	13	52	0.26143	3.99×10^{-6}	15	60	0.38177	8.44×10^{-6}
	x_1	8	32	0.41701	8.28×10^{-6}	14	56	0.58853	5.34×10^{-6}	292	1172	13.59530	9.53×10^{-6}
	x_2	8	32	0.41511	7.96×10^{-6}	14	56	0.58897	5.14×10^{-6}	290	1164	13.30930	9.75×10^{-6}
100,000	x_3	8	32	0.44061	7.01×10^{-6}	14	56	0.57318	4.53×10^{-6}	144	579	6.68150	9.96×10^{-6}
	x_4	8	32	0.43805	4.8×10^{-6}	13	52	0.58712	3.1×10^{-6}	141	567	6.50800	9.92×10^{-6}
	x_5	8	32	0.41147	3.85×10^{-6}	13	52	0.56384	9.56×10^{-6}	70	282	3.30510	8.07×10^{-6}
	x_6	8	32	0.43925	2.27×10^{-6}	13	52	0.53343	5.64×10^{-6}	34	137	1.64510	6.37×10^{-6}

Table 7. Numerical Results for MFRM, ACGD and PDY for problem 7 with given initial points and dimensions.

Dimension	Initial Point	MFRM Iter	Fval	Time	Norm	ACGD Iter	Fval	Time	Norm	PDY Iter	Fval	Time	Norm
	x_1	4	21	0.011834	3.24×10^{-7}	10	42	0.008528	2.46×10^{-6}	14	57	0.00953	5.28×10^{-6}
	x_2	4	21	0.006228	1.43×10^{-7}	9	38	0.008289	3.91×10^{-6}	13	53	0.00896	9.05×10^{-6}
1000	x_3	3	17	0.004096	5.81×10^{-8}	8	34	0.006702	7.43×10^{-6}	3	12	0.00426	8.47×10^{-6}
	x_4	7	34	0.00585	3.89×10^{-6}	11	46	0.009579	5.94×10^{-6}	15	61	0.01169	6.73×10^{-6}
	x_5	7	34	0.006133	6.36×10^{-6}	11	46	0.015328	8.97×10^{-6}	31	126	0.03646	9.03×10^{-6}
	x_6	8	37	0.006106	1.9×10^{-6}	12	49	0.01426	2.87×10^{-6}	15	60	0.01082	3.99×10^{-6}
	x_1	4	21	0.015836	7.25×10^{-7}	10	42	0.023953	5.49×10^{-6}	15	61	0.03215	4.25×10^{-6}
	x_2	4	21	0.014521	3.2×10^{-7}	9	38	0.021065	8.74×10^{-6}	14	57	0.02942	7.40×10^{-6}
5000	x_3	3	17	0.014517	1.3×10^{-7}	9	38	0.025437	4.01×10^{-6}	4	16	0.01107	1.01×10^{-7}
	x_4	7	34	0.028388	8.71×10^{-6}	12	50	0.028607	3.21×10^{-6}	16	65	0.04331	5.43×10^{-6}
	x_5	8	38	0.02787	1.49×10^{-6}	12	50	0.037806	4.84×10^{-6}	33	134	0.09379	7.78×10^{-6}
	x_6	8	37	0.027898	4.26×10^{-6}	12	49	0.029226	6.43×10^{-6}	15	60	0.04077	8.92×10^{-6}
	x_1	4	21	0.028528	1.02×10^{-6}	10	42	0.045585	7.77×10^{-6}	15	61	0.06484	6.01×10^{-6}
	x_2	4	21	0.033782	4.52×10^{-7}	10	42	0.041715	2.98×10^{-6}	15	61	0.07734	3.77×10^{-6}
10,000	x_3	3	17	0.029265	1.84×10^{-7}	9	38	0.036422	5.67×10^{-6}	4	16	0.02707	1.42×10^{-7}
	x_4	8	38	0.043301	1.29×10^{-6}	12	50	0.063527	4.53×10^{-6}	16	65	0.07941	7.69×10^{-6}
	x_5	8	38	0.043741	2.1×10^{-6}	12	50	0.049604	6.85×10^{-6}	34	138	0.14942	6.83×10^{-6}
	x_6	8	37	0.053666	6.02×10^{-6}	12	49	0.050153	9.09×10^{-6}	34	138	0.15224	8.81×10^{-6}
	x_1	4	21	0.10816	2.29×10^{-6}	11	46	0.20624	4.19×10^{-6}	16	65	0.25995	4.89×10^{-6}
	x_2	4	21	0.11969	1.01×10^{-6}	10	42	0.16364	6.67×10^{-6}	15	61	0.24674	8.42×10^{-6}
50,000	x_3	3	17	0.068644	4.11×10^{-7}	10	42	0.1539	3.06×10^{-6}	4	16	0.09405	3.18×10^{-7}
	x_4	8	38	0.16067	2.88×10^{-6}	13	54	0.20728	2.45×10^{-6}	36	146	0.55207	6.39×10^{-6}
	x_5	8	38	0.14484	4.7×10^{-6}	13	54	0.19421	3.69×10^{-6}	35	142	0.54679	9.05×10^{-6}
	x_6	9	41	0.161	1.41×10^{-6}	13	53	0.19386	4.9×10^{-6}	36	146	0.55764	7.59×10^{-6}
	x_1	4	21	0.21825	3.24×10^{-6}	11	46	0.32512	5.93×10^{-6}	17	69	0.52595	5.68×10^{-6}
	x_2	4	21	0.16435	1.43×10^{-6}	10	42	0.30949	9.43×10^{-6}	16	65	0.52102	4.34×10^{-6}
100,000	x_3	3	17	0.13072	5.81×10^{-7}	10	42	0.31031	4.32×10^{-6}	4	16	0.14864	4.50×10^{-7}
	x_4	8	38	0.29012	4.07×10^{-6}	13	54	0.38833	3.46×10^{-6}	36	146	1.05360	9.04×10^{-6}
	x_5	8	38	0.32821	6.65×10^{-6}	13	54	0.3522	5.22×10^{-6}	74	299	2.10730	8.55×10^{-6}
	x_6	9	41	0.43649	1.99×10^{-6}	13	53	0.3561	6.94×10^{-6}	37	150	1.08240	6.66×10^{-6}

Table 8. Numerical results for MFRM, ACGD and PDY for problem 8 with given initial points and dimensions.

Dimension	Initial Point	MFRM Iter	Fval	Time	Norm	ACGD Iter	Fval	Time	Norm	PDY Iter	Fval	Time	Norm
	x_1	8	27	0.1502	1.52×10^{-6}	8	26	0.049826	6.09×10^{-6}	69	279	0.05538	8.95×10^{-6}
	x_2	8	27	0.042248	1.52×10^{-6}	8	26	0.017594	6.09×10^{-6}	270	1085	0.18798	9.72×10^{-6}
1000	x_3	26	114	0.03877	7.85×10^{-6}	8	26	0.010888	6.09×10^{-6}	24	52	0.02439	6.57×10^{-6}
	x_4	26	114	0.017542	7.85×10^{-6}	8	26	0.007873	6.09×10^{-6}	27	58	0.01520	7.59×10^{-6}
	x_5	26	114	0.067692	7.85×10^{-6}	8	26	0.060733	6.09×10^{-6}	28	61	0.04330	9.21×10^{-6}
	x_6	26	114	0.045173	7.85×10^{-6}	8	26	0.006889	6.09×10^{-6}	40	85	0.02116	8.45×10^{-6}
	x_1	6	28	0.023925	8.77×10^{-6}	4	13	0.011005	5.76×10^{-6}	658	2639	1.13030	9.98×10^{-6}
	x_2	15	70	0.043512	7.94×10^{-6}	4	13	0.009131	5.76×10^{-6}	27	58	0.05101	7.59×10^{-6}
5000	x_3	15	70	0.046458	7.94×10^{-6}	4	13	0.011311	5.76×10^{-6}	49	104	0.08035	8.11×10^{-6}
	x_4	15	70	0.044788	7.94×10^{-6}	4	13	0.010475	5.75×10^{-6}	40	85	0.07979	8.15×10^{-6}
	x_5	15	70	0.044639	7.94×10^{-6}	4	13	0.011034	5.77×10^{-6}	18	40	0.09128	9.14×10^{-6}
	x_6	15	70	0.043974	7.94×10^{-6}	4	13	0.00785	5.76×10^{-6}	17	38	0.18528	8.98×10^{-6}
	x_1	11	54	0.06595	6.15×10^{-6}	5	20	0.024232	2.19×10^{-6}	49	104	0.20443	7.62×10^{-6}
	x_2	11	54	0.068125	6.15×10^{-6}	5	20	0.023511	2.19×10^{-6}	40	85	0.15801	8.45×10^{-6}
10,000	x_3	11	54	0.065486	6.15×10^{-6}	5	20	0.023004	2.19×10^{-6}	19	42	0.37880	7.66×10^{-6}
	x_4	11	54	0.064515	6.15×10^{-6}	5	20	0.030435	2.19×10^{-6}	90	187	1.25802	9.7×10^{-6}
	x_5	11	54	0.056261	6.15×10^{-6}	5	20	0.021963	2.19×10^{-6}	988	1988	12.68259	9.93×10^{-6}
	x_6	11	54	0.067785	6.15×10^{-6}	5	20	0.021889	2.21×10^{-6}	27	58	0.32859	7.59×10^{-6}
	x_1	7	38	0.17856	4.5×10^{-6}	5	23	0.087544	2.45×10^{-6}	19	42	0.52291	6.42×10^{-6}
	x_2	7	38	0.17862	4.5×10^{-6}	5	23	0.093227	2.45×10^{-6}	148	304	3.93063	9.92×10^{-6}
50,000	x_3	7	38	0.17746	4.5×10^{-6}	5	23	0.087484	2.45×10^{-6}	937	1886	22.97097	9.87×10^{-6}
	x_4	7	38	0.17392	4.5×10^{-6}	5	23	0.086329	2.4×10^{-6}	27	58	0.68467	7.59×10^{-6}
	x_5	7	38	0.18035	4.5×10^{-6}	5	23	0.08954	2.4×10^{-6}	346	702	8.45043	9.79×10^{-6}
	x_6	7	38	0.17504	4.5×10^{-6}	5	23	0.093203	2.5×10^{-6}	40	85	0.99230	8.45×10^{-6}
	x_1	28	122	0.91448	8.61×10^{-6}	4	20	0.14743	2.71×10^{-6}	-	-	-	-
	x_2	28	122	0.93662	8.61×10^{-6}	4	20	0.14823	2.7×10^{-6}	-	-	-	-
100,000	x_3	28	122	0.90604	8.61×10^{-6}	4	20	0.1497	2.79×10^{-6}	-	-	-	-
	x_4	28	122	0.92351	8.61×10^{-6}	4	20	0.14844	2.37×10^{-6}	-	-	-	-
	x_5	28	122	0.91896	8.61×10^{-6}	4	20	0.12346	1.66×10^{-6}	-	-	-	-
	x_6	28	122	0.91294	8.61×10^{-6}	4	20	0.12522	2.11×10^{-6}	-	-	-	-

Figure 1 reveals that MFRM performed better in terms of number of Iterations, as it solves and wins over 70 percent of the problems with less number of Iterations, while ACGD and PDY solve and win over 40 and almost 10 percent respectively. The story is a little bit different in Figure 2 as ACGD method was very competitive. However, MFRM method performed a little bit better by solving and winning over 50 percent of the problems with less CPU time as against ACGD method which solves and wins less than 50 percent of the problems considered. The PDY method had the least performance with just 10 percent success. The interpretation of Figure 3 was similar to that of Figure 1. Finally, in Table 11 we report numerical results for MFRM, ACGD and PDY for problem 2 with given initial points and dimensions with double float (10^{-16}) accuracy.

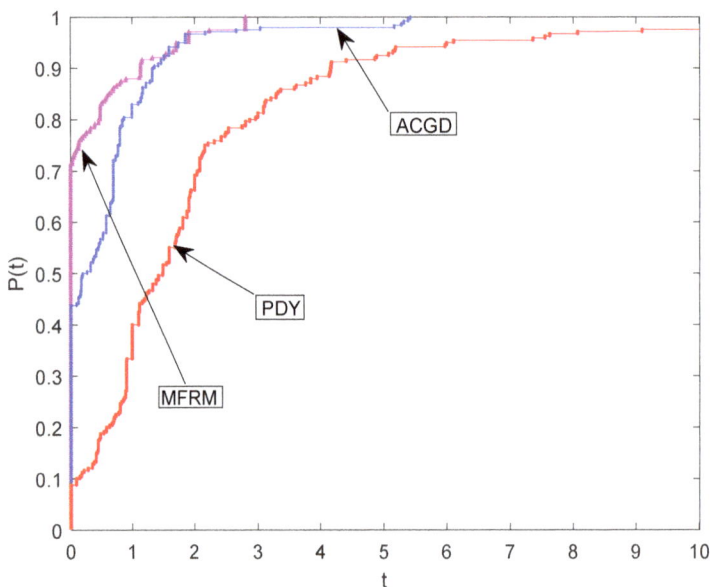

Figure 1. Performance profiles for the number of iterations.

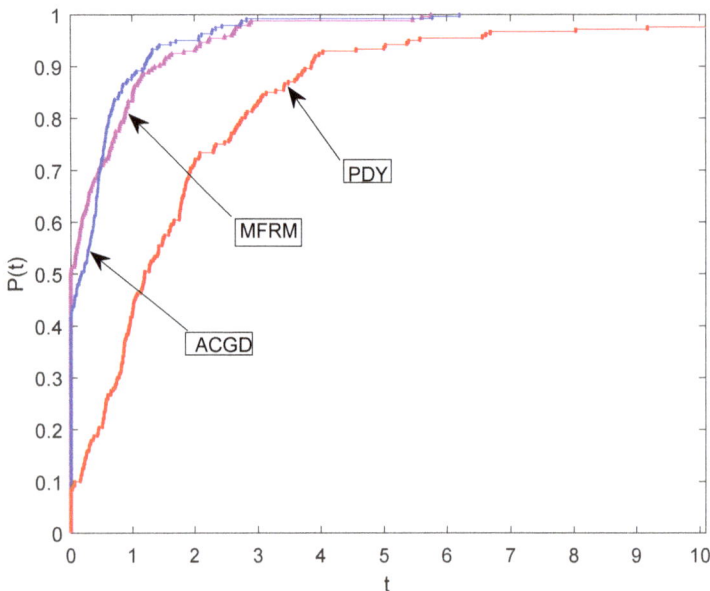

Figure 2. Performance profiles for the CPU time (in seconds).

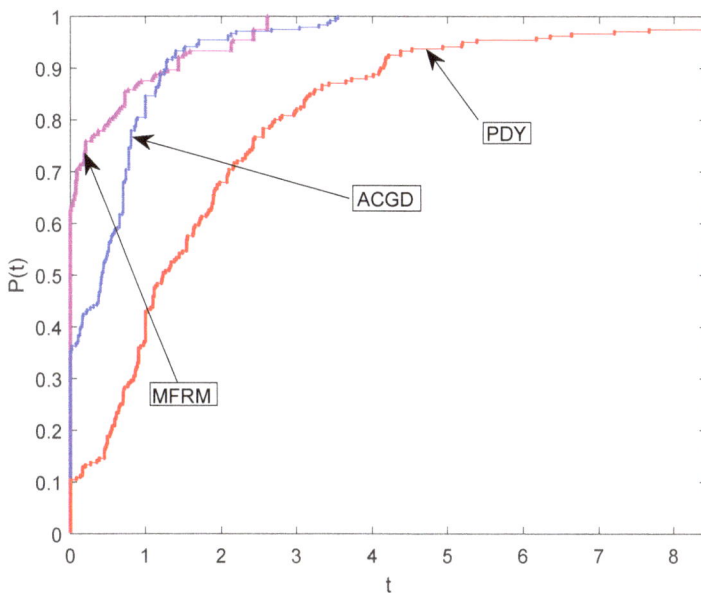

Figure 3. Performance profiles for the number of function evaluations.

4.1. Experiments on Solving Sparse Signal Problems

There were many problems in signal processing and statistical inference involving finding sparse solutions to ill-conditioned linear systems of equations. Among popular approaches was minimizing an objective function which contains quadratic (ℓ_2) error term and a sparse ℓ_1−regularization term, i.e.,

$$\min_x \frac{1}{2}\|y - Bx\|_2^2 + \eta\|x\|_1, \tag{25}$$

where $x \in R^n$, $y \in R^k$ is an observation, $B \in R^{k \times n}$ ($k << n$) is a linear operator, η is a non-negative parameter, $\|x\|_2$ denotes the Euclidean norm of x and $\|x\|_1 = \sum_{i=1}^n |x_i|$ is the ℓ_1−norm of x. It is easy to see that problem (25) is a convex unconstrained minimization problem. Due to the fact that if the original signal is sparse or approximately sparse in some orthogonal basis, problem (25) frequently appears in compressive sensing, and hence an exact restoration can be produced by solving (25).

Iterative methods for solving (25) have been presented in many papers (see [42–45]). The most popular method among these methods is the gradient-based method and the earliest gradient projection method for sparse reconstruction (GPRS) was proposed by Figueiredo et al. [44]. The first step of the GPRS method is to express (25) as a quadratic problem using the following process. Consider a point $x \in \mathbb{R}^n$ such that $x = u - v$, where $u, v \geq 0$. u and v are chosen in such a way that x is splitted into its positive and negative parts as follows $u_i = (x_i)_+$, $v_i = (-x_i)_+$ for all $i = 1, 2, ..., n$, and $(.)_+ = \max\{0, .\}$. By definition of ℓ_1-norm, we have $\|x\|_1 = e_n^T u + e_n^T v$, where $e_n = (1, 1, ..., 1)^T \in R^n$. Now (25) can be written as

$$\min_{u,v} \frac{1}{2}\|y - B(u - v)\|_2^2 + \eta e_n^T u + \eta e_n^T v, \quad u \geq 0, \quad v \geq 0, \tag{26}$$

which is a bound-constrained quadratic program. However, from [44], Equation (26) can be written in standard form as

$$\min_z \frac{1}{2}z^T Dz + c^T z, \quad \text{such that} \quad z \geq 0, \tag{27}$$

where $z = \begin{pmatrix} u \\ v \end{pmatrix}$, $c = \omega e_{2n} + \begin{pmatrix} -b \\ b \end{pmatrix}$, $b = B^T y$, $D = \begin{pmatrix} B^T B & -B^T B \\ -B^T B & B^T B \end{pmatrix}$. Clearly, D is a positive semi-definite matrix, which implies that Equation (27) is a convex quadratic problem.

Xiao et al. [20] translated (27) into a linear variable inequality problem which is equivalent to a linear complementarity problem. Moreover, z is a solution of the linear complementarity problem if and only if it is a solution of the following nonlinear equation:

$$F(z) = \min\{z, Dz + c\} = 0. \tag{28}$$

The function F is a vector-valued function and the "min" was interpreted as component wise minimum. Furthermore, F was proved to be continuous and monotone in [46]. Therefore problem (25) can be translated into problem (1) and thus MFRM method can be applied to solve it.

In this experiment, we consider a simple compressive sensing possible situation, where our goal is to reconstruct a sparse signal of length n from k observations. The quality of recovery is assessed by mean of squared error (MSE) to the original signal $\tilde{x}$,

$$MSE = \frac{1}{n} \|\tilde{x} - x_*\|^2,$$

where x_* is the recovered signal. The signal size is chosen as $n = 2^{11}$, $k = 2^9$ and the original signal contains 2^6 randomly nonzero elements. In addition, the measurement y is distributed with noise, that is, $y = B\tilde{x} + \varrho$, where B is a randomly generated Gaussian matrix and ϱ is the Gaussian noise distributed normally with mean 0 and variance 10^{-4}.

To demonstrate the performance of the MFRM method in signal recovery problems, we compare it with the conjugate gradient descent CGD [20] and projected conjugate gradient PCG [23] methods. The parameters in PCG and CGD methods are chosen as $\gamma = 10$, $\sigma = 10^{-4}$, $\rho = 0.5$. However, we chose $\gamma = 1$, $\sigma = 10^{-4}$, $\rho = 0.9$ and $\mu = 0.01$ in MFRM method. For fairness in comparison, each code was run from the same initial point, same continuation technique on the parameter η, and observed only the behavior of the convergence of each method to have a similar accurate solution. The experiment was initialized with $x_0 = B^T y$ and terminates when

$$\frac{\|f(x_k) - f(x_{k-1})\|}{\|f(x_{k-1})\|} < 10^{-5},$$

where $f(x_k) = \frac{1}{2} \|y - Bx_k\|_2^2 + \eta \|x_k\|_1$.

In Figures 4 and 5, MFRM, CGD and PCG methods recovered the disturbed signal almost exactly. The experiment was repeated for 20 different noise samples (see Table 9). It can be observed that the MFRM is more efficient in terms of the number of Iterations and CPU time than CGD and PCG methods in most cases. Furthermore, MFRM was able to achieve the least MSE in nine (9) out of the twenty (20) experiments. To reveal visually the performance of both methods, two figures were plotted to demonstrate their convergence behavior based on MSE, objective function values, the number of Iterations and CPU time (see Figures 6 and 7). It can also be observed that MFRM requires less computing time to achieve similar quality resolution. This can be seen graphically in Figures 6 and 7 which illustrate that the objective function values obtained by MFRM decrease faster throughout the entire Iteration process.

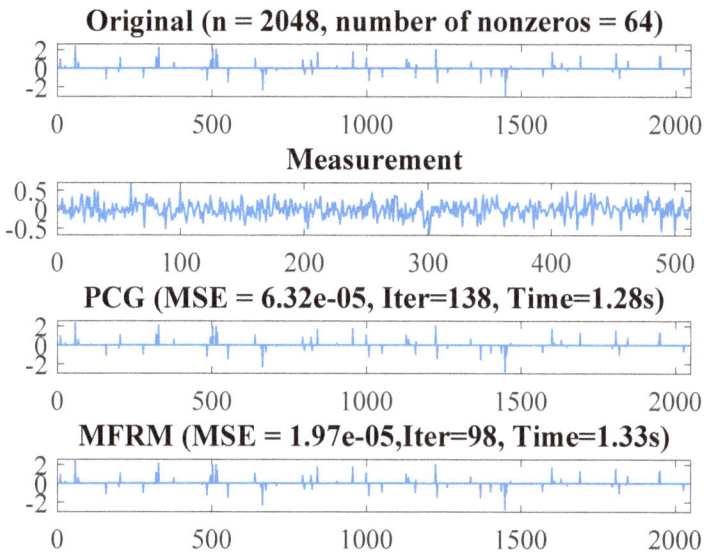

Figure 4. (**top**) to (**bottom**) The original image, the measurement, and the recovered signals by projected conjugate gradient PCG and modified descent Fletcher–Reeves CG method (MFRM) methods.

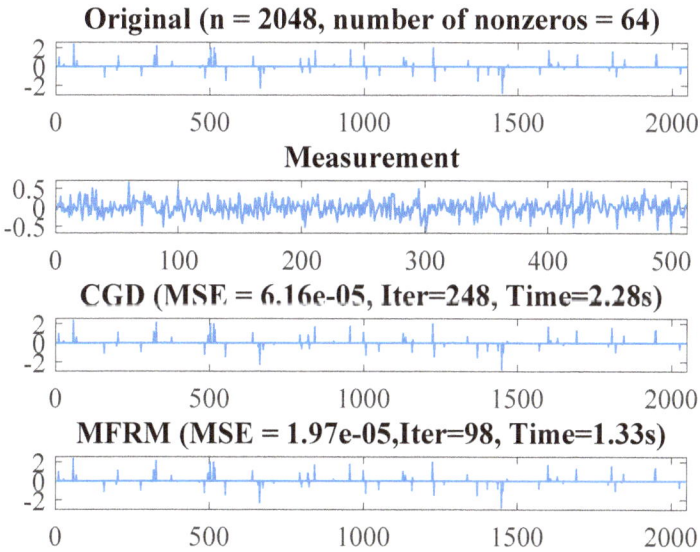

Figure 5. (**top**) to (**bottom**) The original image, the measurement, and the recovered signals by conjugate gradient descent (CGD) and MFRM methods.

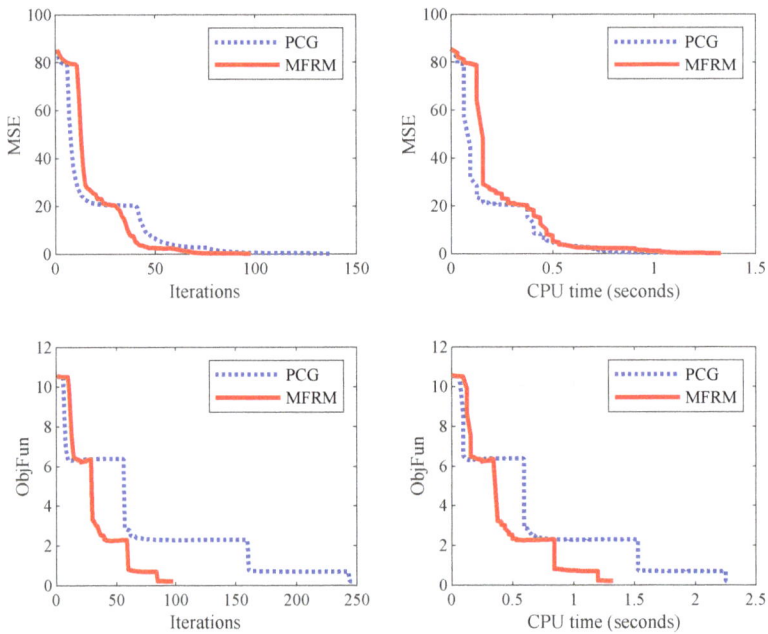

Figure 6. Comparison result of PCG and MFRM. The *x*-axis represent the number of Iterations ((**top left**) and (**bottom left**)) and CPU time in seconds ((**top right**) and (**bottom right**)). The *y*-axis represent the MSE ((**top left**) and (**top right**)) and the objective function values ((**bottom left**) and (**bottom right**)).

Table 9. Twenty experiment results of ℓ_1-norm regularization problem for CGD, PCG and MFRM methods.

S/N	Iter			Time			MSE		
	CGD	**PCG**	**MFRM**	**CGD**	**PCG**	**MFRM**	**CGD**	**PCG**	**MFRM**
1	248	138	98	2.28	1.28	1.33	6.16×10^{-5}	6.32×10^{-5}	1.97×10^{-5}
2	234	138	117	3.37	1.26	1.19	4.08×10^{-5}	3.36×10^{-5}	5.40×10^{-5}
3	224	152	104	1.90	1.29	0.97	2.78×10^{-5}	1.78×10^{-5}	1.02×10^{-5}
4	230	143	117	3.21	2.48	1.17	4.08×10^{-5}	3.36×10^{-5}	5.40×10^{-5}
5	152	119	114	1.65	1.03	1.15	1.23×10^{-5}	2.07×10^{-5}	5.49×10^{-5}
6	223	127	110	1.89	2.56	1.83	3.33×10^{-5}	6.08×10^{-5}	6.50×10^{-6}
7	156	120	125	1.37	1.01	1.20	4.25×10^{-5}	3.26×10^{-5}	1.46×10^{-5}
8	213	89	10	1.90	0.78	1.12	1.86×10^{-5}	3.77×10^{-4}	1.31×10^{-5}
9	227	152	118	2.14	1.53	1.45	2.75×10^{-5}	1.54×10^{-5}	8.11×10^{-6}
10	201	142	101	2.22	1.64	1.01	6.75×10^{-5}	1.86×10^{-5}	1.17×10^{-5}
11	200	151	90	1.70	1.42	0.90	2.36×10^{-5}	1.29×10^{-5}	3.81×10^{-5}
12	202	153	91	1.75	1.34	0.84	6.94×10^{-5}	2.99×10^{-5}	9.21×10^{-5}
13	208	128	125	1.89	1.12	1.26	1.71×10^{-5}	1.42×10^{-5}	9.20×10^{-6}
14	161	145	122	1.47	1.28	1.26	1.15×10^{-5}	8.75×10^{-6}	4.36×10^{-6}
15	227	160	100	1.97	1.42	1.00	3.41×10^{-5}	2.40×10^{-5}	1.54×10^{-5}
16	269	172	88	2.51	1.67	0.98	3.90×10^{-5}	6.59×10^{-5}	2.08×10^{-4}
17	210	129	105	1.84	1.19	1.11	2.11×10^{-5}	1.89×10^{-5}	6.22×10^{-5}
18	225	132	96	1.93	1.15	1.00	3.87×10^{-5}	7.78×10^{-5}	9.49×10^{-5}
19	152	120	92	1.37	1.09	0.87	2.12×10^{-5}	1.32×10^{-5}	4.03×10^{-5}
20	151	128	113	1.31	1.15	1.06	4.48×10^{-5}	1.85×10^{-5}	1.71×10^{-5}

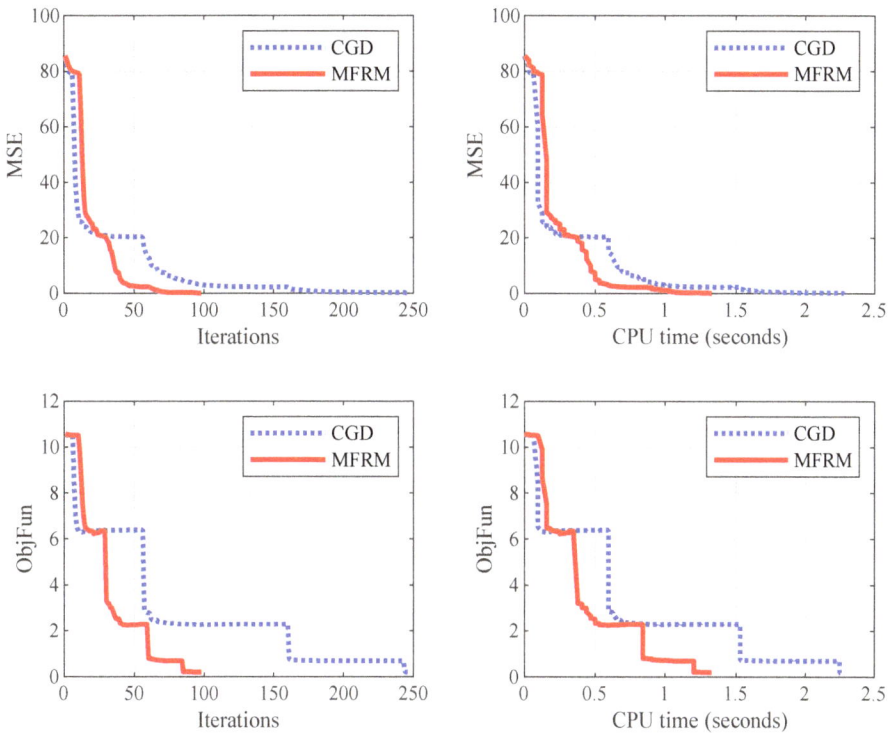

Figure 7. Comparison result of PCG and MFRM. The *x*-axis represent the number of Iterations ((**top left**) and (**bottom left**)) and CPU time in seconds ((**top right**) and (**bottom right**)). The *y*-axis represent the MSE ((**top left**) and (**top right**)) and the objective function values ((**bottom left**) and (**bottom right**)).

4.2. Experiments on Blurred Image Restoration

In this subsection, we test the performance of MFRM in restoring a blurred image. We use the following well-known gray test images; (P1) Cameraman, (P2) Lena, (P3) House and (P4) Peppers for the experiments. We use 4 different Gaussian blur kernels with a standard deviation v to compare the robustness of MFRM method with CGD method proposed in [20].

To assess the performance of each algorithm tested with respect to the metrics that indicate better quality of restoration, in Table 10 we reported the objective function (ObjFun) at the approximate solution, the MSE, the signal-to-noise-ratio (SNR) which is defined as

$$\text{SNR} = 20 \times \log_{10}\left(\frac{\|\bar{x}\|}{\|x - \bar{x}\|}\right),$$

and the structural similarity (SSIM) index that measure the similarity between the original image and the restored image [47] for each of the 16 experiments. The MATLAB implementation of the SSIM index can be obtained at http://www.cns.nyu.edu/~lcv/ssim/.

The original, blurred and restored images by each of the algorithm are given in Figures 8–11. The figures demonstrate that both the two algorithms can restore the blurred images. In contrast to the CGD, the quality of the restored image by MFRM is superior in most cases. Table 11 reported numerical results for MFRM, ACGD and PDY for problem 2.

Table 10. Efficiency comparison based on the value of the objective function (ObjFun) mean-square-error (MSE), SNR and the SSIM index under different Pi(v).

Image	ObjFun		MSE		SNR		SSIM	
	MFRM	**CGD**	**MFRM**	**CGD**	**MFRM**	**CGD**	**MFRM**	**CGD**
P1(1×10^{-4})	1.43×10^6	1.47×10^6	133.90	177.57	21.28	20.05	0.86	0.83
P1(1×10^{-1})	1.43×10^6	1.48×10^6	130.60	177.69	21.39	20.5	0.86	0.83
P1(0.25)	1.47×10^6	1.48×10^6	145.27	177.72	20.93	20.05	0.85	0.83
P1(6.25)	1.58×10^6	1.65×10^6	146.06	183.96	20.9	19.9	0.75	0.79
P2(1×10^{-4})	1.61×10^6	1.65×10^6	36.88	57.55	27.59	25.65	0.88	0.86
P2(1×10^{-1})	1.61×10^6	1.65×10^6	36.85	57.61	27.59	25.65	0.88	0.86
P2(0.25)	1.62×10^6	1.66×10^6	37.78	57.68	27.48	25.64	0.88	0.86
P2(6.25)	1.77×10^6	1.82×10^6	56.65	58.96	25.72	25.55	0.76	0.83
P3(1×10^{-4})	5.74×10^6	5.89×10^6	41.63	44.48	26.26	25.97	0.9	0.88
P3(1×10^{-1})	5.75×10^6	5.90×10^6	42.42	44.54	26.17	25.96	0.89	0.88
P3(0.25)	5.76×10^6	5.91×10^6	43.33	44.65	26.08	25.95	0.88	0.88
P3(6.25)	6.35×10^6	6.60×10^6	106.79	48.47	22.16	25.6	0.63	0.85
P4(1×10^{-4})	1.40×10^6	1.48×10^6	88.81	122.44	22.9	21.5	0.87	0.84
P4(1×10^{-1})	1.41×10^6	1.48×10^6	89.22	122.56	22.88	21.5	0.87	0.84
P4(0.25)	1.41×10^6	1.49×10^6	89.86	122.56	22.85	21.5	0.87	0.84
P4(6.25)	1.56×10^6	1.69×10^6	116.79	138.97	21.71	20.95	0.76	0.82

Table 11. Numerical results for modified Fletcher-Reeves method MFRM, accelerated conjugate gradient descent (ACGD) and projected Dai-Yuan (PDY) methods for problem 2 with given initial points and dimensions with double float (10^{-16}) accuracy.

		MFRM				ACGD				PDY			
Dimension	Initial Point	Iter	Fval	Time	Norm	Iter	Fval	Time	Norm	Iter	Fval	Time	Norm
1000	$x1$	8	27	0.14061	9.47×10^{-19}	12	53	0.030479	3.32×10^{-18}	30	119	0.04027	4.76×10^{-19}
	$x2$	8	36	0.010782	1.49×10^{-18}	7	20	0.013503	1.08×10^{-18}	36	153	0.034454	3.51×10^{-18}
	$x3$	7	20	0.008263	1.21×10^{-18}	13	56	0.021302	3.26×10^{-18}	38	161	0.038168	3.51×10^{-18}
	$x4$	8	23	0.015654	1.80×10^{-19}	12	51	0.02056	3.31×10^{-18}	39	165	0.057793	3.51×10^{-18}
	$x5$	11	38	0.018461	1.59×10^{-18}	14	59	0.088858	3.34×10^{-18}	41	173	0.069756	3.51×10^{-18}
	$x6$	10	34	0.016788	1.07×10^{-18}	10	32	0.012069	5.83×10^{-19}	40	169	0.03311	3.50×10^{-18}
5000	$x1$	9	33	0.028658	7.22×10^{-19}	12	54	0.041685	1.52×10^{-18}	35	149	0.10692	1.57×10^{-18}
	$x2$	7	23	0.024046	2.18×10^{-19}	9	41	0.049194	1.55×10^{-18}	37	157	0.12219	1.57×10^{-18}
	$x3$	6	17	0.03436	3.89×10^{-19}	14	61	0.094129	1.47×10^{-18}	33	131	0.10635	1.06×10^{-19}
	$x4$	8	26	0.03133	7.17×10^{-19}	14	60	0.065147	1.47×10^{-18}	39	165	0.18361	1.57×10^{-18}
	$x5$	9	31	0.036727	5.84×10^{-19}	10	43	0.1165	1.47×10^{-18}	44	2178	0.2178	7.43×10^{-20}
	$x6$	10	34	0.030168	6.41×10^{-19}	12	51	0.038218	1.51×10^{-18}	38	161	0.13144	1.57×10^{-18}
10,000	$x1$	8	28	0.064617	1.89×10^{-19}	11	50	0.068567	1.03×10^{-18}	35	149	0.2253	1.11×10^{-18}
	$x2$	6	19	0.044204	1.90×10^{-19}	14	62	0.15949	1.09×10^{-18}	32	128	0.34325	8.21×10^{-20}
	$x3$	6	17	0.045192	1.45×10^{-19}	18	78	0.10766	1.04×10^{-18}	39	165	0.23899	1.11×10^{-18}
	$x4$	10	35	0.055408	4.99×10^{-19}	12	52	0.061589	1.06×10^{-18}	39	165	0.23162	1.11×10^{-18}
	$x5$	7	20	0.038439	2.06×10^{-19}	14	60	0.087394	1.05×10^{-18}	40	169	0.28998	1.11×10^{-18}
	$x6$	9	29	0.065318	5.27×10^{-19}	16	68	0.09917	1.03×10^{-18}	40	170	0.22564	1.11×10^{-18}
50,000	$x1$	7	26	0.21017	1.93×10^{-19}	23	100	0.51879	4.79×10^{-19}	34	145	0.92896	4.96×10^{-19}
	$x2$	6	21	0.24752	2.09×10^{-19}	25	108	0.64677	4.90×10^{-19}	36	153	0.9954	4.96×10^{-19}
	$x3$	6	17	0.11243	6.27×10^{-20}	23	99	0.50402	4.93×10^{-19}	38	161	0.96768	4.96×10^{-19}
	$x4$	7	20	0.13442	1.02×10^{-19}	24	102	0.63664	4.75×10^{-19}	79	326	1.7542	4.96×10^{-19}
	$x5$	9	30	0.20288	7.25×10^{-20}	25	106	0.51116	4.78×10^{-19}	78	322	1.7246	4.96×10^{-19}
	$x6$	12	52	0.36526	2.28×10^{-19}	23	97	0.56342	4.76×10^{-19}	80	330	1.6812	4.96×10^{-19}
100,000	$x1$	7	27	0.36065	6.53×10^{-20}	23	100	0.88236	3.26×10^{-19}	30	119	1.2102	9.26×10^{-21}
	$x2$	5	14	0.20041	3.91×10^{-20}	25	108	0.90777	3.27×10^{-19}	35	149	1.5699	3.51×10^{-19}
	$x3$	7	24	0.34075	1.47×10^{-19}	25	107	0.95898	3.26×10^{-19}	40	170	1.7126	3.51×10^{-19}
	$x4$	8	31	0.40444	2.09×10^{-20}	24	102	0.83332	3.38×10^{-19}	151	614	5.8306	3.51×10^{-19}
	$x5$	8	26	0.52598	5.03×10^{-20}	25	106	1.0223	3.47×10^{-19}	151	614	5.6777	3.50×10^{-19}
	$x6$	7	20	0.33434	1.45×10^{-19}	23	97	0.87438	3.33×10^{-19}	153	622	5.7906	3.51×10^{-19}

Figure 8. The original image (**top left**), the blurred image (**top right**), the restored image by CGD (**bottom left**) with time = 3.70, signal-to-noise-ratio (SNR) = 20.05 and structural similarity (SSIM) = 0.83, and by MFRM (**bottom right**) with time = 1.97, SNR = 21.28 and SSIM = 0.86.

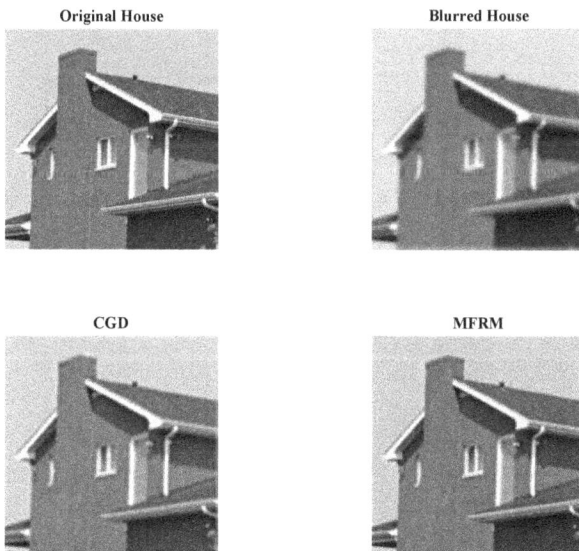

Figure 9. The original image (**top left**), the blurred image (**top right**), the restored image by CGD (**bottom left**) with Time = 1.95, SNR = 25.65 and SSIM = 0.86, and by MFRM (**bottom right**) with Time = 3.59, SNR = 27.59 and SSIM = 0.88.

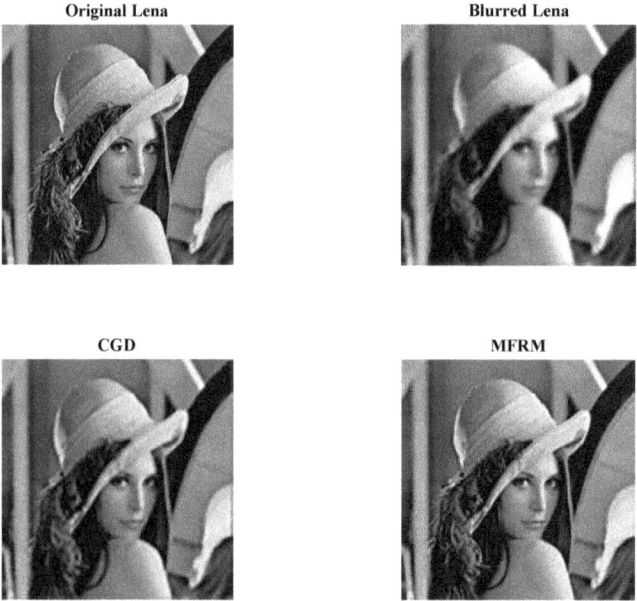

Figure 10. The original image (**top left**), the blurred image (**top right**), the restored image by CGD (**bottom left**) with time = 5.38, SNR = 25.97 and SSIM = 0.88, and by MFRM (**bottom right**) with time = 38.77, SNR = 26.26 and SSIM = 0.90.

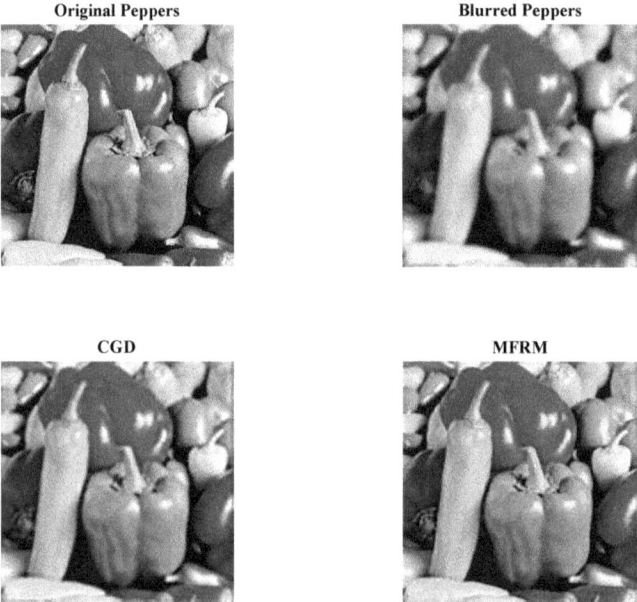

Figure 11. The original image (**top left**), the blurred image (**top right**), the restored image by CGD (**bottom left**) with Time = 2.48, SNR = 21.50 and SSIM = 0.84, and by MFRM (**bottom right**) with Time = 4.93, SNR = 22.90 and SSIM = 0.87.

5. Conclusions

In this paper, a modified conjugate gradient method for solving monotone nonlinear equations with convex constraints was presented which is similar to that in [3]. The proposed method is suitable for non-smooth equations. Under some suitable assumptions, the global convergence of the proposed method was demonstrated. Numerical results were presented to show the effectiveness of the MFRM method compared to the ACGD and PDY methods for the given constrained monotone equation problems. Finally, the MFRM was also shown to be effective in decoding sparse signals and restoration of blurred images.

Author Contributions: conceptualization, A.B.A.; methodology, A.B.A.; software, H.M.; validation, P.K., A.M.A. and K.S.; formal analysis, P.K. and K.S.; investigation, P.K. and H.M.; resources, P.K. and K.S.; data curation, H.M. and A.M.A.; writing–original draft preparation, A.B.A.; writing–review and editing, H.M.; visualization, A.M.A. and K.S.; supervision, P.K.; project administration, P.K. and K.S.; funding acquisition, P.K. and K.S.

Funding: Petchra Pra Jom Klao Doctoral Scholarship for Ph.D. program of King Mongkut's University of Technology Thonburi (KMUTT). This project was partially supported by the Thailand Research Fund (TRF) and the King Mongkut's University of Technology Thonburi (KMUTT) under the TRF Research Scholar Award (Grant No. RSA6080047). Moreover, Kanokwan Sitthithakerngkiet was supported by Faculty of Applied Science, King Mongkuts University of Technology North Bangkok. Contract no. 6242104.

Acknowledgments: We thank Associate Professor Jin Kiu Liu for providing us with the access of the CGD-CS MATLAB codes. The authors acknowledge the financial support provided by King Mongkut's University of Technology Thonburi through the "KMUTT 55th Anniversary Commemorative Fund". The first author was supported by the "Petchra Pra Jom Klao Ph.D. Research Scholarship from King Mongkut's University of Technology Thonburi".

Conflicts of Interest: The authors declare no conflict of interest.

References

1. Abubakar, A.B.; Kumam, P.; Awwal, A.M. A Descent Dai-Liao Projection Method for Convex Constrained Nonlinear Monotone Equations with Applications. *Thai J. Math.* **2018**, *17*, 128–152.
2. Abubakar, A.B.; Kumam, P. A descent Dai-Liao conjugate gradient method for nonlinear equations. *Numer. Algorithms* **2019**, *81*, 197–210. [CrossRef]
3. Abubakar, A.B.; Kumam, P. An improved three-term derivative-free method for solving nonlinear equations. *Comput. Appl. Math.* **2018**, *37*, 6760–6773. [CrossRef]
4. Mohammad, H.; Abubakar, A.B. A positive spectral gradient-like method for nonlinear monotone equations. *Bull. Comput. Appl. Math.* **2017**, *5*, 99–115.
5. Muhammed, A.A.; Kumam, P.; Abubakar, A.B.; Wakili, A.; Pakkaranang, N. A New Hybrid Spectral Gradient Projection Method for Monotone System of Nonlinear Equations with Convex Constraints. *Thai J. Math.* **2018**, *16*, 125–147.
6. Zhou, W.J.; Li, D.H. A globally convergent BFGS method for nonlinear monotone equations without any merit functions. *Math. Comput.* **2008**, *77*, 2231–2240. [CrossRef]
7. Yan, Q.R.; Peng, X.Z.; Li, D.H. A globally convergent derivative-free method for solving large-scale nonlinear monotone equations. *J. Comput. Appl. Math.* **2010**, *234*, 649–657. [CrossRef]
8. DiRksEandM, S.P.; FERRis, C. A collection of nonlinear mixed complementarity problems. *Optim. Methods Softw.* **1995**, *5*, 319–345.
9. Meintjes, K.; Morgan, A.P. A methodology for solving chemical equilibrium systems. *Appl. Math. Comput.* **1987**, *22*, 333–361. [CrossRef]
10. Bellavia, S.; Macconi, M.; Morini, B. STRSCNE: A Scaled Trust-Region Solver for Constrained Nonlinear Equations. *Comput. Optim. Appl.* **2004**, *28*, 31–50. [CrossRef]
11. Kanzow, C.; Yamashita, N.; Fukushima, M. Levenberg–Marquardt methods with strong local convergence properties for solving nonlinear equations with convex constraints. *J. Comput. Appl. Math.* **2004**, *172*, 375–397. [CrossRef]
12. Papp, Z.; Rapajić, S. FR type methods for systems of large-scale nonlinear monotone equations. *Appl. Math. Comput.* **2015**, *269*, 816–823. [CrossRef]

13.   Zhou, W.; Wang, F. A PRP-based residual method for large-scale monotone nonlinear equations. *Appl. Math. Comput.* **2015**, *261*, 1–7. [CrossRef]
14.   Dai, Y.H.; Yuan, Y. A nonlinear conjugate gradient method with a strong global convergence property. *SIAM J. Optim.* **1999**, *10*, 177–182. [CrossRef]
15.   Polak, E.; Ribiere, G. Note sur la convergence de méthodes de directions conjuguées. *Revue Française D'informatique et de Recherche Opérationnelle Série Rouge* 1969, *3*, 35–43. [CrossRef]
16.   Polyak, B.T. The conjugate gradient method in extremal problems. *USSR Comput. Math. Math. Phys.* **1969**, *9*, 94–112. [CrossRef]
17.   Hager, W.W.; Zhang, H. A new conjugate gradient method with guaranteed descent and an efficient line search. *SIAM J. Optim.* **2005**, *16*, 170–192. [CrossRef]
18.   Dai, Y.H.; Liao, L.Z. New conjugacy conditions and related nonlinear conjugate gradient methods. *Appl. Math. Optim.* **2001**, *43*, 87–101. [CrossRef]
19.   Fletcher, R.; Reeves, C.M. Function minimization by conjugate gradients. *Comput. J.* **1964**, *7*, 149–154. [CrossRef]
20.   Xiao, Y.; Zhu, H. A conjugate gradient method to solve convex constrained monotone equations with applications in compressive sensing. *J. Math. Anal. Appl.* **2013**, *405*, 310–319. [CrossRef]
21.   Solodov, M.V.; Svaiter, B.F. A globally convergent inexact Newton method for systems of monotone equations. In *Reformulation: Nonsmooth, Piecewise Smooth, Semismooth and Smoothing Methods*; Springer: Boston, MA, USA, 1998; pp. 355–369.
22.   Liu, S.Y.; Huang, Y.Y.; Jiao, H.W. Sufficient descent conjugate gradient methods for solving convex constrained nonlinear monotone equations. In *Abstract and Applied Analysis*; Hindawi: New York, NY, USA, 2014; Volume 2014.
23.   Liu, J.; Li, S. A projection method for convex constrained monotone nonlinear equations with applications. *Comput. Math. Appl.* **2015**, *70*, 2442–2453. [CrossRef]
24.   Sun, M.; Liu, J. Three derivative-free projection methods for nonlinear equations with convex constraints. *J. Appl. Math. Comput.* **2015**, *47*, 265–276. [CrossRef]
25.   Sun, M.; Liu, J. New hybrid conjugate gradient projection method for the convex constrained equations. *Calcolo* **2016**, *53*, 399–411. [CrossRef]
26.   Ou, Y.; Li, J. A new derivative-free SCG-type projection method for nonlinear monotone equations with convex constraints. *J. Appl. Math. Comput.* **2018**, *56*, 195–216. [CrossRef]
27.   Ding, Y.; Xiao, Y.; Li, J. A class of conjugate gradient methods for convex constrained monotone equations. *Optimization* **2017**, *66*, 2309–2328. [CrossRef]
28.   Liu, J.; Feng, Y. A derivative-free iterative method for nonlinear monotone equations with convex constraints. *Numer. Algorithms* **2018**, 1–18. [CrossRef]
29.   Kabanikhin, S.I. Definitions and examples of inverse and ill-posed problems. *J. Inverse Ill-Posed Probl.* **2008**, *16*, 317–357. [CrossRef]
30.   Belishev, M.I.; Kurylev, Y.V. Boundary control, wave field continuation and inverse problems for the wave equation. *Comput. Math. Appl.* **1991**, *22*, 27–52. [CrossRef]
31.   Beilina, L.; Klibanov, M.V. A Globally Convergent Numerical Method for a Coefficient Inverse Problem. *SIAM J. Sci. Comput.* **2008**, *31*, 478–509. [CrossRef]
32.   Kabanikhin, S.; Shishlenin, M. Boundary control and Gel'fand–Levitan–Krein methods in inverse acoustic problem. *J. Inverse Ill-Posed Probl.* **2004**, *12*, 125–144. [CrossRef]
33.   Lukyanenko, D.; Grigorev, V.; Volkov, V.; Shishlenin, M. Solving of the coefficient inverse problem for a nonlinear singularly perturbed two dimensional reaction diffusion equation with the location of moving front data. *Comput. Math. Appl.* **2019**, *77*, 1245–1254. [CrossRef]
34.   Van, S.B.; Elber, G. Solving piecewise polynomial constraint systems with decomposition and a subdivision-based solver. *Computer-Aided Design* **2017**, *90*, 37–47.
35.   Aizenshtein, M.; Bartoň, M.; Elber, G. Global solutions of well-constrained transcendental systems using expression trees and a single solution test. *Computer Aided Geometric Design* **2012**, *29*, 265–279.
36.   Bartoň, M. Solving polynomial systems using no-root elimination blending schemes. *Computer-Aided Design* **2011**, *43*, 1870–1878.
37.   Wang, X.Y.; Li, S.J.; Kou, X.P. A self-adaptive three-term conjugate gradient method for monotone nonlinear equations with convex constraints. *Calcolo* **2016**, *53*, 133–145. [CrossRef]

38. La Cruz, W.; Martínez, J.; Raydan, M. Spectral residual method without gradient information for solving large-scale nonlinear systems of equations. *Math. Comput.* **2006**, *75*, 1429–1448. [CrossRef]

39. Bing, Y.; Lin, G. An Efficient Implementation of Merrills Method for Sparse or Partially Separable Systems of Nonlinear Equations. *SIAM J. Optim.* **1991**, *1*, 206–221, doi:10.1137/0801015. [CrossRef]

40. Yu, Z.; Lin, J.; Sun, J.; Xiao, Y.H.; Liu, L.Y.; Li, Z.H. Spectral gradient projection method for monotone nonlinear equations with convex constraints. *Appl. Numer. Math.* **2009**, *59*, 2416–2423. [CrossRef]

41. Dolan, E.D.; Moré, J.J. Benchmarking optimization software with performance profiles. *Math. Program.* **2002**, *91*, 201–213. [CrossRef]

42. Hale, E.T.; Yin, W.; Zhang, Y. A fixed-point continuation method for ℓ_1-regularized minimization with applications to compressed sensing. *CAAM TR07-07 Rice Univ.* **2007**, *43*, 44.

43. Beck, A.; Teboulle, M. A fast iterative shrinkage-thresholding algorithm for linear inverse problems. *SIAM J. Imaging Sci.* **2009**, *2*, 183–202. [CrossRef]

44. Figueiredo, M.A.; Nowak, R.D.; Wright, S.J. Gradient projection for sparse reconstruction: Application to compressed sensing and other inverse problems. *IEEE J. Sel. Top. Signal Process.* **2007**, *1*, 586–597. [CrossRef]

45. Birgin, E.G.; Martínez, J.M.; Raydan, M. Nonmonotone spectral projected gradient methods on convex sets. *SIAM J. Optim.* **2000**, *10*, 1196–1211. [CrossRef]

46. Xiao, Y.; Wang, Q.; Hu, Q. Non-smooth equations based method for ℓ_1-norm problems with applications to compressed sensing. *Nonlinear Anal. Theory Methods Appl.* **2011**, *74*, 3570–3577. [CrossRef]

47. Wang, Z.; Bovik, A.C.; Sheikh, H.R.; Simoncelli, E.P. Image quality assessment: From error visibility to structural similarity. *IEEE Trans. Image Process.* **2004**, *13*, 600–612. [CrossRef] [PubMed]

![Sigma] *mathematics*

MDPI

Article

A New Class of Iterative Processes for Solving Nonlinear Systems by Using One Divided Differences Operator

Alicia Cordero [1], **Cristina Jordán** [1], **Esther Sanabria** [2] and **Juan R. Torregrosa** [1,*]

[1] Multidisciplinary Institute of Mathematics, Universitat Politècnica de València, 46022 València, Spain
[2] Department of Applied Mathematics, Universitat Politècnica de València, 46022 València, Spain
*   Correspondence: jrtorre@mat.upv.es

Received: 23 July 2019; Accepted: 19 August 2019; Published: 23 August 2019

check for
updates

Abstract: In this manuscript, a new family of Jacobian-free iterative methods for solving nonlinear systems is presented. The fourth-order convergence for all the elements of the class is established, proving, in addition, that one element of this family has order five. The proposed methods have four steps and, in all of them, the same divided difference operator appears. Numerical problems, including systems of academic interest and the system resulting from the discretization of the boundary problem described by Fisher's equation, are shown to compare the performance of the proposed schemes with other known ones. The numerical tests are in concordance with the theoretical results.

Keywords: nonlinear systems; multipoint iterative methods; divided difference operator; order of convergence; Newton's method; computational efficiency index

1. Introduction

The design of iterative processes for solving scalar equations, $f(x) = 0$, or nonlinear systems, $F(x) = 0$, with n unknowns and equations, is an interesting challenge of numerical analysis. Many problems in Science and Engineering need the solution of a nonlinear equation or system in any step of the process. However, in general, both equations and nonlinear systems have no analytical solution, so we must resort to approximate the solution using iterative techniques. There are different ways to develop iterative schemes such as quadrature formulaes, Adomian polynomials, divided difference operator, weight function procedure, etc. have been used by many researchers for designing iterative schemes to solve nonlinear problems. For a good overview on the procedures and techniques as well as the different schemes developed in the last half century, one refers to some standard texts [1–5].

In this paper, , we want to design Jacobian-free iterative schemes for approximating the solution $\bar{x} = (x_1, x_2, \ldots, x_n)^T$ of a nonlinear system $F(x) = 0$, where $F : D \subseteq \mathbb{R}^n \to \mathbb{R}^n$ is a nonlinear multivariate function defined in a convex set D. The best known method for finding a solution $\bar{x} \in D$ is Newton's procedure,

$$x^{(k+1)} = x^{(k)} - [F'(x^{(k)})]^{-1} F(x^{(k)}), \ k = 0, 1, 2, \ldots,$$

$F'(x^{(k)})$ being the Jacobian of F evaluated in the kth iteration.

Based on Newton-type schemes and by using different techniques, several methods for approximating a solution of $F(x) = 0$ have been published recently. The main objective of all these processes is to speed the convergence or increase their computational efficiency. We are going to recall some of them that we will use in the last section for comparison purposes.

From a variant of Steffensen's method for systems introduced by Samanskii in [6] that replaces the Jacobian matrix $F'(x)$ by the divided difference operator defined as

$$[x, y; F](x - y) = F(x) - F(y),$$

being $x, y \in \mathbb{R}^n$, Wang and Fang in [7] designed a fourth-order scheme, denoted by WF4, whose iterative expression is

$$
\begin{aligned}
r^{(k)} &= x^{(k)} - [a^{(k)}, b^{(k)}; F]^{-1} F(x^{(k)}), \\
x^{(k+1)} &= r^{(k)} - \left(3I - 2[a^{(k)}, b^{(k)}; F]^{-1} [x^{(k)}, r^{(k)}; F]\right) [a^{(k)}, b^{(k)}; F]^{-1} F(r^{(k)}),
\end{aligned}
\tag{1}
$$

where I is the identity matrix of size $n \times n$, $a^{(k)} = x^{(k)} + F(x^{(k)})$ and $b^{(k)} = x^{(k)} - F(x^{(k)})$. Let us observe that this method uses two functional evaluations and two divided difference operators per iteration. Let us remark that Samanskii in [6] defined also a third-order method with the same divided differences operator at the two steps.

Sharma and Arora in [8] added a new step in the previous method obtaining a sixth-order scheme, denoted by SA6, whose expression is

$$
\begin{aligned}
r^{(k)} &= x^{(k)} - [a^{(k)}, b^{(k)}; F]^{-1} F(x^{(k)}), \\
s^{(k)} &= r^{(k)} - \left(3I - 2[a^{(k)}, b^{(k)}; F]^{-1} [x^{(k)}, r^{(k)}; F]\right) [a^{(k)}, b^{(k)}; F]^{-1} F(r^{(k)}), \\
x^{(k+1)} &= s^{(k)} - \left(3I - 2[a^{(k)}, b^{(k)}; F]^{-1} [x^{(k)}, r^{(k)}; F]\right) [a^{(k)}, b^{(k)}; F]^{-1} F(s^{(k)}),
\end{aligned}
\tag{2}
$$

where, as before, $a^{(k)} = x^{(k)} + F(x^{(k)})$ and $b^{(k)} = x^{(k)} - F(x^{(k)})$. In relation with WF4, a new functional evaluation, per iteration, is needed.

By replacing the third step of equation (2), Narang et al. in [9] proposed the following seventh-order scheme that uses two divided difference operators and three functional evaluations per iteration, which is denoted by NM7,

$$
\begin{aligned}
r^{(k)} &= x^{(k)} - Q^{-1} F(x^{(k)}), \\
s^{(k)} &= r^{(k)} - Q^{-1} F(r^{(k)}), \\
x^{(k+1)} &= s^{(k)} - \left(\frac{17}{4} I - Q^{-1} P \left(-\frac{27}{4} I + Q^{-1} P \left(\frac{19}{4} I - \frac{5}{4} Q^{-1} P\right)\right)\right) Q^{-1} F(s^{(k)}),
\end{aligned}
\tag{3}
$$

where $Q = [a^{(k)}, b^{(k)}; F]$, being again $a^{(k)} = x^{(k)} + F(x^{(k)})$, $b^{(k)} = x^{(k)} - F(x^{(k)})$ and $P = [w^{(k)}, t^{(k)}; F]$, with $w^{(k)} = s^{(k)} + F(x^{(k)})$, $t^{(k)} = s^{(k)} - F(x^{(k)})$.

In a similar way, Wang et al. (see [10]) designed a scheme of order 7 that we denote by S7, modifying only the third step of expression (3). Its iterative expression is

$$
\begin{aligned}
r^{(k)} &= x^{(k)} - Q^{-1} F(x^{(k)}), \\
s^{(k)} &= r^{(k)} - \left(3I - 2Q^{-1} [r^{(k)}, x^{(k)}; F]\right) Q^{-1} F(r^{(k)}), \\
x^{(k+1)} &= s^{(k)} - \left(\frac{13}{4} I - Q^{-1} [s^{(k)}, r^{(k)}; F] \left(\frac{7}{2} I - \frac{5}{4} Q^{-1} [s^{(k)}, r^{(k)}; F]\right)\right) Q^{-1} F(s^{(k)}),
\end{aligned}
\tag{4}
$$

where, as in the previous schemes, $Q = [a^{(k)}, b^{(k)}; F]$, with $a^{(k)} = x^{(k)} + F(x^{(k)})$ and $b^{(k)} = x^{(k)} - F(x^{(k)})$.

Different indices can be used to compare the efficiency of iterative processes. For example, in [11], Ostrowski introduced the efficiency index $EI = p^{1/d}$, where p is the convergence order and d is the quantity of functional evaluations at each iteration. Moreover, the matrix inversions appearing in the iterative expressions are in practice calculated by solving linear systems. Therefore, the amount of quotients/products, denoted by op, employed in each iteration play an important role. This is the reason why we presented in [12] the computational efficiency index, CEI, combining EI and the number of operations per iteration. This index is defined as $CEI = p^{1/(d+op)}$.

Our goal of this manuscript is to construct high-order Jacobian-free iterative schemes for solving nonlinear systems involving low computational cost on large systems.

We recall, in Section 2, some basic concepts that we will use in the rest of the manuscript. Section 3 is devoted to describe our proposed iterative methods for solving nonlinear systems and to analyze their convergence. The efficiency indices of our methods are studied in Section 4, as well as a comparative analysis with the schemes presented in the Introduction. Several numerical tests are shown in Section 5, for illustrating the performance of the new schemes. To get this aim, we use a discretized nonlinear one-dimensional heat conduction equation by means of approximations of the derivatives and also some systems of academic interest. We finish the manuscript with some conclusions.

2. Basic Concepts

If a sequence $\{x^{(k)}\}_{k \geq 0}$ in $\mathbb{R}^n$ converges to $\bar{x}$, it is said to be of order of convergence p, being $p \geq 1$, if $C > 0$ ($0 < C < 1$ for $p = 1$) and k_0 exist satisfying

$$\|x^{(k+1)} - \bar{x}\| \leq C\|x^{(k)} - \bar{x}\|^p, \quad \forall k \geq k_0,$$

or

$$\|e^{(k+1)}\| \leq C\|e^{(k)}\|^p, \quad \forall k \geq k_0,$$

being $e^{(k)} = x^{(k)} - \bar{x}$.

Although this notation was presented by the authors in [12], we show it for the sake of completeness. Let $\Phi : D \subseteq \mathbb{R}^n \longrightarrow \mathbb{R}^n$ be sufficiently Fréchet differentiable in D. The qth derivative of Φ at $x \in \mathbb{R}^n$, $q \geq 1$, is the q-linear function $\Phi^{(q)}(x) : \mathbb{R}^n \times \cdots \times \mathbb{R}^n \longrightarrow \mathbb{R}^n$ such that $\Phi^{(q)}(x)(y_1, \ldots, y_q) \in \mathbb{R}^n$. Let us observe that

1. $\Phi^{(q)}(x)(y_1, \ldots, y_{q-1}, \cdot) \in \mathcal{L}(\mathbb{R}^n)$, where $\mathcal{L}(\mathbb{R}^n)$ denotes the set of linear mappings defined from $(\mathbb{R}^n)$ into $(\mathbb{R}^n)$.
2. $\Phi^{(q)}(x)(y_{\sigma(1)}, \ldots, y_{\sigma(q)}) = \Phi^{(q)}(x)(y_1, \ldots, y_q)$, for all permutation σ of $\{1, 2, \ldots, q\}$.

From the above properties, we can use the following notation (let us observe that y^p denotes $(y, \ldots, y)$ p times):

(a) $\Phi^{(q)}(x)(y_1, \ldots, y_q) = \Phi^{(q)}(x)y_1 \ldots y_q$,
(b) $\Phi^{(q)}(x)y^{q-1}\Phi^{(p)}y^p = \Phi^{(q)}(x)\Phi^{(p)}(x)y^{q+p-1}$.

Let us consider $\bar{x} + \varepsilon \in \mathbb{R}^n$ in a neighborhood of $\bar{x}$. By applying Taylor series and considering that $\Phi'(\bar{x})$ is nonsingular,

$$\Phi(\bar{x} + \varepsilon) = \Phi'(\bar{x})\left[\varepsilon + \sum_{q=2}^{p-1} C_q \varepsilon^q\right] + \mathcal{O}(\varepsilon^p), \tag{5}$$

being $C_q = (1/q!)[\Phi'(\bar{x})]^{-1}\Phi^{(q)}(\bar{x})$, $q \geq 2$. Let us notice that $C_q \varepsilon^q \in \mathbb{R}^n$ as $\Phi^{(q)}(\bar{x}) \in \mathcal{L}(\mathbb{R}^n \times \cdots \times \mathbb{R}^n, \mathbb{R}^n)$ and $[\Phi'(\bar{x})]^{-1} \in \mathcal{L}(\mathbb{R}^n)$.

Moreover, we express Φ' as

$$\Phi'(\bar{x} + \varepsilon) = \Phi'(\bar{x})\left[I + \sum_{q=2}^{p-1} qC_q \varepsilon^{q-1}\right] + \mathcal{O}(\varepsilon^{p-1}), \tag{6}$$

the identity matrix being denoted by I. Then, $qC_q \varepsilon^{q-1} \in \mathcal{L}(\mathbb{R}^n)$. From expression (6), we get

$$[\Phi'(\bar{x} + \varepsilon)]^{-1} = \left[I + Y_2 \varepsilon + Y_3 \varepsilon^2 + Y_4 \varepsilon^4 + \cdots\right][\Phi'(\bar{x})]^{-1} + \mathcal{O}(\varepsilon^{p-1}), \tag{7}$$

where

$$
\begin{aligned}
Y_2 &= -2C_2, \\
Y_3 &= 4C_2^2 - 3C_3, \\
Y_4 &= -8C_2^3 + 6C_2C_3 + 6C_3C_2 - 4C_4. \\
&\vdots
\end{aligned}
$$

The equation

$$
e^{(k+1)} = Ke^{(k)^p} + \mathcal{O}(e^{(k)^{p+1}}),
$$

where K is a p-linear operator $K \in \mathcal{L}(\mathbb{R}^n \times \cdots \times \mathbb{R}^n, \mathbb{R}^n)$, known as error equation and p is the order of convergence. In addition, we denote $e^{(k)^p}$ by $(e^{(k)}, e^{(k)}, \cdots, e^{(k)})$.

Divided difference operator of function Φ (see, for example, [2]) is defined as a mapping $[\cdot, \cdot; \Phi]$: $D \times D \subseteq \mathbb{R}^n \times \mathbb{R}^n \rightarrow \mathcal{L}(\mathbb{R}^n)$ satisfying

$$
[x, y; \Phi](x - y) = \Phi(x) - \Phi(y), \quad \text{for all } x, y \in D.
$$

In addition, by using the formula of Gennochi-Hermite [13] and Taylor series expansions around x, the divided difference operator is defined for all $x, x + h \in \mathbb{R}^n$ as follows:

$$
[x + \varepsilon, x; \Phi] = \int_0^1 \Phi'(x + t\varepsilon)dt = \Phi'(x) + \frac{1}{2}\Phi''(x)\varepsilon + \frac{1}{6}\Phi'''(x)\varepsilon^2 + \frac{1}{24}\Phi^{(iv)}(x)\varepsilon^3 + \mathcal{O}(\varepsilon^4). \tag{8}
$$

Being $a^{(k)} = x^{(k)} + \Phi(x^{(k)})$ and $b^{(k)} = x^{(k)} - \Phi(x^{(k)})$, the divided difference operator for points $a^{(k)}$ and $b^{(k)}$ is

$$
\begin{aligned}
[a^{(k)}, b^{(k)}; \Phi] &= \Phi'(b^{(k)}) + \frac{1}{2}\Phi''(b^{(k)})(a^{(k)} - b^{(k)}) + \frac{1}{6}\Phi'''(b^{(k)})(a^{(k)} - b^{(k)})^2 + \mathcal{O}((a^{(k)} - b^{(k)})^3) \tag{9} \\
&= \Phi'(\bar{x})[I + A_1 e^{(k)} + A_2 e^{(k)^2} + A_3 e^{(k)^3}] + \mathcal{O}(e^{(k)^4}),
\end{aligned}
$$

where

$$
\begin{aligned}
A_1 &= 2C_2, \\
A_2 &= C_3(3 + \Phi'(\bar{x})^2), \\
A_3 &= 4C_4(1 + \Phi'(\bar{x})^2 + C_3\Phi'(\bar{x})^2 C_2 + C_3\Phi'(\bar{x})C_2\Phi'(\bar{x})
\end{aligned}
$$

are obtained by replacing the Taylor expansion of the different terms that appear in development (9) and doing algebraic manipulations.

For computational purposes, the following expression (see [2]) is used

$$
[y, x; F]_{i,j} = \frac{F_i(y_1, \ldots, y_{j-1}, y_j, x_{j+1}, \ldots, x_n) - F_i(y_1, \ldots, y_{j-1}, x_j, x_{j+1}, \ldots, x_n)}{y_j - x_j},
$$

where $x = (x_1, \ldots, x_{kj-1}, x_j, x_{j+1}, \ldots, x_n)$ and $y = (y_1, \ldots, y_{j-1}, y_j, y_{j+1}, \ldots, y_n)$ and $1 \leq i, j \leq n$.

3. Proposed Methods and Their Convergence

From a Samanskii-type method and by using the composition procedure "frozening" the divided difference operator (we hold the same divided difference operator in all the steps of the method), we propose the following four-steps iterative class with the aim of reaching order five:

$$
\begin{aligned}
y^{(k)} &= x^{(k)} - \mu [a^{(k)}, b^{(k)}; F]^{-1} F(x^{(k)}), \\
z^{(k)} &= y^{(k)} - \alpha [a^{(k)}, b^{(k)}; F]^{-1} F(y^{(k)}), \\
t^{(k)} &= z^{(k)} - \beta [a^{(k)}, b^{(k)}; F]^{-1} F(z^{(k)}), \\
x^{(k+1)} &= t^{(k)} - \gamma [a^{(k)}, b^{(k)}; F]^{-1} F(t^{(k)}),
\end{aligned} \tag{10}
$$

where μ, α, β and γ are real parameters, $a^{(k)} = x^{(k)} + F(x^{(k)})$ and $b^{(k)} = x^{(k)} - F(x^{(k)})$.

It is possible to prove that, under some assumptions, we can reach order five. We have used different combinations of these steps trying to preserve order 5 and reducing the computational cost. The best result we have been able to achieve is the following:

$$
\begin{aligned}
y^{(k)} &= x^{(k)} - [a^{(k)}, b^{(k)}; F]^{-1} F(x^{(k)}), \\
z^{(k)} &= y^{(k)} - \alpha\, [a^{(k)}, b^{(k)}; F]^{-1} F(y^{(k)}), \\
t^{(k)} &= z^{(k)} - \beta\, [a^{(k)}, b^{(k)}; F]^{-1} F(y^{(k)}), \\
x^{(k+1)} &= z^{(k)} - \gamma\, [a^{(k)}, b^{(k)}; F]^{-1} F(t^{(k)}),
\end{aligned}
\tag{11}
$$

where α, β and γ are real parameters, $a^{(k)} = x^{(k)} + F(x^{(k)})$ and $b^{(k)} = x^{(k)} - F(x^{(k)})$. The convergence of class (11) is presented in the following result.

Theorem 1. *Let us assume $F : D \subseteq \mathbb{R}^n \longrightarrow \mathbb{R}^n$ being a differentiable enough operator at each point of the open neighborhood D of the solution $\bar{x}$ of the system $F(x) = 0$. Let us suppose that $F'(x)$ is continuous and nonsingular in $\bar{x}$ and the initial estimation $x^{(0)}$ is near enough to $\bar{x}$. Therefore, sequence $\{x^{(k)}\}_{k \geq 0}$ calculated from expression (11) converges to $\bar{x}$ with order 4 if $\alpha = 2 - \gamma$, $\beta = \dfrac{(\gamma - 1)^2}{\gamma}$ and for all $\gamma \in \mathbb{R}$, the error equation being*

$$
e^{(k+1)} = \left(\frac{5\gamma - 1}{\gamma} C_2^3 \right) e^{(k)4} + \mathcal{O}(e^{(k)5}).
$$

In addition, if $\gamma = \frac{1}{5}$, the order of convergence is five and the error equation is

$$
e^{(k+1)} = \left(14 C_2^4 - 2C_2 C_3 C_2 + 6C_3 C_2^2 - 2C_2 C_3 F'(\bar{x})^2 C_2 + 2C_3 F'(\bar{x})^2 C_2^2 \right) e^{(k)5} + \mathcal{O}(e^{(k)6}),
$$

where $C_j = \dfrac{1}{j!} [F'(\bar{x})]^{-1} F^{(j)}(\bar{x})$, $j = 2, 3, \ldots$

Proof. By using the Taylor expansion of $F(x^{(k)})$ and its derivatives around $\bar{x}$:

$$
\begin{aligned}
F(x^{(k)}) &= F'(\bar{x}) \left[e^{(k)} + C_2 e^{(k)2} + C_3 e^{(k)3} + C_4 e^{(k)4} + C_5 e^{(k)5} \right] + \mathcal{O}(e^{(k)6}), \\
F'(x^{(k)}) &= F'(\bar{x}) \left[I + 2C_2 e^{(k)} + 3C_3 e^{(k)2} + 4C_4 e^{(k)3} + 5C_5 e^{(k)4} \right] + \mathcal{O}(e^{(k)5}), \\
F''(x^{(k)}) &= F'(\bar{x}) \left[2C_2 + 6C_3 e^{(k)} + 12C_4 e^{(k)2} + 20C_5 e^{(k)3} \right] + \mathcal{O}(e^{(k)4}), \\
F'''(x^{(k)}) &= F'(\bar{x}) \left[6C_3 + 24C_4 e^{(k)} + 60C_5 e^{(k)2} \right] + \mathcal{O}(e^{(k)3}), \\
F^{(iv)}(x^{(k)}) &= F'(\bar{x}) \left[24C_4 + 120C_5 e^{(k)} \right] + \mathcal{O}(e^{(k)2}), \\
F^{(v)}(x^{(k)}) &= F'(\bar{x}) \left[120C_5 \right] + \mathcal{O}(e^{(k)}).
\end{aligned}
$$

From the above expression, by replacing in the first order divided difference operator $[a^{(k)}, b^{(k)}; F]$, the values $a^{(k)} = x^{(k)} + F(x^{(k)})$, $b^{(k)} = x^{(k)} - F(x^{(k)})$, we obtain:

$$
\begin{aligned}
[a^{(k)}, b^{(k)}; F] = F'(\bar{x}) \Big[& I + 2C_2 e^{(k)} + \left(3C_3 + C_3 F'(\bar{x})^2 \right) e^{(k)2} \\
& + \left(4C_4 + 4C_4 F'(\bar{x})^2 + C_3 F'(\bar{x})^2 C_2 + C_3 F'(\bar{x}) C_2 F'(\bar{x}) \right) e^{(k)3} \\
& + \left(5C_5 + C_5 F'(\bar{x})^4 + 10 C_5 F'(\bar{x})^2 + 4C_4 F'(\bar{x})^2 C_2 + 4C_4 F'(\bar{x}) C_2 F'(\bar{x}) \right. \\
& \left. + C_3 F'(\bar{x})^2 C_3 + C_3 F'(\bar{x}) C_2 F'(\bar{x}) C_2 + C_3 F'(\bar{x}) C_3 F'(\bar{x}) \right) e^{(k)4} \Big] + \mathcal{O}(e^{(k)5}).
\end{aligned}
$$

From the above expression, we have

$$
[a^{(k)}, b^{(k)}; F]^{-1} = \left[I + X_2 e^{(k)} + X_3 e^{(k)2} \right] [F'(\bar{x})]^{-1} + \mathcal{O}(e^{(k)3}),
$$

where

$$X_2 = -2C_2,$$
$$X_3 = -3C_3 + 4C_2^2 - C_3F'(\bar{x})^2,$$
$$X_4 = -4C_4 + 6C_2C_3 + 6C_3C_2 - 8C_2^3 + C_3F'(\bar{x})^2C_2 - C_3F'(\bar{x})C_2F'(\bar{x}) + 2C_2C_3F'(\bar{x})^2 - 4C_4F'(\bar{x})^2.$$

Then,

$$\left[a^{(k)}, b^{(k)}; F\right]^{-1}F(x^{(k)}) = e^{(k)} - C_2e^{(k)^2} + \left(-2C_3 + 2C_2^2 - C_3F'(\bar{x})^2\right)e^{(k)^3}$$
$$+ \left(-3C_4 + 4C_2C_3 + 3C_3C_2 - 4C_2^3 - C_3F'(\bar{x})C_2F'(\bar{x}) + 2C_2C_3F'(\bar{x})^2 - 4C_4F'(\bar{x})^2\right)e^{(k)^4} + \mathcal{O}(e^{(k)^5}).$$

Thus,

$$y^{(k)} - \bar{x} = C_2e^{(k)^2} + \left(2C_3 - 2C_2^2 + C_3F'(\bar{x})^2\right)e^{(k)^3}$$
$$+ \left(3C_4 - 4C_2C_3 - 3C_3C_2 + 4C_2^3 + C_3F'(\bar{x})C_2F'(\bar{x}) - 2C_2C_3F'(\bar{x})^2 + 4C_4F'(\bar{x})^2\right)e^{(k)^4} + \mathcal{O}(e^{(k)^5}),$$
$$(y^{(k)} - \bar{x})^2 = C_2^2e^{(k)^4} + \mathcal{O}(e^{(k)^5}),$$

and

$$
\begin{aligned}
F(y^{(k)}) &= F'(\bar{x})\left[(y^{(k)} - \bar{x}) + C_2(y^{(k)} - \bar{x})^2\right] + \mathcal{O}((y^{(k)} - \bar{x})^3) \\
&= F'(\bar{x})\left[C_2e^{(k)^2} + \left(2C_3 - 2C_2^2 + C_3F'(\bar{x})^2\right)e^{(k)^3}\right. \\
&\quad \left. + \left(3C_4 - 4C_2C_3 - 3C_3C_2 + 5C_2^3 + C_3F'(\bar{x})C_2F'(\bar{x}) - 2C_2C_3F'(\bar{x})^2 + 4C_4F'(\bar{x})^2\right)e^{(k)^4}\right] + \mathcal{O}(e^{(k)^5}).
\end{aligned}
$$

From the values of $z^{(k)}$ and $t^{(k)}$ in expression (11), we have

$$t^{(k)} = y^{(k)} - (\alpha + \beta)\left[a^{(k)}, b^{(k)}; F\right]^{-1}F(y^{(k)}).$$

Then,

$$
\begin{aligned}
\left[a^{(k)}, b^{(k)}; F\right]^{-1}F(y^{(k)}) &= C_2e^{(k)^2} + \left(2C_3 - 4C_2^2 + C_3F'(\bar{x})^2\right)e^{(k)^3} \\
&\quad + \left(3C_4 - 8C_2C_3 - 6C_3C_2 + 13C_2^3\right. \\
&\quad \left. + C_3F'(\bar{x})C_2F'(\bar{x}) - 4C_2C_3F'(\bar{x})^2 + 4C_4F'(\bar{x})^2 - C_3F'(\bar{x})^2C_2\right)e^{(k)^4} + \mathcal{O}(e^{(k)^5}).
\end{aligned}
$$

Similarly, we obtain

$$
\begin{aligned}
t^{(k)} - \bar{x} &= \left(1 - (\alpha + \beta)\right)C_2e^{(k)^2} + \left(\left(1 - (\alpha + \beta)\right)\left(2C_3 - 2C_2^2 + C_3F'(\bar{x})^2\right) + 2(\alpha + \beta)C_2^2\right)e^{(k)^3} \\
&\quad + \left(\left(1 - (\alpha + \beta)\right)\left(3C_4 - 4C_2C_3 - 3C_3C_2 + 4C_2^3 + C_3F'(\bar{x})C_2F'(\bar{x}) - 2C_2C_3F'(\bar{x})^2 + 4C_4F'(\bar{x})^2\right)\right. \\
&\quad \left. - (\alpha + \beta)\left(-4C_2C_3 - 3C_3C_2 + 9C_2^3 - 2C_2C_3F'(\bar{x})^2 - C_3F'(\bar{x})^2C_2\right)\right)e^{(k)^4} + \mathcal{O}(e^{(k)^5}), \\
(t^{(k)} - \bar{x})^2 &= \left(1 - (\alpha + \beta)\right)^2C_2^2e^{(k)^4} + \mathcal{O}(e^{(k)^5})
\end{aligned}
$$

and

$$
\begin{aligned}
F(t^{(k)}) &= F'(\bar{x})\left[\left(1 - (\alpha + \beta)\right)C_2e^{(k)^2} + \left(\left(1 - (\alpha + \beta)\right)\left(2C_3 - 2C_2^2 + C_3F'(\bar{x})^2\right) + 2(\alpha + \beta)C_2^2\right)e^{(k)^3}\right. \\
&\quad + \left(\left(1 - (\alpha + \beta)\right)\left(3C_4 - 4C_2C_3 - 3C_3C_2 + 4C_2^3 + C_3F'(\bar{x})C_2F'(\bar{x}) - 2C_2C_3F'(\bar{x})^2 + 4C_4F'(\bar{x})^2\right)\right. \\
&\quad \left.\left. - (\alpha + \beta)\left(-4C_2C_3 - 3C_3C_2 + 9C_2^3 - 2C_2C_3F'(\bar{x})^2 - C_3F'(\bar{x})^2C_2\right) + \left(1 - (\alpha + \beta)\right)^2C_2^3\right)e^{(k)^4}\right] + \mathcal{O}(e^{(k)^5}).
\end{aligned}
$$

Thus,

$$\left[a^{(k)}, b^{(k)}; F\right]^{-1} F(t^{(k)}) = \left(1 - (\alpha + \beta)\right) C_2 e^{(k)^2}$$
$$+ \left(\left(1 - (\alpha + \beta)\right)\left(2C_3 - 4C_2^2 + C_3 F'(\bar{x})^2\right) + 2(\alpha + \beta)C_2^2\right) e^{(k)^3}$$
$$+ \left(\left(1 - (\alpha + \beta)\right)\left(3C_4 - 8C_2 C_3 - 6C_3 C_2 + 12C_2^3 + C_3 F'(\bar{x})C_2 F'(\bar{x}) - 4C_2 C_3 F'(\bar{x})^2\right.\right.$$
$$\left.\left. - C_3 F'(\bar{x})^2 C_2 + 4C_4 F'(\bar{x})^2\right) + \left(1 - (\alpha + \beta)\right)^2 C_2^3\right.$$
$$\left. - (\alpha + \beta)\left(- 4C_2 C_3 - 3C_3 C_2 + 13C_2^3 - C_3 F'(\bar{x})^2 C_2 - 2C_2 C_3 F'(\bar{x})^2\right)\right) e^{(k)^4} + \mathcal{O}(e^{(k)^5}).$$

Therefore, we obtain

$$e^{(k+1)} = e^{(k)} - \left[a^{(k)}, b^{(k)}; F\right]^{-1} F(x^{(k)}) - \alpha \left[a^{(k)}, b^{(k)}; F\right]^{-1} F(y^{(k)}) - \gamma \left[a^{(k)}, b^{(k)}; F\right]^{-1} F(t^{(k)})$$
$$= \left(1 - \alpha - \gamma\left(1 - (\alpha + \beta)\right)\right) C_2 e^{(k)^2} + \left[\left(1 - \alpha - \gamma\left(1 - (\alpha + \beta)\right)\right)\left(2C_3 - 2C_2^2 + C_3 F'(\bar{x})^2\right)\right.$$
$$\left. - 2\left(-\alpha + \gamma(\alpha + \beta) - \gamma\left(1 - (\alpha + \beta)\right)\right) C_2^2\right] e^{(k)^3}$$
$$+ \left[\left(1 - \alpha - \gamma\left(1 - (\alpha + \beta)\right)\right)\left(3C_4 - 4C_2 C_3 - 3C_3 C_2 + 4C_2^3 + C_3 F'(\bar{x})C_2 F'(\bar{x}) - 2C_2 C_3 F'(\bar{x})^2 + 4C_4 F'(\bar{x})^2\right)\right.$$
$$+ \left(-\alpha + \gamma(\alpha + \beta) - \gamma\left(1 - (\alpha + \beta)\right)\right)\left(- 4C_2 C_3 - 3C_3 C_2 + 8C_2^3 - 2C_2 C_3 F'(\bar{x})^2 - C_3 F'(\bar{x})^2 C_2\right)$$
$$\left. \left(-\alpha + 5\gamma(\alpha + \beta) - \gamma\left(1 - (\alpha + \beta)\right)^2\right) C_2^3\right] e^{(k)^4} + \mathcal{O}(e^{(k)^5}).$$

Thus, by requiring that the coefficients of $e^{(k)^2}$ and $e^{(k)^3}$ are null, we get $\alpha = 2 - \gamma$ and $\beta = \dfrac{(\gamma - 1)^2}{\gamma}$ for all $\gamma \in \mathbb{R}$, 4 being its order,

$$e^{(k+1)} = \left(\frac{5\gamma - 1}{\gamma} C_2^3\right) e^{(k)^4} + \mathcal{O}(e^{(k)^5}).$$

By adding the coefficient of $e^{(k)^4}$ to the above system, we get $\gamma = \frac{1}{5}$, 5 being the order of convergence with error equation in this case:

$$e^{(k+1)} = \left(14C_2^4 - 2C_2 C_3 C_2 + 6C_3 C_2^2 - 2C_2 C_3 F'(\bar{x})^2 C_2 + 2C_3 F'(\bar{x})^2 C_2^2\right) e^{(k)^5} + \mathcal{O}(e^{(k)^6}).$$

□

4. Efficiency Indices

As we have mentioned in the Introduction, we use indices $EI = p^{1/d}$ and CEI to compare the different iterative methods.

To evaluate function F, n scalar functions are calculated and $n(n - 1)$ for the first order divided difference $[\cdot, \cdot; F]$. In addition, to calculate an inverse linear operator, an $n \times n$ linear system must be solved; then, we have to do $\frac{1}{3}n^3 + n^2 - \frac{1}{3}n$ quotients/products for getting LU decomposition and solving the corresponding triangular linear systems. Moreover, for solving m linear systems with the same matrix of coefficients, we need to do $\frac{1}{3}n^3 + mn^2 - \frac{1}{3}n$ products-quotients. In addition, we need n^2 products for each matrix-vector multiplication and n^2 quotients for evaluating a divided difference operator.

According to the last considerations, we calculate the efficiency indices EI of methods CJST5, NM7, S7, SA6 and WF4. In case CJST5, for each iteration, we evaluate F three times and once $[\cdot, \cdot; F]$, so $n^2 + 2n$ functional evaluations are needed. Therefore, $EI_{CJST5} = 5^{\frac{1}{n^2 + 2n}}$. The indices obtained for the mentioned methods are also calculated and shown in Table 1.

Table 1. Efficiency indices for different methods.

Method	Order	NFE	EI
NM7	7	$2n^2 + n$	$7^{\frac{1}{2n^2+n}}$
S7	7	$3n^2$	$7^{\frac{1}{3n^2}}$
SA6	6	$2n^2 + n$	$6^{\frac{1}{2n^2+n}}$
WF4	4	$2n^2$	$4^{\frac{1}{2n^2}}$
CJST5	5	$n^2 + 2n$	$5^{\frac{1}{n^2+2n}}$

In Table 2, we present the indices *CEI* of schemes NM7, SA6, S7, WF4 and CJST5. In it, the amount of functional evaluations is denoted by *NFE*, the number of linear systems with the same $[\cdot, \cdot; F]$ as the matrix of coefficients is *NLS1* and $M \times V$ represents the quantity of products matrix-vector. Then, in case CJST5, for each iteration, $n^2 + 2n$ functional evaluations are needed, since we evaluate three times the function F and one divided difference of first order $[a^{(k)}, b^{(k)}, F]$. In addition, we must solve three linear systems with $[a^{(k)}, b^{(k)}, F]$ as coefficients matrix (that is $\frac{1}{3}n^3 + 3n^2 - \frac{1}{3}n$). Thus, the value of *CEI* for CJST5 is

$$CEI_{CJST5} = 5^{\frac{1}{\frac{1}{3}n^3 + 5n^2 + \frac{2}{3}n}}.$$

Analogously, we obtain the indices *CEI* of the other methods. In Figure 1, we observe the computational efficiency index of the different methods of size 5 to 80. The best index corresponds to our proposed scheme.

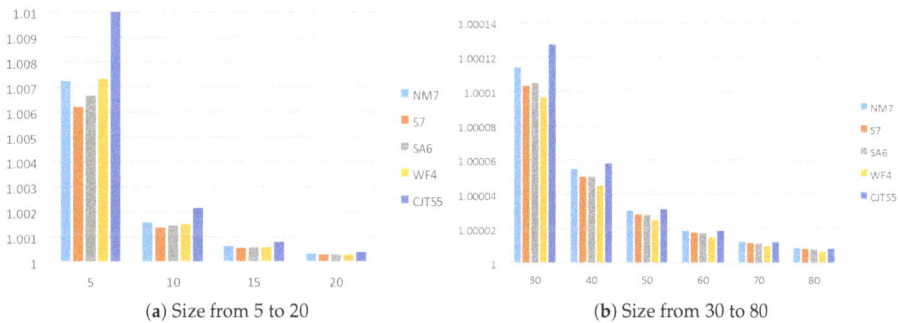

(a) Size from 5 to 20

(b) Size from 30 to 80

Figure 1. *CEI* for several sizes of the system.

Table 2. Computational cost of the procedures.

Method	Order	NFE	NLS1	$M \times V$	CEI
NM7	7	$2n^2 + 3n$	4	3	$7^{\frac{1}{\frac{1}{3}n^3 + 11n^2 + \frac{2}{3}n}}$
S7	7	$3n^2$	6	3	$7^{\frac{1}{\frac{1}{3}n^3 + 14n^2 - \frac{1}{3}n}}$
SA6	6	$2n^2 + n$	4	2	$6^{\frac{1}{\frac{1}{3}n^3 + 11n^2 + \frac{2}{3}n}}$
WF4	4	$2n^2$	3	1	$4^{\frac{1}{\frac{1}{3}n^3 + 8n^2 - \frac{1}{3}n}}$
CJST5	5	$n^2 + 2n$	3	0	$5^{\frac{1}{\frac{1}{3}n^3 + 5n^2 + \frac{2}{3}n}}$

5. Numerical Examples

We begin this section checking the performance of the new method on the resulting system obtained by the discretization of Fisher's partial differential equation. Thereafter, we compare its behavior with that of other known methods on some academic problems. For the computations, we have used *Matlab* R2015a (Natick, Massachusetts, USA) with variable precision arithmetic, with 1000 digits of mantissa. The characteristics of the computer are, regarding the processor, Intel(R) Core(TM) i7-7700 CPU @ 3.6 GHz, 3.601 Mhz, four processors and RAM 16 GB.

We use the estimation of the theoretical order of convergence p, called Computational Order of Convergence (COC), introduced by Jay [14] with the following expression:

$$p \approx COC = \frac{\ln\left(\|F(x^{(k+1)})\|_2 / \|F(x^{(k)})\|_2\right)}{\ln\left(\|F(x^{(k)})\|_2 / \|F(x^{(k-1)})\|_2\right)}, \quad k = 1, 2, \ldots$$

and the Approximated Computational Order of Convergence (ACOC), defined by Cordero and Torregrosa in [15]

$$p \approx ACOC = \frac{\ln\left(\|x^{(k+1)} - x^{(k)}\|_2 / \|x^{(k)} - x^{(k-1)}\|_2\right)}{\ln\left(\|x^{(k)} - x^{(k-1)}\|_2 / \|x^{(k-1)} - x^{(k-2)}\|_2\right)}.$$

Example 1. *Fisher's equation,*

$$u_t = Du_{xx} + ru\left(1 - \frac{u}{p}\right), \quad x \in [a, b] \ t \geq 0, \tag{12}$$

was proposed in [16] by Fisher to model the diffusion process in population dynamics. In it, $D > 0$ is the diffusion constant, r is the level of growth of the species and p is the carrying capacity. Lately, this formulation has proven to be fruitful for many other problems as wave genetics, economy or propagation.

Now, we study a particular case of this equation, when $r = p = 1$ and the spatial interval is $[0, 1]$, $u(x, 0) = sech^2(\pi x)$ and null boundary conditions.

We transform Example 1 in a set of nonlinear systems by applying an implicit method of finite differences, providing the estimated solution in the instant t_k from the estimated one in t_{k-1}. The spacial step $h = 1/n_x$ is selected and the temporal step is $k = T_{max}/n_t$, n_x and n_t being the quantity of subintervals in x and t, respectively, and T_{max} is the final instant. Therefore, a grid of domain $[0, 1] \times [0, T_{max}]$ with points (x_i, t_j), is selected:

$$x_i = 0 + ih, \quad i = 0, 1, \ldots, n_x, \qquad t_j = 0 + jk, \quad j = 0, 1, \ldots, n_t.$$

Our purpose is to estimate the solution of problem (12) at these points, by solving many nonlinear systems, as much as the number of temporal nodes t_j. For it, we use the following finite differences of order $\mathcal{O}(k + h^2)$:

$$u_t(x, t) \approx \frac{u(x, t) - u(x, t - k)}{k},$$

$$u_{xx}(x, t) \approx \frac{u(x + h, t) - 2u(x, t) + u(x - h, t)}{h^2}.$$

By denoting the approximation of the solution at (x_i, t_j) as $u_{i,j}$, and, by replacing it in Example 1, we get the system

$$ku_{i+1,j} + (kh^2 - 2k - h^2)u_{i,j} - kh^2 u_{i,j}^2 + ku_{i-1,j} = -h^2 u_{i,j-1},$$

for $i = 1, 2, \ldots, n_x - 1$ and $j = 1, 2, \ldots, n_t$. The unknowns of this system are $u_{1,j}, u_{2,j}, \ldots, u_{n_x-1,j}$, that is, the approximations of the solution in each spatial node for the fixed instant t_j. Let us remark that, for solving this system, the knowledge of solution in t_{j-1} is required.

Let us observe (Table 3) that the results improve when the temporal step is smaller. In this case, the COC is not a good estimation of the theoretical error. In Figure 2, we show the approximated solution of the problem when $T_{max} = 10$, by taking $n_t = 50$, $n_x = 10$ and using method CJST5.

Table 3. Fisher results by CJST5 and different T_{max}.

T_{max}	n_x	n_t	$\|F(x^1)\|$	$\|F(x^2)\|$	$\|F(x^3)\|$	COC
0.1	10	20	8.033×10^{-9}	2.356×10^{-35}	3.243×10^{-68}	1.2385
0.1	10	200	8.679×10^{-13}	3.203×10^{-44}	7.623×10^{-77}	1.0379
1	10	20	4.158×10^{-5}	3.4×10^{-25}	1.679×10^{-56}	1.5585
1	10	200	8.033×10^{-9}	2.356×10^{-35}	3.243×10^{-68}	1.2385
10	10	20	nc			
10	10	50	0.01945	2.757×10^{-11}	1.953×10^{-38}	3.0683

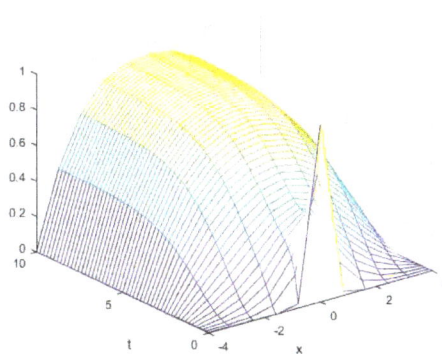

Figure 2. Approximated solution of Example 1.

In the rest of examples, we are going to compare the performance of the proposed method with the schemes presented in the Introduction as well as with the Newton-type method replacing the Jacobian matrix by the divided difference operator, that is, the Samanskii's scheme (see [6]).

Example 2. *Let us define the nonlinear system*

$$\begin{cases} \cos(x_2) - \sin(x_1) = 0, \\ x_3^{x_1} - 1/x_2 = 0, \\ e^{x_1} - x_3^2 = 0. \end{cases}$$

We use, in this example, the starting estimation $x^{(0)} = (1.25, 1.25, 1.25)^T$, the solution being $\bar{x} \approx (0.9096, 0.6612, 1.576)^T$. Table 4 shows the residuals $\|x^{(k)} - x^{(k-1)}\|$ and $\|F(x^{(k)})\|$ for $k = 1, 2, 3$ as well as ACOC and COC. We observe that the COC index is better than the corresponding ACOC of the other methods. In addition, the value $\|x^3 - x^2\|$ is better or similar to that of S7 and NM7 methods, both of them of order 7.

Table 4. Numerical results for Example 2.

	Samanskii	CJST5	WF4	SA6	S7	NM7
$\|x^{(1)} - x^{(0)}\|$	1.415	0.8848	0.8539	0.8934	0.9148	0.9355
$\|x^{(2)} - x^{(1)}\|$	0.5427	0.247	0.2039	0.2689	0.2942	0.3098
$\|x^{(3)} - x^{(2)}\|$	0.1738	0.01159	0.006249	0.005301	0.01069	0.02667
ACOC	1.1875	2.3976	2.433	3.2705	2.9221	4.25
$\|F(x^{(1)})\|$	0.1954	0.1282	0.1038	0.1385	0.1817	0.2098
$\|F(x^{(2)})\|$	0.02369	0.009956	0.004815	0.003584	0.006462	0.01669
$\|F(x^{(3)})\|$	0.00269	2.805×10^{-8}	7.663×10^{-8}	4.009×10^{-11}	8.074×10^{-12}	1.333×10^{-9}
COC	1.0313	5.0015	3.5983	5.0104	6.1445	6.4561

Example 3. *We consider now*

$$\sum_{j=1}^{n} x_j - x_i - e^{-x_i} x_i = 0, \quad i = 1, 2, \dots, n.$$

The numerical results are displayed in Table 5. The initial estimation is $x^{(0)} = (0.25, 0.25, \dots, 0.25)^T$ and the size of the system is $n = 10$, the solution being $\bar{x} = (0, 0, \dots, 0)^T$. We show the same information as in the previous example.

Example 4. *The third example is given by the system:*

$$x_i + 1 - 2 \log \left(1 + \sum_{j=1}^{n} x_j - x_i \right) = 0, \quad i = 1, 2, \dots, n.$$

Its solution is $\bar{x} \approx (9.376, 9.376, \ldots, 9.376)^T$. By using the starting guess $x^{(0)} = (1, 1, \ldots, 1)^T$ with $n = 10$, we obtain the results appearing in Table 6.

Table 5. Numerical results for Example 3.

	Samanskii	CJST5	WF4	SA6	S7	NM7
$\|x^{(1)} - x^{(0)}\|$	1.036	0.8249	0.9116	0.8499	0.7847	0.7897
$\|x^{(2)} - x^{(1)}\|$	0.2552	0.03432	0.121	0.05932	0.00583	0.0008995
$\|x^{(3)} - x^{(2)}\|$	0,009667	1.487×10^{-11}	9.264×10^{-6}	5.367×10^{-10}	2.529×10^{-21}	1.048×10^{-28}
ACOC	2.3361	6.7807	4.6937	6.9572	8.6247	4.25
$\|F(x^{(1)})\|$	1.944	0.2742	0.9634	0.4735	0.04665	0.007196
$\|F(x^{(2)})\|$	0.0773	1.19×10^{-10}	7.411×10^{-5}	4.293×10^{-9}	2.023×10^{-20}	8.381×10^{-28}
$\|F(x^{(3)})\|$	2.65×10^{-5}	2.839×10^{-59}	9.644×10^{-22}	1.3×10^{-57}	9.093×10^{-149}	2.621×10^{-202}
COC	2.4743	5.1932	4.1045	6.0328	6.9895	6.9987

Table 6. Numerical results for Example 4.

	Samanskii	CJST5	WF4	SA6	S7	NM7
$\|x^{(1)} - x^{(0)}\|$	6.013	40.68	67.88	73.31	34.3	39.23
$\|x^{(2)} - x^{(1)}\|$	12.15	13.83	36.07	56.74	17.15	7.369
$\|x^{(3)} - x^{(2)}\|$	10.11	0.0002872	0.06643	0.02688	0.0001485	5.422×10^{-8}
ACOC	-	9.9941	9.9587	29.874	16.819	4.25
$\|F(x^{(1)})\|$	10.84	11.03	30.43	48.74	13.42	5.842
$\|F(x^{(2)})\|$	6.084	0.0002263	0.05234	0.02117	0.000117	4.272×10^{-8}
$\|F(x^{(3)})\|$	1.162	9.286×10^{-28}	6.508×10^{-12}	1.807×10^{-20}	3.105×10^{-40}	7.552×10^{-65}
COC	2.8651	4.9888	3.583	5.3744	7.0313	6.9755

The different methods give us the expected results, according to their order of convergence.

Example 5. *Finally, the last example that we consider is:*

$$\arctan(x_i) + 1 - 2 \left(\sum_{j=1}^{n} x_j^2 - x_i^2 \right) = 0.$$

The solution of it is $\bar{x} \approx (0.1758, 0.1758, \ldots, 0.1758)^T$. By using the initial estimation $x^{(0)} = (0.5, 0.5, \ldots, 0.5)^T$ with $n = 20$, we obtain the numerical results displayed in Table 7.

Table 7. Numerical results for Example 5.

	Samanskii	CJST5	WF4	SA6	S7	NM7
$\|x^{(1)} - x^{(0)}\|$	0.9503	1.323	1.272	1.368	1.394	1.393
$\|x^{(2)} - x^{(1)}\|$	0.3912	0.1266	0.177	0.0821	0.05639	0.05732
$\|x^{(3)} - x^{(2)}\|$	0.1013	4.988×10^{-5}	0.0007407	6.903×10^{-7}	7.214×10^{-9}	6.655×10^{-9}
ACOC	1.5229	3.3404	2.776	4.1543	4.9485	4.25
$\|F(x^{(1)})\|$	8.324	1.706	2.471	1.075	0.7257	0.7381
$\|F(x^{(2)})\|$	1.445	0.0006179	0.009181	8.552×10^{-6}	8.937×10^{-8}	8.245×10^{-8}
$\|F(x^{(3)})\|$	0.0902	1.206×10^{-20}	5.635×10^{-12}	5.437×10^{-36}	8.115×10^{-56}	3.521×10^{-56}
COC	1.5839	4.8559	3.791	5.9219	6.953	6.9577

It is observed in Tables 5–7 that, for the proposed academic problems, the introduced method (CJST5) shows a good performance comparable with higher-order methods. Of course, the worst results are those obtained by Samanskii's method, but it has been included because it is the Jacobian-free version of Newton's scheme and it is also the first step of our proposed scheme. Let us also remark that, when only three iterations are calculated, the index COC gives more reliable information than the ACOC one in all of the examples.

Mathematics **2019**, *7*, 776

6. Conclusions

In this paper, we design a family of iterative methods for solving nonlinear systems with fourth-order convergence. This family does not use Jacobian matrices and one of its elements has order five. The relationship between the proposed method and other known ones in terms of efficiency index and computational efficiency index allows us to see that our method is more efficient than the other ones. In addition, its error bounds are smaller with the same number of iterations in some cases. Thus, our proposal is competitive mostly for big size systems.

Author Contributions: The individual contributions of the authors are as follows: conceptualization, J.R.T.; writing—original draft preparation, C.J. and E.S.; validation, A.C. and J.R.T.; formal analysis, A.C.; numerical experiments, C.J. and E.S.

Funding: This research has been supported partially by Spanish Ministerio de Ciencia, Innovación y Universidades PGC2018-095896-B-C22, PGC2018-094889-B-I00, TEC2016-79884-C2-2-R and also by Spanish grant PROMETEO/2016/089 from Generalitat Valenciana.

Acknowledgments: The authors would like to thank the anonymous reviewers for their useful comments and suggestions that have improved the final version of this manuscript.

Conflicts of Interest: The authors declare that there is no conflict of interest regarding the publication of this paper.

References

1. Traub, J.F. *Iterative Methods for the Solution of Equations*; Prentice-Hall: Englewood Cliffs, NJ, USA, 1964.
2. Ortega, J.M.; Rheinbolt, W.C. *Iterative Solutions of Nonlinears Equations in Several Variables*; Academic Press: New York, NY, USA, 1970.
3. Kelley, C.T. *Iterative Methods for Linear and Nonlinear Equations*; SIAM: Philadelphia, PA, USA, 1995.
4. Petković, M.S.; Neta, B.; Petković, L.D.; Džunić, J. *Multipoint Methods for Solving Nonlinear Equations*; Academic Press: New York, NY, USA, 2013.
5. Amat, S.; Busquier, S. *Advances in Iterative Methods for Nonlinear Equations*; SEMA SIMAI Springer Series; Springer International Publishing: Cham, Switzerland, 2016; Volume 10.
6. Samanskii, V. On a modification of the Newton method. *Ukrain. Mat.* **1967**, *19*, 133–138.
7. Wang, X.; Fang, X. Two Efficient Derivative-Free Iterative Method for Solving Nonlinear Systems. *Algorithms* **2016**, *9*, 14. [CrossRef]
8. Sharma, J.R.; Arora, H. Efficient derivative-free numerical methods for solving systems of nonlinear equations. *Comp. Appl. Math.* **2016**, *35*, 269–284. [CrossRef]
9. Narang, M.; Bathia S.; Kanwar, V. New efficient derivative free family of seventh-order methods for solving systems of nonlinear equations. *Numer. Algorithms* **2017**, *76*, 283–307. [CrossRef]
10. Wang, X.; Zhang, T.; Qian, W.; Teng, M. Seventh-order derivative-free iterative method for solving nonlinear systems. *Numer. Algorithms* **2015**, *70*, 545–558. [CrossRef]
11. Ostrowski, A.M. *Solution of Equations and Systems of Equations*; Prentice-Hall: Englewood Cliffs, NJ, USA, 1964.
12. Cordero, A.; Hueso, J.L.; Martínez, E.; Torregrosa, J.R. A modified Newton-Jarratt's composition. *Numer. Algorithms* **2010**, *55*, 87–99. [CrossRef]
13. Hermite, C. Sur la formule d'interpolation de Lagrange. *Reine Angew. Math.* **1878**, *84*, 70–79. [CrossRef]
14. Jay, L.O. A note of Q-order of convergence. *BIT Numer. Math.* **2001**, *41*, 422–429. [CrossRef]
15. Cordero, A; Torregrosa, J.R. Variants of Newton's method using fifth-order quadrature formulas. *Appl. Math. Comput.* **2007**, *190*, 686–698. [CrossRef]
16. Fisher, R.A. The wave of advance of advantageous genes. *Ann. Eugen.* **1937**, *7*, 353–369. [CrossRef]

$\sum$ *mathematics*

MDPI

Article

An Efficient Iterative Method Based on Two-Stage Splitting Methods to Solve Weakly Nonlinear Systems

Abdolreza Amiri [1], Mohammad Taghi Darvishi [1,*] , Alicia Cordero [2] and Juan Ramón Torregrosa [2]

1   Department of Mathematics, Razi University, Kermanshah 67149, Iran
2   Institute for Multidisciplinary Mathematics, Universitat Politècnica de València, Camino de Vera s/n, 46022 Valencia, Spain
*   Correspondence: darvishi@razi.ac.ir; Tel.: +98-83-3428-3929

Received: 17 July 2019; Accepted: 19 August 2019; Published: 3 September 2019

Abstract: In this paper, an iterative method for solving large, sparse systems of weakly nonlinear equations is presented. This method is based on Hermitian/skew-Hermitian splitting (HSS) scheme. Under suitable assumptions, we establish the convergence theorem for this method. In addition, it is shown that any faster and less time-consuming two-stage splitting method that satisfies the convergence theorem can be replaced instead of the HSS inner iterations. Numerical results, such as CPU time, show the robustness of our new method. This method is easy, fast and convenient with an accurate solution.

Keywords: system of nonlinear equations; Newton method; Newton-HSS method; nonlinear HSS-like method; Picard-HSS method

1. Introduction

For $G : D \subseteq \mathbb{C}^m \longrightarrow \mathbb{C}^m$, we consider the following system of nonlinear equations:

$$G(x) = 0. \tag{1}$$

One may encounter equations like (1) in some areas of scientific computing. In particular, when the technique of finite elements or finite differences are used to discretize nonlinear boundary problems, integral equations and certain nonlinear partial differential equations. Finding the roots of systems like (1) has widespread applications in numerical and applied mathematics. There are many iterative schemes to solve (1). The most common one is the second order classical Newton's scheme, which solves (1) iteratively as

$$x^{(n+1)} = x^{(n)} - G'(x^{(n)})^{-1} G(x^{(n)}), \qquad n = 0, 1, \dots, \tag{2}$$

where $G'(x^{(n)})$ is the Jacobian matrix of G, evaluated in the nth iteration. To avoid computation of inverse of the Jacobian matrix $G'(x)$, Equation (2) is changed to

$$G'(x^{(n)})(x^{(n+1)} - x^{(n)}) = -G(x^{(n)}). \tag{3}$$

Equation (3) is a system of linear equations. Hence, by $s^{(n)} = x^{(n+1)} - x^{(n)}$, we have to solve the following system of equations:

$$G'(x^{(n)})s^{(n)} = -G(x^{(n)}), \tag{4}$$

whence $x^{(n+1)} = x^{(n)} + s^{(n)}$. Thus, by using this approach, we have to solve a system of linear equations such as

$$Ax = b, \tag{5}$$

which we usually use an iterative scheme to solve it.

Furthermore, an inexact Newton method [1–4] is a generalization of Newton's method for solving (1), in which, at the nth iteration, the step-size $s^{(n)}$ from current approximate solution $x^{(n)}$ must satisfy a condition such as

$$\| G(x^{(n)}) + G'(x^{(n)})s^{(n)} \| \le \eta_n \| G(x^{(n)}) \|,$$

for a "forcing term" $\eta_n \in [0,1)$. Let us consider system (1) in which $G(x)$ can be separated into linear and nonlinear terms, Ax and $\varphi(x)$, respectively, that is

$$G(x) = \varphi(x) - Ax \text{ or } Ax = \varphi(x). \tag{6}$$

In (6), the $m \times m$ complex matrix A is a positive definite, large and sparse matrix. In addition, vector-valued function $\varphi : D \subseteq \mathbb{C}^m \longrightarrow \mathbb{C}^m$ is continuously differentiable. Furthermore, x is an m-vector and D is an open set. When the norm of linear part Ax is strongly dominant over the norm of nonlinear part $\varphi(x)$ in a specific norm, system (6) is called a weakly nonlinear system [5,6]. Bai [5] used the separability and strong dominance between the linear and the nonlinear parts and introduced the following iterative scheme

$$Ax^{(n+1)} = \varphi(x^{(n)}). \tag{7}$$

Equation (7) is a system of linear equations. When the matrix A is positive definite, Axelsson et al. [7] solved it by a class of nested iteration methods. To solve linear positive definite systems, Bai et al. [8] applied the Hermitian/skew-Hermitian splitting (HSS) iterative scheme. For solving the large sparse, non-Hermitian positive definite linear systems, Li et al. [9] used an asymmetric Hermitian/skew-Hermitian (AHSS) iterative scheme. Moreover, to improve the robustness of the HSS method, some HSS-based iterative algorithms have been introduced. Bai and Yang [10] presented Picard-HSS and HSS-like methods to solve (7), when matrix A is a positive definite matrix. Based on the matrix multi-splitting technique, block and asynchronous two-stage methods are introduced by Bai et al. [11]. The Picard circulant and skew-circulant splitting (Picard-CSCS) algorithm and the nonlinear CSCS-like iterative algorithm are presented by Zhu and Zhang [12], when the coefficient matrix A is a Teoplitz matrix. A class of lopsided Hermitian/skew-Hermitian splitting (LHSS) algorithms and a class of nonlinear LHSS-like algorithms are used by Zhu [6] to solve the large and sparse of weakly nonlinear systems.

It must be noted that system (6) is a special form of system (1). Generally, system (6) is nonlinear. If we classify Picard-HSS and nonlinear HSS-like iterative methods as Jacobian-free schemes, in many cases, they are not as successful as Jacobian dependent schemes such as the Newton method. Most of the methods for solving nonlinear systems need to compute or approximate the Jacobian matrix in the obtained points at each step of the iterative methods, which is a very time-consuming process, especially when the Jacobian matrices $\varphi'(x^{(n)})$ are dense. Therefore, introducing any scheme that does not need to compute the Jacobian matrix and can solve a wider range of problems than the existing ones is welcome. In fact, Jacobian-free methods to solve nonlinear systems are very important and form an attractive area of research.

In this paper, we present a new iterative method to solve weakly nonlinear systems. Even though the new algorithm uses some notions of mentioned algorithms, but differs from all of them because it has three important characteristics. At the first, the new algorithm is a fully Jacobian-free one. At the second, it is easy to use, and, finally, it is very successful to solve weakly nonlinear systems. The new iterative method is a synergistic combination of high order Newton-like methods and a special splitting of the coefficient matrix A in (5).

The rest of this paper has organized as follows: in the following section, we present our new algorithm. We prove convergence of our algorithm in Section 3. We apply our algorithm to solve some problems in Section 4. In Section 5, we conclude our results and give some comments and discussions.

2. The New Algorithm

In linear system $Ax = b$, we suppose that $A = H + S$, where $H = \frac{1}{2}(A + A^*)$, $S = \frac{1}{2}(A - A^*)$, and A^* is the conjugate transpose of matrix A. Hence, H and S are, respectively, Hermitian and skew-Hermitian parts of A. By an initial guess $x_0 \in \mathbb{C}^n$, and positive constants α and tol, in HSS scheme [8], one computes x_l for $l = 1, 2, \dots$ as

$$\begin{cases} (\alpha I + H)x_{l+\frac{1}{2}} = (\alpha I - S)x_l + b, \\ (\alpha I + S)x_{l+1} = (\alpha I - H)x_{l+\frac{1}{2}} + b, \end{cases} \tag{8}$$

where I is the identity matrix. Stopping criterion for (8) is $\|b - Ax_l\| \le \text{tol}\|b - Ax_0\|$, for known x_0 and tol.

Bai and Guo [13] used an HSS scheme as inner iterations to generate an inexact version of Newton's method as:

(1) Consider the initial guess $x^{(0)}$, α, tol and the sequence $\{l_n\}_{n=0}^{\infty}$ of positive integers.
(2) For $n = 1, 2, \dots$ until $\|G(x^{(n)})\| \le \text{tol}\|G(x^{(0)})\|$ do:

    (2.1) Set $s_0^{(n)} = 0$.
    (2.2) For $l = 1, 2, \dots, l_n - 1$, apply Algorithm HSS as

$$\begin{cases} (\alpha I + H(x^{(n)}))s_{l+\frac{1}{2}}^{(n)} = (\alpha I - S(x^{(n)}))s_l^{(n)} - G(x^{(n)}) \\ (\alpha I + S(x^{(n)}))s_{l+1}^{(n)} = (\alpha I - H(x^{(n)}))s_{l+\frac{1}{2}}^{(n)} - G(x^{(n)}), \end{cases}$$

    and obtain $s_{l_n}^{(n)}$ such that

$$\| G(x^{(n)}) + G'(x^{(n)})s_{l_n}^{(n)} \| \le \eta_n \| G(x^{(n)}) \|, \quad \text{for some } \eta_n \in [0, 1).$$

    (2.3) Set $x^{(n+1)} = x^{(n)} + s_{l_n}^{(n)}$.

In addition, to solve weakly nonlinear problems, one can use a Picard-HSS method as a simple and Jacobian-free method, which is described as follows [10].

2.1. Picard-HSS Iteration Method

Suppose that $\varphi : D \subset \mathbb{C}^n \to \mathbb{C}^n$ is a continuous function and $A \in \mathbb{C}^{n \times n}$ is a positive definite matrix. For an initial guess $x^{(0)}$ and for a positive integer sequence $\{l_n\}_{n=0}^{\infty}$, Picard-HSS iterative method computes $x^{(n+1)}$ for $n = 0, 1, 2, \dots$, by using the following iterative scheme, until the stopping criterion is satisfied [10],

(1) Set $x_l^{(n)} := x^{(n)}$;
(2) For $l = 0, 1, 2, \dots, n - 1$, obtain $x^{(n+1)}$ from solving the following:

$$\begin{cases} (\alpha I + H)x_{l+\frac{1}{2}}^{(n)} = (\alpha I - S)x_l^{(n)} + \varphi(x^{(n)}), \\ (\alpha I + S)x_{l+1}^{(n)} = (\alpha I - H)x_{l+\frac{1}{2}}^{(n)} + \varphi(x^{(n)}). \end{cases}$$

(3) Set $x^{(n+1)} := x^{(n)}_{l_n}$.

The numbers $l_n, n = 0, 1, 2, \ldots$ depend on the problem, so practically they are difficult to be determined in real computations. A modified form of Picard-HSS iteration scheme, called the nonlinear HSS-like method, has been presented [10] to avoid using inner iterations as follows.

2.2. Nonlinear HSS-Like Iteration Method

Obtain $x^{(n+1)}$, $n = 0, 1, 2, \ldots$ from the following [10], for a given $x^{(0)} \in D \subset \mathbb{C}^n$, until the stopping condition is satisfied

$$\begin{cases} (\alpha I + H)x^{(n+\frac{1}{2})} & = (\alpha I - S)x^{(n)} + \varphi(x^{(n)}), \\ (\alpha I + S)x^{(n+1)} & = (\alpha I - H)x^{(n+\frac{1}{2})} + \varphi(x^{(n+\frac{1}{2})}). \end{cases}$$

However, in this method, it is necessary to evaluate the nonlinear term $\varphi(x)$ at each step, which for complicated nonlinear terms $\varphi(x)$ is too costly.

2.3. Our Proposal Iterative Scheme

For solving (6) without computing Jacobian matrices, we present a new algorithm. This algorithm is a strong tool for solving weakly nonlinear problems, as Picard and nonlinear Picard algorithms, but, in comparison with Picard and nonlinear Picard algorithms, it solves a wider range of nonlinear systems. First, we change (7) as

$$Ax^{(n+1)} = Ax^{(n)} - Ax^{(n)} + \varphi(x^{(n)}) \tag{9}$$

and

$$Ax^{(n+1)} - Ax^{(n)} = -Ax^{(n)} + \varphi(x^{(n)}). \tag{10}$$

After computing $x^{(n)}$, set $b^{(n)} = \varphi(x^{(n)})$, $G_n(x) = b^{(n)} - Ax$. Then, by intermediate iterations, obtain $x^{(n+1)}$ as:

- Let $x^{(n)}_0 = x^{(n)}$ and until $\| G(x^{(n)}_k) \| \le \text{tol}_n \| G(x^{(n)}_0) \|$ do:

$$As^{(n)}_k = G(x^{(n)}_k), \tag{11}$$

where $s^{(n)}_k = x^{(n)}_{k+1} - x^{(n)}_k$ (k is the counter of the number of iterations (11)).
- For solving (11), one may use any inner solver; here, we use an HSS scheme. Next, for initial value $x^{(n)}_0$ and $k = 1, 2, \ldots, k_n - 1$ until

$$\| G_n(x^{(n)}_k) \| \le \text{tol}_n \| G_n(x^{(n)}_0) \|, \tag{12}$$

apply the HSS scheme as:

(1) Set $s^{(n)}_{k,0} = 0$.
(2) For $l = 0, 1, 2, \ldots, l_{k_n} - 1$, apply algorithm HSS ($l$ is the counter of the number of HSS iterations):

$$\begin{cases} (\alpha I + H)s^{(n)}_{k,l+\frac{1}{2}} & = (\alpha I - S)s^{(n)}_{k,l} + G_n(x^{(n)}_k), \\ (\alpha I + S)s^{(n)}_{k,l+1} & = (\alpha I - H)s^{(n)}_{k,l+\frac{1}{2}} + G_n(x^{(n)}_k) \end{cases} \tag{13}$$

and obtain $s_{k,l_{k_n}}^{(n)}$ such that

$$\| G_n(x_k^{(n)}) - As_{k,l_{k_n}}^{(n)} \| \leq \eta_k^{(n)} \| G_n(x_k^{(n)}) \|, \quad \eta_k^{(n)} \in [0,1). \tag{14}$$

(3) Set $x_{k+1}^{(n)} = x_k^{(n)} + s_{k,l_{k_n}}^{(n)}$ (l_{k_n} is the required number of HSS inner iterations for satisfying (14)).

- Finally, set $x_0^{(n+1)} = x_{k_n}^{(n)}$ (k_n is the required number of iterations (11) in the nth step, for satisfying (12)), $b^{(n+1)} = \varphi(x_0^{(n+1)})$, $G_{n+1}(x) = b^{(n+1)} - Ax$ and again apply steps 3–14 in Algorithm 1 until to achieve the following stopping criterion:

$$\| Ax^{(n)} - \varphi(x^{(n)}) \| \leq \text{tol} \, \| Ax^{(0)} - \varphi(x^{(0)}) \| \, .$$

Algorithm 1: JFHSS Algorithm

Input: $x^{(0)}$, tol, α, $n \leftarrow 1$
Output: The root of $Ax - \varphi(x) = 0$
1 $root \leftarrow x^{(0)}$
2 **while** $\| Ax^{(n)} - \varphi(x^{(n)}) \| > \text{tol} \, \| Ax^{(0)} - \varphi(x^{(0)}) \|$ **do**

    Input: tol_n
    Set: $x_0^{(n)} = x^{(n)}$, $b^{(n)} = \varphi(x^{(n)})$ and $G_n(x) = b^{(n)} - Ax$, $k = 1$.
3   **while** $\| G_n(x_k^{(n)}) \| > \text{tol}_n \| G_n(x_0^{(n)}) \|$ **do**
4   **Set:** $l = 0$, $s_{k,0}^{(n)} = 0$.
5   **while** $\| G_n(x_k^{(n)}) - As_{k,l_{k_n}}^{(n)} \| > \eta_k^{(n)} \| G_n(x_k^{(n)}) \|$ **do**
6

$$\text{(the HSS algorithm)}$$
$$(\alpha I + H)s_{k,l+\frac{1}{2}}^{(n)} = (\alpha I - S)s_{k,l}^{(n)} + G_n(x_k^{(n)}),$$
$$(\alpha I + S)s_{k,l+1}^{(n)} = (\alpha I - H)s_{k,l+\frac{1}{2}}^{(n)} + G_n(x_k^{(n)}),$$

7   **if** $\| G_n(x_k^{(n)}) - As_{k,l_{k_n}}^{(n)} \| \leq \eta_k^{(n)} \| G_n(x_k^{(n)}) \|$ **then**
8            $l \leftarrow l_{k_n}$, $x_{k+1}^{(n)} = x_k^{(n)} + s_{k,l_{k_n}}^{(n)}$
9   **else**
10            $l \leftarrow l + 1$

11   **if** $\| G_n(x_k^{(n)}) \| \leq \text{tol}_n \| G_n(x_0^{(n)}) \|$ **then**
12            $k \leftarrow k_n$, $x_0^{(n+1)} = x_{k_n}^{(n)}$
13   **else**
14            $k \leftarrow k + 1$

15   **if** $\| Ax^{(n)} - \varphi(x^{(n)}) \| \leq \text{tol} \, \| Ax^{(0)} - \varphi(x^{(0)}) \|$ **then**
16        $root \leftarrow x^{(n)}$
17   **else**
18        $n \leftarrow n + 1$, $x^{(n)} = x_0^{(n+1)}$, $b^{(n)} = \varphi(x^{(n)})$, $G_n(x) = b^{(n)} - Ax$
19   **return** $root$

We call this new method a JFHSS (Jacobian-free HSS) algorithm, and its steps are shown in Algorithm 1.

In addition, we call the intermediate iterations Newton-like iteration because this kind of iteration uses the same procedure as an inexact Newton's method, except, since the function we use here is $b^{(n)} - Ax$ for $n = 1, 2, \cdots$, we don't need to compute any Jacobian and, in fact, the Jacobian is the matrix A. For this reason, we also call this iterative method a "Jacobian-free method".

Since the JFHSS scheme uses many HSS inner iterations, one may use another splitting scheme instead of the HSS method. In fact, if any faster and less time-consuming splitting method is available that satisfies the convergence theorem, presented in the next section, then it can be used instead of the HSS algorithm. One of these methods that is proposed in [14] is GPSS (generalized positive definite and skew-Hermitian splitting) algorithm that uses a positive-definite and skew-Hermitian splitting scheme instead of a Hermitian and skew-Hermitian one. Let H and S be the Hermitian and skew-Hermitian parts of A; then, the GPSS algorithm splits A as $A = P_1 + P_2$ where P_1 and P_2 are, respectively, positive definite and skew-Hermitian matrices. In fact, we have

$$P_1 = \mathcal{D} + 2L_{\mathcal{G}}, \quad P_2 = \mathcal{K} + L_{\mathcal{G}}^* - L_{\mathcal{G}} + S, \tag{15}$$

or

$$P_1 = \mathcal{D} + 2L_{\mathcal{G}}^*, \quad P_2 = \mathcal{K} + L_{\mathcal{G}} - L_{\mathcal{G}}^* + S, \tag{16}$$

where $\mathcal{G}$ and $\mathcal{K}$ are, respectively, Hermitian and Hermitian positive semidefinite matrices of H, that is, $H = \mathcal{G} + \mathcal{K}$; in addition, $\mathcal{D}$ and $L_{\mathcal{G}}$ are the diagonal matrix and the strictly lower triangular matrices of $\mathcal{G}$, respectively (see [14]).

Thus, to solve the system of linear Equation (5) for an initial guess $x_0 \in \mathbb{C}^n$, and positive constants α and tol, the GPSS iteration scheme (until the stopping criterion is satisfied) computes x_l for $l = 1, 2, \ldots$ by

$$\begin{cases} (\bar{\alpha}I + P_1)x_{l+\frac{1}{2}} &= (\bar{\alpha}I - P_2)x_l + b, \\ (\bar{\alpha}I + P_2)x_{l+1} &= (\bar{\alpha}I - P_1)x_{l+\frac{1}{2}} + b, \end{cases} \tag{17}$$

where $\bar{\alpha}$ is a given positive constant and I denotes the identity matrix. In addition, if, in Algorithm 1, we use a GPSS scheme instead of an HSS one, we denote the new method by JFGPSS (Jacobian free GPSS).

3. Convergence of the New Method

As we mentioned in the first section, for solving a nonlinear system, if one can separate (1) into linear and nonlinear terms, Ax and $\phi(x)$, when Ax is strongly dominant over the nonlinear term, Picard-HSS and nonlinear HSS-like methods can solve the problem. However, in many cases, even for weakly nonlinear ones, they may fail to solve the problems. Thus, to obtain a more useful method for solving (6), based on some splitting methods, we presented a new iterative method. Now, we prove that Algorithm 1 converges to the solution of a weakly nonlinear problem (6). In the following theorem, we prove the convergence of the JFHSS scheme.

Theorem 1. *Let $x^{(0)} \in \mathbb{C}^n$ and $\varphi : D \subset \mathbb{C}^n \to \mathbb{C}^n$ be a G-differentiable function on an open set $N_0 \subset D$ on which $\varphi'(x)$ is continuous and $\max \|A^{-1}\varphi'(x)\| = L < 1$. Let us suppose that $H = \frac{1}{2}(A + A^*)$ and $S = \frac{1}{2}(A - A^*)$ are the Hermitian and skew-Hermitian parts of the positive definite matrix A and also that $\mathcal{M}$ is an upper bound for $\|A^{-1}G(x^{(0)})\|$, and l_{k_n} is the number of HSS inner iterations in which the stopping criterion (14) is satisfied,*

$$l_*^n > \left| \frac{\ln\left(\frac{(1-\eta)(1-\eta^{k_{n-1}})}{L} - 1\right)}{\ln(\theta)} \right|, \tag{18}$$

with $l_*^n = \liminf\limits_{k_n \to \infty} l_{k_n}$ for $n = 1, 2, 3, \ldots$, η is the tolerance in Newton-like intermediate iterations with $L < (1 - \eta)^2$ and $\theta = \|T\|$, where T is the HSS inner iteration matrix that can be written as

$$T = (\alpha I + S)^{-1}(\alpha I - H)(\alpha I + H)^{-1}(\alpha I - S).$$

Then, the sequence of iteration $\{x^{(k)}\}_{k=0}^{\infty}$, which is generated by a JFHSS scheme in Algorithm 1, is well-defined and converges to x^*, satisfying $G(x^*) = 0$, and also

$$\|x^{(n+1)} - x^{(n)}\| \le \delta \mathcal{M} \rho^n, \tag{19}$$

$$\|x^{(n+1)} - x^{(0)}\| \le \frac{\delta \mathcal{M}}{1 - \rho}, \tag{20}$$

where $\delta = \limsup\limits_{n \to \infty} \dfrac{(1 + \theta^{l_*^n})}{1 - \eta}$ and $\rho = \limsup\limits_{n \to \infty} \rho_n$ for $\rho_n = \dfrac{(1 + \theta^{l_*^n})}{1 - \eta} L + \eta^{k_n - 1}$.

Proof. Note that $\|T\| \le \max\limits_{\lambda_i \in \lambda(H)} \left| \dfrac{\alpha - \lambda_i}{\alpha + \lambda_i} \right| < 1$ (see [8]), where $\lambda(H)$ is the spectral radius of H and α is a positive constant in HSS inner iterations of JFHSS scheme. Based on Algorithm 1, we can express $x^{(n+1)}$ as

$$
\begin{aligned}
x^{(n+1)} &= x_{k_n}^{(n)} = x_{k_n-1}^{(n)} + (I - T^{l_{k_n}}) G_n'(x_{k_n-1}^{(n)})^{-1} G_n(x_{k_n-1}^{(n)}) \\
&= x_{k_n-1}^{(n)} + (I - T^{l_{k_n}}) A^{-1} G_n(x_{k_n-1}^{(n)}) \\
&= x_{k_n-2}^{(n)} + (I - T^{l_{k_n-1}}) A^{-1} G_n(x_{k_n-2}^{(n)}) + (I - T^{l_{k_n}}) A^{-1} G_n(x_{k_n-1}^{(n)}) \\
&= x_{k_n-3}^{(n)} + (I - T^{l_{k_n-2}}) A^{-1} G_{(n)}(x_{k_n-3}^{(n)}) + (I - T^{l_{k_n-1}}) A^{-1} G_n(x_{k_n-2}^{(n)}) \\
&\quad + (I - T^{l_{k_n}}) A^{-1} G_n(x_{k_n-1}^{(n)}) \\
&= x_0^{(n)} + (I - T^{l_1}) A^{-1} G_n(x_0^{(n)}) + (I - T^{l_2}) A^{-1} G_n(x_1^{(n)}) + \cdots \\
&\quad + (I - T^{l_{k_n-2}}) A^{-1} G_n(x_{k_n-3}^{(n)}) + (I - T^{l_{k_n-1}}) A^{-1} G_n(x_{k_n-2}^{(n)}) \\
&\quad + (I - T^{l_{k_n}}) A^{-1} G_n(x_{k_n-1}^{(n)}) = x^{(n)} + \sum_{j=1}^{k_n}(I - T^{l_j}) A^{-1} G_n(x_{j-1}^{(n)}).
\end{aligned} \tag{21}
$$

In the last equality, we used $x_0^{(n)} = x^{(n)}$. If we set $\eta' = \dfrac{\eta}{\text{cond}(A)}$ in (14) instead of η, where $\text{cond}(A) = \|A\| \|A^{-1}\|$, then $\eta' \le 1$. Because of (14), we have

$$
\begin{aligned}
\|G_n(x_{k_n}^{(n)})\| &\le \|G_n(x_{k_n}^{(n)}) - G_n(x_{k_n-1}^{(n)}) + G_n'(x_{k_n-1}^{(n)})(x_{k_n}^{(n)} - x_{k_n-1}^{(n)})\| \\
&\quad + \|G_n(x_{k_n-1}^{(n)}) - G_n'(x_{k_n-1}^{(n)})(x_{k_n}^{(n)} - x_{k_n-1}^{(n)})\| \\
&= \|G_n(x_{k_n-1}^{(n)}) - A(x_{k_n}^{(n)} - x_{k_n-1}^{(n)})\| \le \eta' \|G_n(x_{k_n-1}^{(n)})\|,
\end{aligned}
$$

so

$$
\begin{aligned}
\|A^{-1} G_n(x_{k_n}^{(n)})\| &\le \|A^{-1}\| \|G_n(x_{k_n}^{(n)})\| \le \eta' \|A^{-1}\| \|G_n(x_{k_n-1}^{(n)})\| \\
&\le \eta' \|A^{-1}\| \|A\| \|A^{-1} G_n(x_{k_n-1}^{(n)})\| = \eta \|A^{-1} G_n(x_{k_n-1}^{(n)})\|.
\end{aligned}
$$

Therefore, by mathematical induction, we can obtain

$$\|A^{-1} G_n(x_{k_n}^{(n)})\| \le \eta^{k_n} \|A^{-1} G_n(x_0^{(n)})\|. \tag{22}$$

Then, from (21), and since $\|I - T^{l_j}\| < 1 + \theta^{l_j} \le 1 + \theta^{l_*^n}$ for $j = 1, 2, \ldots, k_n$, we have

$$
\begin{aligned}
\|x^{(n+1)} - x^{(n)}\| &\leq \sum_{j=1}^{k_n} \|I - T^{l_j}\| \|A^{-1} G_n(x_{j-1}^{(n)})\| \\
&\leq (\|I - T^{l_1}\| + \eta \|I - T^{l_2}\| + \eta^2 \|I - T^{l_3}\| + \cdots \\
&\quad + \eta^{k_n-2}\|I - T^{l_{k_n}-1}\| + \eta^{k_n-1}\|I - T^{l_{k_n}}\|)\|A^{-1} G_n(x_0^{(n)})\| \\
&= (1 + \eta + \eta^2 + \cdots + \eta^{k_n-2} + \eta^{k_n-1})(1 + \theta_*^{l^n})\|A^{-1} G_n(x_0^{(n)})\| \\
&= \frac{1 - \eta^{k_n}}{1 - \eta}(1 + \theta_*^{l^n})\|A^{-1} G_n(x_0^{(n)})\|.
\end{aligned}
\tag{23}
$$

Thus, from the last inequality, since $G_n(x) = b^{(n)} - Ax$, $b^{(n)} = \varphi(x^{(n)})$, we have

$$
\begin{aligned}
\|x^{(n+1)} - x^{(n)}\| &\leq \frac{1 - \eta^{k_n}}{1 - \eta}(1 + \theta_*^{l^n})\|A^{-1}(b^{(n)} - Ax^{(n)})\| \\
&= \frac{1 - \eta^{k_n}}{1 - \eta}(1 + \theta_*^{l^n})(\|A^{-1}(\varphi(x^{(n)}) - \varphi(x^{(n-1)}))\| + \|A^{-1}(\varphi(x^{(n-1)}) - Ax^{(n)})\|).
\end{aligned}
\tag{24}
$$

Then, by using the multivariable Mean Value Theorem (see [15]), we can write

$$
\|A^{-1}(\varphi(x^{(n)}) - \varphi(x^{(n-1)}))\| \leq \max_{x \in S} \|A^{-1}\varphi'(x)\| \|x^{(n)} - x^{(n-1)}\| = L\|x^{(n)} - x^{(n-1)}\|,
$$

where $S = \{x : x = tx^{(n)} + (1-t)x^{(n-1)}, 0 \leq t \leq 1\}$. Thus,

$$
\|A^{-1}(\varphi(x^{(n)}) - \varphi(x^{(n-1)}))\| \leq L\|x^{(n)} - x^{(n-1)}\|.
\tag{25}
$$

From the right-hand side of (24), using (22) for $n-1$, and (25), we have

$$
\begin{aligned}
\|x^{(n+1)} - x^{(n)}\| &\leq \\
&= \frac{1 - \eta^{k_n}}{1 - \eta}(1 + \theta_*^{l^n})(L\|x^{(n)} - x^{(n-1)}\| + \|A^{-1}G_{n-1}(x^n)\|) \\
&\leq \frac{1 - \eta^{k_n}}{1 - \eta}(1 + \theta_*^{l^n})(L\|x^{(n)} - x^{(n-1)}\| + \eta^{k_{n-1}}\|A^{-1}G_{n-1}(x_0^{(n-1)})\|).
\end{aligned}
\tag{26}
$$

If in the last inequality of (26), from (23), we use $\|x^{(n)} - x^{(n-1)}\| \leq \frac{1 - \eta^{k_{n-1}}}{1 - \eta}(1 + \theta_*^{l^{n-1}})\|A^{-1}G_n(x_0^{(n-1)})\|$, then

$$
\begin{aligned}
\|x^{(n+1)} - x^{(n)}\| &\leq \\
&\frac{1 - \eta^{k_n}}{1 - \eta}(1 + \theta_*^{l^n})(L\frac{1 - \eta^{k_{n-1}}}{1 - \eta}(1 + \theta_*^{l^{n-1}})\|A^{-1}G_{n-1}(x_0^{(n-1)})\| + \eta^{k_{n-1}}\|A^{-1}G_{n-1}(x_0^{(n-1)})\|) \\
&\leq \frac{1 - \eta^{k_n}}{1 - \eta}(1 + \theta_*^{l^n})(L\frac{1 - \eta^{k_{n-1}}}{1 - \eta}(1 + \theta_*^{l^{n-1}}) + \eta^{k_{n-1}})\|A^{-1}G_{n-1}(x_0^{(n-1)})\|.
\end{aligned}
$$

As $1 - \eta^{k_n} < 1$, $n = 1, 2, \cdots$ and by the definition of ρ and δ, we have

$$
\|x^{(n+1)} - x^{(n)}\| \leq \delta\rho\|A^{-1}G_{n-1}(x_0^{(n-1)})\|.
\tag{27}
$$

By mathematical induction and since $\|A^{-1}G_0(x_0^{(0)})\| \leq M$,

$$
\|x^{(n+1)} - x^{(n)}\| \leq \delta\rho^n\|A^{-1}G_0(x_0^{(0)})\| \leq \delta M\rho^n,
\tag{28}
$$

which yields (19). By the stopping criterion (18), we must have $\rho < 1$ and then, using (19), it is easy to deduce

$$\|x^{(n+1)} - x^{(0)}\| \le \|x^{(n+1)} - x^{(n)}\| + \|x^{(n)} - x^{(n-1)}\| + \cdots + \|x^{(1)} - x^{(0)}\| \le \frac{\delta \mathcal{M}}{1-\rho},$$

which is the relation (20).

Thus, the sequence $\{x^{(n)}\}$ is in a ball with center $x^{(0)}$ and radius $r = \dfrac{\delta \mathcal{M}}{1-\rho}$. From (28), sequence $\{x^{(n)}\}$ also converges to its limit point x^*. From the following iteration,

$$x_1^{(n)} = x_0^{(n)} + (I - T^{l_1})A^{-1}G_n(x_0^{(n)}),$$

when $n \longrightarrow \infty$, $\|x_0^{(n)} - x^*\| \longrightarrow 0$, $\|x_1^{(n)} - x^*\| \longrightarrow 0$, $l_1 \longrightarrow \infty$. Moreover, as $\|T\| < 1$, then $T^{l_1} \to 0$ and we have

$$G(x^*) = 0,$$

which completes the proof. $\square$

Note that, in some applications, the stopping criterion (18) may be obtained as negative; this shows that, for all $l_* \ge 1$, we must have $\rho < 1$.

In addition, it is easy to deduce from the above theorem that any iterative method that its iteration matrix satisfies in $\|T\| < 1$ can be used instead of the HSS method. For a JFGPSS case, the proof is similar, except, in the inner iteration, the iterative matrix is

$$T = (\bar{\alpha}I + P_2)^{-1}(\bar{\alpha}I - P_1)(\bar{\alpha}I + P_1)^{-1}(\bar{\alpha}I - P_2).$$

The following result shows the convergence of a JFGPSS algorithm:

Theorem 2. *Let $x^{(0)} \in \mathbb{C}^n$ and $\varphi : D \subset \mathbb{C}^n \to \mathbb{C}^n$ be a G-differentiable function on an open set $N_0 \subset D$, on which $\varphi'(x)$ is continuous and $\max \|A^{-1}\varphi'(x)\| = L < 1$. Let us suppose that P_1 and P_2 are generalized positive-definite and skew-Hermitian splitting parts of the positive definite matrix A as (15) and (16) and also that $\mathcal{M}$ is an upper bound for $\|A^{-1}G(x^{(0)})\|$; l_{k_n} is the number of GPSS inner iterations in which the stopping criterion (14) is satisfied,*

$$l_*^n > \left\lceil \frac{\ln\left(\frac{(1-\eta)(1-\eta^{k_n-1})}{L} - 1\right)}{\ln(\theta)} \right\rceil,$$

with $l_*^n = \liminf\limits_{k_n \to \infty} l_{k_n}$ for $n = 1,2,3,\ldots$, η is the tolerance in Newton-like intermediate iterations with $L < (1-\eta)^2$ and $\theta = \|T\|$, where T is the GPSS inner iteration matrix that can be written as

$$T = (\bar{\alpha}I + P_2)^{-1}(\bar{\alpha}I - P_1)(\bar{\alpha}I + P_1)^{-1}(\bar{\alpha}I - P_2).$$

Then, the sequence of iteration $\{x^{(k)}\}_{k=0}^{\infty}$, generated by JFGPSS scheme in Algorithm 1, is well-defined and converges to x^*, satisfying $G(x^*) = 0$, and also

$$\|x^{(n+1)} - x^{(n)}\| \le \delta \mathcal{M} \rho^n,$$

$$\|x^{(n+1)} - x^{(0)}\| \le \frac{\delta \mathcal{M}}{1-\rho},$$

where $\delta = \limsup\limits_{n \to \infty} \dfrac{(1+\theta^{l_*^n})}{1-\eta}$ and $\rho = \limsup\limits_{n \to \infty} \rho_n$ for $\rho_n = \dfrac{(1+\theta^{l_*^n})}{1-\eta}L + \eta^{k_n-1}$.

Proof. Let us note that, in this theorem, we also have $\|T\| < 1$ (for more details, see [16]). The rest of the proof is similar to Theorem 1. $\square$

In the next section, we apply our new iterative method on some weakly nonlinear systems of equations.

4. Application

Now, we use JFHSS and JFGPSS algorithms for solving some nonlinear systems. These examples show that JFHSS and JFGPSS methods perform better than nonlinear HSS-like and Picard-HSS methods.

Example 1. *Consider the following two-dimensional nonlinear convection-diffusion equation*

$$
\begin{aligned}
-(u_{xx} + u_{yy}) + q(u_x + u_y) &= f(x,y), \quad (x,y) \in \Omega \\
u(x,y) &= h(x,y), \quad (x,y) \in \partial\Omega
\end{aligned}
$$

where $\Omega = (0,1) \times (0,1)$, $\partial\Omega$ is its boundary and q is a positive constant for measuring the magnitude of the convection term. We solve this problem for each of the following cases:

Case 1 $f(x,y) = e^{u(x,y)}$, $h(x,y) = 0$.
Case 2 $f(x,y) = -e^{u(x,y)} - \sin(1 + u_x(x,y) + u_y(x,y))$, $h(x,y) = -e^{x+y}$.

To discretize this convection-diffusion equation, for the convective term, we use a central difference method while, for the diffusion term, we use a five-point finite difference method. These yield the following nonlinear system

$$
H(u) = Mu + h^2\psi(u), \tag{29}
$$

where $h = \dfrac{1}{N+1}$ is the equidistance step-size with N as a known natural number and $M = A_N \otimes I_N + A_N \otimes I_N$, $B = C_N \times C_N$ with tridiagonal matrices $A_N = \mathrm{tridiag}(-1 - qh/2, 2, 1 + qh/2)$, $C_N = \mathrm{tridiag}(-1/h, 0, 1/h)$ and I_N is $N \times N$ identity matrix. For case 1, we have $\psi(u) = -\varphi(u)$ and, for case 2, $\psi(u) = \sin(1 + Bu) + \varphi(u)$, where $\varphi(u) = (e^{u_1}, e^{u_2}, ..., e^{u_n})^T$; moreover, $\otimes$ is the Kronecker product symbol, $n = N \times N$ and $\sin(u)$ means $(\sin(u_1), \sin(u_2), \cdots, \sin(u_n))^T$. To apply Picard-HSS, nonlinear HSS-like, JFHSS and JFGPSS methods for solving (29), the stopping criterion for the outer iteration in all methods is chosen as

$$
\frac{\| Mu^{(n)} + h^2\psi(u^{(n)}) \|}{\| Mu^{(0)} + h^2\psi(u^{(0)}) \|} \leq 10^{-12} \tag{30}
$$

Meanwhile, the Newton-like iteration (in JFHSS and JFGPSS methods) is

$$
\frac{\| G_n(u_{k_n}^{(n)}) \|}{\| G_n(u_0^{(n)}) \|} \leq 10^{-1}, \tag{31}
$$

and also the stopping criterion for HSS and GPSS processes in each Newton-like inner iteration is

$$
\| G_n(u_k^{(n)}) - As_{k,l_{k_n}}^{(n)} \| \leq \eta \| G_n(u_k^{(n)}) \|, \tag{32}
$$

where $\{u^{(n)}\}$ is the sequence generated by the JFHSS method. k_n and l_{k_n} are, respectively, the number of Newton-like inner iterations and HSS and GPSS inner iterations, required for satisfying Relations (31) and (32).

Moreover, to avoid computing the Jacobian in Picard-HSS method, we propose the following stopping criterion for inner iterations

$$
\| G(u^{(n)}) + As_{l_n}^{(n)} \| \leq \eta \| G(u^{(n)}) \|. \tag{33}
$$

In order to use a JFGPSS method, we apply the following decomposition on matrix M in Equation (29),

$$P_1 = \mathcal{D} + 2L_\mathcal{G}, \quad P_2 = L_\mathcal{G}^* - L_\mathcal{G} + S. \tag{34}$$

In addition, $\mathcal{K} = 0$, so $\mathcal{G} = H$ is the Hermitian part of M and $S = \frac{1}{2}(M - M^*)$ is the skew Hermitian part of M.

Numerical results for $q = 1000$, $q = 2000$ and initial points $u^{(0)} = \bar{1}$, $u^{(0)} = 4 \times \bar{1}$ for both cases and $u^{(0)} = 12 \times \bar{1}$ for case 1 and $u^{(0)} = 13 \times \bar{1}$ for case 2 and different values of N for JFHSS, JFGPSS, nonlinear HSS-like and Picard-HSS schemes are reported in Tables 1 and 2. Other numerical results such as CPU-time (total CPU time), the number of outer and inner iteration steps (denoted as IT_{out} and IT_{inn}, respectively), and the norm-2 of the function at the last step (denoted by $\|F(u^{(n)})\|$) are also presented in these tables. For JFHSS and JFGPSS algorithms, the values of IT_{int} and IT_{inn} are reported. The former is the obtained number when total inner HSS or GPSS iteration is used in Newton-like iterations, divided by the sum of total Newton-like iterations, while the latter is the total number of intermediate iterations of the Newton-like method.

Except for $u^{(0)} = \bar{1}$, which is relatively close to the solution (in case 1, the real solution u is near zero and, in case 2, almost for all coordinates of the solution, u_i, $i = 1, 2, \cdots, n$, $0 \leq u_i \leq 1$), the nonlinear HSS-like method for other initial points of Tables 1 and 2 could not perform the iterations at all, but JFHSS and JFGPSS methods for all points in both cases could easily solve the problem. Picard-HSS for these three initial points could not solve the problem and, in all cases, fails to solve the problem, especially for $q > 500$.

Numerical results show that the inner iterations for both JFHSS and nonlinear HSS-like are almost the same but for JFGPSS is less than these two methods. For example, in Table 1, for $u^{(0)} = \bar{1}$, $q = 1000$ and $N = 40$, the number of inner iterations for JFHSS and JFGPSS methods are, respectively, 133 and 96 and this number for total iterations in the nonlinear HSS-like method (consider that there is only one kind of iteration in a nonlinear HSS-like method) is 127. However, the nonlinear HSS-like method needs to evaluate a greater number of the nonlinear term $\psi(u)$ than the JFHSS method (for the JFHSS method, only 12 function evaluations are required compared to 254 function evaluations for the nonlinear HSS-like method). Thus, JFHSS and JFGPSS methods can significantly reduce the computational cost of evaluation of the nonlinear term, especially when the nonlinear part is so complicated, e.g., in Example 2, the difference between the computational cost of the nonlinear HSS-like method and the JFHSS method has increased, since the problem has a more complicated nonlinear term.

It must be noted that, in the inner iteration, for solving the linear systems related to the Hermitian part (in HSS scheme) and the skew-Hermitian part (in both HSS and GPSS schemes), we have employed respectively the conjugate gradient (CG) method and the Lanczos method (for more details, see [17]).

In this example, $\eta = tol$ was used for all steps; in most cases, we obtained equal Newton-like and outer iterations at each step; however, in general, choosing equal η and tol does not always lead to equal Newton-like and outer iterations. For example, in cases that nonlinearity increases (e.g., when we choose initial value $u^{(0)} = 12 \times \bar{1}$, in the first steps, the nonlinear term $h^2 \psi(u)$ is so big) result in a different number of Newton-like and outer iterations. In all tables of this paper, a, b denote the number $a \cdot 10^b$.

Table 1. Results for JFHSS, JFGPSS, nonlinear HSS-like and Picard-HSS methods of Example 1, Case 1 ($\eta = \text{tol} = 0.1$).

N			30	40	60	70	80	100
$q = 1000$,	JFHSS	CPU	0.65	1.81	7.46	13.21	24.45	59.32
$u^{(0)} = \bar{1}$		IT_{out}	12	12	12	12	12	12
		IT_{int}	12	12	12	12	12	12
		IT_{inn}	9	11.08	10.75	10.75	10.41	10.91
		$\|F(u^{(n)})\|$	1.86, -11	3.35, -11	1.70, -11	3.41, -11	3.54, -11	2.43, -11
	JFGPSS	CPU	0.63	1.46	5.79	9.84	17.28	44.50
		IT_{out}	12	12	11	11	11	11
		IT_{int}	14	12	11	11	11	11
		IT_{inn}	8.78	8	7.64	7.45	7.90	8.73
		$\|F(u^{(n)})\|$	5.45, -11	1.89, -11	7.69, -11	1.02, -10	9.63, -11	5.09, -11
	Nonlinear HSS-like	CPU	0.82	2.03	8.26	14.60	24.65	61.35
		IT	129	127	123	124	128	126
		$\|F(u^{(n)})\|$	1.45, -10	1.53, -10	1.25, -10	1.10, -10	8.60, -11	8.91, -11
	Picard-HSS		-	-	-	-	-	-
$q = 2000$,	JFHSS	CPU	1.04	2.71	11.32	19.87	31.48	76.13
$u^{(0)} = \bar{1}$		IT_{out}	12	12	12	12	12	12
		IT_{int}	12	12	12	12	12	12
		IT_{inn}	16.08	14.67	14.25	14.17	14	14.08
		$\|F(u^{(n)})\|$	1.47, -10	9.30, -11	7.92, -11	8.80, -11	9.56, -11	6.56, -11
	JFGPSS	CPU	0.85	2.20	8.57	14.26	23.50	54.90
		IT_{out}	12	12	12	12	12	12
		IT_{int}	12	12	12	12	12	12
		IT_{inn}	14.42	12.42	10.58	10	9.84	9.91
		$\|F(u^{(n)})\|$	1.57, -10	8.49, -11	3.33, -11	2.80, -11	2.38, -11	4.23, -11
	Nonlinear HSS-like	CPU	1.32	2.94	12.11	20.51	33.88	80.97
		IT	188	172	167	166	165	165
		$\|F(u^{(n)})\|$	3.24, -10	2.50, -10	2.07, -10	2.32, -10	2.037, -10	1.81, -10
	Picard-HSS		-	-	-	-	-	-
$q = 1000$,	JFHSS	CPU	0.80	2.24	9.34	14.56	23.77	60.21
$u^{(0)} = 4 \times \bar{1}$		IT_{out}	12	12	12	12	12	12
		IT_{int}	12	12	12	12	12	12
		IT_{inn}	11.08	11	10.67	10.75	10.50	11.25
		$\|F(u^{(n)})\|$	1.94, -10	1.70, -10	9.33, -11	9.94, -11	1.15, -10	8.77, -11
	JFGPSS	CPU	0.56	1.51	6.55	12.47	21.03	55.50
		IT_{out}	12	12	11	12	11	11
		IT_{int}	12	12	11	12	11	11
		IT_{inn}	8.92	8.34	8.72	8.75	9.55	10.63
		$\|F(u^{(n)})\|$	9.76, -11	7.68, -11	4.60, -10	6.35, -11	3.73, -10	3.78, -10
	Nonlinear HSS-like		-	-	-	-	-	-
	Picard-HSS		-	-	-	-	-	-
$q = 2000$,	JFHSS	CPU	0.99	2.51	11.20	19.45	32.23	77.58
$u^{(0)} = 4 \times \bar{1}$		IT_{out}	12	12	12	12	12	12
		IT_{int}	12	12	12	12	12	12
		IT_{inn}	16.08	14.67	14.25	14.17	14	14.08
		$\|F(u^{(n)})\|$	5.88, -10	3.71, -10	3.20, -10	3.57, -10	3.75, -10	2.69, -10
	JFGPSS	CPU	0.85	2.20	8.58	14.02	23.22	54.94
		IT_{out}	12	12	12	12	12	12
		IT_{int}	12	12	12	12	12	12
		IT_{inn}	14.42	12.41	10.58	9.92	9.84	9.84
		$\|F(u^{(n)})\|$	6.26, -10	3.44, -10	1.31, -10	1.63, -10	1.08, -10	2.08, -10
	Nonlinear HSS-like		-	-	-	-	-	-
	Picard-HSS		-	-	-	-	-	-
$q = 1000$,	JFHSS	CPU	0.81	2.23	10.41	18.89	31.31	81.74
$u^{(0)} = 12 \times \bar{1}$		IT_{out}	12	12	12	12	12	12
		IT_{int}	14	14	14	14	14	14
		IT_{inn}	10.85	12.83	11.28	11.71	11.64	12.93
		$\|F(u^{(n)})\|$	1.47, -8	1.05, -8	7.50, -9	4.55, -9	3.08, -9	3.29, -9
	JFGPSS	CPU	0.66	1.70	7.95	14.30	25.44	63.80
		IT_{out}	12	12	12	12	12	12
		IT_{int}	14	14	14	14	14	14
		IT_{inn}	8.78	8	7.86	8.64	9.07	9.92
		$\|F(u^{(n)})\|$	8.02, -9	3.11, -8	3.40, -9	2.32, -9	1.61, -9	6.16, -10
	Nonlinear HSS-like		-	-	-	-	-	-
	Picard-HSS		-	-	-	-	-	-

Table 1. *Cont.*

N			30	40	60	70	80	100
$q = 2000$,	JFHSS	CPU	1.06	2.90	13.55	21.72	38.94	87.18
$u^{(0)} = 12 \times \bar{1}$		IT_{out}	12	12	12	12	12	12
		IT_{int}	14	14	14	13	14	13
		IT_{inn}	14.93	14.36	14.86	14.62	14.57	15
		$\|F(u^{(n)})\|$	2.06, -8	1.48, -8	9.76, -9	6.96, -9	6.58, -9	5.72, -9
	JFGPSS	CPU	0.95	2.45	10.03	17.81	29.06	69.57
		IT_{out}	12	12	12	12	12	12
		IT_{int}	14	14	14	14	14	13
		IT_{inn}	13.71	11.64	10.71	10.85	10.64	11.31
		$\|F(u^{(n)})\|$	1.91, -8	1.30, -8	6.08, -9	3.26, -9	6.35, -9	3.23, -9
	Nonlinear HSS-like		-	-	-	-	-	-
	Picard-HSS		-	-	-	-	-	-

Table 2. Results for JFHSS, JFGPSS, nonlinear HSS-like and Picard-HSS methods of Example 1, Case 2 ($\eta = \text{tol} = 0.1$).

N			30	40	60	70	80	100
$q = 1000$,	JFHSS	CPU	0.73	2.03	9.54	16.99	27.27	65.47
$u^{(0)} = \bar{1}$		IT_{out}	11	11	11	12	12	12
		IT_{int}	12	12	12	13	13	13
		IT_{inn}	11.25	11.41	11.92	11.23	10.92	12
		$\|F(u^{(n)})\|$	1.64, -10	1.18, -10	1.43, -11	1.32, -11	1.95, -11	1.56, -11
	JFGPSS	CPU	0.57	1.53	7.19	12.59	19.42	53.59
		IT_{out}	11	11	11	11	11	12
		IT_{int}	12	12	12	12	12	13
		IT_{inn}	9	8.25	8.75	8.92	8.25	9
		$\|F(u^{(n)})\|$	5.45, -11	8.19, -11	8.56, -11	7.6, -11	5.27, -11	4.81, -12
	Nonlinear HSS-like	CPU	0.82	2.30	9.86	14.38	29.31	59.91
		IT	128	128	123	124	121	126
		$\|F(u^{(n)})\|$	1.81, -10	1.43, -10	1.25, -10	1.10, -10	1.15, -10	1.06, -10
	Picard-HSS		-	-	-	-	-	-
$q = 2000$,	JFHSS	CPU	0.98	2.63	11.73	20.51	36.07	77.20
$u^{(0)} = \bar{1}$		IT_{out}	11	11	11	11	12	12
		IT_{int}	12	12	12	12	13	13
		IT_{inn}	16	15	14.50	14.67	14.30	14.62
		$\|F(u^{(n)})\|$	2.49, -10	2.26, -10	2.03, -10	2.30, -10	2.61, -11	1.74, -11
	JFGPSS	CPU	0.88	2.26	8.61	15.08	25.83	60.7
		IT_{out}	11	11	11	11	12	11
		IT_{int}	12	12	12	12	13	12
		IT_{inn}	14.33	12.08	10.58	10.66	9.69	11
		$\|F(u^{(n)})\|$	2.04, -10	2.91, -10	1.91, -10	1.09, -10	1.64, -11	1.72, -10
	Nonlinear HSS-like	CPU	1.15	3.85	12.52	19.61	37.70	79.26
		IT	187	171	166	166	164	164
		$\|F(u^{(n)})\|$	3.68,-10	3.07, -10	2.48, -10	2.17, -10	2.42, -10	2.13, -10
	Picard-HSS		-	-	-	-	-	-
$q = 1000$,	JFHSS	CPU	0.72	2.28	9.39	16.62	28.53	67.23
$u^{(0)} = 4 \times \bar{1}$		IT_{out}	11	11	11	12	11	11
		IT_{int}	12	12	12	12	12	12
		IT_{inn}	11.41	11.33	11.41	11.75	12.08	12.34
		$\|F(u^{(n)})\|$	1.62, -10	2.01, -10	1.64, -10	1.92, 10	2.89, -10	2.47, -10
	JFGPSS	CPU	0.69	1.97	8.85	16.53	26.53	70.80
		IT_{out}	11	11	11	11	12	11
		IT_{int}	12	12	12	12	13	12
		IT_{inn}	10.91	11.16	11	11.42	11.42	12.34
		$\|F(u^{(n)})\|$	2.18, -10	1.22, -10	1.21, -10	8.35, -11	1.15, -10	1.17, -10
	Nonlinear HSS-like		-	-	-	-	-	-
	Picard-HSS		-	-	-	-	-	-

Table 2. *Cont.*

N			30	40	60	70	80	100
$q = 2000$,	JFHSS	CPU	0.97	2.59	11.06	20.24	33.75	79.80
$u^{(0)} = 4 \times \bar{1}$		IT_{out}	11	11	11	11	11	11
		IT_{int}	12	12	12	12	12	12
		IT_{inn}	15.92	15.08	14.50	14.58	14.66	14.92
		$\|F(u^{(n)})\|$	9.65, -10	4.15, -10	4.31, -10	4.40, -10	3.82, -10	3.39, -10
	JFGPSS	CPU	0.88	2.18	8.75	14.60	25.05	64.79
		IT_{out}	11	11	11	11	11	11
		IT_{int}	12	12	12	12	12	12
		IT_{inn}	14.50	12.42	10.66	10.42	10.58	11.08
		$\|F(u^{(n)})\|$	5.06, -10	3.61, -10	2.53, -10	3.32, -10	2.95, -10	1.97, -10
	Nonlinear HSS-like		-	-	-	-	-	-
	Picard-HSS		-	-	-	-	-	-
$q = 1000$,	JFHSS	CPU	0.77	2.33	10.82	19.56	32.13	85.32
$u^{(0)} = 13 \times \bar{1}$		IT_{out}	12	12	12	12	12	12
		IT_{int}	14	14	14	14	14	14
		IT_{inn}	10.85	12.83	11.28	11.71	11.64	12.92
		$\|F(u^{(n)})\|$	1.44, -8	1.36, -8	7.47, -9	4.56, -9	3.8, -9	3.54, -9
	JFGPSS	CPU	0.65	1.74	8.00	14.54	25.02	64.19
		IT_{out}	12	12	12	12	12	12
		IT_{int}	14	14	14	14	14	14
		IT_{inn}	8.78	8	8.28	8.64	9.07	9.86
		$\|F(u^{(n)})\|$	8.03, -9	1.44, -8	3.35, -9	4.76, -9	1.69, -9	1.085, -9
	Nonlinear HSS-like		-	-	-	-	-	-
	Picard-HSS		-	-	-	-	-	-
$q = 2000$,	JFHSS	CPU	1.08	2.97	11.27	22.45	39.98	89.15
$u^{(0)} = 13 \times \bar{1}$		IT_{out}	12	12	12	12	12	12
		IT_{int}	14	14	14	14	14	14
		IT_{inn}	14.93	14.35	14.43	14.62	14.57	15
		$\|F(u^{(n)})\|$	2.01, -8	1.49, -8	8.73, -9	6.97, -9	6.57, -9	5.72, -9
	JFGPSS	CPU	0.99	2.41	10.15	17.98	29.33	67.45
		IT_{out}	12	12	12	12	12	12
		IT_{int}	14	14	14	14	14	13
		IT_{inn}	13.78	11.64	10.71	10.86	10.64	11.31
		$\|F(u^{(n)})\|$	1.70, -8	1.30, -8	6.02, -9	3.23, -9	6.34, -9	3.21, -9
	Nonlinear HSS-like		-	-	-	-	-	-
	Picard-HSS		-	-	-	-	-	-

The optimal value for parameter α that minimizes the boundary of spectral radius of the iteration matrices is important because it also improves the convergence speed of Picard-HSS, nonlinear HSS-like, JFHSS and JFGPSS methods. There are no general results to determine the optimal α and $\bar{\alpha}$, so we need to obtain the optimal values of parameters α and $\bar{\alpha}$ experimentally. However, Bai and Golub [8] proved that spectral radius of HSS iterative matrix that is obtained from the coefficient matrix M in (29) is bounded by $\|T\| \leq \sigma(\alpha) \equiv \max\limits_{\lambda_i \in \lambda(H)} \left| \dfrac{\alpha - \lambda_i}{\alpha + \lambda_i} \right| < 1$, and the minimum of this bound is obtained when

$$\alpha = \alpha^* = \sqrt{\lambda_{min(H)}\lambda_{max(H)}},$$

where $\lambda_{min(H)}$ and $\lambda_{max(H)}$ are, respectively, the smallest and the largest eigenvalues of Hermitian matrix H. Usually, in an HSS scheme, $\alpha_{opt} \neq \alpha^* \equiv \operatorname*{argmin}\limits_{\alpha > 0}\{\sigma(\alpha)\} < 1$ and $\rho(T(\alpha^*)) \geq \rho(T(\alpha_{opt}))$. When q or $qh/2$ is small, $\sigma(\alpha)$ is close to $\rho(T(\alpha))$ and in this case α^* is close to α_{opt} and α^* can be a good estimation for α_{opt}. However, when q or $qh/2$ is large (the skew-Hermitian part is dominant), hence $\sigma(\alpha)$ deviates too much from $\rho(T(\alpha))$, so using α^* is not useful. In this case, $\rho(T(\alpha))$ attains its minimum at α_{opt} that is far from α^*, but close to $qh/2$ (see [8]).

In the GPSS case, a spectral radius of $T(\bar{\alpha})$ is bounded by $\|V(\bar{\alpha})\|$, where $V(\bar{\alpha}) = (\bar{\alpha}I - P_1)(\bar{\alpha}I + P_1)^{-1}$. Since $\|V(\bar{\alpha})\|_2 \leq 1$ (see [18]), GPSS inner iterations unconditionally converge to the exact solution in each inner iteration of a JFGPSS scheme. However, when $P_1 \in \mathbb{C}^{n \times n}$ is a general positive-definite matrix, we do not have any formula to compute $\bar{\alpha}^* \equiv \operatorname*{argmin}\limits_{\bar{\alpha} > 0}\{\|V(\bar{\alpha})\|\}$ that is the value that minimizes the boundary of iteration matrix $T(\bar{\alpha})$, nor do we have a formula for $\bar{\alpha}_{opt}$, the value that minimizes $\|T(\bar{\alpha})\|$.

In Table 3, the optimal values of α_{opt} and $\bar{\alpha}_{opt}$ have been written (tested and optimal α and $\bar{\alpha}_{opt}$) that are determined experimentally by using increments as 0.25. In addition, the corresponding spectral radius of the iteration matrices $T(\alpha)$ and $T(\bar{\alpha})$ for HSS and GPSS algorithms that are used as inner iterations to solve (29) are reported in this table. One can see that the spectral radius of GPSS method in all cases is smaller than HSS scheme, which results in faster convergence.

Table 3. Optimal value of α for HSS and GPSS inner iterations for different values of N and q of Example 1.

N			30	40	50	60	70	80	90	100
$q = 1000$	HSS	α_{opt}	18	15	10.5	9	8	6	5.75	5.75
		$\rho(T(\alpha_{opt}))$	0.7226	0.6930	0.6743	0.6613	0.6513	0.6485	0.6459	0.6467
		α^*	0.4047	0.3062	0.2462	0.2059	0.1769	0.1551	0.1381	0.1244
		$\rho(T(\alpha^*))$	0.8971	0.9211	0.9360	0.9461	0.9535	0.9590	0.9634	0.9669
		$\frac{qh}{2}$	16.1290	12.1951	9.8039	8.1967	7.0423	6.1728	5.4945	4.9505
		$\rho(T(\frac{qh}{2}))$	0.7236	0.6974	0.6783	0.6674	0.6608	0.6574	0.6562	0.6569
	GPSS	$\bar{\alpha}_{opt}$	11.25	9.5	8.5	7.5	7	6.5	6	5.5
		$\rho(T(\bar{\alpha}_{opt}))$	0.5428	0.5140	0.5076	0.4983	0.4959	0.4902	0.4982	0.4983
$q = 2000$	HSS	α_{opt}	26	22	16	13.5	12	10	8.75	8
		$\rho(T(\alpha_{opt}))$	0.7911	0.7663	0.6499	0.7399	0.0.7373	0.7302	0.7302	0.7242
		α^*	0.1638	0.0938	0.0606	0.0424	0.0313	0.0241	0.0191	0.0155
		$\rho(T(\alpha^*))$	0.9579	0.9757	0.9842	0.9889	0.9918	0.9937	0.9950	0.9959
		$\frac{qh}{2}$	32.2581	24.39	19.61	16.3934	14.0845	12.35	10.99	9.9010
		$\rho(T(\frac{qh}{2}))$	0.7953	0.77	0.7512	0.7439	0.7343	0.728	0.7282	0.7270
	GPSS	$\bar{\alpha}_{opt}$	15	13	11	10	9	8	7.5	7
		$\rho(T(\bar{\alpha}_{opt}))$	0.6424	0.6212	0.6144	0.6063	0.6036	0.6028	0.6090	0.6033

Example 2 ([10]). *We consider the two-dimensional nonlinear convection-diffusion equation*

$$-(u_{xx} + u_{yy}) + qe^{x+y}(xu_x + yu_y) = ue^u + \sin(\sqrt{1 + u_x^2 + u_y^2}), \quad (x,y) \in \Omega,$$
$$u(x,y) = 0, \quad (x,y) \in \partial\Omega,$$

where $\Omega = (0,1) \times (0,1)$, $\partial\Omega$ is its boundary and q is a positive constant for measuring magnitude of the convection term. By applying the upwind finite difference scheme on the equidistance discretization grid (stepsize $h = \frac{1}{N+1}$) with the central difference scheme to the convective term, we obtain a system of nonlinear equations in the general form (for more details, see [10])

$$H(x) = Mx - h^2\psi(x). \tag{35}$$

We have selected zero vector $u^{(0)} = \bar{0} = (0, 0, \cdots, 0)^T$ as the initial guess. In addition, again (31) and (32) are used respectively as the stopping criteria for the inner iterations and Newton-like iterations in the JFHSS method and (30) for outer iterations in JFHSS, Picard-HSS and nonlinear HSS-like methods. Moreover, to avoid computing Jacobian in Picard-HSS and nonlinear HSS-like methods, we used (33). Similar to Example 1, one can use other iterative methods instead of HSS in Algorithm 1, for which the spectral radius of its iteration matrix is smaller and thus results in faster convergence.

Numerical results for $N = 32, 48, 64$, optimal α and different values of q for JFNHSS, Picard-HSS and nonlinear HSS-like schemes are reported in Table 4. In addition, we adopted the experimentally optimal parameters α to obtain the least CPU times for these iterative methods. One can see that JFHSS performs better than nonlinear HSS-like and Picard-HSS methods in all cases.

Table 4. Results of JFHSS, nonlinear HSS-like and Picard-HSS methods for Example 2 ($\eta = $ tol $= 0.1$).

q			50	100	200	400	1200	2000
N = 32		α_{opt}	1.4	1.6	2.5	8	21.5	34
	JFHSS	CPU	1.23	1.42	1.29	1.53	1.71	1.86
		IT_{out}	12	12	12	12	12	12
		IT_{int}	12	12	12	12	12	12
		IT_{inn}	11.34	11.67	12	12.75	16.25	20.34
		$\|F(u^{(n)})\|$	1.54, −14	2.2, −14	1.47, −14	6.87, −15	8.22, −15	1.3, −14
	Nonlinear HSS-like	CPU	2.03	2.39	2.25	2.31	2.42	2.45
		IT	129	137	140	146	160	167
		$\|F(u^{(n)})\|$	1.1, −14	2.25, −14	2.31, −14	2.23, −14	2.4, −14	2.3, −14
	Picard-HSS	CPU	7.96	8.31	7.76	8	8.60	8.86
		IT_{out}	12	12	12	12	12	12
		IT_{inn}	121.1	131.91	126.75	145.34	146.34	147
		$\|F(u^{(n)})\|$	1.1, −14	1.24, −14	1.57, −14	1.96, −14	1.84, −14	1.6, −14
N = 48		α_{opt}	0.8	1.4	2.6	4.8	13	20.5
	JFHSS	CPU	5.25	5.31	5.5	5.93	6.21	6.28
		IT_{out}	12	12	12	12	12	12
		IT_{int}	12	12	12	12	12	12
		IT_{inn}	13.66	14.58	15.083	16.08	17.34	17.58
		$\|F(u^{(n)})\|$	2.42, −14	6.04, −15	6.36, −15	1.96, −14	6.15, −15	8.60, −15
	Nonlinear HSS-like	CPU	8.87	11.828	10.02	10.31	11.28	11.85
		IT	161	209	178	186	201	207
		$\|F(u^{(n)})\|$	1.5, −14	1.59, −14	1.46, −14	1.57, −14	1.615, −14	1.46, −14
	Picard-HSS	CPU	50.81	50.01	51.85	53.34	56.32	59.95
		IT_{out}	12	12	12	12	12	12
		IT_{inn}	177.16	179.1	183.50	189.34	202.75	213.25
		$\|F(u^{(n)})\|$	7.7, −15	9.67, −15	1.11, −14	1.23, −14	1.22, −14	1.26, −14
N = 64		α_{opt}	0.7	1	1.8	3.3	8.9	14.2
	JFHSS	CPU	21.68	18.23	18.65	19.156	20.53	21.39
		IT_{out}	12	12	12	12	12	12
		IT_{int}	12	12	12	12	12	12
		IT_{inn}	21	17.39	18.17	18.75	19.91	20.84
		$\|F(u^{(n)})\|$	1.61, −14	6.73, −15	9.15, −15	8.39, −15	7.7, −15	4.71, −15
	Nonlinear HSS-like	CPU	38.57	31.78	33.50	34.65	36.56	37.70
		IT	246	206	213	221	235	242
		$\|F(u^{(n)})\|$	1.17, −14	1.26, −14	1.26, −14	1.16, −14	1.19, −14	1.22, −14
	Picard-HSS	CPU	219.54	217.45	266.83	225.37	228.60	248.35
		IT_{out}	12	12	12	12	12	12
		IT_{inn}	219.54	248.58	230.75	252	258.75	264.50
		$\|F(u^{(n)})\|$	6.12, −15	7.7, −15	8.9, −15	1.0, −14	1.1, −14	1.1, −14

5. Conclusions

In this paper, an iterative method based on two-stage splitting methods has been proposed to solve weakly nonlinear systems and a convergence property of this method has been investigated. This method is a combination of an inexact Newton method, Hermitian and skew-Hermitian splitting (or generalized positive definite and skew-Hermitian splitting) scheme. The advantage of our new method, Picard-HSS and nonlinear HSS-like over the methods like Newton method is that they don't need explicit construction and accurate computation of the Jacobian matrix. Hence, computation works and computer memory may be saved in actual application; however, numerical results show that JFHSS and JFGPSS methods perform better than the two other ones.

Numerical results show that JFHSS and JFGPSS iteration algorithms are effective, robust, and feasible nonlinear solvers for a class of weakly nonlinear systems. Moreover, employing these algorithms to solve nonlinear systems is found to be simple, accurate, fast, flexible, convenient and have small computation cost. In addition, it must be noted that, even though our inner iteration scheme in this paper are HSS and GPSS methods, another inner iteration solver can be used subject to the condition that the iteration matrix satisfies in $\|T\| < 1$.

Author Contributions: The contributions of authors are roughly equal.

Funding: This research received no external funding.

Acknowledgments: The third and fourth authors have been partially supported by the Spanish Ministerio de Ciencia, Innovación y Universidades PGC2018-095896-B-C22 and Generalitat Valenciana PROMETEO/2016/089.

Conflicts of Interest: The authors declare no conflict of interest.

References

1. Shen, W.; Li, C. Kantorovich-type convergence criterion for inexact Newton methods. *Appl. Numer. Math.* **2009**, *59*, 1599–1611. [CrossRef]
2. An, H.-B.; Bai, Z.-Z. A globally convergent Newton-GMRES method for large sparse systems of nonlinear equations. *Appl. Numer. Math.* **2007**, *57*, 235-252. [CrossRef]
3. Eisenstat, S.C.; Walker, H.F. Globally convergent inexact Newton methods. *SIAM J. Optim.* **1994**, *4*, 393–422. [CrossRef]
4. Gomes-Ruggiero, M.A.; Lopes, V.L.R.; Toledo-Benavides, J.V. A globally convergent inexact Newton method with a new choice for the forcing term. *Ann. Oper. Res.* **2008**, *157*, 193–205. [CrossRef]
5. Bai, Z.-Z. A class of two-stage iterative methods for systems of weakly nonlinear equations. *Numer. Algorithms* **1997**, *14*, 295–319. [CrossRef]
6. Zhu, M.-Z. Modified iteration methods based on the Asymmetric HSS for weakly nonlinear systems. *J. Comput. Anal. Appl.* **2013**, *15*, 188–195.
7. Axelsson, O.; Bai, Z.-Z.; Qiu, S.-X. A class of nested iteration schemes for linear systems with a coefficient matrix with a dominant positive definite symmetric part. *Numer. Algorithms* **2004**, *35*, 351–372. [CrossRef]
8. Bai, Z.-Z.; Golub, G.H.; Ng, M.K. Hermitian and skew-Hermitian splitting methods for non-Hermitian positive definite linear systems. *SIAM J. Matrix Anal. Appl.* **2003**, *24*, 603–626. [CrossRef]
9. Li, L.; Huang, T.-Z.; Liu, X.-P. Asymmetric Hermitian and skew-Hermitian splitting methods for positive definite linear systems. *Comput. Math. Appl.* **2007**, *54*, 147–159. [CrossRef]
10. Bai, Z.-Z.; Yang, X. On HSS-based iteration methods for weakly nonlinear systems. *Appl. Numer. Math.* **2009**, *59*, 2923–2936. [CrossRef]
11. Bai, Z.-Z.; Migallón, V.; Penadés, J.; Szyld, D.B. Block and asynchronous two-stage methods for mildly nonlinear systems. *Numer. Math.* **1999**, *82*, 1–20. [CrossRef]
12. Zhu, M.-Z.; Zhang, G.-F. On CSCS-based iteration methods for Toeplitz system of weakly nonlinear equations. *J. Comput. Appl. Math.* **2011**, *235*, 5095–5104. [CrossRef]
13. Bai, Z.-Z.; Guo, X.-P. On Newton-HSS methods for systems of nonlinear equations with positive-definite Jacobian matrices. *J. Comput. Math.* **2010**, *28*, 235–260.
14. Li, X.; Wu, Y.-J. Accelerated Newton-GPSS methods for systems of nonlinear equations. *J. Comput. Anal. Appl.* **2014** , *17*, 245–254.
15. Edwards, C.H. *Advanced Calculus of Several Variables*; Academic Press: New York, NY, USA, 1973.
16. Cao, Y.; Tan, W.-W.; Jiang, M.-Q. A generalization of the positive-definite and skew-Hermitian splitting iteration. *Numer. Algebra Control Optim.* **2012**, *2*, 811–821.
17. Bai, Z.-Z.; Golub, G.H.; Ng, M.K. On inexact Hermitian and skew-Hermitian splitting methods for non-Hermitian positive definite linear systems. *Linear Algebra Appl.* **2008**, *428*, 413–440. [CrossRef]
18. Bai, Z.-Z.; Golub, G.H.; Lu, L.-Z.; Yin, J.-F. Block triangular and skew-Hermitian splitting methods for positive-definite linear systems. *SIAM J. Sci. Comput.* **2005**, *26*, 844–863.

mathematics

MDPI

Article

Higher-Order Iteration Schemes for Solving Nonlinear Systems of Equations

Hessah Faihan Alqahtani [1], Ramandeep Behl [2,*] and Munish Kansal [3]

[1] Department of Mathematics, King Abdulaziz University, Campus Rabigh 21911, Saudi Arabia;
hfalqahtani@kau.edu.sa
[2] Department of Mathematics, King Abdulaziz University, Jeddah 21589, Saudi Arabia
[3] School of Mathematics, Thapar Institute of Engineering and Technology, Patiala 147004, India;
mkmaths@gmail.com
* Correspondence: ramanbehl87@yahoo.in

Received: 23 August 2019; Accepted: 24 September 2019; Published: 10 October 2019

Abstract: We present a three-step family of iterative methods to solve systems of nonlinear equations. This family is a generalization of the well-known fourth-order King's family to the multidimensional case. The convergence analysis of the methods is provided under mild conditions. The analytical discussion of the work is upheld by performing numerical experiments on some application oriented problems. Finally, numerical results demonstrate the validity and reliability of the suggested methods.

Keywords: systems of nonlinear equations; King's family; order of convergence; multipoint iterative methods

---

1. Introduction

System of nonlinear equations (SNEs) finds applications to numerous phenomena in many areas of science and engineering. Given a nonlinear system, $F(X) = 0$, where F is a nonlinear map from $\mathbb{R}^k \to \mathbb{R}^k$, we are interested to compute a vector $X^* = (x_1^*, x_2^*, \cdots, x_k^*)^T$ such that $F(X^*) = 0$, where $F(X) = (f_1(X), f_2(X), \ldots, f_k(X))^T$ is a Fréchet differentiable function and $X = (x_1, x_2, \ldots, x_k)^T \in \mathbb{R}^k$. The classical Newton's method [1] is the most famous procedure to solve SNEs. It is given by

$$X^{(k+1)} = X^{(k)} - \{F'(X^{(k)})\}^{-1} F(X^{(k)}), \ k = 0, 1, 2, \ldots. \tag{1}$$

It converges quadratically if the function F is continuously differentiable and the initial approximation is close enough. In the literature, there are variety of higher-order methods that improve the convergence order of Newton's scheme. For example, several authors have proposed cubically convergent methods [2–5] requiring computation of 2-F' (2-F' stand for F' two times), 1-F (1-F stands for F one time), and two matrix inversions per step. In [6], the authors developed another family of methods of order three, one of which requires one 1-F and 3-F', whereas the other requires 1-F and 4-F' evaluations and two matrix inversions per iteration. In [7], Darvishi and Barati utilized 2-F, 2-F' and two matrix inversions per step to propose a new third-order scheme. Similarly, several third-order methods have been proposed in [8,9] that require 2-F, 1-F', and one matrix inversion. Babajee et al. [10] presented a method having convergence order four which consumes 1-F, 2-F' and two matrix inversions per iteration. Another fourth-order method is developed in [11] using two evaluations of the function and the Jacobian and one matrix inversion, whereas the authors of [12] propose another fourth-order method, utilizing 3-F, 1-F', and one matrix inversion per iteration. Another fifth-order method in [13] requires three evaluations of the function and only one Jacobian evaluation, with the solution of three linear systems with the same matrix of coefficients per iteration.

In pursuit of faster algorithms, researchers have also developed fifth and sixth-order methods, for example, in [6,14–16]. In [15], Narang et al. extended the existing Babajee's fourth-order scheme [17] to solve SNEs and developed a sixth-order convergent family of Chebyshev-Halley type methods. Their scheme requires two F, two F' evaluations, and the solution of two linear systems per iteration. One can notice that although the researchers are making an attempt to improve the order of convergence of an iterative method, it mostly leads to increase in the computational cost per iteration. The computational cost is especially high if the method involves the use of second order Fréchet derivative $F''(X)$. This is a major limitation of the higher-order methods. Thus, although developing new iterative methods, we should try to keep the computational cost low. With this intention, we have made an attempt to develop a family of three-step sixth-order family of methods requiring two F, two F' and one matrix inversion per iteration. This family of methods are compared to be more efficient than existing methods. These have been found to be effective in solving particularly large-scale nonlinear systems.

The outline of the manuscript is as follows. In Section 2, a new class of new sixth-order scheme and its convergence analysis is presented. In Section 3, we present numerous illustrative examples to validate the theoretical results. Finally, Section 4 contains some conclusions.

2. Design of the King's Family for Multidimensional Case

In this section, we proposed a new three-point extension of King's method [18–21] having sixth-order convergence. For this purpose, we consider the well-known fourth-order King's method, which is given by

$$y_k = x_k - \frac{f(x_k)}{f'(x_k)},$$
$$x_{k+1} = y_k - \frac{1 + \alpha u_k}{1 + (\alpha - 2)u_k} \frac{f(y_k)}{f'(x_k)}, \tag{2}$$

where α is a real parameter and $u_k = \frac{f(y_k)}{f(x_k)}$. For $\alpha = 0$, one can obtain the well-known Ostrowski's method [22–24].

Let us now modify the method (2) for SNEs by rewriting the scheme as follows,

$$u_k = \frac{f(y_k) - f(x_k) + f(x_k)}{f(x_k)}$$
$$= \frac{f(y_k) - f(x_k)}{f(x_k)} + 1$$
$$= 1 - \frac{f(y_k) - f(x_k)}{(y_k - x_k)f'(x_k)}$$
$$= 1 - f'(x_k)^{-1}[y_k, x_k; f],$$

where $[y_k, x_k; f] = \frac{f(y_k) - f(x_k)}{y_k - x_k}$. Finally, we can rewrite the above scheme (2) for SNEs with one additional sub-step in the following manner,

$$y^{(k)} = x^{(k)} - F'(x^{(k)})^{-1}F(x^{(k)}),$$
$$z^{(k)} = y^{(k)} - \left(I + (\alpha - 2)U^{(k)}\right)^{-1}\left(I + \alpha U^{(k)}\right)F'(x^{(k)})^{-1}F(y^{(k)}), \tag{3}$$
$$x^{(k+1)} = z^{(k)} - \left([y^{(k)}, z^{(k)}; F]\right)^{-1}F(z^{(k)}),$$

where $[\cdot, \cdot; F]$ is a finite difference of order one and α is a free disposable parameter with $U^{(k)} = I - [x^{(k)}, y^{(k)}; F]F'(x^{(k)})^{-1}$. In addition, $F[Y_n, X_n]$ is a finite difference of order one.

Now, it is necessary to analyze the convergence conditions of this modified King's class of methods. In Theorem 1, we demonstrate the convergence order of the above scheme (3). We have used the following procedures [25] to prove the convergence results.

Let $F : \Omega \subseteq \mathbb{R}^k \longrightarrow \mathbb{R}^k$ be sufficiently differentiable in Ω. Now, we define the qth derivative of F at $\omega \in \Omega$, $q \geq 1$. It can be viewed as a q-linear function $F^{(q)}(\omega) : \mathbb{R}^k \times \cdots \times \mathbb{R}^k \longrightarrow \mathbb{R}^k$, such that $F^{(q)}(\omega)(v_1, \ldots, v_q) \in \mathbb{R}^k$. It is easy to observe that

1. $F^{(q)}(\omega)(v_1, \ldots, v_{q-1}, \cdot) \in \mathcal{L}(\mathbb{R}^k)$.
2. $F^{(q)}(\omega)(v_{\sigma(1)}, \ldots, v_{\sigma(q)}) = F^{(q)}(\omega)(v_1, \ldots, v_q)$, for all permutation σ of $\{1, 2, \ldots, q\}$.

Using the above relations, we can introduce the following notation,

(a) $F^{(q)}(\omega)(v_1, \ldots, v_q) = F^{(q)}(\omega)v_1 \ldots v_q$.
(b) $F^{(q)}(\omega)v^{q-1}F^{(p)}v^p = F^{(q)}(\omega)F^{(p)}(\omega)v^{q+p-1}$.

Now, applying Taylor's expansion for $\xi^* + h \in \mathbb{R}^k$ in the neighborhood of a solution ξ^* of the given linear system, one can get

$$F(\xi^* + h) = F'(\xi^*)\left[h + \sum_{q=2}^{p-1} C_q h^q\right] + O(h^p), \tag{4}$$

where $C_q = (1/q!)[F'(\xi^*)]^{-1}F^{(q)}(\xi^*)$, $q \geq 2$. We note that $C_q h^q \in \mathbb{R}^k$ as $F^{(q)}(\xi^*) \in \mathcal{L}(\mathbb{R}^k \times \cdots \times \mathbb{R}^k, \mathbb{R}^k)$, and $[F'(\bar{x})]^{-1} \in \mathcal{L}(\mathbb{R}^k)$.

Similarly, we can express F' as

$$F'(\xi^* + h) = F'(\xi^*)\left[I + \sum_{q=2}^{p-1} qC_q h^{q-1}\right] + O(h^{p-1}), \tag{5}$$

where I denotes the identity matrix. Therefore, $qC_q h^{q-1} \in \mathcal{L}(\mathbb{R}^k)$. From Equation (5), we obtain

$$[F'(\xi^* + h)]^{-1} = \left[I + X_2 h + X_3 h^2 + X_4 h^4 + \cdots\right][F'(\xi^*)]^{-1} + O(h^p), \tag{6}$$

where

$$X_2 = -2C_2,$$
$$X_3 = 4C_2^2 - 3C_3,$$
$$X_4 = -8C_2^3 + 6C_2C_3 + 6C_3C_2 - 4C_4,$$
$$\vdots$$

Let us denote $e^{(k)} = x^{(k)} - \xi^*$ as the error at the kth iteration. Then, the error equation is given as follows,

$$e^{(k+1)} = M(e^{(k)})^p + O((e^{(k)})^{p+1}),$$

where, M is a p-linear function $M \in \mathcal{L}(\mathbb{R}^k \times \cdots \times \mathbb{R}^k, \mathbb{R}^k)$. Here, p is the *order of convergence* and $(e^{(k)})^p$ is a column vector $\overbrace{(e^{(k)}, e^{(k)}, \cdots, e^{(k)})}^{p}{}^T$.

Theorem 1. *Let $F : \Omega \subseteq \mathbb{R}^k \to \mathbb{R}^k$ be a sufficiently differentiable function defined on a convex set Ω containing the zero ξ^*. Let us assume that $F'(x)$ is continuous and non-vanishing at ξ^*. If the initial guess $x^{(0)}$ is close enough to ξ^*, the iterative scheme (3) attains sixth-order convergence for each α.*

Proof. Let $e^{(k)} = x^{(k)} - \xi^*$ be the error at the kth-iteration. Now, expanding $F(x^{(k)})$ and $F'(x^{(k)})$ using Taylor's expansion in a neighborhood of ξ^*, we get

$$F(x^{(k)}) = F'(\xi^*) \left[e^{(k)} + C_2(e^{(k)})^2 + C_3(e^{(k)})^3 + C_4(e^{(k)})^4 + C_5(e^{(k)})^5 + C_6(e^{(k)})^6 \right] + O((e^{(k)})^7) \qquad (7)$$

and

$$F'(x^{(k)}) = F'(\xi^*) \left[I + 2C_2 e^{(k)} + 3C_3(e^{(k)})^2 + 4C_4(e^{(k)})^3 + 5C_5(e^{(k)})^4 + 6C_6(e^{(k)})^5 \right] + O((e^{(k)})^6), \qquad (8)$$

where I is the identity matrix of size $n \times n$ and $C_k = \frac{1}{k!} F'(\xi^*)^{-1} F^{(k)}(\xi^*)$, $k \geq 2$.

With the help of above expression (8), we have

$$F'(x^{(k)})^{-1} = \left[I - 2C_2 e^{(k)} + \Delta_0(e^{(k)})^2 + \Delta_1(e^{(k)})^3 + \Delta_2(e^{(k)})^4 + \Delta_3(e^{(k)})^5 + \Delta_4(e^{(k)})^6 + O((e^{(k)})^7) \right] F'(\xi^*)^{-1} \qquad (9)$$

where $\Delta_i = \Delta_i(C_2, C_3, \ldots, C_6)$, for example, $\Delta_0 = 4C_2^2 - 3C_3$, $\Delta_1 = -(8C_2^3 - 6C_2C_3 - 6C_3C_2 + 4C_4)$, $\Delta_2 = 8C_2C_4 + 9C_3^2 + 8C_4C_2 - 12C_2^2C_3 - 12C_2C_3C_2 - 12C_3C_2^2 + 16C_2^4 - 5C_5$, $\Delta_3 = 10C_2C_5 + 12C_3C_4 + 12C_4C_3 + 10C_5C_2 - 16C_2^2C_4 - 18C_2^2C_3^2 - 16C_2C_4C_2 - 18C_3C_2C_3 - 18C_3^2C_2 - 16C_4C_2^2 + 24C_2^3C_3 + 24C_2^2C_3C_2 + 24C_2C_3C_2^2 + 24C_3C_2^3 - 32C_2^5 - 6C_6$, etc.

From expressions (7) and (9), we yield

$$F'(x^{(k)})^{-1} F(x^{(k)}) = e^{(k)} + \Theta_0(e^{(k)})^2 + \Theta_1(e^{(k)})^3 + \Theta_2(e^{(k)})^4 + \Theta_3(e^{(k)})^5 + \Theta_4(e^{(k)})^6 + O((e^{(k)})^7). \qquad (10)$$

where $\Theta_j = \Theta_j(C_2, C_3, \ldots, C_6)$, for example, $\Theta_0 = -C_2$, $\Theta_1 = 2C_2^2 - 2C_3$, $\Theta_2 = -(4C_2^3 - 4C_2C_3 - 3C_3C_2 + 3C_4)$, $\Theta_3 = 6C_2C_4 + 6C_3^2 + 4C_4C_2 - 8C_2^2C_3 - 6C_2C_3C_2 - 6C_3C_2^2 + 8C_2^4 - 4C_5$, $\Theta_4 = 8C_2C_5 + 9C_3C_4 + 8C_4C_3 + 5C_5C_2 - 12C_2^2C_4 - 12C_2C_3^2 - 8C_2C_4C_2 - 12C_3C_2C_3 - 9C_3^2C_2 - 8C_4C_2^2 + 16C_2^3C_3 + 12C_2^2C_3C_2 + 12C_2C_3C_2^2 + 12C_3C_2^3 - 16C_2^5 - 5C_6$, etc.

By inserting the expression (10) in the first substep of (3), we obtain

$$y^{(k)} - \xi^* = -\Theta_0(e^{(k)})^2 - \Theta_1(e^{(k)})^3 - \Theta_2(e^{(k)})^4 - \Theta_3(e^{(k)})^5 - \Theta_4(e^{(k)})^6 + O((e^{(k)})^7). \qquad (11)$$

which further produces

$$F(y^{(k)}) = F'(\xi^*) \left[-\Theta_0(e^{(k)})^2 - \Theta_1(e^{(k)})^3 + (C_2\Theta_0^2 - \Theta_2)(e^{(k)})^4 + (2C_2\Theta_0\Theta_1 - \Theta_3)(e^{(k)})^5 + \right.$$
$$\left. - \left(C_3\Theta_0^3 - C_2(\Theta_1^2 + 2\Theta_0\Theta_2) + \Theta_4 \right)(e^{(k)})^6 + O((e^{(k)})^7) \right] \qquad (12)$$

and

$$U^{(k)} = I - F'(x^{(k)})^{-1}[x^{(k)}, y^{(k)}; F] = C_2 e^{(k)} + (C_2\Theta_0 - 2C_2^2 + 2C_3)(e^{(k)})^2 + \left(-2C_2^2\Theta_0 + C_2(\Theta_1 - 7C_3) + C_3\Theta_0 + 4C_2^3 \right.$$
$$+ 3C_4 \big)(e^{(k)})^3 + \left(4C_2^3\Theta_0 + C_2^2(20C_3 - 2\Theta_1) + C_2(-5C_3\Theta_0 - 10C_4 + \Theta_2) + C_4\Theta_0 \right.$$
$$+ C_3(\Theta_1 - \Theta_0^2) - 8C_2^4 - 6C_3^2 + 4C_5 \big)(e^{(k)})^4 + \left[-8C_2^4\Theta_0 + C_2^3(4\Theta_1 - 52C_3) \right.$$
$$+ 2C_2^2(8C_3\Theta_0 + 14C_4 - \Theta_2) + C_2 \left(C_3(2\Theta_0^2 - 5\Theta_1) - 6C_4\Theta_0 + 33C_3^2 - 13C_5 + \Theta_3 \right)$$
$$- C_4\Theta_0^2 - 3C_3^2\Theta_0 + C_5\Theta_0 + C_4\Theta_1 + C_3(-17C_4 - 2\Theta_0\Theta_1 + \Theta_2) + 16C_2^5 + 5C_6 \big](e^{(k)})^5$$
$$+ O((e^{(k)})^6). \qquad (13)$$

By using expressions (9), (12), and (13), we obtain

$$\left(I + (\alpha - 2)U^{(k)} \right)^{-1} \left(I + \alpha U^{(k)} \right) F'(x^{(k)})^{-1} F(y^{(k)}) = -\Theta_0(e^{(k)})^2 - \Theta_1(e^{(k)})^3 + (2\alpha C_2^2\Theta_0 - C_2\Theta_0^2$$
$$- C_3\Theta_0 - \Theta_2)(e^{(k)})^4 + O((e^{(k)})^5) \qquad (14)$$

By adopting expressions (11)–(14) in the scheme (3), we have

$$z^{(k)} - \xi^* = \tau_1(e^{(k)})^4 + \tau_2(e^{(k)})^5 + \tau_3(e^{(k)})^6 + O((e^{(k)})^7). \tag{15}$$

where $\tau_j = \tau_j(\Theta_0, \Theta_1, \ldots, \Theta_4, C_2, C_3, \ldots, C_6, \alpha), j = 1, 2, 3$, for example, $\tau_1 = \Theta_0(-2\alpha C_2^2 + C_2\Theta_0 + C_3)$. Now, expanding the $F(z^{(k)})$ in a neighborhood of ξ^*, we have

$$F(z^{(k)}) = F'(\xi^*)\left[\tau_1(e^{(k)})^4 + \tau_2(e^{(k)})^5 + \tau_3(e^{(k)})^6 + O((e^{(k)})^7)\right], \tag{16}$$

which further produces with the help of expression (12)

$$[z^{(k)}, y^{(k)}; F]^{-1}F(z^{(k)}) = \tau_1(e^{(k)})^4 + \tau_2(e^{(k)})^5 + (C_2\Theta_0\tau_1 + \tau_3)(e^{(k)})^6 + O((e^{(k)})^7). \tag{17}$$

By using equation (17) in the final substep of (3), we have

$$\begin{aligned} e^{(j+1)} &= \left(\tau_1(e^{(k)})^4 + \tau_2(e^{(k)})^5 + \tau_3(e^{(k)})^6\right) - \left(\tau_1(e^{(k)})^4 + \tau_2(e^{(k)})^5 + (C_2\Theta_0\tau_1 + \tau_3)(e^{(k)})^6\right) + O((e^{(k)})^7) \\ &= -(C_2\Theta_0\tau_1)(e^{(k)})^6 + O((e^{(k)})^7) \\ &= C_2^3((2\alpha+1)c_2^2 - c_3)(e^{(k)})^6 + O((e^{(k)})^7). \end{aligned} \tag{18}$$

Therefore, the scheme (3) has sixth-order convergence. $\square$

3. Numerical Experiments

Here, we checked the efficiency and effectiveness of our scheme on real-life and standard academic test problems. Therefore, we consider five number of the examples' details, as seen in the examples (1)–(5). Further, we also depicted the starting points and zeros of the considered nonlinear system in the examples (1)–(5). Now, we employ our sixth-order scheme (3), called (PM), to verify the computational performance of them with existing methods considered in the previous section.

Now, we compare (3) with a sixth-order family proposed by Abbasbandy et al. [26] and Hueso et al. [27]. We choose their best expressions (8) and (14–15) (for $t_1 = -\frac{9}{4}$ and $s_2 = \frac{9}{8}$), respectively denoted by (AM_6) and (HM_6). Moreover, a comparison of a newly proposed scheme has been done with the sixth-order family of iterative method proposed by Sharma and Arora [28] and Wang and Li [29], out these works we choose two methods, namely, (13) and (6), respectively, called (SM_6) and (WM_6). Finally, we compare (3) with sixth-order methods suggested by Mona et al. [15] and Lotfi et al. [16], we consider methods (3.1) (for $\lambda = 1$, $\beta = 2$, $p = 1$ and $q = \frac{3}{2}$) and (5), respectively, called by (MM_6) and (LM_6). All the iterative expressions are mentioned below.

Method AM_6:

$$y^{(k)} = x^{(k)} - \frac{2}{3}F'(x^{(k)})^{-1}F(x^{(k)}),$$

$$z^{(k)} = x^{(k)} - \left[I + \frac{21}{8}F'(x^{(k)})^{-1}F'(y^{(k)}) - \frac{9}{2}(F'(x^{(k)})^{-1}F'(y^{(k)}))^2 + \frac{15}{8}(F'(x^{(k)})^{-1}F'(y^{(k)}))^3\right]F'(x^{(k)})^{-1}F(x^{(k)}), \tag{19}$$

$$x^{(k+1)} = z^{(k)} - \left[3I - \frac{5}{2}F'(x^{(k)})^{-1}F'(y^{(k)}) + \frac{1}{2}(F'(x^{(k)})^{-1}F'(y^{(k)}))^2\right]F'(x^{(k)})^{-1}F(z^{(k)}).$$

Method HM_6:

$$y^{(k)} = x^{(k)} - F'(x^{(k)})^{-1}F(x^{(k)}),$$
$$H(x^{(k)}, y^{(k)}) = F'(x^{(k)})^{-1}F(y^{(k)}),$$
$$G_s(x^{(k)}, y^{(k)}) = s_1I + s_2H(y^{(k)}, x^{(k)}) + s_3H(x^{(k)}, y^{(k)}) + s_4H(y^{(k)}, x^{(k)})^2, \tag{20}$$
$$z^{(k)} = x^{(k)} - G_s(x^{(k)}, y^{(k)})F'(x^{(k)})^{-1}F(x^{(k)}),$$
$$x^{(k+1)} = z^{(k)}.$$

where s_1 s_2, s_3, and s_4 are real numbers.

Method SM_6:

$$y^{(k)} = x^{(k)} - \frac{2}{3}F'(x^{(k)})^{-1}F(x^{(k)}),$$
$$z^{(k)} = \phi_4^{(4)}(x^k, y^k),$$
$$x^{(k+1)} = z^{(k)} - \left[sI + tF'(x^{(k)})^{-1}F'(y^{(k)})\right]F'(x^{(k)})^{-1}F(z^{(k)}),$$

(21)

where s and t are real parameters.

Method WM_6:

$$y^{(k)} = x^{(k)} - F'(x^{(k)})^{-1}F(x^{(k)}),$$
$$z^{(k)} = y^{(k)} - \left[2I - F'(x^{(k)})^{-1}F'(y^{(k)})\right]F'(x^{(k)})^{-1}F(y^{(k)}),$$
$$x^{(k+1)} = z^{(k)} - \left[2I - F'(x^{(k)})^{-1}F'(y^{(k)})\right]F'(x^{(k)})^{-1}F(z^{(k)}).$$

(22)

Method MM_6:

$$y^{(k)} = x^{(k)} - \frac{2}{3}F'(x^{(k)})^{-1}F(x^{(k)}),$$
$$z^{(k)} = x^{(k)} - \left[I + \frac{1}{2\beta}\left(I - \frac{\lambda}{\beta}G(x^{(k)})\right)H(G(x^{(k)}))u(x^{(k)})\right],$$
$$x^{(k+1)} = z^{(k)} - \left[pI + qG(x^{(k)})\right]F'(x^{(k)})^{-1}F(z^{(k)}),$$

(23)

where λ, β, p, and q are real numbers.

Method LM_6:

$$y^{(k)} = x^{(k)} - F'(x^{(k)})^{-1}F(x^{(k)}),$$
$$z^{(k)} = x^{(k)} - 2(F'(x^{(k)} + F'(y^{(k)})^{-1}F(x^{(k)}),$$
$$x^{(k+1)} = z^{(k)} - \left[\frac{7}{2}I - 4F'(x^{(k)})^{-1}F'(y^{(k)} + \frac{3}{2}F'(x^{(k)})^{-1}F'(y^{(k)})^2\right]F'(x^{(k)})^{-1}F(z^{(k)}).$$

(24)

The abscissas t_j and the weights w_j are known and depicted in the Table 1 when $t = 8$(for the more details please have a look at Example 1). In Tables 2–6, we mention the number of iteration indexes (n), residual errors $(\|F(x^{(k)})\|)$, error $\|x^{(k+1)} - x^{(k)}\|$ and computational convergence order $\rho^* \approx \dfrac{\log\left[\|x^{(k+1)} - x^{(k)}\| / \|x^{(k)} - x^{(k-1)}\|\right]}{\log\left[\|x^{(k)} - x^{(k-1)}\| / \|x^{(k-1)} - x^{(k-2)}\|\right]}$. In addition, the value of η is the last calculated value of $\dfrac{\|x^{(k+1)} - x^{(k)}\|}{\|x^{(k)} - x^{(k-1)}\|^6}$.
Finally, the comparison on the basis of number of iterations taken by different methods on numerical Examples 1–5 is also depicted in Table 7.

All the computations have been done with multiple precision arithmetic with 1000 digits of mantissa, which minimize round-off errors in Mathematica-9. Here, a $(\pm b)$ is $a \times 10^{(\pm b)}$ in all the tables. The stopping criteria for the programming is defined as follows,

$$\text{(i) } \|F(x^{(k)})\| < 10^{-100} \text{ and (ii) } \|x^{(k+1)} - x^{(k)}\| < 10^{-100}.$$

Example 1. *Let us consider the Hammerstein integral equation (see [1], pp. 19–20) given as follows,*

$$y(s) = 1 + \frac{1}{5}\int_0^1 F(s,t)y(t)^3dt,$$

where $y \in C[0,1]$, $s, t \in [0,1]$, and the kernel F is

$$F(s,t) = \begin{cases} (1-s)t, t \le s, \\ s(1-t), s \le t. \end{cases}$$

Now, using the Gauss Legendre formula, we transform the above equation into a finite-dimensional problem, which is given as $\int_0^1 f(t)dt \simeq \sum_{j=1}^{8} w_j f(t_j)$, where t_j and w_j denote abscissas and weights, respectively. Now $t_j^{'s}$ and $w_j^{'s}$ are determined for $t = 8$ by Gauss Legendre quadrature formula. Let us call $y(t_i)$ by $y_i(i = 1, 2, \ldots, 8)$, then we get the following nonlinear system,

$$5y_i - 5 - \sum_{j=1}^{8} a_{ij} y_j^3 = 0, \text{ where } i = 1, 2, \ldots, 8,$$

where $a_{ij} = \begin{cases} w_j t_j (1 - t_i), j \leq i, \\ w_j t_i (1 - t_j), i < j. \end{cases}$

Here, the abscissas t_j and the weights w_j are known and depicted in the Table 1 when $t = 8$.
The convergence of the methods towards the root

$$\zeta^* = (1.00209\ldots, 1.00990\ldots, 1.01972\ldots, 1.02643\ldots, 1.02643\ldots, 1.01972\ldots, 1.00990\ldots, 1.00209\ldots)^T,$$

is tested in the Table 2 on the choice of the initial guess $x^{(0)} = \left(\frac{1}{2}, \frac{1}{2}, \frac{1}{2}, \frac{1}{2}, \frac{1}{2}, \frac{1}{2}, \frac{1}{2}, \frac{1}{2}\right)^T$. The numerical results show that the proposed methods $PM1_6$ and $PM2_6$ have better residual errors in comparison with the existing ones. In addition, smaller asymptotic error constants also belong to our methods $PM1_6$ and $PM2_6$.

Table 1. $t_j^{'s}$ and $w_j^{'s}$ of Gauss–Legendre formula for $t = 8$.

j	t_j	w_j
1	0.01985507175123188415821957...	0.05061426814518812957626567...
2	0.10166676129318663020422303...	0.11119051722668723527217800...
3	0.23723379504183550709113047...	0.15685533229389436436898110...
4	0.40828267875217509753026193...	0.18134189168918099148257522...
5	0.59171732124782490246973807...	0.18134189168918099148257522...
6	0.76276620495816449290886952...	0.15685533229389436436898110...
7	0.89833323870681336979577696...	0.11119051722668723527217800...
8	0.98014492824876811584178043...	0.05061426814518812957626567...

Table 2. Comparison of methods on Hammerstein integral equation in Example 1.

Methods	k	$\|F(x^{(k)})\|$	$\|x^{(k+1)} - x^{(k)}\|$	ρ^*	$\frac{\|x^{(k+1)} - x^{(k)}\|}{\|x^{(k)} - x^{(k-1)}\|^6}$
	1	1.1(−5)	2.4(−6)		
AM_6	2	5.4(−39)	1.2(−39)		6.596956919(−6)
	3	8.0(−239)	1.7(−239)	5.9991	7.072478176(−6)
	1	3.0(−5)	6.5(−6)		
HM_6	2	6.9(−31)	1.5(−31)		1.994799598
	3	4.4(−159)	9.4(−160)	4.9991	9.220736175(+25)
	1	9.4(−6)	2.0(−6)		
SM_6	2	1.5(−39)	3.2(−40)		4.964844066(−6)
	3	2.7(−242)	5.7(−243)	5.9991	5.324312398(−6)

Table 2. *Cont.*

Methods	k	$\|F(x^{(k)})\|$	$\|x^{(k+1)} - x^{(k)}\|$	ρ^*	$\frac{\|x^{(k+1)} - x^{(k)}\|}{\|x^{(k)} - x^{(k-1)}\|^6}$
WM_6	1	1.6(−6)	3.3(−7)		
	2	9.0(−45)	1.9(−45)		1.377597868(−6)
	3	3.3(−274)	7.1(−275)	5.9996	1.431074607(−6)
MM_6	1	3.9(−6)	8.2(−7)		
	2	5.8(−36)	1.2(−36)		3.943508142
	3	4.6(−185)	9.9(−186)	4.9992	2.773373608(+30)
LM_6	1	8.6(−6)	1.8(−6)		
	2	8.4(−37)	1.7(−37)		4.445532921(−3)
	3	1.4(−189)	3.0(−190)	4.9224	1.232404905(+31)
$PM1_6$	1	1.7(−7)	1.5(−7)		
	2	1.9(−47)	4.1(−48)		3.291248720(−7)
	3	7.6(−291)	1.6(−291)	5.9998	3.358306951(−7)
$PM2_6$	1	6.6(−7)	1.4(−7)		
	2	1.1(−47)	2.3(−48)		2.820724919(−7)
	3	2.0(−292)	4.3(−293)	5.9998	2.880288646(−7)

Example 2. *Let us consider the Van der Pol equation [30], which governs the flow of current in a vacuum tube, defined as follows,*

$$y'' - \mu(y^2 - 1)y' + y = 0, \ \mu > 0. \tag{25}$$

Here, boundary conditions are given by $y(0) = 0$, $y(2) = 1$. Further, we divide the given interval $[0, 2]$ as follows,

$$x_0 = 0 < x_1 < x_2 < x_3 < \cdots < x_n, \text{ where } x_i = x_0 + ih, \ h = \frac{2}{n}.$$

Moreover, we assume that

$$y_i = y(x_i), \quad i = 0, 1, \ldots, n.$$

Now, if we discretize the above problem (25) by using the finite-difference formula for the first and second derivatives, which are given by

$$y_k' = \frac{y_{k+1} - y_{k-1}}{2h}, \ y_k'' = \frac{y_{k-1} - 2y_k + y_{k+1}}{h^2}, \ k = 1, 2, \ldots, n-1,$$

then, we obtain a SNEs of order $(n-1) \times (n-1)$.

$$2h^2 y_k - h\mu \left(y_k^2 - 1\right)(y_{k+1} - y_{k-1}) + 2(y_{k-1} + y_{k+1} - 2y_k) = 0, \ k = 1, 2, \ldots, n-1.$$

Let us consider $\mu = \frac{1}{2}$ and initial approximation $y_k^{(0)} = \log\left(\frac{1}{k^2}\right)$, $k = 1, 2, \ldots, n-1$. In this example, we solve 9×9 SNEs by taking $n = 10$. The solution of this problem is

$$\xi^* = (-0.4795\ldots, -0.9050\ldots, -1.287\ldots, -1.641\ldots, -1.990\ldots, -2.366\ldots, -2.845\ldots, -3.673\ldots, -6.867\ldots,)^T.$$

The numerical results are displayed in Table 3. It is found that the newly proposed methods perform better in all aspects, whereas the existing methods show larger residual errors.

Table 3. Comparisons of methods on Van der Pol equation in Example 2.

Methods	k	$\|F(x^{(k)})\|$	$\|x^{(k+1)} - x^{(k)}\|$	ρ^*	$\frac{\|x^{(k+1)} - x^{(k)}\|}{\|x^{(k)} - x^{(k-1)}\|^6}$
	1	9.8(+16)	1.2(+6)		
AM_6	2	9.3(+15)	5.3(+5)		2.207185393(−31)
	3	8.9(+14)	2.4(+5)	1.0000	1.115653208(−29)
	1	2.0(+2)	3.4(+1)		
HM_6	2	2.6(+0)	1.4(+3)		8.575063031(−7)
	3	6.9(+7)	8.5(+2)	0.13787	1.044481693(−16)
	1	4.7(+10)	9.4(+3)		
SM_6	2	3.9(+9)	4.1(+3)		6.141757152(−21)
	3	3.3(+8)	1.8(+3)	0.99993	3.757585542(−19)
	1	3.8(+10)	8.7(+3)		
WM_6	2	3.3(+9)	3.9(+3)		8.901990983(−21)
	3	2.9(+8)	1.7(+3)	0.99992	5.216543771(−19)
	1	1.8(+9)	3.2(+3)		
MM_6	2	1.4(+8)	1.4(+3)		1.233926627(−18)
	3	1.1(+7)	6.0(+2)	0.99947	8.435924608(−17)
	1	7.8(+3)	5.1(+1)		
LM_6	2	6.0(+2)	1.9(+1)		1.090110495(−9)
	3	3.9(+1)	5.0(0)	1.3904	9.718708834(−8)
	1	2.6(0)	7.2(−1)		
$PM1_6$	2	1.5(−4)	2.8(−5)		2.099986524(−4)
	3	6.7(−27)	1.6(−27)	5.0546	3.049589936
	1	2.6(−1)	1.2(−1)		
$PM2_6$	2	8.3(−9)	1.5(−9)		4.804752944(−4)
	3	2.8(−48)	7.4(−49)	4.9654	7.413843150(+4)

Example 3. *The 2D Bratu problem [31,32] is defined as*

$$u_{xx} + u_{tt} + Ce^u = 0, on$$
$$\Omega : (x,t) \in 0 \le x \le 1, 0 \le t \le 1, \quad (26)$$
$$with\ boundary\ conditions\ u = 0\ on\ \Omega.$$

Using finite difference discretization, a given nonlinear PDE can be reduced to a SNEs. Let $\Theta_{i,j} = u(x_i, t_j)$ be the numerical solution at the grid points of the mesh. Let M_1 and M_2 be the number of steps in x and t directions, respectively, and h and k be the respective step sizes. To solve the given PDE, we apply the central difference formula to u_{xx} and u_{tt} in the following way,

$$u_{xx}(x_i, t_j) = \frac{\Theta_{i+1,j} - 2\Theta_{i,j} + \Theta_{i-1,j}}{h^2}, \quad C = 0.1, \ t \in [0,1] \quad (27)$$

By using expression (27) in (26) and after some simplification, we have

$$\Theta_{i,j+1} + \Theta_{i,j-1} - \Theta_{i,j} + \Theta_{i+1,j} + \Theta_{i-1,j} + h^2 C \exp\left(\Theta_{i,j}\right) \quad i = 1,2,3,\ldots,M_1, j = 1,2,3,\ldots,M_2 \quad (28)$$

By choosing $M_1 = M_2 = 11$, $C = 0.1$, and $h = \frac{1}{11}$, we obtain a system of nonlinear equations of size 10×10, with the initial guess $x_0 = 0.1(\sin(\pi h)\sin(\pi k), \sin(2\pi h)\sin(2\pi k),\ldots,\sin(10\pi h)\sin(10\pi k))^T$. Numerical estimations are given in Table 4. Numerical results demonstrate that the new methods have much improved error estimations and computational order of convergence in comparison to its competitors.

Table 4. Comparisons of different methods on 2D Bratu problem in Example 3.

Methods	k	$\|F(x^{(k)})\|$	$\|x^{(k+1)} - x^{(k)}\|$	ρ^*	$\frac{\|x^{(k+1)}-x^{(k)}\|}{\|x^{(k)}-x^{(k-1)}\|^6}$
AM_6	1	4.4(−15)	2.4(−14)		
	2	6.9(−95)	3.5(−94)		1.428095547(−12)
	3	7.9(−574)	3.9(−573)	5.9994	1.973434769(−12)
HM_6	1	2.1(−13)	1.2(−12)		
	2	2.1(−71)	1.2(−70)		7.368055345(−11)
	3	1.7(−361)	9.3(−361)	4.9997	3.495510769(+1)
SM_6	1	4.4(−15)	2.4(−14)		
	2	7.1(−95)	3.6(−94)		1.433541371(−12)
	3	9.2(−574)	4.5(−573)	5.9994	1.433541371(−12)
WM_6	1	8.1(−2)	5.0(−1)		
	2	5.0(−19)	2.9(−18)		1.754949400(−16)
	3	1.7(−122)	1.0(−121)	5.9999	1.666475363(−16)
MM_6	1	7.5(−15)	4.1(−14)		
	2	1.2(−80)	6.4(−80)		1.375781477(+1)
	3	1.2(−409)	6.3(−409)	4.9997	9.148606524(+66)
LM_6	1	6.4(−16)	2.5(−15)		
	2	2.9(−87)	5.3(−87)		1.447342836(−13)
	3	9.1(−445)	3.3(−444)	4.9853	2.808772449(+1)
$PM1_6$	1	5.7(−18)	7.0(−18)		
	2	4.9(−117)	5.9(−117)		5.229619528(−14)
	3	1.9(−711)	2.3(−711)	5.9999	5.367043263(−14)
$PM2_6$	1	5.7(−18)	6.9(−18)		
	2	4.7(−117)	5.6(−117)		1.698178527(−6)
	3	1.4(−711)	1.6(−711)	5.9999	1.854013522(−6)

Example 4. *Consider another typical nonlinear problem, that is, Fisher's equation [33] with homogeneous Neumann's BC's, the diffusion coefficient H is*

$$
\begin{aligned}
u_t &= H u_{xx} + u(1 - u) = 0, \\
u(x,0) &= 1.5 + 0.5\cos(\pi x), \quad 0 \le x \le 1, \\
u_x(0,t) &= 0, \quad \forall\, t \ge 0, \\
u_x(1,t) &= 0, \quad \forall\, t \ge 0.
\end{aligned}
\tag{29}
$$

Again using finite difference discretization, the equation (29) reduces to a SNEs. Consider $\Theta_{i,j} = u(x_i, t_j)$ be its approximate solution at the grid points of the mesh. Let M_1 and M_2 be the number of steps in x and t directions, and h and k be the respective step size. Applying central difference formula to u_{xx}, backward difference for $u_t(x_i, t_j)$, and forward difference for $u_x(x_i, t_j)$, respectively, in the following way,

$$
\begin{aligned}
u_{xx}(x_i, t_j) &= \frac{\Theta_{i+1,j} - 2\Theta_{i,j} + \Theta_{i-1,j}}{h^2}, \\
u_t(x_i, t_j) &= \frac{\Theta_{i,j} - \Theta_{i,j-1}}{k} \\
\text{and} \\
u_x(x_i, t_j) &= \frac{\Theta_{i+1,j} - \Theta_{i,j}}{h},
\end{aligned}
\tag{30}
$$

where $h = \frac{1}{M_1}, k = \frac{1}{M_2}, t \in [0,1]$.

By adopting expression (30) *in* (29), *after some simplification, we get*

$$\frac{\Theta_{1,j} - \Theta_{i,j-1}}{k} - \Theta_{i,j}\left(1 - \Theta_{i,j}\right) - H\frac{\Theta_{i+1,j} - 2\Theta_{i,j} + \Theta_{i-1,j}}{h^2}, \quad i = 1, 2, 3, \ldots, M_1, j = 1, 2, 3, \ldots, M_2 \quad (31)$$

For the system of nonlinear equations, we considered $M_1 = M_2 = 5$, $h = \frac{1}{5}$, $k = \frac{1}{5}$ *and* $H = 1$, *which reduces to nonlinear system of size* 5×5, *with the initial guess* $x_0 = \left(1 + \frac{i}{25}\right)^T$, $i = 1, 2, \ldots, 25$. *All the numerical results are shown in Table* 5. *Numerical results show that our methods have better computational efficiency than the already existing schemes, in terms of residual errors and difference between consecutive approximations.*

Table 5. Comparisons of different methods on Fisher's equation in Example 4.

Methods	k	$\|F(x^{(k)})\|$	$\|x^{(k+1)} - x^{(k)}\|$	ρ^*	$\frac{\|x^{(k+1)} - x^{(k)}\|}{\|x^{(k)} - x^{(k-1)}\|^6}$
AM_6	1	6.9(−3)	1.5(−3)		
	2	4.2(−21)	6.5(−22)		7.021836209(−5)
	3	4.5(−131)	7.0(−132)	5.9941	9.022505886(−5)
HM_6	1	6.0(−3)	1.3(−3)		
	2	1.3(−18)	2.0(−19)		5.236602433(−2)
	3	1.8(−97)	2.8(−98)	4.9940	4.016099735(+14)
SM_6	1	4.8(−3)	1.0(−3)		
	2	2.9(−22)	4.5(−23)		4.391559871(−5)
	3	3.0(−138)	4.7(−139)	5.9945	5.610610203(−5)
WM_6	1	4.8(−3)	1.0(−3)		
	2	2.9(−22)	4.5(−23)		4.391559871(−5)
	3	3.0(−138)	4.7(−139)	5.9945	5.610610203(−5)
MM_6	1	2.7(−3)	5.5(−4)		
	2	9.5(−21)	1.5(−21)		5.507288873(−2)
	3	1.6(−108)	2.5(−109)	4.9964	2.350947999(+16)
LM_6	1	4.4(−3)	1.0(−3)		
	2	4.6(−20)	7.4(−21)		5.815894917(−3)
	3	2.2(−107)	1.2(−108)	5.1204	7.024389352(+12)
$PM1_6$	1	5.7(−18)	7.0(−18)		
	2	4.9(−117)	5.9(−117)		5.229619528(−14)
	3	1.9(−711)	2.3(−711)	5.9999	5.367043263(−14)
$PM2_6$	1	5.7(−18)	6.9(−18)		
	2	4.7(−117)	5.6(−117)		5.194329469(−14)
	3	1.4(−711)	1.6(−711)	5.9999	5.331099808(−14)

Example 5. *Let us consider the following nonlinear system,*

$$F(X) = \begin{cases} x_j^2 x_{j+1} - 1 = 0, \ 1 \le j \le n-1, \\ x_n^2 x_1 - 1 = 0. \end{cases} \quad (32)$$

To obtain a large SNEs 200×200, *we choose* $n = 200$ *and the initial approximation* $x^{(0)} = (1.25, \ 1.25, \ 1.25, \ \cdots, \ 1.25(200times))^T$ *for this problem. The required solution of this problem is* $\xi^* = (1, \ 1, \ 1, \ \cdots, \ 1(200times))^T$. *The obtained results can be observed in Table* 6. *It can be easily seen that the proposed scheme performs well; in this case, in terms of error estimates as compared to the available methods of the same nature.*

Table 6. Comparisons of different methods on Example 5.

Methods	k	$\|F(x^{(k)})\|$	$\|x^{(k+1)} - x^{(k)}\|$	ρ^*	$\frac{\|x^{(k+1)}-x^{(k)}\|}{\|x^{(k)}-x^{(k-1)}\|^6}$
	1	5.5(−2)	1.8(−2)		
AM_6	2	8.2(−15)	2.7(−15)		7.599151595(−5)
	3	9.5(−92)	3.2(−92)	5.9993	7.695242316(−5)
	1	3.0(−2)	1.0(−2)		
HM_6	2	2.0(−14)	6.7(−15)		6.625126396(−3)
	3	2.7(−75)	8.9(−76)	4.9997	9.962407479(+9)
	1	3.9(−2)	1.3(−2)		
SM_6	2	6.4(−16)	2.1(−16)		4.636076578(−5)
	3	1.3(−98)	4.3(−99)	6.0000	4.674761498(−5)
	1	4.2(−2)	1.4(−2)		
WM	2	1.2(−15)	3.9(−16)		5.254428593(−5)
	3	5.7(−97)	1.9(−97)	5.9995	5.303300859(−5)
	1	2.7(−2)	8.9(−3)		
MM_6	2	5.3(−15)	1.8(−15)		3.463613165(−3)
	3	1.6(−78)	5.4(−79)	4.9993	1.781568229(+10)
	1	3.4(−2)	1.1(−2)		
LM_6	2	2.3(−16)	7.7(−17)		3.721130070(−5)
	3	2.3(−101)	7.6(−102)	5.9996	3.746683848(−5)
	1	5.5(−3)	1.8(−3)		
$PM1_6$	2	3.4(−22)	1.1(−22)		2.944381059(−6)
	3	1.8(−137)	5.9(−138)	6.0000	2.946278255(−6)
	1	2.8(−3)	9.5(−4)		
$PM2_6$	2	2.5(−24)	8.5(−25)		1.178221976(−6)
	3	1.3(−150)	4.4(−151)	6.0000	1.178511302(−6)

Table 7. Number of iterations taken by different methods on Examples 1–5.

Methods	Ex.1	Ex.2	Ex.3	Ex.4	Ex. 5	Total	Average
AM_6	3	20	3	3	4	33	6.6
HM_6	3	D	3	3	4	13 *	3.25 *
SM_6	3	13	3	3	4	26	5.2
WM_6	3	13	2	3	4	25	5
MM_6	3	12	3	3	4	25	5
LM_6	3	7	3	3	3	19	3.8
$PM1_6$	3	4	2	2	3	14	2.8
$PM2_6$	3	4	2	2	3	14	2.8

* means, the total number of iteration calculated on all examples except Example 2, because HM_6 is divergent in Example 2.

4. Conclusions

In this work, we have developed new family of sixth-order iterative methods for solving SNEs, numerically. To check their effectiveness, the proposed scheme is applied on some large-scale systems arising from various academic problems. Further, the numerical results show that the proposed methods perform better than already existing schemes in the scientific literature.

Author Contributions: R.B. and M.K.: Conceptualization; Methodology; Validation; Writing–Original Draft Preparation; Writing–Review & Editing. H.F.A.: Review & Editing.

Funding: Deanship of Scientific Research (DSR), King Abdulaziz University, Jeddah, under grant No. G-1349-665-1440.

Acknowledgments: This project was supported by the Deanship of Scientific Research (DSR), King Abdulaziz University, Jeddah, under grant No. G-1349-665-1440. The authors, therefore, gratefully acknowledge with thanks DSR for technical and financial support.

Conflicts of Interest: The authors declare no conflicts of interest.

References

1. Ortega J.M.; Rheinboldt, W.C. *Iterative Solution of Nonlinear Equations in Several Variables*; Academic Press: New York, NY, USA, 1970.
2. Cordero, A.; Torregrosa, J.R. Variants of Newton's method for functions of several variables. *Appl. Math. Comput.* **2006**, *183*, 199–208. [CrossRef]
3. Frontini M.; Sormani, E. Third-order methods from quadrature formulae for solving systems of nonlinear equations. *Appl. Math. Comput.* **2004**, *149*, 771–782. [CrossRef]
4. Grau-Sánchez, M.; Grau, À; Noguera, M. On the computational efficiency index and some iterative methods for solving systems of non-linear equations. *Comput. Appl. Math.* **2011**, *236*, 1259–1266.
5. Homeier, H.H.H. A modified Newton method with cubic convergence: The multivariable case. *Comput. Appl. Math.* **2004**, *169*, 161–169. [CrossRef]
6. Cordero, A.; Torregrosa, J.R. Variants of Newton's method using fifth-order quadrature formulas. *Appl. Math. Comput.* **2007**, *190*, 686–698. [CrossRef]
7. Darvishi, M.T.; Barati, A. Super cubic iterative methods to solve systems of nonlinear equations. *Appl. Math. Comput.* **2007**, *188*, 1678–1685. [CrossRef]
8. Darvishi, M.T.; Barati, A. A third-order Newton-type method to solve systems of non-linear equations. *Appl. Math. Comput.* **2007**, *187*, 630–635.
9. Potra, F.A.; Pták, V. *Nondiscrete Induction and Iterarive Processes*; Pitman Publishing: Boston, MA, USA, 1984.
10. Babajee, D.K.R.; Madhu, K.; Jayaraman, J. On some improved harmonic mean Newton-like methods for solving systems of nonlinear equations. *Algorithms* **2015**, *8*, 895–909. [CrossRef]
11. Cordero, A.; Martínez, E.; Torregrosa, J.R. Iterative methods of order four and five for systems of nonlinear equations. *Comput. Appl. Math.* **2009**, *231*, 541–551. [CrossRef]
12. Cordero, A.; Hueso, J.L.; Martínez, E.; Torregrosa, J.R. Efficient high-order methods based on golden ratio for nonlinear systems. *Appl. Math. Comput.* **2011**, *217*, 4548–4556. [CrossRef]
13. Arroyo, V.; Cordero, A.; Torregrosa, J.R. Approximation of artificial satellites' preliminary orbits: The efficiency challenge. *Math. Comput. Modell.* **2011**, *54*, 1802–1807. [CrossRef]
14. Alzahrani, A.K.H.; Behl, R.; Alshomrani, A. Some higher-order iteration functions for solving nonlinear models, *Appl. Math. Comput.* **2018**, *334*, 80–93. [CrossRef]
15. Narang, M.; Bhatia, S.; Kanwar, V. New two parameter Chebyshev-Halley like family of fourth and sixth-order methods for systems of nonlinear equations. *Appl. Math. Comput.* **2016**, *275*, 394–403. [CrossRef]
16. Lotfi, T.; Bakhtiari, P.; Cordero, A.; Mahdiani, K.; Torregrosa, J.R. Some new efficient multipoint iterative methods for solving nonlinear systems of equations. *Int. J. Comput. Math.* **2015**, *92*, 1921–1934. [CrossRef]
17. Babajee, D.K.R. On a two-parameter Chebyshev-Halley like family of optimal two-point fourth order methods free from second derivatives. *Afrika Matematika* **2015**, *26*, 689–695. [CrossRef]
18. Campos, B.; Cordero, A.; Torregrosa, J.R.; Vindel, P. Stability of King's family of iterative methods with memory. *J. Comput. Appl. Math.* **2017**, *318*, 504–514. [CrossRef]
19. Cordero, A.; Maimó, J.G.; Torregrosa, J.R.; Vassileva, M.P.; Vindel, P. Chaos in King's iterative family. *Appl. Math. Lett.* **2013**, *26*, 842–848. [CrossRef]
20. Chicharro, F.; Cordero, A.; Torregrosa, J.R. Dynamics of iterative families with memory based on weight functions procedure. *J. Comput. Appl. Math.* **2019**, *354*, 286–298. [CrossRef]
21. King, R.F. A family of fourth-order methods for nonlinear equations. *SIAM J. Numer. Anal.* **1973**, *10*, 876–879. [CrossRef]
22. Argyros, I.K. *Convergence and Application of Newton-Type Iterations*; Springer: Berlin/Heidelberg, Germany, 2008.
23. Petković, M.S.; Neta, B.; Petković, L.D.; Džunić, J. *Multipoint Methods for Solving Nonlinear Equations*; Academic Press: Amsterdam, The Netherlands, 2012.
24. Traub, J.F. *Iterative Methods for the Solution of Equations*; Prentice-Hall: Englewood Cliffs, NJ, USA, 1964.

25. Cordero, A.; Hueso, J.L.; Martínez, E.; Torregrosa, J.R. A modified Newton-Jarratt's composition. *Numer. Algor.* **2010**, *55*, 87–99. [CrossRef]
26. Abbasbandy, S.; Bakhtiari, P.; Cordero, A.; Torregrosa, J.R.; Lotfi, T. New efficient methods for solving nonlinear systems of equations with arbitrary even order. *Appl. Math. Comput.* **2016**, *287*, 287–288. [CrossRef]
27. Hueso, J.L.; Martínez, E.; Teruel, C. Convergence, efficiency and dynamics of new fourth and sixth order families of iterative methods for nonlinear systems. *J. Comput. Appl. Math.* **2015**, *275*, 412–420. [CrossRef]
28. Sharma, J.R.; Arora, H. Efficient Jarratt-like methods for solving systems of nonlinear equations. *Calcolo* **2014**, *51*, 193–210. [CrossRef]
29. Wang, X.; Li, Y. An Efficient Sixth Order Newton Type Method for Solving Nonlinear Systems. *Algorithms* **2017**, *10*, 45. [CrossRef]
30. Burden, R.L.; Faires, J.D. *Numerical Analysis*; PWS Publishing Company: Boston, MA, USA, 2001.
31. Simpson, R.B. A method for the numerical determination of bifurcation states of nonlinear systems of equations. *SIAM J. Numer. Anal.* **1975**, *12*, 439–451. [CrossRef]
32. Kapania, R.K. A pseudo-spectral solution of 2-parameter Bratu's equation. *Comput. Mech.* **1990**, *6*, 55–63. [CrossRef]
33. Sauer, T. *Numerical Analysis*, 2nd ed.; Pearson: Upper Saddle River, NJ, USA, 2012.

mathematics

MDPI

Article

A Seventh-Order Scheme for Computing the Generalized Drazin Inverse

Dilan Ahmed [1], Mudhafar Hama [2], Karwan Hama Faraj Jwamer [2] and Stanford Shateyi [3,*]

[1] Department of Mathematics, College of Education, University of Sulaimani, Kurdistan Region, Sulaimani 46001, Iraq
[2] Department of Mathematics, College of Science, University of Sulaimani, Kurdistan Region, Sulaimani 46001, Iraq
[3] Department of Mathematics, University of Venda, Private Bag X5050, Thohoyandou 0950, South Africa
[*] Correspondence: stanford.shateyi@univen.ac.za

Received: 26 May 2019; Accepted: 22 June 2019; Published: 12 July 2019

Abstract: One of the most important generalized inverses is the Drazin inverse, which is defined for square matrices having an index. The objective of this work is to investigate and present a computational tool in the form of an iterative method for computing this task. This scheme reaches the seventh rate of convergence as long as a suitable initial matrix is chosen and by employing only five matrix products per cycle. After some analytical discussions, several tests are provided to show the efficiency of the presented formulation.

Keywords: drazin inverse; generalized inverse; iterative methods; higher order; efficiency index

JEL Classification: 15A09; 65F30.

---

1. Introduction

Drazin, in the pioneering work in [1], proposed and generalized a different type of outer inverse in associative rings and semigroups that does not possess the reflexivity feature but commutes with the element. The significance of this type of inverse and its calculation was then discussed wholly in [2]. Accordingly, several authors attempted to propose procedures for the calculation of generalized inverses. See, e.g., [3–5].

It is recalled that the smallest non-negative integer k such that [6]

$$\text{rank}(A^k) = \text{rank}(A^{k+1}), \tag{1}$$

is named as the matrix A's index and is shown by $\text{ind}(A)$. Furthermore, assume that A is an $N \times N$ matrix with complex entries. The Drazin inverse of A, shown by A^D, is the unique matrix X reading the following identities [7]:

1. $A^k X A = A^k$,
2. $X A X = X$,
3. $A X = X A$.

Throughout the paper, with A^*, $\mathcal{R}(A)$, $\mathcal{N}(A)$, and $\text{rank}(A)$, we show the conjugate transpose, the range, the null space, and the rank of $A \in \mathbb{C}^{N \times N}$, respectively [8]. It is remarked that if $\text{ind}(A) = 1$, X is named as the g-inverse or group inverse of A. In addition, if A is nonsingular, then it is easily seen that

$$\text{ind} A = 0, \text{ and } A^D = A^{-1}. \tag{2}$$

For the square system $Ax = b$, the general solution can be represented using the concept of the Drazin inverse as follows [7]:

$$x = A^D b + (I - AA^D)z, \tag{3}$$

wherein $z \in \mathcal{R}(A^{k-1}) + \mathcal{N}(A)$.

Methods in the category of iteration schemes for computing generalized inverses like the Drazin inverse are quite sensitive to the choice of the initial approximation. In fact, the convergence of the Schulz-type method can only be observed if the initial value is chosen correctly [9,10]. This selection can be done under special care on some criteria, as already discussed, and given in the literature for different types of outer inverses. For some in-depth discussions about this, one may refer to [11].

Authors in [12] showed that iterative Schulz-type schemes can be used for finding the Drazin inverse of square matrices both possessing real or complex spectra. Actually, the authors investigated the initial matrix below:

$$X_0 = \alpha A^l, \ l \geq \text{ind}(A) = k, \tag{4}$$

wherein the parameter α should be selected such that the criterion $\|I - AX_0\| < 1$ is read. Employing the starting value (4) yields an iterative scheme for computing the famous Drazin inverse with second-order convergence.

It is in fact necessary to apply an appropriate initial matrix when calculating the Drazin inverse. One way is as follows [12,13]:

$$X_0 = \frac{2}{\text{Tr}(A^{k+1})} A^k, \tag{5}$$

wherein $\text{Tr}(\cdot)$ stands for the trace of an arbitrary square matrix. Another fruitful initial matrix which could lead in converging sequence of matrices for computing the generalized Drazin inverse can be written as

$$X_0 = \frac{1}{2\|A\|_2^{k+1}} A^k. \tag{6}$$

The Schulz method of the quadratic convergence rate for doing this task can be defined by [14]

$$X_{n+1} = X_n(2I - AX_n), \ n = 0, 1, 2, \cdots, \tag{7}$$

where I is the identity matrix and requires only two matrix products to achieve this rate per cycle.

Authors in [15] re-studied Chebyshev's method for calculating $A_{MN}^{\dagger}$ using a suitable initial value as follows:

$$X_{n+1} = X_n(3I - AX_n(3I - AX_n)), \ n = 0, 1, 2, \cdots, \tag{8}$$

with a third-order of convergence having three matrix products per cycle. Another scheme, having a cubic rate of convergence and a greater number of matrix products, was given by the same authors as follows:

$$X_{n+1} = X_n \left[I + \frac{1}{2}(I - AX_n)(I + (2I - AX_n)^2) \right], \ n = 0, 1, 2, \cdots. \tag{9}$$

A general procedure for having p-order methods with a p number of matrix–matrix products was given in [16] (Chapter 5). For instance, the authors presented the following fourth-order scheme:

$$X_{n+1} = X_n(I + B_n(I + B_n(I + B_n))), \ n = 0, 1, 2, \cdots, \tag{10}$$

in which $B_n = I - AX_n$.

The main goal and motivation for investigating novel or useful matrix schemes for computing the Drazin inverse is not only because of the applications of such solvers in different kinds of mathematical problems [17,18] but also to improve the computational efficiency index. In fact, the hyperpower structure as given for a sample case in (10) requires a p number of matrix–matrix products to achieve a

p rate of convergence. This leads to more and more inefficient methods from this class of methods, particularly when the order increases.

As such, motivated by extending efficient methods of higher orders for calculating generalized inverses, here we focus on a seventh-order scheme and discuss how we can reach this higher rate by employing only five matrix by matrix products. This will reveal a higher efficiency index for the discussed scheme in computing the Drazin inverse. For more studies and investigations in this field and related issues of generalized matrix inverses, interested readers are guided to [19–22].

The organization of this paper is as follows. In Section 1, we quickly review the definition, literature, and the need for development of higher order schemes. Section 2 is devoted to extending and proposing an efficient iterative expression for the Drazin inverse. It is derived that the scheme requires only five matrix–matrix products to achieve this rate.

Theoretical discussions with some concrete proofs are provided in Section 3, while Section 4 is oriented on the application of this scheme for computing the Drazin inverses. Results will reveal the effectiveness of this scheme for calculating the Drazin inverse. Lastly, some concluding comments are given in Section 5.

2. Derivation of an Efficient Formulation

Here, the aim is to present a competitive formulation for a member of the hyperpower family of iterations, so as to not only gain a high rate of convergence but also improve the computational efficiency index. In fact, we must factorize the formulation so as to gain the same high convergence rate but a lower number of matrix products.

Toward this objective, let us take into account the following seventh-order method from the family of hyperpower iteration schemes [23]:

$$X_{n+1} \;=\; X_n(I + B_n + B_n^2 + B_n^3 + B_n^4 + B_n^5 + B_n^6). \tag{11}$$

It is necessary to emphasize that we are looking for a seventh-order scheme and not a higher-order one, since we wish to hit several targets at the same time. First, the derived scheme for the Drazin inverse must be efficient, viz., it must improve the computational efficiency index of the existing solvers, as will be shown at the end of this section. Second, very high-order schemes might occasionally become hard for coding purposes, and this limits their application, so we aim to have this order be high but not so high. Besides, higher-order schemes mean that fewer stopping criteria (the computation of matrix norms) should be calculated per cycle, which is useful in terms of the elapsed time.

Now, to improve the performance of (11), we factorize (11) in what follows:

$$X_{n+1} = X_n \left(I + B_n(I + B_n)(I - B_n + B_n^2)(I + B_n + B_n^2) \right). \tag{12}$$

However, the formulation (12) needs six matrix–matrix multiplications, which is still not that useful for improving the computational index of efficiency theoretically. As such, doing more factorization would yield the following scheme:

$$X_{n+1} = X_n \left(I + (B_n + B_n^2)(I - B_n + B_n^2)(I + B_n + B_n^2) \right). \tag{13}$$

The scheme (13) requires five matrix products per cycle to achieve the seventh order of convergence. Noting that one reason for the need to propose and have an efficient higher scheme in the category of matrix Schulz-type methods is also in the fact that Schulz-type schemes of lower orders are quite slow at the initial stage of iterates, and this could yield a greater computational burden for finding the Drazin inverse [24]. In fact, it sometimes takes several iterates for the scheme to arrive at its convergence phase, and due to imposing some stopping termination based on matrix norms, this might add some more elapsed time for the application of lower-order schemes.

It is recalled that the definition of index of efficiency is given by [25]:

$$EI = \rho^{\frac{1}{\kappa}}, \tag{14}$$

wherein ρ and κ are the convergence rate and the total cost per cycle, respectively. Hence, the efficiency indexes of different methods are reported by

$$EI(7) = 2^{\frac{1}{2}} \simeq 1.41421, \ EI(8) = 3^{\frac{1}{3}} \simeq 1.44225, \tag{15}$$

$$EI(9) = 4^{\frac{1}{4}} \simeq 1.41421, \ EI(12) = 7^{\frac{1}{6}} \simeq 1.38309, \ EI(13) = 7^{\frac{1}{5}} \simeq 1.47577. \tag{16}$$

This shows that we have achieved our motivation by improving the efficiency index for calculating the Drazin inverse via a competitive formulation.

3. Seventh Rate of Convergence

Let us now recall some of the well-known lemmas we need in the rest of this section.

Proposition 1 ([26]). *Assume that $M \in \mathbb{C}^{n \times n}$ and $\varepsilon > 0$ are given. There is at least one matrix norm $\| \cdot \|$ such that*

$$\rho(M) \leq \|M\| \leq \rho(M) + \varepsilon, \tag{17}$$

wherein $\rho(M)$ shows the collection of all of M's eigenvalues (in the maximum of absolute value sense).

Proposition 2 ([26]). *If $P_{L,M}$ shows the projector on a space L on space M, then*
(i) $P_{L,M} Q = Q$ if and only if $\mathcal{R}(Q) \subseteq L$,
(ii) $Q P_{L,M} = Q$ if and only if $\mathcal{N}(Q) \supseteq M$.

The proof of the main theorem concerning the convergence as well as its rate of (13) for calculating the generalized Drazin inverse is now addressed as follows.

Theorem 1. *Consider that $A \in \mathbb{C}^{N \times N}$ is a square singular matrix. In addition, let the initial value X_0 be selected via (4) or (5). Thence, the matrices $\{X_n\}_{n=0}^{n=\infty}$ generated via (13) satisfy the following error estimate for calculating the Drazin inverse:*

$$\|A^D - X_n\| \leq \|A^D\| \|I - AX_0\|^{7^n}. \tag{18}$$

In addition, the convergence order is seven.

Proof. To prove that the sequence is converging, we first take into consideration that

$$\begin{aligned}
R_{n+1} &= I - AX_{n+1} \\
&= I - A(X_n \left(I + (B_n + B_n^2)(I - B_n + B_n^2)(I + B_n + B_n^2) \right)) \\
&= I - A(X_n \left(I + B_n(I + B_n)(I - B_n + B_n^2)(I + B_n + B_n^2) \right)) \\
&= I - A(X_n \left(I + B_n + B_n^2 + B_n^3 + B_n^4 + B_n^5 + B_n^6 \right)) \\
&= (I - AX_n)^7 \\
&= R_n^7,
\end{aligned} \tag{19}$$

wherein $R_n = I - AX_n$. Employing a matrix norm on (19), we obtain that

$$\|R_{n+1}\| \leq \|R_n\|^7. \tag{20}$$

Since X_0 is selected as in (4) or (5), we have

$$\mathcal{R}(X_0) \subseteq \mathcal{R}(A^k). \tag{21}$$

This could now be stated as

$$\mathcal{R}(X_n) \subseteq \mathcal{R}(X_{n-1}). \tag{22}$$

Thus, we can conclude that

$$\mathcal{R}(X_n) \subseteq \mathcal{R}(A^k), \; n \geq 0. \tag{23}$$

In a similar way, by defining the scheme by the left-multiplying of X_n, we can state that

$$\mathcal{N}(X_n) \supseteq \mathcal{N}(A^k), \; n \geq 0. \tag{24}$$

Now, an application of the definition of the Drazin inverse yields

$$AA^D = A^D A = P_{\mathcal{R}(A^k),\mathcal{N}(A^k)}. \tag{25}$$

Proposition 2, along with (23), (24), and (25) could lead to

$$X_n AA^D = X_n = A^D A X_n, \; n \geq 0. \tag{26}$$

To complete the proof, we proceed in what follows. The error matrix $\delta_n = A^D - X_n$ satisfies

$$\begin{aligned}
\delta_n &= A^D - X_n \\
&= A^D - A^D A X_n \\
&= A^D (I - A X_n) \\
&= A^D R_n.
\end{aligned} \tag{27}$$

Using (20), we obtain the following inequality

$$\|\delta_n\| = \|A^D\| \|R_n\| \leq \|A^D\| \|R_0\|^{7^n}, \tag{28}$$

which is an affirmation of (18). Employing (28) and Proposition 2, one gets that

$$\begin{aligned}
A\delta_{n+1} &= AA^D - A X_{n+1} \\
&= AA^D - I + I - A X_{n+1} \\
&= AA^D - I + R_{n+1}.
\end{aligned} \tag{29}$$

Note that the idempotent matrix AA^D is the projector on $\mathcal{R}(A^k)$ along $\mathcal{N}(A^k)$, where $\mathcal{R}(A^k)$ denotes the range of A^k, and $\mathcal{N}(A^k)$ is the null space of A^k. Considering (19) and applying several simplifications, one obtains

$$A\delta_{n+1} = AA^D - I + R_n^7. \tag{30}$$

Now, by taking into account the following feature,

$$(I - AA^D)^t = (I - AA^D), \; t \geq 1, \tag{31}$$

we can get that

$$\begin{aligned}
(I - AA^D)A\delta_n &= (I - AA^D)A(A^D - X_n) \\
&= X_n - AA^D V_n \\
&= 0.
\end{aligned} \tag{32}$$

We obtain, for each $t \geq 1$, that (here we use (32) in simplifications)

$$
\begin{aligned}
(R_n)^t + AA^D - I &= (I - AV_n)^t + AA^D - I \\
&= (I - AA^D + AA^D - AV_n)^t + AA^D - I \\
&= \left((I - AA^D) + A\delta_n \right)^t + AA^D - I \\
&= I - AA^D + (A\delta_n)^t + AA^D - I \\
&= (A\delta_n)^t.
\end{aligned}
\tag{33}
$$

From (33) and (30), we have

$$
A\delta_{n+1} = (A\delta_n)^7.
\tag{34}
$$

Taking matrix norms from both sides yields

$$
\|A\delta_{n+1}\| \leq \|A\delta_n\|^7.
\tag{35}
$$

Considering (35) and the second criterion of (1), we obtain that

$$
\begin{aligned}
\|\delta_{n+1}\| &= \|X_{n+1} - A^D\| \\
&= \|A^D AV_{n+1} - A^D AA^D\| \\
&= \|A^D(AV_{n+1} - AA^D)\| \\
&\leq \|A^D\|\|A\delta_{n+1}\| \\
&\leq \|A^D\|\|\delta_n\|^7.
\end{aligned}
\tag{36}
$$

The relations in (36) yield the point that $X_n \to A^D$ as $n \to +\infty$ with the seventh order of convergence. The proof is complete. $\square$

4. Computational Tests

The purpose of this section is to investigate the efficiency of our competitive formulation for computing the Drazin inverse, both theoretically and numerically. For such a task, we employ the challenging schemes (7), (8), and (13).

Here, we have simulated the tests in Mathematica 11.0, [27,28] and the time shown is in seconds. Noting that the compared methods are programmed in the same environment using the hardware CPU Intel Core i5 2430–M, 16 GB RAM in Windows 7 Ultimate with an SSD hard disk.

Test Problem 1. *The aim of this test is to testify the computation of the Drazin inverse for the following input [12]:*

$$
A = \begin{bmatrix}
2 & 0.4 & 0 & 0 & 0 & 0 & 0 & 0 & 0 & 0 & 0 & 0 \\
-2 & 0.4 & 0 & 0 & 0 & 0 & 0 & 0 & 0 & 0 & 0 & 0 \\
-1 & -1 & 1 & -1 & 0 & 0 & 0 & 0 & -1 & 0 & 0 & 0 \\
-1 & -1 & -1 & 1 & 0 & 0 & 0 & 0 & 0 & 0 & 0 & 0 \\
0 & 0 & 0 & 0 & 1 & 1 & -1 & -1 & 0 & 0 & -1 & 0 \\
0 & 0 & 0 & 0 & 1 & 1 & -1 & -1 & 0 & 0 & 0 & 0 \\
0 & 0 & 0 & -1 & -2 & 0.4 & 0 & 0 & 0 & 0 & 0 & 0 \\
0 & 0 & 0 & 0 & 2 & 0.4 & 0 & 0 & 0 & 0 & 0 & 0 \\
0 & -1 & 0 & 0 & 0 & 0 & 0 & 0 & 1 & -1 & -1 & -1 \\
0 & 0 & 0 & 0 & 0 & 0 & 0 & 0 & -1 & 1 & -1 & -1 \\
0 & 0 & 0 & 0 & 0 & 0 & 0 & 0 & 0 & 0 & 0.4 & -2 \\
0 & 0 & 0 & 0 & 0 & 0 & 0 & 0 & 0 & 0 & 0.4 & 2
\end{bmatrix},
$$

at which $k = \text{ind}(A) = 3$. Here, the Drazin inverse is expressed by

$$A^D = \begin{bmatrix}
0.25 & -0.25 & 0. & 0. & 0. & 0. & 0. & 0. & 0. & 0. & 0. & 0. \\
1.25 & 1.25 & 0. & 0. & 0. & 0. & 0. & 0. & 0. & 0. & 0. & 0. \\
-1.66406 & -0.992187 & 0.25 & -0.25 & 0. & 0. & 0. & 0. & -0.0625 & -0.0625 & 0. & 0.15625 \\
-1.19531 & -0.679687 & -0.25 & 0.25 & 0. & 0. & 0. & 0. & -0.0625 & 0.1875 & 0.6875 & 1.34375 \\
-2.76367 & -1.04492 & -1.875 & -1.25 & -1.25 & 1.25 & 1.25 & 1.25 & 1.48438 & 2.57813 & 3.32031 & 6.64063 \\
-2.76367 & -1.04492 & -1.875 & -1.25 & -1.25 & 1.25 & 1.25 & 1.25 & 1.48438 & 2.57813 & 4.57031 & 8.51563 \\
14.1094 & 6.30078 & 6.625 & 3.375 & 5. & -3. & -5. & -5. & -4.1875 & -8.5 & -10.5078 & -22.4609 \\
-19.3242 & -8.50781 & -9.75 & -5.25 & -7.5 & 4.5 & 7.5 & 7.5 & 6.375 & 12.5625 & 15.9766 & 33.7891 \\
-0.625 & -0.3125 & 0. & 0. & 0. & 0. & 0. & 0. & 0.25 & -0.25 & -0.875 & -1.625 \\
-1.25 & -0.9375 & 0. & 0. & 0. & 0. & 0. & 0. & -0.25 & 0.25 & -0.875 & -1.625 \\
0. & 0. & 0. & 0. & 0. & 0. & 0. & 0. & 0. & 0. & 1.25 & 1.25 \\
0. & 0. & 0. & 0. & 0. & 0. & 0. & 0. & 0. & 0. & -0.25 & 0.25
\end{bmatrix}.$$

The results are obtained by applying the stop termination

$$||X_{n+1} - X_n||_1 \le 10^{-6}, \tag{37}$$

and now by employing the definition in Section 1, we have for (13),

$$||A^{k+1}X_{n+1} - A^k||_\infty \simeq 3.69638 \times 10^{-12},$$
$$||X_{n+1}AX_{n+1} - X_{n+1}||_\infty \simeq 8.43992 \times 10^{-10}, \tag{38}$$
$$||AX_{n+1} - X_{n+1}A||_\infty \simeq 3.75205 \times 10^{-10}.$$

It is also necessary to mention that the domain of validity for the proposed formulation (13) is not only limited to the Drazin inverse, and if a suitable initial approximation is used, under some assumptions we can construct a converging sequence of matrix iterates for other types of generalized inverses.

Test Problem 2. *In this test, we compare the results of various schemes for computing the regular inverse using the initial matrix*

$$X_0 = \frac{1}{||A||_F} A^*, \tag{39}$$

the stopping condition (37), and the following complex matrices constructed in Mathematica:

```
N = 5000; no = 25;
ParallelTable[
  A[j] = SparseArray[
    {Band[{-100, 1100}] -> RandomReal[20], Band[{1, 1}] -> 2.,
     Band[{1000, -50}, {N - 20, N - 25}] -> {2.8, RandomReal[] + I},
     Band[{600, 150}, {N - 100, N - 400}] -> {-RandomReal[3], 3. + 3 I}
    },
    {N, N}, 0.],
  {j, no}
];
```

The plot of the large sparse matrices in Test Problem 2 is plotted in Figure 1, showing the sparsity pattern of these matrices, while the pattern of sparsity for the inverse matrix is provided in Figure 2. Figure 3 shows the clear superiority of the proposed formulation in computing the inverse of large sparse matrices.

Here, we give a simple Mathematica implementation of (13) in solving Test Problem 2:

```
For[j = 1, j <= number, j++,
  {
    X = A[j]/(Norm[A[j], "Frobenius"]^2);
    k = 1;
    X1 = 20 X;
    Time[j] = Part[
      While[k <= 75 && N[Norm[X - X1, 1]] >= 10^(-6),
        X1 = SparseArray[X];
        XX = Id - A[j].X1;
        X2 = XX.XX;
        X =
        Chop@
          SparseArray[
            X1.(Id + (XX + X2).(Id - XX + X2).(Id + XX + X2))];
        k++]; // AbsoluteTiming,
      1];
  }];
```

To apply our scheme in modern applications of numerical linear algebra getting involved with sparse large matrices, one may use some commands such as `SparseArray[]` for tackling matrices in sparse forms and subsequently reduce the computational effort and time for preserving the sparsity pattern and finding an approximate inverse. Such applications may occur in various types of problems like the ones in [29,30].

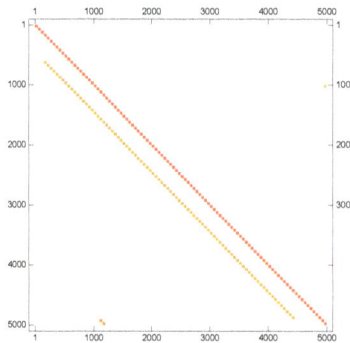

Figure 1. The sparsity pattern of matrices in Test Problem 2.

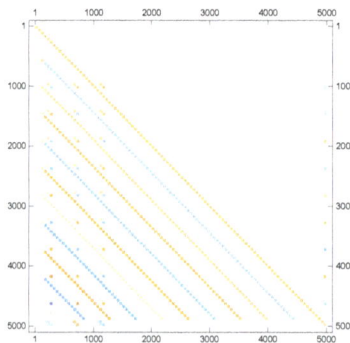

Figure 2. The sparsity pattern of the inverse matrix $X = A_{25}^{-1}$ in Test Problem 2.

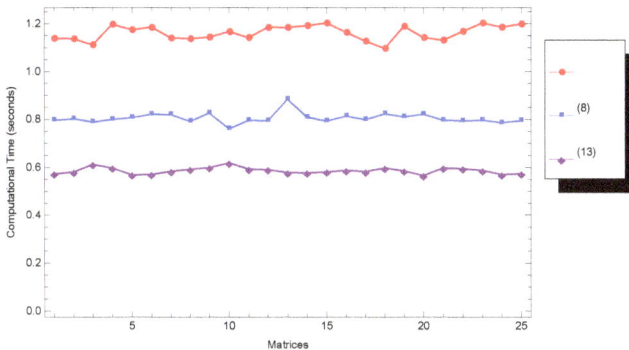

Figure 3. The CPU time required for different matrices in Test Problem 2.

5. Conclusions

Following the motivation of proposing and extending efficient iteration schemes for computing generalized inverses, and particularly for Drazin inverses, in this work we have extended and discussed theoretically how we could achieve a seventh rate in a hyperpower structure for an iterative method. The scheme is a matrix product method and employs only five products to reach this rate. Clearly, the efficiency index will hit the bound $7^{1/5} \simeq 1.47577$.

Several computational tests for calculating the Drazin inverses of several randomly generated matrices were provided to show the superiority and stability of the scheme in doing this task. Other computational problems of different sizes for different matrices were also done and showed similar behavior and the superiority of (13) for the Drazin inverse.

Author Contributions: All authors contributed equally in preparing and writing this work.

Funding: This manuscript receives no funding.

Acknowledgments: We are grateful to four anonymous referees for several comments which improved the readability of this work.

Conflicts of Interest: The authors declare no conflict of interest.

References

1. Drazin, M.P. Pseudoinverses in associative rings and semigroups. *Am. Math. Mon.* **1958**, *65*, 506–514. [CrossRef]
2. Wilkinson, J.H. *Note on the Practical Significance of the Drazin Inverse*; Campbell, S.L., Ed.; Recent Applications of Generalized Inverses, Pitman Advanced Publishing Program, Research Notes in Mathematics, No. 66, Boston; NASA: Washington, DC, USA, 1982; pp. 82–99.
3. Kyrchei, I. Explicit formulas for determinantal representations of the Drazin inverse solutions of some matrix and differential matrix equations. *Appl. Math. Comput.* **2013**, *219*, 7632–7644. [CrossRef]
4. Liu, X.; Zhu, G.; Zhou, G.; Yu, Y. An analog of the adjugate matrix for the outer inverse $A_{T,S}^{(2)}$. *Math. Prob. Eng.* **2012**, *2012*, 591256.
5. Moghani, Z.N.; Khanehgir, M.; Karizaki, M.M. Explicit solution to the operator equation $AXD + FX^*B = C$ over Hilbert C*-modules. *J. Math. Anal.* **2019**, *10*, 52–64.
6. Ben-Israel, A.; Greville, T.N.E. *Generalized Inverses: Theory and Applications*, 2nd ed.; Springer: New York, NY, USA, 2003.
7. Wei, Y. Index splitting for the Drazin inverse and the singular linear system. *Appl. Math. Comput.* **1998**, *95*, 115–124. [CrossRef]
8. Ma, H.; Li, N.; Stanimirović, P.S.; Katsikis, V.N. Perturbation theory for Moore–Penrose inverse of tensor via Einstein product. *Comput. Appl. Math.* **2019**, *38*, 111. [CrossRef]
9. Soleimani, F.; Soleymani, F.; Shateyi, S. Some iterative methods free from derivatives and their basins of attraction. *Discret. Dyn. Nat. Soc.* **2013**, *2013*, 301718. [CrossRef]

10. Soleymani, F. Efficient optimal eighth-order derivative-free methods for nonlinear equations. *Jpn. J. Ind. Appl. Math.* **2013**, *30*, 287–306. [CrossRef]

11. Pan, V.Y. *Structured Matrices and Polynomials: Unified Superfast Algorithms*; BirkhWauser: Boston, MA, USA; Springer: New York, NY, USA, 2001.

12. Li, X.; Wei, Y. Iterative methods for the Drazin inverse of a matrix with a complex spectrum. *Appl. Math. Comput.* **2004**, *147*, 855–862. [CrossRef]

13. Stanimirović, P.S.; Ciric, M.; Stojanović, I.; Gerontitis, D. Conditions for existence, representations, and computation of matrix generalized inverses. *Complexity* **2017**, *2017*, 6429725. [CrossRef]

14. Schulz, G. Iterative Berechnung der Reziproken matrix. *Z. Angew. Math. Mech.* **1933**, *13*, 57–59. [CrossRef]

15. Li, H.-B.; Huang, T.-Z.; Zhang, Y.; Liu, X.-P.; Gu, T.-X. Chebyshev-type methods and preconditioning techniques. *Appl. Math. Comput.* **2011**, *218*, 260–270. [CrossRef]

16. Krishnamurthy, E.V.; Sen, S.K. *Numerical Algorithms: Computations in Science and Engineering*; Affiliated East-West Press: New Delhi, India, 1986.

17. Ma, J.; Gao, F.; Li, Y. An efficient method to compute different types of generalized inverses based on linear transformation. *Appl. Math. Comput.* **2019**, *349*, 367–380. [CrossRef]

18. Soleymani, F.; Stanimirović, P.S.; Khaksar Haghani, F. On Hyperpower family of iterations for computing outer inverses possessing high efficiencies. *Linear Algebra Appl.* **2015**, *484*, 477–495. [CrossRef]

19. Qin, Y.; Liu, X.; Benítez, J. Some results on the symmetric representation of the generalized Drazin inverse in a Banach algebra. *Symmetry* **2019**, *11*, 105. [CrossRef]

20. Wang, G.; Wei, Y.; Qiao, S. *Generalized Inverses: Theory and Computations*; Science Press: Beijing, China; New York, NY, USA, 2004.

21. Xiong, Z.; Liu, Z. The forward order law for least Squareg-inverse of multiple matrix products. *Mathematics* **2019**, *7*, 277. [CrossRef]

22. Zhao, L. The expression of the Drazin Inverse with rank constraints. *J. Appl. Math.* **2012**, *2012*, 390592. [CrossRef]

23. Sen, S.K.; Prabhu, S.S. Optimal iterative schemes for computing Moore-Penrose matrix inverse. *Int. J. Syst. Sci.* **1976**, *8*, 748–753. [CrossRef]

24. Soleymani, F. An efficient and stable Newton–type iterative method for computing generalized inverse $A_{T,S}^{(2)}$. *Numer. Algorithms* **2015**, *69*, 569–578. [CrossRef]

25. Ostrowski, A.M. Sur quelques transformations de la serie de LiouvilleNewman. *C.R. Acad. Sci. Paris* **1938**, *206*, 1345–1347.

26. Jebreen, H.B.; Chalco-Cano, Y. An improved computationally efficient method for finding the Drazin inverse. *Discret. Dyn. Nat. Soc.* **2018**, *2018*, 6758302. [CrossRef]

27. Sánchez León, J.G. *Mathematica Beyond Mathematics: The Wolfram Language in the Real World*; Taylor & Francis Group: Boca Raton, FL, USA, 2017.

28. Wagon, S. *Mathematica in Action*, 3rd ed.; Springer: Berlin, Germany, 2010.

29. Soleymani, F. Efficient semi-discretization techniques for pricing European and American basket options. *Comput. Econ.* **2019**, *53*, 1487–1508. [CrossRef]

30. Soleymani, F.; Barfeie, M. Pricing options under stochastic volatility jump model: A stable adaptive scheme. *Appl. Numer. Math.* **2019**, *145*, 69–89. [CrossRef]

![Σ mathematics logo] *mathematics*

MDPI

Article

Calculating the Weighted Moore–Penrose Inverse by a High Order Iteration Scheme

Haifa Bin Jebreen[ORCID]

Mathematics Department, College of Science, King Saud University, Riyadh 11451, Saudi Arabia;
hjebreen@ksu.edu.sa

Received: 16 July 2019; Accepted: 7 August 2019; Published: 10 August 2019

Abstract: The goal of this research is to extend and investigate an improved approach for calculating the weighted Moore–Penrose (WMP) inverses of singular or rectangular matrices. The scheme is constructed based on a hyperpower method of order ten. It is shown that the improved scheme converges with this rate using only six matrix products per cycle. Several tests are conducted to reveal the applicability and efficiency of the discussed method, in contrast with its well-known competitors.

Keywords: iteration scheme; Moore–Penrose; rectangular matrices; rate of convergence; efficiency index

MSC: 15A09; 65F30

1. Introduction

1.1. Background

Constructing and discussing different features of iterative schemes for the calculation of outer inverses is an active topic of current research in Applied Mathematics (for more details, refer to [1–3]). Many papers have been published in the field of outer inverses over the past few decades, each having their own domain of validity and usefulness. In fact, in 1920, Moore was a pioneer of this field and published seminal works about the outer inverse [4,5]. However, several deep works were published during the 1950s (as reviewed and observed in [4]). It is also noted that pseudo-inverse operator was first introduced by Fredholm in [6].

The method of partitioning (due to Greville) was a pioneering work in computing generalized inverses, which was re-introduced and re-investigated in [4,7]. This scheme requires a lot of operations and is subject to cancelation and rounding errors. Among the generalized inverses, the weighted Moore–Penrose (WMP) inverse is important, as it can be simplified to a pseudo-inverse, as well as a regular inverse. Several applications of computing the WMP inverse can be observed, with some discussion, in the recent literature [8,9]; including applications to the solution of matrix equations. See [10–13] for further discussions and applications.

Furthermore, for large matrices, or as long as the weight matrices in the process of computing the WMP inverse are ill-conditioned, symbolic computation of the current algorithms may not properly work due to several reasons, such as time consumption, requiring higher memory space, or instability. On the other hand, several numerical methods for the weighted Moore–Penrose (WMP) inverse are not stabie or possess slow convergence rates. Hence, it is necessary to investigate and extend novel and useful iterative matrix methods for such an objective; see, also, the discussions in [14,15].

1.2. Definition

Let us consider that M and N are two square Hermitian positive definite (HPD) matrices of sizes m and n ($m \leq n$) and $A \in \mathbb{C}^{m \times n}$. Then, there is a unique matrix X satisfying the following identities [16]:

1. $AXA = A$,
2. $XAX = X$,
3. $(MAX)^* = MAX$,
4. $(NXA)^* = NXA$.

Then, $X \in \mathbb{C}^{n \times m}$ is called the WMP inverse of A, and is shown by $A_{MN}^{\dagger}$. Noting that, as long as $M = I_{m \times m}$ and $N = I_{n \times n}$, then X is the Moore–Penrose (MP) inverse, or simply the pseudo-inverse of A, and we show it by $A^{\dagger}$ [17]. Furthermore, when the matrix A is non-singular, then the pseudo-inverse will be simplified to the regular inverse.

The weighted singular value decomposition (WSVD), first introduced in [18], is normally applied to define this generalized inverse. Consider that the rank of A is r. Then, we have $U \in \mathbb{C}^{m \times m}$ and $V \in \mathbb{C}^{n \times n}$, satisfying the following relations:

$$U^* M U = I_{m \times m},\tag{1}$$

and

$$V^* N^{-1} V = I_{n \times n},\tag{2}$$

such that

$$A = U \begin{pmatrix} D & 0 \\ 0 & 0 \end{pmatrix} V^*.\tag{3}$$

Thus, $A_{MN}^{\dagger}$ is furnished as follows:

$$A_{MN}^{\dagger} = N^{-1} V \begin{pmatrix} D^{-1} & 0 \\ 0 & 0 \end{pmatrix} U^* M,\tag{4}$$

where we have a diagonal matrix $D = \mathrm{diag}(\sigma_1, \sigma_2, \ldots, \sigma_r)$, for $\sigma_1 \geq \sigma_2 \geq \ldots \geq \sigma_r > 0$, while σ_i^2 is the non-zero eigenvalue of $N^{-1} A^* M A$. In addition,

$$\|A\|_{MN} = \sigma_1, \quad \|A_{MN}^{\dagger}\|_{NM} = \frac{1}{\sigma_r}.\tag{5}$$

In this work, $A^{\#} = N^{-1} A^* M$ is used as the weighted matrix of the conjugate transpose of A. See [19] for more details.

1.3. Literature

Schulz-type methods for the calculation of the WMP inverse are sensitive to the choice of the initial value; that is, the initial choice of matrix must be close enough to the generalized inverse so as to guarantee the scheme to converge [20]. More precisely, convergence can only be observed if the starting matrix is chosen carefully. However, this starting value can be chosen simply for the case of the WMP inverse. The pioneering work in [21] gave several suggestions, along with deep discussions, about how to make such a choice quickly.

Let us, now, briefly provide some of the pioneering and most important matrix iterative methods for computing the WMP inverse.

The second–order Schulz scheme for finding the WMP inverse, requiring only two matrix products per computing cycle, is given by [22]:

$$X_{k+1} = X_k(2I - AX_k), \ k = 0, 1, 2, \cdots . \tag{6}$$

Throughout the work, I stands for the identity matrix, unless clearly stated otherwise.

An improvement of (6) with third-order convergence, known as Chebyshev's method, was discussed in [23] for computing $A_{MN}^\dagger$ as follows:

$$X_{k+1} = X_k(3I - AX_k(3I - AX_k)), \ k = 0, 1, 2, \cdots . \tag{7}$$

The authors in [23] proposed another third-order iterative formulation, having one more matrix multiplication, as follows:

$$X_{k+1} = X_k \left[I + \frac{1}{2}(I - AX_k)(I + (2I - AX_k)^2) \right], \ k = 0, 1, 2, \cdots . \tag{8}$$

It is necessary to recall that a general class of iteration schemes for computing the WMP inverse and some other kinds of other generalized inverses was discussed and investigated in [24] (Chapter 5) to have p-th order using a total of p matrix products. An example could be the following fourth-order iteration:

$$X_{k+1} = X_k(I + B_k(I + B_k(I + B_k))), \ k = 0, 1, 2, \cdots , \tag{9}$$

where $B_k = I - AX_k$. As another instance, a tenth-order matrix method could be furnished as follows [25]:

$$\begin{aligned} X_{k+1} &= X_k(I + B_k(I + B_k(I + B_k(I + B_k(I \\ &+ B_k(I + B_k(I + B_k(I + B_k)))))))), \ k = 0, 1, 2, \cdots . \end{aligned} \tag{10}$$

1.4. Motivation and Organization

The main motivation behind proposing and extending new iterative methods for the WMP inverse is to apply them in practical large scale problems [26], as well as to improve the computational efficiency, which is directly linked to the concept of numerical analysis for designing new iterative expressions which are economically useful, in being able to reduce computational complexity and time requirements.

Hence, with this motivation at hand, to increase the computational efficiency index as well as to contribute in this field, the main focus of this work is to investigate a tenth-order method requiring only six matrix multiplications per cycle. We prove that this can provide an improvement of the computational efficiency index in calculating the WMP inverse.

The paper is organized as follows. Section 1 discusses the preliminaries and literature of this topic very briefly, to prepare the reader for the analytical discussions of Section 2, in which we describe an effective iteration formulation for the WMP inverse. It is investigated that the method needs only six matrix multiplications to reach its tenth order of convergence.

Concrete proofs of convergence are furnished in Section 3. Section 4 discusses the application of our formulation to the WMP inverses of many randomly generated matrices of various dimensions. Numerical evidence demonstrates the usefulness of this method for computing the WMP inverse, in terms of the elapsed computation time. Finally, several concluding remarks and comments are given in Section 5.

2. A High Order Scheme for the WMP Inverse

For the use of iterative methods, such as the ones described in Section 1, it is required to employ a starting value when computing the WMP inverse. As in [27], one general procedure to find this starting matrix is of the following form:

$$X_0 = \lambda A^\#, \tag{11}$$

where $A^\# = N^{-1}A^*M$ is the matrix of weighted conjugate transpose (WCT) for the input matrix A and

$$\lambda = \frac{1}{\sigma_1^2}. \tag{12}$$

Recall that, in (12), σ_1 is the the largest eigenvalue of $N^{-1}A^*MA$.

2.1. Derivation

Another reason for proposing a higher order method is that methods based on improvements of the Schulz iteration scheme are slow in the initial phase of iteration. This means that the convergence order cannot be observed at the beginning, it can be seen only after performing several iterates. On the other hand, by incorporating a stop condition using matrix norms, we can increase the elapsed time of executing the written programs for finding the WMP inverse.

Accordingly, to contribute and extend a high order matrix iteration scheme in this context, we first take into account a tenth-order scheme having ten matrix multiplications per cycle, as follows:

$$X_{k+1} = X_k(I + B_k + B_k^2 + \cdots + B_k^9). \tag{13}$$

Now, to develop the performance of (13), we factorize (13) to reduce the number of products. So, we can write

$$X_{k+1} = X_k (I + B_k) [(I - B_k + B_k^2 - B_k^3 + B_k^4)(I + B_k + B_k^2 + B_k^3 + B_k^4)]. \tag{14}$$

This formulation for the matrix iteration requires seven matrix products. However, it is possible to reduce this number of products by considering a more tight formulation for (14). Hence, we write

$$X_{k+1} = X_k (I + B_k) M_k, \tag{15}$$

where

$$M_k = [(I + \chi B_k^2 + B_k^4)(I + \kappa B_k^2 + B_k^4)]. \tag{16}$$

To find the unknown weighting coefficients in (15) and, more specifically, in (16), we need to solve a symbolic problem. As such, a Mathematica code [28] was employed to do such a task, as follows:

```
ClearAll["Global'*"];
fact1 = (1 + a B^2 + B^4);
fact2 = (1 + b B^2 + B^4);
sol = fact1*fact2 + (c B^2) // Expand
S = Table[
s[i] = Coefficient[sol, B^i], {i, 2, 6, 2}
] // Simplify
Solve[
s[2] == 1 && s[4] == 1 && s[6] == 1, {a, b, c}
] // Simplify
{a, b, c} = {a, b, c} /. %[[1]] // Simplify
Chop@sol // Simplify
```

This was given only to ease understanding of the procedure of obtaining the coefficient. Now, we obtain:

$$\chi = \frac{1}{2}\left(1 - \sqrt{5}\right), \qquad \kappa = \frac{1}{2}\left(1 + \sqrt{5}\right). \tag{17}$$

This means that (15) requires only six matrix products per cycle to hit a convergence speed of ten.

2.2. Several Lemmas

Before providing the main results concerning the convergence analysis of the proposed scheme, we furnish the following lemmas, inspired by [29], which reveal how the iterates generated by (15) have some specific important relations and, then, show a relation between (4) and (15).

Lemma 1. *For* $\{X_k\}_{k=0}^{k=\infty}$ *produced by* (15) *using the starting matrix* (11), *for any* $k \geq 0$, *it holds that*

$$
\begin{aligned}
(MAX_k)^* &= MAX_k, \\
(NX_kA)^* &= NX_kA, \\
X_kAA^\dagger_{MN} &= X_k, \\
A^\dagger_{MN}AX_k &= X_k.
\end{aligned}
\tag{18}
$$

Proof. The proof can be done by employing mathematical induction. When $k = 0$ and X_0 is the suitable initial matrix, the first two relations in (18) are straightforward. Hence, we discuss the last two relations by applying the following identities:

$$(AA^\dagger_{MN})^{\#} = AA^\dagger_{MN}, \tag{19}$$

and

$$(A^\dagger_{MN}A)^{\#} = A^\dagger_{MN}A. \tag{20}$$

Accordingly, we have:

$$
\begin{aligned}
X_0AA^\dagger_{MN} &= \lambda A^{\#}AA^\dagger_{MN} \\
&= \lambda A^{\#}(AA^\dagger_{MN})^{\#} \\
&= \lambda A^{\#}(A^\dagger_{MN})^{\#}A^{\#} \\
&= \lambda(AA^\dagger_{MN}A)^{\#} \\
&= \lambda A^{\#} \\
&= X_0,
\end{aligned}
\tag{21}
$$

and also

$$
\begin{aligned}
A^\dagger_{MN}AX_0 &= \lambda A^\dagger_{MN}AA^{\#} \\
&= \lambda(A^\dagger_{MN}A)^{\#}A^{\#} \\
&= \lambda(A^{\#}(A^\dagger_{MN})^{\#}A^{\#}) \\
&= \lambda(A(A^\dagger_{MN}A)^{\#} \\
&= \lambda A^{\#} \\
&= X_0.
\end{aligned}
\tag{22}
$$

Subsequently, now the relation is valid for $k > 0$, then we discuss that it will still be true for $k + 1$. Taking our matrix iteration (15) into consideration, we have:

$$
\begin{aligned}
(MAX_{k+1})^* &= (MA(X_k(I + B_k)[(I + \chi B_k^2 + B_k^4)(I + \kappa B_k^2 + B_k^4)]))^* \\
&= [MAX_k(I + B_k + B_k^2 + B_k^3 + B_k^4 + B_k^5 + B_k^6 + B_k^7 + B_k^8 + B_k^9)]^* \\
&= MA[X_k(I + B_k + B_k^2 + B_k^3 + B_k^4 + B_k^5 + B_k^6 + B_k^7 + B_k^8 + B_k^9)] \\
&= MAX_{k+1},
\end{aligned}
$$

using that

$$(M(AX_k))^* = MAX_k, \tag{23}$$

M is a Hermitian positive definite matrix ($M^* = M$), and similar facts, such as:

$$
\begin{aligned}
(M(AX_k)^2)^* &= (M(AX_k)(AX_k))^* \\
&= (AX_k)^*(M(AX_k))^* \\
&= (AX_k)^*(M(AX_k)) \\
&= (AX_k)^*M^*(AX_k) \\
&= (M(AX_k))^*(AX_k) \\
&= M(AX_k)(AX_k) \\
&= M(AX_k)^2.
\end{aligned}
\tag{24}
$$

Hence, the first relation in (18) is true for $k+1$, and the 2nd relation could be investigated similarly. For the other relation in (18), by employing the assumption that

$$
X_k AA^\dagger_{MN} = X_k,
\tag{25}
$$

and (15), we have:

$$
\begin{aligned}
X_{k+1}AA^\dagger_{MN} &= (X_k(I+B_k)[(I+\chi B_k^2+B_k^4)(I+\kappa B_k^2+B_k^4)])AA^\dagger_{MN} \\
&= (X_k + X_kB_k + X_kB_k^2 + X_kB_k^3 + X_kB_k^4 + X_kB_k^5 + X_kB_k^6 \\
&\quad +X_kB_k^7 + X_kB_k^8 + X_kB_k^9)AA^\dagger_{MN} \\
&= X_kAA^\dagger_{MN} + X_kB_kAA^\dagger_{MN} + X_kB_k^2AA^\dagger_{MN} + X_kB_k^3AA^\dagger_{MN} + X_kB_k^4AA^\dagger_{MN} \\
&\quad +X_kB_k^5AA^\dagger_{MN} + X_kB_k^6AA^\dagger_{MN} + X_kB_k^7AA^\dagger_{MN} + X_kB_k^8AA^\dagger_{MN} + X_kB_k^9AA^\dagger_{MN} \\
&= (X_k + X_kB_k + X_kB_k^2 + X_kB_k^3 + X_kB_k^4 + X_kB_k^5 \\
&\quad +X_kB_k^6 + X_kB_k^7 + X_kB_k^8 + X_kB_k^9) \\
&= X_{k+1}.
\end{aligned}
$$

Therefore, the third relation in (18) is valid for $k+1$. The final relation could be investigated in a similar way, and the result follows. The proof is, thus, complete. □

Lemma 2. *Employing the assumptions of Lemma 1 and (3), then for (15) we have:*

$$
(V^{-1}N)X_k(M^{-1}(U^*)^{-1}) = diag(T_k,0),
\tag{26}
$$

where T_k is a diagonal matrix, $V^*N^{-1}V = I_{n\times n}$, $U^*MU = I_{m\times m}$, $V \in \mathbb{C}^{n\times n}$, $U \in \mathbb{C}^{m\times m}$, and $A = U\Sigma V^*$.

Proof. Assume that $T_0 = \lambda D$ and that σ_i^2 are the non-zero eigenvalues of the matrix $N^{-1}A^*MA$, while $D = diag(\sigma_1, \sigma_2, \ldots, \sigma_r)$, $\sigma_i > 0$ for any i. Thus, we can write that:

$$
\begin{aligned}
T_{k+1} := \varphi(T_k) &= T_k(I+(I-DT_k))[(I+\chi(I-DT_k)^2 + (I-DT_k)^4) \\
&\quad \times (I+\kappa(I-DT_k)^2 + (I-DT_k)^4)].
\end{aligned}
\tag{27}
$$

Applying mathematical induction, one can write that

$$
\begin{aligned}
(V^{-1}N)X_0(M^{-1}(U^*)^{-1}) &= \lambda(V^{-1}N)A^\#(M^{-1}(U^*)^{-1}) \\
&= \lambda(V^{-1}N)N^{-1}A^*(MM^{-1}(U^*)^{-1}) \\
&= \lambda(V^{-1}N)N^{-1}Vdiag(D,0)U^*(MM^{-1}(U^*)^{-1}) \\
&= diag(\lambda D,0).
\end{aligned}
\tag{28}
$$

In addition, when (31) is satisfied, then using (15), one can get that:

$$
\begin{aligned}
(V^{-1}N)X_{k+1}(M^{-1}(U^*)^{-1}) &= (V^{-1}N)X_k(M^{-1}(U^*)^{-1}) \\
&\times (2I - (V^{-1}N)AX_k(M^{-1}(U^*)^{-1})) \\
&\times [(I + \chi(V^{-1}N)(I - AX_k)^2(M^{-1}(U^*)^{-1}) \\
&+ (V^{-1}N)(I - AX_k)^4(M^{-1}(U^*)^{-1})) \\
&\times (I + \kappa(V^{-1}N)(I - AX_k)^2(M^{-1}(U^*)^{-1}) \\
&+ (V^{-1}N)(I - AX_k)^4)(M^{-1}(U^*)^{-1})].
\end{aligned}
\tag{29}
$$

Using the fact that $A = U^*MU\text{diag}(D,0) = V^*NV$, one attains

$$
(V^{-1}N)X_{k+1}(M^{-1}(U^*)^{-1}) = \text{diag}(\varphi(T_k),0),
\tag{30}
$$

which shows that (27) is a diagonal matrix. This completes the proof. □

3. Error Analysis

The objective of this section is to provide a matrix analysis for the convergence of the iteration scheme (15).

Theorem 1. *Let us consider that A is an $m \times n$ matrix whose WSVD is provided by (4). Furthermore, assume that the starting value is given by (11). Thus, the matrix sequence from (15) tends to $A_{MN}^\dagger$.*

Proof. In light of (4), to prove our convergence for the WMP inverse, we now just need to prove that

$$
\lim_{k \to \infty} (V^{-1}N)X_k(M^{-1}(U^*)^{-1}) = \text{diag}(D^{-1},0).
\tag{31}
$$

It is obtained, using Lemmas 1 and 2, that

$$
T_k = \text{diag}(\tau_1^{(k)}, \tau_2^{(k)}, \ldots, \tau_r^{(k)}),
\tag{32}
$$

where

$$
\tau_i^{(0)} = \lambda \sigma_i
\tag{33}
$$

and

$$
\begin{aligned}
\tau_i^{(k+1)} &= \tau_i^{(k)} \left(2I + \sigma_i \tau_i^{(k)}\right) [(I \\
&+ \chi(\sigma_i \tau_i^{(k)})^2 + (\sigma_i \tau_i^{(k)})^4)(I \\
&+ \kappa(\sigma_i \tau_i^{(k)})^2 + (\sigma_i \tau_i^{(k)})^4)].
\end{aligned}
\tag{34}
$$

The sequence produced by (34) is the result of employing (15) in calculating the zero σ_i^{-1} of the function

$$
\phi(\tau) = \sigma_i - \tau^{-1},
\tag{35}
$$

using the starting condition $\tau_i^{(0)}$.

We observe that convergence to σ_i^{-1} can be achieved, as long as

$$
0 < \tau_i^{(0)} < \frac{2}{\sigma_i},
\tag{36}
$$

which results in a criterion on λ (the selection in formula (12) has now been shown). Hence, $\{T_k\} \to \Sigma^{-1}$, and (31) is satisfied. It is now clear that $\{X_k\}_{k=0}^{k=\infty} \to A_{MN}^\dagger$ when $k \to \infty$. This concludes the proof. □

4. Computational Tests

In this section, our aim is to study the efficiency of the proposed approach for calculating the WMP inverse computationally and analytically. To do this, we considered several competitors from the literature in our comparisons, such as those from (6), (7), (10), and (15), denoted by "SM2", "CM3", "KMS10", and "PM10", respectively.

Note that all computations were done in Mathematica 11.0 [30] and the time is reported in seconds. The hardware used was a CPU Intel Core i5 2430-M with 16 GB of RAM.

We know that the efficiency index is expressed by [31]:

$$EI = \rho^{\frac{1}{\kappa}}, \tag{37}$$

where ρ and κ stand for the speed and the whole cost in each cycle, respectively.

As such, the efficiency index of different methods (6–10) and (15) are reported by: $2^{\frac{1}{2}} \simeq 1.414$, $3^{\frac{1}{3}} \simeq 1.442$, $3^{\frac{1}{4}} \simeq 1.316$, $4^{\frac{1}{4}} \simeq 1.414$, $10^{\frac{1}{10}} \simeq 1.258$, and $10^{\frac{1}{6}} \simeq 1.467$, respectively. Clearly, our investigated iterative expression has better a index and can be more useful in finding the WMP inverse.

Example 1. *[29] The purpose of this experiment was to examine the calculation of WMP inverses for 10 uniform randomly provided* $m1 \times n1 = 200 \times 210$ *matrices, as follows:*

```
SeedRandom[12]; no = 10; m1 = 200; n1 = 210;
ParallelTable[A[k] = RandomReal[{1}, {m1, n1}];, {k, no}];
```

where the ten various HPD matrices M and N were given by:

```
ParallelTable[MM[k] = RandomReal[{2}, {m1, m1}];, {k, no}];
ParallelTable[MM[k] = Transpose[MM[k]].MM[k];, {k, no}];
ParallelTable[NN[k] = RandomReal[{3}, {n1, n1}];, {k, no}];
ParallelTable[NN[k] = Transpose[NN[k]].NN[k];, {k, no}];
```

The results by applying the stop termination

$$||X_{k+1} - X_k||_2 \leq 10^{-10}, \tag{38}$$

are reported in Tables 1 and 2, based on the number of iterations, elapsed CPU time (in seconds), and $X_0 = \frac{1}{\sigma_1^2} A^\#$. As can be observed from the results, the best scheme in terms of number of iterations and time was (15).

Table 1. Comparison based on the number of iterations and the required mean in Experiment 1.

Methods	SM2	CM3	KMS10	PM10
A_1	68	43	22	22
A_2	69	44	22	22
A_3	67	43	21	21
A_4	71	46	23	23
A_5	72	46	23	23
A_6	72	46	23	23
A_7	66	42	21	21
A_8	78	50	25	25
A_9	63	41	20	20
A_{10}	69	44	22	22
Mean	69.5	44.5	22.2	22.2

Table 2. Comparison based on the elapsed CPU time and its mean in Experiment 1.

Methods	SM2	CM3	KMS10	PM10
A_1	1.4954	1.04826	0.996155	0.755317
A_2	1.4563	1.08006	0.984057	0.767785
A_3	1.37301	1.03847	0.967427	0.720294
A_4	1.53201	1.10927	1.0365	0.789994
A_5	1.50908	1.10164	0.998853	0.794098
A_6	1.51215	1.11421	1.03361	0.823177
A_7	1.39481	1.00116	0.915244	0.743779
A_8	1.62742	1.24438	1.12434	0.87007
A_9	1.32916	0.999683	0.903072	0.709523
A_{10}	1.49738	1.05736	0.985084	0.764156
Mean	1.47267	1.07945	0.994434	0.773819

Example 2. *The iterative methods were compared for five randomly generated dense* $m1 \times n1 = 500 \times 500$ *matrices produced in Mathematica environment by the following piece of code:*

```
m1 = 500; n1 = 500; no = 5; SeedRandom[12];
ParallelTable[A[k] = RandomReal[{0, 1}, {m1, n1}];, {k, no}];
ParallelTable[MM[k] = RandomReal[{0, 1}, {m1, m1}];, {k, no}];
ParallelTable[MM[k] = Transpose[MM[k]].MM[k];, {k, no}];
ParallelTable[NN[k] = RandomReal[{0, 1}, {n1, n1}];, {k, no}];
ParallelTable[NN[k] = Transpose[NN[k]].NN[k];, {k, no}];
```

Here, we applied the stopping condition

$$||X_{k+1} - X_k||_\infty \leq 10^{-10}, \tag{39}$$

with a change in the initial approximation as $X_0 = \frac{1.5}{\sigma_1^2} A^\#$. Noting that the weights M and N were very ill-conditioned, as we had produced them to be. We report the results in Tables 3 and 4, which reveal that the novel approach was superior to the existing solvers.

Table 3. Comparison based on the number of iterations and the required mean in Experiment 2.

Methods	SM2	CM3	KMS10	PM10
A_1	98	61	30	30
A_2	86	55	27	27
A_3	83	53	26	26
A_4	85	54	27	27
A_5	81	52	26	26
Mean	86.6	55.	27.2	27.2

Table 4. Comparison based on the elapsed time and its mean in Experiment 2.

Methods	SM2	CM3	KMS10	PM10
A_1	7.89745	7.20885	12.1801	7.11963
A_2	6.90346	6.6397	10.933	6.50042
A_3	2.34013	2.23622	3.75341	2.20977
A_4	2.23133	2.15679	3.78848	2.23819
A_5	2.44316	2.26733	3.79153	2.2391
Mean	4.36311	4.10178	6.88929	4.06142

One other application of (15), aside from computing the WMP inverse, is in finding good approximate inverse pre-conditioners for Krylov methods when tackling large sparse linear system of equations (see, e.g., [29]). In fact, to apply our scheme in such environments, we can employ several commands, such as `SparseArray[]` for handling sparse matrices.

The main advantage of the proposed method is the improvement of convergence order obtained by improving the computational efficiency index. Although this computational efficiency index improvement was not observed to be drastic, in solving practical problems in higher dimensions it leads to a clear reduction of computation time.

5. Ending Notes

We have investigated a tenth order iterative method for computing the WMP inverse requiring only six matrix products. The WMP inverse has many applications, from the numerical solution of non-linear equations (those involving singular linear systems [32]) to direct engineering applications. Clearly, the efficiency index will reach $10^{1/6} \simeq 1.46$, which is better than the Newton–Schulz and Chebyshev methods for calculating the WMP inverse. The convergence order of the scheme was supported and upheld analytically. The extension of this improved version of the hyperpower family for computing other types generalized inverses, such as outer and inner inverses, under special criteria and initial matrices provides a direction for future works in this active topic of research.

Funding: This research project was supported by a grant from the "Research Center of the Female Scientific and Medical Colleges", Deanship of Scientific Research, King Saud University.

Conflicts of Interest: The authors declare no conflict of interest.

References

1. Bin Jebreen, H.; Chalco-Cano, Y. An improved computationally efficient method for finding the Drazin inverse. *Disc. Dyn. Nat. Soc.* **2018**, *2018*, 6758302. [CrossRef]
2. Niazi Moghani, Z.; Khanehgir, M.; Mohammadzadeh Karizaki, M. Explicit solution to the operator equation $AXD + FX^*B = C$ over Hilbert C^*-modules. *J. Math. Anal.* **2019**, *10*, 52–64.
3. Stanimirović, P.S.; Katsikis, V.N.; Srivastava, S.; Pappas, D. A class of quadratically convergent iterative methods. *RACSAM* **2019**, 1–22. [CrossRef]
4. Ben-Israel, A.; Greville, T.N.E. *Generalized Inverses: Theory and Applications*, 2nd ed.; Springer: New York, NY, USA, 2003.
5. Godunov, S.K.; Antonov, A.G.; Kiriljuk, O.P.; Kostin, V.I. *Guaranteed Accuracy in Numerical Linear Algebra*; Springer: Dordrecht, The Netherlands, 1993.
6. Fredholm, I. Sur une classe d'équations fonctionnelles. *Acta Math.* **1903**, *27*, 365–390. [CrossRef]
7. Wang, G.R. A new proof of Grevile's method for computing the weighted M-P inverse. *J. Shangai Norm. Univ.* **1985**, *3*, 32–38.
8. Bakhtiari, Z.; Mansour Vaezpour, S. Positive solutions to the system of operator equations $T_iX = U_i$ and $T_iXV_i = U_i$. *J. Math. Anal.* **2016**, *7*, 102–117.
9. Xia, Y.; Chen, T.; Shan, J. A novel iterative method for computing generalized inverse. *Neural Comput.* **2014**, *26*, 449–465. [CrossRef]
10. Courriee, P. Fast computation of Moore-Penrose inverse matrices. *arXiv preprint* **2008**, arXiv:0804.4809.
11. Lu, S.; Wang, X.; Zhang, G.; Zhou, X. Effective algorithms of the Moore-Penrose inverse matrices for extreme learning machine. *Intell. Data Anal.* **2015**, *19.4*, 743–760. [CrossRef]
12. Sheng, X.; Chen, G. The generalized weighted Moore-Penrose inverse. *J. Appl. Math. Comput.* **2007**, *25*, 407–413. [CrossRef]
13. Soleymani, F.; Soheili, A.R. A revisit of stochastic theta method with some improvements. *Filomat* **2017**, *31*, 585–596. [CrossRef]
14. Söderström, T.; Stewart, G.W. On the numerical properties of an iterative method for computing the Moore-Penrose generalized inverse. *SIAM J. Numer. Anal.* **1974**, *11*, 61–74. [CrossRef]
15. Stanimirović, P.S.; Ciric, M.; Stojanović, I.; Gerontitis, D. Conditions for existence, representations, and computation of matrix generalized inverses. *Complexity* **2017**, *2017*, 6429725. [CrossRef]

16. Gulliksson, M.E.; Wedin, P.A.; Wei, Y. Perturbation identities for regularized Tikhonov inverse and weighted pseudo inverse. *BIT* **2000**, *40*, 513–523. [CrossRef]

17. Roy, F.; Gupta, D.K.; Stanimirović, P.S. An interval extension of SMS method for computing weighted Moore-Penrose inverse. *Calcolo* **2018**, *55*, 15. [CrossRef]

18. Van Loan, C.F. Generalizing the singular value decomposition. *SIAM J. Numer. Anal.* **1976**, *13*, 76–83. [CrossRef]

19. Zhang, N.; Wei, Y. A note on the perturbation of an outer inverse. *Calcolo* **2008**, *45*, 263–273. [CrossRef]

20. Ghorbanzadeh, M.; Mahdiani, K.; Soleymani, F.; Lotfi, T. A class of Kung-Traub-type iterative algorithms for matrix inversion. *Int. J. Appl. Comput. Math.* **2016**, *2*, 641–648. [CrossRef]

21. Pan, V.Y. *Structured Matrices and Polynomials: Unified Superfast Algorithms*; BirkhWauser: Boston, MA, USA; Springer: New York, NY, USA, 2001.

22. Schulz, G. Iterative Berechnung der Reziproken matrix. *Z. Angew. Math. Mech.* **1933**, *13*, 57–59. [CrossRef]

23. Li, H.-B.; Huang, T.-Z.; Zhang, Y.; Liu, X.-P.; Gu, T.-X. Chebyshev-type methods and preconditioning techniques. *Appl. Math. Comput.* **2011**, *218*, 260–270. [CrossRef]

24. Krishnamurthy, E.V.; Sen, S.K. *Numerical Algorithms—Computations in Science and Engineering*; Affiliated East-West Press: New Delhi, India, 1986.

25. Sen, S.K.; Prabhu, S.S. Optimal iterative schemes for computing Moore-Penrose matrix inverse. *Int. J. Syst. Sci.* **1976**, *8*, 748–753. [CrossRef]

26. Grevile, T.N.E. Some applications of the pseudo-inverse of matrix. *SIAM Rev.* **1960**, *3*, 15–22. [CrossRef]

27. Huang, F.; Zhang, X. An improved Newton iteration for the weighted Moore-Penrose inverse. *Appl. Math. Comput.* **2006**, *174*, 1460–1486. [CrossRef]

28. Sánchez León, J.G. *Mathematica Beyond Mathematics: The Wolfram Language in the Real World*; Taylor & Francis Group: Boca Raton, FL, USA, 2017.

29. Zaka Ullah, M.; Soleymani, F.; Al-Fhaid, A.S. An efficient matrix iteration for computing weighted Moore-Penrose inverse. *Appl. Math. Comput.* **2014**, *226*, 441–454. [CrossRef]

30. Trott, M. *The Mathematica Guide-Book for Numerics*; Springer: New York, NY, USA, 2006.

31. Ostrowski, A.M. Sur quelques transformations de la serie de LiouvilleNewman. *CR Acad. Sci. Paris* **1938**, *206*, 1345–1347.

32. Soheili, A.R.; Soleymani, F. Iterative methods for nonlinear systems associated with finite difference approach in stochastic differential equations. *Numer. Algor.* **2016**, *71*, 89–102. [CrossRef]

![Sigma mathematics logo]

mathematics

MDPI

Article

An Improved Curvature Circle Algorithm for Orthogonal Projection onto a Planar Algebraic Curve

Zhinan Wu [1,†] **and Xiaowu Li** [2,*,†]

[1] School of Mathematics and Computer Science, Yichun University, Yichun 336000, China; zhi_nan_7@163.com
[2] College of Data Science and Information Engineering, Guizhou Minzu University, Guiyang 550025, China
*   Correspondence: lixiaowu002@126.com; Tel.:+86-187-8613-2431
†   These authors contributed equally to this work.

Received: 10 August 2019; Accepted: 25 September 2019; Published: 1 October 2019

Abstract: Point orthogonal projection onto planar algebraic curve plays an important role in computer graphics, computer aided design, computer aided geometric design and other fields. For the case where the test point **p** is very far from the planar algebraic curve, we propose an improved curvature circle algorithm to find the footpoint. Concretely, the first step is to repeatedly iterate algorithm (the Newton's steepest gradient descent method) until the iterated point could fall on the planar algebraic curve. Then seek footpoint by using the algorithm (computing footpoint **q**) where the core technology is the curvature circle method. And the next step is to orthogonally project the footpoint **q** onto the planar algebraic curve by using the algorithm (the hybrid tangent vertical foot algorithm). Repeatedly run the algorithm (computing footpoint **q**) and the algorithm (the hybrid tangent vertical foot algorithm) until the distance between the current footpoint and the previous footpoint is near 0. Furthermore, we propose Second Remedial Algorithm based on Comprehensive Algorithm B. In particular, its robustness is greatly improved than that of Comprehensive Algorithm B and it achieves our expected result. Numerical examples demonstrate that Second Remedial Algorithm could converge accurately and efficiently no matter how far the test point is from the plane algebraic curve and where the initial iteration point is.

Keywords: point projection; intersection; planar algebraic curve; Newton's iterative method; the improved curvature circle algorithm

1. Introduction

Reconstructing curve/surface is an important work in the field of computer aided geometric design, especially in geometric modeling and processing where it is crucial to fit curve/surface in high accuracy and reduce the error of representation curve/surface. The representation of the four curve types are the explicit-type, implicit-type, parametric-type and subdivision-type. Because implicit representation has unique advantage in the process of computer aided geometric design, it has wide and far-reaching applications. From scattered and unorganized three-dimensional data, Bajaj et al. [1] reconstructed surface and functions on surfaces. They [2,3] have constructed the algebraic B-spline surfaces with least-squares fitting feature using tensor product technique. Schulz et al. [4] constructed an enveloping algebraic surface using gradually approximate algebraization method. Kanatani et al. [5] applied the algebraic curve to construct geometric ellipse fitting using unified strict maximum likelihood estimation method. Mullen et al. [6] reconstructed robust and accurate algebraic surface functions to sign the unsigned from scattered and unorganized three-dimensional data point sets. Upreti et al. [7] used a technique to sign algebraic level sets on NURBS surface and algebraic Boolean level sets on NURBS surfaces. Rouhani et al. [8] applied the algebraic function for polynomial representation system. And L.G. Zagorchev et al. [9] applied the algebraic function for general algebraic surface.

Up to now, there are three main types of methods to solve the problem of point orthogonal projection onto planar algebraic curve: local method, global method and compromise method between these two methods. Here are three typical approaches.

According to the most basic geometric characteristic, orthogonal projection of test point $\mathbf{p}$ onto the planar algebraic curve is actually the point $\mathbf{x}$ on the curve such that cross product of vectors $\overrightarrow{\mathbf{xp}}$ and $\nabla f(\mathbf{x})$ is 0.

$$\nabla f(\mathbf{x}) \times (\mathbf{p} - \mathbf{x}) = 0. \tag{1}$$

Equation (1) can be transformed into Newton's iteration formula (3). Furthermore, Sullivan et al. [10] adopted a hybrid method with Lagrange multiplier and Newton's iterative method to compute the closest point on the planar algebraic curve for each test point. Some orthogonal projection problems can be transformed into solving system of nonlinear equations. The common characteristic of methods [10,11] is that they converge locally and fast, while methods [10,11] are dependent on the initial points.

The first global method of solving system of nonlinear equations is the Homotopy continuous method [12,13]. They constructed Homotopy continuous formula.

$$H(\mathbf{x}, t) = (1 - t)P(\mathbf{x}) + tQ(\mathbf{x}), \quad t \in [0, 1] \tag{2}$$

where t is a parameter of continuous transformation from 0 to 1, $P(\mathbf{x}) = 0$ is the original system of nonlinear equations to be solved, $Q(\mathbf{x})$ is the objective solution of system of nonlinear equations $P(\mathbf{x}) = 0$. All isolated solutions of system of nonlinear equations $P(\mathbf{x}) = 0$ can be computed by the numerical continuous Homotopy methods [12,13]. So the Homotopy methods [12,13] are global convergence. The Homotopy methods' robustness is proved by [14] and their high time-consuming property is verified in [15]. Of course, the Homotopy methods [12,13] are ideal in theory, but it is difficult to find or construct the objective system of nonlinear equations $Q(\mathbf{x}) = 0$ in practical engineering applications.

The second global resultant methods convert system of nonlinear equations into the expression of the resultants and then solve the resultants [16–19]. According to classical elimination theory, system of two nonlinear equations with two variables can be turned into a resultant polynomial with one variable, which is equivalent to the two simultaneous equations. The Sylvester's resultant and Cayley's statement of Bézout's method are the most famous resultant methods [16–19]. Because the resultant methods [16–19] can solve all roots if the degree of the planar algebraic curve is less than 4, they are good global methods. However, if the degree of the planar algebraic curve is more than quintic, it becomes harder and harder with increasing degree to solve two-polynomial system with the resultant methods.

The third global method is the adoption of the Bézier clipping technique [20–22]. In the first step, solving the nonlinear system of Equation (1) is transformed into solving all roots of Bernstein-Bézier representation with convex hull property. In the second step, if the parts of the domains do not include the solution, we clip the parts of the domains by using convex hull box with Bernstein-Bézier form such that the discarded parts of the region has no solution and all the solutions are in the retained parts of the region. In the third step, the de Casteljau subdivision rule is used to segment the remaining part of the curve obtained by elimination in step 2. Repeat steps 2 and 3 until we can find all the solutions to Equation (1). The advantage of this method is that all solutions of Equation (1) can be found. But this global clipping method has one difficulty: sometimes Equation (1) is difficult or even impossible to convert into Bernstein-Bézier form. For example, specific Equation (1) $\nabla f(\mathbf{x}) \times (\mathbf{p} - \mathbf{x}) = -36p_1y^{17} + 36xy^{17} + 6p_2x^5 - 6x^5y - 4p_1x + 4p_2y + 4x^2 - 4y^2$ is impossible to convert into Bernstein-Bézier form where $f(\mathbf{x}) = x^6 + 4xy + 2y^{18} - 1 = 0$.

The compromise method is between local and global methods. Consisting of the geometric property with computing the nearest point is proposed by Hartmann [23,24] named as the first compromise method. Repeatedly run the Newton's steepest gradient descent method (3) until the iterative point falls on the planar algebraic curve, where the initial iterative point is unrestricted.

$$\mathbf{x}_{n+1} = \mathbf{x}_n - (f(\mathbf{x}_n) / \langle \nabla f(\mathbf{x}_n), \nabla f(\mathbf{x}_n) \rangle) \nabla f(\mathbf{x}_n). \tag{3}$$

$$\mathbf{q} = \mathbf{p} - (\langle \mathbf{p} - \mathbf{y}_n, \nabla f(\mathbf{y}_n) \rangle / \langle \nabla f(\mathbf{y}_n), \nabla f(\mathbf{y}_n) \rangle) \nabla f(\mathbf{y}_n). \tag{4}$$

Running the iterative formula (4) one time, the method [23,24] can obtain the vertical foot point $\mathbf{q}$ where the iterative point $\mathbf{y}_n$ of the formula (4) is the final iterative point obtained by formula (3). Continuously iterate the above two steps until the vertical foot point $\mathbf{q}$ is on the planar algebraic curve $f(\mathbf{x})$. Unluckily, progressive geometric tangent approximation iteration method with computing vertical foot point $\mathbf{q}$ fails for some planar algebraic curves $f(\mathbf{x})$.

The second compromise method is developed by Nicholas [25] who adopted the osculating circle technique to realize orthogonal projection onto the planar algebraic curve. Calculate the corresponding curvature at one point on the planar algebraic curve, and then the radius and center of the curvature circle. The line segment formed by the test point and the center of the curvature circle intersects the curvature circle at footpoint $\mathbf{q}$. Approximately take the footpoint $\mathbf{q}$ as a point on the planar algebraic curve. For the new point on the planar algebraic curve, repeat the above procedure to get a new footpoint and corresponding new approximate point on the planar algebraic curve. Repeat the above behavior until the footpoint $\mathbf{q}$ is the orthogonal projection point $\mathbf{p}_\Gamma$. Because the planar algebraic curve does not have parametric control like parametric curve, taking the footpoint as an approximate point on the planar algebraic curve will bring about large errors. So it makes the operation of the whole algorithm unstable.

The third compromise method is the circle shrinking technique [26]. Repeatedly run the iterative formula (3) such that the final iterative point $\mathbf{p}_c$ falls on the planar algebraic curve as far as possible, where the selection of initial iterative point is arbitrary. The next iterative point on the planar algebraic curve is obtained through a series of combined operations of circle and the planar algebraic curve, where the center and radius of the circle are test point $\mathbf{p}$ and $\|\mathbf{p} - \mathbf{p}_c\|$, respectively. A series of combined operations include the two most important steps: Find a point $\mathbf{p}^+$ on the circle by means of the mean value theorem; Seek the intersection of the line segment $\overline{\mathbf{pp}^+}$ and the circle where we call this intersection as the current intersection point $\mathbf{p}_c$. Repeatedly run this series of combined operations until the distance between the current point $\mathbf{p}_c$ and the previous point $\mathbf{p}_c$ is 0. The circle shrinking technique [26] takes a lot of time to seek point $\mathbf{p}^+$ each time. The algorithm has one difficulty: if the degree of the planar algebraic curve is higher than 5, the intersection point $\mathbf{p}_c$ of line segment $\overline{\mathbf{pp}^+}$ and the planar algebraic curve cannot be solved directly by formula or the iterative methods to find the intersection $\mathbf{p}_c$ will lead to instability.

The four compromise method is a circle double-and-bisect algorithm [27]. The circle doubling algorithm begins with a very small circle where the center is the test point $\mathbf{p}$ and the radius is very small r_1. Keep the same center of the circle, take the radius r_2 twice of r_1 to draw a new circle. If there is no intersection between the new circle and the curve, draw a new circle with radius twice of r_2. Continuously repeat the above process until new circle can intersect with the planar algebraic curve and the former circle does not. Naturally, the former circle and the current circle are called interior circle and exterior circle, respectively. Moreover, the bisecting technology implements the rest of the process. Continue to draw a new circle with new radius $r = (r_1 + r_2)/2$. If the new current circle whose radius is r intersects with the curve, substitute r for r_2, else for r_1. Repeatedly run the above progress until the difference between the two radii is approximate zero($|r_1 - r_2| < \varepsilon$). But this method is very difficult to judge whether the exterior circle intersects the planar algebraic curve or not [27].

The fifth compromise method is the integrated hybrid second order algorithm [28]. It includes two sub-algorithms: the hybrid second order algorithm and the initial iterative value estimation algorithm. They mainly exploit three ideas: (1) the tangent orthogonal vertical foot method coupled with calibration method; (2) Newton's steepest gradient descent iterative method to impel the iteration point to be on the planar implicit curve; (3) Newton's iterative method to speed up the whole iteration process. Before running the hybrid second order algorithm, the initial iterative value estimation algorithm is used to force the initial iterative value of the formula (17) of the hybrid second order algorithm and the orthogonal projection point $\mathbf{p}_\Gamma$ as close as possible. After a lot of tests, if the distance between the test point $\mathbf{p}$ and the curve is not very far, the advantages of this algorithm are obvious in term of robustness and efficiency. But when the test point is very far from the curve, the integrated hybrid second order algorithm is invalid.

2. The Improved Curvature Circle Algorithm

In Reference [28], when the test point $\mathbf{p}$ is not particularly far away, the integrated hybrid algorithm can have ideal result. But if the test point $\mathbf{p}$ is very far from the curve, the algorithm is invalid where the test point $\mathbf{p}$ can not be robustly and effectively orthogonally projected onto the planar algebraic curve. In order to overcome this difficulty, we propose an improved curvature circle algorithm to ensure robustness and effective convergence with the test point $\mathbf{p}$ being arbitrarily far away. No matter how far the test point $\mathbf{p}$ is from the planar algebraic curve, if the initial iteration point $\mathbf{x}_0$ is very close to the orthogonal projection point of the test point $\mathbf{p}$, the preconceived algorithm can converge well. So we attempt to construct an algorithm to find an initial iterative point very close to the orthogonal projection point $\mathbf{p}_\Gamma$ of the test point $\mathbf{p}$. The general idea is the following. Repeatedly iterate the formula (3) by utilizing the Newton's steepest gradient descent method until the iteration point fall on the planar algebraic curve as far as possible, written as $\mathbf{p}_c$. This time, the distance between the iteration point $\mathbf{p}_c$ and the orthogonal projection point $\mathbf{p}_\Gamma$ is much smaller than that between the original iteration point $\mathbf{x}_0$ and the orthogonal projection point $\mathbf{p}_\Gamma$. The iteration point $\mathbf{p}_c$ is closer to the orthogonal projection point $\mathbf{p}_\Gamma$. In order to further promote the iteration point $\mathbf{p}_c$ and the orthogonal projection point $\mathbf{p}_\Gamma$ to be closer, we introduce a key step with curvature circle algorithm. Draw a curvature circle through point $\mathbf{p}_c$ on the planar algebraic curve with the radius R determined by the curvature k and the center $\mathbf{m}$ being a normal direction point of point $\mathbf{p}_c$ on the planar algebraic curve. Line segment $\overline{\mathbf{mp}}$ determined by the test point $\mathbf{p}$ and the center $\mathbf{m}$ intersects curvature circle at point $\mathbf{q}$. We take the intersection point $\mathbf{q}$ as the next iteration point for the iteration point $\mathbf{p}_c$. Of course, the distance between the intersection point $\mathbf{q}$ and the orthogonal projection point $\mathbf{p}_\Gamma$ is much smaller than the previous one. We use the intersection point $\mathbf{q}$ as the new test point, and run the hybrid algorithm again where the initial iterative point at this moment can be set as $\mathbf{q} - (0.1, 0.1)$. Repeatedly iterate until the iteration point falls on the planar algebraic curve $f(\mathbf{x})$, written as $\mathbf{p}_c$. We repeat the last two key steps in this procedure until the iteration point $\mathbf{p}_c$ and the orthogonal projection point $\mathbf{p}_\Gamma$ overlap (See Figure 1).

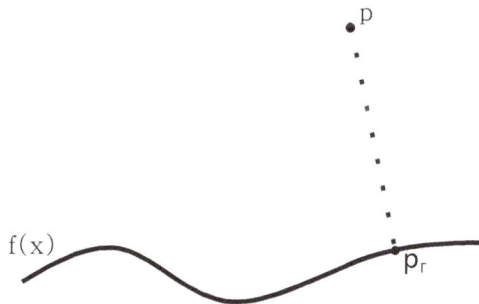

Figure 1. Test point $\mathbf{p}$ orthogonal projection onto planar algebraic curve $f(\mathbf{x})$.

Let's elaborate on the general idea. Let $\mathbf{p}$ be a test point on the plane. There is an planar algebraic curve Γ on the plane.

$$f(x, y) = 0. \tag{5}$$

The plane algebraic curve (5) can be simply written as

$$f(\mathbf{x}) = 0, \tag{6}$$

where $\mathbf{x} = (x, y)$. The goal of this paper is to find a point $\mathbf{p}_\Gamma$ on the planar algebraic curve $f(\mathbf{x})$ to satisfy the basic relationship

$$\|\mathbf{p} - \mathbf{p}_\Gamma\| = \min_{\mathbf{x} \in \Gamma} \|\mathbf{p} - \mathbf{x}\|. \tag{7}$$

The above problem can be written as

$$\begin{cases} f(\mathbf{p}_\Gamma) = 0, \\ \nabla f(\mathbf{p}_\Gamma) \times (\mathbf{p} - \mathbf{p}_\Gamma) = 0, \\ \|\mathbf{p} - \mathbf{p}_\Gamma\| = \min\limits_{\mathbf{x} \in \Gamma} \|\mathbf{p} - \mathbf{x}\|, \end{cases} \tag{8}$$

where $\nabla f = \left[\dfrac{\partial f}{\partial x}, \dfrac{\partial f}{\partial y}\right]$ is Hamiltonian operator and symbol $\times$ is cross product. We take s as the arc length parameter of the planar algebraic curve $f(\mathbf{x})$ and $\mathbf{t} = \left[\dfrac{dx}{ds}, \dfrac{dy}{ds}\right]$ is unit tangent vector along the planar algebraic curve $f(\mathbf{x})$. Take derivative of Equation (6) with respect to arc length parameter s and combine with unit tangent vector condition $\|\mathbf{t}\| = 1$, we obtain the following simultaneous system of nonlinear equations,

$$\begin{cases} \langle \mathbf{t}, \nabla f \rangle = 0, \\ \|\mathbf{t}\| = 1. \end{cases} \tag{9}$$

It is easy to get the solution of Equation (9).

$$\mathbf{t} = \left[-\frac{\partial f}{\partial y}, \frac{\partial f}{\partial x}\right] / \|\nabla f\|. \tag{10}$$

Repeatedly iterate Equation (3) called as the Newton's steepest gradient descent method until until the iterative termination criteria $|f(\mathbf{x}_{n+1})| < \varepsilon$, where the initial iterative point is $\mathbf{x}_0 = \mathbf{p} - (0.1, 0.1)$ and refer to the iterative point $\mathbf{x}_{n+1}$ as $\mathbf{p}_c$. The first advantage of the Newton's steepest gradient descent method (3) is to make the iteration point fall on the planar algebraic curve $f(\mathbf{x})$ as far as possible. Its second advantage of the Newton's steepest gradient descent method (3) is that the iteration point fallen on the planar algebraic curve is relatively close to the orthogonal projection point $\mathbf{p}_\Gamma$, and it brings great convenience to implementation of the subsequent sub-algorithms. The Newton's steepest gradient descent method (Algorithm 1) can be specifically described as (See Figure 2).

Algorithm 1: The Newton's steepest gradient descent method.

Input: The test point $\mathbf{p}$ and the planar algebraic curve $f(\mathbf{x}) = 0$

Output: The iterative point $\mathbf{p}_c$ fallen on planar algebraic curve $f(\mathbf{x}) = 0$

Description:

Step 1:

  $\mathbf{x}_{n+1} = \mathbf{p} - (0.1, 0.1)$;

  Do {

    $\mathbf{x}_n = \mathbf{x}_{n+1}$;

    Update $\mathbf{x}_{n+1}$ according to the iterative Equation (3);

  }while ($|f(\mathbf{x}_{n+1})| > \varepsilon$ && $\|\mathbf{x}_{n+1} - \mathbf{x}_n\| > \varepsilon$);

Step 2:

  $\mathbf{p}_c = \mathbf{x}_{n+1}$;

  Return $\mathbf{p}_c$;

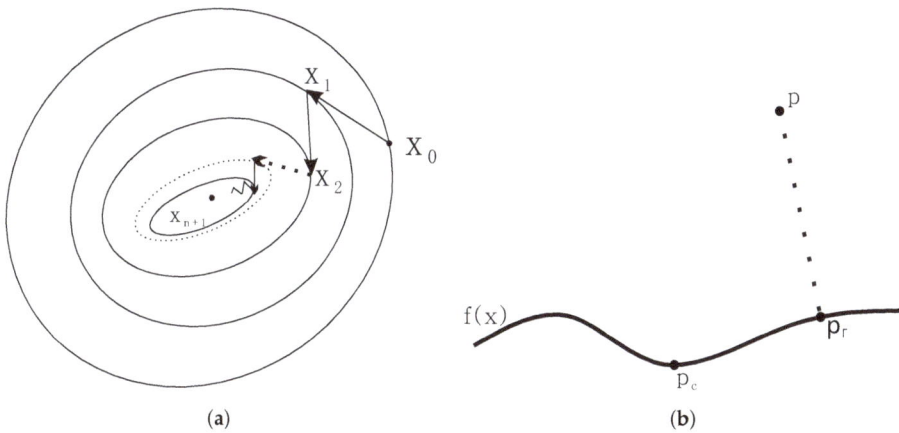

Figure 2. The entire graphic demonstration of Algorithm 1. (**a**) The whole iterative process of the Newton's steepest gradient descent method; (**b**) The last step of the iterative point $\mathbf{p}_c$ fallen on the planar algebraic curve $f(\mathbf{x})$ through the Newton's steepest gradient descent method.

In this case, if the iterative point $\mathbf{p}_c$ fallen on the planar algebraic curve $f(\mathbf{x})$ is taken as the initial iterative point of the hybrid algorithm, convergence or divergence may occur where divergence can not improve the algorithm. As for divergence, it can not achieve the purpose of improving the algorithm. From another point of view, the distance between iteration point $\mathbf{p}_c$ and orthogonal projection point $\mathbf{p}_\Gamma$ of the test point $\mathbf{p}$ should be closer. It lays a good foundation for the implementation of subsequent sub-algorithms. In order to get the iteration point and the orthogonal projection point $\mathbf{p}_\Gamma$ closer, we adopt curvature circle way to promote the iteration point and the orthogonal projection point $\mathbf{p}_\Gamma$ being closer. Because the iterative point is on the planar algebraic curve, the curvature k at the iterative point $\mathbf{p}_c$ fallen on the planar algebraic curve $f(\mathbf{x})$ is defined as [29],

$$k = k(x, y) = \frac{\left[-f_y, f_x\right] \mathbf{G} \left[-f_y, f_x\right]^T}{\|\nabla f\|^3}, \tag{11}$$

where $\mathbf{G} = \begin{pmatrix} f_{xx} & f_{xy} \\ f_{yx} & f_{yy} \end{pmatrix}$. The radius R and the center $\mathbf{m}$ of the curvature circle $\bigcirc$ directed by the curvature k are

$$R = |1/k|, \tag{12}$$

and

$$\mathbf{m} = \mathbf{p}_c + \frac{\overrightarrow{\mathbf{n}}}{k}, \tag{13}$$

where the unit normal vector $\overrightarrow{\mathbf{n}}$ is $\overrightarrow{\mathbf{n}} = \frac{\nabla f}{\|\nabla f\|}$. The line segment $\overline{\mathbf{mp}}$ determined by the test point $\mathbf{p}$ and the center $\mathbf{m}$ of the curvature circle $\bigcirc$ intersects the curvature circle $\bigcirc$ at point $\mathbf{q}$ which is named as footpoint $\mathbf{q}$. From elementary geometric knowledge, the parametric equation of the line segment $\overline{\mathbf{mp}}$ can be expressed as

$$\mathbf{x} = \mathbf{p} + (\mathbf{m} - \mathbf{p})w, \tag{14}$$

where parametric $0 \leq w \leq 1$ is undetermined. In addition, the equation of the curvature circle $\bigcirc$ can be written as

$$\|\mathbf{m} - \mathbf{x}\| = R. \tag{15}$$

By solving Equation (14) and Equation (15) together, the analytic expression of the intersection **q** is obtained

$$\mathbf{q} = \mathbf{p} + (\mathbf{m} - \mathbf{p})w, \tag{16}$$

where the undetermined parameter w is accurately identified as $w = 1 - \dfrac{R}{\|\mathbf{m} - \mathbf{p}\|}$. The computation of the footpoint **q** can be realized through Algorithm 2 (See Figure 3).

Algorithm 2: Computing footpoint **q** via the curvature circle ○ and the line segment $\overline{\mathbf{mp}}$.

Input: The test point **p**, the planar algebraic curve $f(\mathbf{x}) = 0$ and the iterative point $\mathbf{p}_c$ on the planar algebraic curve $f(\mathbf{x}) = 0$.
Output: The footpoint **q**.
Description:
Step 1:
　　Compute the curvature k of the iterative point $\mathbf{p}_c$ fallen on the planar algebraic curve $f(\mathbf{x}) = 0$ by the curvature calculation formula (11).
Step 2:
　　Calculate the radius R and the center **m** of the curvature circle ○ through the formulas (12) and (13), respectively.
Step 3:
　　Compute the footpoint **q** by the formula (16).
　　Return **q**;

Remark 1. *The important formula for computing the curvature k is the formula (11). If the denominator of the curvature k with the formula (11) is 0, the whole iteration process will degenerate. In order to solve this special degeneration, we adopt a small perturbation of the curvature k of the formula (11) in programming implementation of Algorithm 2. Namely, the denominator of the curvature k with the formula (11) could be incremented by a small positive constant ε, the denominator of the curvature k is the denominator of the curvature k +ε, and Algorithm 2 continues to calculate the center and the radius of the curvature circle corresponding to the curvature after disturbance. Of course, in all subsequent formulas or iterative formulas, we also do the same denominators perturbation treatment for the case of the zero denominators of the formulas or the iterative formulas.*

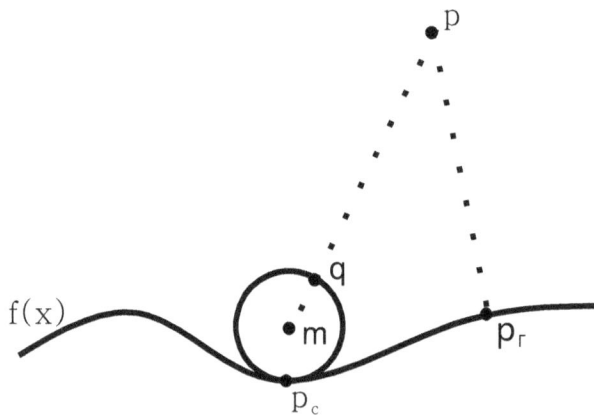

Figure 3. Graphic demonstration for Algorithm 2.

Under this circumstance, if the footpoint point **q** at this moment is taken as the initial iteration point of the hybrid algorithm, the convergence probability of the hybrid algorithm is much greater than that of using the point $\mathbf{p}_c$ in Algorithm 1 as the initial iterative point of the hybrid algorithm. The reason is that the distance $\|\mathbf{q} - \mathbf{p}_\Gamma\|$ is smaller than the distance $\|\mathbf{p}_c - \mathbf{p}_\Gamma\|$. But divergence may happen in this case. In order to further guarantee the robustness,we orthogonally project the footpoint q onto the planar algebraic curve $f(\mathbf{x})$ by using the hybrid algorithm, instead of directly using the footpoint q as the initial iterative point. At this time we still call the orthogonal projection point of the footpoint q as the point $\mathbf{p}_c$ which is just fallen on the planar algebraic curve $f(\mathbf{x})$. Because at this time the footpoint q is close to the planar algebraic curve $f(\mathbf{x})$, the algorithm can ensure complete convergence. The distance between the iterative point $\mathbf{p}_c$ and the orthogonal projection point $\mathbf{p}_\Gamma$ of the test point p becomes smaller again. The core iterative formula (17) of the hybrid algorithm is as follows (See [28]).

$$\begin{cases} \mathbf{y}_n = \mathbf{x}_n - (f(\mathbf{x}_n)/\langle \nabla f(\mathbf{x}_n), \nabla f(\mathbf{x}_n)\rangle)\, \nabla f(\mathbf{x}_n), \\ \mathbf{z}_n = \mathbf{y}_n - (\mathbf{F}(\mathbf{y}_n)/\langle \nabla \mathbf{F}(\mathbf{y}_n), \nabla \mathbf{F}(\mathbf{y}_n)\rangle)\, \nabla \mathbf{F}(\mathbf{y}_n), \\ \mathbf{Q} = \mathbf{q} - (\langle (\mathbf{q} - \mathbf{z}_n), \nabla f(\mathbf{z}_n)\rangle / \langle \nabla f(\mathbf{z}_n), \nabla f(\mathbf{z}_n)\rangle)\nabla f(\mathbf{z}_n), \\ \mathbf{u}_n = \mathbf{z}_n + sign(\langle \mathbf{q} - \mathbf{z}_n, \mathbf{t}_0\rangle)\mathbf{t}_0 \Delta s, \\ \mathbf{v}_n = \mathbf{u}_n - (\mathbf{F}(\mathbf{u}_n)/\langle \nabla \mathbf{F}(\mathbf{u}_n), \nabla \mathbf{F}(\mathbf{u}_n)\rangle)\, \nabla \mathbf{F}(\mathbf{u}_n), \\ \mathbf{x}_{n+1} = \mathbf{v}_n + [-\Delta e, 0]\left[\nabla f^T, (\Delta \mathbf{v}_n)^T\right]^{-1} \ (if \, |[\nabla f^T, (\Delta \mathbf{v}_n)^T]| = 0, \mathbf{x}_{n+1} = \mathbf{v}_n), \end{cases} \tag{17}$$

where $\mathbf{F}_0(\mathbf{x}) = [(\mathbf{q} - \mathbf{x}) \times \nabla f(\mathbf{x})] = 0$, $\mathbf{F}(\mathbf{x}) = \dfrac{\mathbf{F}_0(\mathbf{x})}{\sqrt{\langle \nabla f(\mathbf{x}), \nabla f(\mathbf{x})\rangle}}$, $\mathbf{t}_0 = \dfrac{[-f_y, f_x]}{\|\Delta f\|}$, $\Delta s = \|\mathbf{Q} - \mathbf{z}_n\|$,

$f(\mathbf{v}_n) = \Delta e$, $\Delta \mathbf{v}_n = -(\mathbf{F}(\mathbf{u}_n)/\langle \nabla \mathbf{F}(\mathbf{u}_n), \nabla \mathbf{F}(\mathbf{u}_n)\rangle)\nabla \mathbf{F}(\mathbf{u}_n)$.

The iterative formula (17) mainly contains four techniques. The core technology is the tangent foot vertical method with the third step and the fourth step of the iterative formula (17). Draw a tangent line **L** from a point on a plane algebraic curve $f(\mathbf{x})$. Through the footpoint q (The footpoint q at this time is as the test point of iterative formula (17)), make a vertical line of the tangent **L** and get its corresponding vertical foot point **Q**, which is equivalent to the third step in the formula (17). From the fourth step of the iterative formula (17), we get the next iteration point of particular importance for the initial iteration point. When the next iteration point is not very close to the planar algebraic curve $f(\mathbf{x})$, we adopt the second important technique with the iteration point correction method, equivalent to the sixth step of the iterative formula (17). The iteration point is to move to the plane algebra curve as close as possible such that the distance between the correction point of the iteration point and the plane algebra curve $f(\mathbf{x})$ is as close as possible. These two techniques are pure geometric techniques. When the distance between the test point and the planar algebraic curve is very close, the effect of convergence is obvious. Of course, when the distance between the test point and the planar algebraic curve is relatively long, sometimes there will be non-convergence. In order to improve the robustness of convergence, we add the Newton's steepest gradient descent method before the first technique with the third step and the fourth step of the iterative formula (17). Its first aim is to bring the initial iteration point closer to the planar algebraic curve $f(\mathbf{x})$. Its second aim is to promote the accuracy of subsequent iterations. In order to accelerate the whole iteration process of the iterative formula (17), we once again incorporate the fourth technology of Newton's iterative method which is closely related to the footpoint q. This technique not only accelerates the convergence rate of the whole iteration process but also improves the iteration robustness. Furthermore, the accuracy of the whole iteration process can be improved by the fourth technique. So we add Newton's iterative method after the first step with the second technique, and then add it again before the last step with the third technique. Based on the above explanation and illustration, we get the following the hybrid tangent vertical foot algorithm (Algorithm 3).

Algorithm 3: The hybrid tangent vertical foot algorithm (See Figure 4).

Input: The footpoint **q** and the planar algebraic curve $f(\mathbf{x}) = 0$.
Output: The point $\mathbf{p}_c$ fallen on the planar algebraic curve $f(\mathbf{x}) = 0$.
Description:
Step 1:

$\mathbf{x}_{n+1} = \mathbf{q} - (0.1, 0.1)$;
Do {

$\mathbf{x}_n = \mathbf{x}_{n+1}$;
Execute $\mathbf{x}_{n+1}$ according to the iterative Equation (17);
}while $(\|\mathbf{x}_{n+1} - \mathbf{x}_n\|^2 > \varepsilon \&\& |f(\mathbf{x}_{n+1})| > \varepsilon)$

Step 2:

$\mathbf{p}_c = \mathbf{x}_{n+1}$;
Return $\mathbf{p}_c$;

With the description of the above three algorithms, we propose a comprehensive and complete algorithm (Algorithm 4) closely related to Algorithm 2 (See Figure 4).

Algorithm 4: The first improved curvature circle algorithm (Comprehensive Algorithm A).

Input: Test point **p** and the planar algebraic curve $f(\mathbf{x})$.
Output: Orthogonal projection point $\mathbf{p}_\Gamma$ of the test point **p** which orthogonally projects the test point **p** onto a planar algebraic curve $f(\mathbf{x})$.
Description:
Step 1: Starting from the neighbor point of the test point **p**, calculate the point $\mathbf{p}_c$ fallen on the $f(\mathbf{x})$ via Algorithm 1.
Do{

Step 2: Compute the footpoint **q** via Algorithm 2.
Step 3: Project footpoint **q** onto the planar algebraic curve $f(\mathbf{x})$ via Algorithm 3, then get the new iterative point $\mathbf{p}_c$ fallen on the $f(\mathbf{x})$.
}while (distance (the current $\mathbf{p}_c$, the previous $\mathbf{p}_c$)> ϵ).

$\mathbf{p}_\Gamma = \mathbf{p}_c$;
Return $\mathbf{p}_\Gamma$;

Through a series of rigorous deductions, Comprehensive Algorithm A is the important algorithm of our paper. No matter how far the test point **p** is from the planar algebraic curve $f(\mathbf{x})$, test point **p** could very robustly orthogonally projects onto the planar algebraic curve $f(\mathbf{x})$. This has achieved our desired result. After a lot of testing and observation, when the point on the curve is close to the orthogonal projection point, we find that Comprehensive Algorithm A presents two characteristics: (1) difference between the first distance and the second distance decreases slower and slower, where the first distance and the second distance are the one between the previous iterative point $\mathbf{p}_c$ on the planar algebraic curve and the orthogonal projection point $\mathbf{p}_\Gamma$, and the one between the current iterative point $\mathbf{p}_c$ on the planar algebraic curve and the orthogonal projection point $\mathbf{p}_\Gamma$, respectively; (2) the rate goes even slower at which the absolute value of the inner product gradually approaches zero. These two characteristics are what we don't want to obtain because they are contrary to the efficiency of computer systems. On the premise of ensuring robustness, we try our best to improve and excavate a certain degree of efficiency for the problem of point orthogonal projection onto planar algebraic curve. We have an ingenious discovery. After each running of Algorithm 3, we run the Newton's iterative method associated with the original test point **p**, which can improve the convergence and ensure the orthogonality. Namely, that is to add this step after the last step of the formula (17). Thus the iterative formula (18) is obtained.

$$\begin{cases}
\mathbf{y}_n = \mathbf{x}_n - (f(\mathbf{x}_n)/\langle \nabla f(\mathbf{x}_n), \nabla f(\mathbf{x}_n)\rangle)\,\nabla f(\mathbf{x}_n), \\
\mathbf{z}_n = \mathbf{y}_n - (\mathbf{F}(\mathbf{y}_n)/\langle \nabla \mathbf{F}(\mathbf{y}_n), \nabla \mathbf{F}(\mathbf{y}_n)\rangle)\,\nabla \mathbf{F}(\mathbf{y}_n), \\
\mathbf{Q} = \mathbf{q} - (\langle\langle(\mathbf{q}-\mathbf{z}_n),\nabla f(\mathbf{z}_n)\rangle / \langle \nabla f(\mathbf{z}_n), \nabla f(\mathbf{z}_n)\rangle\rangle)\nabla f(\mathbf{z}_n), \\
\mathbf{u}_n = \mathbf{z}_n + sign(\langle \mathbf{q}-\mathbf{z}_n, \mathbf{t}_0\rangle)\mathbf{t}_0\Delta s, \\
\mathbf{v}_n = \mathbf{u}_n - (\mathbf{F}(\mathbf{u}_n)/\langle \nabla \mathbf{F}(\mathbf{u}_n), \nabla \mathbf{F}(\mathbf{u}_n)\rangle)\,\nabla \mathbf{F}(\mathbf{u}_n), \\
\mathbf{w}_n = \mathbf{v}_n + [-\Delta e, 0]\left[\nabla f^T, (\Delta \mathbf{v}_n)^T\right]^{-1} (if\,|\left[\nabla f^T, (\Delta \mathbf{v}_n)^T\right]| = 0, then\ \ \mathbf{w}_n = \mathbf{v}_n), \\
\mathbf{x}_{n+1} = \mathbf{w}_n - (G(\mathbf{w}_n)/\langle \nabla G(\mathbf{w}_n), \nabla G(\mathbf{w}_n)\rangle)\,\nabla G(\mathbf{w}_n),
\end{cases} \qquad (18)$$

where $\mathbf{F}_0(\mathbf{x}) = [(\mathbf{q}-\mathbf{x}) \times \nabla f(\mathbf{x})] = 0$, $\mathbf{F}(\mathbf{x}) = \dfrac{\mathbf{F}_0(\mathbf{x})}{\sqrt{\langle \nabla f(\mathbf{x}), \nabla f(\mathbf{x})\rangle}}$, $\mathbf{t}_0 = \dfrac{[-f_y, f_x]}{\|\Delta f\|}$, $\Delta s = \|\mathbf{Q}-\mathbf{z}_n\|$,

$f(\mathbf{v}_n) = \Delta e$, $\Delta \mathbf{v}_n = -(\mathbf{F}(\mathbf{u}_n)/\langle \nabla \mathbf{F}(\mathbf{u}_n),\nabla \mathbf{F}(\mathbf{u}_n)\rangle)\nabla \mathbf{F}(\mathbf{u}_n)$, $G_0(\mathbf{x}) = [(\mathbf{p}-\mathbf{x}) \times \nabla f(\mathbf{x})] = 0$, $G(\mathbf{x}) = \dfrac{G_0(\mathbf{x})}{\sqrt{\langle \nabla f(\mathbf{x}), \nabla f(\mathbf{x})\rangle}}$. Because the iterative formula (17) of Algorithm 3 naturally becomes the iterative

formula (18), so Algorithm 3 naturally becomes the following Algorithm 5.

Algorithm 5: The hybrid tangent vertical foot algorithm.

Input: The footpoint $\mathbf{q}$ and the planar algebraic curve $f(\mathbf{x}) = 0$.

Output: The point $\mathbf{p}_c$ fallen on planar algebraic curve $f(\mathbf{x}) = 0$.

Description:

Step 1:

$\mathbf{x}_{n+1} = \mathbf{q} - (0.1, 0.1)$;

Do {

$\mathbf{x}_n = \mathbf{x}_{n+1}$;

Execute $\mathbf{x}_{n+1}$ according to the iterative Equation (18);

}while $(\|\mathbf{x}_{n+1} - \mathbf{x}_n\|^2 > \varepsilon \&\&|f(\mathbf{x}_{n+1})| > \varepsilon)$

Step 2:

$\mathbf{p}_c = \mathbf{x}_{n+1}$;

Return $\mathbf{p}_c$;

Now let's replace Algorithm 3 of Comprehensive Algorithms A with Algorithm 5. We get the following Comprehensive Algorithm B (Algorithm 6).

Algorithm 6: The second improved curvature circle algorithm (Comprehensive Algorithm B).

Input: Test point $\mathbf{p}$ and the planar algebraic curve $f(\mathbf{x})$.

Output: Orthogonal projection point $\mathbf{p}_\Gamma$ of the test point $\mathbf{p}$ which orthogonally projects the test point $\mathbf{p}$ onto the planar algebraic curve $f(\mathbf{x})$.

Description:

Step 1: Starting from the neighbor point of the test point $\mathbf{p}$, calculate the point $\mathbf{p}_c$ fallen on the $f(\mathbf{x})$ via Algorithm 1.

Do{

Step 2: Compute the footpoint $\mathbf{q}$ via Algorithm 2.

Step 3: Project footpoint $\mathbf{q}$ onto the planar algebraic curve $f(\mathbf{x})$ via Algorithm 5, then get new point $\mathbf{p}_c$ fallen on the $f(\mathbf{x})$.

}while(distance(the current $\mathbf{p}_c$, the previous $\mathbf{p}_c$)$> \epsilon$).

$\mathbf{p}_\Gamma = \mathbf{p}_c$;

Return $\mathbf{p}_\Gamma$;

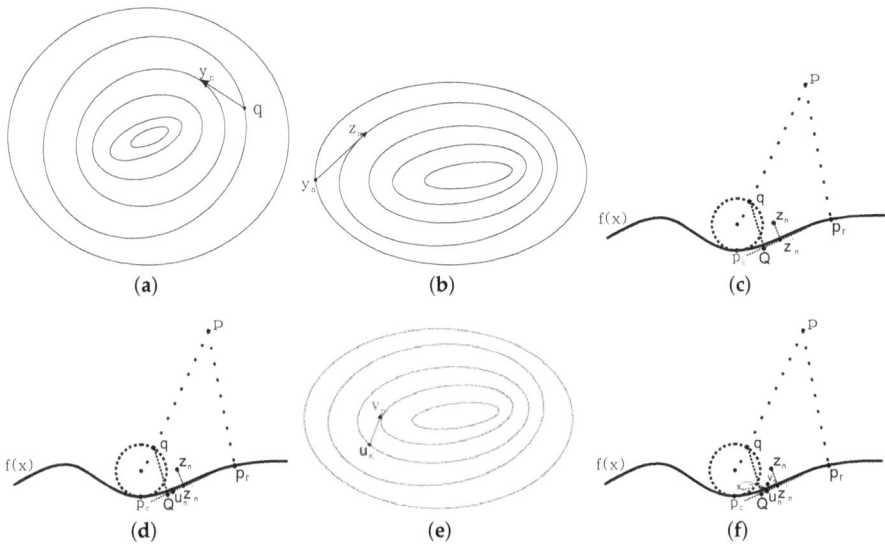

Figure 4. Graphic interpretation of the whole iteration process in Algorithm 3. (**a**) Newton's steepest gradient descent method in the first step; (**b**) The Newton's iteration method related to the test point in the second step; (**c**) The vertical foot point **Q** being the footpoint **q** orthogonal projection onto tangent line induced by the iterative point z_n on the planar algebraic curve in the third step; (**d**) Calculating line incremental iterative value in the fourth step; (**e**) Once again running the Newton's iteration method related to the test point in the fifth step; (**f**) Correcting the previous iteration value to improve the robustness of iteration in the last step.

Comprehensive Algorithm A and Comprehensive Algorithm B share common advantage: the robustness of the two algorithms is substantially improved than that of the existing algorithms because our algorithms are not subject to any restrictions on test points and initial iteration points. By comparison, Comprehensive Algorithm B has four advantages over Comprehensive Algorithm A. (1) The last step of the iterative formula (18) in Comprehensive Algorithm B can make corrections continuously; (2) The last step of the iterative formula (18) in Comprehensive Algorithm B accelerates the whole Comprehensive Algorithm B; (3) The last step of the iterative formula (18) in Comprehensive Algorithm B accelerates the inner product of two vectors to 0, where the first vector refers to the vector connecting the test point **p** and the iteration point z_{n+1} of Comprehensive Algorithm B and the second vector $\left[-\frac{\partial f}{\partial y}, \frac{\partial f}{\partial x} \right]_{|x=x_{n+1}}$ is the tangent vector derived from the iteration point x_{n+1} on the planar algebraic curve, respectively; (4) Comprehensive Algorithm B overcomes two shortcomings of Comprehensive Algorithm A.

Of course, when the test point is not too far from the plane algebra curve, Comprehensive Algorithm is also convergent for any initial iterative point. However, Comprehensive Algorithm A takes more time than directly using the hybrid second order algorithm. In practical applications such as computer graphics, it's hard to know if the test point **p** is close to or far from a planar algebraic curve. Because the main reason is that the degree and the type of the planar algebraic curve restrict the relative distance between the test point **p** and the planar algebraic curve. In order to optimize time efficiency, we take advantage of Comprehensive Algorithm A and the hybrid second order algorithm such that no matter where the test point **p** is located, it can be orthogonally projected onto the planar algebraic curve efficiently and robustly. First, the hybrid second order algorithm is iterated. If it does not converge after 100 iterations, it will be changed to Comprehensive Algorithm A to iterate until the iteration point reaches the orthogonal projection point $\mathbf{p}_\Gamma$. Specific algorithm implementation is the following Comprehensive Integrated Algorithm A (Algorithm 7).

Algorithm 7: The first comprehensive integrated improved curvature circle algorithm (Comprehensive Integrated Algorithm A).

Input: Test point $\mathbf{p}$ and the planar algebraic curve $f(\mathbf{x})$.
Output: Orthogonal projection point $\mathbf{p}_\Gamma$ of the test point $\mathbf{p}$.
Description:
Step 1:

$\quad\quad x_{n+1} = \mathbf{p} - (0.1, 0.1);$
$\quad\quad \text{for}(i = 0; i < N; i++) \{$
$\quad\quad\quad x_n = x_{n+1};$
$\quad\quad\quad x_{n+1} = \text{Hybrid second order algorithm}(f, \mathbf{p}, x_n);$
$\quad\quad\quad \text{if}(\|x_{n+1} - x_n\| < \varepsilon) \quad\quad \text{break};$
$\quad\quad \}$

Step 2:

$\quad\quad \text{if}(i \geq N \&\& d \geq 1e - 15) \{$
$\quad\quad\quad x_n = x_{n+1};$
$\quad\quad\quad x_{n+1} = \text{Comprehensive Algorithm A}(f, \mathbf{p}, x_n);$
$\quad\quad \}$
$\quad\quad \mathbf{p}_\Gamma = x_{n+1};$
$\quad\quad \text{Return } \mathbf{p}_\Gamma;$

Number N is an empirical value of the iterative times where the value N is specified as 5 or 6.

Similar to Comprehensive Algorithm A, by replacing Algorithm 3 with Algorithm 5, the following Comprehensive Integrated Algorithm B (Algorithm 8) can be obtained naturally.

Algorithm 8: The second comprehensive integrated improved curvature circle algorithm (Comprehensive Integrated Algorithm B).

Input: The test point $\mathbf{p}$ and the planar algebraic curve $f(\mathbf{x})$.
Output: Orthogonal projection point $\mathbf{p}_\Gamma$ of the test point $\mathbf{p}$.
Description:
Step 1:

$\quad\quad x_{n+1} = \mathbf{p} - (0.1, 0.1);$
$\quad\quad \text{for}(i = 0; i < N; i++) \{$
$\quad\quad\quad x_n = x_{n+1};$
$\quad\quad\quad x_{n+1} = \text{Hybrid second order algorithm}(f, \mathbf{p}, x_n);$
$\quad\quad\quad \text{if}(\|x_{n+1} - x_n\| < \varepsilon) \quad\quad \text{break};$
$\quad\quad \}$

Step 2:

$\quad\quad \text{if}(i \geq N \&\& d \geq 1e - 15) \{$
$\quad\quad\quad x_n = x_{n+1};$
$\quad\quad\quad x_{n+1} = \text{Comprehensive Algorithm B}(f, \mathbf{p}, x_n);$
$\quad\quad \}$
$\quad\quad \mathbf{p}_\Gamma = x_{n+1};$
$\quad\quad \text{Return } \mathbf{p}_\Gamma;$

Number N is an empirical value of the iterative times where the value N is specified as 5 or 6.

To sum up, we have presented four synthesis algorithms altogether. After analysis and judgment, Comprehensive Algorithm B and Comprehensive Integrated Algorithm B are the most robust and efficient. On the problem of orthogonal projection of point onto planar algebraic curve, if the distance between the test point and the planar algebraic curve is close, we recommend the hybrid second order algorithm, if the distance between the test point and the planar algebraic curve is not close, we recommend Comprehensive Algorithm B. Of course, if the distance between the test point and the planar algebraic curve cannot be known to be very far or close, Comprehensive Integrated Algorithm B is the best choice.

Remark 2. *In sum, Comprehensive Algorithm B has strong superiority over existing algorithms [10–28]. If the distance between the test point and the planar algebraic curve is very far away, the test point can be ideally orthogonally projected onto the planar algebraic curve. But when there are singular points $\frac{\partial f}{\partial x} \cdot \frac{\partial f}{\partial x} + \frac{\partial f}{\partial y} \cdot \frac{\partial f}{\partial y} = 0$ in the planar algebraic curve, this case will seriously hinder the correct execution and implementation of Comprehensive Algorithm B. In order to solve the problem in the case of singularities in the planar algebraic curves, we propose a remedy to Comprehensive Algorithm B (Algorithm 9). The specific description is as follows (See Figure 5).*

Algorithm 9: The first remedial algorithm of Comprehensive Algorithm B.

Input: Test point $\mathbf{p}$ and the planar algebraic curve $f(\mathbf{x})$.
Output: Orthogonal projection point $\mathbf{p}_\Gamma$ of the test point $\mathbf{p}$.
Description:
Step 1.
   Starting from the neighbor point of the test point $\mathbf{p}$, calculate the iterative point $\mathbf{p}_c$ fallen on the planar algebraic curve $f(\mathbf{x})$ via Algorithm 1.
Step 2.
   Judge whether to use curvature circle method or tangent method in the next step.
Step 3.
   Find the left endpoint $\mathbf{L}_0$ on the other side of $f(\mathbf{x})$ relative to the test point $\mathbf{p}$. According to the result of **step 2**, if use curvature circle method, then the left endpoint $\mathbf{L}_0$ is equal to the intersection point $\mathbf{q}$ which is computed by the curvature circle method with the formula (16). If not, then the left endpoint $\mathbf{L}_0$ is equal to the vertical foot $\mathbf{Q}$ which is computed by the tangent method with the third step of the formula (17).
Step 4.
   Calculate the intersection point $\mathbf{p}_c$ of the line segment $\overline{\mathbf{L}_0\mathbf{p}}$ connecting the current left endpoint $\mathbf{L}_0$ and the test point $\mathbf{p}$ and the planar algebraic curve $f(\mathbf{x})$ by the hybrid method of combining Newton's iterative method and binary search method. The intersection point $\mathbf{p}_c$ is called as the current iterative point $\mathbf{p}_c$;
Step 5.
   Repeat **Step 2**, **Step 3** and **Step 4** until the distance between the current iterative point $\mathbf{p}_c$ and the previous iterative point $\mathbf{p}_c$ is near zero;
Step 6.
   $\mathbf{p}_\Gamma = \mathbf{p}_c$;
   Return $\mathbf{p}_\Gamma$;

Now let's describe the hybrid method of combining Newton's iterative method and binary search method in detail. The parameter equation of the line segment $\overline{\mathbf{L}_0\mathbf{p}}$ can be expressed as

$$\begin{cases} x = L_1 + (p_1 - L_1)w, \\ y = L_2 + (p_2 - L_2)w, \end{cases} \tag{19}$$

where $\mathbf{L}_0 = (L_1, L_2)$, $\mathbf{p} = (p_1, p_2)$, and $0 \leq w \leq 1$ is a parameter of Equation (19). Substitute Equation (19) into Equation (6) of the planar algebraic curve to get a equation on the parameter w,

$$K(w) = f(x, y) = 0, \tag{20}$$

where the x and y of Equation (20) are completely determined by the x and y of Equation (19). So the most basic Newton's iterative formula corresponding to Equation (20) is not difficult to write as,

$$w_{n+1} = w_n - \frac{K(w_n)}{DK(w_n)}, \tag{21}$$

where $DK(w)$ is the first derivative of $K(w)$ about the parameter w. Now we start to iterate the Newton's iterative formula (21) with the initial iterative value $w_0 = 0.0$. Based on the actual situation, the intersection of the line segment $\overline{\mathbf{L}_0 \mathbf{p}}$ and the planar algebraic curve is much closer to the left endpoint $\mathbf{L}_0$ and much farther from the original test point $\mathbf{p}$, therefore, the initial interval of the binary search method can be specified as $[a, b] = [0.0, 0.5]$. The detailed description of the hybrid method of combining Newton's iterative method and binary search method is as following Algorithm 10.

Algorithm 10: The hybrid method of combining Newton's iterative method and binary search method.

    Input: The planar algebraic curve $f(\mathbf{x})$, the original test point $\mathbf{p} = (p_1, p_2)$, the iterative point $\mathbf{p}_c$ via Algorithm 1.

    Output: The intersection $\mathbf{p}_c$ between the line segment $\overline{\mathbf{L}_0 \mathbf{p}}$ and the planar algebraic curve $f(\mathbf{x})$.

    Description:

    Step 1:

        The initial interval of the binary search method $[a, b] = [0.0, 0.5]$, the initial iterative value $w = 0.0$;

    Step 2:

        $w = w - K(w) / DK(w)$;

        kmin=min($K(a), K(b)$);

        kmax=max($K(a), K(b)$);

        if ($K(w) <$ kmin or $K(w) >$ kmax)

            $w = (a + b) / 2$;

        sa=sign($K(a)$);

        sw=sign($K(w)$);

        if(sa $==$ sw)

            $a = w$;

        else

            $b = w$;

    Step 3:

        Repeatedly iterate **Step 2** until $|a - b| < \varepsilon$;

    Step 4:

        $\mathbf{p}_c = \mathbf{L}_0 + (\mathbf{p} - \mathbf{L}_0)w$;

        Return $\mathbf{p}_c$;

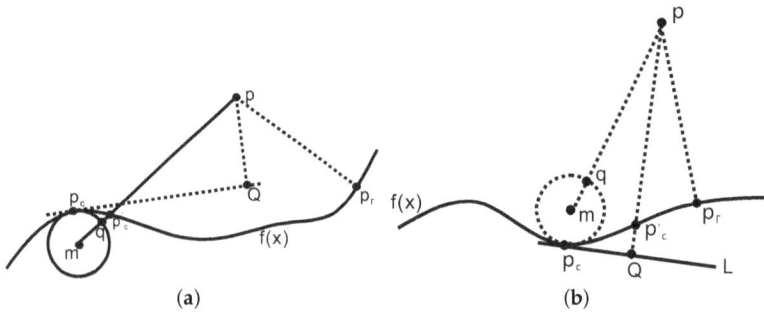

Figure 5. Graphical interpretation for the first remedial algorithm of Comprehensive Algorithm B. (a) The intersection point **q** between the line segment $\overline{mp}$ and the curvature circle and the test point **p** on the opposite side of the planar algebraic curve $f(\mathbf{x})$; (b)The vertical foot **Q** of the tangential line **L** and the test point **p** on the opposite side of the planar algebraic curve $f(\mathbf{x})$.

The robustness of the first remedial algorithm of Comprehensive Algorithm B is much better than that of Comprehensive Algorithm B while the first remedial algorithm of Comprehensive Algorithm B takes much more time than Comprehensive Algorithm B. The hybrid method of combining Newton's iterative method and binary search method is a hybrid method which binary search method ensures global convergence and the Newton's iterative method plays an accelerating role. In order to ensure robustness and improve efficiency, we have fully excavated Comprehensive Algorithm B. We have developed Second Remedial Algorithm (Algorithm 11). The specific description is as follows (See Figure 6).

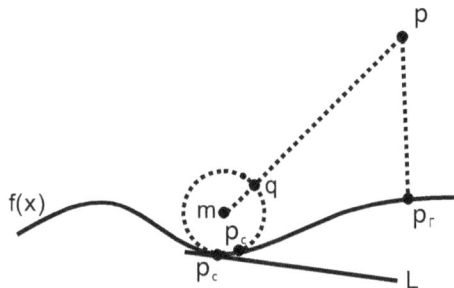

Figure 6. Graphic demonstration for Second Remedial Algorithm.

Algorithm 11: Second Remedial Algorithm.

Input: Test point **p** and the planar algebraic curve $f(\mathbf{x})$.

Output: Orthogonal projection point $\mathbf{p}_\Gamma$ of the test point **p** which orthogonally projects the test point **p** onto the planar algebraic curve $f(\mathbf{x})$

Description:

Step 1: Starting from a certain percentage of the test point **p**, calculate the point $\mathbf{p}_c$ fallen on the $f(\mathbf{x})$ via Algorithm 1.

Do{

    Step 2: Compute the footpoint **q** via Algorithm 2.

    Step 3: Starting from the footpoint **q**, compute the iterative point $\mathbf{p}_c$ fallen on the $f(\mathbf{x})$ via Algorithm 1.

}while(distance(the current $\mathbf{p}_c$, the previous $\mathbf{p}_c$)> ϵ).

Step 4: Compute the orthogonal projection point $\mathbf{p}_\Gamma$ of the test point **p** via Algorithm 12.

    Return $\mathbf{p}_\Gamma$;

Algorithm 12: The hybrid Newton-type iterative algorithm.

Input: The current iterative point $\mathbf{p}_c$ fallen on the planar algebraic curve $f(\mathbf{x})$ and the planar algebraic curve $f(\mathbf{x})$.

Output: Orthogonal projection point $\mathbf{p}_\Gamma$ of the test point $\mathbf{p}$ which orthogonally projects the test point $\mathbf{p}$ onto the planar algebraic curve $f(\mathbf{x})$.

Description:

Step 1:

$$\mathbf{x}_{n+1} = \mathbf{p}_c;$$
Do {

$$\mathbf{x}_n = \mathbf{x}_{n+1};$$
Compute $\mathbf{x}_{n+1}$ according to the iterative formula (22);
}while $(\|\mathbf{x}_{n+1} - \mathbf{x}_n\|^2 > \varepsilon \&\& |f(\mathbf{x}_{n+1})| > \varepsilon)$

Step 2:

$$\mathbf{p}_\Gamma = \mathbf{x}_{n+1};$$
Return $\mathbf{p}_\Gamma;$

The expression of the iterative formula (22) is as follow,

$$\begin{cases} \mathbf{y}_n = \mathbf{x}_n - (f(\mathbf{x}_n)/\langle \nabla f(\mathbf{x}_n), \nabla f(\mathbf{x}_n) \rangle) \nabla f(\mathbf{x}_n), \\ \mathbf{z}_n = \mathbf{y}_n - (\mathbf{F}(\mathbf{y}_n)/\langle \nabla \mathbf{F}(\mathbf{y}_n), \nabla \mathbf{F}(\mathbf{y}_n) \rangle) \nabla \mathbf{F}(\mathbf{y}_n), \\ \mathbf{x}_{n+1} = \mathbf{z}_n + [-\Delta e, 0] \left[\nabla f^T, (\Delta \mathbf{z}_n)^T \right]^{-1} (if \ |[\nabla f^T, (\Delta \mathbf{z}_n)^T]| = 0, then \ \mathbf{x}_{n+1} = \mathbf{z}_n), \end{cases} \tag{22}$$

where $\mathbf{F}_0(\mathbf{x}) = [(\mathbf{p} - \mathbf{x}) \times \nabla f(\mathbf{x})] = 0$, $\mathbf{F}(\mathbf{x}) = \dfrac{\mathbf{F}_0(\mathbf{x})}{\sqrt{\langle \nabla f(\mathbf{x}), \nabla f(\mathbf{x}) \rangle}}$, $f(\mathbf{z}_n) = \Delta e$, $\Delta \mathbf{z}_n = -(\mathbf{F}(\mathbf{y}_n)/\langle \nabla \mathbf{F}(\mathbf{y}_n), \nabla \mathbf{F}(\mathbf{y}_n) \rangle) \nabla \mathbf{F}(\mathbf{y}_n)$.

Remark 3. *In this remark, we present the geometric interpretation of Second Remedial Algorithm. The purpose of the first step is to make the iterative point $\mathbf{p}_c$ fall on the planar algebraic curve as much as possible through Newton's steepest gradient descent method of Algorithm 1, where the coordinates of the initial iterative point take proportional value of that of the test point $\mathbf{p}$ to ensure that Algorithm 1 converges successfully. Otherwise, the distance between the initial iterative point and the planar algebraic curve is very large, which easily leads to the divergence of Algorithm 1. The purpose of **Do** ... **While** cycle body in Second Remedial Algorithm is to continuously and gradually move the iterative point $\mathbf{p}_c$ to fall on the planar algebraic curve to the orthogonal projection point $\mathbf{p}_\Gamma$. The second step in **Do**... **While** cycle body in Second Remedial Algorithm has two characteristics. Since the footpoint $\mathbf{q}$ is the unique intersection point of the curvature circle and the straight line segment $\overline{mp}$ connecting the centre $\mathbf{m}$ of the curvature circle and the test point $\mathbf{q}$, the footpoint $\mathbf{q}$ is always closely related to the iterative point $\mathbf{p}_c$ fallen on the planar algebraic curve and the test point $\mathbf{p}$. The first characteristic is that the footpoint $\mathbf{q}$ can guarantee the global convergence of the whole algorithm (Second Remedial Algorithm). The second characteristic is that the distance between the footpoint $\mathbf{q}$ and the planar algebraic curve is much smaller than the distance between the test point $\mathbf{p}$ and the planar algebraic curve. So the third step with Algorithm 1 in **Do** ... **While** cycle body can very robustly iterate the footpoint $\mathbf{q}$ onto the planar algebraic curve. The core thought of **Do** ... **While** cycle body in Second Remedial Algorithm is to keep the iterative point $\mathbf{p}_c$ to fall on the planar algebraic curve and to move towards the orthogonal projection point $\mathbf{p}_\Gamma$ such that the distance $\|\mathbf{p}_c - \mathbf{p}_\Gamma\|$ between the iterative point $\mathbf{p}_c$ and the orthogonal projection point $\mathbf{p}_\Gamma$ becomes smaller and smaller. As the distance $\|\mathbf{p}_c - \mathbf{p}_\Gamma\|$ gets smaller and smaller, we have found that there is a defect in **Do** ... **While** cycle body in Second Remedial Algorithm. The decreasing speed of the distance $\|\mathbf{p}_c - \mathbf{p}_\Gamma\|$ is getting slower and slower. Especially the second formula of the formula (8) is very difficult to be satisfied. Namely, it is difficult to orthogonalize the vector $\overrightarrow{pp_c}$ and the vector $\nabla f(\mathbf{p}_c)$. In order to overcome the difficulty, we add Algorithm 12 behind **Do** ... **While** cycle body in Second Remedial Algorithm.*

411

Algorithm 12 includes three components: (1) The Newton's steepest gradient descent method in the first step; (2) The Newton's iterative method closely associated with the test point p in the second step; (3) Correcting method in the third step. Algorithm 12 has four advantages and important roles: (1) Algorithm 12 plays a role for accelerating the whole algorithm (Second Remedial Algorithm); (2) The first step plays a role for making the iteration point fall on the planar algebraic curve as far as possible; (3) The second step plays a role for accelerating orthogonalization between the vector $\overrightarrow{pp_c}$ and the vector $\nabla f(p_c)$; (4) The third step plays a dual role for the accelerating orthogonalization and the promoting the iterative point to fall on the planar algebraic curve. The numerical tests of Second Remedial Algorithm achieve our expected results. No matter how far the test point p is from the planar algebraic curve $f(x)$, Second Remedial Algorithm can converge very robustly and efficiently. Second Remedial Algorithm is the best one in our paper. Of course, the robustness and the efficiency of Second Remedial Algorithm are better than that of the existing algorithms. We are very happy about that.

Remark 4. *In order to further improve the efficiency of the test point p orthogonal projecting onto plane algebraic curve $f(x)$, we construct a Comprehensive Integrated Algorithm C which includes two parts: the hybrid second order algorithm in [28] and Second Remedial Algorithm. Firstly run the hybrid second order algorithm in [28]. If the hybrid second order algorithm converges, then it means that Comprehensive Integrated Algorithm C is finished. Otherwise, Second Remedial Algorithm runs until it converges successfully. That is the end of Comprehensive Integrated Algorithm C. The specific description of Comprehensive Integrated Algorithm C is similar to that of Comprehensive Integrated Algorithm B. Here, we are not giving a detailed description of Comprehensive Integrated Algorithm C. When the distance between the test point p and the planar algebraic curve $f(x)$ is not far, the hybrid second order algorithm in [28] has very high robustness and efficiency. When the distance between the test point p and the planar algebraic curve $f(x)$ is particularly far, the hybrid second order algorithm does not converge and fails, while Second Remedial Algorithm converges particularly robustly and successfully. To sum up, Comprehensive Integrated Algorithm C absorbs the advantages of two sub-algorithms and overcomes their respective shortcomings such that the robustness and the efficiency of Comprehensive Integrated Algorithm C are maximized.*

3. Convergence Analysis

This section proves that several Comprehensive Algorithms do not depend on the initial iteration points.

Theorem 1. *Comprehensive Algorithm A is independent of the initial iterative point.*

Proof. Firstly, we state the whole operation process of Comprehensive Algorithm A. Comprehensive Algorithm A contains three sub-algorithms (Algorithms 1–3). The role of Algorithm 1 is to repeatedly iterate the iterative formula (3) through Newton's steepest gradient descent method such that the final iteration point x_{n+1} could fall on the planar algebraic curve where the final iteration point x_{n+1} is denoted as p_c. The function of Algorithm 2 is to seek the footpoint q. The curvature circle $\bigcirc$ determined by the point p_c is obtained from the iterative point p_c on the planar algebraic curve $f(x)$ of Algorithm 1, where the curvature k, the radius R and the center m are determined by formulas (11)–(13), respectively. The intersection of the line segment $\overline{mp}$ connecting the center m and the test point p and the curvature circle $\bigcirc$ is foot point q. The footpoint q could be orthogonally projected onto the planar algebraic curve $f(x)$ by repeated iteration of Algorithm 3 where at this moment the test point is not the original test point p but the footpoint point q solved by Algorithm 2. Repeatedly run Algorithm 2 and Algorithm 3 bound together until the distance between the current footpoint q and the previous footpoint q is near zero.

Secondly, the Comprehensive Algorithm A is independent of the initial iterative point. No matter how far the original test point p is from the planar algebraic curve $f(x)$, no matter where the initial iterative point x_0 is located, Algorithms 1 can ensure that the final iterative point x_{n+1} or p_c of the initial iterative point can fall on the planar algebraic curve $f(x)$. It is obvious that the distance $\|p_c - p_\Gamma\|$

between the iteration point $\mathbf{p}_c$ and the orthogonal projection point $\mathbf{p}_\Gamma$ is much smaller than the distance $\|\mathbf{p} - \mathbf{p}_\Gamma\|$ between the orthogonal projection point $\mathbf{p}_\Gamma$ and the original test point $\mathbf{p}$. From the iterative point $\mathbf{p}_c$ fallen on the planar algebraic curve $f(\mathbf{x})$, we can calculate the corresponding curvature k and its center $\mathbf{m}$ and radius R. The intersection point $\mathbf{q}$ of the curvature circle $\bigcirc$ and the line segment $\overline{mp}$ connecting the original test point $\mathbf{p}$ and the center $\mathbf{m}$ of the curvature circle $\bigcirc$ is just the footpoint $\mathbf{q}$. That is to say, the footpoint $\mathbf{q}$ is directly generated by the curvature circle $\bigcirc$ and the line segment $\overline{mp}$, and the curvature circle $\bigcirc$ is controlled by the iterative point $\mathbf{p}_c$ fallen on the planar algebraic curve $f(\mathbf{x})$. So the footpoint $\mathbf{q}$ is directly controlled by the original test point $\mathbf{p}$ and the iterative point $\mathbf{p}_c$, while the current footpoint $\mathbf{q}$ is between the orthogonal projection point $\mathbf{p}_\Gamma$ and the current iterative point $\mathbf{p}_c$. It also shows that Algorithm 2 plays a decisive role in the convergence robustness of Comprehensive Algorithm. In addition, we can also know that the distance between the footpoint point $\mathbf{q}$ and the planar algebraic curve $f(\mathbf{x})$ is much smaller than the distance between the original test point $\mathbf{p}$ and the planar algebra curve $f(\mathbf{x})$. At this point, we keep running Algorithm 3 with the footpoint point $\mathbf{q}$ as the current test point until the current test point can be orthogonally projected onto the plane algebraic curve $f(\mathbf{x})$ with guaranteed convergence of Algorithm 3. And now we can call the orthogonal projection point of the footpoint point $\mathbf{q}$ as also the iterative point $\mathbf{p}_c$ fallen on the planar algebraic curve $f(\mathbf{x})$. The first reason is the distance between the current iterative point $\mathbf{p}_c$ and the orthogonal projection point $\mathbf{p}_\Gamma$ of the original test point point $\mathbf{p}$ is smaller than the one between the previous iterative point $\mathbf{p}_c$ and the orthogonal projection point $\mathbf{p}_\Gamma$ of the original test point $\mathbf{p}$. The second reason is that it establishes a solid foundation for the convergence robustness of the subsequent sub-algorithms implementation. Then according to the requirements of Comprehensive Algorithm A, the second step and third step of Comprehensive Algorithm A are executed once per cycle, the distance $\|\mathbf{p}_c - \mathbf{p}_\Gamma\|$ between the current iterative point $\mathbf{p}_c$ on the planar algebraic curve and the orthogonal projection point $\mathbf{p}_\Gamma$ of the original test point $\mathbf{p}$ of the execution result is smaller than that between the previous iterative point $\mathbf{p}_c$ on the planar algebraic curve $f(\mathbf{x})$ and the orthogonal projection point $\mathbf{p}_\Gamma$ of the original test point $\mathbf{p}$. The distance $\|\mathbf{p}_c - \mathbf{p}_\Gamma\|$ between the current iterative point $\mathbf{p}_c$ and the orthogonal projection point $\mathbf{p}_\Gamma$ of the original test point $\mathbf{p}$ is becoming smaller. So repeatedly iterate the second step and the third step of Comprehensive Algorithm A until the distance $\|\mathbf{p}_c - \mathbf{p}_\Gamma\|$ between the current iterative point $\mathbf{p}_c$ and the orthogonal projection point $\mathbf{p}_\Gamma$ of the original test point $\mathbf{p}$ is becoming smaller and smaller. Ultimately, the distance $\|\mathbf{p}_c - \mathbf{p}_\Gamma\|$ between the current iterative point $\mathbf{p}_c$ and the orthogonal projection point $\mathbf{p}_\Gamma$ of the original test point $\mathbf{p}$ is becoming zero. It also demonstrates that Comprehensive Algorithm A is completely convergent. This further proves that Comprehensive Algorithm A can completely converge no matter how far away the original test point $\mathbf{p}$ is from the planar algebraic curve and no matter where the initial iterative point $\mathbf{x}_0$ of Comprehensive Algorithm A is on the plane. This means Comprehensive Algorithm A is independent of the initial iterative point. $\square$

Theorem 2. *Comprehensive Algorithm B is independent of the initial iterative point.*

Proof. In the last step of the iterative formula (18) in Algorithm 5, Newton's iteration method, which is closely related to the original test point $\mathbf{p}$, is added. In this way, the iterative formula (17) is transformed into the iterative formula (18) in Algorithm 5. Algorithm 5 has several advantages over Algorithm 3. It can speed up the iteration, improve its accuracy and promote the orthogonalization of the tangent vector derived from the iteration point on the planar algebraic curve and the tangent vector connecting the test point and the iterative point. Replace Algorithm 3 of Comprehensive Algorithm A with Algorithm 5 to get Comprehensive Algorithm B. Since Comprehensive Algorithm A is independent of the initial iterative point, so Comprehensive Algorithm B is naturally independent of the initial iterative point. $\square$

Theorem 3. *Comprehensive Integrated Algorithm A is independent of the initial iterative point.*

Proof. Comprehensive Integrated Algorithm A consists of two parts: the hybrid second order algorithm and Comprehensive Algorithm A. Whether the test point is very far or very close to the planar algebraic curve, the hybrid second order algorithm is executed several times. If this algorithm converges, then it represents that the execution of Comprehensive integrated Algorithm A is over. So Comprehensive Integrated Algorithm A is independent of the initial iterative point. If the hybrid second order algorithm does not converge, then run Comprehensive Algorithm A of the second step of Comprehensive Integrated Algorithm A. Because whether the test point is very far from or very close to the planar algebraic curve, we know from Theorem 1 that Comprehensive Algorithm A is independent of the initial iterative point. To sum up, Comprehensive Integrated Algorithm A is independent of the initial iterative point. □

In a similar way to the proof of Theorem 3, we can state the following result.

Theorem 4. *Comprehensive Integrated Algorithm B is independent of the initial iterative point.*

Theorem 5. *The first remedial algorithm of Comprehensive Algorithm B is independent of the initial iterative point.*

Proof. From the Figure 5, for any initial iterative point, the final iterative point $\mathbf{p}_c$ of Algorithm 1 in the first step of the first remedial algorithm of Comprehensive Algorithm B can ensure that it falls on the planar algebraic curve $f(\mathbf{x})$. The left endpoint $\mathbf{L}_0$ is the only one that can be determined through third step of the first remedial algorithm of Comprehensive Algorithm B. Graphic display shows that the left endpoint $\mathbf{L}_0$ and the original test point $\mathbf{p}$ are on both sides of the planar algebraic curve. Namely, there is only one intersection point (also written as $\mathbf{p}_c$) between the line segment $\overline{\mathbf{L}_0\mathbf{p}}$ and the planar algebraic curve $f(\mathbf{x})$. Because the hybrid method of combining Newton's iterative method and binary search method is global convergence method, the intersection $\mathbf{p}_c$ of the line segment $\overline{\mathbf{L}_0\mathbf{p}}$ and the planar algebraic curve $f(\mathbf{x})$ can be accurately and uniquely solved by this method. Then repeatedly iterate and run Step 2, Step 3 and Step 4, the distance $\|\mathbf{p}_c - \mathbf{p}_\Gamma\|$ between the current intersection point $\mathbf{p}_c$ and the orthogonal projection point $\mathbf{p}_\Gamma$ of the original test point $\mathbf{p}$ continues to shrink to zero. So we have this conclusion that the first remedial algorithm of Comprehensive Algorithm B is independent of the initial iterative point. □

Theorem 6. *Second Remedial Algorithm is independent of the initial iterative point.*

Proof. In Remark 3, we give a detailed description of the geometric interpretation of Second Remedial Algorithm. In this proof, we only elaborate on the most important geometric significance of Second Remedial Algorithm. The first step of Second Remedial Algorithm is to let the initial iteration point fall on the planar algebraic curve as much as possible through Newton's steepest gradient descent method of Algorithm 1. Moreover, there is few restriction on the selection of the initial iterative point. The purpose of **Do** ... **While** cycle body in Second Remedial Algorithm is to continuously and gradually move the iterative point $\mathbf{p}_c$ to fall on the planar algebraic curve to the orthogonal projection point $\mathbf{p}_\Gamma$. The second step in **Do**... **While** cycle body in Second Remedial Algorithm has two characteristics. Since the footpoint $\mathbf{q}$ is the unique intersection point of the curvature circle and the straight line segment $\overline{\mathbf{mp}}$ connecting the centre $\mathbf{m}$ of the curvature circle and the test point $\mathbf{p}$, the footpoint $\mathbf{q}$ is always closely related to the iterative point $\mathbf{p}_c$ fallen on the planar algebraic curve and the test point $\mathbf{p}$. The first characteristic is that the footpoint $\mathbf{q}$ can guarantee the global convergence of the whole algorithm (Second Remedial Algorithm). The second characteristic is that the distance between the footpoint $\mathbf{q}$ and the planar algebraic curve is much smaller than the distance between the test point $\mathbf{p}$ and the planar algebraic curve. So the third step with Algorithm 1 in **Do** ... **While** cycle body can very robustly iterate the footpoint $\mathbf{q}$ onto the planar algebraic curve. The core thought of **Do** ... **While** cycle body in Second Remedial Algorithm is to keep the iterative point $\mathbf{p}_c$ fallen on the planar algebraic curve moving towards the orthogonal projection point $\mathbf{p}_\Gamma$ such that the distance $\|\mathbf{p}_c - \mathbf{p}_\Gamma\|$ between the iterative point $\mathbf{p}_c$ and the orthogonal projection point $\mathbf{p}_\Gamma$ becomes smaller and smaller. If the distance $\|\mathbf{p}_c - \mathbf{p}_\Gamma\|$ gets smaller and smaller, we have found that the decreasing speed of the distance $\|\mathbf{p}_c - \mathbf{p}_\Gamma\|$ is getting slower and slower. Especially the second formula of the formula (8) is very difficult to be satisfied. Algorithm 12 behind the loop body has four advantages and important roles: (1) Algorithm 12 plays a role for accelerating the whole algorithm (Second Remedial Algorithm); (2) The first step plays a role for making the iteration point fall on the planar algebraic curve as far as possible; (3) The second step plays a role for accelerating orthogonalization between the vector $\overrightarrow{\mathbf{p}\mathbf{p}_c}$ and the vector $\nabla f(\mathbf{p}_c)$; (4) The third step plays a dual role for the accelerating orthogonalization and the promoting the iterative point to fall on the planar algebraic curve. No matter how far the test point is from the planar algebraic curve, Second Remedial Algorithm converges very robustly and efficiently. By adding this step, the efficiency and the robustness for Algorithm 12 of Second Remedial Algorithm is further improved. Then the robustness and the efficiency of Second Remedial Algorithm is also further improved. So Second Remedial Algorithm is independent of the initial iterative point. In addition, in a similar way to the proof of Theorem 3, it is not difficult to know that Comprehensive Integrated Algorithm C is also independent of the initial iterative point. □

4. Numerical Comparison Results

We now present some examples to illustrate the efficiency and the comparison of the newly developed method of Comprehensive Algorithm B and Second Remedial Algorithm to show the two algorithms' high robustness and efficiency for very remote test points. We have three examples to represent closed planar algebraic curve, two sub-closed planar algebraic curves, two branches but not closed planar algebra curves and a single branch not closed the planar algebra curve, respectively. All computations were done using VC++6.0. We used $\varepsilon = 10^{-16}$. The following stopping criteria is used for Comprehensive Algorithm B and Second Remedial Algorithm . In Tables 1–3, the four symbols $\mathbf{p}$, $\mathbf{p}_\Gamma$, $|f(\mathbf{p}_\Gamma)|$ and $|\langle \mathbf{V}_1, \mathbf{V}_2 \rangle|$ are the original test point, the orthogonal projection point of the original test point, the deviation degree of the orthogonal projection point on the planar algebraic curve and the absolute value of the inner product of two vectors $\mathbf{V}_1$ and $\mathbf{V}_2$, respectively, where $\mathbf{V}_1$ is $\overrightarrow{\mathbf{p}\mathbf{p}_\Gamma}$ and $\mathbf{V}_2$ is the tangent vector $\left[-\frac{\partial f}{\partial y}, \frac{\partial f}{\partial x} \right]$ of the orthogonal projection point $\mathbf{p}_\Gamma$ on the planar algebraic curve $f(\mathbf{x})$. Thanks to the suggestions by the reviewers, the fourth quadrant result values of the three tables are implemented by Second Remedial Algorithm in Maple 18 environment.

Example 1 (Reference to [28]). *Suppose a planar algebraic curve* $f(x,y) = x^6 + 2x^5y - 2x^3y^2 + x^4 - y^3 + 2y^8 - 4 = 0$ *(See Figure 7). In each of the four quadrants, randomly select four distant test points. We calculate the corresponding orthogonal projection point for each test point via computation by Comprehensive Algorithm B and Second Remedial Algorithm. The specific results are shown in Table 1).*

Example 2. *Suppose a planar algebraic curve* $f(x,y) = x^{10} + 6xy + 2y^{18} - 2 = 0$(*See Figure 8). In each of the four quadrants, randomly select four distant test points. We calculate the corresponding orthogonal projection point for each test point via computation by Comprehensive Algorithm B and Second Remedial Algorithm. The specific results are shown in Table 2.*

Example 3. *Suppose a planar algebraic curve* $f(x,y) = x^{10} + 6xy + 2y^{16} + 2 = 0$(*See Figure 9). In each of the four quadrants, randomly select four distant test points. We calculate the corresponding orthogonal projection point for each test point via computation by Comprehensive Algorithm B and Second Remedial Algorithm. The specific results are shown in Table 3.*

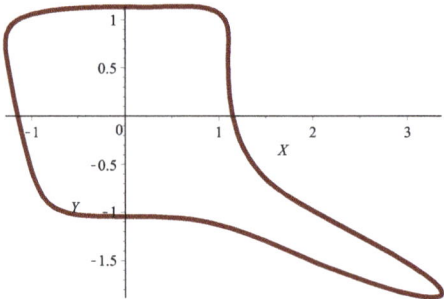

Figure 7. Graphical representation of the planar algebraic curve for Example 1.

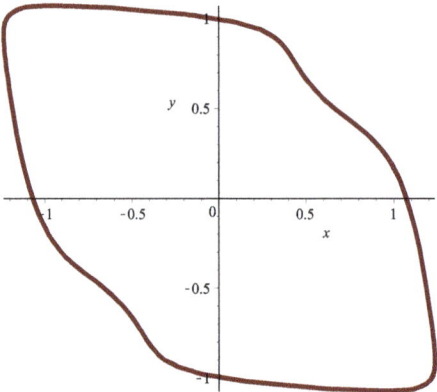

Figure 8. Graphical representation of the planar algebraic curve for Example 2.

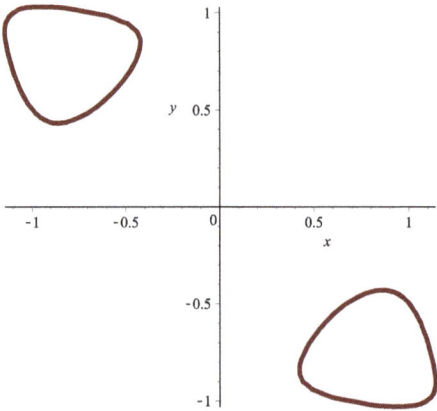

Figure 9. Graphical representation of the planar algebraic curve for Example 3.

Table 1. Test results of Comprehensive Algorithm B for Example 1.

p	(1325, 7447)	(779, 325)	(990, 1375)	(−0.5, 8623)	(−16, 598)	(−3, 231)	(−21, 247)
$\mathbf{p}_\Gamma$	(3.1087, −1.8666)	(3.3647, −1.8184)	(3.3695, −1.8380)	(−0.01582, 1.13368)	(−0.3353, 1.13074)	(−0.2268, 1.1328)	(−0.6071, 1.1166)
$|f(\mathbf{p}_\Gamma)|$	$3.9\cdot10^{-10}$	$7.9\cdot10^{-11}$	$1.4\cdot10^{-11}$	$2.3\cdot10^{-12}$	$4.0\cdot10^{-12}$	$1.9\cdot10^{-12}$	$2.3\cdot10^{-12}$
$|\langle\mathbf{V}_1,\mathbf{V}_2\rangle|$	$2.3\cdot10^{-10}$	$4.4\cdot10^{-11}$	$2.9\cdot10^{-09}$	$2.1\cdot10^{-13}$	$1.5\cdot10^{-12}$	$1.4\cdot10^{-11}$	$9.6\cdot10^{-12}$
$\mathbf{p}$	(−42, −127)	(−5, −38)	(−9, −579)	(78, −123)	(168, −12)	(537, −31)	(91, −221)
$\mathbf{p}_\Gamma$	(0.6633, −0.9941)	(0.4951, −1.0292)	(−0.2183, −1.0445)	(3.3519, −1.8685)	(3.3694, −1.8417)	(3.3694, −1.8414)	(3.3415, −1.8736)
$|f(\mathbf{p}_\Gamma)|$	$7.4\cdot10^{-12}$	$1.4\cdot10^{-12}$	$1.7\cdot10^{-12}$	$4.9\cdot10^{-12}$	$1.8\cdot10^{-13}$	$5.4\cdot10^{-13}$	$8.6\cdot10^{-12}$
$|\langle\mathbf{V}_1,\mathbf{V}_2\rangle|$	$1.7\cdot10^{-13}$	$1.2\cdot10^{-12}$	$1.1\cdot10^{-12}$	$2.3\cdot10^{-13}$	$5.7\cdot10^{-14}$	$2.8\cdot10^{-13}$	$4.5\cdot10^{-13}$

Table 2. Test results of Comprehensive Algorithm B for Example 2.

p	(565, 945)	(979, 325)	(375, 405)	(1959, 1377)	(−9356, 8623)	(−816, 798)	(−3987, 1231)	(−4821, 647)
$\mathbf{p}_\Gamma$	(0.2978, −0.9108)	(1.0773, −0.0164)	(0.6094, 0.5450)	(0.4780, 0.6961)	(−1.2055, 1.0209)	(−1.2035, 1.0230)	(−1.2288, 0.9794)	(−1.2341, 0.9554)
$|f(\mathbf{p}_\Gamma)|$	$1.1\cdot10^{-14}$	$6.0\cdot10^{-13}$	$8.0\cdot10^{-18}$	$9.2\cdot10^{-15}$	$5.8\cdot10^{-15}$	$1.1\cdot10^{-14}$	$2.5\cdot10^{-15}$	$1.1\cdot10^{-14}$
$|\langle\mathbf{V}_1,\mathbf{V}_2\rangle|$	0	$1.8\cdot10^{-12}$	$2.3\cdot10^{-13}$	$4.5\cdot10^{-12}$	$6.9\cdot10^{-10}$	$2.1\cdot10^{-11}$	$1.7\cdot10^{-10}$	$1.1\cdot10^{-10}$
$\mathbf{p}$	(−7942, −275)	(−598, −98)	(−3709, −1979)	(−2937, −1391)	(9708, −323)	(2608, −1912)	(7347, −931)	(5091, −1921)
$\mathbf{p}_\Gamma$	(−1.2367, 0.8930)	(−1.1847, 0.4851)	(−1.0035, −0.1601)	(−1.0225, 0.1224)	(1.1427, −0.9100)	(1.1224, −0.9787)	(1.1414, −0.9273)	(1.1344, −0.9563)
$|f(\mathbf{p}_\Gamma)|$	$2.3\cdot10^{-15}$	$1.3\cdot10^{-11}$	$9.5\cdot10^{-15}$	$1.4\cdot10^{-13}$	$4.7\cdot10^{-15}$	$3.2\cdot10^{-15}$	$4.4\cdot10^{-15}$	$1.7\cdot10^{-17}$
$|\langle\mathbf{V}_1,\mathbf{V}_2\rangle|$	$4.0\cdot10^{-11}$	$3.6\cdot10^{-12}$	$8.7\cdot10^{-11}$	$7.2\cdot10^{-12}$	$5.8\cdot10^{-11}$	$8.7\cdot10^{-11}$	$7.2\cdot10^{-12}$	$2.1\cdot10^{-10}$

Table 3. Test results of Comprehensive Algorithm B for Example 3.

p	(1387, 645)	(1879, 395)	(3075, 205)	(1956, 777)	(−9256, 4603)	(−836, 1798)	(−5987, 1031)	(−4181, 1247)
$\mathbf{p}_\Gamma$	(−1.0629, 0.6023)	(0.4232, −0.8650)	(0.4214, −0.8515)	(0.4276, −0.8795)	(−1.1305, 0.9655)	(−1.0823, 1.0103)	(−0.4232, 0.8220)	(−1.1369, 0.9490)
$|f(\mathbf{p}_\Gamma)|$	$1.7\cdot10^{-13}$	$1.3\cdot10^{-16}$	$8.1\cdot10^{-17}$	$1.0\cdot10^{-16}$	$5.3\cdot10^{-17}$	$1.8\cdot10^{-15}$	$6.1\cdot10^{-17}$	$1.0\cdot10^{-17}$
$|\langle\mathbf{V}_1,\mathbf{V}_2\rangle|$	$3.6\cdot10^{-12}$	$3.2\cdot10^{-12}$	$2.3\cdot10^{-13}$	$6.3\cdot10^{-12}$	$1.7\cdot10^{-10}$	$4.3\cdot10^{-11}$	$1.8\cdot10^{-12}$	$6.5\cdot10^{-11}$
$\mathbf{p}$	(−7342, −1275)	(−5098, −918)	(−3217, −2079)	(−2337, −1251)	(9508, −375)	(6608, −712)	(2347, −931)	(1491, −1321)
$\mathbf{p}_\Gamma$	(−0.4226, 0.8617)	(−0.4227, 0.8623)	(−1.0274, 0.5370)	(−1.0471, 0.5706)	(1.2367, −0.9267)	(1.2353, −0.9460)	(1.2255, −0.9888)	(1.2068, −1.0194)
$|f(\mathbf{p}_\Gamma)|$	$5.7\cdot10^{-17}$	$1.3\cdot10^{-16}$	$4.4\cdot10^{-14}$	$1.1\cdot10^{-13}$	$2.2\cdot10^{-15}$	$2.8\cdot10^{-15}$	$1.8\cdot10^{-15}$	$1.1\cdot10^{-15}$
$|\langle\mathbf{V}_1,\mathbf{V}_2\rangle|$	$1.0\cdot10^{-11}$	$5.5\cdot10^{-12}$	$7.3\cdot10^{-12}$	$9.1\cdot10^{-12}$	$2.2\cdot10^{-11}$	$3.6\cdot10^{-12}$	$5.4\cdot10^{-11}$	$2.2\cdot10^{-11}$

Remark 5. *Besides all test points of the three examples mentioned above are tested by Comprehensive Algorithm B, we have tested them again with the Second Remedial Algorithm. All the test results are consistent with those of Comprehensive Algorithm B and convergent. In addition, in the region* $[-3000, 3000] \times [-3000, 3000]$ *of each example, we randomly select a large number of test points, the probability of non-convergence is particularly low by Second Remedial Algorithm. Further, we use Second Remedial Algorithm other examples with test points in a very large area, and the probability of non-convergence is also very low. Second Remedial Algorithm is verified to be the best one again in our paper. Of course, the robustness and the efficiency of Second Remedial Algorithm is better than that of the existing algorithms.*

5. Conclusions and Future Work

In this paper, we have constructed a Comprehensive Algorithm which is an improved curvature circle algorithm for orthogonal projecting onto planar algebraic curve. Based on an integrated hybrid second-order algorithm [28], the Comprehensive Algorithm (the improved curvature circle algorithm) has also incorporated the curvature circle technique and Newton's gradient steepest descent method such that it can converge robustly and efficiently no matter how far the test point is from the planar algebraic curve and no matter where the initial iterative point is located. Furthermore, we propose Second Remedial Algorithm based on Comprehensive Algorithm B. In particular, its robustness and efficiency is greatly improved than that of Comprehensive Algorithm B and it achieves our expected result. The numerical examples show that the improved curvature circle algorithm is superior to the existing ones. In future work, we try to refine the idea of Comprehensive Algorithm and Second Remedy Algorithm. And the idea is applied to point orthogonal projecting onto spatial algebraic curve and algebraic surface.

Author Contributions: The contribution of all the authors is the same. All of the authors team up to develop the current draft. Z.W. is responsible for investigating, providing resources and methodology, the original draft, writing, reviewing, validation, editing and supervision of this work. X.L. is responsible for software, algorithm, program implementation, formal analysis, visualization, writing, reviewing, editing and supervision of this work.

Funding: This research was funded by the National Natural Science Foundation of China Grant No. 61263034, the Feature Key Laboratory for Regular Institutions of Higher Education of Guizhou Province Grant No. 2016003, the Key Laboratory of Advanced Manufacturing Technology, Ministry of Education, Guizhou University Grant No. 2018479, the National Bureau of Statistics Foundation Grant No. 2014LY011, the Shandong Provincial Natural Science Foundation of China Grant No.ZR2016GM24.

Acknowledgments: We take the opportunity to thank the anonymous reviewers for their thoughtful and meaningful comments.

Conflicts of Interest: The authors declare no conflict of interest.

References

1. Bajaj, C.L.; Bernardini, F.; Xu, G. Reconstructing surfaces and functions on surfaces from unorganized three-dimensional data. *Algorithmica* **1997**, *19*, 243–261. [CrossRef]
2. Jüttler, B.; Felis, A. Least-squares fitting of algebraic spline surfaces. *Adv. Comput. Math.* **2002**, *17*, 135–152.
3. Song, X.; Jüttler, B. Modeling and 3D object reconstruction by implicitly defined surfaces with sharp features. *Comput. Graph.* **2009**, *33*, 321–330. [CrossRef]
4. Schulz, T.; Jüttler, B. Envelope computation in the plane by approximate implicitization. *Appl. Algebra Eng. Commun. Comput.* **2011**, *22*, 265–288. [CrossRef]
5. Kanatani, K.; Sugaya, Y. Unified computation of strict maximum likelihood for geometric fitting. *J. Math. Imaging Vis.* **2010**, *38*, 1–13. [CrossRef]
6. Mullen, P.; De Goes, F.; Desbrun, M.; Cohen-Steiner, D.; Alliez, P. Signing the unsigned: Robust surface reconstruction from raw pointsets. *Comput. Graph. Forum* **2010**, *29*, 1733–1741. [CrossRef]
7. Upreti, K.; Subbarayan, G. Signed algebraic level sets on NURBS surfaces and implicit Boolean signed algebraic level sets on NURBS surfaces and implicit Boolean. *Comput. Aided Des.* **2017**, *82*,112–126. [CrossRef]
8. Rouhani, M., Sappa, A.D. Implicit polynomial representation through a fast fitting error estimation. *IEEE Trans. Image Process.* **2012**, *21*, 2089–2098. [CrossRef]

9. Zagorchev, L.G.; Goshtasby, A.A. A curvature-adaptive implicit surface reconstruction for irregularly spaced points. *IEEE Trans. Vis. Comput. Graph.* **2012**, *18*, 1460–1473. [CrossRef]

10. William, H.P.; Brian, P.F.; Teukolsky, S.A.; William, T.V. *Numerical Recipes in C: The Art of Scientific Computing*, 2nd ed.; Cambridge University Press: Cambridge, UK, 1992.

11. Steve, S.; Sandford, L.; Ponce, J. Using geometric distance fits for 3-D object modeling and recognition. *IEEE Trans. Pattern Anal. Mach. Intell.* **1994**, *16*, 1183–1196.

12. Morgan, A.P. Polynomial continuation and its relationship to the symbolic reduction of polynomial systems. In *Symbolic and Numerical Computation for Artificial Intelligence*; Academic Press: Cambridge, MA, USA, 1992; pp. 23–45.

13. Layne, T.W.; Billups, S.C.; Morgan, A.P. Algorithm 652: HOMPACK: A suite of codes for globally convergent homotopy algorithms. *ACM Trans. Math. Softw.* **1987**, *13*, 281–310.

14. Berthold, K.P.H. Relative orientation revisited. *J. Opt. Soc. Am. A* **1991**, *8*, 1630–1638.

15. Dinesh, M.; Krishnan, S. Solving algebraic systems using matrix computations. *ACM Sigsam Bull.* **1996**, *30*, 4–21.

16. Chionh, E.-W. Base Points, Resultants, and the Implicit Representation of Rational Surfaces. Ph.D. Thesis, University of Waterloo, Waterloo, ON, Canada, 1990.

17. De Montaudouin, Y.; Tiller, W. The Cayley method in computer aided geometric design. *Comput. Aided Geom. Des.* **1984**, *1*, 309–326. [CrossRef]

18. Albert, A.A. *Modern Higher Algebra*; D.C. Heath and Company: New York, NY, USA, 1933.

19. Thomas, W.; David, S.; Anderson, C.; Goldman, R.N. Implicit representation of parametric curves and surfaces. *Comput. Vis. Graph. Image Proc.* **1984**, *28*, 72–84.

20. Nishita, T.; Sederberg, T.W.; Kakimoto, M. Ray tracing trimmed rational surface patches. *ACM SIGGRAPH Comput. Graph.* **1990**, *24*, 337–345. [CrossRef]

21. Elber, G.; Kim, M.-S. Geometric Constraint Solver Using Multivariate Rational Spline Functions. In Proceedings of the 6th ACM Symposium on Solid Modeling and Applications, Ann Arbor, MI, USA, 4–8 June 2001; pp. 1–10.

22. Sherbrooke, E.C.; Patrikalakis, N.M. Computation of the solutions of nonlinear polynomial systems. *Comput. Aided Geom. Des.* **1993**, *10*, 379–405. [CrossRef]

23. Hartmann, E. The normal form of a planar curve and its application to curve design. In *Mathematical Methods for Curves and Surfaces II*; Vanderbilt University Press: Nashville, TN, USA, 1997; pp. 237–244.

24. Hartmann, E. On the curvature of curves and surfaces defined by normal forms. *Comput. Aided Geom. Des.* **1999**, *16*, 355–376. [CrossRef]

25. Nicholas, J.R. Implicit polynomials, orthogonal distance regression, and the closest point on a curve. *IEEE Trans. Pattern Anal. Mach. Intell.* **2000**, *22*, 191–199.

26. Martin, A.; Bert, J. Robust computation of foot points on implicitly defined curves. In *Mathematical Methods for Curves and Surfaces: Troms?* Nashboro Press: Brentwood, TN, USA, 2004; pp. 1–10.

27. Hu, M.; Zhou, Y.; Li, X. Robust and accurate computation of geometric distance for Lipschitz continuous implicit curves. *Vis. Comput.* **2017**, *33*, 937–947. [CrossRef]

28. Li, X.; Pan, F.; Cheng, T.; Wu, Z.; Liang, J.; Hou, L. Integrated hybrid second order algorithm for orthogonal projection onto a planar implicit curve. *Symmetry* **2018**, *10*, 164. [CrossRef]

29. Goldman R. Curvature formulas for implicit curves and surfaces. *Comput. Aided Geom. Des.* **2005**, *22*, 632–658. [CrossRef]

![Σ mathematics logo]

mathematics

MDPI

Article

Nonlinear Operators as Concerns Convex Programming and Applied to Signal Processing

Anantachai Padcharoen [1] and **Pakeeta Sukprasert** [2,*]

[1] Department of Mathematics, Faculty of Science and Technology, Rambhai Barni Rajabhat University, Chanthaburi 22000, Thailand; anantachai.p@rbru.ac.th
[2] Department of Mathematics and Computer Science, Faculty of Science and Technology, Rajamangala University of Technology Thanyaburi (RMUTT), Thanyaburi, Pathumthani 12110, Thailand
* Correspondence: pakeeta_s@rmutt.ac.th

Received: 16 August 2019; Accepted: 13 September 2019; Published: 19 September 2019

Abstract: Splitting methods have received a lot of attention lately because many nonlinear problems that arise in the areas used, such as signal processing and image restoration, are modeled in mathematics as a nonlinear equation, and this operator is decomposed as the sum of two nonlinear operators. Most investigations about the methods of separation are carried out in the Hilbert spaces. This work develops an iterative scheme in Banach spaces. We prove the convergence theorem of our iterative scheme, applications in common zeros of accretive operators, convexly constrained least square problem, convex minimization problem and signal processing.

Keywords: convexity; least square problem; accretive operators; signal processing

1. Introduction

Let $\mathcal{E}$ be a real Banach space. The zero point problem is as follows:

$$\text{find} \quad x \in \mathcal{E} \quad \text{such that} \quad 0 \in \mathcal{A}u + \mathcal{B}u, \tag{1}$$

where $\mathcal{A} : \mathcal{E} \to \mathcal{E}$ is an operator and $\mathcal{B} : \mathcal{E} \to 2^{\mathcal{E}}$ is a set-valued operator. This problem includes, as special cases, convex programming, variational inequalities, split feasibility problem and minimization problem [1–7]. To be more precise, some concrete problems in machine learning, image processing [4,5], signal processing and linear inverse problem can be modeled mathematically as the form in Equation (1).

Signal processing and numerical optimization are independent scientific fields that have always been mutually influencing each other. Perhaps the most convincing example where the two fields have met is compressed sensing (CS) [2]. Several surveys dedicated to these algorithms and their applications in signal processing have appeared [3,6–8]

Fixed point iterations is an important tool for solving various problems and is known in a Banach space $\mathcal{E}$. Let $\mathcal{K}$ be a nonempty closed convex subset of $\mathcal{E}$ and $\mathcal{S} : \mathcal{K} \to \mathcal{K}$ is the operator with at least one fixed point. Then, for $u_1 \in \mathcal{K}$:

1. The Picard iterative scheme [9] is defined by:

$$u_{n+1} = \mathcal{S}u_n, \quad \forall n \in \mathbb{N}.$$

2. The Mann iterative scheme [10] is defined by:

$$u_{n+1} = (1 - \eta_n)u_n + \eta_n \mathcal{S}u_n, \quad \forall n \in \mathbb{N},$$

where $\{\eta_n\}$ is a sequence in $(0, 1)$.

3. The Ishikawa iterative scheme [11] is defined by:

$$u_{n+1} = (1 - \eta_n)u_n + \eta_n S[(1 - \vartheta_n)u_n + \vartheta_n Su_n], \quad \forall n \in \mathbb{N},$$

where $\{\eta_n\}$ and $\{\vartheta_n\}$ are sequences in $(0,1)$.

4. The S-iterative scheme [12] is defined by:

$$u_{n+1} = (1 - \eta_n)Su_n + \eta_n S[(1 - \vartheta_n)u_n + \vartheta_n Su_n], \quad \forall n \in \mathbb{N},$$

where $\{\eta_n\}$ and $\{\vartheta_n\}$ are sequences in $(0,1)$.

Recently, Sahu et al. [13] and Thakur et al. [14] introduced the following same iterative scheme for nonexpansive mappings in uniformly convex Banach space:

$$\begin{cases} w_n = (1 - \xi_n)u_n + \xi_n Su_n, \\ z_n = (1 - \vartheta_n)w_n + \vartheta_n Sw_n, \\ u_{n+1} = (1 - \eta_n)Sw_n + \eta_n Sz_n, \quad \forall n \in \mathbb{N}, \end{cases}$$

where $\{\eta_n\}$, $\{\vartheta_n\}$ and $\{\xi_n\}$ are sequences in $(0,1)$. The authors proved that this scheme converges to a fixed point of contraction mapping, faster than all known iterative schemes. In addition, the authors provided an example to support their claim.

In this paper, we first develop an iterative scheme for calculating common solutions and using our results to solve the problem in Equation (1). Secondly, we find common solutions of convexly constrained least square problems, convex minimization problems and applied to signal processing.

2. Preliminaries

Let $\mathcal{E}$ be a real Banach space with norm $\|\cdot\|$ and $\mathcal{E}^*$ be its dual. The value of $f \in \mathcal{E}^*$ at $u \in \mathcal{E}$ ia denoted by $\langle u, f \rangle$. A Banach space $\mathcal{E}$ is called strictly convex if $\frac{\|u+v\|}{2} < 1$, for all $u, v \in \mathcal{E}$ with $\|u\| = \|v\| = 1$. It is called uniformly convex if $\lim_{n \to \infty} \|u_n - v_n\| = 0$ for any two sequences $\{u_n\}, \{v_n\}$ in $\mathcal{E}$ such that $\|u\| = \|v\| = 1$ and $\lim_{n \to \infty} \frac{\|u+v\|}{2} = 1$.

The (normalized) duality mapping $\mathcal{J}$ from $\mathcal{E}$ into the family of nonempty (by Hahn Banach theorem) weak-star compact subsets of its dual $\mathcal{E}$ is defined by

$$\mathcal{J}(u) = \{f \in \mathcal{E}^* : \langle u, f \rangle = \|u\|^2 = \|f\|^2\}$$

for each $u \in \mathcal{E}$, where $\langle \cdot, \cdot \rangle$ denotes the generalized duality pairing.

For an operator $A : \mathcal{E} \to 2^{\mathcal{E}}$, we denote its domain, range and graph as follows:

$$\mathcal{D}(A) = \{u \in \mathcal{E} : Au \neq \varnothing\}$$
$$\mathcal{R}(A) = \cup\{Ap : p \in \mathcal{D}(A)\},$$

and

$$\mathcal{G}(A) = \{(u,v) \in \mathcal{E} \times \mathcal{E} : u \in \mathcal{D}(A), v \in Au\},$$

respectively. The inverse A^{-1} of A is defined by $u \in A^{-1}v$, if and only if $v \in Au$. If $\forall u_i \in \mathcal{D}(A)$ and $v_i \in Au_i$ $(i = 1, 2)$, and there is $j \in \mathcal{J}(u_1 - u_2)$ such that $\langle v_1 - v_2, j \rangle \geq 0$, then A is called accretive.

An accretive operator A in a Banach space $\mathcal{E}$ is said to satisfy the range condition if $\overline{\mathcal{D}(A)} \subset \mathcal{R}(I + \mu A)$ for all $\mu > 0$, where $\overline{\mathcal{D}(A)}$ denotes the closure of the domain of A. We know that for an accretive operator A which satisfies the range condition, $A^{-1}0 = \mathcal{F}ix(\mathcal{J}_\mu^A)$ for all $\mu > 0$.

A point $u \in \mathcal{K}$ is a fixed point of S provided $Su = u$. Denote by $\mathcal{F}ix(S)$ the set of fixed points of S, i.e., $\mathcal{F}ix(S) = \{u \in \mathcal{K} : Su = u\}$.

1. The mapping S is called $\mathcal{L}-$Lipschitz, $\mathcal{L} > 0$, if

$$\|Su - Sv\| \le \mathcal{L}\|u - v\|, \quad \forall u, v \in \mathcal{K}.$$

2. The mapping S is called nonexpansive if

$$\|Su - Sv\| \le \|u - v\|, \quad \forall u, v \in \mathcal{K}.$$

3. The mapping S is called quasi-nonexpansive if $Fix(S) \ne \emptyset$ and

$$\|Su - v\| \le \|u - v\|, \quad \forall u \in \mathcal{K}, v \in Fix(S).$$

In this case, $\mathcal{H}$ is a real Hilbert space. If $A : \mathcal{E} \to 2^{\mathcal{E}}$ is an $m-$accretive operator (see [15–17]), then A is called maximal accretive operator [18], and for all $\mu > 0$, $\mathcal{R}(I + \mu A) = \mathcal{H}$ if and only if A is called maximal monotone [19]. Denote by $\text{dom}(h)$ the domain of a function $h : \mathcal{H} \to (-\infty, \infty]$, i.e.,

$$\text{dom}(h) = \{u \in \mathcal{H} : h(u) < \infty\}.$$

The subdifferential of $h \in \Gamma_0(\mathcal{H})$ at $u \in \mathcal{H}$ is the set

$$\partial h(u) = \{z \in \mathcal{H} : h(u) \le h(v) + \langle z, u - v \rangle, \quad \forall v \in \mathcal{H}\},$$

where $\Gamma_0(\mathcal{H})$ denotes the class of all *l.s.c.* functions from $\mathcal{H}$ to $(-\infty, \infty]$ with nonempty domains.

Lemma 1 ([20]). *Let $h \in \Gamma_0(\mathcal{H})$. Then, ∂h is maximal monotone.*

We denote by $B_\lambda[v]$ the closed ball with the center at v and radius λ :

$$B_\lambda[v] = \{u \in \mathcal{E} : \|v - u\| \le \lambda\}.$$

Lemma 2 ([21]). *Let $\mathcal{E}$ be a Banach space, and $p > 1$ and $R > 0$ be two fixed numbers. Then, $\mathcal{E}$ is uniformly convex if and only if there exists a continuous, strictly increasing, and convex function $\varphi : [0, \infty) \to [0, \infty)$ with $\varphi(0) = 0$ such that*

$$\|\alpha u + (1 - \alpha)v\|^p \le \|u\|^p + (1 - \alpha)\|v\|^p - \alpha(1 - \alpha)\varphi(\|u - v\|),$$

for all $u, v \in B_R[0]$ and $\alpha \in [0, 1]$.

Definition 1 ([22]). *A vector space $\mathcal{H}$ is said to satisfy Opial's condition, if for each sequence $\{u_n\}$ in $\mathcal{H}$ which converges weakly to point $u \in \mathcal{H}$,*

$$\liminf_{n \to \infty} \|u_n - u\| < \liminf_{n \to \infty} \|u_n - v\|, \quad v \in \mathcal{H}, v \ne u.$$

Lemma 3 ([23]). *Let $\mathcal{K}$ be a nonempty subset of a Banach space $\mathcal{E}$, let $S : \mathcal{K} \to \mathcal{E}$ be a uniformly continuous mapping, and let $\{u_n\} \subset \mathcal{K}$ an approximating fixed point sequence of S. Then, $\{v_n\}$ is an approximating fixed point sequence of S whenever $\{v_n\}$ is in $\mathcal{K}$ such that $\lim_{n \to \infty} \|u_n - v_n\| = 0$.*

Lemma 4 ([16]). *Let $\mathcal{K}$ be a nonempty closed convex subset of a uniformly convex Banach space $\mathcal{E}$. If $S : \mathcal{K} \to \mathcal{E}$ is a nonexpansive mapping, then $I - S$ has the demiclosed property with respect to 0.*

A subset $\mathcal{K}$ of Banach space $\mathcal{E}$ is called a retract of $\mathcal{E}$ if there is a continuous mapping $\mathcal{Q}$ from $\mathcal{E}$ onto $\mathcal{K}$ such that $\mathcal{Q}u = u$ for all $u \in \mathcal{K}$. We call such $\mathcal{Q}$ a retraction of $\mathcal{E}$ onto $\mathcal{K}$. It follows that, if a mapping $\mathcal{Q}$ is a retraction, then $\mathcal{Q}v = v$ for all v in the range of $\mathcal{Q}$. A retraction $\mathcal{Q}$ is called a sunny if

$\mathfrak{Q}(\mathfrak{Q}u + \lambda(u - \mathfrak{Q}u)) = \mathfrak{Q}u$ for all $u \in \mathcal{E}$ and $\lambda \geq 0$. If a sunny retraction $\mathfrak{Q}$ is also nonexpansive, then $\mathcal{K}$ is called a sunny nonexpansive retract of $\mathcal{E}$ [24].

Let $\mathcal{E}$ be a strictly convex reflexive Banach space and $\mathcal{K}$ be a nonempty closed convex subset of $\mathcal{E}$. Denote by $\mathcal{P}_{\mathcal{K}}$ the (metric) projection from $\mathcal{E}$ onto $\mathcal{K}$, namely, for $u \in \mathcal{E}$, $\mathcal{P}_{\mathcal{K}}(u)$ is the unique point in $\mathcal{K}$ with the property

$$\inf\{\|u - v\| : v \in \mathcal{K}\} = \|u - \mathcal{P}_{\mathcal{K}}(u)\|.$$

Let an inner product $\langle \cdot, \cdot \rangle$ and the induced norm $\| \cdot \|$ are specified with a real Hilbert space $\mathcal{H}$. Let $\mathcal{K}$ is a nonempty subset of $\mathcal{H}$, we have the nearest point projection $\mathcal{P}_{\mathcal{K}} : \mathcal{H} \to \mathcal{K}$ is the unique sunny nonexpansive retraction of $\mathcal{H}$ onto $\mathcal{K}$. It is also known that $\mathcal{P}_{\mathcal{K}}(u) \in \mathcal{K}$ and

$$\langle u - \mathcal{P}_{\mathcal{K}}(u), \mathcal{P}_{\mathcal{K}}(u) - v \rangle \geq 0, \quad \forall u \in \mathcal{H}, v \in \mathcal{K}.$$

3. Main Results

Let $\mathcal{K}$ be a nonempty closed convex subset of a Banach space $\mathcal{E}$ with $\mathfrak{Q}_{\mathcal{K}}$ as a sunny nonexpansive retraction. We denote by $\Psi := \mathcal{F}ix(\mathcal{S}) \cap \mathcal{F}ix(\mathcal{T})$.

Lemma 5. *Let $\mathcal{K}$ be a nonempty closed convex subset of a Banach space $\mathcal{E}$ with $\mathfrak{Q}_{\mathcal{K}}$ as the sunny nonexpansive retraction, let $\mathcal{S}, \mathcal{T} : \mathcal{K} \to \mathcal{E}$ be quasi-nonexpansive mappings which $\Psi \neq \varnothing$, and let $\{\eta_n\}, \{\vartheta_n\}$ and $\{\xi_n\}$ be sequences in $(0, 1)$ for all $n \in \mathbb{N}$. Let $\{u_n\}$ be defined by Algorithm 1. Then, for each $\bar{u} \in \Psi$, $\lim_{n \to \infty} \|u_n - \bar{u}\|$ exists and*

$$\|w_n - \bar{u}\| \leq \|u_n - \bar{u}\|, \quad \text{and} \quad \|z_n - \bar{u}\| \leq \|u_n - \bar{u}\|, \quad \forall n \in \mathbb{N}. \tag{2}$$

Algorithm 1: Three-step sunny nonexpansive retraction

---

initialization: $\eta_n, \vartheta_n, \xi_n \in (0, 1)$, $u_1 \in \mathcal{K}$ and $n = 1$.
while *stopping criterion not met* **do**
$\quad w_n = \mathfrak{Q}_{\mathcal{K}}[(1 - \xi_n)u_n + \xi_n \mathcal{S}u_n]$,
$\quad z_n = \mathfrak{Q}_{\mathcal{K}}[(1 - \vartheta_n)w_n + \vartheta_n \mathcal{T}w_n]$,
$\quad u_{n+1} = \mathfrak{Q}_{\mathcal{K}}[(1 - \eta_n)\mathcal{S}w_n + \eta_n \mathcal{T}z_n]$.
end

Proof. Let $\bar{u} \in \Psi$. Then, we have

$$\begin{aligned}
\|w_n - \bar{u}\| &= \|\mathfrak{Q}_{\mathcal{K}}[(1 - \xi_n)u_n + \xi_n \mathcal{S}u_n] - \bar{u}\| \\
&\leq \|(1 - \xi_n)(u_n - \bar{u}) + \xi_n(\mathcal{S}u_n - \bar{u})\| \\
&\leq (1 - \xi_n)\|u_n - \bar{u}\| + \xi_n\|\mathcal{S}u_n - \bar{u}\| \\
&\leq (1 - \xi_n)\|u_n - \bar{u}\| + \xi_n\|u_n - \bar{u}\| \\
&= \|u_n - \bar{u}\|,
\end{aligned} \tag{3}$$

$$\begin{aligned}
\|z_n - \bar{u}\| &= \|\mathfrak{Q}_{\mathcal{K}}[(1 - \vartheta_n)w_n + \vartheta_n \mathcal{T}w_n] - \bar{u}\| \\
&\leq \|(1 - \vartheta_n)(w_n - \bar{u}) + \vartheta_n(\mathcal{T}w_n - \bar{u})\| \\
&\leq (1 - \vartheta_n)\|w_n - \bar{u}\| + \vartheta_n\|\mathcal{T}w_n - \bar{u}\| \\
&\leq (1 - \vartheta_n)\|w_n - \bar{u}\| + \vartheta_n\|w_n - \bar{u}\| \\
&= \|w_n - \bar{u}\| \\
&\leq \|u_n - \bar{u}\|,
\end{aligned} \tag{4}$$

and

$$\begin{aligned}
\|u_{n+1} - \bar{u}\| &= \|\mathcal{Q}_{\mathcal{K}}[(1 - \eta_n)\mathcal{S}w_n + \eta_n \mathcal{T}z_n] - \bar{u}\| \\
&\leq \|(1 - \eta_n)(\mathcal{S}w_n - \bar{u}) + \eta_n(\mathcal{T}z_n - \bar{u})\| \\
&\leq (1 - \eta_n)\|\mathcal{S}w_n - \bar{u}\| + \eta_n\|\mathcal{T}z_n - \bar{u}\| \\
&\leq (1 - \eta_n)\|w_n - \bar{u}\| + \eta_n\|z_n - \bar{u}\| \\
&\leq (1 - \eta_n)\|u_n - \bar{u}\| + \eta_n\|u_n - \bar{u}\| \\
&= \|u_n - \bar{u}\|.
\end{aligned} \tag{5}$$

Therefore,

$$\|u_{n+1} - \bar{u}\| \leq \|u_n - \bar{u}\| \leq \cdots \leq \|u_1 - \bar{u}\|, \quad \forall n \in \mathbb{N}. \tag{6}$$

Since $\{\|u_n - \bar{u}\|\}$ is monotonically decreasing, we have that the sequence $\{\|u_n - \bar{u}\|\}$ is convergent. $\square$

From Lemma 5, we have results:

Theorem 1. *Let* $\mathcal{K}$ *be a nonempty closed convex subset of a Banach space* $\mathcal{E}$ *with* $\mathcal{Q}_{\mathcal{K}}$ *as the sunny nonexpansive retraction, let* $\mathcal{S}, \mathcal{T} : \mathcal{K} \to \mathcal{E}$ *be quasi-nonexpansive mappings with* $\Psi \neq \varnothing$, *and let* $\{\eta_n\}, \{\vartheta_n\}$ *and* $\{\zeta_n\}$ *be sequences of real numbers, for which* $0 < c_1 \leq \eta_n \leq \hat{c}_1 < 1$, $0 < c_2 \leq \vartheta_n \leq \hat{c}_2 < 1$, $0 < c_3 \leq \zeta_n \leq \hat{c}_3 < 1$ *for all* $n \in \mathbb{N}$. *Let* $u_1 \in \mathcal{K}$, $\mathcal{P}_\Psi(u_1) = u_*$ *and* $\{u_n\}$ *is defined by Algorithm 1. Then, we have the following:*

(i) $\{u_n\}$ *is in a closed convex bounded set* $B_\lambda[u_*] \cap \mathcal{K}$, *where* λ *is a constant in* $(0, \infty)$ *such that* $\|u_1 - u_*\| \leq \lambda$.

(ii) *If* $\mathcal{S}$ *is uniformly continuous, then* $\lim_{n \to \infty} \|u_n - \mathcal{S}u_n\| = 0$ *and* $\lim_{n \to \infty} \|u_n - \mathcal{T}u_n\| = 0$.

(iii) *If* $\mathcal{E}$ *fulfills the Opial's condition and* $I - \mathcal{S}$ *and* $I - \mathcal{T}$ *are demiclosed at 0, then* $\{u_n\}$ *converges weakly to an element of* $\Psi \cap B_\lambda[u_*]$.

Proof. (i) Since $u_* \in \Psi$, from Equation (6), we obtain

$$\|u_{n+1} - u_*\| \leq \|u_n - u_*\| \leq \cdots \leq \|u_1 - u_*\| \leq \lambda, \quad \forall n \in \mathbb{N}. \tag{7}$$

Therefore, $\{u_n\}$ is in the closed convex bounded set $B_\lambda[u_*] \cap \mathcal{K}$.

(ii) Suppose that $\mathcal{S}$ is uniformly continuous. Using Lemma 5, we get that $\{u_n\}$, $\{z_n\}$ and $\{w_n\}$ are in $B_\lambda[u_*] \cap \mathcal{K}$, and hence, from Equation (2), we obtain

$$\|\mathcal{T}w_n - u_*\| \leq \lambda, \quad \|\mathcal{S}w_n - u_*\| \leq \lambda \quad \text{and} \quad \|\mathcal{S}u_n - u_*\| \leq \lambda, \quad \forall n \in \mathbb{N}.$$

Using Lemma 2 for $p = 2$ and $R = \lambda$, from Equation (5), we obtain

$$\begin{aligned}
\|u_{n+1} - u_*\|^2 &\leq \|(1 - \eta_n)(\mathcal{S}w_n - u_*) + \eta_n(\mathcal{T}z_n - u_*)\|^2 \\
&\leq (1 - \eta_n)\|\mathcal{S}w_n - u_*\|^2 + \eta_n\|\mathcal{T}z_n - u_*\|^2 \\
&\quad - \eta_n(1 - \eta_n)\varphi(\|\mathcal{S}w_n - \mathcal{T}z_n\|) \\
&\leq (1 - \eta_n)\|w_n - u_*\|^2 + \eta_n\|z_n - u_*\|^2 \\
&\quad - \eta_n(1 - \eta_n)\varphi(\|\mathcal{S}w_n - \mathcal{T}z_n\|) \\
&\leq (1 - \eta_n)\|u_n - u_*\|^2 + \eta_n\|u_n - u_*\|^2 \\
&\quad - \eta_n(1 - \eta_n)\varphi(\|\mathcal{S}w_n - \mathcal{T}z_n\|) \\
&= \|u_n - u_*\|^2 - \eta_n(1 - \eta_n)\varphi(\|\mathcal{S}w_n - \mathcal{T}z_n\|),
\end{aligned} \tag{8}$$

which implies that

$$\eta_n(1 - \eta_n)\varphi(\|\mathcal{S}w_n - \mathcal{T}z_n\|) = \|u_n - u_*\| - \|u_{n+1} - u_*\|^2. \tag{9}$$

Note that: $c_1(1 - \hat{c}_1) \leq \eta_n(1 - \eta_n)$. Thus,

$$c_1(1 - \hat{c}_1) \sum_{i=1}^{n} \varphi(\|\mathcal{S}w_i - \mathcal{T}z_i\|) = \|u_1 - u_*\| - \|u_{n+1} - u_*\|^2, \quad \forall n \in \mathbb{N}. \tag{10}$$

In the same way, we obtain

$$c_1(1 - \hat{c}_1) \sum_{n=1}^{\infty} \varphi(\|\mathcal{S}w_n - \mathcal{T}z_n\|) \leq \|u_1 - u_*\| < \infty. \tag{11}$$

Therefore, we have $\lim_{n \to \infty} \|\mathcal{S}w_n - \mathcal{T}z_n\| = 0$. From the relations in Algorithm 1, we obtain

$$
\begin{aligned}
\|w_n - u_*\|^2 &\leq (1 - \xi_n)\|u_n - u_*\|^2 + \xi_n\|\mathcal{S}u_n - u_*\|^2 \\
&\quad - \xi_n(1 - \xi_n)\varphi(\|u_n - \mathcal{S}u_n\|) \\
&\leq (1 - \xi_n)\|u_n - u_*\|^2 + \xi_n\|u_n - u_*\|^2 \\
&\quad - \xi_n(1 - \xi_n)\varphi(\|u_n - \mathcal{S}u_n\|) \\
&= \|u_n - u_*\|^2 - \xi_n(1 - \xi_n)\varphi(\|u_n - \mathcal{S}u_n\|)
\end{aligned}
\tag{12}
$$

and

$$
\begin{aligned}
\|z_n - u_*\|^2 &\leq \|(1 - \vartheta_n)(w_n - u_*) + \vartheta_n(\mathcal{T}w_n - u_*)\|^2 \\
&\leq (1 - \vartheta_n)\|w_n - u_*\|^2 + \vartheta_n\|\mathcal{T}w_n - u_*\|^2 \\
&\quad - \vartheta_n(1 - \vartheta_n)\varphi(\|w_n - \mathcal{T}w_n\|) \\
&\leq (1 - \vartheta_n)\|w_n - u_*\|^2 + \vartheta_n\|w_n - u_*\|^2 \\
&= \|w_n - u_*\|^2 - \vartheta_n(1 - \vartheta_n)\varphi(\|w_n - \mathcal{T}w_n\|) \\
&\leq \|u_n - u_*\|^2 - \vartheta_n(1 - \vartheta_n)\varphi(\|w_n - \mathcal{T}w_n\|).
\end{aligned}
\tag{13}
$$

From Equations (8), (13) and (12), we obtain

$$
\begin{aligned}
\|u_{n+1} - u_*\|^2 &\leq \|(1 - \eta_n)(\mathcal{S}w_n - u_*) + \eta_n(\mathcal{T}z_n - u_*)\|^2 \\
&\leq (1 - \eta_n)\|\mathcal{S}w_n - u_*\|^2 + \eta_n\|\mathcal{T}z_n - u_*\|^2 \\
&\quad - \eta_n(1 - \eta_n)\varphi(\|\mathcal{S}w_n - \mathcal{T}z_n\|) \\
&\leq (1 - \eta_n)\|w_n - u_*\|^2 + \eta_n\|z_n - u_*\|^2 \\
&\quad - \eta_n(1 - \eta_n)\varphi(\|\mathcal{S}w_n - \mathcal{T}z_n\|) \\
&\leq (1 - \eta_n)[\|u_n - u_*\|^2 - \xi_n(1 - \xi_n)\varphi(\|u_n - \mathcal{S}u_n\|)] \\
&\quad + \eta_n[\|u_n - u_*\|^2 - \vartheta_n(1 - \vartheta_n)\varphi(\|w_n - \mathcal{T}w_n\|)] \\
&\quad - \eta_n(1 - \eta_n)\varphi(\|\mathcal{S}w_n - \mathcal{T}z_n\|) \\
&= \|u_n - u_*\|^2 - (1 - \eta_n)\xi_n(1 - \xi_n)\varphi(\|u_n - \mathcal{S}u_n\|) \\
&\quad - \eta_n\vartheta_n(1 - \vartheta_n)\varphi(\|w_n - \mathcal{T}w_n\|) \\
&\quad - \eta_n(1 - \eta_n)\varphi(\|\mathcal{S}w_n - \mathcal{T}z_n\|).
\end{aligned}
\tag{14}
$$

Note that: $(1 - \hat{c}_1)c_3(1 - \hat{c}_3) \leq (1 - \eta_n)\xi_n(1 - \xi_n)$ and $c_1c_2(1 - \hat{c}_2) \leq \eta_n\vartheta_n(1 - \vartheta_n)$. Thus,

$$(1 - \hat{c}_1)c_3(1 - \hat{c}_3) \sum_{i=1}^{n} \varphi(\|u_i - \mathcal{S}u_i\|) \leq \|u_1 - u_*\|^2 - \|u_{n+1} - u_*\|^2, \quad \forall n \in \mathbb{N}.$$

It follows that $\lim_{n \to \infty} \|u_n - \mathcal{S}u_n\| = 0$. Note that:

$$
\begin{aligned}
\|w_n - u_n\| &= \|\mathcal{Q}_{\mathcal{K}}[(1 - \xi_n)u_n + \xi_n\mathcal{S}u_n] - \mathcal{Q}_{\mathcal{K}}[u_n]\| \\
&\leq \|\mathcal{S}u_n - u_n\| \to 0 \quad \text{as} \quad n \to \infty.
\end{aligned}
$$

Since $\mathcal{S}$ is uniformly continuous, it follows from Lemma 3 that $\lim_{n\to\infty}\|w_n - \mathcal{S}w_n\| = 0$. Thus, from $\lim_{n\to\infty}\|\mathcal{S}w_n - \mathcal{T}z_n\| = 0$, we obtain $\lim_{n\to\infty}\|u_n - \mathcal{T}u_n\| = 0$.

(iii) By assumption, $\mathcal{E}$ satisfies the Opial's condition. Let $w^* \in \Psi$ such that $w^* \in B_\lambda[u_*] \cap \mathcal{K}$. From Lemma 5, we have $\lim_{n\to\infty}\|u_n - w^*\|$ exists. Suppose there are two subsequences $\{u_{n_q}\}$ and $\{u_{m_l}\}$ which converge to two distinct points u^* and v^* in $B_\lambda[u_*] \cap \mathcal{K}$, respectively. Then, since both $I - \mathcal{S}$ and $I - \mathcal{T}$ have the demiclosed property at 0, we have $\mathcal{S}u^* = \mathcal{T}u^* = u^*$ and $\mathcal{S}v^* = \mathcal{T}v^* = v^*$. Moreover, using the Opial's condition:

$$\lim_{n\to\infty}\|u_n - u^*\| = \lim_{q\to\infty}\|u_{n_q} - u^*\| < \lim_{l\to\infty}\|u_{m_l} - v^*\| = \lim_{n\to\infty}\|u_n - v^*\|.$$

Similarly, we obtain

$$\lim_{n\to\infty}\|u_n - v^*\| < \lim_{n\to\infty}\|u_n - u^*\|,$$

which is a contradiction. Therefore, $u^* = v^*$. Hence, the sequence $\{u_n\}$ converges weakly to an element of $\Psi \cap B_\lambda[u_*] \cap \mathcal{K}$. □

Theorem 2. *Let $\mathcal{K}$ be a nonempty closed convex subset of a Banach space $\mathcal{E}$ with $\mathcal{Q}_\mathcal{K}$ as the sunny nonexpansive retraction, let $\mathcal{S}, \mathcal{T} : \mathcal{K} \to \mathcal{E}$ be nonexpansive mappings with $\Psi \neq \emptyset$, and let $\{\eta_n\}, \{\vartheta_n\}$ and $\{\xi_n\}$ be sequences of real numbers, for which $0 < c_1 \leq \eta_n \leq \hat{c}_1 < 1, 0 < c_2 \leq \vartheta_n \leq \hat{c}_2 < 1, 0 < c_3 \leq \xi_n \leq \hat{c}_3 < 1$ for all $n \in \mathbb{N}$. Let $u_1 \in \mathcal{K}$, $\mathcal{P}_\Psi(u_1) = u_*$ and $\{u_n\}$ is defined by Algorithm 1. Then, we have the following:*

(i) *$\{u_n\}$ is in a closed convex bounded set $B_\lambda[u_*] \cap \mathcal{K}$, where λ is a constant in $(0,\infty)$ such that $\|u_1 - u_*\| \leq \lambda$.*
(ii) *$\lim_{n\to\infty}\|u_n - \mathcal{S}u_n\| = 0$ and $\lim_{n\to\infty}\|u_n - \mathcal{T}u_n\| = 0$.*
(iii) *If $\mathcal{E}$ fulfills the Opial's condition, then $\{u_n\}$ converges weakly to an element of $\Psi \cap B_\lambda[u_*]$.*

Proof. It follows from Theorem 1. □

Corollary 1. *Let $\mathcal{K}$ be a nonempty closed convex subset of a real Hilbert space $\mathcal{H}$, let $\mathcal{S}, \mathcal{T} : \mathcal{K} \to \mathcal{E}$ be nonexpansive mappings with $\Psi \neq \emptyset$, and let $\{\eta_n\}, \{\vartheta_n\}$ and $\{\xi_n\}$ be sequences of real numbers, for which $0 < c_1 \leq \eta_n \leq \hat{c}_1 < 1, 0 < c_2 \leq \vartheta_n \leq \hat{c}_2 < 1, 0 < c_3 \leq \xi_n \leq \hat{c}_3 < 1$ for all $n \in \mathbb{N}$. Let $\{u_n\}$ be defined by*

$$\begin{cases} w_n = (1 - \xi_n)u_n + \xi_n \mathcal{S}u_n, \\ z_n = (1 - \vartheta_n)w_n + \vartheta_n \mathcal{T}w_n, \\ u_{n+1} = (1 - \eta_n)\mathcal{S}w_n + \eta_n \mathcal{T}z_n, \quad \forall n \in \mathbb{N}. \end{cases} \tag{15}$$

Then, $\{u_n\}$ converges weakly to an element of Ψ.

Proof. It follows from Theorem 1. □

4. Applications

4.1. Common Zeros of Accretive Operators

From Equation (15), we set $\mathcal{S} = J_\mu^A$ and $\mathcal{T} = J_\mu^B$, and inherit the convergence analysis for solving Equation (1).

Theorem 3. *Let $\mathcal{K}$ be a nonempty closed convex subset of a r.u.c. Banach space $\mathcal{E}$ satisfying the Opial's condition. Let $A : \mathcal{D}(A) \subseteq \mathcal{K} \to 2^\mathcal{E}, B : \mathcal{D}(B) \subseteq \mathcal{K} \to 2^\mathcal{E}$ be accretive operators, for which $\mathcal{D}(A) \subseteq \mathcal{K} \subseteq \cap_{\mu>0}\mathcal{R}(I + \mu A), \mathcal{D}(B) \subseteq \mathcal{K} \subseteq \cap_{\mu>0}\mathcal{R}(I + \mu B)$ and $A^{-1}(0) \cap B^{-1}(0) \neq \emptyset$. Let $\{\eta_n\}, \{\vartheta_n\}$ and $\{\xi_n\}$ be sequences of real numbers, for which $0 < c_1 \leq \eta_n \leq \hat{c}_1 < 1, 0 < c_2 \leq$*

$\vartheta_n \leq \hat{c}_2 < 1, 0 < c_3 \leq \xi_n \leq \hat{c}_3 < 1$ for all $n \in \mathbb{N}$. Let $\mu > 0$, $u_1 \in \mathcal{K}$ and $\mathcal{P}_{A^{-1}(0) \cap B^{-1}(0)}(u_1) = u_*$. Let $\{u_n\}$ be defined by

$$
\begin{cases}
w_n = (1 - \xi_n)u_n + \xi_n J_\mu^A u_n, \\
z_n = (1 - \vartheta_n)w_n + \vartheta_n J_\mu^B w_n, \\
u_{n+1} = (1 - \eta_n)J_\mu^A w_n + \eta_n J_\mu^B z_n, \quad \forall n \in \mathbb{N}.
\end{cases}
\tag{16}
$$

Then, we have the following:

(i) $\{u_n\}$ *is in a closed convex bounded set* $B_\lambda[u_*] \cap \mathcal{K}$, *where* λ *is a constant in* $(0, \infty)$ *such that* $\|u_1 - u_*\| \leq \lambda$.

(ii) $\lim_{n \to \infty} \|u_n - J_\mu^A u_n\| = 0$ *and* $\lim_{n \to \infty} \|u_n - J_\mu^B u_n\| = 0$.

(iii) $\{u_n\}$ *converges weakly to an element of* $A^{-1}(0) \cap B^{-1}(0) \cap B_\lambda[u_*]$.

Proof. By assumption $\mathcal{D}(A) \subseteq \mathcal{K} \subseteq \cap_{\mu > 0} R(I + \mu A)$, we known that $J_\mu^A, J_\mu^B : \mathcal{K} \to \mathcal{K}$ be nonexpansive. Note that $\mathcal{D}(A) \cap \mathcal{D}(B) \subseteq \mathcal{K}$ and hence

$$
\begin{aligned}
u_* \in A^{-1}(0) \cap B^{-1}(0) &\Rightarrow u_* \in \mathcal{D}(A) \cap \mathcal{D}(B) \text{ with } 0 \in Au_* \text{ and } 0 \in Bu_* \\
&\Rightarrow u_* \in \mathcal{K} \text{ with } J_\mu^A u_* = u_* \text{ and } J_\mu^B u_* = u_* \\
&\Rightarrow u_* \in Fix(J_\mu^A, J_\mu^B) \cap \mathcal{K}.
\end{aligned}
$$

Next, set $\mathcal{S} = J_\mu^A$ and $\mathcal{T} = J_\mu^B$. Hence, Theorem 3 is the same way as Theorem 2. $\square$

4.2. Convexly Constrained Least Square Problem

We provide applications of Theorem 2 for finding solutions to common problems with two convexly constrained least square problems. We consider the following problem:

Let $A, B \in \mathcal{B}(\mathcal{H})$, and $y, z \in \mathcal{H}$. Define $\varphi, \psi : \mathcal{H} \to \mathbb{R}$ by

$$
\varphi = \|Au - y\|^2 \quad \text{and} \quad \psi = \|Bu - z\|^2, \quad \forall u \in \mathcal{H},
$$

where $\mathcal{H}$ is a real Hilbert space.

Let $\mathcal{K}$ be a nonempty closed convex subset of $\mathcal{H}$. The objective is to find $b \in \mathcal{K}$ such that

$$
b \in \arg\min_{u \in \mathcal{K}} \varphi(u) \cap \arg\min_{u \in \mathcal{K}} \psi(u),
\tag{17}
$$

where

$$
\arg\min_{u \in \mathcal{K}} \varphi(u) := \{\bar{u} \in \mathcal{K} : \varphi(u_*) = \inf_{u \in \mathcal{K}} \varphi(u)\}.
$$

Proposition 1 ([8]). *Let* $\mathcal{H}$ *be a real Hilbert space,* $A \in \mathcal{B}(\mathcal{H})$ *with the adjoint* A^* *and* $y \in \mathcal{H}$. *Let* $\mathcal{K}$ *be a nonempty closed convex subset of* $\mathcal{H}$. *Let* $b \in \mathcal{H}$ *and* $\delta \in (0, \infty)$. *Then, the following statements are equivalent:*

(i) *b solves the following problem:*

$$
\min_{u \in \mathcal{K}} \|Au - y\|^2.
$$

(ii) $b = \mathcal{P}_{\mathcal{K}}(b - \delta A^*(Ab - y))$.

(iii) $\langle Av - Ab, y - Ab \rangle \leq 0$, *for all* $v \in \mathcal{K}$.

Theorem 4. *Let* $\mathcal{K}$ *be a nonempty closed convex subset of a real Hilbert space* $\mathcal{H}$, $y, z \in \mathcal{H}$ *and* $A, B \in \mathcal{B}(\mathcal{H})$, *for which the solution set of the problem in Equation* (17) *is nonempty. Let* $\{\eta_n\}$, $\{\vartheta_n\}$ *and* $\{\xi_n\}$ *be sequences of real numbers, for which* $0 < c_1 \leq \eta_n \leq \hat{c}_1 < 1, 0 < c_2 \leq \vartheta_n \leq \hat{c}_2 < 1, 0 < c_3 \leq \xi_n \leq \hat{c}_3 < 1$ *for all*

$n \in \mathbb{N}$. Let $u_1 \in \mathcal{H}$, $\mathcal{P}_{\arg\min_{u \in \mathcal{K}} \varphi(u) \cap \arg\min_{u \in \mathcal{K}} \psi(u)}(u_1) = u_*$, $\delta \in (0, 2\min\{\frac{1}{\|A\|^2}, \frac{1}{\|B\|^2}\})$, $u_1 \in \mathcal{K}$ and $\{u_n\}$ is defined by

$$\begin{cases} w_n = (1 - \xi_n)u_n + \xi_n \mathcal{S}u_n, \\ z_n = (1 - \vartheta_n)w_n + \vartheta_n \mathcal{T}w_n, \\ u_{n+1} = (1 - \eta_n)\mathcal{S}w_n + \eta_n \mathcal{T}z_n, \quad \forall n \in \mathbb{N}. \end{cases} \tag{18}$$

where $\mathcal{S}, \mathcal{T} : \mathcal{K} \to \mathcal{K}$ defined by $\mathcal{S}u = \mathcal{P}_{\mathcal{K}}(u - \delta A^*(Au - y))$ and $\mathcal{T}u = \mathcal{P}_{\mathcal{K}}(u - \delta B^*(Bu - z))$ for all $u \in \mathcal{K}$. Then, we have the following:

(i) $\{u_n\}$ is in the closed ball $B_\lambda[u_*]$, where λ is a constant in $(0, \infty)$ such that $\|u_1 - u_*\| \le \lambda$.
(ii) $\lim_{n \to \infty} \|u_n - \mathcal{S}u_n\| = 0$ and $\lim_{n \to \infty} \|u_n - \mathcal{T}u_n\| = 0$.
(iii) $\{u_n\}$ converges weakly to an element of $\arg\min_{u \in \mathcal{K}} \varphi(u) \cap \arg\min_{u \in \mathcal{K}} \psi(u) \cap B_\lambda[u_*]$.

Proof. Note that: $\nabla\varphi(u) = A^*(Au - y)$, for all $u \in \mathcal{H}$; we obtain that $\|\nabla\varphi(u) - \nabla\varphi(v)\| = \|A^*(Au - y) - A^*(Av - y)\| \le \|A\|^2\|u - v\|$, for all $u, v \in \mathcal{H}$. Thus, $\nabla\varphi$ is $\frac{1}{\|A\|^2}$-ism and hence $(I - \delta\nabla\varphi)$ is nonexpansive from $\mathcal{K}$ into $\mathcal{H}$ for $\sigma \in (0, \frac{2}{\|A\|^2})$. Therefore, $\mathcal{S} = \mathcal{P}_{\mathcal{K}}(I - \sigma\nabla\varphi)$ and $\mathcal{T} = \mathcal{P}_{\mathcal{K}}(I - \tau\nabla\varphi)$ are nonexpansive mappings from $\mathcal{K}$ into itself for $\sigma \in (0, \frac{2}{\|A\|^2})$ and $\tau \in (0, \frac{2}{\|B\|^2})$, respectively. Hence, Theorem 4 is the same way as Theorem 2. $\square$

4.3. Convex Minimization Problem

We give an application to common solutions to convex programming problems in a Hilbert space $\mathcal{H}$. We consider the following problem:

Let $g_1, g_2 : \mathcal{H} \to (-\infty, \infty]$ be proper *l.s.c.* functions. The objective is to find $x \in \mathcal{H}$ such that:

$$x \in \partial g_1^{-1}(0) \cap g_2^{-1}(0). \tag{19}$$

Note that: $J_\mu^{\partial g_1} = prox_{\mu g_1}$.

Theorem 5. *Let $\mathcal{K}$ be a nonempty closed convex subset of a real Hilbert space $\mathcal{H}$. Let $g_1, g_2 \in \Gamma_0(\mathcal{H})$, for which the solution set of the problem in Equation (19) is nonempty. Let $\{\eta_n\}$, $\{\vartheta_n\}$ and $\{\xi_n\}$ be sequences of real numbers, for which $0 < c_1 \le \eta_n \le \hat{c}_1 < 1$, $0 < c_2 \le \vartheta_n \le \hat{c}_2 < 1$, $0 < c_3 \le \xi_n \le \hat{c}_3 < 1$ for all $n \in \mathbb{N}$. Let $\mu > 0$, $u_1 \in \mathcal{H}$ and $\mathcal{P}_{\partial g_1^{-1}(0) \cap g_2^{-1}(0)}(u_1) = u_*$. Let $u_1 \in \mathcal{K}$ and $\{u_n\}$ is defined by*

$$\begin{cases} w_n = (1 - \xi_n)u_n + \xi_n prox_{\mu g_1}(u_n), \\ z_n = (1 - \vartheta_n)w_n + \vartheta_n prox_{\mu g_2}(w_n), \\ u_{n+1} = (1 - \eta_n)prox_{\mu g_1}(w_n) + \eta_n prox_{\mu g_2}(z_n), \quad \forall n \in \mathbb{N}. \end{cases} \tag{20}$$

Then, we have the following:

(i) $\{u_n\}$ is in the closed ball $B_\lambda[u_*]$, where λ is a constant in $(0, \infty)$ such that $\|u_1 - u_*\| \le \lambda$.
(ii) $\lim_{n \to \infty} \|u_n - prox_{\mu g_1}(u_n)\| = 0$ and $\lim_{n \to \infty} \|u_n - prox_{\mu g_2}(u_n)\| = 0$.
(iii) $\{u_n\}$ converges weakly to an element of $\partial g_1^{-1}(0) \cap g_2^{-1}(0) \cap B_\lambda[u_*]$.

Proof. Using Lemma 1, we have that ∂g_1 is maximal monotone. We know that $\mathcal{R}(I + \mu\partial f) = \mathcal{H}$ and using the maximal monotonicity of ∂g_1. Thus, $J_\mu^{\partial g_1} = prox_{\mu g_1} : \mathcal{H} \to \mathcal{H}$ is nonexpansive. Similarly, $J_\mu^{\partial g_2} = prox_{\mu g_2} : \mathcal{H} \to \mathcal{H}$ is nonexpansive. Hence, Theorem 5 is the same way as Theorem 2. $\square$

4.4. Signal Processing

We consider some applications of our algorithm to inverse problems occurring from signal processing. For example, we consider the following underdeterminated linear equation system:

$$y = \mathcal{A}u + e, \tag{21}$$

where $u \in \mathbb{R}^N$ is recovered, $y \in \mathbb{R}^M$ is observations or measured data with noisy e, and $\mathcal{A} : \mathbb{R}^N \to \mathbb{R}^M$ is a bounded linear observation operator. It determines a process with loss of information. For finding solutions of the linear inverse problems in Equation (21), a successful one of some models is the convex unconstrained minimization problem:

$$\min_{u \in \mathcal{R}^N} \frac{1}{2}\|\mathcal{A}u - y\|^2 + d\|u\|_1, \tag{22}$$

where $d > 0$ and $\|\cdot\|_1$ is the l_1-norm. Thus, we can find solution to Equation (22) by applying our method in the case $g_1(u) = \frac{1}{2}\|\mathcal{A}u - y\|^2$ and $g_2(u) = d\|u\|_1$. For any $\alpha \in (0, \frac{2}{\mathcal{L}}]$, the corresponding forward-backward operator $J_\alpha^{g_1, d\|\cdot\|_1}$ as follows:

$$J_\alpha^{g_1, d\|\cdot\|_1}(u) = prox_{\alpha d\|\cdot\|_1}(u - \alpha\nabla g_1(u)), \tag{23}$$

where g_1 is the squared loss function of the Lasso problem in Equation (22). The proximity operator for l_1-norm is defined as the shrinkage operator as follows:

$$prox_{\alpha d\|\cdot\|_1}(u) = \max(|u_i| - \alpha d, 0) \cdot \text{sgn}(u_i), \tag{24}$$

where $\text{sgn}(\cdot)$ is the signum function. We apply the algorithm to the problem in Equation (22) follow as Algorithm 2:

Algorithm 2: Three-step forward-backward operator

---

initialization: $\eta_n, \vartheta_n, \xi_n \in (0, 1)$, $\alpha, d \in (0, 1)$ $u_1 \in \mathcal{K}$ and $n = 1$.
while *stopping criterion not met* **do**

$\quad w_n = (1 - \xi_n)u_n + \xi_n J_\alpha^{g_1, d\|\cdot\|_1}(u_n),$

$\quad z_n = (1 - \vartheta_n)w_n + \vartheta_n J_\alpha^{g_1, d\|\cdot\|_1}(w_n),$

$\quad u_{n+1} = (1 - \eta_n)J_\alpha^{g_1, d\|\cdot\|_1}(w_n) + \eta_n J_\alpha^{g_1, d\|\cdot\|_1}(z_n).$

end

In our experiment, we set the hits of a signal $u \in \mathbb{R}^N$. The matrix $\mathcal{A} \in \mathbb{R}^{M \times N}$ was generated from a normal distribution with mean zero and one invariance. The observation y is generated by Gaussian noise distributed normally with mean 0 and variance 10^{-4}. We compared our Algorithm 2 with SPGA [12]. Let $\eta_n = \vartheta_n = \xi_n = 0.5, \alpha = 0.1$ and $d = 0.01$ in both Algorithm 2 and SPGA. The experiment was initialized by $u_1 = A^*y$ and terminated when $\frac{\|u_{n+1} - u_n\|}{\|u_n\|} < 10^{-4}$. The restoration accuracy was measured by means of the mean squared error: MSE $= \frac{\|u_* - u\|^2}{N}$, where u_* is an estimated signal of u. All codes were written in Matlab 2016b and run on Dell i-5 Core laptop. We present the numerical comparison of the results in Figures 1–6.

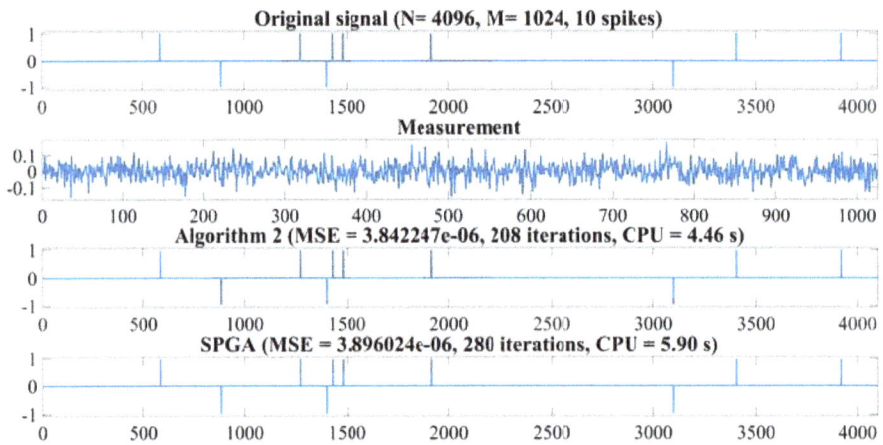

Figure 1. From top to bottom: Original signal, observation data, recovered signal by Algorithm 2 and SPGA with $N = 4096$, $M = 1024$ and 10 spikes, respectively.

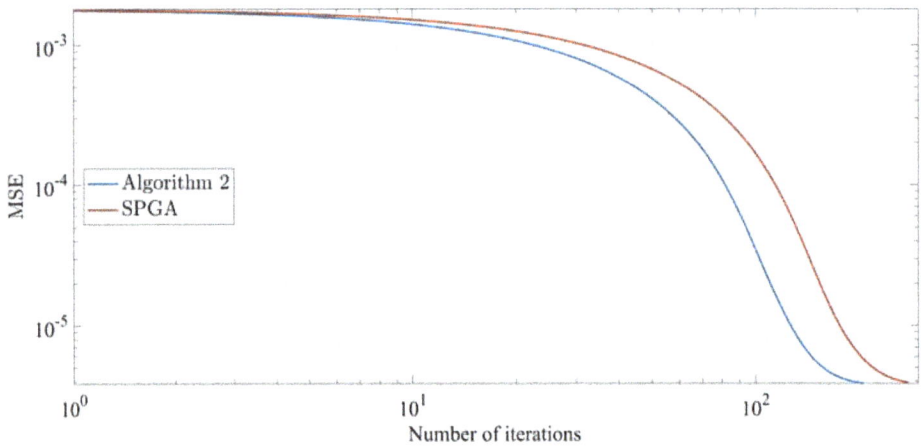

Figure 2. Comparison MSE of two algorithms for recovered signal with $N = 4096$, $M = 1024$ and 10 spikes, respectively.

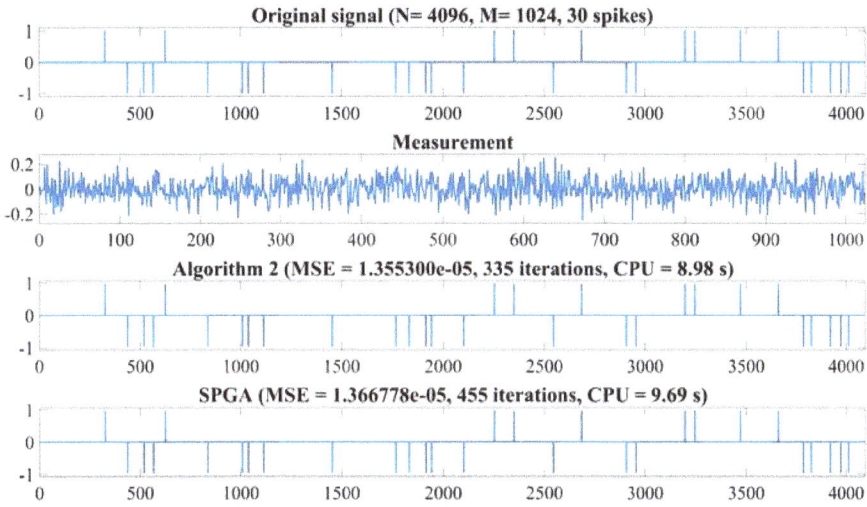

Figure 3. From top to bottom: Original signal, observation data, recovered signal by Algorithm 2 and SPGA with $N = 4096$, $M = 1024$ and 30 spikes, respectively.

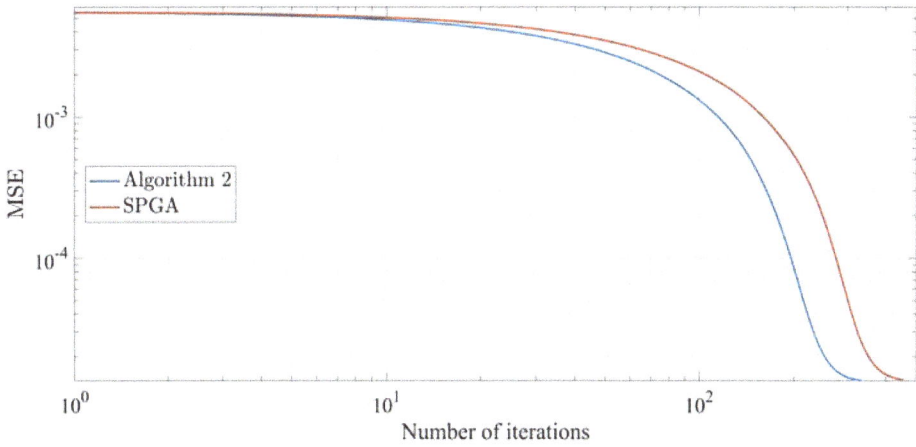

Figure 4. Comparison MSE of two algorithms for recovered signal with $N = 4096$, $M = 1024$ and 30 spikes, respectively.

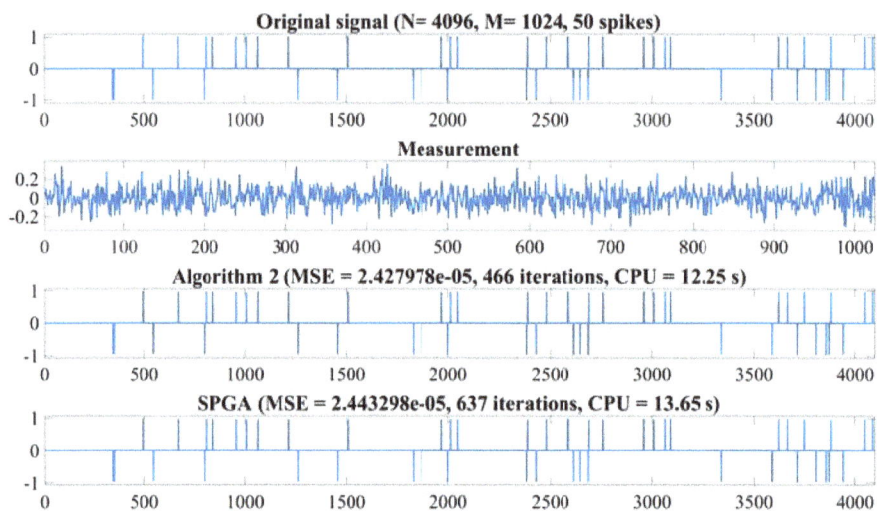

Figure 5. From top to bottom: Original signal, observation data, recovered signal by Algorithm 2 and SPGA with $N = 4096$, $M = 1024$ and 50 spikes, respectively.

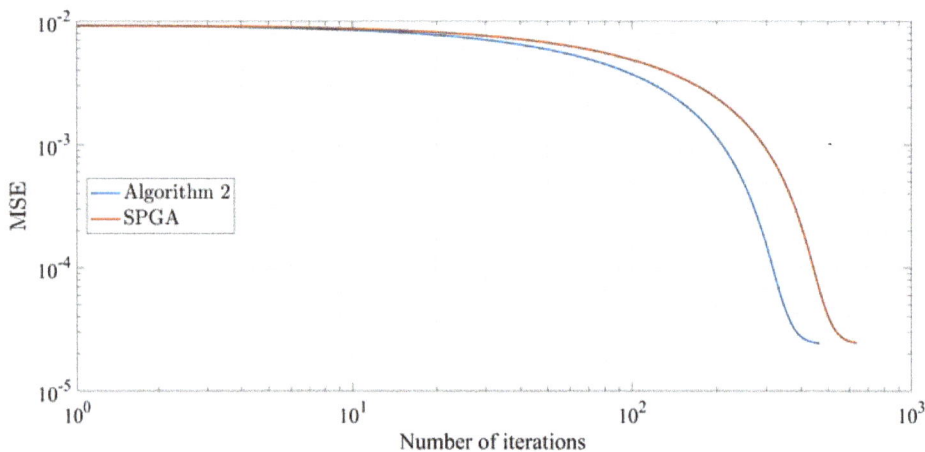

Figure 6. Comparison MSE of two algorithms for recovered signal with $N = 4096$, $M = 1024$ and 50 spikes, respectively.

5. Conclusions

In this work, we introduce a modified iterative scheme in Banach spaces and solve common zeros of accretive operators, convexly constrained least square problem, convex minimization problem and signal processing. In the case of signal processing, all results are compared with the forward-backward method in Algorithm 2 and SPGA, as proposed in [12]. The numerical results show that Algorithm 2 has a better convergence behavior than SPGA when using the same step sizes for both.

Author Contributions: A.P. and P.S.; writing original draft, A.P. and P.S.; data analysis, A.P. and P.S.; formal analysis and methodology

Funding: This research was funded by Rajamangala University of Technology Thanyaburi (RMUTT).

Acknowledgments: The first author thanks Rambhai Barni Rajabhat University for the support. Pakeeta Sukprasert was financially supported by Rajamangala University of Technology Thanyaburi (RMUTT).

Mathematics **2019**, *7*, 866

Conflicts of Interest: The authors declare no conflict of interest.

Abbreviations

The following abbreviations are used in this manuscript:

Symbols	Display
l.s.c.	lower semicontinuous, convex
$\mathcal{B}(\mathcal{H})$	the set of all bounded and linear operators from $\mathcal{H}$ into itself
r.u.c.	real uniformly convex

References

1. Kankam, K.; Pholasa, N.; Cholamjiak, P. On convergence and complexity of the modified forward-backward method involving new linesearches for convex minimization. *Math. Meth. Appl. Sci.* **2019**, *42*, 1352–1362. [CrossRef]
2. Candès, E.J.; Wakin, M.B. An introduction to compressive sampling. *IEEE Signal Process. Mag.* **2008**, *25*, 21–30. [CrossRef]
3. Suantai, S.; Kesornprom, S.; Cholamjiak, P. A new hybrid CQ algorithm for the split feasibility problem in Hilbert spaces and Its applications to compressed Sensing. *Mathematics* **2019**, *7*, 789; doi:10.3390/math7090789. [CrossRef]
4. Kitkuan, D.; Kumam, P.; Padcharoen, A.; Kumam, W.; Thounthong, P. Algorithms for zeros of two accretive operators for solving convex minimization problems and its application to image restoration problems. *J. Comput. Appl. Math.* **2019**, *354*, 471–495. [CrossRef]
5. Padcharoen, A.; Kumam, P.; Cho, Y.J. Split common fixed point problems for demicontractive operators. *Numer. Algorithms* **2019**, *82*, 297–320. [CrossRef]
6. Cholamjiak, P.; Shehu, Y. Inertial forward-backward splitting method in Banach spaces with application to compressed sensing. *Appl. Math.* **2019**, *64*, 409–435. [CrossRef]
7. Jirakitpuwapat, W.; Kumam, P.; Cho, Y.J.; Sitthithakerngkiet, K. A general algorithm for the split common fixed point problem with its applications to signal processing. *Mathematics* **2019**, *7*, 226. [CrossRef]
8. Combettes, P.L.; Wajs, V.R. Signal recovery by proximal forward-backward splitting. *Multiscale Model Simul.* **2005**, *4*, 1168–1200. [CrossRef]
9. Picard, E. Memoire sur la theorie des equations aux d'erives partielles et la methode des approximations successives. *J. Math Pures Appl.* **1890**, *231*, 145–210.
10. Mann, W.R. Mean value methods in iteration. *Proc. Am. Math. Soc.* **1953**, *4*, 506–510. [CrossRef]
11. Ishikawa, S. Fixed points by a new iteration method. *Proc. Am. Math. Soc.* **1974**, *44*, 147–150. [CrossRef]
12. Agarwal, R.P.; O'Regan, D.; Sahu, D.R. Iterative construction of fixed points of nearly asymptotically nonexpansive mappings. *J. Nonlinear Convex Anal.* **2007**, *8*, 61–79.
13. Sahu, V.K.; Pathak, H.K.; Tiwari, R. Convergence theorems for new iteration scheme and comparison results. *Aligarh Bull. Math.* **2016**, *35*, 19–42.
14. Thakur, B.S.; Thakur, D.; Postolache, M. New iteration scheme for approximating fixed point of non-expansive mappings. *Filomat* **2016**, *30*, 2711–2720. [CrossRef]
15. Chang, S.S.; Wen, C.F.; Yao, J.C. Zero point problem of accretive operators in Banach spaces. *Bull. Malays. Math. Sci. Soc.* **2019**, *42*, 105–118. [CrossRef]
16. Browder, F.E. Nonlinear mappings of nonexpansive and accretive type in Banach spaces. *Bull. Am. Math. Soc.* **1967**, *73*, 875–882. [CrossRef]
17. Browder, F.E. Semicontractive and semiaccretive nonlinear mappings in Banach spaces. *Bull. Am. Math. Soc.* **1968**, *7*, 660–665. [CrossRef]
18. Cioranescu, I. *Geometry of Banach Spaces, Duality Mapping and Nonlinear Problems*; Kluwer: Amsterdam, The Netherlands, 1990.
19. Takahashi, W. *Nonlinear Functional Analysis, Fixed Point Theory and Its Applications*; Yokohama Publishers: Yokohama, Japan, 2000.
20. Rockafellar, R.T. On the maximal monotonicity of subdifferential mappings. *Pac. J. Math.* **1970**, *33*, 209–216. [CrossRef]

21. Xu, H.K. Inequalities in Banach spaces with applications. *Nonlinear Anal.* **1991**, *16*, 1127–1138. [CrossRef]
22. Opial, Z. Weak convergence of the sequence of successive approximations for nonexpansive mappings. *Bull. Am. Math. Soc.* **1967**, *73*, 591–597. [CrossRef]
23. Sahu, D.R.; Pitea, A.; Verma, M. A new iteration technique for nonlinear operators as concerns convex programming and feasibility problems. *Numer. Algorithms* **2019**. [CrossRef]
24. Goebel, K.; Reich, S. *Uniform Convexity, Hyperbolic Geometry and Non Expansive Mappings*; Marcel Dekker: New York, NY, USA; Basel, Switzerland, 1984 .

![Sigma mathematics logo]

mathematics

MDPI

Article

Hybrid Second Order Method for Orthogonal Projection onto Parametric Curve in n-Dimensional Euclidean Space

Juan Liang [1,2,†], **Linke Hou** [3,†,*], **Xiaowu Li** [4,†,*] , **Feng Pan** [4,†], **Taixia Cheng** [5,†] and **Lin Wang** [4,†]

[1] Data Science and Technology, North University of China, Taiyuan 030051, Shanxi, China; liangjuan76@126.com
[2] Department of Science, Taiyuan Institute of Technology, Taiyuan 030008, Shanxi, China
[3] Center for Economic Research, Shandong University, Jinan 250100, Shandong, China
[4] College of Data Science and Information Engineering, Guizhou Minzu University, Guiyang 550025, Guizhou, China; panf@vip.163.com (F.P.); wanglin@gzmu.edu.cn (L.W.)
[5] Graduate School, Guizhou Minzu University, Guiyang 550025, Guizhou, China; lissacheng@163.com
[*] Correspondence: abram75@163.com (L.H.); lixiaowu002@126.com (X.L.); Tel.: +86-135-0640-1186 (L.H.); +86-187-8613-2431 (X.L.)
[†] These authors contributed equally to this work.

Received: 16 October 2018; Accepted: 28 November 2018; Published: 5 December 2018

Abstract: Orthogonal projection a point onto a parametric curve, three classic first order algorithms have been presented by Hartmann (1999), Hoschek, et al. (1993) and Hu, et al. (2000) (hereafter, H-H-H method). In this research, we give a proof of the approach's first order convergence and its non-dependence on the initial value. For some special cases of divergence for the H-H-H method, we combine it with Newton's second order method (hereafter, Newton's method) to create the hybrid second order method for orthogonal projection onto parametric curve in an n-dimensional Euclidean space (hereafter, our method). Our method essentially utilizes hybrid iteration, so it converges faster than current methods with a second order convergence and remains independent from the initial value. We provide some numerical examples to confirm robustness and high efficiency of the method.

Keywords: point projection; intersection; parametric curve; n-dimensional Euclidean space; Newton's second order method; fixed point theorem

1. Introduction

In this research, we will discuss the minimum distance problem between a point and a parametric curve in an n-dimensional Euclidean space, and how to gain the closest point (footpoint) on the curve as well as its corresponding parameter, which is termed as the point projection or inversion problem of a parametric curve in an n-dimensional Euclidean space. It is an important issue in the themes such as geometric modeling, computer graphics, computer-aided geometry design (CAGD) and computer vision [1,2]. Both projection and inversion are fundamental for a series of techniques, for instance, the interactive selection of curves and surfaces [1,3], the curve fitting [1,3], reconstructing curves [2,4,5] and projecting a space curve onto a surface [6]. This vital technique is also used in the ICP (iterative closest point) method for shape registration [7].

The Newton-Raphson algorithm is deemed as the most classic one for orthogonal projection onto parametric curve and surface. Searching the root of a polynomial by a Newton-Raphson algorithm can be found in [8–12]. In order to solve the adaptive smoothing for the standard finite unconstrained minimax problems, Polak et al. [13] have presented a extended Newton's algorithm where a new feedback precision-adjustment rule is used in their extended Newton's algorithm.

Once the Newton-Raphson method reaches its convergence, two advantages emerge and it converges very fast with high precision. However, the result relies heavily on a good guess of initial value in the neighborhood of the solution.

Meanwhile, the classic subdivision method consists of several procedures: Firstly, subdivide NURBS curve or surface into a set of Bézier sub-curves or patches and eliminate redundancy or unnecessary Bézier sub-curves or Bézier patches. Then, get the approximation candidate points. Finally, get the closest point through comparing the distances between the test point and candidate points. This technique is reflected in [1]. Using new exclusion criteria within the subdivision strategy, the robustness for the projection of points on NURBS curves and surfaces in [14] has been improved than that in [1], but this criterion is sometimes too critical. Zou et al. [15] use subdivision minimization techniques which rely on the convex hull characteristic of the Bernstein basis to impute the minimum distance between two point sets. They transform the problem into solving of n-dimensional nonlinear equations, where n variables could be represented as the tensor product Bernstein basis. Cohen et al. [16] develop a framework for implementing general successive subdivision schemes for nonuniform B-splines to generate the new vertices and the new knot vectors which are satisfied with derived polygon. Piegl et al. [17] repeatedly subdivide a NURBS surface into four quadrilateral patches and then project the test point onto the closest quadrilateral until it can find the parameter from the closest quadrilateral. Using multivariate rational functions, Elber et al. [11] construct a solver for a set of geometric constraints represented by inequalities. When the dimension of the solver is greater than zero, they subdivide the multivariate function(s) so as to bind the function values within a specified domain. Derived from [11] but with more efficiency, a hybrid parallel method in [18] exploits both the CPU and the GPU multi-core architectures to solve systems under multivariate constraints. Those GPU-based subdivision methods essentially exploit the parallelism inherent in the subdivision of multivariate polynomial. This geometric-based algorithm improves in performance compared to the existing subdivision-based CPU. Two blending schemes in [19] efficiently remove no-root domains, and hence greatly reduce the number of subdivisions. Through a simple linear combination of functions for a given system of nonlinear equations, no-root domain and searching out all control points for its Bernstein-Bézier basic with the same sign must be satisfied with the seek function. During the subdivision process, it can continuously create these kinds of functions to get rid of the no-root domain. As a result, van Sosin et al. [20] efficiently form various complex piecewise polynomial systems with zero or inequality constraints in zero-dimensional or one-dimensional solution spaces. Based on their own works [11,20], Bartoň et al. [21] propose a new solver to solve a non-constrained (piecewise) polynomial system. Two termination criteria are applied in the subdivision-based solver: the no-loop test and the single-component test. Once two termination criteria are satisfied, it then can get the domains which have a single monotone univariate solution. The advantage of these methods is that they can find all solutions, while their disadvantage is that they are computationally expensive and may need many subdivision steps.

The third classic methods for orthogonal projection onto parametric curve and surface are geometry methods. They are mainly classified into eight different types of geometry methods: tangent method [22,23], torus patch approximating method [24], circular or spherical clipping method [25,26], culling technique [27], root-finding problem with Bézier clipping [28,29], curvature information method [6,30], repeated knot insertion method [31] and hybrid geometry method [32]. Johnson et al. [22] use tangent cones to search for regions with satisfaction of distance extrema conditions and then to solve the minimum distance between a point and a curve, but it is not easy to construct tangent cones at any time. A torus patch approximatively approaches for point projection on surfaces in [24]. For the pure geometry method of a torus patch, it is difficult to achieve high precision of the final iterative parametric value. A circular clipping method can remove the curve parts outside a circle with the test point being the circle's center, and the radius of the elimination circle will shrink until it satisfies the criteria to terminate [26]. Similar to the algorithm [26], a spherical clipping technique for computing the minimum distance with clamped B-spline surface is provided by [25]. A culling technique to remove superfluous curves and surfaces containing no projection from the given point is proposed in [27], which is in line with the idea in [1]. Using Newton's method for the last step [1,25–27], the special case of non-convergence may happen. In view of the convex-hull property of Bernstein-Bézier representations, the problem to be solved can be formulated as a univariate root-finding problem. Given a C^1 parametric curve $c(t)$ and a point p, the projection constraint problem can be formulated as a univariate root-finding problem $\langle c'(t), c(t) - p \rangle = 0$ with a metric induced by the Euclidean scalar product in R^n. If the curve is parametrized by a (piece-wise) polynomial, then the fast root-finding schemes as a Bézier clipping [28,29] can be used. The only issue is the C^1 discontinuities that can be checked in a post-process. One advantage of these methods is that they do not need any initial guess on the parameter value. They adopt the key technology of degree reduction via clipping to yield a strip bounded of two quadratic polynomials. Curvature information is found for computing the minimum distance between a point and a parameter curve or surface in [6,30]. However, it needs to consider the second order derivative and the method [30] is not fit for n-dimensional Euclidean space. Hu et al. [6] have not proved the convergence of their two algorithms. Li et al. [33] have strictly proved convergence analysis for orthogonal projection onto planar parametric curve in [6]. Based on repeated knot insertion, Mørken et al. [31] exploit the relationship between a spline and its control polygon and then present a simple and efficient method to compute zeros of spline functions. Li et al. [32] present the hybrid second order algorithm which orthogonally projects onto parametric surface; it actually utilizes the composite technology and hence converges nicely with convergence order being 2. The geometric method can not only solve the problem of point orthogonal projecting onto parametric curve and surface but also compute the minimum distance between parametric curves and parametric surfaces. Li et al. [23] have used the tangent method to compute the intersection between two spatial curves. Based on the methods in [34,35], they have extended to compute the Hausdorff distance between two B-spline curves. Based on matching a surface patch from one model to the other model which is the corresponding nearby surface patch, an algorithm for solving the Hausdorff distance between two freeform surfaces is presented in Kim et al. [36], where a hierarchy of Coons patches and bilinear surfaces that approximate the NURBS surfaces with bounding volume is adopted. Of course, the common feature of geometric methods is that the ultimate solution accuracy is not very high. To sum up, these algorithms have been proposed to exploit diverse techniques such as Newton's iterative method, solving polynomial equation roots methods, subdividing methods, geometry methods. A review of previous algorithms on point projection and inversion problem is obtained in [37].

More specifically, using the tangent line or tangent plane with first order geometric information, a classical simple and efficient first order algorithm which orthogonally project onto parametric curve and surface is proposed in [38–40] (H-H-H method). However, the proof of the convergence for the H-H-H method can not be found in this literature. In this research, we try to give two contributions. Firstly, we give proof that the algorithm is first order convergent and it does not depend on the initial value. We then provide some numerical examples to show its high convergence rate. Secondly, for several special cases where the H-H-H method is not convergent, there are two methods (Newton's method and the H-H-H method) to combine our method. If the H-H-H method's iterative parametric value is satisfied with the convergence condition of the Newton's method, we then go to Newton's method to increase the convergence process. Otherwise, we go on the H-H-H method until its iterative parametric value is satisfied with the convergence condition of the Newton's method, and we then turn to it as above. This algorithm not only ensures the robustness of convergence, but also improves the convergence rate. Our hybrid method can go faster than the existing methods and ensures the independence to the initial value. Some numerical examples verify our conclusion.

The rest of this paper is arranged as follows. In Section 2, convergence analysis of the H-H-H method is presented. In Section 3, for several special cases where the H-H-H method is not convergent, an improved our method is provided. Convergence analysis for our method is also provided in this section. In Section 4, some numerical examples for our method are verified. In Section 5, conclusions are provided.

2. Convergence Analysis of the H-H-H Method

In this part, we will prove that the algorithm defined by Equations (2) or (3) is of first order convergence and its convergence does not rely on the initial value. Suppose a C^2 curve $c(t) = (f_1(t), f_2(t), \ldots, f_n(t))$ in an n-dimensional Euclidean space $R^n (n \geq 2)$ and a test point $p = (p_1, p_2, \ldots, p_n)$. The first order geometric method to compute the footpoint q of test point p can be implemented as below. Projecting test point p onto the tangent line of the parametric curve $c(t)$ in an n-dimensional Euclidean space at $t = t_m$ gets a point q determined by $c(t_m)$ and its derivative $c'(t_m)$. The footpoint can be approximated as

$$q = c(t_m) + \Delta t c'(t_m). \tag{1}$$

Then,

$$\Delta t = \frac{\langle c'(t_m), p - c(t_m) \rangle}{\langle c'(t_m), c'(t_m) \rangle}, \tag{2}$$

where $\langle x, y \rangle$ is the scalar product of vectors $x, y \in R^n$. Equation (2) can also be expressed as

$$K_1(t_m) = t_m + \frac{\langle c'(t_m), p - c(t_m) \rangle}{\langle c'(t_m), c'(t_m) \rangle}. \tag{3}$$

Let $t_m \leftarrow K_1(t_m)$, and repeatedly iterate the above process until $|K_1(t_m) - t_m|$ is less than an error tolerance ε. This method is addressed as H-H-H method [38–40]. Furthermore, convergence of this method will not depend on the choice of the initial value. According to many of our test experiments, when the iterative parametric value approaches the target parametric value α, the iteration step size becomes smaller and smaller, while the corresponding number of iterations becomes bigger and bigger.

Theorem 1. *The convergence order of the method defined by Equations (2) or (3) is one, and its convergence does not depend on the initial value.*

Proof. We adopt the numerical analysis method which is equivalent to those in the literature [41,42]. Firstly, we deduce the expression of footpoint q. Suppose that parameter curve $c(t)$ is a C^2 curve in an n-dimensional Euclidean space $R^n (n \geq 2)$, where the corresponding projecting point with parameter α is orthogonal projecting of the test point $p = (p_1, p_2, \ldots, p_n)$ onto the parametric curve $c(t)$. It is easy to indicate a relational expression

$$\langle p - h, \mathbf{n} \rangle = 0, \tag{4}$$

where $h = c(\alpha)$ and tangent vector $\mathbf{n} = c'(\alpha)$. In order to solve the intersection (footpoint q) between the tangent line, which goes through the parametric curve $c(t)$ at $t = t_m$, and the perpendicular line, which is determined by the test point p, we try to express the equation of the tangent line as:

$$x = c(t_m) + c'(t_m) \cdot s, \tag{5}$$

where $x = (x_1, x_2, \ldots, x_n)$ and s is a parameter. In addition, the vector of line segment both going through the test point p and the point $c(t_m)$ will be

$$y = p - x, \tag{6}$$

where $y = (y_1, y_2, \ldots, y_n)$. Because the vector (6) and the tangent vector $c'(t_m)$ of Equation (5) are orthogonal to each other, the current parameter value s of Equation (5) is

$$s_0 = \frac{\langle p - c(t_m), c'(t_m) \rangle}{\langle c'(t_m), c'(t_m) \rangle}. \tag{7}$$

Substituting (7) into (5), we have

$$q = c(t_m) + c'(t_m) \cdot s_0. \tag{8}$$

Thus, the footpoint $q = (q_1, q_2, \ldots, q_n)$ is determined by Equation (8).

Secondly, we deduce that the convergence order of the method defined by (2) or (3) is first order convergent. Our proof method absorbs the idea of [41,42]. Substituting (8) into (2), and simplifying, we get the relationship,

$$\Delta t = \frac{\langle p - c(t_m), c'(t_m) \rangle}{\langle c'(t_m), c'(t_m) \rangle}. \tag{9}$$

Using Taylor's expansion, we get

$$c(t_m) = B_0 + B_1 e_m + B_2 e_m^2 + o(e_m^3), \tag{10}$$

$$c'(t_m) = B_1 + 2B_2 e_m + o(e_m^2), \tag{11}$$

where $e_m = t_m - \alpha$, and $B_i = (1/i!)c^{(i)}(\alpha), i = 0, 1, 2, \ldots$ From (10) and (11) and combining with (4), the numerator of Equation (9) can be transformed into the following one:

$$\begin{aligned} &\langle p - c(t_m), c'(t_m) \rangle \\ &= L_1 e_m + L_2 e_m^2 + o(e_m^3), \end{aligned} \tag{12}$$

where $L_1 = 2\langle p - B_0, B_2 \rangle - \langle B_1, B_1 \rangle, L_2 = -3\langle B_1, B_2 \rangle$. By (11), the denominator of Equation (9) can be changed as follows:

$$\begin{aligned} &\langle c'(t_m), c'(t_m) \rangle \\ &= M_1 + M_2 e_m + M_3 e_m^2 + o(e_m^3), \end{aligned} \tag{13}$$

where $M_1 = \langle B_1, B_1 \rangle, M_2 = 4\langle B_1, B_2 \rangle, M_3 = 4\langle B_2, B_2 \rangle$. Substituting Equations (12) and (13) into the right-hand side of Equation (9), we get

$$\begin{aligned} \Delta t &= \frac{\langle p - c(t_m), c'(t_m) \rangle}{\langle c'(t_m), c'(t_m) \rangle} \\ &= \frac{L_1 e_m + L_2 e_m^2 + o(e_m^3)}{M_1 + M_2 e_m + M_3 e_m^2 + o(e_m^3)}. \end{aligned} \tag{14}$$

Using Taylor's expansion by Maple 18, and through simplification, we get

$$\begin{aligned} K_1(t_m) &= \alpha + (\frac{L_1}{M_1} + 1)e_m + \frac{L_2 M_1 - L_1 M_2}{M_1^2} e_m^2 + o(e_m^3), \\ &= \alpha + (\frac{L_1}{M_1} + 1)e_m + o(e_m^2), \\ &= \alpha + C_0 e_m + o(e_m^2), \end{aligned} \tag{15}$$

where the symbol C_0 is the coefficient of the first order error e_m of Equation (15). The result implies the iterative Equations (2) or (3) is of first order convergence.

Now, we try to interpret that Equations (2) or (3) do not depend on the initial value.

Our proof method absorbs the idea of references [43,44]. Without loss of generality, we only prove that convergence of Equations (2) or (3) does not depend on the initial value in two-dimensional case. As to convergence of Equations (2) or (3) not being dependent on the initial value in general n-dimensional Euclidean space case, it is completely equivalent to the two-dimensional case.

Firstly, we interpret Figure 1. For a horizontal axis t, there are two points are on the planar parametric curve $c(t)$. For the first point $c(t_m)$ on the horizontal axis, the test point p orthogonal projects it onto the planar parametric curve $c(t)$ and yields the second point and its corresponding parameter value α on the horizontal axis. Then, by the iterative methods (2) or (3), the line segment connected by the point p and the point $c(\alpha)$ is perpendicular to the tangent line of the planar parametric curve $c(t)$ at $t = \alpha$. The footpoint q is determined by the tangent line of the planar parametric curve $c(t)$ through the point $c(t_m)$. Evidently, the parametric value t_{m+1} of footpoint q can be used as the next iterative value. M is the corresponding parametric value of the middle point of the point $c(t_m)$ and the footpoint q.

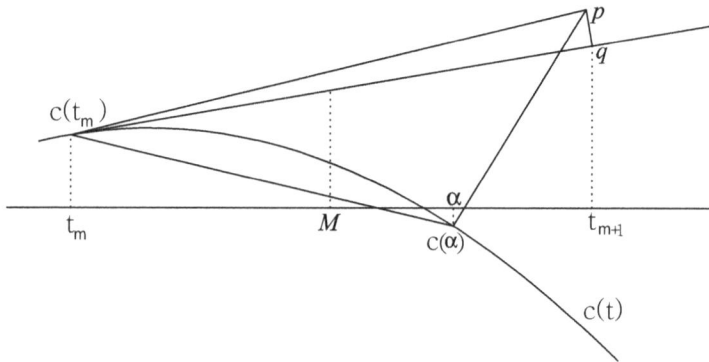

Figure 1. Geometric illustration for convergence analysis.

Secondly, we prove the argument whose convergence of Equations (2) or (3) does not depend on the initial value. It is easy to know that t denotes the corresponding parameter for the first dimensional of the planar parametric curve on the two-dimensional plane. When the iterative Equations (2) or (3) start to run, we suppose that the iterative parameter value is satisfied with the inequality relationship $t_m < \alpha$ and the corresponding parameter of the footpoint q is t_{m+1}, as shown in Figure 1. The middle point of two points $(t_{m+1}, 0)$ and $(t_m, 0)$ is $(M, 0)$, i.e., $M = \dfrac{t_m + t_{m+1}}{2}$, and, because of $0 < \Delta t = t_{m+1} - t_m$, then there exists an inequality $t_m < M < \alpha$. Equivalently, $t_m - \alpha < t_{m+1} - \alpha < \alpha - t_m = -(t_m - \alpha)$, which can be expressed as $|e_{m+1}| < |e_m|$, where $e_m = t_m - \alpha$. If $t_m > \alpha$, we can get the same result through the same method. Thus, an iterative error expression $|e_{m+1}| < |e_m|$ in a two-dimensional plane is demonstrated. Thus, it is known that convergence of the iterative Equations (2) or (3) does not depend on the initial value in two-dimensional planes (see Figure 1). Furthermore, we could get the argument that convergence of the iterative Equations (2) or (3) does not depend on the initial value in an n-dimensional Euclidean space. The proof is completed. □

3. The Improved Algorithm

3.1. Counterexamples

In Section 2, convergence of the H-H-H method does not depend on the initial value. For special cases with non-convergence by the H-H-H method, we then enumerate nine counterexamples.

Counterexample 1. *There are a parametric curve* $c(t) = (t, 1 + t^2)$ *and a test point* $p = (0,0)$. *The projection point and parametric value of the test point* p *are* $(0,1)$ *and* $\alpha = 0$, *respectively. As to many initial values, the H-H-H method fails to converge to* α. *When the initial values are* $t = -3, -2, -1.5, 1.5, 2, 3$, *respectively, there repeatedly appear alternating oscillatory iteration values of* 0.412415429665, -0.412415429665. *Furthermore, for a parametric curve* $c(t) = (t, 1 + a_1t^2 + a_2t^4 + a_3t^6 + a_4t^8 + a_5t^{10})$, $a_1 \neq 0, a_2 \neq 0, a_3 \neq 0, a_4 \neq 0, a_5 \neq 0$, *about* $p = (0,0)$ *and many initial values, the H-H-H method fails to converge to* α *(see Figure 2).*

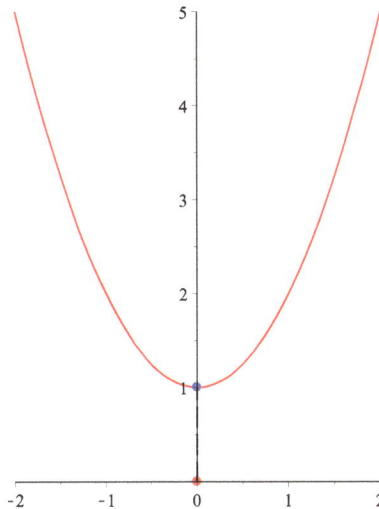

Figure 2. Geometric illustration for counterexample 1.

Counterexample 2. *There are a parametric curve* $c(t) = (t, t^2, t^4, t^6, 1 + t^2 + t^4 + t^6 + t^8)$ *and a test point* $p = (0, 0, 0, 0, 0)$. *The projection point and parametric value of the test point* p *are* $(0, 0, 0, 0, 1)$ *and* $\alpha = 0$, *respectively. For any initial value, the H-H-H method fails to converge to* α. *When the initial values are* $t = -5, -4, -3, -2, -1, 1, 2, 3, 4, 5$, *respectively, there repeatedly appear alternating oscillatory iteration values of* 0.304949569175, -0.304949569175. *Furthermore, for a parametric curve* $c(t) = (a_0t, a_1t^2, a_2t^4, a_3t^6, 1 + a_4t^2 + a_5t^4 + a_6t^6 + a_7t^8 + a_8t^{10} + a_9t^{28})$, $a_0 \neq 0, a_1 \neq 0, a_2 \neq 0, a_3 \neq 0, a_4 \neq 0, a_5 \neq 0, a_6 \neq 0, a_7 \neq 0, a_8 \neq 0, a_9 \neq 0$, *about point* $p = (0, 0, 0, 0, 0)$ *and any initial value, the H-H-H method fails to converge to* α.

Counterexample 3. *There are a parametric curve* $c(t) = (t, \sin(t)), t \in [0, 3]$ *and a test point* $p = (4, 9)$. *The projection point and parametric value of the test point* p *are* $(1.842576, 0.9632946)$ *and* $\alpha = 1.842576$, *respectively. For point* p *and any initial value, the H-H-H method fails to converge to* α. *When the initial values are* $t = -5, -4, -3, -2, -1, 1, 2, 3, 4, 5$, *respectively, there repeatedly appear alternating oscillatory iteration values of* 2.165320, 0.0778704, 6.505971, 9.609789. *In addition, for a parametric curve* $c(t) = (t, \sin(at))$, $a \neq 0$, *for any test point* p *and any initial value, the H-H-H method fails to converge to* α *(see Figure 3).*

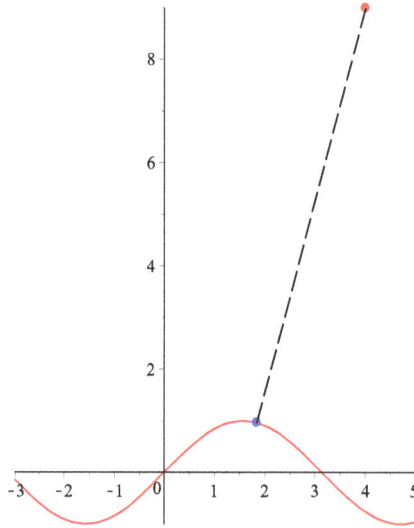

Figure 3. Geometric illustration of counterexample 3.

Counterexample 4. *There are a parametric curve $c(t) = (t, \cos(t)), t \in [0, 3]$ and a test point $p = (2, 6)$. The projection point and parametric value of the test point p are $(0.3354892, 0.9442493)$ and $\alpha = 0.3354892$, respectively. For test point p and any initial value, the H-H-H method fails to converge to α. When the initial value is $t = -5$, alternating oscillatory iteration values of 5.18741299662, 3.59425803253, -0.507188248308, 1.6901041247, 3.82746208506 repeatedly appear. When the initial value is $t = 2$, very irregular oscillatory iteration values of 0.652526561595, -0.720371663877, -2.39555359952, 0.365881194752, 2.06880954777, 3.18725085474, 1.71447110647, etc. appear In addition, for a parametric curve $c(t) = (t, \cos(at)), a \neq 0$, for any test point p and any initial value, the H-H-H method fails to converge to α (see Figure 4).*

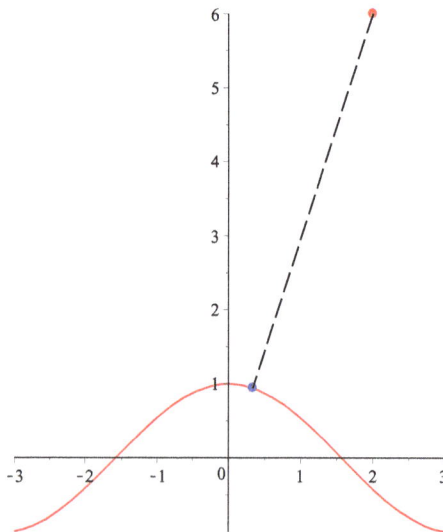

Figure 4. Geometric illustration of counterexample 4.

Counterexample 5. *There are a parametric curve* $c(t) = (t, t, t, t, \sin(t)), t \in [6, 9]$ *and a test point* $p = (3, 5, 7, 9, 11)$. *The projection point and parametric value of the test point* p *are* $(7.310786, 7.310786, 7.310786, 7.310786, 0.8560612)$ *and* $\alpha = 7.310786$, *respectively. For point* p *and any initial value, the H-H-H method fails to converge to* α. *When the initial values are* $t = -9, -7, -5, 6, 8$, *respectively, there repeatedly appear alternating oscillatory iteration values of* $7.24999006346, 6.37363460615$. *In addition, for a parametric curve* $c(t) = (t, t, t, t, \sin(at)), t \in [6, 9], a \neq 0$ *with a test point* $p = (3, 5, 7, 9, 11)$, *for any initial value, the H-H-H method fails to converge to* α.

Counterexample 6. *There are a parametric curve* $c(t) = (t, t, t, t, \cos(t)), t \in [4, 8]$ *and a test point* $p = (2, 4, 6, 8, 10)$. *The projection point and parametric value of the test point* p *are* $(5.883406, 5.883406, 5.883406, 5.883406, 0.9211469)$ *and* $\alpha = 5.883406$, *respectively. For point* p *and any initial value, the H-H-H method fails to converge to* α. *When the initial values are* $t =, -4, -3, -2, 4, 5, 6, 7$, *respectively, there repeatedly appear alternating oscillatory iteration values of* $4.17182145828, 7.80116702003$. *In addition, about a parametric curve* $c(t) = (t, t, t, t, \cos(at)), t \in [4, 8], a \neq 0$ *with a point* $p = (2, 4, 6, 8, 10)$, *for any initial value, the H-H-H method fails to converge. The non-convergence explanation of the three counterexamples below are similar to the preceding six ones and omitted to save space.*

Counterexample 7. *There are a parametric curve* $c(t) = (t^4 + 2t^2 + 1, t^2 + 1, t^4 + 2, t^2, 3t^6 + t^4 + 2t^2)$ *in five-dimensional Euclidean space and a test point* $p = (0, 0, 0, 0, 0)$. *The projection point and parametric value of the test point* p *are* $(1, 1, 2, 0, 0)$ *and* $\alpha = 0$, *respectively. For any initial value* t_0, *the H-H-H method fails to converge. We also test many other examples, such as when parametric curve is completely symmetrical and the point is on the symmetrical axis of parametric curve. For any initial value* t_0, *the same results remain.*

Counterexample 8. *There are a parametric curve* $c(t) = (t, \sin(t), t, \sin(t), \sin(t)), t \in [-5, 5]$ *in five-dimensional Euclidean space and a test point* $p = (3, 4, 5, 6, 7)$. *The corresponding orthogonal projection parametric value* α *are* $-3.493548, -2.280571, 1.875969, 4.791677$, *respectively. For any initial value* t_0, *the H-H-H method fails to converge.*

Counterexample 9. *There is a parametric curve* $c(t) = (\sin(t), \cos(t), t, \sin(t), \cos(t)), t \in [-5, 5]$ *in five-dimensional Euclidean space and a test point* $p = (3, 4, 5, 6, 7)$. *The corresponding orthogonal projection parametric value* α *are* $-4.833375, -3.058735, 0.9730030, 3.738442$, *respectively. For any initial value* t_0, *the H-H-H method fails to converge.*

3.2. The Improved Algorithm

Due to the H-H-H method's non-convergence for some special cases, the improved algorithm is presented to ensure the converge for any parametric curve, test point and initial value. The most classic Newton's method can be expressed as

$$t_{m+1} = t_m - \frac{f(t_m)}{f'(t_m)}, \tag{16}$$

where $f(t) = <T_1, V_1> = 0$, $T_1 = c'(t)$, $V_1 = p - c(t)$. It converges faster than the H-H-H method. However, the convergence of this depends on the chosen initial value. Only when the local convergence condition for the Newton's method is satisfied, the method can acquire high effectiveness. In order to improve the robustness and rate of convergence, based on the the H-H-H method, our method is proposed. Combining the respective advantage of their two methods, if the iterative parametric value of the H-H-H method is satisfied with the convergence condition of the Newton's method, we then go to the method to increase the convergence process. Otherwise, we continue the H-H-H method until it can generate iterative parametric value while satisfying the convergence condition by the Newton's method, and we then go to the iterative process mentioned above. Thus, we run to the end of the whole process. The procedure not only ensures the robustness of convergence, but also improves the convergence rate. Using a hybrid strategy, our method is faster than current methods and independent from the initial value. Some numerical examples verify our conclusion. Our method can be realized as follows (see Figure 5).

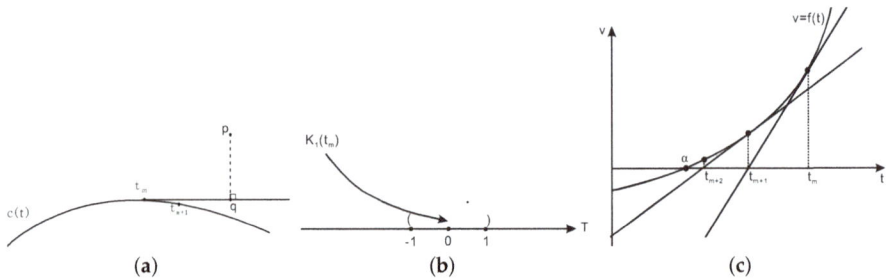

Figure 5. Geometric illustration for our method. (**a**) Running the H-H-H method; (**b**) Judging the H-H-H method whether being satisfied the convergence condition of fixed point theorem for the Newton's iterative method; (**c**) Running the Newton's iterative method.

Hybrid second order method
Input: Initial iterative value t_0, test point p and parametric curve $c(t)$ in an n-dimensional Euclidean space.
Output: The corresponding parameter α determined by orthogonal projection point.
Step 1. Initial iterative parametric value t_0 is input.
Step 2. Using the iterative Equation (3), calculate the parametric value $K_1(t_0)$, and update $K_1(t_0)$ to t_1, namely, $t_1 = K_1(t_0)$.
Step 3. Determine whether absolute value of difference between the current t_0 and the new t_1 is near 0. If so, this algorithm is ended.
Step 4. Substitute the new t_1 into $\left|\frac{f(t)f''(t)}{f'(t)^2}\right|$, determine if $\left|\frac{f(t_1)f''(t_1)}{f'(t_1)^2}\right| < 1$.

If ($\left|\frac{f(t_1)f''(t_1)}{f'(t_1)^2}\right| < 1$) {

Using Newton's iterative Equation (16), compute $t_0 = t_1 - \frac{f(t_1)}{f'(t_1)}$ until absolute value of difference between the current t_1 and the new t_0 is near 0; then, this algorithm ends.

}
Else {

    turn to **Step 2.**

}

Remark 1. *Firstly, a geometric illustration of our method in Figure 5 would be presented. Figure 5a illustrates the second step of our method where the next iterative parameter value $t_{m+1} = K_1(t_m) = t_m + \frac{\langle c'(t_m), p - c(t_m)\rangle}{\langle c'(t_m), c'(t_m)\rangle}$ is determined by the iterative Equation (3). During the iterative process, the step Δt will become smaller and smaller. Thus, the next iterative parameter value t_{m+1} comes close to parameter value t_m but far from the footpoint q. If the third step of our method is not over, then our method goes into the fourth step. Figure 5b is judging condition of a fixed point theorem of the fourth step of our method. If $T = \left|\frac{f(t)f''(t)}{f'(t)^2}\right| < 1$, then it turns to the Newton's method in Figure 5c until it runs to the end of the whole process of Newton's second order iteration; otherwise, it goes to the second step in Figure 5a.*

Secondly, we give an interpretation for the singularity case of the iterative Equation (16). As to some special cases where the H-H-H method is not convergent in Section 3.1, our method still converges. We test many examples for arbitrary initial value, arbitrary test point and arbitrary parametric curve and find that our method remains more robust to converge than the H-H-H method. If the first order derivative $f'(t_m)$ of the iterative Equation (16) develops into 0, i.e., $f'(t_m) = 0$ about some non-negative integer m, we use a perturbed method to solve the special problem, which adopts the idea in [23,45]. Namely, the function $f'(t_m) = 0$ could be *increased by a very small positive number ε, i.e., $f'(t_m) = f'(t_m) + \varepsilon$, and then the iteration by Equation (16) is continued in order to calculate the parameter value. On the other hand, if the curve can be parametrized by a (piece-wise) polynomial, then the fast root-finding schemes such as Bézier clipping [28,29] are efficient ones. The only issue is the C^1 discontinuities that can be checked in a post-process. One then does not need any initial guess on the parameter value.*

Thirdly, if the curve is only C^0 continuous, and the closest point can be exactly such a point, then the derivative is not well defined and our method may fail to find such a point. Namely, there are singular points on the parametric curve. We adopt the following technique to solve the problem of singularity. We use the methods [46–48] to find all singular points on the parametric curve and the corresponding parametric value of each singular point as many as possible. Then, the hybrid second order method comes into work. If the current iterative parametric value t_m is the corresponding parametric value of a singular point, we make a very small perturbation ε to the current iterative parametric value t_m, i.e., $t_m = t_m + \varepsilon$. The purpose of this behavior is to enable the hybrid second order method to run normally. Then, from all candidate points (singular points and orthogonal projection points), a corresponding point is selected so that the distance between the corresponding point and the test point is the minimum one. When the entire program terminates, the minimum distance and its corresponding parameter value are found.

3.3. Convergence Analysis of the Improved Algorithm

In this subsection, we prove the convergence analysis of our method.

Theorem 2. In Reference [49] (Fixed Point Theorem)
If $\phi(x) \in C[c,d]$, $\phi(x) \in [c,d]$ for all $x \in [c,d]$; furthermore, if $\phi'(x)$ exists on (c,d) and a positive constant $L < 1$ exists with $|\phi'(x)| \leq L$ for all $x \in (c,d)$, then there exists exactly one fixed point in $[c,d]$.

In addition, if $\phi(t) = t - \dfrac{f(t)}{f'(t)}$, the corresponding fixed point theorem of Newton's method is as follows:

Theorem 3. *Let $f : [c,d] \to [c,d]$ be a differentiable function, if for all $t \in [c,d]$, there is*

$$\left| \frac{f(t)f''(t)}{f'^2(t)} \right| < 1. \tag{17}$$

Then, there is a fixed point $l_0 \in [c,d]$ in Newton's iteration expression (16) such that $l_0 - l_0 - \dfrac{f(l_0)}{f'(l_0)}$. Meanwhile, the iteration sequence $\{t_m\}$ been from expression (16) can converge to the fixed point when $\forall t_0 \in [c,d]$.

Theorem 4. *Our method is second order convergent.*

Proof: Let α be a simple zero for a nonlinear function $f(t) = < T_1, V_1 > = 0$, where $T_1 = c'(t)$, $V_1 = p - c(t)$. Using Taylor's expansion, we have

$$f(t_m) = f'(\alpha)[e_m + b_2 e_m^2 + b_3 e_m^3 + o(e_m^4)], \tag{18}$$

$$f'(t_m) = f'(\alpha)[2b_2 e_m + 3b_3 e_m^2 + o(e_m^3)], \tag{19}$$

where $b_k = \dfrac{f^{(k)}(\alpha)}{k! f'(\alpha)}$, $k = 2, 3, \ldots$, and $e_m = t_m - \alpha$. Combining with (15), we then have

$$y_m = \phi(t_m) = t_m - \frac{f(t_m)}{f'(t_m)} = \alpha + b_2 C_0^2 e_m^2 + o(e_m^3). \tag{20}$$

This means that the convergence order of our method is 2. The proof is completed. □

Theorem 5. *Convergence of our method does not depend on the initial value.*

Proof. According to the description of our method, if the iterative parametric value of the H-H-H method is satisfied with the convergence condition of the Newton's method, we then go to the Newton's method. Otherwise, we steadily adopt the H-H-H method until its iterative parametric value is satisfied with the convergence condition of the Newton's method, and we go to Newton's method. Then, we run to the end of the whole process. Theorem 1 ensures that it does not depend on the initial value. If our method goes to the fourth step and if it is appropriate to the condition of the fixed point theorem (Theorem 3), Newton's method is realized by our method. Then, the fourth step of our method being also independent of the initial value can be confirmed by Theorem 3. In brief, convergence of our method does not depend on the initial value via the whole algorithm execution process. The proof is completed. □

4. Numerical Experiments

In order to illustrate the superiority of our method to other algorithms, we provide five numerical examples to confirm its robustness and high efficiency. From Tables 1–14, the iterative termination criteria is satisfied such that $|t_m - \alpha| < 10^{-17}$ and $|t_{m+1} - t_n| < 10^{-17}$. All numerical results were computed through g++ in a Fedora Linux 8 environment. The approximate zero α reached up to the 17th decimal place is reflected. These results of our five examples are obtained from computer hardware configuration with T2080 1.73 GHz CPU and 2.5 GB memory.

Example 1. *There is a parametric curve $c(t) = (f_1(t), f_2(t), f_3(t)) = (6t^7 + t^5, 5t^8 + 3t^6, 10t^{12} + 8t^8 + 6t^6 + 4t^4 + 2t^2 + 3), t \in [-2, 2]$ in three-dimensional Euclidean space and a test point $p = (p_1, p_2, p_3) = (2.0, 4.0, 2.0)$. Using our method, the corresponding orthogonal projection parametric value is $\alpha = 0.0$, the initial values t_0 are 0,2,4,5,6,8,9,10, respectively. For each initial value, the iteration process runs 10 times and then 10 different iteration times in nanoseconds, respectively. In Table 1, the average run time of our method for eight different initial values are 536,142, 77,622, 101,481, 119,165, 126,502, 142,393, 150,801, 156,413 nanoseconds, respectively. Finally, the overall average running time is 176,315 nanoseconds (see Figure 6). If test point p is (2.0, 2.0, 2.0), the corresponding orthogonal projection parametric value is $\alpha = 0.0$, we replicate the procedure using our method and report the results in Table 2. In Table 2, the average running time of our method for 8 different initial values are 627,996, 89,992, 119,241, 139,036, 148,269, 167,364, 167,364, 178,554 nanoseconds, respectively. Finally, the overall average running time is 205,228 nanoseconds (see Figure 7). However, for the above two cases, the H-H-H method does not converge for any initial iterative value.*

Because of a singular point on the parametric curve, we have also added some pre-processing steps before our method. (1) Find the singular point (0,0,3) and the corresponding parametric value 0 by using the methods [21,46–48]. (2) Using our method, the orthogonal projection points of test points (2,4,2) and (2,2,2) and their corresponding parameter values 0 and 0 are calculated, respectively. (3) From all candidate points(singular point and orthogonal projection point), corresponding point is selected so that the distance between the corresponding point and the test point is the minimum one. In Figure 6, the blue point denotes singular point (0,0,3), which is also the orthogonal projecting point of the test point (2,4,2). This is the same for the blue point in Figure 7.

Table 1. Running time for different initial values of Example 1 by our method with test point $p = (2.0, 4.0, 2.0)$.

t_0	0	2	4	5	6	8	9	10
α	0.0	0.0	0.0	0.0	0.0	0.0	0.0	0.0
1	498,454	75,487	105,563	116,470	123,031	134,941	154,253	156,872
2	555,709	81,629	108,064	117,762	125,946	140,940	153,468	155,830
3	509,173	82,824	100,744	111,206	134,367	141,705	150,013	158,715
4	564,222	77,465	96,721	114,757	129,128	173,027	150,320	158,580
5	502,986	81,028	97,142	118,535	120,668	132,856	155,335	149,437
6	553,198	79,520	104,307	120,795	129,351	150,085	151,073	143,065
7	576,814	74,268	100,231	115,002	132,322	139,919	154,754	159,014
8	524,848	81,982	99,604	115,263	122,401	139,345	143,568	175,169
9	528,848	71,228	103,186	140,023	122,040	135,006	145,434	154,016
10	547,161	70,789	99,247	121,834	125,766	136,103	149,790	153,435
Average	536,142	77,622	101,481	119,165	126,502	142,393	150,801	156,413
Total Average	176,315							

Table 2. Running time for different initial values of Example 1 by our method with test point $p = (2.0, 2.0, 2.0)$.

t_0	0	2	4	5	6	8	9	10
α	0.0	0.0	0.0	0.0	0.0	0.0	0.0	0.0
1	595,515	92,371	119,904	135,660	148,751	162,758	171,535	177,355
2	648,825	91,746	119,348	135,284	148,531	162,541	171,333	176,431
3	595,772	91,633	119,248	135,322	148,222	162,240	171,095	176,501
4	648,472	91,565	119,139	135,355	148,165	191,884	171,366	176,395
5	595,856	91,556	119,168	135,406	148,144	162,224	171,417	176,507
6	648,305	91,532	119,018	135,316	148,169	183,342	171,413	176,473
7	647,406	91,587	119,069	135,283	148,197	162,291	171,282	176,397
8	595,423	91,617	119,247	135,140	14,8101	162,116	171,342	196,529
9	646,551	83,167	119,135	172,412	148,149	162,148	171,313	176,390
10	657,838	83,147	119,131	135,179	148,259	162,094	171,609	176,557
Average	627,996	89,992	119,241	139,036	148,269	167,364	171,371	178,554
Total Average	205,228							

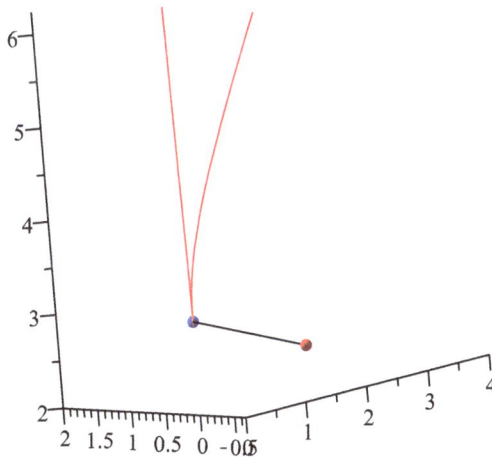

Figure 6. Geometric illustration for the test point $p = (2.0, 4.0, 2.0)$ of Example 1.

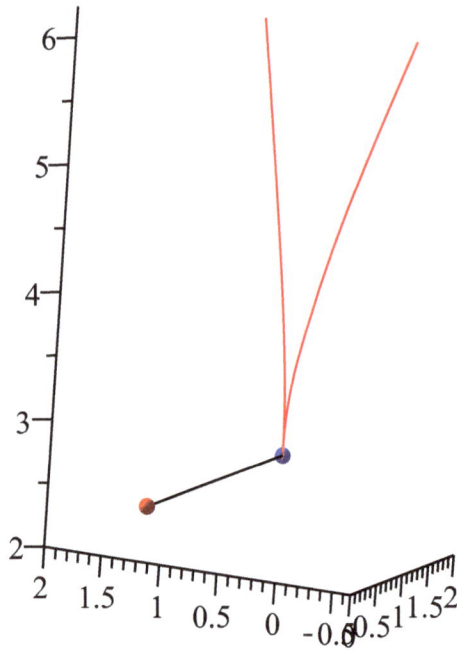

Figure 7. Geometric illustration for the test point *p* = (2.0, 2.0, 2.0) of Example 1.

Example 2.   *There is a spatial quartic quasi-rational Bézier curve $c(t) = (f_1(t), f_2(t), f_3(t)) = (\frac{u(t)}{a(t)}, \frac{v(t)}{a(t)}, \frac{w(t)}{a(t)})$, where $u(t) = 2t^4 + 3t^3 + 3t^2 + 12t + 1$, $v(t) = 4t^4 + 3t^3 + 7t^2 + 7t + 21$, $w(t) = 5t^4 + t^3 + 9t^2 + 11t + 13$, $a(t) = 4t^4 + 8t^3 + 17t^2 + 15t + 6$, $t \in [-2, 2]$ and a test point $p = (p_1, p_2, p_3) = (1.0, 3.0, 5.0)$. The corresponding orthogonal projection parametric value α are -1.4118250062741212, -0.61917136491841674, -0.059335038305820650, 1.8493434997820080, respectively. Using our method, the initial values t_0 are $-2.4, -2.1, -2.0, -1.8, -1.6, -1.2, -1.0, -0.8$, respectively. For each initial value, the iteration process runs 10 times and then 10 different iteration times in nanoseconds, respectively. From Table 3, the average running time of our method for eight different initial values are 85,344, 93,936, 79,424, 62,643, 54,482, 22,982, 25,654, 26,868 nanoseconds, respectively. Finally, the overall average running time is 56,417 nanoseconds (see Figure 8). If test point p is $(2.0, 4.0, 8.0)$, the corresponding orthogonal projection parametric value α are -1.2589948653798823, -0.62724968160147096, -0.14597283439336865, 1.8584532894110559, respectively. We firstly replicate the procedure using our method and report the results in Table 4. From Table 4, the average running time of our method for eight different initial iterative values are 101,436, 109,001, 95,061, 77,563, 62,366, 27,054, 29,587, 32,501 nanoseconds, respectively. Finally, the overall average running time is 66,821 nanoseconds (see Figure 9). We then replicate the procedure using the algorithm [26] and report the results in Table 5. From Table 5, the average running time of the algorithm [26] for eight different initial values are 619,772, 654,281, 584,653, 467,856, 384,393, 163,225, 183,257, 195,013 nanoseconds, respectively. Finally, the overall average running time is 406,556 nanoseconds. However, for the above two cases, the H-H-H method does not converge for any initial value.*

Table 3. Running time for different initial values of Example 2 by our method with test point $p = (1.0, 3.0, 5.0)$.

t_0	-2.4	-2.1	-2	-1.8	-1.6	-1.2	-1	-0.8
α	-1.4118	0.61917	-1.4118	0.61917	-0.059	-0.059	1.84934	1.84934
1	88,695	90,501	75,137	68,499	52,014	24,731	26,295	28,444
2	89,958	91,254	79,411	64,563	54,321	22,014	26,278	28,024
3	83,956	95,063	79,553	63,237	54,683	22,733	24,813	28,760
4	83,623	96,033	82,022	68,075	51,098	23,270	24,573	26,707
5	83,368	95,700	76,197	63,518	51,752	22,321	24,644	26,586
6	83,631	97,303	80,984	62,608	53,473	21,658	24,009	28,209
7	87,286	94,655	78,483	66,844	52,277	23,502	25,554	28,725
8	87,150	96,316	79,215	64,333	51,554	23,217	26,234	28,295
9	86,300	89,399	94,487	66,665	50,279	23,190	25,791	26,160
10	89,761	96,377	82,362	64,371	50,367	22,332	23,929	27,273
Average	85,344	93,936	79,424	62,643	54,482	22,982	25,654	26,868
Total Average	56,417							

Table 4. Running time for different initial values of Example 2 by our method with test point $p = (2.0, 4.0, 8.0)$.

t_0	-2.4	$--2.1$	$--2$	-1.8	-1.6	-1.2	-1	-0.8
α	-0.6272	-0.1459	-0.6272	-1.2589	-0.1459	1.858	-1.2589	1.858
1	101,366	109,667	92,799	77,983	62,865	29,460	29,755	32,649
2	102,027	108,844	92,709	77,477	62,269	27,177	29,555	32,458
3	101,526	109,010	92,709	77,587	62,284	26,885	29,619	32,538
4	101,266	108,909	92,724	77,441	62,374	26,785	29,557	32,478
5	101,346	108,944	92,714	77,386	62,214	26,691	29,559	32,505
6	101,315	108,990	92,764	77,557	62,334	26,731	29,564	32,497
7	101,415	108,834	92,614	77,582	62,415	26,720	29,573	32,512
8	101,306	108,945	92,528	77,461	62,309	26,715	29,548	32,493
9	101,562	108,954	116,107	77,542	62,284	26,684	29,549	32,429
10	101,235	108,910	92,939	77,616	62,314	26,690	29,595	32,451
Average	101,436	109,001	95,061	77,563	62,366	27,054	29,587	32,501
Total Average	66,821							

Table 5. Running time for different initial values of Example 2 by the algorithm [26].

t_0	-2.4	-2.1	-2.0	-1.8	-1.6	-1.2	-1.0	-0.8
α	-0.6272	-0.1459	-0.6272	-1.2589	-0.1459	1.858	-1.2589	1.858
1	633,173	660,734	566,675	470,236	391,687	171,352	175,965	198,543
2	597,065	628,741	565,012	485,368	367,539	161,649	185,457	197,798
3	652,494	675,268	600,951	463,899	396,359	163,879	188,682	187,128
4	649,281	653,066	573,597	460,967	385,325	156,876	182,979	195,214
5	622,109	687,282	568,766	472,217	402,669	170,876	189,508	202,540
6	633,737	627,667	562,864	490,735	374,340	165,445	175,457	191,037
7	584,705	637,608	563,523	468,230	395,411	163,631	175,676	187,539
8	607,439	693,001	585,948	449,706	400,728	161,467	189,216	187,433
9	637,036	639,359	671,613	444,834	359,918	157,235	188,119	195,867
10	580,678	640,082	587,577	472,368	369,954	159,834	181,510	207,033
Average	619,772	654,281	584,653	467,856	384,393	163,225	183,257	195,013
Total Average	406,556							

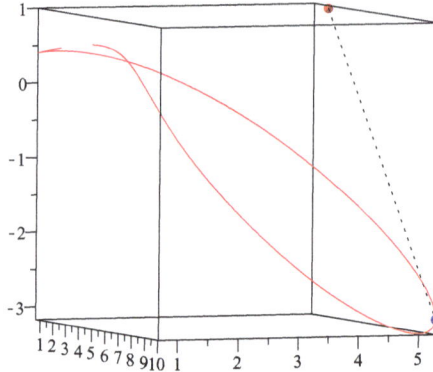

Figure 8. Geometric illustration for the first case of Example 2.

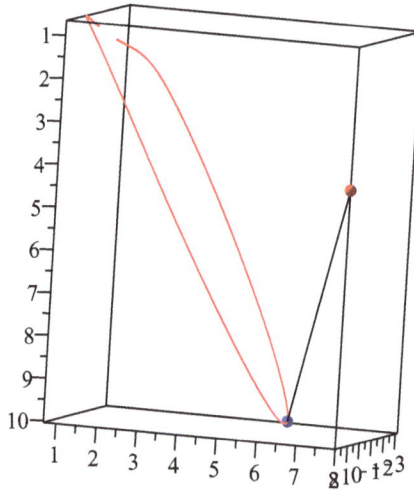

Figure 9. Geometric illustration for the second case of Example 2.

Example 3. *There is a parametric curve* $c(t) = (f_1(t), f_2(t), f_3(t), f_4(t), f_5(t)) = (\cos(t), \sin(t), t, \cos(t), \sin(t)), t \in [-2, 2]$ *in five-dimensional Euclidean space and a test point* $p = (p_1, p_2, p_3, p_4, p_5) = (3.0, 4.0, 5.0, 6.0, 7.0)$. *Using our method, the corresponding orthogonal projection parametric value is* $\alpha = 1.1587403612284800$, *the initial values* t_0 *are* $-10, -8, -6, -4, 4, 8, 12, 16$, *respectively. For each initial value, the iteration process runs 10 times and then 10 different iteration times in nanoseconds, respectively. In Table 6, the average running time of our method for eight different initial values are 391,013, 424,444, 391,092, 249,376, 115,617, 170,212, 179,465, 196,912 nanoseconds, respectively. Finally, the overall average running time is 264,766 nanoseconds. If test point* p *is* $(30.0, 40.0, 50.0, 60.0, 70.0)$, *the corresponding orthogonal projection parametric value* α *is* 1.2352898417860202. *We then replicate the procedure using our method and report the results in Table 7. In Table 7, the average running time of our method for eight different initial values are* 577,707, 485,417, 460,913, 289,232, 133,661, 199,470, 211,915, 229,398 *nanoseconds, respectively. Finally, the overall average running time is 323,464 nanoseconds. However, for the above parametric curve and many test points, the H-H-H method does not converge for any initial value.*

Table 6. Running time for different initial values of Example 3 by our method with test point $p = (3, 4, 5, 6, 7)$.

t_0	−10	−8	−6	−4	4	8	12	16
α	1.15874	1.15874	1.15874	1.15874	1.15874	1.15874	1.15874	1.15874
1	407,427	425,388	387,337	306,115	110,887	161,079	187,144	184,119
2	417,729	446,171	398,801	341,895	121,148	169,115	169,954	194,671
3	420,894	390,507	383,308	260,183	115,033	165,103	171,989	198,884
4	383,836	421,365	427,391	242,641	109,521	161,121	179,152	195,714
5	373,696	421,551	373,171	266,584	120,844	187,930	179,184	186,309
6	374,791	445,114	373,974	242,889	119,449	183,082	180,269	201,487
7	381,353	408,011	402,073	216,762	109,054	162,402	172,013	188,206
8	398,662	442,008	373,328	194,821	119,236	192,990	180,472	197,299
9	364,491	417,139	396,843	230,070	110,243	164,273	204,410	196,163
10	387,246	427,188	394,694	191,799	120,759	155,029	170,059	226,270
Average	391,013	424,444	391,092	249,376	115,617	170,212	179,465	196,912
Total Average	264,766							

Table 7. Running time for different initial values of Example 3 by our method with test point $p = (30, 40, 50, 60, 70)$.

t_0	−10	−8	−6	−4	4	8	12	16
α	1.235289	1.235289	1.235289	1.235289	1.235289	1.235289	1.235289	1.235289
1	1,190,730	475,499	453,879	369,551	133,651	191,093	208,202	223,695
2	1,031,760	500,975	486,534	380,881	133,638	190,959	208,490	236,637
3	482,018	475,395	450,480	297,272	133,674	199,528	208,292	223,312
4	428,081	475,588	475,100	277,356	133,635	186,919	208,438	223,802
5	455,282	475,033	448,776	296,510	133,535	220,570	208,139	223,471
6	428,321	499,776	448,617	277,353	133,590	220,625	208,046	223,213
7	428,246	474,978	474,667	247,245	133,620	192,326	208,101	230,791
8	453,374	502,500	448,503	235,415	133,594	220,635	208,087	223,183
9	426,949	474,816	474,167	275,526	133,546	198,204	245,226	223,213
10	452,306	499,605	448,409	235,207	134,128	173,843	208,127	262,661
Average	577,707	485,417	460,913	289,232	133,661	199,470	211,915	229,398
Total Average	323,464							

Example 4. (Reference to [6]) *There is a parametric curve $c(t) = (f_1(t), f_2(t)) = (t^2, \sin(t)), t \in [-3, 3]$ in two-dimensional Euclidean space and a test point $p = (p_1, p_2) = (1.0, 2.0)$. The corresponding orthogonal projection parametric value is $\alpha = 1.1063055095030472$. Using our method, the initial values t_0 are $-100, -4, 5, 7, 8, 10, 11, 100$, respectively. For each initial value, the iteration process runs 10 times and then 10 different iteration times in nanoseconds, respectively. In Table 8, the average running time of our method for eight different initial iterative values are 62,816, 35,042, 27,648, 43,122, 21,625, 38,654, 21,518, 72,917 nanoseconds, respectively. Finally, the overall average running time is 40,418 nanoseconds (see Figure 10). Implementing the same procedure, the overall average running time given by the H-H-H method is 231,613 nanoseconds in Table 9, while the overall average running time given by the second order method [6] is 847,853 nanoseconds in Table 10. Thus, our method is faster than the H-H-H method [38–40] and the second order method [6].*

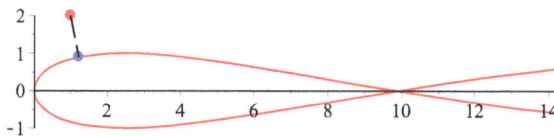

Figure 10. Geometric illustration for Example 4.

Table 8. Running time for different initial values of Example 4 by our method.

t_0	-100	-4	5	7	8	10	11	100
α	1.106305	1.106305	1.106305	1.106305	1.106305	1.106305	1.106305	1.106305
1	63,345	35,580	27,069	41,551	22,304	36,858	21,478	72,257
2	63,192	36,203	28,160	41,733	20,042	38,680	20,338	71,620
3	61,306	33,833	27,400	44,198	23,078	37,704	23,757	73,108
4	66,627	34,502	26,014	44,160	21,147	39,374	22,530	70,154
5	62,583	35,053	29,275	42,800	20,817	39,339	23,046	73,189
6	63,957	34,398	25,650	42,282	22,184	37,376	20,070	75,872
7	60,865	35,929	28,944	42,134	19,964	40,078	21,943	71,608
8	63,522	35,427	27,578	41,688	23,650	39,456	21,076	76,283
9	60,551	35,508	28,563	44,542	20,280	38,463	20,596	71,781
10	62,216	33,987	27,830	46,130	22,781	39,209	20,349	73,296
Average	62,816	35,042	27,648	43,122	21,625	38,654	21,518	72,917
Total Average	40,418							

Table 9. Running time for different initial values of Example 4 by the H-H-H method.

t_0	-100	-4	5	7	8	10	11	100
α	1.106305	1.106305	1.106305	1.106305	1.106305	1.106305	1.106305	1.106305
1	424,579	357,276	443,858	179,583	176,984	175,859	175,249	178,445
2	425,680	358,510	179,137	177,849	182,701	176,665	176,463	207,164
3	359,794	356,912	180,000	180,472	177,867	179,743	178,929	179,372
4	371,119	357,214	179,567	179,804	184,542	177,675	177,854	179,651
5	358,128	358,119	232,337	179,285	179,113	175,632	177,690	181,976
6	358,470	357,893	179,985	179,941	178,600	178,289	178,565	181,868
7	358,083	359,391	178,815	177,857	177,613	178,014	177,385	179,361
8	477,393	357,011	178,029	179,525	175,684	176,000	175,413	180,966
9	356,254	359,356	176,148	178,581	176,351	177,024	185,103	180,013
10	356,801	359,773	213,327	177,252	176,993	178,060	177,655	181,427
Average	384,630	358,146	214,120	179,015	178,645	177,296	178,031	183,024
Total Average	231,613							

Table 10. Running time for different initial values of Example 4 by the Algorithm [6].

t_0	-100	-4	5	7	8	10	11	100
α	1.106305	1.106305	1.106305	1.106305	1.106305	1.106305	1.106305	1.106305
1	681,353	107,102	119,083	120,328	122,504	115,181	113,566	542,116
2	725,571	124,514	136,810	121,111	116,824	111,116	117,466	5,250,481
3	669,249	111,052	122,151	125,261	124,865	116,105	120,309	5,523,805
4	713,982	112,146	131,494	118,104	121,099	111,410	118,658	5,407,166
5	699,433	111,347	118,830	121,003	118,694	115,182	124,917	5,259,412
6	693,396	113,323	116,046	109,176	108,194	111,420	117,342	5,508,049
7	691,375	114,667	115,748	123,330	127,812	118,635	119,208	5,348,517
8	663,125	107,484	127,493	120,134	116,818	111,717	117,079	5,446,703
9	731,148	128,918	122,897	120,947	120,985	113,777	125,463	5,251,580
10	676,286	128,567	130,775	118,031	116,725	111,095	108,275	5,356,125
Average	694,492	115,912	124,133	119,743	119,452	113,564	118,228	5,377,300
Total Average	847,853							

Example 5. (Reference to [6]) *There is a parametric curve $c(t) = (f_1(t), f_2(t)) = (t, sin(t)), t \in [-3, 3]$ in two-dimensional Euclidean space and a test point $p = (p_1, p_2) = (1.0, 2.0)$, the corresponding orthogonal projection parametric value is $\alpha = 1.2890239979093887$. Using our method, the initial values t_0 are $-100, -4, 5, 7, 8, 10, 11, 100$, respectively. For each initial value, the iteration process runs 10 times and then 10 different iteration time in nanoseconds, respectively. In Table 11, the average running time of our method for eight different initial values are $50,579, 28,238, 22,687, 34,974,17,781, 31,186, 17,210, 59,116$ nanoseconds, respectively. Finally, the overall average running time is 32,721 nanoseconds (see Figure 11). We then replicate the procedure using the second order method [6] and report the results in Table 12. In Table 12, the average running time of the second order method [6] for 8 different initial values are $320,035,182,451, 147,031, 235,779, 112,090, 200,431, 113,284, 369,294$ nanoseconds, respectively. Finally, the overall average*

running time is 210,049 nanoseconds. In addition, we compare the iterations by different methods where the NC denotes non-convergence in Table 13.

Table 11. Running time for different initial values of Example 5 by our method.

t_0	−100	−4	5	7	8	10	11	100
α	1.28902	1.28902	1.28902	1.28902	1.28902	1.28902	1.28902	1.28902
1	52,010	27,426	21,791	33,323	18,399	29,551	16,486	58,995
2	50,335	29,269	23,949	32,810	15,820	30,342	16,066	58,080
3	49,047	26,841	23,061	37,063	19,611	31,569	19,756	57,458
4	52,651	29,124	21,403	33,838	17,472	33,295	18,583	54,566
5	49,871	29,814	25,062	35,870	16,655	32,949	18,304	61,860
6	53,651	28,678	19,550	35,731	18,373	31,429	16,342	59,570
7	47,275	28,115	24,177	35,456	16,933	30,510	18,010	59,042
8	49,982	27,896	22,639	34,292	19,927	30,959	16,449	63,652
9	49,704	29,359	22,502	34,164	17,274	30,391	16,044	61,373
10	51,268	25,859	22,736	37,190	17,342	30,864	16,060	56,564
Average	50,579	28,238	22,687	34,974	17,781	31,186	17,210	59,116
Total Average	32,721							

Table 12. Running time for different initial values of Example 5 by the Algorithm [6].

t_0	−100	−4	5	7	8	10	11	100
α	1.28902	1.28902	1.28902	1.28902	1.28902	1.28902	1.28902	1.28902
1	308,942	191,002	152,199	235,287	114,568	199,404	110,512	379,771
2	348,554	175,800	146,728	232,260	102,698	190,860	101,754	352,834
3	311,680	190,863	148,384	242,131	118,602	207,376	125,517	408,978
4	332,421	166,849	145,131	234,536	102,795	198,956	113,523	370,826
5	319,660	185,059	160,358	235,072	108,557	211,429	119,911	350,188
6	329,882	177,252	132,242	233,702	120,945	199,978	107,366	363,299
7	304,977	200,038	151,398	229,166	102,315	220,162	122,013	354,466
8	326,645	171,624	137,588	228,181	113,627	195,782	108,512	369,899
9	291,369	191,878	156,871	247,614	108,418	189,534	112,319	363,905
10	326,221	174,148	139,415	239,836	128,377	190,831	111,411	378,781
Average	320,035	182,451	147,031	235,779	112,090	200,431	113,284	369,294
Total Average	210,049							

Table 13. Comparison of iterations by different methods in Example 5.

t_0	−100.0	−4.0	5.0	7.0	8.0	10.0	11.0	100.0
α	1.28902	1.28902	1.28902	1.28902	1.28902	1.28902	1.28902	1.28902
H-H-H method [38–40]	NC	NC	NC	NC	NC	NC	NC	NC
Second order method [6]	75	30	32	32	33	29	31	101
Newton's method	NC	NC	NC	NC	NC	NC	NC	NC
Our method	15	19	17	17	15	17	15	23

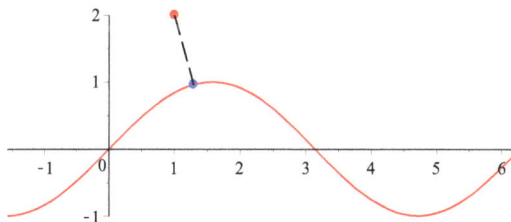

Figure 11. Geometric illustration for Example 5.

Remark 2. *From the results of five examples, the overall average running time of our method is 145.5 μs. From the results of Table 9, the overall average running time of the H-H-H method is 231.6 μs. From results of six examples in [26], the overall average running time of the algorithm [1] is 680.8 μs. From results of* six examples in [26], the overall average running time of the algorithm [14] is 1270.8 μs. From results of Table 5, the overall average running time of the algorithm [26] is 406.6 μs. From results of Tables 10 and 12, the overall average running time of the algorithm [6] is 528.9 μs. Table 14 displays time comparison for these algorithms. In short, the robustness and efficiency of our method are more superior to those of the existing algorithms [1,6,14,26,38–40].

Table 14. Time comparison of various algorithms.

Algorithms	Ours	H-H-H	Algorithm [1]	Algorithm [14]	Algorithm [6]	Algorithm [26]
Time (μs)	145.5	231.6	680.8	1270.8	528.9	406.6

Remark 3. *For general parametric curve containing the elementary functions, such as* $\sin(t), \cos(t), e^t$, $\ln t$, $\arcsin t$, $\arccos t$, *etc., it is very difficult to transform general parametric curve into Bézier-type curve. In contrast, our method can deal with the general parametric curve containing the elementary functions. Furthermore, the convergence of our method does not depend on the initial value. From Table 13, only the H-H-H method or the Newton's method can not ensure convergence, while our method can ensure convergence. For multiple solutions of orthogonal projection, our approach works as follows:*
(1) The parameter interval $[a, b]$ *of parametric curve* $c(t)$ *is divided into M identical subintervals.*
(2) An initial value is selected randomly in each interval.
(3) Using our method and using each initial parametric value, do iterations, respectively. Suppose that the iterative parametric values are $\alpha_1, \alpha_2, \ldots, \alpha_M$, *respectively.*
(4) Calculate the local minimum distances $d_1, d_2, \ldots, d_M$, *where* $d_i = \|p - c(\alpha_i)\|$.
(5) Seek the global minimum distance $d = \|p - c(\alpha)\|$ *from* $\{\|p - c(a)\|, d_1, d_2, \ldots, d_M, \quad \|p - c(b)\|\}$.
If we are to solve all solutions as far as possible, we urge the positive integer M to be as large as possible.

We use Example 2 to illustrate how the procedure works, where, for $t \in [-2, 2]$, three parameter values are -1.4118250062741212, -0.61917136491841674, 1.8493434997820080, respectively. It is easy to find that the projection point with the parameter value -0.61917136491841674 will be the one with minimum distance, whereas other projection points without these parameter values can not be the one with minimum distance. Thus, only the orthogonal projection point with minimum distance remains after the procedure to select multiple orthogonal projection points.

Remark 4. *We have done many test examples including five test examples. In the light of these test results, our method has good convergent properties for different initial values, namely, if initial value is* t_0, *then the corresponding orthogonal projection parametric value* α *for the orthogonal projection point of the test point* p *is suitable for one inequality relationship*

$$|\langle p - c(\alpha), c'(\alpha) \rangle| < 10^{-17}. \tag{21}$$

This indicates that the inequality relationship satisfies requirements of Equation (4). This shows that convergence of our method does not depend on the initial value. Furthermore, our method is robust and efficient, which is satisfied with the previous two of ten challenges proposed by [50].

5. Conclusions

This paper discusses the problem related to a point orthogonal projection onto a parametric curve in an *n*-dimensional Euclidean space on the basis of the H-H-H method, combining with a fixed point theorem of Newton's method. Firstly, we run the H-H-H method. If the current iterative parametric value from the H-H-H method is satisfied with the convergence condition of the Newton's method, we then go to the method to increase the convergence rate. Otherwise, we continue the H-H-H method to generate the iterative parametric value with satisfaction of the local convergence condition by the Newton's method, and we then go to the previous step. Then, we run to the end of the whole process. The presented procedures ensure the convergence of our method and it does not depend on the initial value. Analysis of convergence demonstrates that our method is second order convergent. Some numerical examples confirm that our method is more efficient and performs better than other methods, such as the algorithms [1,6,14,26,38–40].

In this paper, our discussion focuses the algorithms in the parametric curve C^2. For the parametric curve being C^0, C^1, piecewise curve or having singular points, we only present a preliminary idea. However, we have not completely implemented an algorithm for this kind of spline with low continuity. In the future, we will try to construct several brand new algorithms to handle the kind of spline with low continuity such that they can ensure very good robustness and efficiency. In addition, we also try to extend this idea to handle point orthogonal projecting onto implicit curves and implicit surfaces that include singularity points. Of course, the realization of these ideas is of great challenge. However, it is of great value and significance in practical engineering applications.

Author Contributions: The contribution of all the authors is the same. All of the authors team up to develop the current draft. J.L. is responsible for investigating, providing methodology, writing, reviewing and editing this work. X.L. is responsible for formal analysis, visualization, writing, reviewing and editing of this work. F.P. is responsible for software, algorithm and program implementation to this work. T.C. is responsible for validation of this work. L.W. is responsible for supervision of this work. L.H. is responsible for providing resources, writing, and the original draft of this work.

Funding: This research was funded by the National Natural Science Foundation of China Grant No. 61263034, the Feature Key Laboratory for Regular Institutions of Higher Education of Guizhou Province Grant No. 2016003, the Training Center for Network Security and Big Data Application of Guizhou Minzu University Grant No. 20161113006, the Key Laboratory of Advanced Manufacturing Technology, Ministry of Education, Guizhou University Grant No. 2018479, the National Bureau of Statistics Foundation Grant No. 2014LY011, the Key Laboratory of Pattern Recognition and Intelligent System of Construction Project of Guizhou Province Grant No. 20094002, the Information Processing and Pattern Recognition for Graduate Education Innovation Base of Guizhou Province, the Shandong Provincial Natural Science Foundation of China Grant No.ZR2016GM24, the Scientific and Technology Key Foundation of Taiyuan Institute of Technology Grant No. 2016LZ02, the Fund of National Social Science Grant No. 14XMZ001 and the Fund of the Chinese Ministry of Education Grant No. 15JZD034.

Acknowledgments: We take the opportunity to thank the anonymous reviewers for their thoughtful and meaningful comments.

Conflicts of Interest: The authors declare no conflict of interest.

References

1. Ma, Y.L.; Hewitt, W.T. Point inversion and projection for NURBS curve and surface: Control polygon approach. *Comput. Aided Geom. Des.* **2003**, *20*, 79–99. [CrossRef]
2. Piegl, L.; Tiller, W. Parametrization for surface fitting in reverse engineering. *Comput.-Aided Des.* **2001**, *33*, 593–603. [CrossRef]
3. Yang, H.P.; Wang, W.P.; Sun, J.G. Control point adjustment for B-spline curve approximation. *Comput.-Aided Des.* **2004**, *36*, 639–652. [CrossRef]
4. Johnson, D.E.; Cohen, E. A Framework for efficient minimum distance computations. In Proceedings of the IEEE International Conference on Robotics & Automation, Leuven, Belgium, 20 May 1998.
5. Pegna, J.; Wolter, F.E. Surface curve design by orthogonal projection of space curves onto free-form surfaces. *J. Mech. Des. ASME Trans.* **1996**, *118*, 45–52. [CrossRef]
6. Hu, S.M.; Wallner, J. A second order algorithm for orthogonal projection onto curves and surfaces. *Comput. Aided Geom. Des.* **2005**, *22*, 251–260. [CrossRef]
7. Besl, P.J.; McKay, N.D. A method for registration of 3-D shapes. *IEEE Trans. Pattern Anal. Mach. Intell.* **1992**, *14*, 239–256. [CrossRef]
8. Mortenson, M.E. *Geometric Modeling*; Wiley: New York, NY, USA, 1985.
9. Limaien, A.; Trochu, F. Geometric algorithms for the intersection of curves and surfaces. *Comput. Graph.* **1995**, *19*, 391–403. [CrossRef]
10. Press, W.H.; Teukolsky, S.A.; Vetterling, W.T.; Flannery, B.P. Numerical recipes. In *C: The Art of Scientific Computing*, 2nd ed.; Cambridge University Press: New York, NY, USA, 1992.

11. Elber, G.; Kim, M.S. Geometric Constraint solver using multivariate rational spline functions. In Proceedings of the 6th ACM Symposiumon Solid Modeling and Applications, Ann Arbor, MI, USA, 4–8 June 2001; pp. 1–10.

12. Patrikalakis, N.; Maekawa, T. *Shape Interrogation for Computer Aided Design and Manufacturing*; Springer: Berlin, Germany, 2001.

13. Polak, E.; Royset, J.O. Algorithms with adaptive smoothing for finite minimax problems. *J. Optim. Theory Appl.* **2003**, *119*, 459–484. [CrossRef]

14. Selimovic, I. Improved algorithms for the projection of points on NURBS curves and surfaces. *Comput. Aided Geom. Des.* **2006**, 439–445. [CrossRef]

15. Zhou, J.M.; Sherbrooke, E.C.; Patrikalakis, N. Computation of stationary points of distance functions. *Eng. Comput.* **1993**, *9*, 231–246. [CrossRef]

16. Cohen, E.; Lyche, T.; Riesebfeld, R. Discrete B-splines and subdivision techniques in computer-aided geometric design and computer graphics. *Comput. Graph. Image Process.* **1980**, *14*, 87–111. [CrossRef]

17. Piegl, L.; Tiller, W. *The NURBS Book*; Springer: New York, NY, USA, 1995.

18. Park, C.-H.; Elber, G.; Kim, K.-J.; Kim, G.-Y.; Seong, J.-K. A hybrid parallel solver for systems of multivariate polynomials using CPUs and GPUs. *Comput.-Aided Des.* **2011**, *43*, 1360–1369. [CrossRef]

19. Bartoň, M. Solving polynomial systems using no-root elimination blending schemes. *Comput.-Aided Des.* **2011**, *43*, 1870–1878.

20. van Sosin, B.; Elber, G. Solving piecewise polynomial constraint systems with decomposition and a subdivision-based solver. *Comput.-Aided Des.* **2017**, *90*, 37–47. [CrossRef]

21. Bartoň, M.; Elber, G.; Hanniel, I. Topologically guaranteed univariate solutions of underconstrained polynomial systems via no-loop and single-component tests. *Comput.-Aided Des.* **2011**, *43*, 1035–1044.

22. Johnson, D.E.; Cohen, E. Distance extrema for spline models using tangent cones. In Proceedings of the 2005 Conference on Graphics Interface, Victoria, Canada, 9–11 May 2005.

23. Li, X.W.; Xin, Q.; Wu, Z.N.; Zhang, M.S.; Zhang, Q. A geometric strategy for computing intersections of two spatial parametric curves. *Vis. Comput.* **2013**, *29*, 1151–1158. [CrossRef]

24. Liu, X.-M.; Yang, L.; Yong, J.-H.; Gu, H.-J.; Sun, J.-G. A torus patch approximation approach for point projection on surfaces. *Comput. Aided Geom. Des.* **2009**, *26*, 593–598. [CrossRef]

25. Chen, X.-D.; Xu, G.; Yong, J.-H.; Wang, G.Z.; Paul, J.-C. Computing the minimum distance between a point and a clamped B-spline surface. *Graph. Models* **2009**, *71*, 107–112. [CrossRef]

26. Chen, X.-D.; Yong, J.-H.; Wang, G.Z.; Paul, J.-C.; Xu, G. Computing the minimum distance between a point and a NURBS curve. *Comput.-Aided Des.* **2008**, *40*, 1051–1054. [CrossRef]

27. Oh, Y.-T.; Kim, Y.-J.; Lee, J.; Kim, Y.-S. Gershon Elber, Efficient point-projection to freeform curves and surfaces. *Comput. Aided Geom. Des.* **2012**, *29*, 242–254. [CrossRef]

28. Sederberg, T.W.; Nishita, T. Curve intersection using Bézier clipping. *Comput.-Aided Des.* **1990**, *22*, 538–549. [CrossRef]

29. Bartoň, M.; Jüttler, B. Computing roots of polynomials by quadratic clipping. *Comput. Aided Geom. Des.* **2007**, *24*, 125–141.

30. Li, X.W.; Wu, Z.N.; Hou, L.K.; Wang, L.; Yue, C.G.; Xin, Q. A geometric orthogonal pojection strategy for computing the minimum distance between a point and a spatial parametric curve. *Algorithms* **2016**, *9*, 15. [CrossRef]

31. Mørken, K.; Reimers, M. An unconditionally convergent method for computing zeros of splines and polynomials. *Math. Comput.* **2007**, *76*, 845–865. [CrossRef]

32. Li, X.W.; Wang, L.; Wu, Z.N.; Hou, L.K.; Liang, J.; Li, Q.Y. Hybrid second-order iterative algorithm for orthogonal projection onto a parametric surface. *Symmetry* **2017**, *9*, 146. [CrossRef]

33. Li, X.W.; Wang, L.; Wu, Z.N.; Hou, L.K.; Liang, J.; Li, Q.Y. Convergence analysis on a second order algorithm for orthogonal projection onto curves. *Symmetry* **2017**, *9*, 210.

34. Chen, X.-D.; Ma, W.Y.; Xu, G.; Paul, J.-C. Computing the Hausdorff distance between two B-spline curves. *Comput.-Aided Des.* **2010**, *42*, 1197–1206. [CrossRef]

35. Chen, X.-D.; Chen, L.Q.; Wang, Y.G.; Xu, G.; Yong, J.-H.; Paul, J.-C. Computing the minimum distance between two Bézier curves. *J. Comput. Appl. Math.* **2009**, *229*, 294–301. [CrossRef]

36. Kim, Y.J.; Oh, Y.T.; Yoon, S.H.; Kim, M.S.; Elber, G. Efficient Hausdorff distance computation for freeform geometric models in close proximity. *Comput.-Aided Des.* **2013**, *45*, 270–276. [CrossRef]

37. Sundar, B.R.; Chunduru, A.; Tiwari, R.; Gupta, A.; Muthuganapathy, R. Footpoint distance as a measure of distance computation between curves and surfaces. *Comput. Graph.* **2014**, *38*, 300–309. [CrossRef]

38. Hoschek, J.; Lasser, D. *Fundamentals of Computer Aided Geometric Design*; A. K. Peters: Natick, MA, USA, 1993.

39. Hu, S.M.; Sun, J.G.; Jin, T.G.; Wang, G.Z. Computing the parameter of points on NURBS curves and surfaces via moving affine frame method. *J. Softw.* **2000**, *11*, 49–53.

40. Hartmann, E. On the curvature of curves and surfaces defined by normal forms. *Comput. Aided Geom. Des.* **1999**, *16*, 355–376. [CrossRef]

41. Li, X.W.; Mu, C.L.; Ma, J.W.; Wang, C. Sixteenth-order method for nonlinear Equations. *Appl. Math. Comput.* **2010**, *215*, 3754–3758. [CrossRef]

42. Liang, J.; Li, X.W.; Wu, Z.N.; Zhang, M.S.; Wang, L.; Pan, F. Fifth-order iterative method for solving multiple roots of the highest multiplicity of nonlinear equation. *Algorithms* **2015**, *8*, 656–668. [CrossRef]

43. Melmant, A. Geometry and Convergence of Euler's and Halley's Methods. *SIAM Rev.* **1997**, *39*, 728–735. [CrossRef]

44. Traub, J.F. A Class of Globally Convergent Iteration Functions for the Solution of Polynomial Equations. *Math. Comput.* **1966**, *20*, 113–138. [CrossRef]

45. Śmietański, M.J. A perturbed version of an inexact generalized Newton method for solving nonsmooth equations. *Numer. Algorithms* **2013**, *63*, 89–106. [CrossRef]

46. Chen, F.; Wang, W.-P.; Liu, Y. Computing singular points of plane rational curves. *J. Symb. Comput.* **2008**, *43*, 92–117. [CrossRef]

47. Jia, X.-H.; Goldman, R. Using Smith normal forms and μ-bases to compute all the singularities of rational planar curves. *Comput. Aided Geom. Des.* **2012**, *29*, 296–314. [CrossRef]

48. Shi, X.-R.; Jia, X.-H.; Goldman, R. Using a bihomogeneous resultant to find the singularities of rational space curves. *J. Symb. Comput.* **2013**, *53*, 1–25. [CrossRef]

49. Burden, R.L.; Faires, J.D. *Numerical Analysis*, 9th ed.; Brooks/Cole Cengage Learning: Boston, MA, USA, 2011.

50. Piegl, L.A. Ten challenges in computer-aided design. *Comput.-Aided Des.* **2005**, *37*, 461–470. [CrossRef]

mathematics

MDPI

Article

How to Obtain Global Convergence Domains via Newton's Method for Nonlinear Integral Equations

José Antonio Ezquerro *[ID] and **Miguel Ángel Hernández-Verón**

Department of Mathematics and Computation, University of La Rioja, Calle Madre de Dios, 53, 26006 Logrono, Spain; mahernan@unirioja.es
* Correspondence: jezquer@unirioja.es

Received: 3 May 2019; Accepted: 14 June 2019; Published: 17 June 2019

Abstract: We use the theoretical significance of Newton's method to draw conclusions about the existence and uniqueness of solution of a particular type of nonlinear integral equations of Fredholm. In addition, we obtain a domain of global convergence for Newton's method.

Keywords: Fredholm integral equation; Newton's method; global convergence

1. Introduction

Integral equations are very common in physics and engineering, since a lot of problems of these disciplines can be reduced to solve an integral equation. In general, we cannot solve integral equations exactly and are forced to obtain approximate solutions. For this, different numerical methods can be used. So, for example, iterative schemes based on the homotopy analysis method in [1], adapted Newton-Kantorovich schemes in [2] and schemes based on a combination of the Newton-Kantorovich method and quadrature methods in [3]. Besides, techniques based on using iterative methods are also interesting, since the theoretical significance of the methods allows drawing conclusions about the existence and uniqueness of solution of the equations. The use of an iterative method allows approximating a solution and, by analysing the convergence, proving the existence of solution, locating a solution and even separating such solution from other possible solutions by means of results of uniqueness. The theory of fixed point plays an important role in the development of iterative methods for approximating, in general, a solution of an equation and, in particular, for approximating a solution of an integral equation.

In this work, we pay attention to the study of nonlinear Fredholm integral equations with nonlinear Nemytskii operators of type

$$x(s) = \ell(s) + \lambda \int_a^b \mathcal{K}(s,t)\mathcal{H}(x)(t)\,dt, \quad s \in [a,b], \quad \lambda \in \mathbb{R}, \tag{1}$$

where $\ell(s) \in \mathcal{C}[a,b]$, kernel $\mathcal{K}(s,t)$ of integral equation is a known function in $[a,b] \times [a,b]$, $\mathcal{H}$ is a Nemytskii operator [4] given by $\mathcal{H} : \Omega \subseteq \mathcal{C}[a,b] \to \mathcal{C}[a,b]$, such that $\mathcal{H}(x)(t) = H(x(t))$ and $H : \mathbb{R} \to \mathbb{R}$ is a derivable real function, and $x(s) \in \mathcal{C}[a,b]$ is the unknown function to find.

It is common to use the Banach Fixed Point Theorem [5–7] to prove the existence of a unique fixed point of an operator and approximate it by the method of successive approximations. Moreover, global convergence for the method is obtained in the full space. For this, we use that the operator involved is a contraction.

Our main aim of this work is to do a study of integral Equation (1) from Newton's method,

$$x_{n+1} = x_n - [F'(x_n)]^{-1}F(x_n), \quad n \geq 0, \quad \text{with } x_0 \text{ given,}$$

that has quadratic convergence, superior to the convergence of the method of successive approximations, which is linear. This study is similar to that of the Fixed Point Theorem for the method of successive approximations. In addition, we obtain a domain of global convergence, $B(\widetilde{x}, R) = \{x \in \mathcal{C}[a,b] : \|x - \widetilde{x}\| < R\}$, with $\widetilde{x} \in \mathcal{C}[a,b]$, for Newton's method. Also, we obtain a result of uniqueness of solution that separate the approximate solution from other possible solutions. To carry out this study, we develop a technique based on the use of auxiliary points, which allows obtaining domains of global convergence, locating solutions of (1) and domains of uniqueness of these solutions.

On the other hand, if $\mathcal{H}(x) = x$, integral Equation (1) is linear and well-known, it is a Fredholm integral equation of the second kind, which is connected with the eigenvalue problem represented by the homogeneous equation

$$x(s) = \lambda \int_a^b \mathcal{K}(s,t) x(t)\, dt, \quad s \in [a,b],$$

and has non-trivial solutions $x(s) \not\equiv 0$ for the characteristic values or eigenvalues λ (the latter term is sometimes reserved to the reciprocals $\nu = 1/\lambda$) of kernel $\mathcal{K}(s,t)$ and every non-trivial solution of (1) is called characteristic function or eigenfunction corresponding to characteristic value λ. If Equation (1) is nonlinear, our results allow doing a study of the equation based on the values of parameter λ, which is another important aim of our work.

2. Global Convergence and Uniqueness of Solution

If we are interested in proving the convergence of an iteration, we can usually follow three ways to do it: local convergence, semilocal convergence and global convergence. First, from some conditions on the operator involved, if we require conditions to the solution x^*, we establish a local analysis of convergence and obtain a ball of convergence of the iteration, which, from the initial approximation x_0 lying in the ball, shows the accessibility to x^*. Second, from some conditions on the operator involved, if we require conditions to the initial iterate x_0, we establish a semilocal analysis of convergence and obtain a domain of parameters, which corresponds to the conditions required to the initial iterate, so that the convergence of iteration is guaranteed to x^*. Third, from some conditions on the operator involved, the convergence of iteration to x^* in a domain, and independently of the initial approximation x_0, is established and global convergence is called. Observe that the three studies require conditions on the operator involved and requirement of conditions to the solution, to the initial approximation, or to none of these, is what determines the way of analysis.

The local analysis of the convergence has the disadvantage that it requires conditions on the solution and this is unknown. The global analysis of convergence, as a consequence of the absence of conditions on the initial approximations and the solution, is very specific for the operators involved.

In this paper, we focus our attention on the analysis of the global convergence of Newton's method and, as a consequence, we obtain domains of global convergence for nonlinear integral Equation (1) and also locate a solution. For this, we obtain a ball of convergence, by using an auxiliary point, that contains a solution and guarantees the convergence of Newton's method from any point of the ball.

Solving Equation (1) is equivalent to solving the equation $\mathcal{F}(x) = 0$, where $\mathcal{F} : \Omega \subseteq \mathcal{C}[a,b] \longrightarrow \mathcal{C}[a,b]$ and

$$[\mathcal{F}(x)](s) = x(s) - \ell(s) - \lambda \int_a^b \mathcal{K}(s,t)\mathcal{H}(x)(t)\, dt, \quad s \in [a,b], \quad \lambda \in \mathbb{R}, \quad n \in \mathbb{N}. \tag{2}$$

Then,

$$[\mathcal{F}'(x)y](s) = y(s) - \lambda \int_a^b \mathcal{K}(s,t)[\mathcal{H}'(x)y](t)\, dt = \lambda \int_a^b \mathcal{K}(s,t)H'(x(t))y(t)\, dt.$$

As a consequence,

$$\|\mathcal{F}'(x) - \mathcal{F}'(y)\| \le K\|x - y\|,$$

where $K = |\lambda|\mathcal{S}L$, L is such that $\|\mathcal{H}'(x) - \mathcal{H}'(y)\| \le L\|x - y\|$, for all $x, y \in \Omega$, and $\mathcal{S} = \left\|\int_a^b \mathcal{K}(s, t)\, dt\right\|$. From the Banach lemma on invertible operators, it follows

$$\|\tilde{\Gamma}\| = \|[\mathcal{F}'(\tilde{x})]^{-1}\| \le \frac{1}{1 - |\lambda|\mathcal{S}\|\mathcal{H}'(\tilde{x})\|} = \beta, \qquad \|\tilde{\Gamma}\mathcal{F}(\tilde{x})\| \le \frac{\|\tilde{x} - u\| + |\lambda|\mathcal{S}\|\mathcal{H}(\tilde{x})\|}{1 - |\lambda|\mathcal{S}\|\mathcal{H}'(\tilde{x})\|} = \eta.$$

provided that

$$|\lambda|\mathcal{S}\|\mathcal{H}'(\tilde{x})\| < 1. \tag{3}$$

Next, we give some properties that are used later.

Lemma 1. *For operator* (2), *we have:*

(a) $\quad \tilde{\Gamma}\mathcal{F}(x) = \tilde{\Gamma}\mathcal{F}(\tilde{x}) + (x - \tilde{x}) + \int_0^1 \tilde{\Gamma}\left(\mathcal{F}'(\tilde{x} + t(x - \tilde{x})) - \mathcal{F}'(\tilde{x})\right)(x - \tilde{x})\, dt$, *with* $x \in \Omega$.

(b) $\quad \mathcal{F}(x_n) = \int_0^1 \left(\mathcal{F}'(x_{n-1} + t(x_n - x_{n-1})) - \mathcal{F}'(x_{n-1})\right)(x_n - x_{n-1})\, dt$, *with* $x_{n-1}, x_n \in \Omega$.

As a consequence of item (b) of Lemma 1, it follows, for $x_{n-1}, x_n \in \Omega$,

$$\|\mathcal{F}(x_n)\| \le \frac{K}{2}\|x_n - x_{n-1}\|^2.$$

From the last result, and taking into account the parameters obtained previously, we analyze the first iteration of Newton's method, what leads us to the convergence of the method.

If $x_0 \in B(\tilde{x}, R)$, then

$$\|\Gamma_0\| = \|[\mathcal{F}'(x_0)]^{-1}\| \le \frac{\beta}{1 - K\beta R} = \alpha, \qquad \|\Gamma_0\mathcal{F}'(\tilde{x})\| \le \frac{1}{1 - K\beta R}.$$

provided that

$$K\beta R < 1. \tag{4}$$

Moreover, from item (a) of Lemma 1, it follows

$$\|x_1 - x_0\| \le \|\Gamma_0\mathcal{F}'(\tilde{x})\|\|\tilde{\Gamma}\mathcal{F}(x_0)\| < \frac{\eta + R + K\beta R^2/2}{1 - K\beta R} = \delta,$$

and, from item (b) of Lemma 1, we have

$$\|x_1 - \tilde{x}\| = \left\|-\Gamma_0\left(\mathcal{F}(x_0) + \mathcal{F}'(x_0)(\tilde{x} - x_0)\right)\right\| \le \|\Gamma_0\mathcal{F}'(\tilde{x})\|\|\tilde{\Gamma}\mathcal{F}(\tilde{x})\| + \frac{K\beta R^2/2}{1 - K\beta R} \le \frac{2\eta + K\beta R^2}{2(1 - K\beta R)},$$

so that $x_1 \in B(\tilde{x}, R)$, provided that

$$\frac{2\eta + K\beta R^2}{2(1 - K\beta R)} \le R. \tag{5}$$

Observe now that condition (5) holds if

$$K\beta\eta \le 1/6 \qquad \text{and} \qquad R \in [R_-, R_+],$$

where $R_- = \frac{1 - \sqrt{1 - 6K\beta\eta}}{3K\beta}$ and $R_+ = \frac{1 + \sqrt{1 - 6K\beta\eta}}{3K\beta}$ are the two real positive roots of quadratic equation

$$2\eta - 2R + 3K\beta R^2 = 0.$$

After that, if we assume that

$$\|x_n - x_{n-1}\| < \gamma^{2^{n-2}} \|x_{n-1} - x_{n-2}\|, \tag{6}$$

$$\|x_n - \tilde{x}\| < \frac{2\eta + K\beta R^2}{2(1 - K\beta R)} \leq R, \tag{7}$$

where $\gamma = K\alpha\delta/2$, for all $n \geq 2$, and provided that condition (5) holds, it follows in the same way that

$$\|x_{n+1} - x_n\| < \gamma^{2^{n-1}} \|x_n - x_{n-1}\|, \qquad \|x_{n+1} - \tilde{x}\| < \frac{2\eta + K\beta R^2}{2(1 - K\beta R)} \leq R,$$

so that (6) and (7) are true for all positive integers n by mathematical induction.

In addition, $\gamma < 1$ if

$$3(K\beta R)^2 - 10(K\beta R) + 2(2 - K\beta\eta) > 0, \tag{8}$$

which is satisfied provided that

$$K\beta\eta \leq 1/6 \quad \text{and} \quad R < \frac{5 - \sqrt{13 + 6K\beta\eta}}{3K\beta}.$$

As a consequence, condition (4) holds. More precisely, we can establish the following result.

Lemma 2. *There always exists $R > 0$, such that inequalities (4), (5) and (8) hold, if*

(a) $K\beta\eta \leq 0.1547\ldots$ *and $R \in \left[R_-, \frac{5-\sqrt{13+6K\beta\eta}}{3K\beta} \right)$,*

(b) $K\beta\eta \in [0.1547\ldots, 1/6)$ *and $R \in [R_-, R_+]$,*

where $R_- = \frac{1-\sqrt{1-6K\beta\eta}}{3K\beta}$ and $R_+ = \frac{1+\sqrt{1-6K\beta\eta}}{3K\beta}$.

Proof. First, we prove item (a) of Lemma 2. Observe that $R_- < \frac{5-\sqrt{13+6K\beta\eta}}{3K\beta}$, since $K\beta\eta \leq 0.1547\ldots$, so that $\left[R_-, \frac{5-\sqrt{13+6K\beta\eta}}{3K\beta} \right) \neq \emptyset$. Moreover, as $K\beta\eta \leq 0.1547\ldots$, we have $3(K\beta\eta)^2 + 6(K\beta\eta) - 1 \leq 0$ and, as a consequence, $R_+ > \frac{5-\sqrt{13+6K\beta\eta}}{3K\beta}$ and $R \in \left[R_-, \frac{5-\sqrt{13+6K\beta\eta}}{3K\beta} \right) \subset [R_-, R_+]$, so that (5) and (8) hold.

Second, if $K\beta\eta \in [0.1547\ldots, 1/6)$, then $3(K\beta\eta)^2 + 6(K\beta\eta) - 1 \geq 0$ and $R_+ < \frac{5-\sqrt{13+6K\beta\eta}}{3K\beta}$, so that $R \in [R_-, R_+]$. Then, (5) and (8) hold.

Third, in both cases, $K\beta R < 1$ follows immediately, since $R < \frac{5-\sqrt{13+6K\beta\eta}}{3K\beta}$ in items (a) and (b) of Lemma 2. $\square$

2.1. Convergence

Now, we can establish the following result.

Theorem 1. *Suppose that $K\beta\eta \leq 1/6$ and consider $R > 0$ satisfying item (a) or item (b) of Lemma 2 and such that $B(\tilde{x}, R) \subset \Omega$. If condition (3) holds, then Newtons's method is well-defined and converges to a solution x^* of $\mathcal{F}(x) = 0$ in $\overline{B(\tilde{x}, R)}$ from every point $x_0 \in B(\tilde{x}, R)$.*

Proof. From (6) and $\gamma < 1$, we have $\|x_{n+1} - x_n\| < \|x_n - x_{n-1}\|$, for all $n \in \mathbb{N}$, so that sequence $\{\|x_{n+1} - x_n\|\}$ is strictly decreasing for all $n \in \mathbb{N}$ and, therefore, sequence $\{x_n\}$ is convergent. If $x^* = \lim_{n \to \infty} x_n$, then $\mathcal{F}(x^*) = 0$, by the continuity of $\mathcal{F}$ and $\|\mathcal{F}(x_n)\| \to 0$ when $n \to \infty$. $\square$

From Theorem 1, the convergence of Newton's method to a solution of equation $\mathcal{F}(x) = 0$ is guaranteed. Moreover, the best ball of location of the solution is $B(\tilde{x}, R_-)$ and the biggest ball of

convergence is $B(\tilde{x}, R_+)$ or $B\left(\tilde{x}, \frac{5-\sqrt{13+6K\beta\eta}}{3K\beta}\right)$, depending on the value of $K\beta\eta$ is: $K\beta\eta \leq 0.1547\ldots$
for the former and $K\beta\eta \in [0.1547\ldots, 1/6)$ for the latter.

2.2. Uniqueness of Solution

For uniqueness of solution, we establish the following result, where uniqueness of solution is proved in $\overline{B(\tilde{x}, R)}$.

Theorem 2. *Under conditions of Theorem 1, solution x^* of $\mathcal{F}(x) = 0$ is unique in $\overline{B(\tilde{x}, R)}$.*

Proof. Assume that w^* is another solution of $\mathcal{F}(x) = 0$ in $\overline{B(\tilde{x}, R)}$ such that $w^* \neq x^*$. If operator $Q = \int_0^1 \mathcal{F}'(w^* + t(x^* - w^*))\, dt$ is invertible, we have $x^* = w^*$, since $Q(w^* - x^*) = \mathcal{F}(w^*) - \mathcal{F}(x^*)$. Then, as

$$\|I - \tilde{\Gamma}Q\| \leq \|\tilde{\Gamma}\| \int_0^1 \|\mathcal{F}'(\tilde{x}) - \mathcal{F}'(w^* + t(x^* - w^*))\|\, dt$$

$$\leq \beta K \int_0^1 \|\tilde{x} - (w^* + t(x^* - w^*))\|\, dt \tag{9}$$

$$= \beta K R$$

$$< 1,$$

it follows that Q is invertible by the Banach lemma on invertible operators and uniqueness follows immediately. □

Notice that, from Theorems 1 and 3, the best ball of location of a solution of (1) is $B(\tilde{x}, R_-)$ and the best ball of uniqueness of solution and the biggest ball of convergence is $B\left(\tilde{x}, \frac{5-\sqrt{13+6K\beta\eta}}{3K\beta}\right)$ or $B(\tilde{x}, R_+)$, depending on the value of $K\beta\eta$ lies.

Once given the uniqueness of solution in the domain of existence of solution $B(\tilde{x}, R)$, we enlarge such domain from the following theorem.

Theorem 3. *Under conditions of Theorem 1, we have that the solution x^* is unique in the domain $B(\tilde{x}, \varrho) \cap \Omega$, where $\varrho = \frac{2}{K\beta} - R$.*

Proof. Assume that w^* is another solution of $\mathcal{F}(x) = 0$ in $B(\tilde{x}, \varrho) \cap \Omega$ such that $w^* \neq x^*$. Then, from (9), it follows

$$\|I - \tilde{\Gamma}Q\| < \beta K \int_0^1 ((1-t)\varrho + tR)\, dt = 1.$$

and Q is again invertible by the Banach lemma on invertible operators. □

Note that $\varrho > 0$, since $\beta K R < 1$, and uniqueness of solution is obtained in the ball of global convergence given in Theorem 1, since $\varrho = \frac{2}{K\beta} - R \geq \frac{5-\sqrt{13+6K\beta\eta}}{3K\beta}, R_+$.

3. Example

Now, we apply the last result to the following nonlinear integral equation:

$$x(s) = s^3 + \frac{18}{25} \int_0^1 s^3 t^3 x(t)^2\, dt, \quad s \in [0,1]. \tag{10}$$

For Equation (10), we have $\lambda = 18/25$ and $\mathcal{S} = \left\|\int_0^1 s^3 t^3\, dt\right\| = 1/4$ with the max-norm. As $\mathcal{H}(\tilde{x})(t) = \tilde{x}(t)^2$, then $L = 2$. If we choose $\tilde{x}(s) = s^3$, then condition (3) holds, since $|\lambda|\mathcal{S}\|H'(\tilde{x})\| = 9/25 < 1$. Moreover, $\beta = 25/16$ and $\eta = 9/32$ and $K = |\lambda|\mathcal{S}L = 9/25$, so that $K\beta\eta = 0.1582\ldots \in$

$[R_-, R_+]$, where $R_- = 0.4590\ldots$ and $R_+ = 0.7261\ldots$. Therefore, from Theorem 1, the convergence of Newton's method to a solution of Equation (10) is guaranteed and the best ball of location of the solution is $B(s^3, 0.4590\ldots)$ and the biggest ball of convergence is $B(s^3, 0.7261\ldots)$. Furthermore, from Theorem 3, it follows that the domain of uniqueness of solution is $B(s^3, 3.0965\ldots)$.

Next, we approximate a solution of Equation (10) by Newton's method. After four iterations with stopping criterion $\|x_n - x_{n-1}\|_\infty < 10^{-18}$, $n \in \mathbb{N}$, we obtain solution shown in Table 1, where errors $\|x^* - x_n\|$ and sequence $\|\mathcal{F}(x_n)\|$ are also shown. Observe from the last sequence that solution shown in Table 1 is a good approximation of the solution of Equation (10). Finally, we observe in Figure 1 that solution shown in Table 1 lies within the domain of location of solution found above.

Table 1. Approximated solution $x^*(s)$ of (10), absolute errors and $\{\|\mathcal{F}(x_n)\|\}$.

n	$x_n(s)$	$\|x^* - x_n\|$	$\|\mathcal{F}(x_n)\|$
0	s^3	$8.4715\ldots \times 10^{-2}$	7.2×10^{-2}
1	$(1.0841121495327102\ldots)s^3$	$6.0365\ldots \times 10^{-4}$	$5.0938\ldots \times 10^{-4}$
2	$(1.0847157717628998\ldots)s^3$	$3.1090\ldots \times 10^{-8}$	$2.6233\ldots \times 10^{-8}$
3	$(1.0847158028530592\ldots)s^3$	$8.2478\ldots \times 10^{-17}$	$6.9595\ldots \times 10^{-17}$
4	$(1.0847158028530593\ldots)s^3$		

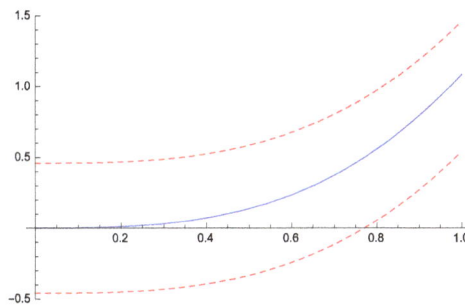

Figure 1. Approximated solution $x^*(s)$ of (10) and domain of location of solution.

4. Study of the Integral Equation from Parameter λ

Next, we study the integral Equation (1) from the values of parameter λ.

First, we observe that $K\beta\eta \leq 1/6$ if

$$6|\lambda|\mathcal{S}L \left(\|\tilde{x} - \ell\| + |\lambda|\mathcal{S}\|\mathcal{H}(\tilde{x})\|\right) \leq \left(1 - |\lambda|\mathcal{S}\|\mathcal{H}'(\tilde{x})\|\right)^2 \tag{11}$$

and condition (3) holds.

Now, we analyze condition (11). Observe that (11) is satisfied if

- $\|\mathcal{H}'(\tilde{x})\|^2 < 6L\|\mathcal{H}(\tilde{x})\|$ and $|\lambda| \in [0, \mu_+]$, where

$$\mu_+ = \frac{-(3L\|\tilde{x} - \ell\| + \|\mathcal{H}'(\tilde{x})\|) + \sqrt{\Delta}}{\mathcal{S}(6L\|\mathcal{H}(\tilde{x})\| - \|\mathcal{H}'(\tilde{x})\|^2)}$$

and $\Delta = 3L(3L\|\tilde{x} - \ell\|^2 + 2\|\tilde{x} - \ell\|\|\mathcal{H}'(\tilde{x})\| + 2\|\mathcal{H}(\tilde{x})\|)$.
- $\|\mathcal{H}'(\tilde{x})\|^2 > 6L\|\mathcal{H}(\tilde{x})\|$ and $|\lambda| \in [0, \mu_+] \cup [\mu_-, +\infty)$, where

$$\mu_- = \frac{-(3L\|\tilde{x} - \ell\| + \|\mathcal{H}'(\tilde{x})\|) - \sqrt{\Delta}}{\mathcal{S}(6L\|\mathcal{H}(\tilde{x})\| - \|\mathcal{H}'(\tilde{x})\|^2)}.$$

- $\|\mathcal{H}'(\tilde{x})\|^2 = 6L\|\mathcal{H}(\tilde{x})\|$ and $|\lambda| \leq \dfrac{1}{2\mathcal{S}(3L\|\tilde{x} - \ell\| + \|\mathcal{H}'(\tilde{x})\|)}$.

Second, once $\tilde{x}$ is fixed, we have two chances: $K\beta\eta \leq 0.1547\ldots$ or $K\beta\eta \in [0.1547\ldots, 1/6)$. If first holds, then $R \in \left[R_-, \frac{5-\sqrt{13+6K\beta\eta}}{3K\beta}\right)$ and, if second does, then $R \in [R_-, R_+]$.

Finally, as condition (3) is satisfied, then Newtons's method is well-defined and converges to a solution x^* of $\mathcal{F}(x) = 0$ in $\overline{B(\tilde{x}, R)}$ from every point $x_0 \in B(\tilde{x}, R)$ by Theorem 1.

5. Application

Now, we apply the last study to the following particular Davis-type integral Equation [8]:

$$x(s) = s + \lambda \int_0^1 G(s,t)x(t)^2 dt, \quad \lambda \in \mathbb{R}, \quad s \in [0,1], \tag{12}$$

where the kernel of (12) is a Green's function defined as follows:

$$G(s,t) = \begin{cases} (1-s)t, & t \leq s, \\ s(1-t), & s \leq t. \end{cases}$$

One can show that the function $x(s)$ that satisfied Equation (12) is any solution of the differential equation

$$x''(s) + \lambda x(s)^2 = 0,$$

that also satisfies the two-point boundary condition: $x(0) = 0$, $x(1) = 1$.

For Equation (12), we have $S = \left\|\int_0^1 G(s,t)\, dt\right\| = 1/8$ with the max-norm and $\mathcal{H}(\tilde{x})(t) = \tilde{x}(t)^2$. Therefore, $L = 2$ and condition (3) is reduced to $|\lambda| < 4/\|\tilde{x}\|$. In addition,

$$\|\mathcal{H}'(\tilde{x})\|^2 = 4\|\tilde{x}\|^2 < 12\|\tilde{x}\|^2 = 6L\|\mathcal{H}(\tilde{x})\|$$

and, as a consequence,

$$|\lambda| \leq \mu_+ = \frac{-(6\|\tilde{x}-s\| + 2\|\tilde{x}\|) + \sqrt{12\left(3\|\tilde{x}-s\|^2 + 2\|\tilde{x}-s\|\|\tilde{x}\| + \|\tilde{x}\|^2\right)}}{\|\tilde{x}\|^2}.$$

After that, we choose $\tilde{x}(s) = s$ and hence $\mu_+ = 2(-1+\sqrt{3}) = 1.4641\ldots$, so that $|\lambda| \leq 1.4641\ldots$, that satisfies condition (3). In this case, from Theorem 1, we can guarantee the convergence of Newton's method to a solution of Equation (12) with λ such that $|\lambda| \leq 1.4641\ldots$ Moreover, once λ is fixed, depending on the value of $K\beta\eta$, we can obtain the best ball of location of solution and the biggest ball of convergence.

Observe that we cannot apply Newton's method directly, since we do not know the inverse operator that is involved in the algorithm of Newton's method. Then, we use a process of discretization to transform (12) into a finite dimensional problem. For this, we use a Gauss–Legendre quadrature formula to approximate the integral of (12),

$$\int_0^1 \phi(t)\, dt \simeq \sum_{j=1}^m w_j \phi(t_j),$$

where the m nodes t_j and weights w_j are known.

Next, we denote the approximations $x(t_i)$ by x_i, with $i = 1, 2, \ldots, m$, so that (12) is equivalent to the nonlinear system given by

$$x_j = t_j + \lambda \sum_{k=1}^m a_{jk} x_k^2, \quad k = 1, 2, \ldots, m, \tag{13}$$

where

$$a_{jk} = \begin{cases} w_k \left(1 - t_j\right) t_k, & k \le j, \\ w_k \left(1 - t_k\right) t_j, & k > j. \end{cases}$$

After that, we write system (13) compactly in matrix form as

$$F(\mathbf{x}) \equiv \mathbf{x} - \mathbf{v} - \lambda A \mathbf{y} = 0, \qquad F : \mathbb{R}^m \longrightarrow \mathbb{R}^m, \tag{14}$$

where

$$\mathbf{x} = (x_1, x_2, \dots, x_m)^t, \quad \mathbf{v} = (t_1, t_2, \dots, t_m)^t, \quad A = \left(a_{jk}\right)_{j,k=1}^m, \quad \mathbf{y} = \left(x_1^2, x_2^2, \dots, x_m^2\right)^t.$$

Choose $m = 8$, $\lambda = 7/5$, $\tilde{\mathbf{x}} = \mathbf{v}$ and hence $K = 0.2471\ldots$, $\beta = 1.2179\ldots$, $\eta = 0.0665\ldots$ and $K\beta\eta = 0.0200\ldots$ As $K\beta\eta < 0.1547\ldots$, it follows, from Theorem 1, that the best ball of location of solution is $\overline{B}(\mathbf{v}, 0.0686\ldots)$ and the biggest ball of convergence is $B(\mathbf{v}, 1.5259\ldots)$.

If the starting point for Newton's method is $x_0 = \mathbf{v}$, the method converges to the solution $x^* = (x_1^*, x_2^*, \dots, x_8^*)^t$ of system (14), which is shown in Table 2, after four iterations with stopping criterion $\|x_n - x_{n-1}\|_\infty < 10^{-18}$, $n \in \mathbb{N}$.

Table 2. Numerical solution x^* of system (14) with $\lambda = 7/5$.

i	x_i^*	i	x_i^*
1	0.02267000...	5	0.65888692...
2	0.11607746...	6	0.82291926...
3	0.27057507...	7	0.93276524...
4	0.46275932...	8	0.98797444...

Moreover, errors $\|x^* - x_n\|$ and sequence $\{\|F(x_n)\|\}$ are shown in Table 3. Observe then that vector shown in Table 2 is a good approximation of a solution of (14).

Table 3. Absolute errors and $\{\|F(x_n)\|\}$.

n	$\|x^* - x_n\|$	$\|F(x_n)\|$
0	$6.7169\ldots \times 10^{-2}$	$9.6624\ldots \times 10^{-1}$
1	$6.4734\ldots \times 10^{-4}$	$5.3956\ldots \times 10^{-4}$
2	$6.0570\ldots \times 10^{-8}$	$5.0516\ldots \times 10^{-8}$
3	$5.4063\ldots \times 10^{-16}$	$4.5077\ldots \times 10^{-16}$

Furthermore, as a solution of (12) satisfies $x(0) = 0$ and $x(1) = 1$, if values of Table 2 are interpolated, an approximated solution is obtained, which is painted in Figure 2. Notice that this approximated solution lies in the domain of location of solution $\overline{B}(\mathbf{v}, 0.0686\ldots)$ which is obtained from Theorem 1.

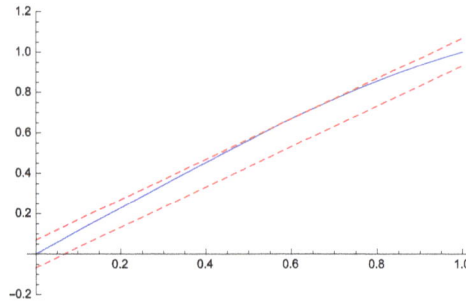

Figure 2. Solution $\mathbf{x}^*$ of system (14) and domain of location of solution.

6. Conclusions

Following the idea of the Fixed Point Theorem for the method of successive approximations, we do an analysis for Newton's method, use the theoretical significance of the method to prove the existence and uniqueness of solution of a particular type of nonlinear integral equations of Fredholm and, in addition, obtain a domain of global convergence for the method that allows locating a solution and separating it from other possible solutions. For this, we use a technique based on using auxiliary points. Moreover, we present a study of the nonlinear equations which is based on the real parameter involved in the equation.

Author Contributions: The contributions of the two authors have been similar. Both authors have worked together to develop the present manuscript.

Funding: This research was partially supported by Ministerio de Ciencia, Innovación y Universidades under grant PGC2018-095896-B-C21.

Conflicts of Interest: The authors declare no conflict of interest.

References

1. Awawdeh, F.; Adawi, A.; Al-Shara, S. A numerical method for solving nonlinear integral equations. *Int. Math. Forum* **2009**, *4*, 805–817.
2. Nadir, M.; Khirani, A. Adapted Newton-Kantorovich method for nonlinear integral equations. *J. Math. Stat.* **2016**, *12*, 176–181. [CrossRef]
3. Saberi-Nadja, J.; Heidari, M. Solving nonlinear integral equations in the Urysohn form by Newton-Kantorovich-quadrature method. *Comput. Math. Appl.* **2010**, *60*, 2018–2065.
4. Matkowski, J. Functional Equations and Nemytskii Operators. *Funkc. Ekvacioj* **1982**, *25*, 127–132.
5. Berinde, V. *Iterative Approximation of Fixed Point*; Springer: New York, NY, USA, 2005.
6. Ragusa, M.A. Local Hölder regularity for solutions of elliptic systems. *Duke Math. J.* **2002**, *13*, 385–397. [CrossRef]
7. Wang, H.; Zhang, L.; Wang, X. Fixed point theorems for a class of nonlinear sum-type operators and application in a fractional differential equation. *Bound. Value Probl.* **2018**, *2018*, 140. [CrossRef]
8. Davis, H.T. *Introduction to Nonlinear Differential and Integral Equations*; Dover: New York, NY, USA, 1962.

![mathematics logo] *mathematics*

MDPI

Article

Numerical Solution of Heston-Hull-White Three-Dimensional PDE with a High Order FD Scheme

Malik Zaka Ullah

Department of Mathematics, King Abdulaziz University, Jeddah 21589, Saudi Arabia; zmalek@kau.edu.sa

Received: 27 June 2019; Accepted: 1 August 2019; Published: 6 August 2019

check for
updates

Abstract: A new numerical method for tackling the three-dimensional Heston–Hull–White partial differential equation (PDE) is proposed. This PDE has an application in pricing options when not only the asset price and the volatility but also the risk-free rate of interest are coming from stochastic nature. To solve this time-dependent three-dimensional PDE as efficiently as possible, high order adaptive finite difference (FD) methods are applied for the application of method of lines. It is derived that the new estimates have fourth order of convergence on non-uniform grids. In addition, it is proved that the overall procedure is conditionally time-stable. The results are upheld via several numerical tests.

Keywords: heston model; Hull–White; option pricing; PDE; finite difference (FD)

MSC: 41A25; 65M22

1. Introduction

To model different types of derivatives in finance, a common approach is to investigate the connections of these factors to each other, formulated as a stochastic differential equation (SDEs). The factors could be the underlying asset, the volatility, domestic and foreign interest rates, etc., [1,2]. As such, the important action of pricing option under different payoffs can be modeled and simulated via the SDEs or their corresponding partial differential equation (PDE) formulation.

However, a frequently occurring issue is that whatever the model becomes complicated and more realistic, the procedure of having and representing its exact solution becomes harder, see, e.g., [3–5].

To discuss more and from the beginning, the classical model of Black–Scholes in pricing contracts does not cover and illustrate all the aspects of an option in a complete market, such as market risks, stochastic volatility (SV), and asymmetries seen in data of market, [6]. Some remedies to this well-known model are via non-lognormal hypothesis for a SDE, that indicates some modifications of the volatility and the underlying asset. We recall that Heston in [7] extended and improved the behavior of the Black–Scholes model by involving more risky factor into the model, i.e., by considering the volatility to be stochastic as well. Further discussions can be found at [6,8].

On the other hand, as long as the foreign exchange (FX) products are involved and a trader encounters a situation in which the interest rate is not anymore constant during the lifetime of an option, then investigating and proposing an improved model, having stochastic rates of interest, such as the power-reverse dual-currency and the Heston–Cox–Ingersoll–Ross (HCIR) problems (refer to [9] and the references therein for more background).

1.1. Problem Formulation

The option pricing problem under the 3D Heston–Hull–White (HHW) model as a PDE model is defined by [10]:

$$
\begin{aligned}
\frac{\partial u(s,v,r,t)}{\partial t} =& \frac{1}{2}s^2 v \frac{\partial^2 u(s,v,r,t)}{\partial s^2} + \frac{1}{2}\sigma_1^2 v \frac{\partial^2 u(s,v,r,t)}{\partial v^2} + \frac{1}{2}\sigma_2^2 \frac{\partial^2 u(s,v,r,t)}{\partial r^2} \\
&+ \rho_{12}\sigma_1 s v \frac{\partial^2 u(s,v,r,t)}{\partial s \partial v} + \rho_{13}\sigma_2 s \sqrt{v} \frac{\partial^2 u(s,v,r,t)}{\partial s \partial r} \\
&+ \rho_{23}\sigma_1\sigma_2 \sqrt{v} \frac{\partial^2 u(s,v,r,t)}{\partial v \partial r} \\
&+ rs \frac{\partial u(s,v,r,t)}{\partial s} + \kappa(\eta - v) \frac{\partial u(s,v,r,t)}{\partial v} \\
&+ a(b(T-t)-r) \frac{\partial u(s,v,r,t)}{\partial r} \\
&- ru(s,v,r,t).
\end{aligned}
\tag{1}
$$

Here $\kappa > 0$ shows the volatility adjustment speed to the analytical mean $\eta > 0$, while σ_1, σ_2, a are some parameters. In addition, the correlation parameters are $\rho_{12}, \rho_{13}, \rho_{23} \in [-1,1]$, b is a time function. In pricing under call options, the (terminal/)initial condition is given by [11,12]:

$$
u(s,v,r,0) = (s - E)^+,
\tag{2}
$$

where the strike price is E. In a similar way, for a put option, it is given as follows:

$$
u(s,v,r,0) = (E - s)^+.
\tag{3}
$$

As discussed in [13,14], the fair pricing procedure should be carried out by computational schemes since the corresponding high-dimensional PDEs, constructed for such options, do not admit any analytical or semi-analytical solutions, see [15,16] for further background.

1.2. Novelties and Motivation

The contribution of this article reads in proposing a solution method via an un-equally spaced grid having a focus on the hot area in option pricing under the HHW PDE problem. Studying and coding multi dimensional problems with discretization methods while the grid of points are non-uniform is a challenging and intensive task, but could clearly increase the accuracy of the approximate solution by applying fewer numbers of grid nodes in contrasts to the uniform discretization. This reduces the size of the discretized problem and is useful in practice.

To this aim, (1) is tackled by employing high order fourth-order finite difference approximations. We apply fourth order discretizations on a stencil having five and six non-equidistant nodes. Derivation and construction of fourth-order compact FD method for HHW PDE is new and useful in practice.

In fact, the method-of-lines technique is considered to build a set of ODEs with time-varying system matrix. All the side conditions are imposed therein as well. Thence, a method to march along time for the set of ODEs is provided in Section 3 and it is analytically illustrated that the presented numerical procedure is conditionally time-stable when b is not changing by time.

Recalling that here adaptive FD formulas are constructed to hit some features simultaneously, viz., to be effective, results in sparse operators and being able to handling non-uniform grids.

Motivated by recent works in this field (see e.g., [17]), we aim at proposing higher order schemes for the HHW equation on non-uniform meshes so as to increase the accuracy of obtained option prices without increasing the computational load so much. The novelties and contributions of our work are given below:

- We propose fourth-order adaptive discretizations for the spatial variables.
- The beauty of our scheme is the use of non-uniform grid of nodes with an adaptation on the hotzone.
- We provide a new stability bound for the resulting fully discretized set of equations when pricing under HHW PDE using high order discretization methods along the spatial as well as the temporal variables.

1.3. Grid Generation

The option pricing problem (1) is considered in the unbounded area

$$(s, v, r, t) \in \Omega \times (0, T], \tag{4}$$

wherein $\Omega = [0, +\infty) \times [0, +\infty) \times [0, +\infty)$. For tackling the financial model numerically, one can take into account the following domain [18]:

$$\Omega = [0, s_{\max}] \times [0, v_{\max}] \times [-r_{\max}, r_{\max}], \tag{5}$$

wherein $s_{\max}, v_{\max}, r_{\max}$ are three positive real constants and assumed to be large enough.

Since the PDE model is coercive (sometimes called degenerate) at $v = 0$, its payoff is non-smooth at $s = E$, and the working domain has large width, thus it is requisite to use non-uniform meshes, at which the location of the nodes are not equally-spaced. This helps in producing results of higher accuracy with adapting to the hotzone of the problem.

Let $\{s_i\}_{i=1}^m$ be a set of non-uniform nodes along s as follows [13,19]:

$$s_i = \varphi(\xi_i), \qquad 1 \le i \le m, \tag{6}$$

where $m > 1$ and $\xi_{\min} = \xi_1 < \xi_2 < \cdots < \xi_m = \xi_{\max}$ are m equi–distant points with the following characteristics: $\xi_{\min} = \sinh^{-1}\left(\frac{s_{\min} - s_{\text{left}}}{d_1}\right)$, $\xi_{\text{int}} = \frac{s_{\text{right}} - s_{\text{left}}}{d_1}$, $\xi_{\max} = \xi_{\text{int}} + \sinh^{-1}\left(\frac{s_{\max} - s_{\text{right}}}{d_1}\right)$, wherein $s_{\min} = 0$. Here $d_1 > 0$ controls the density of the nodes around $s = E$. We also have:

$$\varphi(\xi) = \begin{cases} s_{\text{left}} + d_1 \sinh(\xi), & \xi_{\min} \le \xi < 0, \\ s_{\text{left}} + d_1 \xi, & 0 \le \xi \le \xi_{\text{int}}, \\ s_{\text{right}} + d_1 \sinh(\xi - \xi_{\text{int}}), & \xi_{\text{int}} < \xi \le \xi_{\max}. \end{cases} \tag{7}$$

Throughout this work, we used the same value for $d_1 = \frac{E}{20}$ while $s_{\text{left}} = \max\{0.5, \exp\{-0.25T\}\} \times E$, $[s_{\text{left}}, s_{\text{right}}] \subset [0, s_{\max}]$, $s_{\text{right}} = E$ and $s_{\max} = 14E$.

The nodes along v, i.e., $\{v_j\}_{j=1}^n$ are defined by:

$$v_j = d_2 \sinh(\varsigma_j), \qquad 1 \le j \le n, \tag{8}$$

where $d_2 > 0$ gives the concentration around $v = 0$. In this work, we used $d_2 = \frac{v_{\max}}{500}$, where $v_{\max} = 10$. In addition, ς_j are equally spaced points given by:

$$\varsigma_j = (j-1)\Delta\varsigma, \Delta\varsigma = \frac{1}{n-1} \sinh^{-1}\left(\frac{v_{\max}}{d_2}\right), \tag{9}$$

for any $1 \le j \le n$. The non-uniform nodes along r are defined as follows:

$$r_k = d_3 \sinh(\zeta_k), \qquad 1 \le k \le o, \tag{10}$$

whereas $d_3 = \frac{r_{\max}}{500}$ is a positive parameter and $r_{\max} = 1$. We also have $\zeta_k = (k-1)\Delta\zeta$, $\Delta\zeta = \frac{1}{o-1} \sinh^{-1}\left(\frac{r_{\max}}{d_3}\right)$. Note that denser mesh points in the important area could circumvent the problems happening in solving (1), like non-smoothness of payoffs (2) and (3) at $s = E$, and the degeneracy at $v = r = 0$.

We state that a detailed study into possibly better choices for the involved parameters in mesh generating may be interesting, but this is beyond the scope of the current research. Furthermore, the non-smoothness arising in the payoff would ruin the convergence rate of most derivative approximation particularly on uniform meshes and due to this, the application of non-uniform nodes is indispensable for efficient numerical solution of (1).

1.4. Manuscript Organization

The remaining parts of this work are organized as follows. In Section 2, the weights of the FD scheme over non-uniform grids (here we also call adaptive grids with special emphasis on the hot zone) are derived to attain the higher rate of convergence four.

Section 3 is devoted to the application of a sixth order Runge–Kutta time stepping method to advance along time when semi-discretize the HHW PDE. We prove that the new procedure is time-stable conditionally based on the largest eigenvalue of the system matrix. Section 4 shows that numerical performances are more useful than the earlier schemes with quicker convergence behavior. Finally, some conclusions are drawn in Section 5.

2. Calculating the Weights of the High Order FD Scheme

In this section, by applying a methodology as in ([20], Chapters 3–4) or [21], but with more Taylor expansion terms, we can construct fourth-order FD approximations on (non-uniform) grids.

Five points are required in estimating the first derivative as well as six points in approximating the second derivative in order to obtain a consistent fourth-order scheme throughout the discretized mesh of points.

Without losing the generality, let us construct the weights in the one dimensional case. Then, the concept of tensors using Kronecker product may be applied easily to transfer the weights to the appropriate dimensions. To this objective, consider a sufficiently smooth function $g(s)$ and a grid as follows:

$$\{s_1, s_2, \cdots, s_{m-1}, s_m\}. \tag{11}$$

Consider the following five adjacent nodes:

$$\{\{s_{i-2}, g(s_{i-2})\}, \{s_{i-1}, g(s_{i-1})\}, \{s_i, g(s_i)\}, \{s_{i+1}, g(s_{i+1})\}, \{s_{i+2}, g(s_{i+2})\}\}, \tag{12}$$

and calculate the interpolation polynomial $p(z)$ going via the nodes and then its first derivative $p'(z)$.

At this moment, by employing a computer algebra system to do some symbolic computations and setting $z = s_i$, we attain the FD estimate for the first derivative as follows:

$$g'(s_i) = \alpha_{i-2}g(s_{i-2}) + \alpha_{i-1}g(s_{i-1}) + \alpha_i g(s_i) + \alpha_{i+1}g(s_{i+1}) + \alpha_{i+2}g(s_{i+2}) + \mathcal{O}\left(h^4\right), \tag{13}$$

where the maximum local grid spacing is h and we have

$$\alpha_{i-2} = -\frac{\Gamma_{i-1,i}\Gamma_{i,i+1}\Gamma_{i,i+2}}{\Gamma_{i-2,i-1}\Gamma_{i-2,i}\Gamma_{i-2,i+1}\Gamma_{i-2,i+2}},$$

$$\alpha_{i-1} = \frac{\Gamma_{i-2,i}\Gamma_{i,i+1}\Gamma_{i,i+2}}{\Gamma_{i-2,i-1}\Gamma_{i-1,i}\Gamma_{i-1,i+1}\Gamma_{i-1,i+2}},$$

$$\alpha_i = \frac{1}{\Gamma_{i-2,i}\Gamma_{i,i-1}\Gamma_{i,i+1}\Gamma_{i,i+2}} \Xi_1, \tag{14}$$

$$\alpha_{i+1} = \frac{\Gamma_{i-2,i}\Gamma_{i,i-1}\Gamma_{i,i+2}}{\Gamma_{i-2,i+1}\Gamma_{i+1,i-1}\Gamma_{i+1,i}\Gamma_{i+1,i+2}},$$

$$\alpha_{i+2} = \frac{\Gamma_{i-2,i}\Gamma_{i,i-1}\Gamma_{i,i+1}}{\Gamma_{i-2,i+2}\Gamma_{i+2,i-1}\Gamma_{i+2,i}\Gamma_{i+2,i+1}},$$

using $\Gamma_{l,q} = s_l - s_q$ and

$$\begin{aligned}
\Xi_1 =& s_{i-2}(s_{i-1}(\Gamma_{i+1,i} + \Gamma_{i+2,i}) \\
& + 3s_i^2 - 2(s_{i+1} + s_{i+2})s_i + s_{i+1}s_{i+2}) + s_i(-4s_i^2 + 3(s_{i+1} + s_{i+2})s_i \\
& - 2s_{i+1}s_{i+2}) + s_{i-1}(3s_i^2 - 2(s_{i+1} + s_{i+2})s_i + s_{i+1}s_{i+2}).
\end{aligned} \tag{15}$$

Recalling that the above procedure should be similarly done for the nodes $\{s_1, s_2, s_{m-1}, s_m\}$, viz, to find the weighting coefficients with fourth order of convergence for such nodes, we should consider the five adjacent points and then calculate the interpolating polynomial at that specific point. In this way, the sided FD formulas are constructed and used.

Similarly, FD estimates for the second derivative terms can be obtained applying a similar methodology as above. To this objective, we consider a set of points as follows:

$$\begin{aligned}
& \{\{s_{i-3}, g(s_{i-3})\}, \{s_{i-2}, g(s_{i-2})\}, \{s_{i-1}, g(s_{i-1})\}, \\
& \{s_i, g(s_i)\}, \{s_{i+1}, g(s_{i+1})\}, \{s_{i+2}, g(s_{i+2})\}\},
\end{aligned} \tag{16}$$

and compute the second-derivative interpolating polynomial $p''(z)$ based on z. Now by taking into account $z = s_i$ in Mathematica [22], one obtains that:

$$\begin{aligned}
g''(s_i) =& \beta_{i-3} g(s_{i-3}) + \beta_{i-2} g(s_{i-2}) + \beta_{i-1} g(s_{i-1}) \\
& + \beta_i g(s_i) + \beta_{i+1} g(s_{i+1}) + \beta_{i+2} g(s_{i+2}) + \mathcal{O}\left(h^4\right),
\end{aligned} \tag{17}$$

where

$$\beta_{i-3} = \frac{\Xi_2}{\Gamma_{i-3,i-2}\Gamma_{i-3,i-1}\Gamma_{i-3,i}\Gamma_{i-3,i+1}\Gamma_{i-3,i+2}},$$

$$\beta_{i-2} = \frac{\Xi_3}{\Gamma_{i-3,i-2}\Gamma_{i-2,i-1}\Gamma_{i-2,i}\Gamma_{i-2,i+1}\Gamma_{i-2,i+2}}, \tag{18}$$

$$\beta_{i-1} = \frac{\Xi_4}{\Gamma_{i-2,i-1}\Gamma_{i-1,i-3}\Gamma_{i-1,i}\Gamma_{i-1,i+1}\Gamma_{i-1,i+2}},$$

$$\beta_i = \frac{\Xi_5}{\Gamma_{i-3,i}\Gamma_{i,i-2}\Gamma_{i,i-1}\Gamma_{i,i+1}\Gamma_{i,i+2}},$$

$$\beta_{i+1} = \frac{\Xi_6}{\Gamma_{i-3,i+1}\Gamma_{i+1,i-2}\Gamma_{i+1,i-1}\Gamma_{i+1,i}\Gamma_{i+1,i+2}},$$

$$\beta_{i+2} = \frac{\Xi_7}{\Gamma_{i-3,i+2}\Gamma_{i+2,i-2}\Gamma_{i+2,i-1}\Gamma_{i+2,i}\Gamma_{i+2,i+1}}.$$

Here, we have

$$
\begin{aligned}
\Xi_2 =& s_{i-1}(-6s_i^2 + 4(s_{i+1}+s_{i+2})s_i - 2s_{i+1}s_{i+2}) + 2s_i(4s_i^2 \\
& - 3(s_{i+1}+s_{i+2})s_i + 2s_{i+1}s_{i+2}) + s_{i-2}(-6s_i^2 \\
& + 4(s_{i+1}+s_{i+2})s_i - 2s_{i+1}s_{i+2} + s_{i-1}(4s_i - 2(s_{i+1}+s_{i+2}))), \\
\Xi_3 =& 2(s_{i-3}(s_{i-1}(\Gamma_{i+1,i}+\Gamma_{i+2,i}) + 3s_i^2 - 2(s_{i+1} \\
& + s_{i+2})s_i + s_{i+1}s_{i+2}) + s_i(-4s_i^2 + 3(s_{i+1}+s_{i+2})s_i - 2s_{i+1}s_{i+2}) \\
& + s_{i-1}(3s_i^2 - 2(s_{i+1}+s_{i+2})s_i + s_{i+1}s_{i+2})), \\
\Xi_4 =& 2(s_{i-3}(s_{i-2}(\Gamma_{i+1,i}+\Gamma_{i+2,i}) + 3s_i^2 - 2(s_{i+1} \\
& + s_{i+2})s_i + s_{i+1}s_{i+2}) + s_i(-4s_i^2 + 3(s_{i+1}+s_{i+2})s_i - 2s_{i+1}s_{i+2}) \\
& + s_{i-2}(3s_i^2 - 2(s_{i+1}+s_{i+2})s_i + s_{i+1}s_{i+2})),
\end{aligned}
\tag{19}
$$

$$
\begin{aligned}
\Xi_5 =& 2(s_{i-2}(s_{i-1}(\Gamma_{i+1,i}+\Gamma_{i+2,i} - s_i) + 6s_i^2 - 3(s_{i+1}+s_{i+2})s_i + s_{i+1}s_{i+2}) + s_{i-3}(s_{i-1}(\Gamma_{i+1,i} \\
& + \Gamma_{i+2,i} - s_i) + s_{i-2}(\Gamma_{i+1,i}+\Gamma_{i+2,i} + s_{i-1} - s_i) + 6s_i^2 - 3s_{i+1}s_i - 3s_{i+2}s_i + s_{i+1}s_{i+2}) \\
& + s_i(2s_i(3s_{i+1} - 5s_i) + s_{i-1}(6s_i - 3s_{i+1})) + (3s_i(2s_i - s_{i+1}) + s_{i-1}(s_{i+1} - 3s_i))s_{i+2}),
\end{aligned}
$$

$$
\begin{aligned}
\Xi_6 =& 2(s_i(s_{i-1}(2\Gamma_{i,i+2} + s_i) + s_i(3s_{i+2} - 4s_i)) + s_{i-2}(s_i(2\Gamma_{i,i+2} + s_i) \\
& + s_{i-1}(\Gamma_{i+2,i} - s_i)) + s_{i-3}(2s_i\Gamma_{i,i+2} \\
& + s_{i-1}(\Gamma_{i+2,i} - s_i) + s_{i-2}(\Gamma_{i-1,i}+\Gamma_{i+2,i}) + s_i^2)), \\
\Xi_7 =& 2(s_i(s_{i-1}(2\Gamma_{i,i+1} + s_i) + s_i(3s_{i+1} - 4s_i)) + s_{i-2}(s_i(2\Gamma_{i,i+1} \\
& + s_i) + s_{i-1}(\Gamma_{i+1,i} - s_i)) + s_{i-3}(2s_i\Gamma_{i,i+1} \\
& + s_{i-1}(\Gamma_{i+1,i} - s_i) + s_{i-2}(\Gamma_{i-1,i}+\Gamma_{i+1,i}) + s_i^2)).
\end{aligned}
$$

Summarizing the following theorem has been established.

Theorem 1. *As long as the function g is sufficiently smooth, the first and second derivative of the this function can be approximated by five and six adjacent points respectively on non-uniform meshes, via the formulas (13) and (17).*

Proof. The proof can be investigated by Taylor expansions as in the derivation in this section. It is hence omitted. □

The procedure for obtaining the weights for the points $\{s_1, s_2, s_3, s_{m-1}, s_m\}$ to keep the fourth convergence order should be investigated by the six adjacent points as described above but for that specific node.

It is noted that the formulations derived in (13) and (17) can be used for both uniform and nonuniform distribution of the discretization nodes, and can be simplified to more simpler formulations if the nodes are equidistant.

3. Application to Option Pricing under 3D HHW PDE

Considering the non-uniform nodes discussed in Section 1 along with the high order FD formulations calculated in Section 2, one is able to derive the differentiation matrices corresponding to the first and second derivatives of the function. These derivative matrices contains the weights of the fourth order approximations and are sparse in general since they are banded matrices whose zero elements are much more than their non-zero elements. This feature would help us in solving the financial model (1) as would be observed later.

For multi-dimensional derivatives, a matrix is constructed such that this is done on the flattened data, and subsequently the Kronecker product of the matrices for the derivatives (in one-dimension) are being considered.

One way for imposing the impact of (13)–(17) is with matrices including the weights of (13)–(17), i.e., the non-equidistant second-order FD weights, as their elements. A matrix which shows an estimation to the differential operator is called as a matrix of differentiation [20]. Forming and implementing the proposed scheme based on these matrices are invaluable aids for analysis.

Taking all the weights into consideration, the PDE (1) can be semi discretized to obtain:

$$\frac{\partial U(t)}{\partial t} = A(t)U(t), \qquad 0 \le t \le T, \tag{20}$$

at which $U(t) = \underbrace{(u_{1,1,1}(t), u_{1,1,2}(t), \ldots, u_{m,n,o-1}(t), u_{m,n,o}(t))^*}_{N \text{ elements}},$ is the unknowns vector

and $N = m \times n \times o$. Noting that $A_{N \times N}(t)$ is the coefficient of the problem (1) at which the boundaries have not yet been imposed inside.

Here the boundaries along s are defined as follows [13]:

$$u(s, v, r, t) = 0, \qquad s = 0, \tag{21}$$
$$u_s(s, v, r, t) = 1, \qquad s = s_{\max}. \tag{22}$$

For $v = v_{\max}$, the following Dirichlet condition is prescribed:

$$u(s, v, r, t) = s, \qquad v = v_{\max}. \tag{23}$$

Remarking that the nodes which are located on the boundary $v = 0$ are considered as interior nodes and we take a fact into consideration that they must read the PDE model. That is to say, we incorporate the semi-discretized equations at this boundary.

At last, for $r = \pm r_{\max}$, we impose:

$$u_r(s, v, r, t) = 0, \qquad r = r_{\max}, \tag{24}$$
$$u_r(s, v, r, t) = 0, \qquad r = -r_{\max}. \tag{25}$$

By incorporating the above mentioned conditions, we obtain the following system of semi-discretized ODEs as follows:

$$\dot{U}(t) = \bar{A}(t)U(t), \tag{26}$$

where $\bar{A}(t)$ is the coefficient matrix including the boundaries.

Integrator

For discretizing in temporal variable t, many schemes are existing, for example refer to [23]. Explicit methods are basically straightforward to implement, but suffer from stability problems. Implicit schemes are unconditionally stable, but only exhibit low convergence or very time-consuming because of solving nonlinear system of algebraic equations per step.

Consider $\mathbf{u}_\iota$ to be the computational solution for the exact solution $U(t_\iota)$ and choose $k + 1$ temporal nodes with the step size $\Delta t = \frac{T}{k}$.

At the moment, we use the δ–stage Runge–Kutta scheme [23] at $t_{\iota+1} = t_\iota + \Delta t$, $(0 \le \iota \le k)$ by:

$$
\begin{aligned}
g_i &= \mathbf{u}_\iota + \Delta t \sum_{j=1}^{\delta} k_j a_{i,j}, \\
k_i &= f\left(\Delta t c_i + t_\iota, g_i\right), \\
\mathbf{u}_{\iota+1} &= \mathbf{u}_\iota + \Delta t \sum_{i=1}^{\delta} b_i k_i,
\end{aligned}
\tag{27}
$$

wherein f is defined based on the right hand side of (26). It is generally assumed that the row-sum conditions hold:

$$c_i = \sum_{j=1}^{\delta} a_{i,j}, \qquad i = 1, 2, \ldots, \delta. \tag{28}$$

Now we consider a sixth-order explicit Runge–Kutta scheme (RK6) below [24]:

$$\Lambda = \left(
\begin{array}{ccccccc}
0 & 0 & 0 & 0 & 0 & 0 & 0 \\
1 & 0 & 0 & 0 & 0 & 0 & 0 \\
\frac{3}{8} & \frac{1}{8} & 0 & 0 & 0 & 0 & 0 \\
\frac{8}{27} & \frac{2}{27} & \frac{8}{27} & 0 & 0 & 0 & 0 \\
\frac{3(3p_1-7)}{392} & \frac{(p_1-7)}{49} & \frac{-6(p_1-7)}{49} & \frac{3(p_1-21)}{392} & 0 & 0 & 0 \\
\frac{-3(17p_1+77)}{392} & \frac{(-p_1-7)}{49} & -\frac{(8p_1)}{49} & \frac{3(121p_1+21)}{1960} & \frac{(p_1+6)}{5} & 0 & 0 \\
\frac{(7p_1+22)}{12} & \frac{2}{3} & \frac{2(7p_1-5)}{9} & \frac{-7(3p_1-2)}{20} & \frac{-7(9p_1+49)}{90} & \frac{-7(p_1-7)}{18} & 0
\end{array}
\right), \tag{29}$$

with $b = (9/180, 0, 64/180, 0, 49/180, 49/180, 9/180)$, $C = (1, 1/2, 2/3, (7 - p_1)/14, (7 + p_1)/14, 1)$, and $p_1 = 21^{1/2}$.

Notice that a consequence of explicitness is $c_1 = 0$ in (28), so that the function is sampled at the beginning of the current integration step. Here, the sixth-order time-stepping solver consists of seven stages and reaches sixth order of convergence. The sixth order says that the error of local truncation is on the order of $\mathcal{O}(\Delta t^7)$, while the total accumulated error is on the order of $\mathcal{O}(\Delta t^6)$.

In the sequel, we study that under what criteria the numerical discretized solution does not blow up. The following theorem is one of the contributions of this work. This is given for the time-independent case, i.e., when $\bar{A}(t) = \bar{A}$.

Theorem 2. *If the system of ODEs (26) reads the condition of Lipschitz, then the time-stepping method (27)–(29) has conditional stability.*

Proof. To find a stability conditions, we proceed as follows. Incorporating the time-stepping solver (27) on the system of ODEs (26) yields:

$$
\begin{aligned}
\mathbf{u}_{l+1} = &\left(I + \Delta t \bar{A} + \frac{(\Delta t \bar{A})^2}{2!} \right. \\
&+ \frac{(\Delta t \bar{A})^3}{3!} + \frac{(\Delta t \bar{A})^4}{4!} \\
&+ \frac{(\Delta t \bar{A})^5}{5!} + \frac{(\Delta t \bar{A})^6}{6!} \\
&\left. - \frac{(\Delta t \bar{A})^7}{2160} \right) \mathbf{u}_l.
\end{aligned}
\tag{30}
$$

Thus, the numerical stability is reduced to:

$$
\left| 1 + \Delta t \omega_i + \frac{(\Delta t \omega_i)^2}{2} + \frac{(\Delta t \omega_i)^3}{6} + \frac{(\Delta t \omega_i)^4}{24} \right.
$$
$$
\left. + \frac{(\Delta t \omega_i)^5}{120} + \frac{(\Delta t \omega_i)^6}{720} - \frac{(\Delta t \omega_i)^7}{2126} \right| \leq 1,
\tag{31}
$$

which is due to (30) for any ω_i as the eigenvalue of $\bar{A}$. Now by considering: $\omega_{max}\left(\bar{A}\right)$, the stability condition can now be represented as follows:

$$\left| 1 + \Delta t\omega_{max} + \frac{(\Delta t\omega_{max})^2}{2} + \frac{(\Delta t\omega_{max})^3}{6} + \frac{(\Delta t\omega_{max})^4}{24} \right.$$
$$\left. + \frac{(\Delta t\omega_{max})^5}{120} + \frac{(\Delta t\omega_{max})^6}{720} - \frac{(\Delta t\omega_{max})^7}{2126} \right| \leq 1. \tag{32}$$

Noting that the negative semi-definiteness of $\bar{A}$ makes all its eigenvalues to have negative real parts. Thus, the proposed scheme has numerical stability if the temporal step size Δt satisfy (32). Noting that this can be computed in the language Mathematica [22] via the command

$$\texttt{Eigenvalues[matrix, 1]}. \tag{33}$$

The proof is ended. □

4. Experiments

In this section, some tests were given for our proposed method showed via Adaptive Finite Difference Method (AFDM) to price at the money call options, when $T = 1$ year and $E = 100\$$. A comparison was done by the standard uniform FD scheme [4], which by second order FD approximations and the Euler's scheme as a temporal solver shown by FDM. We also compare with the method provided in [13] shown by Haentjens-In't Method (HIM).

Mathematica 11.0 is used for the simulations [25]. Time is also reported in second while we employ the following stopping condition:

$$\text{Error} = \left| \frac{u_{\text{approx}}(s, v, r, t) - u_{\text{ref}}(s, v, r, t)}{u_{\text{ref}}(s, v, r, t)} \right|, \tag{34}$$

wherein u_{ref} and u_{approx} are the exact and numerical results.

To increase the computational efficiency for very large scale semi-discrete systems that we are dealing with, here we set $\texttt{AccuracyGoal} \to 5$, $\texttt{PrecisionGoal} \to 5$.

Here, we consider more number of discretization nodes along s rather than v and r, since its working interval is larger than the others and the non-smoothness of the initial condition occurs along this spatial variable.

The non-constant b is defined as follows:

$$b(\tau) = c_1 - c_2 \exp\left(-c_3\tau\right), \qquad \tau \geq 0, \tag{35}$$

where c_1, c_2, c_3 are constants, and $\tau = T - t$. The following two test cases are considered:

1. $\kappa = 3.0$, $\eta = 0.12$, $a = 0.20$, $\sigma_1 = 0.80$, $\sigma_2 = 0.03$, $\rho_{12} = 0.6$, $\rho_{13} = 0.2$, $\rho_{23} = 0.4$, $c_1 = 0.05$, $c_2 = 0$, $c_3 = 0$, where the reference value is $u_{\text{ref}}(100, 0.04, 0.1, 1) \simeq 16.176$.
2. $\kappa = 0.5$, $\eta = 0.8$, $a = 0.16$, $\sigma_1 = 0.90$, $\sigma_2 = 0.03$, $\rho_{12} = -0.5$, $\rho_{13} = 0.2$, $\rho_{23} = 0.1$, $c_1 = 0.055$, $c_2 = 0$, $c_3 = 0$, where the reference value is $u_{\text{ref}}(100, 0.04, 0.1, 1) \simeq 20.994$.

The results are brought forward in Tables 1 and 2 showing the stable and efficient valuations of options under HHW PDE via the new high-order procedure. Furthermore, to reveal the positivity and stability of the numerical results, in Experiment 2, and by considering $m = 30$, $n = 18$ and $o = 18$ discretization nodes, the results based on AFDM are plotted in Figures 1 and 2.

Table 1. Error results in Heston–Hull–White (HHW) Option 1.

Procedure	m	n	o	Size	Δt	Price	Error	CPU Timing
FDM								
	10	8	6	480	0.002	25.492	5.7×10^{-1}	0.39
	14	10	10	1400	0.001	11.098	3.1×10^{-1}	0.62
	18	12	12	2592	0.0005	17.203	6.3×10^{-2}	1.22
	24	14	14	4704	0.00025	18.731	1.5×10^{-1}	3.39
	28	16	16	7168	0.0002	13.329	1.7×10^{-1}	6.07
	45	22	22	21,780	0.00005	14.636	9.4×10^{-2}	70.54
HIM								
	10	8	6	480	0.001	14.472	1.0×10^{-1}	0.44
	14	10	10	1400	0.0005	15.300	5.3×10^{-2}	0.85
	18	12	12	2592	0.00025	15.615	3.4×10^{-2}	2.09
	24	14	14	4704	0.0001	15.806	2.2×10^{-2}	8.04
	28	16	16	7168	0.0001	15.871	1.8×10^{-2}	11.84
	50	22	22	24,200	0.000025	16.006	5.9×10^{-3}	186.77
AFDM								
	10	8	6	480	0.005	15.123	6.5×10^{-2}	0.41
	14	10	10	1400	0.002	15.986	1.1×10^{-2}	0.82
	18	12	12	2592	0.001	16.059	7.2×10^{-3}	1.99
	24	14	14	4704	0.000625	16.136	2.4×10^{-3}	5.56
	28	16	16	7168	0.0005	16.160	9.8×10^{-4}	9.02
	50	22	22	24,200	0.0001	16.179	1.8×10^{-4}	103.26

Table 2. Error results in HHW option 2.

Procedure	m	n	o	Size	Δt	Price	Error	CPU Timing
FDM								
	20	10	10	2000	0.00025	22.022	4.9×10^{-2}	1.77
	24	12	12	3456	0.0002	21.436	2.1×10^{-2}	3.14
	26	14	14	5096	0.0001	19.678	6.1×10^{-2}	8.67
	28	16	16	7168	0.0001	17.376	1.7×10^{-1}	11.72
	30	18	18	9720	0.00005	17.404	1.7×10^{-1}	35.50
	36	20	20	14,400	0.000025	20.510	2.2×10^{-2}	107.89
	38	22	22	18,392	0.000025	20.275	3.3×10^{-2}	161.16
	42	22	22	20,328	0.00002	18.370	1.2×10^{-1}	244.32
HIM								
	20	10	10	2000	0.00025	20.631	1.6×10^{-2}	1.69
	24	12	12	3456	0.0002	20.709	1.2×10^{-2}	3.40
	26	14	14	5096	0.0001	20.729	1.1×10^{-2}	8.64
	28	16	16	7168	0.0001	20.748	1.0×10^{-2}	12.36
	30	18	18	9720	0.00005	20.767	9.9×10^{-3}	36.54
	36	20	20	14,400	0.000025	20.810	7.8×10^{-3}	108.92
	38	22	22	18,392	0.000025	20.818	7.5×10^{-3}	166.08
	42	22	22	20,328	0.00002	20.833	6.7×10^{-3}	250.67
AFDM								
	20	10	10	2000	0.0005	20.832	7.7×10^{-3}	0.89
	24	12	12	3456	0.0004	20.899	4.5×10^{-3}	3.36
	26	14	14	5096	0.00025	20.910	4.0×10^{-3}	6.62
	28	16	16	7168	0.0002	20.926	3.2×10^{-3}	11.27
	30	18	18	9720	0.0001	20.951	2.0×10^{-3}	34.55
	36	20	20	14,400	0.0000625	20.972	1.0×10^{-3}	98.22
	38	22	22	18,392	0.00005	20.980	6.6×10^{-4}	165.27
	42	22	22	20,328	0.00004	20.999	2.3×10^{-4}	240.25

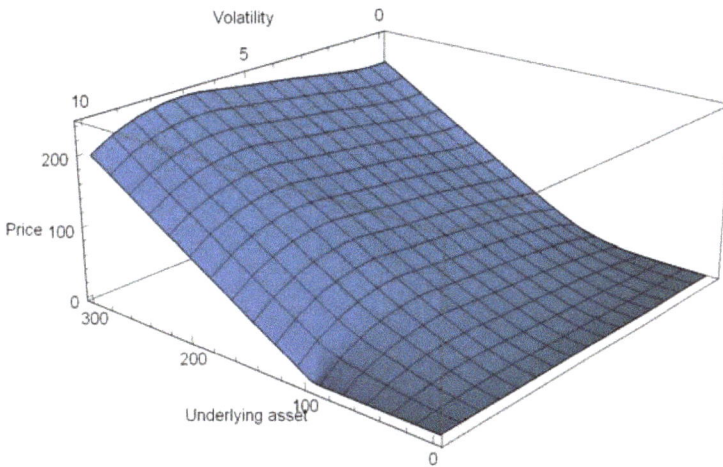

Figure 1. A numerical solution based on AFDM in Heston–Hull–White (HHW) option 2.

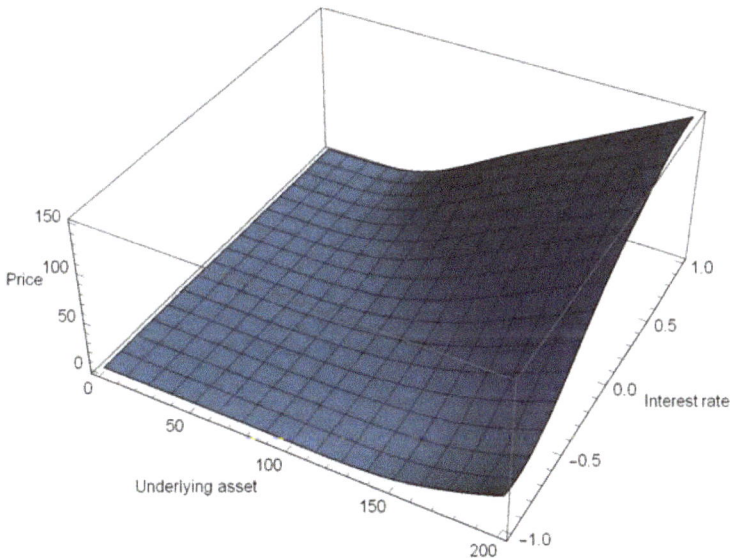

Figure 2. A numerical solution based on AFDM in HHW option 2.

5. Ending Comments

In financial engineering, it is famous that the Black–Scholes PDE could not be useful in real application due to several restrictions. Several ideas to observe the market's reality are models based upon the stochastic volatility and interest rate models. The resulted PDE problem in this way is hard to be solved theoretically due to higher involved dimensions and so numerical methods are required.

In this paper, we have proposed a new discretized numerical method based on adaptive FD methodology on non-uniform grids in order to tackle an important problem in computational finance known as HHW PDE (1). It was proved that the new procedure has conditional stability and shown to be efficient in practice.

Further discussions can be investigated to extend the results of this work for other types of options defined on HHW model such as digital (binary) options, at which the initial condition is not only non-smooth at the strike but also discontinues.

Funding: This research received no external funding.

Acknowledgments: This work was supported by the Deanship of Scientific Research (DSR), King Abdulaziz University, Jeddah, under grant No. (D-235-130-1439). The authors, therefore, gratefully acknowledge the DSR technical and financial support.

Conflicts of Interest: The author declares no conflict of interest.

References

1. Brigo, D.; Mercurio, F. *Interest Rate Models-Theory and Practice: With Smile, Inflation and Credit*, 2nd ed.; Springer Finance: Berlin, Germany, 2007.
2. Cakici, N.; Chatterjee, S.; Chen, R.-R. Default risk and cross section of returns. *J. Risk Financ. Manag.* **2019**, *12*, 95. [CrossRef]
3. Ballestra, L.V.; Cecere, L. A numerical method to estimate the parameters of the CEV model implied by American option prices: Evidence from NYSE. *Chaos Solitons Fractals* **2016**, *88*, 100–106. [CrossRef]
4. Duffy, D.J. *Finite Difference Methods in Financial Engineering: A Partial Differential Equation Approach*; Wiley: Chichester, UK, 2006.
5. Magoulès, F.; Gbikpi-Benissan, G.; Zou, Q. Asynchronous iterations of parareal algorithm for option pricing models. *Mathematics* **2018**, *6*, 45. [CrossRef]
6. Fouque, J.-P.; Papanicolaou, G.; Sircar, K.R. *Derivatives in Financial Markets with Stochastic Volatility*; Cambridge Univ. Press: Cambridge, UK, 2000.
7. Heston, S.L. A closed-form solution for options with stochastic volatility with applications to bond and currency options. *Rev. Finan. Stud.* **1993**, *6*, 327–343. [CrossRef]
8. Hull, J.; White, A. Using Hull-White interest rate trees. *J. Deriv.* **1996**, *4*, 26–36. [CrossRef]
9. Schöbel, R.; Zhu, J. Stochastic volatility with an Ornstein-Uhlenbeck process: An extension. *Eur. Financ. Rev.* **1999**, *3*, 23–46. [CrossRef]
10. Guo, S.; Grzelak, L.A.; Oosterlee, C.W. Analysis of an affine version of the Heston-Hull-White option pricing partial differential equation. *Appl. Numer. Math.* **2013**, *72*, 143–159. [CrossRef]
11. Sargolzaei, P.; Soleymani, F. A new finite difference method for numerical solution of Black-Scholes PDE. *Adv. Diff. Equat. Control Process.* **2010**, *6*, 49–55.
12. Soleymani, F.; Barfeie, M. Pricing options under stochastic volatility jump model: A stable adaptive scheme. *Appl. Numer. Math.* **2019**, *145*, 69–89. [CrossRef]
13. Haentjens, T.; In't Hout, K.J. Alternating direction implicit finite difference schemes for the Heston-Hull-White partial differential equation. *J. Comput. Fin.* **2012**, *16*, 83–110. [CrossRef]
14. Soleymani, F.; Akgül, A. Asset pricing for an affine jump-diffusion model using an FD method of lines on non-uniform meshes. *Math. Meth. Appl. Sci.* **2019**, *42*, 578–591. [CrossRef]
15. Itkin, A.; Carr, P. Jumps without tears: A new splitting technology for Barrier options. *Int. J. Numer. Anal. Model.* **2011**, *8*, 667–704.
16. Sumei, Z.; Jieqiong, Z. Efficient simulation for pricing barrier options with two-factor stochastic volatility and stochastic interest rate. *Math. Prob. Eng.* **2017**, *2017*, 3912036. [CrossRef]
17. Soleymani, F.; Saray, B.N. Pricing the financial Heston-Hull-White model with arbitrary correlation factors via an adaptive FDM. *Comput. Math. Appl.* **2019**, *77*, 1107–1123. [CrossRef]
18. Kwok, Y.K. *Mathematical Models of Financial Derivatives*, 2nd ed.; Springer: Heidelberg, Germany, 2008.
19. Ballestra, L.V.; Sgarra, C. he evaluation of American options in a stochastic volatility model with jumps: An efficient finite element approach. *Comput. Math. Appl.* **2010**, *60*, 1571–1590. [CrossRef]
20. Fornberg, B. *A Practical Guide to Pseudospectral Methods*; Cambridge University Press: Cambridge , UK, 1996.
21. Soleymani, F.; Barfeie, M.; Khaksar Haghani, F. Inverse multi-quadric RBF for computing the weights of FD method: Application to American options. *Commun. Nonlinear Sci. Numer. Simul.* **2018**, *64*, 74–88. [CrossRef]
22. Wellin, P.R.; Gaylord, R.J.; Kamin, S.N. *An Introduction to Programming with Mathematica*, 3rd ed.; Cambridge University Press: Cambridge, UK, 2005.

23. Sofroniou, M.; Knapp, R. *Advanced Numerical Differential Equation Solving in Mathematica, Wolfram Mathematica, Tutorial Collection*; Wolfram Research, Inc.: Champaign, IL, USA, 2008.
24. Luther, H.A. An explicit sixth-order Runge-Kutta formula. *Math. Comput.* **1967**, *22*, 434–436. [CrossRef]
25. Mangano, S. *Mathematica Cookbook*; O'Reilly Media: Sevastopol, CA, USA, 2010.

MDPI

St. Alban-Anlage 66

4052 Basel

Switzerland

Tel. +41 61 683 77 34

Fax +41 61 302 89 18

www.mdpi.com

Mathematics Editorial Office

E-mail: mathematics@mdpi.com

www.mdpi.com/journal/mathematics